MICROBIAL PHYSIOLOGY

2nd Revised and Enlarged Edition

Preface (2nd Edition)

The first edition of this book was brought out more than a decade ago. It embodied the core aspects of microbial physiology and metabolism, and was well received by teachers and students. However, since then there has been an unprecedented accumulation of information on unknown aspects of microbial life that has relevance to microbial technology and human welfare, and as such there is a need to translocate this latest information to all stake-holders. Further, many teacher and research cleeagues while commending the first edition suggested some improvement and additions to be made in future editions.

We, therefore, thought of bringing out this thoroughly revised new edition keeping in view of latest information on the subject and to fill the gaps in already described topics. Several new chapters like aminoacid metabolism, nucleotide metabolism, lipid metabolism are added. Similarly, some chapters on applied aspects of microbial physiology like microbial toxins, antimicrobials and signal transduction and quorum sensing are added. With the additions of new chapters the new addition becomes comprehensive providing information on all aspects of microbial metabolism and caters the needs of all sections. Hopefully, as in the case of first edition, the researchers, teachers and students would appreciate our efforts and patronize this edition also.

We thank all our teacher colleagues for providing the necessary inputs for the improvement of this edition. We gratefully acknowledge the sincere efforts put up by Scientific Publishers (India), Jodhpur in bringing out this edition.

12 April, 2021

S. Ram Reddy
S.M. Reddy

Preface (1st Edition)

Microbial physiology is a field that is experiencing rapid growth and making inroads to allied disciplines like biochemistry, biophysics, genetics and molecular biology. Physiology and metabolism are twins so much so it is difficult to decide where physiology leaves off and metabolism begins. Similarly, it is difficult to draw a borderline between physiology and related subjects like nutrition, biochemistry and genetics. The difficulties in understanding the microbial physiology are confound probably due to the fact that microorganisms display a bewildering physiological and metabolic diversity. Further, a particular organelle like cell membrane carry a multitude of physiological and biochemical functions. Similarly, microbial physiology is not properly understood without substantial consideration of genetic make up and of the basic principles of molecular biology.

The role of microorganisms in nature and in human life activities is a matter that has been reckoned so poignantly than ever before. Microbial physiology is at the heart of different disciplines of microbiology. All the activities of microbes are the manifestations of basic physiological activities. Thus, the understandings of the physiology of the microbes enable us in manipulating these organisms in a desired direction.

Though there are several books available dealing with microbial physiology to different depths, a comprehensive book covering all aspects of physiology and catering the needs of students, teachers and researchers is a long felt need. The present book is basically designed to fill this vacuum and it may serve users as a textbook or reference book as it is written in a flexible manner covering the basic aspects and at the same time projecting the expanding areas of the subject.

The layout of the chapters and contents of the chapters are planned in such way to provide the principles of microbial physiology in a right perspective. Chapters I to VI are intended to provide basic information on cell structure, nutrition, growth, bioenergetics and environmental adaptations. Chapters VII to X provide a picture of how different microorganisms obtain energy and carbon for synthetic or anabolic processes. Chapters XI–XIII deal with the catabolic processes under diverse conditions. The diversity of both catabolic and anabolic processes of microorganisms are also discussed. The last three chapters explain about the biological nitrogen fixation, secondary metabolism and sporulation which are specific to microorganisms.

It is hoped that the contents of this book fulfill the needs of students of undergraduate and post-graduate as well as teachers of microbiology of all Indian Universities. An attempt has been made to present the subject matter in a simple and lucid manner avoiding too much of quantitative information. Suggested readings are provided at the end of each chapter for the benefit of those who intend in depth of the subject. Authors have taken every care to present the facts of the subject matter. However, they are conscious of their limitations and invite suggestions, which will go a along way for improving the subject matter and the book in future.

S. Ram Reddy
S.M. Reddy

Contents

Chapter - 1

Cell Structure and Function

1.1. INTRODUCTION

The cell is the basic unit of all living organisms. Many organisms are unicellular and microscopic carrying all the life functions and that too near efficiency, whereas others are multicellular, macroscopic with cell specialization. Whether one is a microbe or large life form, the essential processes of life are mediated by structures. All life processes essential to life are accomplished by a microbe within a small space. Microbes compensate for morphological simplicity by physiological and metabolic complexity. Though microbes and large life forms differ physiologically in many respects, the former display all the physiological possibilities of nature at the cellular level.

Study of cellular phenomenon in microbial systems is essential for understanding the physiology of larger forms. Comparative study of cells in microbial systems allows us to the critical differences between the cell types from which all life is made and at the same time to understand essentials of life. Thus microbes serve as a model for understanding of nutrition, metabolism and biochemistry, protein and nucleic acid synthesis, energy generation, enzyme regulation, membrane transport, motility, differentiation, cellular communication and the behavior of the higher life forms. Subtle differences in the life processes with in the microbes exist. Understanding these differences allows us to exploit them for the benefit of mankind.

1.2. PROKARYOTES vs. EUKARYOTES

At macro level, the prokaryotes and eukaryotes perform the same life functions. However, there are both similarities and differences in the nature of the processes and in the structures that accomplish them. Fig.1.1(a) and Fig. 1.1(b), and Table 1.1 provide the main structural features of prokaryotic and eukaryotic cells.

Table 1.1. Structures and processes of prokaryotic and eukaryotic cells

		Eukaryotes			
Structure/ Process	**Prokaryote**	**Protozoa**	**Algae**	**Fungi**	**Metazoan**
Membrane-bound Nucleus	Absent	Present	Present	Present	Present
Histones •	Absent	Present	Present	Present	Present

Structure/ Process	Prokaryote	Eukaryotes			
		Protozoa	**Algae**	**Fungi**	**Metazoan**
Mitochondria	Absent	Present	Present	Present	Present
Chloroplasts	Absent	Absent	Present	Absent	Absent
Phagocytosis	Absent	May occur	Absent	Absent	May occur
Pinocytosis	Absent	May occur	Absent	Absent	May occur
Protoplasmic streaming	Absent	May occur	Absent	Absent	Absent
Endoplasmic reticulum	Absent	Present	Present	Present	Present
Mesosomes	Present	Absent	Absent	Absent	Absent
Electron transport	In cytoplasmic membrane	In mitochondrion	In mitochondria and chloroplast	In mitochondrion	In mitochondrion
Rigid cell wall	Present	Absent	Present	Present	Absent
Ribosomes	Cytoplasmic 70S	Cytoplasmic 80S	Cytoplasmic 80S	Cytoplasmic 80S	Cytoplasmic 80S
		Mitochondrion 70S	Mitochondrion 70S	Mitochondrion 70S	Mitochondrion 70S
Golgi apparatus	Absent	Present	Dicytosomes	Present	Present

Woese *et al.* (1990) divided all the life forms into three domains based on 16S RNA *viz,* **Bacteria** (eubacteria), **Eucarya** (eukaryotes) and **Archaea** (archaebacteria) that diverged in the distant past from a common ancestor (Fig. 1.2).

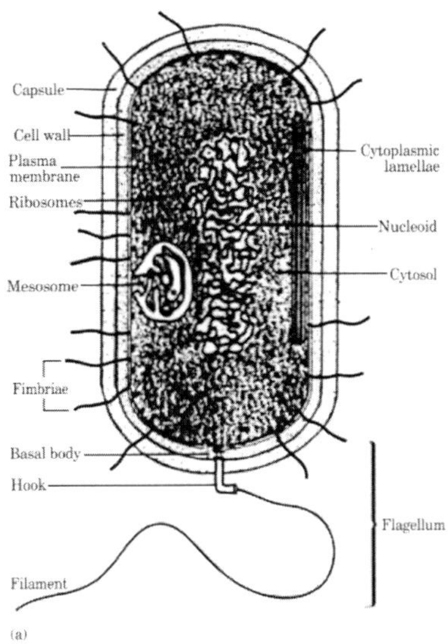

Fig. 1.1(a). The structure of a typical prokaryotic cell

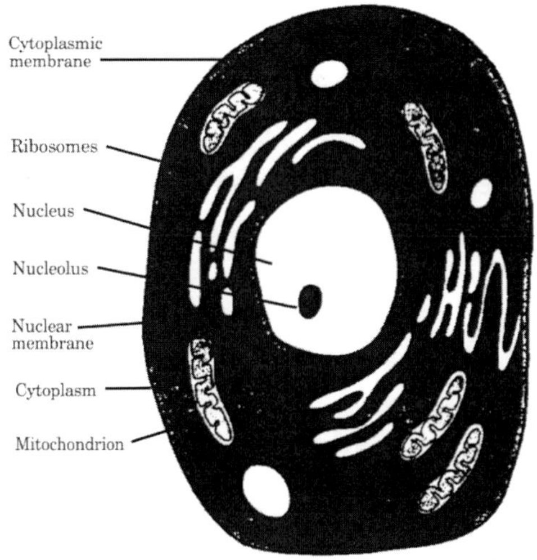

Fig. 1.1(b). Structure of a eukaryotic (animal) cell

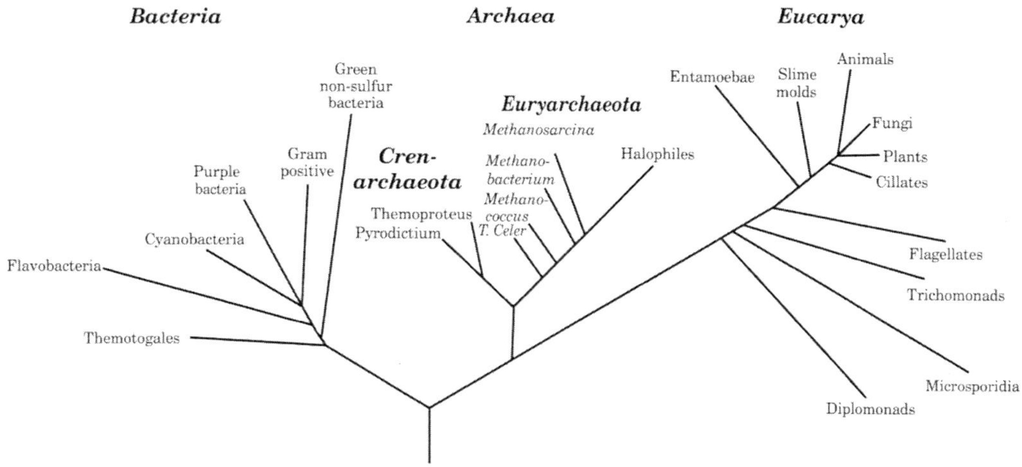

Fig. 1.2. Phylogenetic relationships among life forms based upon rRNA sequences.

1.3. PROKARYOTIC CELL STRUCTURE

Prokaryotic cell contains of a variety of structures to perform different life functions. The major cell organelles and their functions are summarized in Table 1.2. However, it should be remembered that not all structures are found in all genera of bacteria. Further, gram-negative and gram-positive cells differ with respect to their cell walls. Inspite of minor variations, prokaryotes are consistent in their fundamental structure and most important components.

Table 1.2. Functions of prokaryotic structures

Cell organelle	Function
Plasma membrane	Selectively permeable barrier, mechanical boundary of cell, nutrient and waste transport, location of many metabolic processes (respiration, photosynthesis), detection of environmental cues for chemotaxis
Gas vacuoles	Buoyancy for floating in aquatic environments
Ribosomes	Protein synthesis
Inclusion bodies	Storage of carbon, phosphate, and other substances
Nucleoid	Localization of genetic material (DNA)
Periplasmic space	Contains hydrolytic enzymes and binding proteins for nutrients processing and uptake
Cell wall	Gives bacteria shape and protection from lysis in dilute solutions
Capsules and slime layers	Resistance to phagocytosis, adherence to surfaces
Fimbriae and pili	Attachment to surfaces, bacterial mating
Flagella	Movement
Endospore	Survival under harsh environmental conditions

1.3.1. Cell Wall

All the prokaryotic cells are surrounded by a definite cell wall. Mycoplasmas and some archaebacteria are exceptions. Cell wall is one of the most important part of prokaryotic cells because it is associated with many accessory structures, such as glycocalyx, flagella, fimbriae and pili. It protects the cell from osmolysis. Pathogenicity of several bacteria can be attributed to the cell wall nature. The cell wall can protect the cell from toxic substances and it is the target for action of many antibiotics.

Christian Gram, in the nineteenth century, developed a staining procedure by which it now became possible to divide the bacteria into two broad groups Gram-positive and Gram-negative on the basis of cell wall structure. The difference in the gram staining is also related to outer wall layer of Gram-negative bacteria which is rich in phospholipids which made leaky by the lipid solvents, alcohol and acetone. Archaea can stain either Gram-positive or Gram-negative, but their cell wall composition is different from that of bacteria. No archeal cell wall contains peptidoglycan, instead it contains pseudo peptidoglycan, polysaccharide or protein. Because of the importance of cell wall in prokaryotic cell structure and function, and its composition in different groups is discussed in detail separately.

1.3.1.1. Glycocalyx

The term glycocalyx is often used to describe all extracellular material to the cell wall. All bacteria are probably surrounded by glycocalyces as they grow in their natural habitat. However, they tend to loose these external layers when cultivated in the laboratory. The extracellular materials are usually polymers and may be in the form of S-layer, capsules and slimes.

Glycocalyx helps the bacterium to adhere to the surfaces of other cells or to inanimate objects which is necessary for colonization of solid surfaces or growth in biofilms. Still, another role of the glycocalyx is protection from phagocytosis.

1.3.1.2. S-layers

These are an array of protein or glycoprotein molecules on the cell wall. S-layers are found on the cell wall surface of both gram-positive and gram-negative bacteria. In wall less archaea, the S-layer is found over the cell membrane and serves as the cell wall itself.

1.3.1.3. Capsules

The capsules are composed of fibrous material at the cell surface. They may be rigid, flexible or closely associated and tightly bound to the cell surface. Loosely bound material called slime or slime capsule, is generally shed into the medium.

1.3.2. Appendages

At least three types of appendages, *viz.*, flagella, fimbriae and sex pili, each with a specific designed function extend from the surface of bacterial cell.

1.3.2.1. Flagella

Flagella are organelles of locomotion that protrude from the cell surface of motile bacteria. Bacterial genera differ with regard to number and location of flagella. Bacterial flagella are unrelated in composition, structure and mechanism of action to the eukaryotic flagellum. The flagellum of *E. coli* and *Salmonella typhimurium* serves as the prototype for bacterial flagella (Fig. 1.3). Much less is known about the flagella of archaea but they appear to be similar in morphology to bacterial flagella.

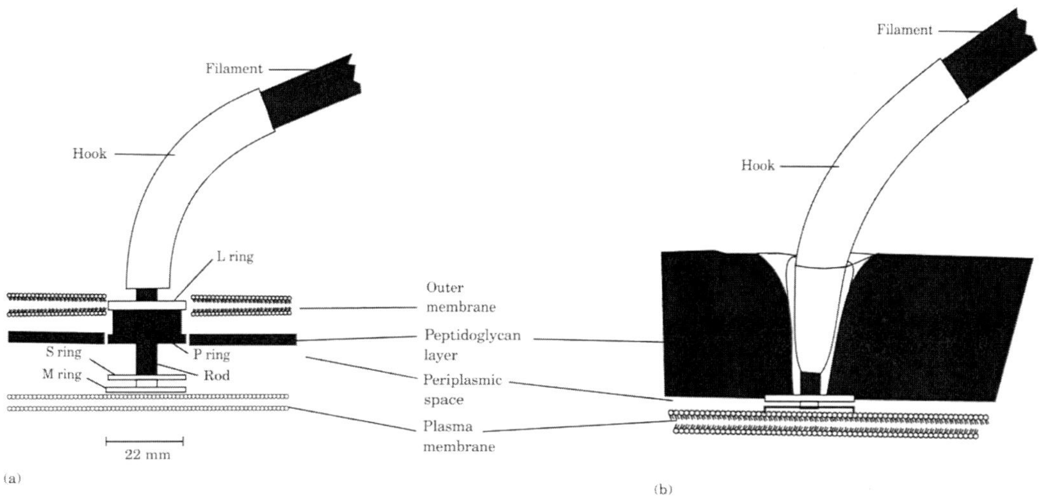

Fig. 1.3. The ultrastructure of bacterial flagella. Flagellar basal bodies and hooks in (a) Gram-negative and (b) Gram- positive bacteria.

The flagellum consists of three parts: a basal body, a hook and a filament plus additional proteins. The basal body consists of four rings M, S, P and L. M and S rings are embedded in the cell membrane. These two rings along with switch, Mot A and Mot B proteins function as a motor. P ring is embedded in the peptidoglycan layer and L in the outer membrane. Gram-positive bacteria lack the two outer rings (L and P rings). A central rod passes through these four rings and at the end of which a curved hook is attached. The hook in turn attached to long filament. The motor is actually electrochemical machine. The energy to drive the motor comes from current of protons that moves down a proton potential gradient through the flagellar motor from the outside of the membrane to the inside. When the motor turns, the attached filament rotates and propels the bacterium through the medium. The motor can rotate either clock-wise or counter clock-wise which determines the directionality of swimming.

In addition to flagella, most gram-negative bacteria and a few gram-positive bacteria possess protein fibrils extending from the cell surface. They have been called as fimbriae, pili, filaments and fibrils. All such appendages are not alike nor their functions known in all cases. Their sizes vary considerably. Freshly isolated strains frequently have fibrils, they are often lost during sub-culturing in the laboratory.

1.3.2.2. Fimbriae

It has been proposed that fibrils which mediate attachment be called fimbriae. Fibrils mediate attachment of the bacteria to other cells, for example to other bacteria, animal, plant or fungal cells. They are, therefore, important for colonization because they help the bacteria stick to surfaces. Fimbriae possess adhesive proteins (called **adhesins**) that recognize and bind to specific receptors on the surface of cells. Adhesins of fimbriae have medical importance also. For example, pathogenic bacteria identify their host cells through fimbriae. Adhesins are specialized for attachment to specific receptors and can account for the specificity of bacterial attachment to hosts and tissues.

1.3.2.3. Sex pili

The fimbriae that mediate attachment between mating cells are different from other fimbriae and are called sex pili. Sex pili are found in enteric bacteria such as *E. coli* and *Pseudomonas*. Sex pili seem to be helpful in cell suspension where the bacteria are not in intimate contact.

1.3.3. Periplasm

Periplasm is the space between cell membrane and the outer membrane. This aqueous compartment containing proteins, oligosaccharides, salts and peptidoglycan is found only in Gram-negative bacteria. Periplasm is often considered as a cellular compartment with specialized activities. These activities include oxidation – reduction reactions, osmotic regulation, solute transport, protein secretion and hydrolytic activities.

Oligosaccharides of the periplasm are presumed to be involved in osmotic regulation. Some proteins of the periplasm function as solute binding proteins and assist in solute transport. Some of the enzymes in the periplasm are cytochrome C that oxidise organic compounds, and inorganic compounds and deliver the electrons to electron transport chain in cell membrane. Periplasmic oxidations are important for energy metabolism in many Gram-negative bacteria. Hydrolytic enzymes present in the periplasm degrade nutrients to smaller molecules that can be transported across the cell membrane by specific transporters. For example, periplasmic amylase degrades starch to simple sugars. Still, some enzymes of periplasm are detoxifying agents. For instance, β - lactamase that degrades the penicillin is a periplasmic enzyme. Ton B, a periplasmic protein of *E.coli* is required for the uptake of several solutes that do not diffuse through the porins, but require specific transport systems. Examples of solutes that require Ton B for uptake are iron (siderophores) and vitamin B_{12}. This protein is thought to couple the energy in the cytoplasm or the cell membrane to the uptake of certain solutes through the outer envelope and into the periplasm.

1.3.4. Cell Membrane

Prokaryotic cells, like all other living cells, are encompassed by a thin, complex and functionally dynamic membrane called cell membrane (CW). The cell membranes are very thin structures, about 5 to 10 nm thick and can only be seen with electron microscope by freeze – etching technique. This membrane is the chief point of contact of the cells with environment and thus is responsible for much of its relationship with the outside world. Cell membrane is responsible for a broad range of physiological activities including solute transport, electron transport, photosynthetic electron transport, the establishment of electrochemical gradients, motility, ATP synthesis, and biosynthesis of lipids, biosynthesis of cell wall polymers, secretion of proteins, intercellular signaling and responses to environmental signals.

Bacterial cell membranes consist primarily of phospholipids and proteins in a fluid mosaic structure in which the phospholipids form a bilayer. It is a complex mosaic part whose structure and interactions at the molecular level are not well understood. Although the lipids are very important for determining the basic functions of the cell membrane the membrane proteins confer most of the impotent functions of biological membranes. There are many different types of membrane proteins, reflecting their very different functions. More than 100 different types of proteins are reported in the cell membrane. Many of these proteins are clustered in functional aggregates, such as the proton tanslocating ATP as the flagellar motor, electron transport complexes, and solute transports. There are two classes of proteins in membranes, integral and peripheral. Integral proteins are embedded in the membrane and bound to the fatty acids of the phospholipids via hydrophobic bonding. They can be removed only with detergents or solvents. Peripheral proteins are attached to the membrane surface to the phospholipids by ionic interactions and can be removed by washing the membrane, with salt solutions. The lipids and proteins diffuse laterally in the plane of the membrane but do not move across the membrane. Many archaebacterial membranes differ from other bacterial membranes in having a monolayer with lipid molecules spanning the whole membrane. Archaeal membrane also possesses certain unusual proteins like bacteriorhodopsin and halorhodopsin. Bacteriorhodopsin is a pigment protein pump responsible for the light driven electrogenic efflux of protons. Halorhodopsin is also a light-driven electrogenic ion pump, which is used to accumulate intracellularly in order to maintain osmotic stability.

The phospholipids are fatty acids esterified to two of the hydroxyl groups of phosphoglycerides. The third hydroxyl group in the glycerol backbone of the phospholipid is covalently bound to a substituted phosphate group. Since the phospholipids are polar at one end and nonpolar at the other end, they are said to be amphipathic. They spontaneously aggregate with their nonpolar fatty acid regions interacting with each other by hydrophobic bonding. The polar phosphorylated regions face the aqueous phase where ionic interactions with cations, water and polar groups on protein occur. The phospholipids in *E. coli* are approximately 75% phosphatidyl ethanolamine (PE) and 20% phosphatidyl glycerol (PG) and the remainder is cardiolipin phosphatidyl serine and trace amounts of other phospholipids. Some bacteria possess phosphatidyl choline, or lecithin, which is characteristic of higher organisms other bacteria produce glycolipids, such as monogalactosyl diglyceride. The membrane lipids from archaca are quite different from those in bacteria and eukarya. Their hydrocarbon chains are based on isoprenoid units and these are linked to the glycerol backbone in an other, rather than ester linkage. In some archaea, a glycerol backbone and head group are attached to both ends of a pair of isoprenoid units, steroids are not commonly found in bacteria and archaea except for the cell wall-less **Mycoplasma**. Very complex lipids, including the very long, branched mycolic acids, are common in **Mycobacteria,** but occur in a very thick and rigid outer layer than in the cytoplasmic membrane.

Cell membrane is permeable to water molecules, gases and small hydrophobic molecules. However, it acts as a permeability barrier to virtually all water soluble molecules so that most solutes diffuse or carried across the membrane through or on special protein transporters.

1.3.5. Cytoplasm

Cytoplasm is the substance lying between the plasma membrane and the nucleoid. It is a viscous material containing largely water (about 70% of the bacterial mass). The liquid or soluble portion of the cytoplasm is called cytosol. Suspended in the cytoplasm are ribosomes, DNA, different membranes and large protein aggregates.

Cytosol contains a large number and variety of enzymes which catalyse a major portion of the biochemical reactions in the cell, such as the enzymes of the central pathways for the metabolism of carbohydrates (glycolysis, PP pathway and ED pathway) as well as the central pathway for the organic acid metabolism, TCA cycle and glyoxylate pathways, and enzymes for other pathways such as the biosynthesis and degradation of amino acids, lipids and nucleotides. Extensive protein – protein interactions occur among enzymes of the cytosol.

1.3.6. Intracytoplasmic Membranes

Many bacterial cells possess intracytoplasmic membranes that perform specialized physiological functions (Table 1.3). These membranes are very often connected to the cell membrane and are generally believed to be derived from the invagination of chemically modified areas of the cell membrane. However, in some cases like thylakoids of cell membrane their exact origin is not known.

Table 1.3. Bacteria containing intracytoplasmic membrane (ICM)

Prokaryotic group	Function of ICM
Methanotrophs	Methane oxidation
Nitrogen fixers	Cellular respiratory activity to provide more ATP
Nitrifiers	Ammonia and nitrite oxidation
Phototrophs	Sites of photosynthetic apparatus

1.3.7. Inclusion Bodies

Some bacteria contain specialized organelles in the cytoplasm. These differ from eukaryotic organelles in not being surrounded by a lipo-protein membrane. They are referred by some researchers as inclusion bodies rather than organelles.

1.3.7.1. Gas vesicles

Gas vesicles are found in aquatic bacteria, such as cyanobacteria, photosynthetic and non-photosynthetic bacteria. These are surrounded by protein coat and filled by a gas that is in equilibrium with gases dissolved in the cytosol. The gas vesicles enable the bacteria to float in lakes and ponds at depths that support growth because of favourable light, temperature or nutrients. Halophilic bacteria such as *Halobacterium* and a few marine cyanobacteria like *Trichodesmium* also have gas vesicles.

1.3.7.2. Carboxysomes

These are large polyhedral inclusion bodies found in obligate autotrophs that fix atmospheric **carbondioxide**. The carboxysomes are observed in nitrifying bacteria, sulfur oxidizers and cyanobacteria. The presence of carboxysomes appears to be species specific. These inclusion bodies store ribulose biphosphate carboxylase (Rubp carboxylase) the enzyme that incorporates carbon of CO_2 to organic carbon in Calvin cycle.

1.3.7.3. Chlorosomes

Chlorosomes are ellipsoid inclusion bodies found in green sulfur photosynthetic bacteria and photosynthetic gliding bacteria. They lie immediately underneath the cytoplasmic membrane and are surrounded by a nonunit membrane of galactolipid with some proteins. The major light harvesting pigments are located in the chlorosomes. However, it is interesting to note that these bacteria reaction centers are located in the cell membrane. In those photosynthetic bacteria that donot have chlorosomes, the light harvesting pigments surround the reaction centers in the cell membrane.

1.3.7.4. Granules and Globules

Bacteria often contain cytoplasmic granules whose contents vary with the bacterium. In many bacteria, these granules store poly–β–hydroxybutyric acid (PHB) as a carbon and energy reserve. Some bacteria store glycogen for the same purpose. Other granules found in bacterial cells include polyphosphate or volutin granules and in some sulfur oxidizing bacteria elemental sulfur globules. Table 1.4 presents a section of bacteria that accumulate one or more of the typical storage materials of microorganisms.

Table 1.4. Bacteria containing granules and globules

Organism	Glycogen	PHB	Polyphosphate
Enterobacter aerogenes	+	–	+
Alcaligenes eutrophus	–	+	+
Azotobacter vienlandii	–	+	+
Bacillus megaterium	+	+	–
Escherichia coli	+	–	+
Mycobacterium phlei	+	–	+
Pseudomonas aeruginosa	–	–	–
P. multivorans	–	+	?
Rhodospirillum rubrum	+	+	?
Sphaerotilus natans	–	+	?

1.3.8. Ribosomes

Ribosomes are the only cell organelles that are shared by both prokaryotes and eukaryotes. However, there are many fundamental differences between the ribosomes of two groups. Prokaryotic ribosomes sediment in a centrifugal field at the velocity of 70 svedberg units (i.e, they are 70S) in contrast to eukaryotic cytosolic ribosomes that are 80S. Bacterial ribosomes are similar regardless of the bacterium. Archaeal ribosomes are also 70S type. Needless to mention, ribosomes are the sites for protein synthesis.

1.3.9. Nucleoid

The most striking difference between prokaryotes and eukaryotes is the way in which their genetic material is organized. Prokaryotes lack a membrane delimited nucleus. The prokaryotic chromosome is a single circle of double stranded DNA. It is located in an irregularly shaped region called the nucleoid. The nucleoid is visible in the light microscope after staining with the Feulgen stain, which specifically reacts with DNA. A cell can have more than one nucleoid especially when it is in actively dividing stage. There is some evidence available that bacterial DNA is attached to cell membrane and membrane may be involved in the separation of DNA into daughter cells during division. In *E. coli,* the closed circular DNA measures about 144 μm about 500 times longer than the cell. The DNA is efficiently packed to fit with in the nucleoid. The looped and extensively coiled DNA is believed to be packed by nucleoid proteins, other than histone proteins.

Many bacteria possess plasmids in addition to chromosome. These are circular, double stranded DNA molecules that can exist and replicate independently of the chromosome. Plamids are not required for growth and reproduction of host. They carry genes that confer bacterial host a selective advantage. Plasmid genes can render bacteria drug resistant, give them new metabolic abilities, make them pathogenic or endow them with a number of other properties.

1.4. FURTHER READING

Costerton, J.W., R.T. Irwin and K.J. Cheng (1981), The bacterial glycocalyx in nature and disease *Annu. Rev. Microbiol*. **35:** 299 – 324.

Ferguson, S.J. (1992), The periplasm. In : Prokaryotic Structure and Function : A New Perspective. S. Mohan, C. Dow and J.A. Cole (eds). Society for General Microbiology, Symposium 47, Cambridge University Press.

Karwuec, S. and M. Riley (1990), Organization of the bacterial chromosome. *Microbiological Reviews* **54:** 502-539.

Kurland, C.G. (1972). Structure and function of the bacterial ribosome. *Annu. Rev. Biochem.,* **41** : 377-408.

Manson, M.D. (1992). Bacterial motility and chemotoxis. *Advances in Microbial Physiology,* **33** : 277-346.

Nikaido, H. and M. Vaara (1987). Outer membrane In: *Eschericha coli* and *Salmonella typhimurium.* Cellular and Molecular Biology Vol. 1. F.C. Neidhardt (ed) ASM Press, Washington DC.

Rose, A.H. and D.W. Tempest (eds.) (1988). *Advances in Microbial Physiology.* Academic Press, New York.

Salton, M.R. J. and P. Owen (1976). Bacterial membrane structure. *Annu. Rev. Microbiol.*, **30:** 451- 482.

Shockman, G.D. and J. Barratt (1983). Structure, function and assembly of cell walls of gram-positive bacteria. *Ann. Rev. Microbiol.,* **37:** 501 – 528.

Wilson, D.R. and T.J. Boveridge (1993). Bacterial flagellar filaments and their components of flagellins *Can. J. Microbiol.,* **39 :** 451-472.

Woese, C.R., O. Kandler and M.L. Wheelis (1990). Towards a natural system of organisms. Proposals for the domains Archaea, Bacteria and Eucarya. *Proc. Natl. Acad. Sci. USA,* **87 :** 4576-4579.

■■■■

Chapter - 2

Cell Wall and its Biosynthesis

2.1. INTRODUCTION

A great majority of microbes possess rigid exterior layers known as cell walls or envelopes. In eukaryotic microbes (algae and fungi), the cell wall composition is almost similar to higher plants. Typically, they contain cellulose, hemicellulose, lignin, mannoprotein and in the case of certain fungi chitin. The chemical composition of prokaryotic cell walls differs substantially from that of their eukaryotic counterparts.

The bacterial cell wall or envelope (in case of Gram-negative bacteria) is a complex structure consisting of different layers that have to perform many critical functions for the cell. Besides being a protective shield, it allows the organisms to maintain their shape. It allows communication with the surroundings in order to find optimal conditions for growth. This highly developed cell organelle is mechanically stabilized to withstand high intracellular osmotic pressure of approximately 2-25 atm. Organisms devoid of walls or envelopes are extremely sensitive to osmotic pressure changes and undergo osmotic lysis. Cell wall stabilizes the cytoplasmic membrane and at the same time precludes certain membrane-dependent processes, particularly pinocytosis and phagocytosis. The cell walls of many pathogens have components that contribute to their pathogenicity. The wall can protect a cell from toxic substances and is also the site of action of several antibiotics.

2.2. TYPES OF CELL WALLS

Christian Gram developed a stain in 1884. Very soon, it became evident that bacteria could be divided into two major groups based on their response to this stain and named it as Gram-stain. Bacteria possessing one type of wall can be stained using this stain and are called Gram-positive, while bacteria possessing the second type wall do not stain and are called Gram-negative. Substantial differences are also found both in the chemical composition and in the organization of the components of Gram–positive wall and the Gram-negative envelope (Fig. 2.1). Furthermore, there are also differences between the walls and envelopes found in both Gram-positive and Gram-negative bacteria and those found in the archaebacteria.

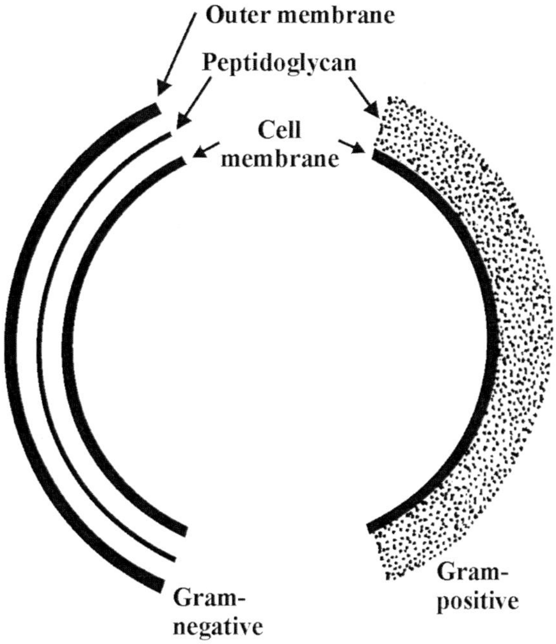

Fig. 2.1. Bacterial cell walls. Gram-negative and a Gram-positive walls.

2.3. PEPTIDOGLYCAN

The strength and rigidity of bacterial cell walls are due to a glycopeptide called peptidoglycan or murein. Peptidoglycan is an enormous polymer composed of many identical subunits. The glycan polymer contains alternating two sugar derivatives, N- acetylglucosamine (Glc Nac or G) and N-acetylmuramic acid (Mur Nac or M) (the lactyl ether of N-acetylglucosamine) attached to each other via β,1–4 linkages. These polymers are interconnected by a short peptide consisting of L- and D-amino acids in an alternating sequence (Fig. 2.2 a & b). The presence of LD and DD peptide bonds renders the peptides resistant to the majority of peptidases. Cross linkage of two peptides is made possible by the presence of a diamino acid, such as L-lysine, mesodiaminopimelic acid, or L-ornithine. This allows the formation of a tail to tail bridging peptide bond between the terminal carboxyl group of the diamino acid at position 3 of another peptide side chain. In many Gram-positive bacteria, the cross linkage is not a direct one but is mediated by an additional peptide called 'Intervening-peptide'. A pentaglycine, for example, is inserted between the D-alanine and L-lysine in case of *Staphylococcus aureus*. In contrast to the types of cross linkages referred to as group A, in which the cross-linkage is between positions 4 and 3 of two peptides, the linkage in group B is via an intervening peptide. The linkage in group B is via an intervening peptide containing one diamino acid that allows for a cross linkage between the α-carboxyl group of the D-glutamic acid in position 2 of one peptide to the carboxyl group of D-alanine in position 4 of another peptide. In general, there are only a few modifications of the sugar part. The glycan can be O_6-acetylated, which makes the murein resistant to lysozyme or they can be de- N- acetylated as in the case of *Bacillus cereus*.

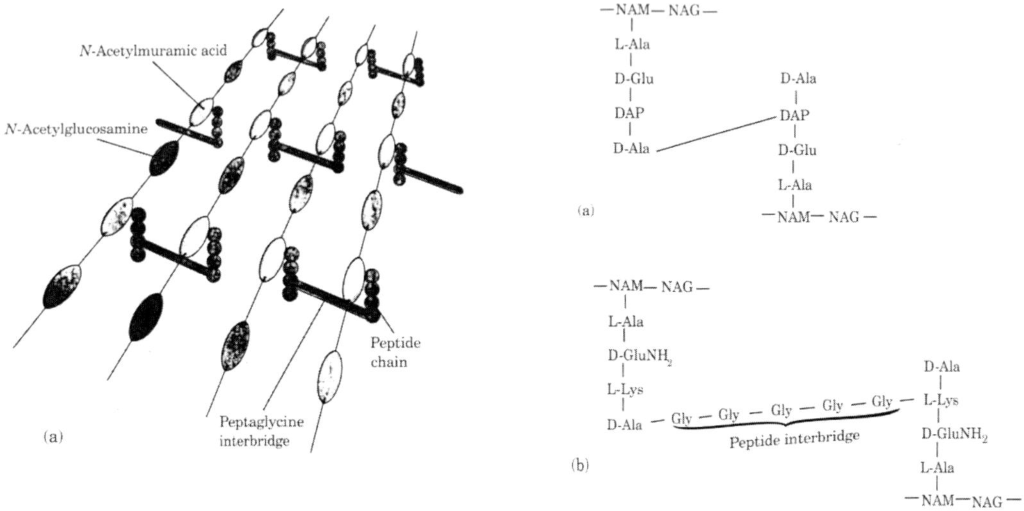

Fig. 2.2. Peptidoglycan cross-links (a) *E. coli* peptidoglycan direct cross-linking, typical of many Gram-negative bacteria. (b) *Staphylococcus aureus* peptidoglycan. *S. aureus* is a Gram-positive bacterium. *NAM* is N-acetylmuramic acid. *NAG* is N-acetylglucosamine Gly is glycine.

The peptidoglycan of Gram-negative bacteria can be isolated as a sac of pure peptidoglycan which surrounds the cell membrane in the living cell. It is called 'murein sacculus'. The sacculus is elastic and believed to be under stress *in vivo* because of the expansion due to turgor pressure against the cell membrane. The peptidoglycan in these bacteria is attached non-covalently to the outer envelope via lipoprotein. Evidences available to suggest that peptidoglycans are also attached to cell membrane by zones of adhesion.

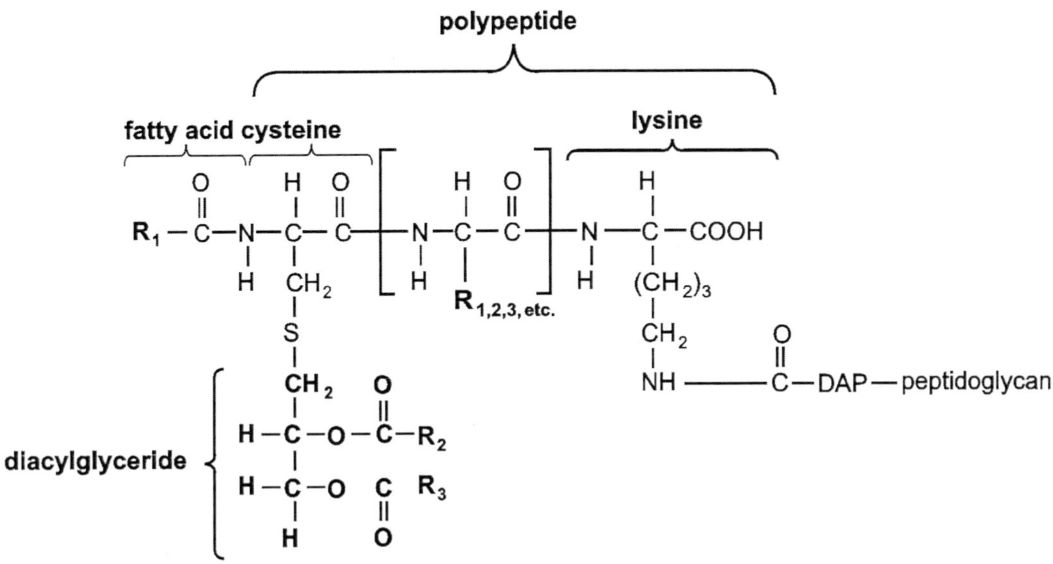

Fig. 2.3. Murein lipoprotein.

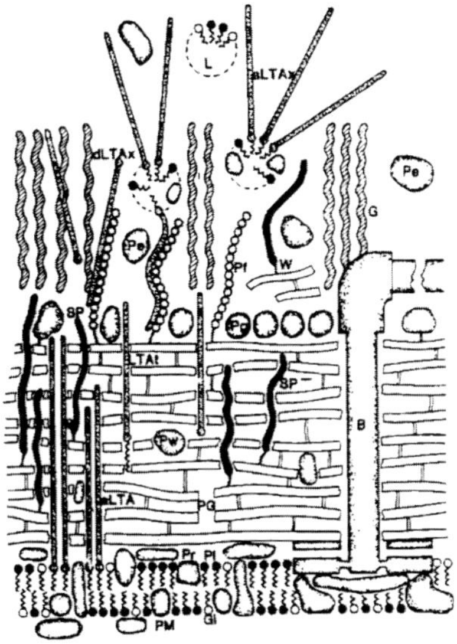

Fig. 2.4a. Gram + ve cell wall PM- plasma membrane, pr-protein Pl- phospholipid, PG-peptidoglycon, SP-polysaccharide, LTA-acetylated lipoteichoic acid, L-lipids, Pg-globular protein, G-glycocalyx, B-body of the flagellum

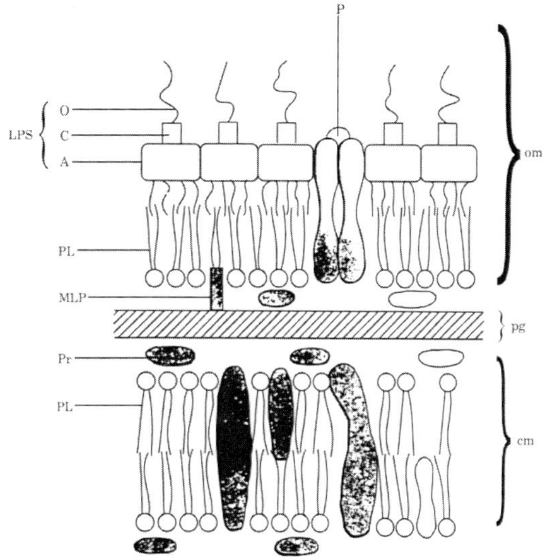

Fig. 2.4b. Gram-negative envelope. CM- cell membrane, PG-peptidoglycan, OM- outer membrane, Pl-phospholipid, pr-protein, MLP-murein lipoprotein, A-lipid A , C- core, LPS-lipopolysaccharide

2.3.1. Murein Lipoprotein

The murein of Gram-negative bacteria is substituted by lipoprotein. On an average, lipoprotein is attached to every tenth of muramic acid residue. The linkage is from the epsilon amino group of the lysine at the carboxyl terminus to the carboxyl group at the L-center of meso-diaminopimelic acid (Fig. 2.3). The murein lipoprotein is a member of the super family of lipoproteins that are characterized by a typical consensus processing site (LLLAGCSSNS). The murein lipoprotein is believed to play a structural role in keeping the outer membrane attached to the cell surface.

Some of the major differences between the Gram-positive and Gram-negative bacterial cell walls are shown in Fig. 2.4 a & b and presented Table 2.1.

Table 2.1. Differences between Gram-positive and Gram-negative bacterial cell walls

Character	Gram-positive	Gram-negative
Peptidoglycan	Thick layer	Thin layer
Peptidoglycan tetrapeptide	Most contain lysine	All contain diaminopimelic acid
Peptidoglycan cross linkage	Generally pentapeptide, entirely glycine	Direct bonding of diaminopimelic acid of one chain to the terminal D-alanine of another chain
Teichoic acid	Present	Absent

Character	Gram-positive	Gram-negative
Teichuronic acid	Present	Absent
Lipoproteins	Absent	Present
LPS (Lipopolysaccharide endotoxin)	Absent	Present
Outer membrane	Absent	Present
Periplasmic space	Absent	Present

2.4. GRAM-POSITIVE CELL WALL

The Gram-positive cell wall in bacteria is a thick structure, approximately 15–30 nm wide and consists of several polymers, the major one being peptidoglycan. The kind and amount of the other polymers in the wall vary with the genus of bacterium. The non-peptidoglycan polymers can comprise up to 50% of the dry weight of the wall. Several polymers are covalently bound to the glycan chain of peptidoglycan, which include.

i. Teichoic acid: Teichoic acid is a polymer of either ribitol phosphate or glycerol phosphate (Fig. 2.5 a & b)

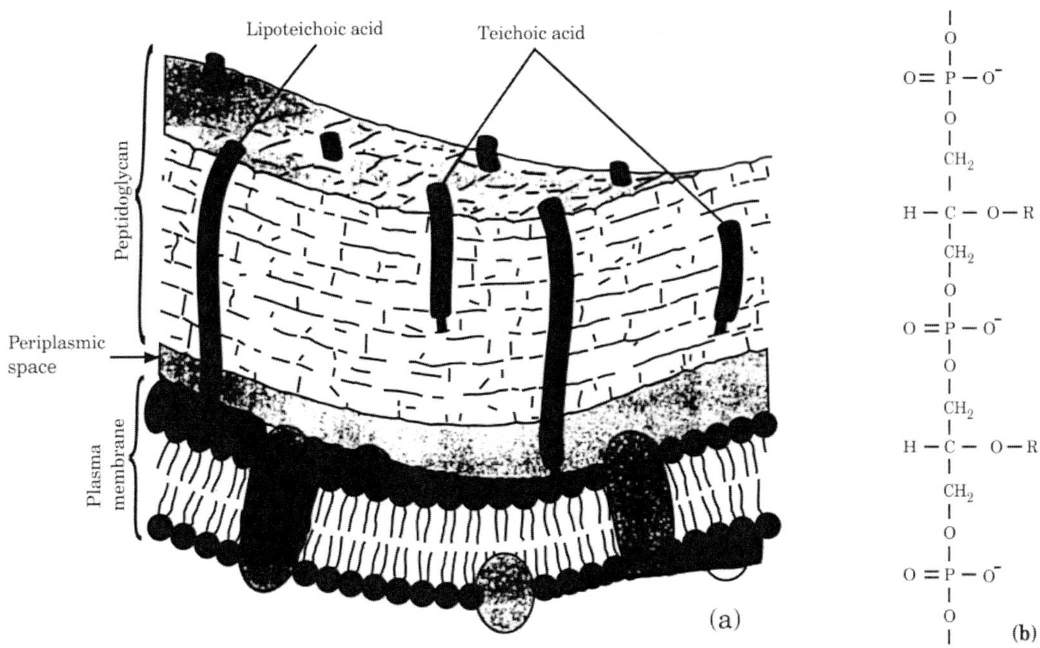

Fig. 2.5a. The Gram-positive envelope with lipoteichoic acid and teichoic acid
Fig. 2.5b. Teichoic acid structure. The segment of a teichoic acid made of phosphate, glycerol, and a side chain, *R*. *R* may represent D-alanine, glucose or other molecules.

ii. Teichuronic acids : These are acidic polysaccharides containing uronic acids.

iii. Neutral polysaccharides: These are particularly important for the classification of streptococci and lactobacilli, where they are used to divide the bacteria into serological groups.

iv. Lipoteichoic acid: A teichoic acid connected to the lipids of the plasma membrane is lipoteichoic acid (LTA)

v. Other glycolipids : In many bacteria, like *Micrococcus, Streptococcus, Mycobacterium, Corynebacterium* and *Propionibacterium*, LTA is replaced by other amphiphilic glycolipids. These are also called lipoglycans.

vi. Mycolic acid: It is a polymer found in the cell walls of *Mycobacteria*. Mycolic acid has a strong affinity for basic fuchsin, which is not removed by dilute HCl. They are, therefore, called acid fast bacteria.

The functions of most of the above polymers are largely unknown, although some may act as adhesins and/or presumably effect the permeability characteristic of the cell wall.

2.5. GRAM-NEGATIVE CELL WALL

Gram-negative cell walls are structurally and chemically more complex than Gram-positive cell walls. This cell wall consists of an outer membrane composed of lipopolysaccharides, phospholipids and proteins, and an underlying peptidoglycan layer. Between the outer and cell membrane is a compartment called periplasm where in the peptidoglycan lies.

i. Lipopolysaccharide : The best characterized lipopolysaccharides are those of *E. coli* and *Salmonella typhimurium*. A lipopolysaccharide is composed of three parts, (i) a hydrophobic region called lipid A. It is composed of a backbone of two glucosamine residues linked by 1,6 and esterified via hydroxyl groups to fatty acids, (ii) a core polysaccharide, whose composition is similar in all the Enterobacteriaceae and is connected to lipid A via 3-deoxy-D-mannooctusonate (KDO) and (iii) the distal polysaccharide region (O-antigen). It is made up of four to six sugars that vary considerably in composition between different strains of bacteria. The arrangement of the LPS in the outer envelope is shown in Fig 2.6. Thus, it is evident that LPS is anchored in the outer envelope by the hydrophobic lipid A region, while the repeating oligosaccharide protrudes in to the surroundings.

   LPS provides a permeability barrier to hydrophobic compounds. Loss of most of the core and oligosaccharide in enteric bacteria is associated with increased sensitivity to hydrophobic compounds (e.g. antibiotics, bile salts, and hydrophobic dyes such as eosin and methylene blue). Mutants devoid of oligosaccharide loose the virulence and it is believed that LPS can increase pathogenicity.

ii. Murein lipoprotein: Explained in section 2.3.1

iii. Porin and other proteins: The major proteins in the outer envelope are called porins (Fig. 2.6). The porins form small non-specific hydrophilic channels through the outer envelope allowing the diffusion of neutral and charged solutes of molecular weight less than 600D. The channels are necessary to allow passage of small molecules into and out of the cell. *E. coli* has three major porins : OmpF, OmpC and PhoE. Each porin makes a separate channel. Since, the porins exclude molecules with molecular weights larger than 600D, there are other proteins in the outer membrane that facilitate the translocation of larger solutes across the outer membrane. One such protein in *E. coli* is Lam B protein which forms channels for maltose and maltodextrins.

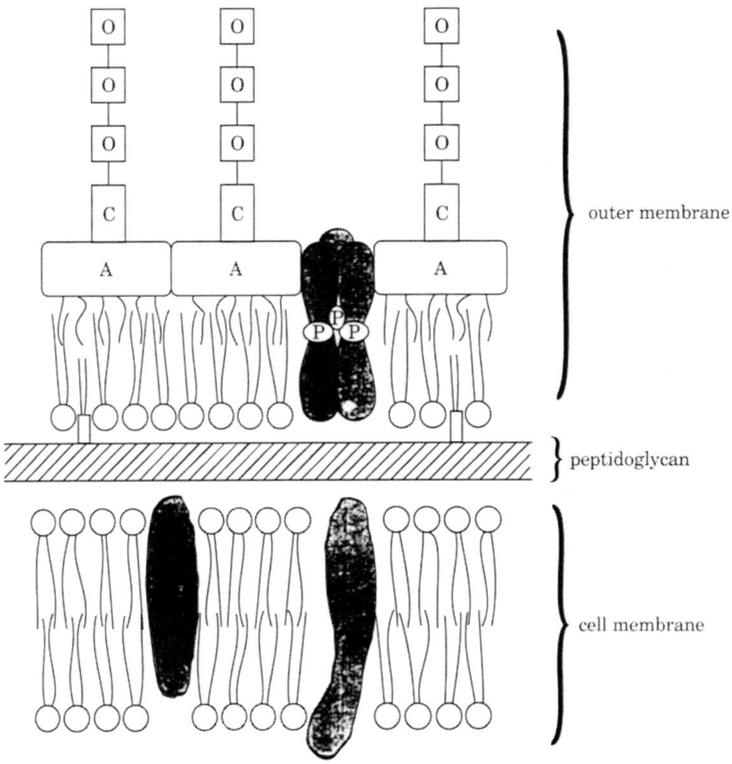

Fig. 2.6. The arrangement of the LPS in the outer membrane of gram-negative bacteria. A, lipid A; C, core; O, oligosaccharide; P, porin.

2.6. ARCHAEAL CELL WALL

Cell walls of archaebacteria are quite different from bacterial cell walls. No archaeal cell wall contains peptidoglycan. The cell walls of these bacteria, instead, contain pseudopeptido-glycan (pseudomurein) or protein (the S layer). Pseudopeptidoglycan resembles peptidoglycan in that it has the glycan chains cross linked by peptides (Fig. 2.7). However, the glycan chain strands consist of N-acetyl glucosamine or N-acetylgalactosamine (G) and N-acetyltalosaminuronic acid (T). Futher, the sugars are linked by a β, 1-3 glycosidic linkage rather than a β, 1- 4. The peptide bridges contain L-amino acids including lysine, glutamic acid, alanine, threonine and serine, rather than D-amino acids. As a consequence of this different chemical set up, the synthesis of pseudomurein is not inhibited by D-cycloserine, vancomycin or penicillin, which are typical inhibitors of murein synthesis.

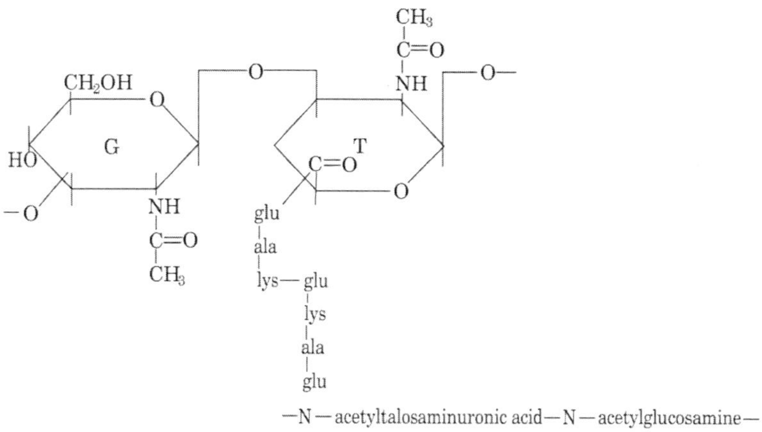

Fig. 2.7. Pseudopeptidoglycan in archaea.

Table 2.2. Differences between Gram-positive and archaeal cell walls

Character	Gram-positive	Archaebacteria
Major cell wall constituent	Peptidoglycan	Pseudopeptidoglycan
Monomers in peptidoglycan	N-acetyl muramicacid N-acetyl glucosamine	N-acetylglucosamine, N-acetylgalactosamine N-acetyltalosaminuronic acid
Linkage of sugars	β, 1-4 linkages	β, 1-3 linkages
Peptide bridges	D-amino acids	L-amino acids

The structural differences between Gram-positive and Gram-negative cell walls have significant physiological consequences. Murein constitutes the major portion of gram-positive cell walls and largely exposed to the external environment. As a consequence, lysozyme is generally more destructive to the structure of Gram-positive walls than for Gram-negative envelopes. In addition, antibiotics that interfere with murein synthesis are generally more effective with Gram-positive cells than with Gram-negative walls. Another significant difference between Gram-positive and Gram-negative cell walls is the degree to which the two structures participate in transport and selective permeability. The Gram-negative envelope is much more involved in these processes than the Gram-positive wall. The various receptors in the outer membrane of the Gram-negative bacterium facilitate selective accumulation of materials at the cell exterior, and combined with porins, hydrolytic enzymes and receptors in the periplasmic space allow the entrance of materials into the cell that might otherwise be excluded.

2.7. PROTOPLASTS AND SPHEROPLASTS

Investigations carried out on the physiological properties of Gram-positive and Gram-negative bacterial cells from which the walls are removed, and comparison with normal cells yielded much useful information in understanding the physiological functions of the wall and of the cell interior. Osmotically sensitive structures from Gram-positive cell are usually referred to as

protoplasts, while those obtained from Gram-negative cells are termed **spheroplasts**. Alternatively, when cell wall material is absent, the structure is known as protoplast, whereas osmotically sensitive spherical structures from which some, but not all, of the wall has been removed are referred to as spheroplasts. As discussed earlier, and because of distinctive nature of Gram-positive cells, it is easier to obtain complete wall removal with Gram-positive cells than with Gram-negative cells.

Properties of intact cells, protoplasts and spheroplasts are given in Table 2.3.

Table 2.3. Properties of intact cells, spheroplasts and protoplasts

Property	Intact cells	Spheroplasts	Protoplasts
Cell division	Possible	Possible in some cases	Limited or not possible when possible normal cells are formed
Cell wall regeneration	Possible	Possible in some cases	Impossible
Phage attachment	Possible	Some times possible	Impossible
Phage maturation	Possible	Possible	Possible
Spore induction	Possible	Possible	Impossible
Spore maturation	Possible	Possible	Possible once induced
DNA replication	Yes	Yes	Yes
Protein synthesis	Yes	Yes	Yes
Transport and relative permeability	Yes	Yes	Yes

2.8. SYNTHESIS OF GRAM-POSITIVE WALL

Gram-positive cell wall, a homogenous thick layer, predominantly consists of peptidoglycan and covalently bound teichoic acid or teichuronic acids. Biosynthesis of these two components is discussed separately here under.

2.8.1. Synthesis of Peptidoglycan

Biosynthesis of peptidoglycan is accomplished in four stages and in three cellular compartments

i. Synthesis of precursors of peptidoglycan in cytosol
ii. Transfer of precursors to a lipid carrier across the cell membrane
iii. Polymerization of precursors into peptidoglycan on the outer surface of the membrane
iv. Cross-linking of peptidoglycan strands by transpeptidation.

2.8.1.1 Synthesis of precursors

The two amino sugar precursors to the peptidoglycan are N-acetylglucosamine and N-acetylmuramic acid. Both these amino sugars are made from fructose-6-phosphate, an intermediate of glycolysis (Fig. 2.8). In step 1, glutamine donates an amino group to fructose-6-phosphate, converting it to glucosamine-6-phosphate. Then in step 2, a transacylase transfers an

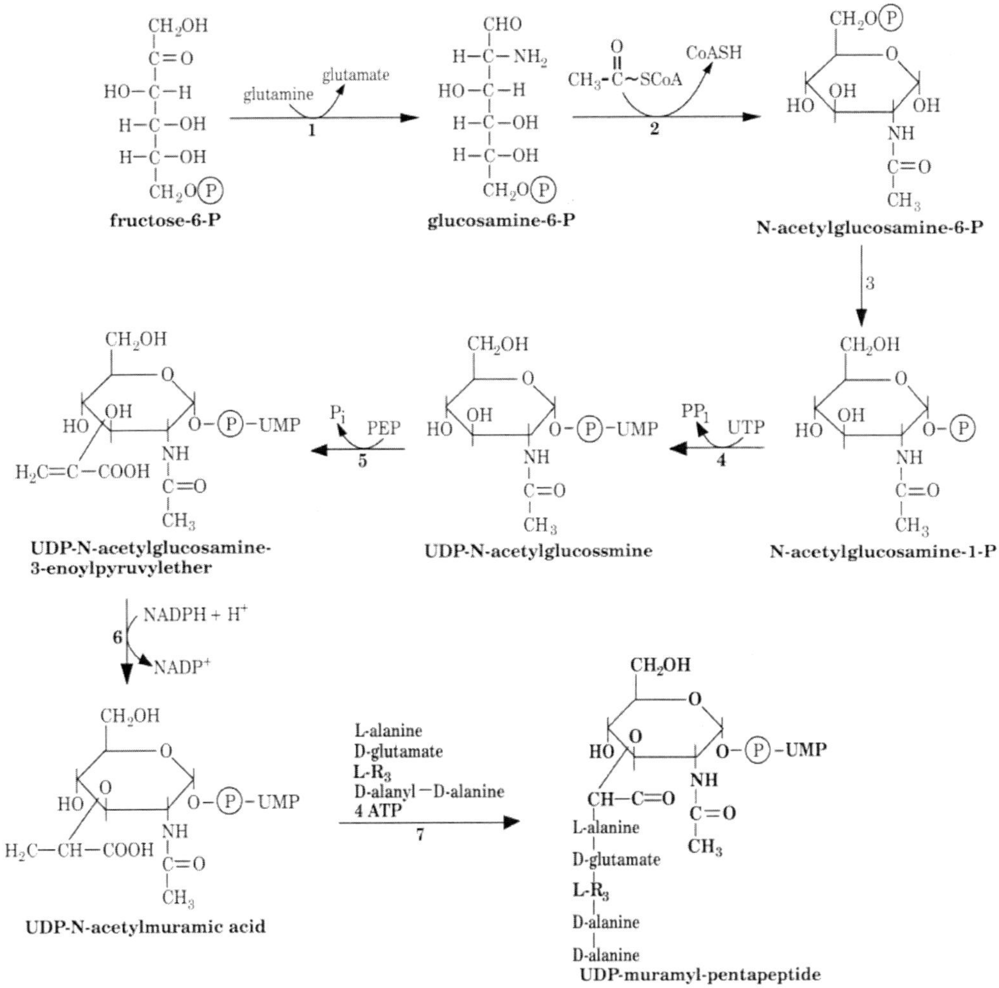

Fig. 2.8. Synthesis of N-acetylglucosamine and N-acetylmuramyl-pentapeptide. Enzymes: (1) glutamine: fructose-6-phosphate aminotransferase; (2) glucosamine phosphate transacetylase; (3) N-acetylglucosamine phosphomutase; (4) UDP-N-acetylglucosamine pyrophosphorylase; (5) enoylpyruvate transferase; (6) UDP-N-acetylenoylpyruvylglucosamine reducatse. Step 7: The UDP-N-acetylmuramic acid is converted to the pentapeptide derivative by the sequential additions of L-alanine, D-glutamate, L-R, and D-alanyl-D-alanine by separate enzymes.

acetyl group from acetyl-CoA to the amino group on glucosamine-6-phosphate to make N-acetyl glucosamine-6-phosphate, which is isomerized in step 3 to N-acetylglucosamine-1-phosphate. The latter attacks UTP displacing pyrophosphate in step 4 to form UDP-N-acetylglucosamine. The reaction is driven to completion by a pyrophosphate. Here, some of the UDP-N-acetylglucosamine (UDP-GlcNAC) is used as the precursor to the N-acetylglucosamine in peptidoglycan and some is converted to UDP-N-acetylmuramic acid (UDP-MurNAC). The UDP-GlcNAC is converted to UDP-MurNAC by the addition of a lactyl group to the sugar in step 5. In this reaction, the C_3 OH of the sugar displaces the phosphate from the α carbon of phosphoenol pyruvate forming the enol pyruvate ether derivative of the UDP-N-acetylmuramic acid. Then, in step 6, the enol derivative is reduced to the lactyl moiety by NADPH. The UDP-MurNAC is converted to UDP-MurNAC-pentapeptide by the sequential addition of five amino acids, L-alanine, D-glutamate, L-R_3 (residue 3), and the dipeptide D-alanyl-D-alanine in step 7. Each reaction is catalyzed by a separate enzyme and requires ATP to activate the carboxyl group of the amino acid. The activated carboxyl is probably an acyl phosphate. The products are ADP and inorganic phosphate (The fifth D-alanine is removed during the cross-linking). The D-alanine-D-alanine is made by separate enzymes viz., racemase and synthetase which are inhibited by the antibiotic cycloserine.

2.8.1.2. Transfer of precursors to lipid carrier

In the second stage, the MurNAC-pentapeptide is transferred to the lipid carrier in a membrane. The lipid carrier is a undecaprenyl phosphate or bactoprenol. Bactoprenol not only serves as a carrier for peptidoglycan precursors but also serves as a carrier for the precursors of other cell wall polymers (lipopolysaccharide and teichoic acids). The nucleotide sugars (UDP-M and UDP-G) diffuse to the membrane where undecaprenyl phosphate (lipid-P) attaches the UDP-MurNAC-pentapeptide displacing UMP (Fig. 2.9). The product is lipid-P-murNAC-pentapeptide. The GlcNAC is then transferred from UDP-GlcNAC to the MurNAC on the lipid carrier. The product is the disaccharide precursor to the peptidoglycan, lipid-pp-murNAC pentapeptide-GlcNAC. Now, the lipid disaccharide moves to the other side of the membrane.

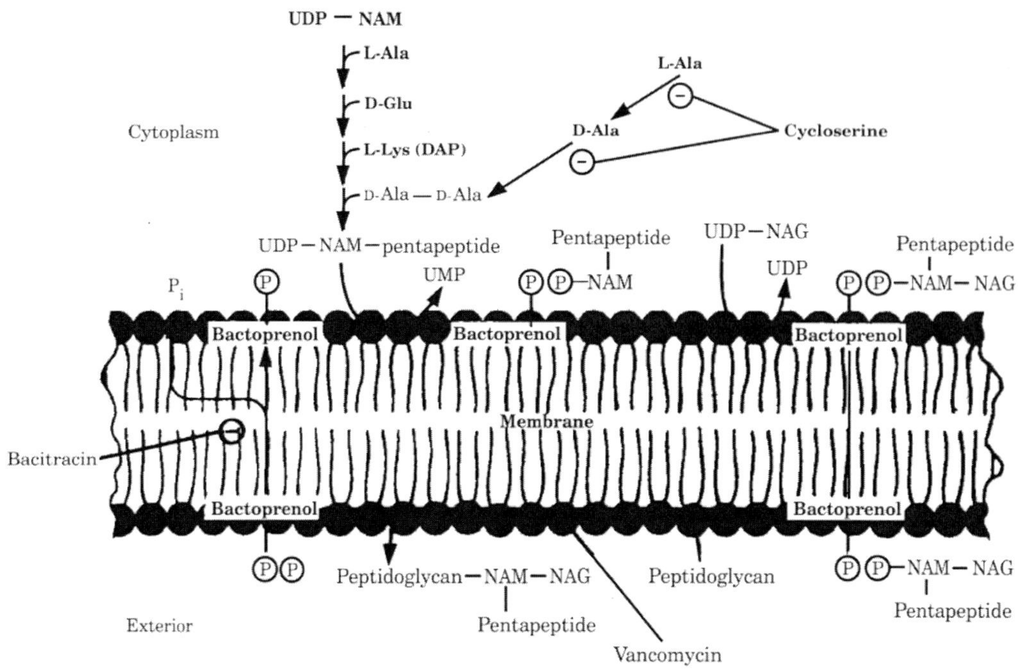

Fig. 2.9. Peptidogylcan synthesis. NAM is N-acetylmuramic acid and NAG is N-acetylglucosamine. The pentapeptide contains L-lysine in *S. aureus* peptidoglycan, and diaminopimelic acid (DAP) in *E. coli*.

2.8.1.3. Polymerisation of precursors into glycan

On the outer surface of the membrane, the lipid disaccharide is transferred to the growing end of the acceptor glycan chain. This transfer is brought about by a transglycoylation where the C_4 hydroxyl of the incoming GlcNAC attacks C_1 of the MurNAC in the glycan, displacing the lipid-pp from growing glycan chain. This reaction is catalyzed by a membrane bound enzyme called transglycosylase. The translocation of lipid linked disaccharide across the cytoplasmic membrane is inhibited by antibiotic **vancomycin**, which binds to the D-ala-D-alanine terminus of the precursor. The lipid-pp released from the growing glycan chain is hydrolyzed by a membrane bound **pyrophosphatase** to lipid-p and pi. This reaction is considered very important because it helps to drive the transglycosylation reaction to completion, since the hydrolysis of the phospho-diester results in the release of substantial energy. This hydrolysis also regenerates the lipid-p that is necessary for continued growth of the peptidoglycan as well as other cell wall polymers. The hydrolysis is inhibited by the antibiotic **bacitracin**.

2.8.1.4. Cross-linking of peptidoglycan strands by transpeptidation

Cross-linking of peptidoglycan strands is brought about by transpeptidation (Fig. 2.10). In this reaction, an-NH_2 group from the diamino acid in the 3 position (e.g. DAP) form bond with the carboxyl carbon in the peptide bond holding the two D-alanine residues together in the pentapeptide and displaces the terminal D-alanine. The result is a new peptide bond. The transpeptidation reaction is inhibited by **penicillin**.

2.8.2. Synthesis of Teichoic Acid and Teichuronic Acid

Teichoic acids are the second major components of the Gram-positive cell wall. As explained earlier, they are composed of polymers of ribitol or glycerol connected by phosphodiester linkages. Ribitol containing polymers are connected to the wall through covalent linkages, whereas glycerol containing polymers are attached to glycolipid in the membrane.

A general picture regarding the synthesis of teichoic acid and teichuronic acid is slowly emerging out. However, controversy exists regarding the precise details of both processes.

In step-I, sugar or sugar acid moiety is converted to sugar or sugar acid 6-phosphate. The later is isomerized to a 1-phosphate moiety (step-2). In the next step, sugar-1-phosphate moiety is transferred to uridine triphosphate with the release of pyrophoshate (step-3). In the fourth step the sugar phosphate is transferred to undecaprenyl diphosphate (UDPRPP). This results in the release of uridine monophosphate. In the fifth step, sugar residues from uridine diphosphate sugar moiety are transferred sequentially to the UDPRPP-bound building block with concomitant release of uridine diphopshate in every step. These events lead to the formation of UDRPP-bound teichoic acid polymer. Now this teichoic acid polymer is transferred to N-acetylmuramic acid polypeptide (NAMPP) with release of UDPRPP (step-6), In some cases UDPRPP is converted to undecaprenol monophosphate (UDPRP), which is recycled to accept the sugar phosphate. Most of the teichoic acids are modified by the addition of amino acids, notably D-alanine, and sugars to hydroxyl groups of the ribitol or glycerol residues. Although modification of teichoic acids is very common, great variation is found in nature and extent to which modification occurs.

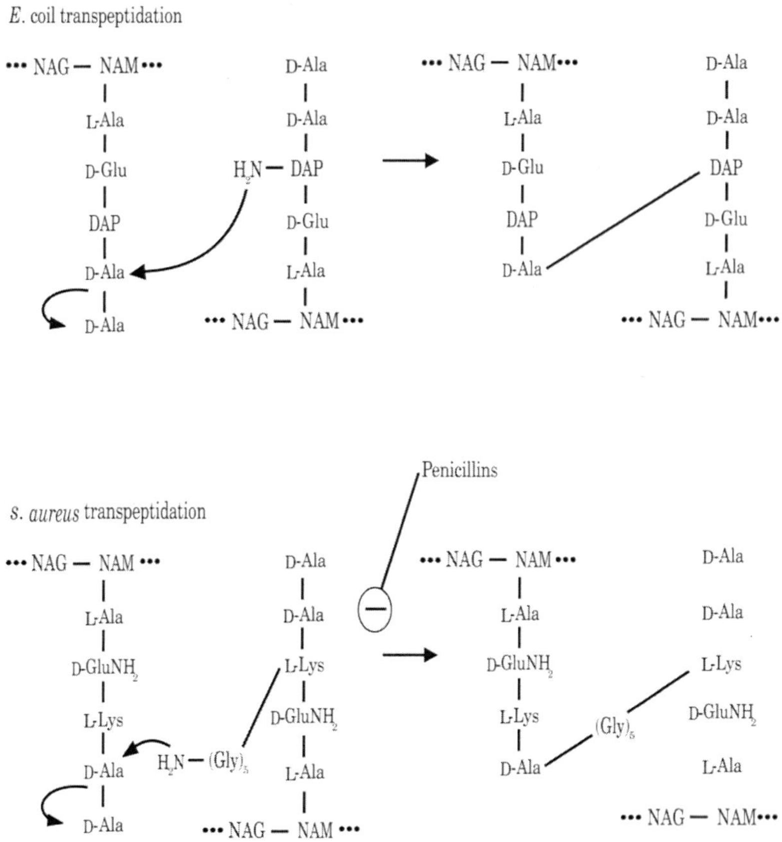

Fig. 2.10. Transpeptidation in *Escherichia coli* and *Staphylococcus aureus*.

2.9. SYNTHESIS OF GRAM-NEGATIVE CELL WALL

The relative morphological complexity of the Gram-negative envelope is accompanied by more complicated biosynthetic and assembly mechanisms than those required for Gram-positive wall synthesis. Synthesis of the Gram-negative envelope requires all of the processes necessary for a Gram-positive bacteria and quite a few more. Unlike Gram-positive organisms, the Gram-negative organisms additionally must form lipopolysaccharides. However, the accomplishment of certain tasks is similar in both the organisms. For example, both types of organisms synthesize precursors of polymers in the cytoplasm and have the mechanism to transport polymer building blocks to their addition sites. The Gram-positive organisms transport only murein precursors, while Gram-negative organisms transport both murein and lipopolysaccharide precursors.

Synthesis of murein precursors and their transport across the membrane has been described earlier. Synthesis of lipopolysaccharide components, transport and their assembly are discussed in this section.

Synthesis of lipopolysaccharide may be divided into three major sub-processes:

i. Synthesis of lipid A

ii. Synthesis of core polysaccharide, and

iii. Synthesis of the O-antigen

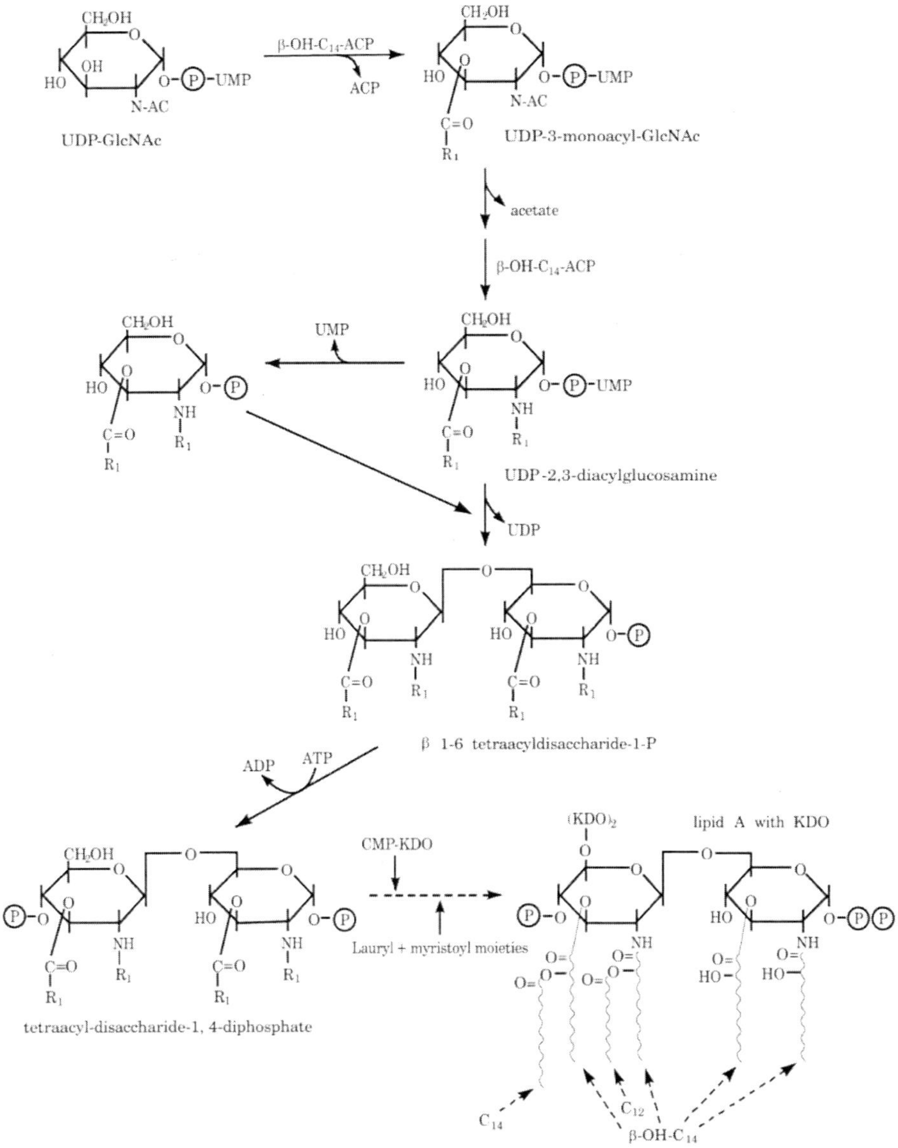

Fig. 2.11. Model for synthesis of lipid A in *E. coli*.

2.9.1 Synthesis of Lipid A

The lipid A portion of the lipopolysaccharide is synthesized from UDP-GlcNAC, which inturn is made from fructose-6-phosphate. A model for lipid A synthesis in *E. coli* as proposed by Raetz is presented in Fig. 2.11. It is believed that lipid A is made in the cytoplasmic membrane by peripheral membrane proteins. β-OH myristic acid is transferred from an ACP derivative to C_3 of UDP-GlcNAC to form the monoacyl derivative. Then, the acetate from the nitrogen on C_2 is removed and a second molecule of β-OH myristic is transferred to the nitrogen to form the 2,3-diacyl derivative. Some of the latter looses UMP and is converted to 2,3-diacylglucosamine-1-P which condenses with UDP-2, 3-diacylglucosamine to form the disaccharide linked in β-1,6 linkage. A phosphate is added to form the 1,4-diphosphate derivative and this is then modified by the addition of KDO and the esterification of fatty acids to the OH of the β-hydroxymyristic moieties. However, these final modifications are not well characterized.

2.9.2 Synthesis of Core Polysaccharide

The core in the Enterobacteriaceae members contains an inner region consisting of KDO, heptose, ethanolamine and phosphate and an outer region that consists of hexoses. The biosynthesis of the inner region is not fully understood. The outer core region grows as hexose units which are donated one at a time from nucleoside diphosphate derivatives to the non-reducing end of the growing glycan chain attached to the KDO. The addition of each sugar is catalyzed by a specific glycosyl transferase, which is membrane bound.

2.9.3. Synthesis of O-antigen

The O-antigen region is synthesized by a mechanism that is different from that of core synthesis (Fig. 2.12). The O-antigen is synthesized as a separate polymer on a lipid carrier and then transferred as a unit to the core. The lipid carrier is undecaprenyl phosphate (bactoprenol). First, the repeat unit of the O-antigen is synthesized on the lipid carrier. This is done by a series of consecutive reactions in which a sugar moiety is transferred from a nucleoside diphosphate carrier to the non-reducing end of the growing repeating unit. Each of these reactions is catalyzed by a different enzyme. Then the repeat unit is transferred to the growing oligosaccharide chain. The lipid-pp of the acceptor oligosaccharide chain is displaced and enters the lipid pyrophosphate pool where it is hydrolyzed to lipid-p by a bacitracin-sensitive enzyme. Finally, the completed oligosacc-haride chain is transferred to the lipid A-core displacing lipid pyrophosphate.

2.9.4. Assembly of the Three Components and Formation of Lipopolysaccharide

All the three individual components of the lipopolysaccharide are synthesized on the cytoplasmic side of the membrane. Then they are transferred to periplasmic side where they are assembled. A model for lipopolysaccharide assembly is depicted in Fig. 2.13. The O-antigen tetrasaccharide subunit is probably synthesized on the lipid carrier on the cytoplasmic side of the membrane and then moves to the periplasmic side, where it is added to the growing O-antigen anchored to the membrane by its lipid carrier. The lipid carrier on the growing oligosaccharide is displaced. The core-lipid A region may also be assembled on the cytoplasmic surface and translocated to the periplasmic surface. The final completion of the lipopolysaccharide would take place on the periplasmic surface, where the O-antigen is transferred to the core-lipid A, displacing lipid-pp. The lipid is regenerated via a phosphatase and enters a common pool of lipid-p also used in the biosynthesis of peptidoglycan. The details of movement of lipopolysaccharide through periplasm into the outer envelope are not known.

2.10. EXOPOLYSACCHARIDES

Exopolysaccharides are produced by the great majority of bacteria and also by variety of eukaryotes. These, exopolysaccharides may form capsules or slime layers. When morphologically discrete they are referred to as **capsules**. However, when they are extruded into the extracellular environment and are not cell associated, they are considered **slime layers**. Though many workers do not consider exopolysaccharides as a part of the cell, but none can deny their physiological importance.

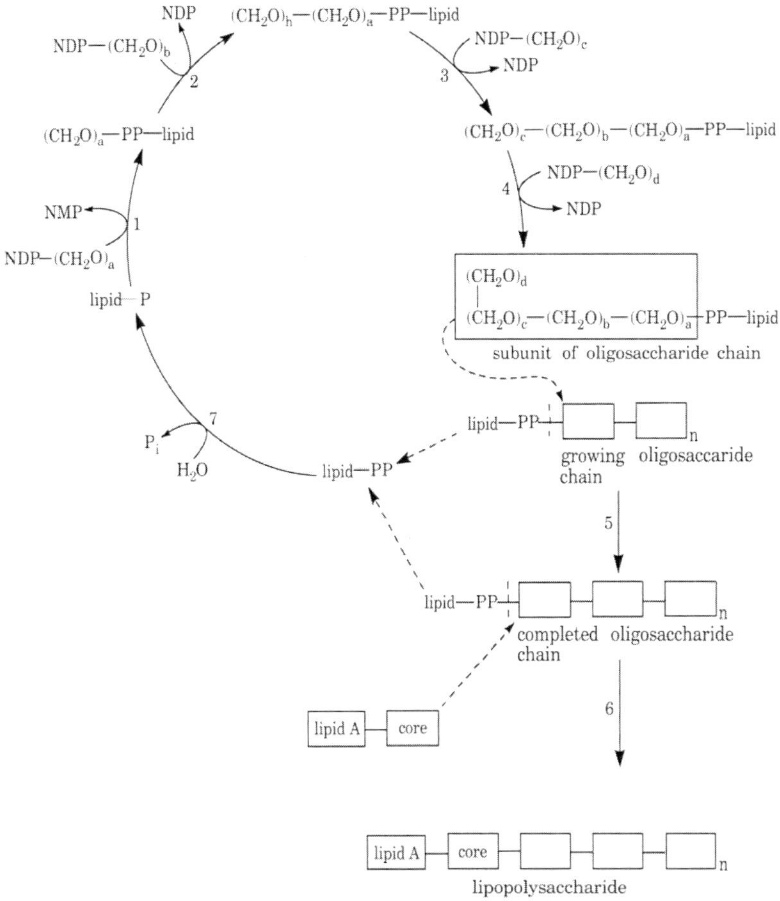

Fig. 2.12. Synthesis of O-antigen and attachment to core in *S. typhimurium.* The O-antigen is polymerized on the same phospholipid carrier that is used for peptidoglycan synthesis. Each sugar is added from a nucleotide diphosphate derivative using a specific sugar transferase (reactions 1-4). The nucleotide that is used depends upon the sugar and the specific glycosly transferase. When the O-antigen tetrasaccharide subunit is finished, it is transferred to the growing O-antigen chain by an enzyme called O-antigen polymerase (reaction 5). The completed O-antigen is transferred from its lipid carrier to core-lipid A by an enzyme called O-antigen: lipopolysaccharide ligase (reaction 6). The lipid PP that is displaced is hydrolyzed by a bacteria sensitive phosphatase (reaction 7). All of the reactions take place in the cell membrane.

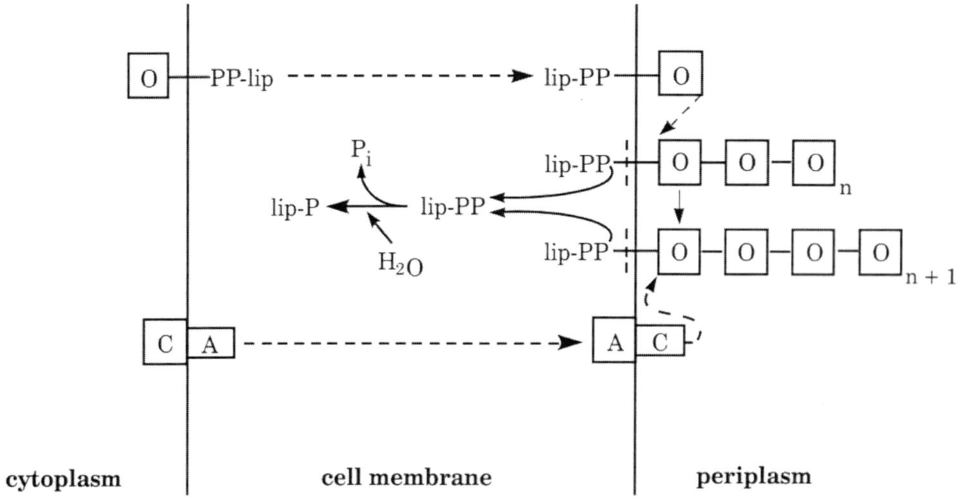

Fig. 2.13. A model for LPS assembly. O,O-antigen; C, core; lip, lipid carrier; A, lipid A.

2.10.1 Synthesis of Exopolysaccharides

Exopolysaccharide synthesis is a process that involves both the external and internal environment. It involves uptake of materials from environment, intracellular polymer synthesis, and return of the completed polymer to the cell surface. Exopolysaccharide synthesis thus, serves as a model for understanding transport of both micro-and macromolecules.

Exopolysaccharide synthesis and formation is influenced by a number of physiological factors and also by the external conditions. In general, microbes form exopolysaccharides from a variety of starting materials like carbohydrates, amino acids and hydrocarbons. It is an interesting and noteworthy fact that exopolysaccharides formation may occur in organisms that do not use carbohydrates efficiently.

Five modes of exopolysaccharide synthesis are reported. In four methods, synthesis occurs intracellularly using externally supplied precursors. They require nucleotide-activated sugar precursors. Some organisms produce intracellularly synthesized hetero-polysaccharide from a single specific extracellular substrate. Others produce homogenous intracellularly formed polymers from a variety of extracellular substrates. A third mode of nucleotide-sugar-mediated exopolysaccharide formation produces an intracellular heteropolymer with a repeating unit from a variety of extracellular precursors. The fourth of nucleotide-sugar-mediated exopolysaccharide formation produces intracellular polymers composed of two monomeric units, but with no repeating units.

In the fifth mode, synthesis occurs at the cell surface. It does not require the transport of either precursors to the intracellular environment or of the completed polymer to the cell surface. This type of synthesis is exhibited by species of *Streptococcus* and *Leuconostoc*. The organisms require a specific carbohydrate, sucrose, and produce exopolysaccharide with the aid of two enzymes viz., dextran sucrase and levan sucrase. With dextran sucrase, the sucrose molecule is cleaved to a dextran (glucose) polymer and free fructose is liberated. With levan sucrase, the process is inverse with the formation of a polymer of fructose (levulose) and free glucose. This mode of exopolysaccharide synthesis requires neither nucleotide activated sugars nor lipid carrier molecules.

As explained earlier, in most cases a nucleotide diphosphate is required for polymer formation. UDP and GDP derivatives are most commonly involved. In prokaryotes, UDP-glucose serves as the precursor for both exopolysaccharide and lipopolysaccharide synthesis. Glucose-1-phosphate from which UDP-glucose is formed, is the most important carbohydrate in prokaryotes. Exopolysaccharide synthesis also requires participation of lipid carrier molecules as intermediates. Studies carried out on the nature of lipid carrier revealed that more than one type of lipid might be required for exopolysaccharide synthesis in different organisms. Still, in a number of systems lipid involvement may not be necessary for exopolysaccharide formation.

Polymers, from the building blocks, are formed by two mechanisms.

i. Addition of the building block to the non-reducing end of the growing polymer e.g., *E. coli.*

ii. Addition of the building block to the reducing end of the existing growing polymer e.g., *Salmonella* species

2.10.2 Excretion of the Exopolysaccharides

The exact mechanism by which intracellularly synthesized exopolysaccharides are excreted to the cell surface or to external environment is not known. At present, it is presumed that Bayer's junctions are substantially involved in Gram-negative bacteria, but other mechanisms must exist in Gram-positive species. Perhaps, lipid carrier mediated and proton motive force facilitated processes occur in certain organisms.

Lastly, the question remains to be answered is how these excreted polysaccharides are attached to the cell surface. Three locations are attributed for polymer attachment: (i) attachment to the cell wall surface, (ii) attachment within the cell wall and (iii) attachment interior to the cell wall. However, the details of precise mechanism of their attachment and site are still to be elucidated.

2.11 FURTHER READING

Beveridge, T.J. (1995), The periplasmic space and the periplasm in gram-positive and gram-negative bacteria. *ASM News*, 61(3) : 125-130.

Ghuysen, J.M. and R. Hackenbeck (ed) (1994), Bacterial cell wall. New York: Elsevier.

Hancock, R.E.W. (1991), Bacterial outer membranes. Evolving concepts. ASM News, 57(4) : 175-182.

Holtje, J. (2000), Bacterial cell walls. In Encyclopedia of Microbiology Vol. I 2nd edn. J. Lederberg (ed.) 759-771. San Diego Academic Press.

Jennings, D.R. (1988), Turnover of cell walls in microorganisms. Microbiological Reviews, 52: 554-567.

Michelis. S., and Beckwith (1982), Mechanisms of incorporation of cell envelope proteins in *Escherichia coli*. *Annu. Rev. Microbiol.*, 36 : 435-465.

Navare, W.W. and Schneewind, O. (1999), Surface proteins of gram-positive bacteria and mechani-sms of their targeting to the cell wall envelope. *Microbiol Mol. Biol. Rev.*, 63 (1) : 174-229.

Osborne, M.J. and H.C.P. Wu (1980), Proteins of the outer membrane of gram-negative bacteria. *Annu. Rev. Microbiol.* 34 : 369-422.

Raetz, C.R.H. (1987), Structure and biosynthesis of lipid A in *Escherichia coli*. In *Escherichia coli* and *Salmonella typhimurium* : Cellular and Molecular Biology. F.C. Neidhardt, J.L. Ingraham, K.B. Low, M.Magasanik, M. Schaechter and H.E. Umbarger (eds.). American Society for Microbiology, Washington, DC.

Schockman, G.D. and J. Barret (1983), Structure, function and assembly of cell walls of gram-positive bacteria. *Annu. Rev. Microbiol.,* 37: 501-528.

Ward, J.B. (1983), Teichoic and Teichuronic acids : Biosynthesis, assembly and location. *Microbiological Reviews,* 45 : 211-243.

■■■■

Chapter - 3

Cell Metabolism — An Overview

3.1. METABOLISM

Metabolism is the total of all chemical reactions occurring in the cell. It is a highly coordinated cellular activity in which many multienzyme systems (metabolic pathways) cooperate to perform four basic functions:

i. Obtain chemical energy by capturing solar energy or degrading energy – rich nutrients from the environment.

ii. Convert nutrient molecules into the cell's own characteristic molecules, including the precursors of macromolecules.

iii. Polymerize monomeric precursors into macromolecules: proteins, nucleic acids, lipids and polysaccharides.

iv. Synthesize and degrade molecules required in specialized cellular functions, such as membrane lipids, intracellular messengers and pigments.

The above four functions can be broadly categorized into two types:

i. Catabolism

ii. Anabolism

Catabolism is the degradative phase of metabolism in which complex organic molecules (carbohydrates, fats, proteins) are broken down into smaller, simpler end products. Catabolic pathways release energy. Some of this energy is trapped in the formation of ATP and reduced electron carriers ($NADH_2$, $NADPH_2$ and $FADH_2$) and used for various cellular functions, including to drive the anabolic reactions. The remainder of energy is dissipated as heat.

Anabolism is the synthetic phase of metabolism in which small, simplest precursors are built up into larger and more complex molecules like lipids, polysaccharides, proteins and nucleic acids. Anabolic reactions require an input of energy, generally in the form of phosphoryl group transfer potentials of ATP and the reducing power ($NADH_2$, $NADPH_2$ and $FADH_2$). The differences between anabolism and catabolism are précised in Table 3.1.

A simplified view of cell metabolism is shown in the Fig. 3.1, which depicts how catabolic degradative reactions supply energy needed for cell functions and how anabolic reactions bring about the synthesis of cell components from nutrients. In anabolism, nutrients from the environment are absorbed and converted into cell components. In catabolism energy stored in macromolecules is released to drive anabolic reactions and other cell functions. Simultaneously, some waste products are also produced which are released into the environment.

Table 3.1. Differences between anabolism and catabolism

	Anabolism	Catabolism
(i)	Synthesis of macromolecules from simpler molecules in a living cell.	Breakdown of macromolecules into micromolecules
(ii)	Inorganic compounds are converted into simple organic compounds. Simple organic compounds are converted into complex organic compounds	Complex organic compounds are broken down into simple organic compounds. Simple organic compounds are converted into inorganic compounds.
(iii)	It is an endergonic process and requires input of energy	It is an exergonic process and associated with release of energy
(iv)	Photosynthesis, protein synthesis, nucleic acid synthesis.	Respiration

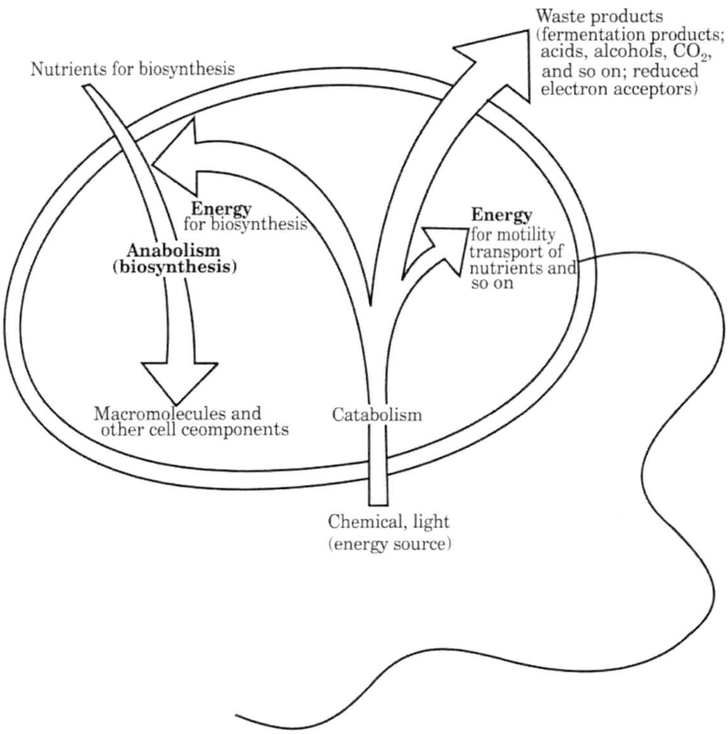

Fig. 3.1. A simplified view of cell metabolism

Metabolic pathways may be linear or branched and even cyclic. In some, a single precursor yields many end products and in others many precursors yield a single end product. Still some pathways, like TCA cycle are cyclic. In general catabolic pathways are convergent and anabolic pathways are divergent. In cyclic metabolic pathways, for example TCA cycle, the starting component of the pathway is regenerated through a series of reactions. In amphibolic pathways, that function both catabolically and anabolically, both are reciprocally regulated, *i.e.,* when one occurs the other is suppressed. Catabolic and anabolic pathways that connect the same two end points may employ many of the same enzymes, but invariably atleast one of the steps is catalysed by different enzymes (in the catabolic and anabolic directions) and these enzymes are the sites of separate regulation. Moreover, for both anabolic and catabolic pathways to be essentially irreversible, the reactions unique to each direction must include at least one that is thermodynamically very favourable (Fig. 3.2). As a further contribution in the separate regulation of catabolic and anabolic reaction sequences, paired anabolic and catabolic pathways commonly occur in different cellular component compartments. The concentration of intermediates, enzymes and regulators can be maintained at different levels in these compartments. Finally, anabolic and catabolic pathways often use different cofactors.

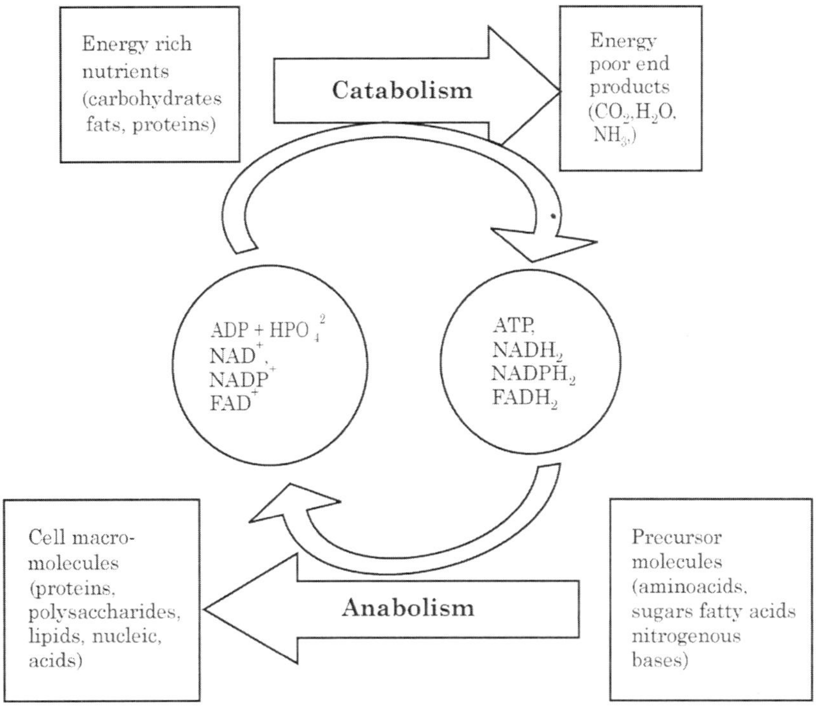

Fig. 3.2 : Energy relation between anabolic and catabolic pathways

Catabolism may be further divided into three stages (Fig 3.3). In the first of the catabolism, larger nutrient molecules (proteins, polysaccharides and lipid) are hydrolysed or broken down into their constituent parts. The chemical reactions occurring during this stage do not release much energy. In the second stage, amino acids, monosaccharides, fatty acids, glycerol and other products of the first stage are degraded to a few simpler molecules. During this stage metabolites like pyruvate, acetyl coenzyme A, and intermediates of tricarboxylic acid are formed. This process can be operated either aerobically or anaerobically. Finally, nutrient carbon is fed into the

TCA cycle during the third stage of catabolism and molecules are oxidised completely to CO_2 with the production of ATP, $NADH_2$ and $FADH_2$. This stage is responsible for the generation of most of the energy. Microorganisms exhibit a great diversity in the catabolic process. Most pathways closely resemble their counterparts in higher organisms. The uniqueness of microbial metabolism lies in the diversity of sources from which it generates ATP and $NADH_2$.

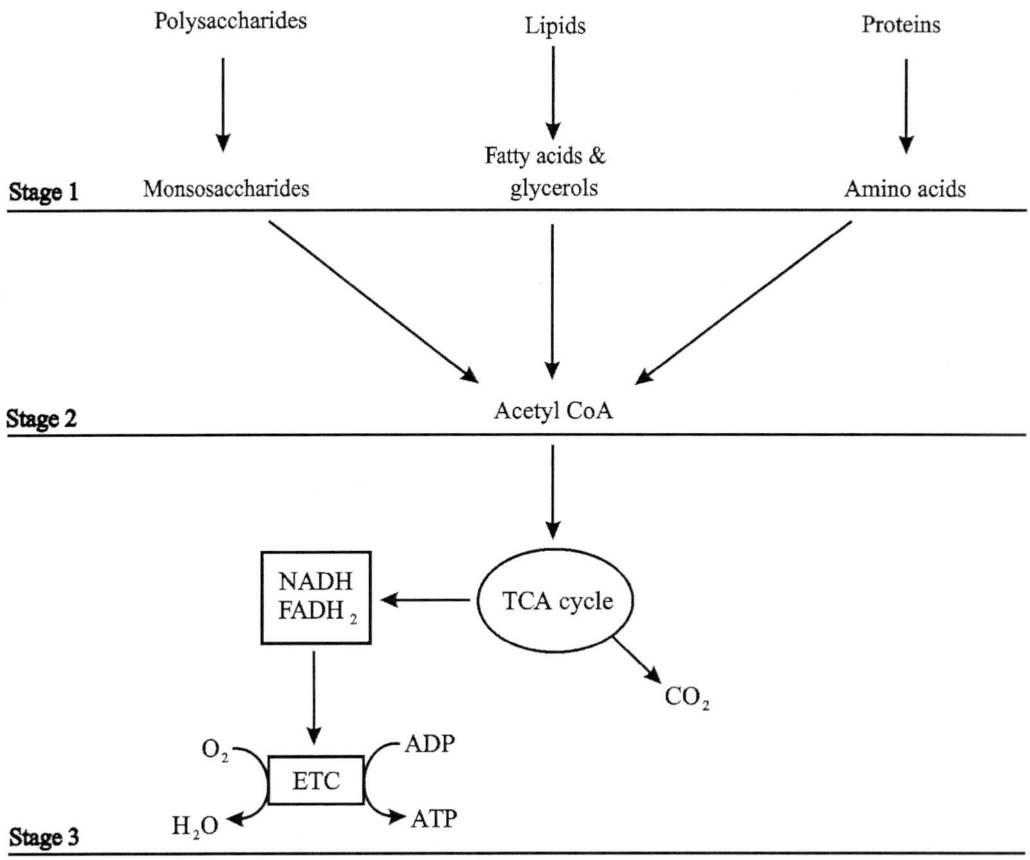

Fig. 3.3. Three stages of catabolism

Anabolism in a cell is highly ordered process and immensely complex. As such much energy is required for this process. A rapidly growing microbial cell like *E. coli* has an astonishing rate of biosynthesis (Table 3.2). The order of anabolism in living organisms is presented in Fig 3.4. The simple organic molecules or monomers are synthesized from inorganic molecules. This stage, as in photosynthesis and chemosynthesis, needs input of a lot of external energy. In the second stage macromolecules are constructed from their respective monomers. Macromolecules, in turn are assembled into larger, more complex structures such as supramolecular systems and organelles. Macromolecules normally contain the necessary information to form supramolecular system or organelle spontaneously in a process known as self-assembly. For example, ribosomes are large assemblages of many proteins and RNA molecules. They arise by the self assembly of their components without the involvement of external factors.

Table 3.2. Biosynthesis of different macromolecules in *E. coli.*

Cell constituent	Number of molecules per cell	Molecules synthesized per second	Molecules of ATP required per second for synthesis
DNA	1	0.00083	60,000
RNA	15,000	12.5	75,000
Polysaccharides	39,000	32.5	65,000
Lipids	15,000,000	12,500	87.000
Proteins	1,700,000	1,400	2,120,000

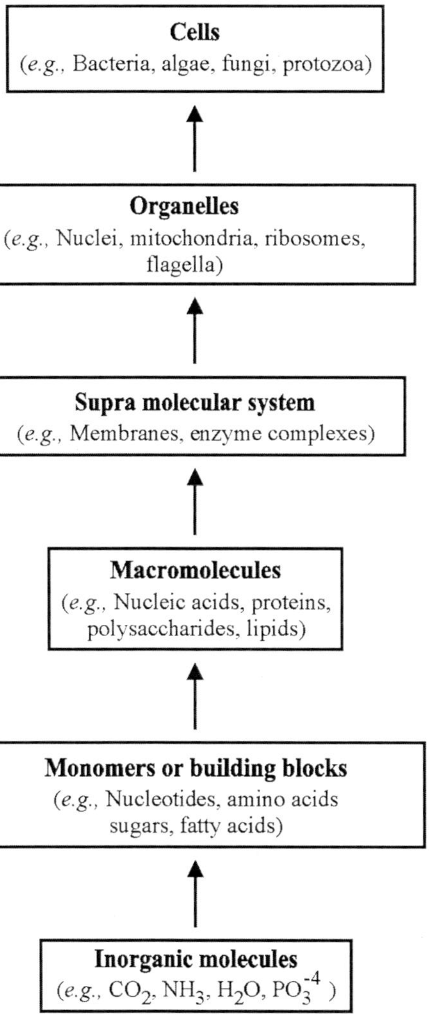

Fig. 3.4. The order of the anabolism for the construction of the cell

3.2. ROLE OF CENTRAL METABOLIC PATHWAYS IN METABOLISM

There are major metabolic pathways (both anabolic and catabolic) in living organisms including microorganisms each one with a specific purpose. However, there are certain metabolic pathways that serve a central role for many of these pathways, in that the intermediates of these pathways are either exit or entry points for many metabolic pathways. The central metabolic pathways (CMP) are those pathways that provide the precursor metabolites to all of the other pathways and also many of the intermediates serve as the entry points for catabolic end products for energy production or synthetic process (Table 3.3). They are the pathways for metabolism of mainly carbohydrates and to certain extent some carboxylic acids. The major carbohydrate pathways are Embden-Meyerhof Parnas pathway (EMP pathway or glycolysis), the pentose phosphate pathway (PPP) and the Entner - Doudoroff pathway(ED). These pathways are conserved in all the living organisms. However, the ED pathway has been found only among the prokaryotes. In aerobic organisms citric acid or tricarboxylic acid (TCA) cycle is common to all these pathways. Hence, TCA cycle is also treated as a part of central metabolic pathways.

Table 3.3. Twelve critical intermediate compounds of central metabolic pathway

Compound	Pathway(s)
Glucose-6-phosphate	EMP, PP, ED
Fructose-6-phosphate	EMP, PP, ED
Ribose-5-phosphate	PP
Erytrose-4-phosphate	PP
Triose phosphate	EMP, PP, ED
3-phosphoglycerate	EMP, PP, ED
Phosphoenol pyruvate	EMP, PP, ED
Pyruvate	EMP, PP, ED
Acetyl-CoA	Connecting link between EMP, PP, ED and TCA
Alpha-ketoglutaric acid	TCA
Oxaloacetic acid	TCA
Succinyl-CoA	TCA

Central metabolic pathways are important both for anabolic and catabolic pathways, in that many of intermediates serve as precursors for various anabolic pathways and similarly many end products of catabolic processes end up with the intermediates of CMP.

3.2.1. Anabolic Pathways Linked to Central Metabolic Pathways

Many intermediates of EMP, PP and ED pathways directly or indirectly serve as the precursors for various synthetic processes that include: polysaccharide synthesis, cell wall synthesis, lipid synthesis, amino acid synthesis, nucleic acid synthesis and many more.

Polysaccharide synthesis: Glucose-6- phosphate of glycolysis is a starting material for the synthesis of many polysaccharides, most importantly glycogen, starch, cellulose etc.

Synthesis of cell wall materials: Two amino sugars, N-acetyl muramic acid and N-acetyl glucosamine, which are the main constituents of peptidoglycan are synthesized utilizing fructose-6-phosphate as a precursor. Further, in the synthesis of these compounds phosphoenol pyruvate is source for lactyl portion of muramic acid.

Synthesis of assimilatory power: ATP and $NAPH^+$ together known as assimilatory power. $NAPH^+$ synthesized from PP pathway serves as a source of electrons for the reduction of molecules during biosynthesis as in C_3 and C_4 cycles. ATPS are produced both in glycolysis and TCA cycle through substrate level phosphorylation or in electron transport chain by oxidative phosphorylation.

Lipid synthesis: Both fatty acids and glycerol-3-phosphate, the basic constituents of lipids and phospholipids are synthesized from the intermediates of CMP. Dihydroxyacetone phosphate (DHAP) is the precursor for glycerol-3-P which participates in phospholipids biosynthesis. Acetyl -CoA, a connecting link between glycolysis and TCA is the precursor for synthesis of fatty acids and thus lipids. It can also be source of cell carbon when it is assimilated via glyoxylate cycle.

Amino acid synthesis: For many amino acids the intermediates of glycolysis serve as the precursors. 3-phosphoglycerate is the precursor for serine from which glycine, cysteine and tryptophan are synthesized. Synthesis of pyruvate family amino acids, leucine, isoleucine and valine is initiated from pyruvate. The precursors of the aromatic amino acids, phenyl alanine, tyrosine and tryptophan are glycolytic intermediate phosphoenol pyruvate (PEP) and erythrose-4-phosphate, an intermediate in the pentose phosphate pathway. Biosynthesis pathway of histidine utilizes 5-phosphoryl-α- pyrophosphate (PRPP) an intermediate also involved in the biosynthesis of tryptophan. Aspartate served from oxaloacetate is common precursor of asparagine lysine, methionine and threonine and isoleucine; α-ketoglutarate is the precursor for glutamate, glutamine, arginine, proline and ornithine.

Nucleic acid synthesis: The pentose, ribose 5-phosphate, one of the intermediates of PP-pathway is a major component of nucleic acids both RNA and DNA. Similarly, important component of nucleic acids, purines and pyrimidines are indirectly derived from the glycolysis and TCA cycle intermediates. Purines are derived from 3—phosphoglycerate through serine and glycine. Pyrimidines, on the other hand, are derived from oxaloacetate via aspartate.

Porphyrin synthesis: Succinyl-CoA, a central intermediate in the TCA cycle is the precursor for the synthesis of the porphyrin ring of heme groups which serve as oxygen carriers (in hemoglobin and myoglobin) and electron carriers (in cytochromes).

Anaplerotic reactions: The pathways of glycolysis and TCA are basically seen as oxidative and decarboxylate in nature. However, when in demanding situations to replenish various intermediates, different intermediates like pyruvate, PEP, function as CO_2 acceptors to produce products like oxaloacetate, malate, which are later converted to glucose by gluconeogenesis. In fact, the reverse TCA cycle reactions, which are basically reductive in nature, are dependent on such carboxylations.

Synthesis of secondary metabolites: Notwithstanding the physiological role, a plethora of secondary metabolites are produced by different microorganisms. All these secondary metabolites are synthesized, directly or indirectly, from the intermediates of central metabolic pathways (CMP), directly or indirectly as precursors for their biosynthesis (Chapter 23).

3.2.2. Catabolic Pathways Linked to Central Metabolic Pathways

The entire central metabolic pathways involving breakdown of glucose to pyruvate (EMP pathway, PP pathway and ED pathway) and subsequent TCA cycle reactions are basically catabolic in nature. However, the glucose molecule is enzymatically sequentially broken down. At first, glucose broken down to two pyruvates and then pyruvate is first oxidized to acetyl CoA and CO_2. In the last and final stage acetyl CoA is oxidized to CO_2 in TCA cycle. During these reactions reducing power $NADP^+$, FAD^+ and $NADPH^+_2$ is generated which is used for either generation of energy or biosynthesis. In addition ATPs are also produced.

It is not only carbohydrates that undergo catabolism through central metabolic pathways, but also many other cell constituent molecules like proteins, lipids, nucleic acids and many other compounds are degraded and their end products are channeled to central metabolic pathways at various levels. The end products that are fed into central metabolic pathway are utilized for generation of cellular energy or assimilated into cellular constituents.

Proteins: Proteins are first broken down into oligopeptides by proteolytic enzymes and later degraded to amino acids by peptidases. The resulting amino acids are degraded to one of seven common metabolic pathways. Glucogenic amino acids are degraded to pyruvate, α-ketoglutarate, succinyl-CoA, fumarate or oxaloacetate. Ketogenic amino acids are broken down to acetyl CoA or acetoacetate and can thus be converted to fatty acids or ketone bodies.

Lipids: On hydrolysis, lipids yield fatty acids and glycerol. Fatty acids are fed into the CMP at acetyl CoA level, whereas glycerols enter the CMP through 3-phosphoglyceraldehyde or dihydroxyacetone phosphate.

Nucleic acids: RNA and DNA are made up of purine (ATP, GTP) and pyrimidine (CTP, TTP, UTP) nucleotides. Purine nucleotides are degraded to D-ribose and hypoxanthine. The D-ribose enters the central metabolic pathways through PP-pathway which may be utilized for either in anabolic or catabolic processes. Degradation of pyrimidine nucleotides leads to production of D-ribose and NH_4^+. Thymine degraded through propionyl CoA and methyl malonyl CoA to succinyl CoA, an intermediate of TCA cycle.

■■■■

Chapter - 4

Principles of Bioenergetics

4.1. INTRODUCTION

In the study of physiology and biochemistry, cells are often thought of as tiny machines in which all events may be explained in terms of either chemical reactions, fluid dynamics, electrical fluxes across partitions, or the absorption or emission of light. Thus, all cellular activities are ultimately based on the known laws of physics and chemistry. These laws of physics and chemistry are fundamental to an understanding of cellular activities, especially the production, transduction and consumption of energy during cell metabolism.

The metabolism of a cell is characterized by myriad of chemical reactions in which energy is either consumed, produced or transduced from one form to another. All the metabolic reactions (both anabolic and catabolic) that proceed in cells are accompanied by energy changes. **The study of energy changes accompanying the metabolic reactions is known as bioenergetics**. More precisely, bioenergetics is the quantitative study of the energy transductions that occur in living cells and of the nature and function of the chemical processes underlying these transductions.

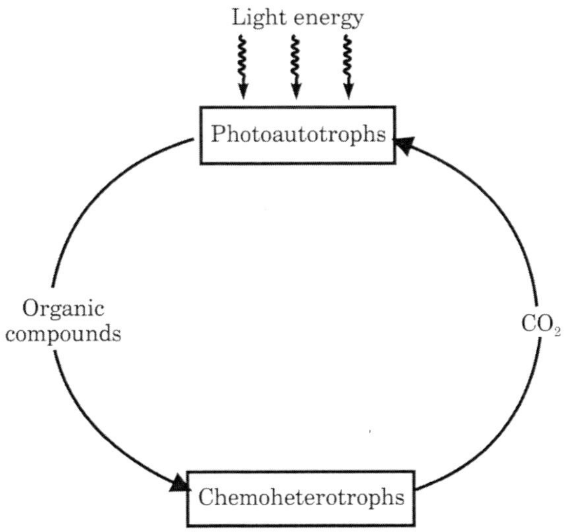

Fig. 4.1. The flow of carbon and energy in an ecosystem.

All living cells and organisms need energy. All cellular (physical and chemical) changes or processes are the result of application or movement of energy. Broadly speaking, energy is needed for three types of work, *viz.,* chemical, transport and mechanical. However, from cell physiology point of view, energy is needed for (i) construction of physical parts of the cell, (ii) synthesis of macromolecules, (iii) repair, (iv) maintenance, (v) growth and multiplication, (vi) accumulation of certain nutrients, (vii) efflux and (viii) motility.

The energy required for living organisms must ultimately be obtained from the cell's environment in some form. Photoautotrophs utilize the light as energy source, whereas chemoautotrophs obtain their energy from the oxidation of inorganic substances. Hetero- trophs meet their energy requirements from the oxidation of organic compounds. The flow of energy in an ecosystem and in living organisms is shown diagrammatically in Fig. 4.1.

Energy transformations in a living cell are shown in Fig. 4.2 a & b. Biological energy transformations obey the same physical laws that govern all the natural processes. It is, therefore, essential to understand the laws of thermodynamics and how they apply to the flow of energy, especially free energy, enthalpy and entropy.

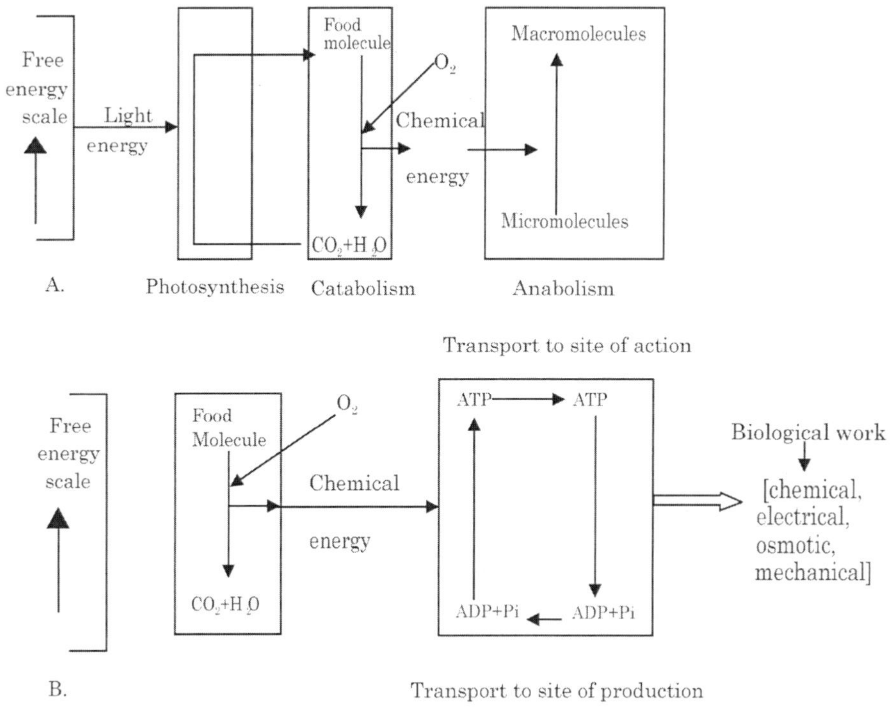

Fig. 4.2. The energy flow in living organisms. **A.** Energy flow between anabolic and catabolic reactions, **B.** Energy transduction to carry to cell functions.

4.2. LAWS OF THERMODYNAMICS

The science of thermodynamics analyses energy changes in a collection of matter called 'system'. All other matter in the universe is called surroundings. Thermodynamics focuses on the energy differences between the initial state and final state in a system, but it is not concerned with how the system gets the energy and with what rate. Energy changes in the system obey the first two laws of thermodynamics.

4.2.1 The First Law of Thermodynamics

The first law is concerned with the conversion of energy within a system where a system is defined as a body (a cell or an organism) and its surroundings. It states that energy cannot be created nor destroyed but can be converted from one form into another. During such a conversion, the total amount of the energy of the system remains constant.

This law applies to all levels of organization in the living world. It applies to organisms, cells, organelles and to the individual chemical reactions that characterize metabolism.

From a biological viewpoint, the first law of thermodynamics indicates that at any given moment a cell possesses a specific quantity of energy. This energy takes several forms which include (i) potential energy (ii) electrical energy and (iii) thermal energy. According to the first law, these forms of energy may be interconverted. For example, some of the cell's potential energy can be converted into electrical or thermal energy, but the cell cannot create or destroy energy.

4.2.2 The Second Law of Thermodynamics

The first law of thermodynamics tells us that the total energy of an isolated system consisting of a cell and its surroundings is the same before and after a series of events or chemical reactions have taken place. The first law does not tell us the direction in which the reactions proceed. To understand the spontaneity of a reaction and predict its direction, one must take into account two functions called **entropy** and **enthalpy**.

Entropy is a measure of the degree of randomness or disorder of a system. The entropy increases with increasing disorder. It is expressed as **S. Enthalpy** is the heat content of the reacting system. It reflects the number and kinds of chemical bonds in the reactants and products. It is the tendency to achieve the most stable bonding state and expressed as '**H**'.

The second law of thermodynamics states that in all processes the entropy of the system increases until an equilibrium state is attained.

4.2.3 Free Energy

The amount of free energy liberated or taken up during the course of a reaction is referred to as the free energy change. It is represented by ΔG and expressed in terms of calories. If the ΔG value is measured under standard conditions, it is referred to as standard free energy change and is represented by $\Delta G^{\circ'}$.

The first and second laws of thermodynamics, now can be combined in a useful equation

$$\Delta G = \Delta H - T.\Delta S.$$

Where

T = Temperature on Kelvin scale

ΔS = Change of entropy (increase in randomness)

ΔH = Change of enthalpy

ΔG = Change in free energy

The change in free energy has a definite concrete relationship to the direction of chemical reactions.

For the reaction,

$$A + B \longrightarrow C + D$$

the equilibrium constant is defined as

$$K_{eq} = \frac{[C][D]}{[A][B]} \text{ at equilibrium}$$

The equilibrium constant of a reaction is directly related to its change in free energy. The $\Delta G^{\circ'}$ can be calculated from equilibrium constant K_{eq}, of a reaction using the relationship.

$$\Delta G^{\circ'} = -\ 2.303\ RT\ .\ \log K_{eq}$$

where R is the gas constant (8.314 J/mol/degree), T is the absolute temperature (in degree Kelvin) and K_{eq} is the equilibrium constant. Table 4.1 lists standard free energy of hydrolysis of some energy rich biological compounds. From the foregoing discussion certain valid conclusions can be drawn. If the equilibrium constant is greater than 1.0, then the $\Delta G^{\circ'}$ value is negative, and the reaction is said to be exergonic *i.e.*, energy releasing. When the K_{eq} value is less than 1.0, the $\Delta G^{\circ'}$ value is positive and the reaction is said to be endergonic, *i.e.*, energy consuming. If the equilibrium constant is 1, then the $\Delta G^{\circ'}$ value equals zero.

Table 4.1. Standard free energy of hydrolysis of some energy rich biological molecules

Compounds	**$\Delta G^{\circ'}$ at pH 7.0 (k cal/mol)**
Phosphoenol pyruvate (PEP)	- 14.8
Cyclic AMP (cAMP)	- 12.0
1,3-diphosphoglyceric acid	- 11.9
Acetyl phosphate	-10.5
Acetoacetyl CoA	-10.5
Phosphocreatine	-10.3
S-adenosylmethionine	-10.0
Pyrophosphate	-8.0
Acetyl CoA	-7.7
ATP $\rightarrow$ AMP + PPi	-8.6
ATP $\rightarrow$ ADP + Pi	-7.3

Compounds	ΔG°' at pH 7.0 (k cal/mol)
UDP – glucose	-7.6
Sucrose	-6.6
Glucose-1-phosphate	-5.0
Fructose-6-phosphate	-3.8
Glucose-6-phosphate	-3.3
Lactose	-3.0
3-phosphoglyceric acid	-2.4
Sn-Glycerol-3-phosphate	-2.2

(i) Exergonic reactions

$$A + B \rightleftharpoons C + D$$

$$K_{eq} = \frac{[C][D]}{[A][B]} > 1.0$$

$\Delta G^{\circ'}$ is negative.

(ii) Endergonic reactions

$$A + B \rightleftharpoons C + D$$

$$K_{eq} = \frac{[C][D]}{[A][B]} < 1.0$$

$\Delta G^{\circ'}$ is positive.

4.3. COUPLED REACTIONS

Living cells commonly employ either of two mechanisms to cause endergonic reactions to take place. One mechanism is to create reactant and product concentrations that are markedly different from equilibrium values. This can be brought about either through the production of additional reactants or through removal of products by drawing them into other cellular reactions. A second type of coupling occurs when exergonic and endergonic reactions are catalyzed by the same enzyme. For example, the two reactions in the cell are catalyzed (and also coupled) by the enzyme glucokinase

$$ATP + H_2O \longrightarrow ADP + \text{phosphate};\ \Delta G^{\circ'} = -\ 30.65\ \text{kJ/mole}$$

$$\text{Glucose} + \text{phosphate} \longrightarrow \text{glucose-6-(P)} + H_2O;\ \Delta G^{\circ'} = +\ 13.86\ \text{kJ/mole}$$

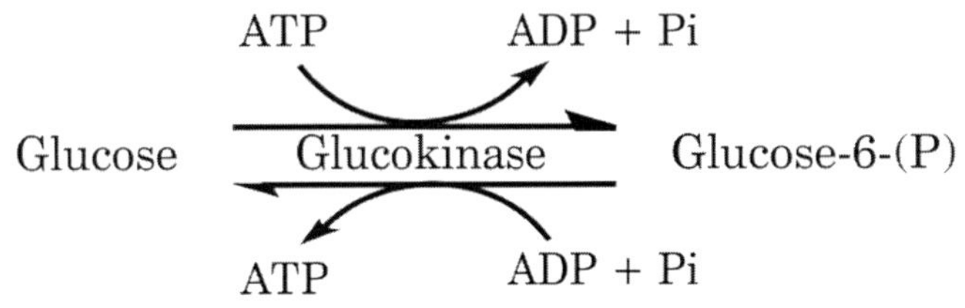

The exergonic reaction is thus coupled to the endergonic reaction by glucokinase, and some of the free energy yielded by exergonic reaction is consumed to drive the endergonic reaction.

4.4. TRANSPORT OF ENERGY: THE ROLE OF ATP AND OTHER ENERGY RICH MOLECULES

In a living cell, catabolic (or exergonic) and anabolic (endergonic) reactions are both spatially and temporally separated. Hence, it is necessary to carry the energy released from the place of exergonic reactions to the place of endergonic reactions (Fig. 4.3). Living systems possess a variety of such high-energy-transfer compounds which faithfully carry the energy from place of energy generation to place of utilization. ATP, a high energy molecule, is the major connecting link between the catabolic and anabolic components. Hydrolysis of ATP into ADP and Pi is strongly exergonic and is associated with free energy $\Delta G^{\circ'}$ of – 7.3 K cal/mol

$$ATP + H_2O \longrightarrow ADP + Pi;\ \Delta G^{\circ'} = -\ 7.3\ K\ cals/mol$$

ATP is ideally suited for its role as an energy currency. It is formed in energy trapping and generating processes, such as photosynthesis, fermentation and aerobic respiration. In the cell's economy, exergonic breakdown of ATP is coupled with various endergonic reactions to promote their completion. In other words, ATP links energy generating reactions, which liberate free energy, with energy using reactions, which require free energy input to proceed towards completion.

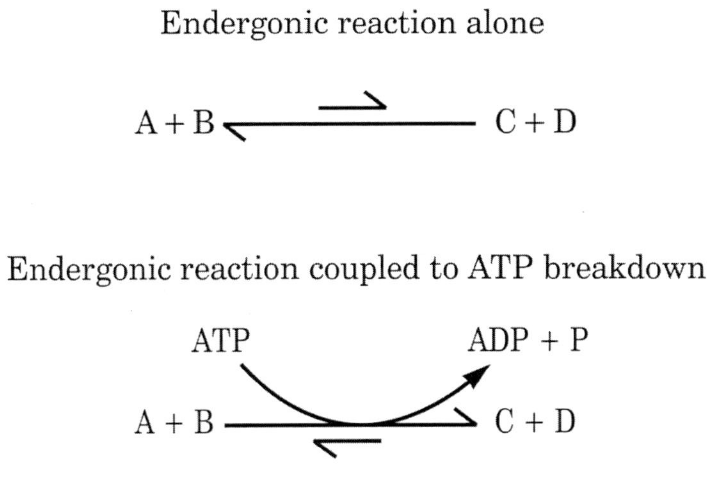

Fig. 4.3. ATP as a coupling agent. The use of ATP to make endergonic reactions more favourable. It is formed by exergonic reactions and then used to drive endergonic reactions.

Undoubtedly, ATP is by far the most important high-energy-transfer compound in all the living organisms. However, there are many other high energy transfer compounds whose hydrolysis yields a large amount of free energy. Such high-energy-transfer compounds found in the cell and their standard free energy changes upon hydrolysis are presented in Table 4.1.

4.5. OXIDATION-REDUCTION REACTIONS

Free energy changes are not only involved in the equilibria of regular chemical reactions, but also in the equilibria of oxidation-reduction reactions. Oxidation is the loss of electrons and reduction is the gain of electrons. An oxidising agent or oxidant will accept electrons and in the process it gets reduced. Similarly, a reducing agent (reductant) donates the electrons and in the process it gets oxidised. Oxidation – reduction reactions are mutually dependant and occur simultaneously.

$$\text{Oxidant} + ne^- \rightleftharpoons \text{reductant}$$

Thus, in each oxidation – reduction reactions, a pair of substances is involved : one is the reduced form, the other is oxidised form. The oxidant and reductant pair is referred to as redox-couple or oxidation-reduction (O/R) system.

The equilibrium constant for the O/R system reaction is called the **'standard reduction potential (E_o)** and is a measure of the tendency of the reducing agent to loose electrons. The reference standard for reduction potential is the hydrogen system with an E_o' (the reduction potential at pH 7.0) of –0.72 volts or – 720 millivolts).

The reduction potential value of redox couples has some thing to convey. Redox couples with more negative reduction potentials will donate electrons to couples with more positive potentials. Redox potentials of some biologically important redox couples are presented in Table 4.2.

Table 4.2. Redox potentials of some selected biologically important redox couples.

Redox couple	**E_o' (volts)**
$2H^+ + 2e^- \longrightarrow H_2$	– 0.42
Ferredoxin (Fe^{3+}) + e^- $\longrightarrow$ Ferredoxin (Fe^{+2})	– 0.42
$NAD(P)^+ + H^+ + 2e^- \longrightarrow NAD(P)H$	– 0.32
$S + 2H^+ + 2e^- \longrightarrow H_2S$	– 0.274
Acetaldehyde + $2H^+ + 2e^-$ $\longrightarrow$ ethanol	– 0.197
Pyruvate + $2H^+ + 2e^-$ $\longrightarrow$ lactate $^{2-}$	– 0.185
$FAD + 2H^+ + 2e^- \longrightarrow FADH_2$	– 0.18
Oxaloacetate^{2-} + $2H^+ + 2e^-$ $\longrightarrow$ malate	– 0.166
Furmarate^{2-} + $2H^+ + 2e^-$ $\longrightarrow$ succinate	– 0.031
Cytochrome b(Fe^{3+}) + e^- $\longrightarrow$ cytochrome b(Fe^{2+})	+ 0.075

Redox couple	E_o' (volts)
Ubiquinone + $2H^+ + 2e^-$ ⟶ ubiquinone H_2	+ 0.10
Cytochrome c(Fe^{3+}) + e^- ⟶ cytochrome c(Fe^{2+})	+ 0.254
$NO_3^- + 2H^+ + 2e^-$ ⟶ $NO_2^- + H_2O$	+ 0.421
$Fe^{3+} + e^-$ ⟶ Fe^{2+}	+ 0.771
$O_2 + 4H^+ + 4e^-$ ⟶ $2H_2o$	+ 0.815

Thus, electrons will tend to move from reductants at the top of the list in Table 4.2 to oxidants at the bottom because they have more positive potentials. Take the example of electron carrier nicotinamide adenine dinucleotide (NAD^+). The NAD^+/NADH couple has a very negative Eo' and can, therefore, give electrons to many acceptors, including O_2.

$$NAD^{2+} + 2H^+ + 2e^- \rightleftharpoons NADH + H;\ E_o = -0.32 \text{ volts}$$

$$\tfrac{1}{2}O_2 + 2H^+ + 2e^- \rightleftharpoons H_2O;\ E_o = +0.82 \text{ volts}$$

Because NAD^+/NADH is more negative than ½ O_2/H_2O, electrons will flow from NADH (the reductant) to O_2 (the oxidant)

$$NADH + H^+ + \tfrac{1}{2}O_2 \longrightarrow H_2O + NAD^+$$

When electrons move from a reductant to an acceptor with more positive redox potential, the free energy is released. The $\Delta G°'$ of the reaction is directly related to the magnitude of the difference between the reduction potentials of the two couples. The larger $\Delta E_o'$, the greater the amount of free energy made available. The relationship between an oxidation reduction potential difference and the standard free energy change is

$$\Delta G°' = -\,nFE'$$

where n is the number of moles of electrons transferred in the reaction, F is the Faraday constant (23,061 cals/V per equivalent) and E_o is the standard oxidation-reduction potential difference.

For example, let us calculate $\Delta\ G°'$ for the reaction in which cytochrome c is oxidised by an oxygen from the ferrous to the ferric state with $E_o' = +0.56V$

$$2 \text{ Cytochrome c} - Fe^{+2} \text{(reduced)} + \tfrac{1}{2}O_2 \rightleftharpoons 2 \text{ cytochrome c} - Fe^{3+} \text{(oxidized)} + O^{2-}$$

$$\Delta G°' = -\,2\,(23{,}061)\,(0.56)$$

$$\Delta G°' = -\,25.828 \text{ cal/mol}$$

4.6 TRANSDUCTION OF ENERGY

The conversion of energy from one form to another is called transduction. A good example to transduction is the conversion of light energy into potential chemical bond energy in photosynthesis. The conversion of chemical bond energy into light energy occurs in bioluminescence. Similarly, during the flagellar movement chemical bond energy is converted to mechanical energy. Energy transductions are carried out by transducers, which are nothing but different cell membranes like cristae of mitochondria, lamellae of chloroplasts and cell membranes of both prokaryotes and eukaryotes. Cell membranes are flexible, self-sealing and selectively permeable to polar solutes. Membranes are not merely passive barriers, they are complex dynamic structures. They include an array of proteins specialized for promoting or catalysing various cellular processes and organise complex reaction sequences and are central to both biological energy conservation and transduction.

The principal method by which energy is generated for physiological processes such as biosynthesis and solute transport in living organisms is by coupling the flow of electrons in membranes to the creation of electrochemical proton gradient. The electrons flow from primary electron donors to terminal electron acceptors occur through a series of electron carrier proteins and a class of lipids called quinones. These electron carriers are located in cell membranes, as mentioned earlier, such as cristae of mitochondria, lamellae of chloroplasts and cell membrane of prokaryotes. Interestingly, in these membranes electron carriers are located one after the other as per their standard electrode potentials, so that the electrons can pass on sequentially from negative redox potential couple to positive redox potential couple. Let us, first try to understand types, nature and arrangement of electron carries.

4.6.1 Electron Carriers

There are four principal types of electron carriers

1. Flavoproteins
2. Quinones
3. Iron sulfur proteins
4. Cytochromes

1. Flavoproteins: Flavoproteins are electron carriers that have their prosthetic group an organic molecule called flavin. There are two flavins, flavin mononucleotide (FMN) and flavin adenine nucleotide (FAD). Reduced flavins carry 2H (two electrons and two hydrogens). Flavoproteins catalyse diverse oxidation and reduction reactions in the cytoplasm, not merely those of the electron transport chain in the membranes. Although all the flavoproteins have FMN or FAD as their prosthetic group, they differ in which oxidations they catalyze as well as in their redox potentials. These differences are due to differences in the protein component of the enzyme, not in the flavin itself.

2. Quinones: Quinones are lipid electron carriers. Due to their hydrophobic nature, some are believed to be highly mobile in the lipid phase of the membrane and carry hydrogen and electrons to and from the complexes of protein electron carriers which are not mobile. Bacteria possess two types of quinones: **ubiquinone (UQ),** a quinone also found in mitochondria, and **menaquinones (MQ or MK).** The latter is a derivative of vitamin K. Menaquinones have a much lower electrode potential than ubiquinones and are used predominantly during anaerobic respiration, where the electron acceptor has a low potential. A third type of quinone occurs in chloroplasts and cyanobacteria, and functions in photosynthetic electron transport.

3. **Iron – sulfur proteins**: Iron – sulfur proteins contain nonheme iron and usually acid-labile sulfur as coenzyme. Generally, the proteins contain clusters in which iron and acid – labile sulfur

are present in the ratio of 1:1. However, there may be more than one iron sulfur cluster per protein. The Fe-S clusters have different Eh values and the electron travels from one FeS cluster to the next towards the higher Eh. These proteins cover a wide range of potentials from approximately –400 mV to + 350 mV. They, therefore, carry out oxidation – reduction reactions at both, the low potential end and the high potential end of the electron transport chain and are found in several locations.

4. Cytochromes: Cytochromes are electron carriers that have heme as the prosthetic group. Heme consists of four pyrrole rings attached to each other by methane bridges. Because hemes have four pyrroles, they are called tetrapyrroles. The substituted tetrapyrroles are called **porphyrins**. Therefore, hemes are also called porphyrins. The iron, in the center of the heme, is the electron carrier and is oxidised to ferric or reduced to ferrous ion during electron transport. The cytochromes are, therefore, one electron carriers. There are four classes of cytochromes (*i.e.,* a, b, c and d) which have hemes a,b,c and d. Hemes vary according to the side group and distinguished spectrophotometrically.

4.6.2 Arrangement of Electron Carriers in Mitochondria

The electron carriers are arranged as an electron transport chain that transfers from electron donors at a low electrode potential to electron acceptors at higher electrode potential. The carriers are organised in the membrane as individual complexes. Four complexes can be recognized in mitochondria. They are complex I (NADH – ubiquinone oxidoreductase), complex II (succinate dehydrogenase), complex III (Ubiquinol-cytochrome c oxidoreductase, also called the bc_1 complex) and complex IV (cytochrome c oxidase, which is cytochrome aa_3). Each complex can have several proteins. The arrangement of the elctron carriers in mitochondria is shown in Fig. 4.4(a).

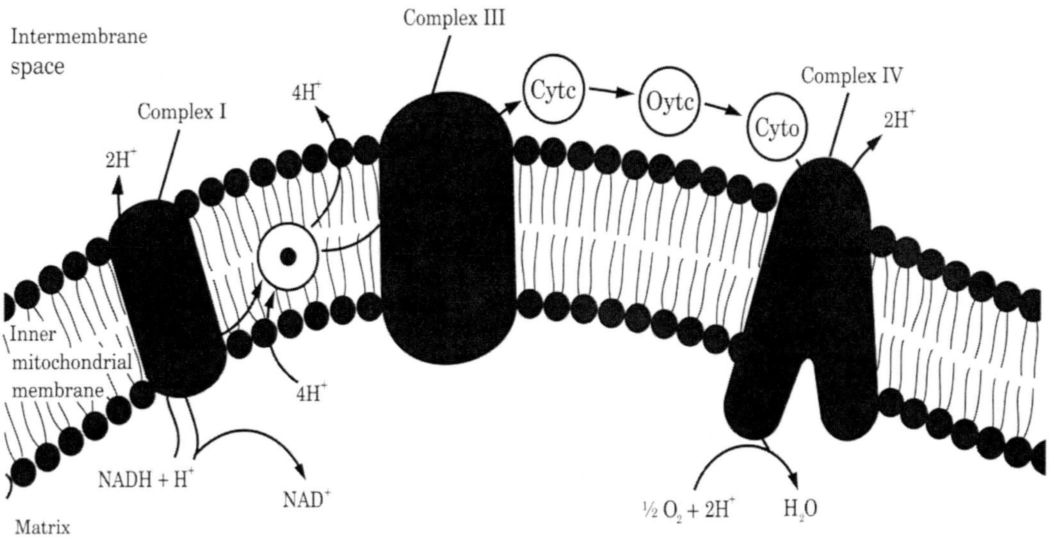

Fig. 4.4. (a) The arrangement of electron carriers in mitochondria.

4.6.3 Arrangement of Electron Carriers in Bacteria

Bacterial electron transport chains vary among different bacteria and also according to growth conditions. Bacterial chains are organised into dehydrogenases and oxidase complexes connected by quinones. The quinones accept electrons from dehydrogenase and transfer them to oxidase complexes that reduce the terminal electron acceptor (Fig. 4.4b). Some of the bacterial electron carrier complexes like complex I and complex II are similar to mitochondria. Nevertheless, they differ in many respects, from the mitochondria because they are capable of using electron acceptors other than oxygen and due to the diversity of substrates oxidised by them (Fig. 4.5). In aerobic respiration, where the terminal electron acceptor is oxygen, dehydrogenase complex removes electrons from an electron donor and transfers them to a quinone. The electrons are transferred to an oxidase complex via branched pathway. Depending upon the bacterium, the pathway may branch at the quinone or at cytochrome. Many bacteria have bc_1, cytochrome c and cytochrome aa_3 in one of the branches. Other bacteria do not have a bc_1 complex and may or may not have cytochrome aa_3. In anaerobic respiration, where the electron acceptor is other than oxygen (*e.g.*, sulfate, nitrate, fumarate), the electrons are transferred to reductase complexes which are synthesized anaerobically. Several reductases exist, each one specific for the electron acceptor. More than one reductase can simultaneously exist in a bacterium.

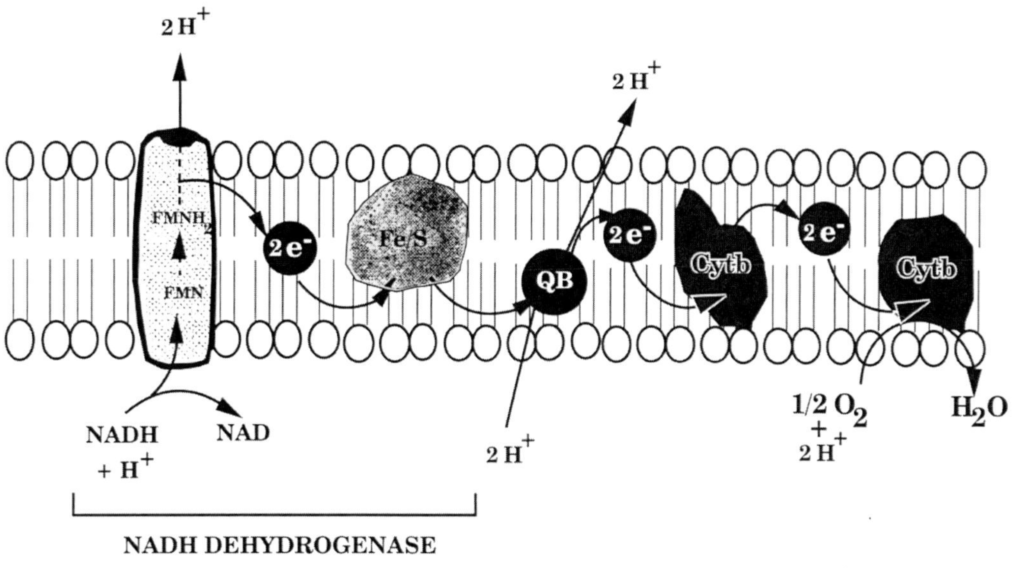

Fig. 4.4 (b). An electron-transport chain of *E. coli* FMN = flavin mononucleotide; Fe-S = iron - sulfur group; Q8 = ubiquinone.

Two major differences between mitochondrial and bacterial electron transport chains are that (i) the routes to the terminal electron acceptors in bacteria are branched, the branch point being at the quinone or cytochrome, and (ii) many bacteria can alter their electron transport chains depending upon the growth conditions under aerobic conditions. There are often two or three branches leading to different oxidases, whereas under anaerobic conditions, they do not make the oxidase complexes, but instead synthesize reductase.

4.7. CHEMIOSMOTIC THEORY

Chemiosmotic theory is central to energy metabolism. It lies at the foundation of all bacterial physiology. Chemiosmotic theory brings together principles of physics and thermodynamics in explaining the membrane bioenergetics. The student should be familiar with the principles of chemiosmotic theory for a deeper understanding of how the bacterial cell uses ion gradients to couple energy yielding (exergonic) reactions to energy requiring (endergonic) reactions.

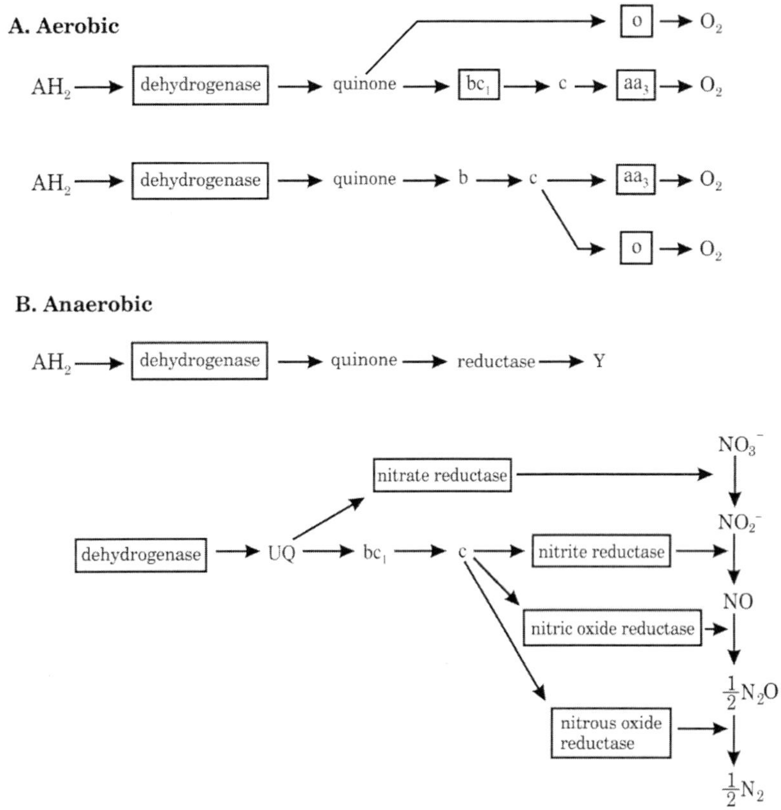

Fig. 4.5. A model for electron transport pathways in bacteria.
A. Aerobic bacteria B. Anaerobic bacteria

The purpose of electron transport is to generate ATP, but electron transport to oxygen proceeds very readily in broken up mitochondria without any ATP generation at all. This situation perplexed biochemists for many decades. If electron transport generates ATP in the cell, why it cannot occur in broken up mitochondria. In other words, ATP synthesis occurs in intact healthy mitochondria but not in broken mitochondria. A solution was obtained as a result of theoretical ideas of Peter Mitchel (1961) who proposed the same in the form of **Chemiosmotic theory**. His idea explained how electron flow causes ATP synthesis, that is so novel that it was at first hardly taken seriously by most. It took a long time to be accepted and finally he was awarded Nobel prize in 1978. The concept is based on the simple notion that gradients have the ability to do work. For example, a gradient of water pressure can be used to generate electricity, a gradient of air pressure to drive a windmill etc. Similarly, a chemical gradient can be utilized to do useful work if suitable energy harnessing device is interposed.

Chemiosmotic theory states that energy transducing membranes (*i.e.,* bacterial cell membranes, mitochondria and chloroplast membranes) pump protons across the membrane, thereby generating an electrochemical gradient of protons across the membrane (the proton potential) that can be used to do useful work when the protons return across the membrane to the lower potential. Mitchell discovered that electron flow caused protons to be ejected from inside the mitochondrion to the outside, thus creating a proton gradient (ΔP) across the membrane. Proton translocation across the membrane occurs through the coupling sites. These are the sites in electron transport pathway where redox reactions are coupled to proton translocation. In mitochondria, there are three coupling sites and they are called sites 1 (complex I) 2, (complex III) and 3 (complex IV). This process creates an electrochemical gradient (designated $\Delta\ \mu H^+$) across the membrane consisting of a negative – inside membrane potential ($\Delta\ \psi$) and an outside acid gradient of H^+(Δ pH). In any case, this proton expulsion or translocation, a membrane potential, negative inside and positive outside, is generated which contributes to the total energy gradient or proton motive force (PMF or ΔP). The accumulated protons against gradient (electrical, chemical and concentration), tend to return to the inside. As the membrane is impermeable to protons, certain proton conducting channels are interpolated in the membrane to facilitate the proton flow. It is, indeed, during the back flow of protons through specialised proton–conducting channels, useful cellular work is done including the synthesis of ATP.

The proton motive force (PMF) is an interconvertible form of free energy. As shown in the Figure 4.6, it can be used for ATP synthesis, NADPH synthesis (reverse electron flow), flagellar movement, active transport and heat production.

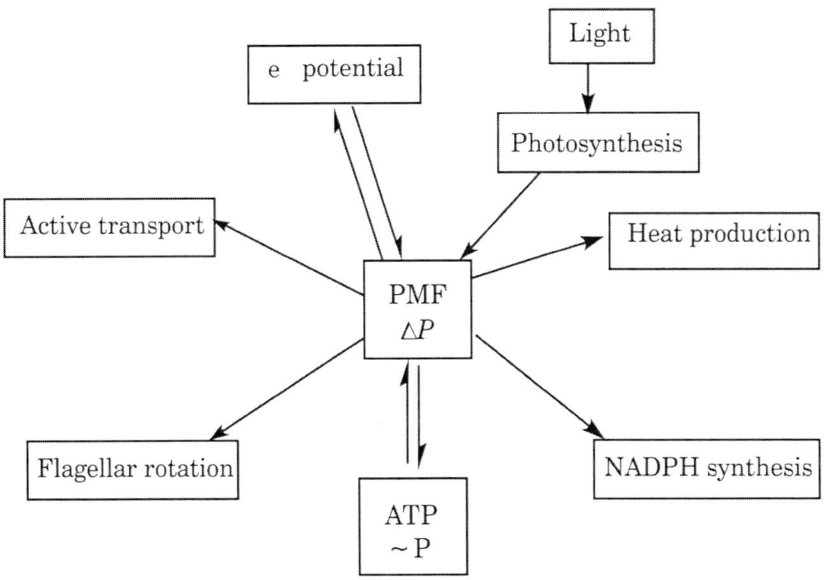

Fig. 4.6 : The central role of proton motive force (PMF)

4.8. ATP SYNTHASE

There are many transmembrane protein proton conductors, such as solute transporters, ATP synthases, flagellar rotation motors. ATP synthases or proton translocating ATP synthases are enzyme complexes built in the cell membranes of prokaryotes and mitochondrial, and chloroplast membranes that couple the translocation of protons down a proton potential gradient to the phosphorylation of ADP to make ATP.

Structural and functional details of ATP synthase are worked out by Paul Boyer of UCLA, who won a Nobel prize for this contribution. He referred it to as 'splendid molecular machine' and added 'all enzymes are beautiful, but ATP synthase is one of the most beautiful as well as one of the most unusual and important. The enzyme complex is found in aerobic organisms wherever the energy derived from electron transport has to be trapped as ATP. It is located in inner membrane of mitochondria, lamellae of chloroplasts and cell membrane of prokaryotes (Fig. 4.7). ATP synthase has been highly conserved during evolution and, therefore, the enzymes from different sources differ slightly in detail. Experiments on a variety of bacteria, chloroplasts and mitochondria have contributed to our current understanding of the structure and function of ATP synthase.

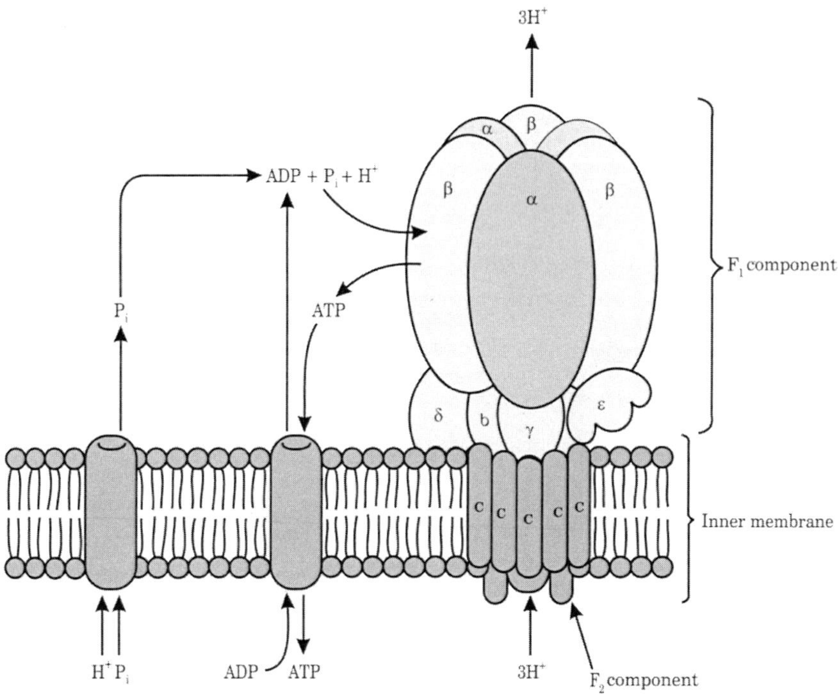

Fig. 4.7. ATP synthase.

The ATP synthase consists of two regions *i.e.,* the F_0 and F_1 (Fig. 4.7). The F_0 region spans the membrane and serves as a proton channel through which the protons are pumped. The F_1 region is located in the inner surface of the membrane and is the catalytic sub-unit, responsible for the synthesis and hydrolysis of ATP. The F_1 unit is also called coupling factor. The F_1 subunit from *E. coli* is made from five different polypeptides with the following stoichiometry, α_3, β_3, γ, δ and ε. The β polypeptides each contain a single catalytic site for ATP synthesis and hydrolysis. The F_0 portion has three different polypeptides, which in *E. coli* are a_1, b_2 and c_{10}, all of which occur in multiples. In all *E. coli* uses eight different types of polypeptides to construct a 22-polypeptide machine that acts as a reversible pump driven by proton potential.

4.9. MECHANISM OF ATP SYNTHESIS

ATP synthesis is brought about by the ATP synthase complex. The model that is favoured by most investigators is called the **'binding change mechanism'**. It is based upon the finding that purified F_1 will synthesize tightly bound ATP. The model proposes that when protons move down the electrochemical gradient through the F_0-F_1 complex, a conformational change in F_1 occurs that results in the release of newly synthesized ATP from high affinity site. The β subunit of F_1 unit has a catalytic site which synthesizes ATP from ADP + Pi. The sequence of events in the synthesis of ATP as proposed by Boyer is also follows (Fig. 4.8):

1. The site is open and nothing is bound to it (the O state)
2. A conformational change in the subunit converts a low affinity state; ADP and Pi now bind to it loosely, but there is no catalytic activity (the L state)
3. A further conformational change produces a tight binding state the ADP and Pi become tightly bound. This is now catalytically active and ATP is formed (the T state)
4. The conformational change opens up the site, ATP escapes and the site is back to the original open state.

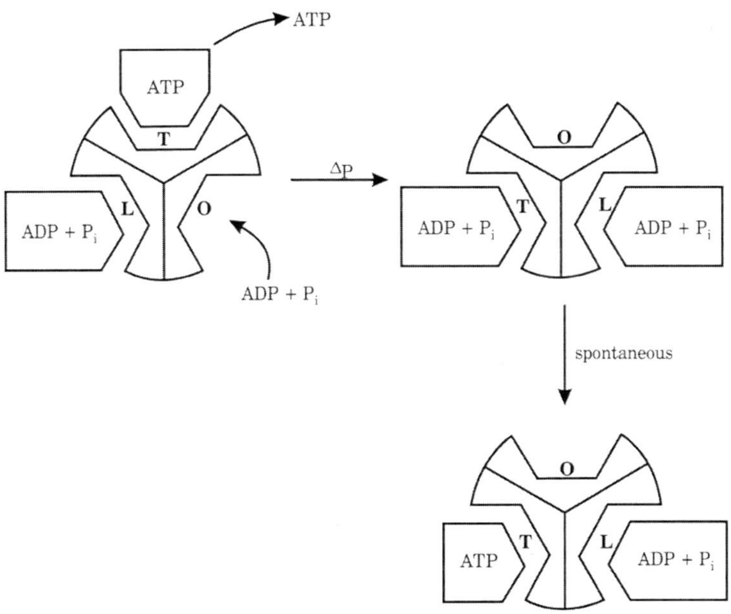

Fig. 4.8. Binding-change mechanism for ATP synthesis.

The model postulates that each site progresses sequentially through the three conformations. ATP synthase is a most unusual enzyme in that catalylic activity is dependent on cooperation between the subunits. At any time one of the three is in O state, one in the L state and one in the T state. It is the conformational changes in the F_1 subunits that constitute the mechanism by which energy is transmitted into ATP production. Results of several studies ultimately yielded the picture that ATP synthase is the smallest known rotary motor, driven by proton rather than electron. Though a clear picture is emerging out on the mechanism of rotary motion, still some issues are in a diffused state. For instance, we do not completely understand how the proton flow through the Fo induces rotary motion, nor how shaft rotation interacts with the F_1 subunits to induce the sequence of conformational changes believed to occur in the β subunits during the synthesis of ATP.

4.10. BIOENERGETICS IN CYTOSOL

Cellular reactions are driven by two different kinds of energy. One kind of energy is electrochemical energy which is due to ion gradient across the cell membrane. The second type of energy is chemical energy in the form of high energy molecules (e.g. ATP) in the soluble part of the cell. High energy molecules are important because they drive anabolic pathways such as synthesis of nucleic acids, proteins, lipids polysaccharides in the cytoplasm. Uptake of certain solutes also requires high energy molecules rather than electrochemical energy. The high energy molecules include ATP, other nucleotide derivatives, phosphoenol pyruvate, acylphosphates and acyl-CoA derivatives. These molecules have bonds that have free energy of hydrolysis. These bonds often referred as high energy bonds and represented by squiggle (~). However, it is a misnomer, as these so called energy rich bonds, infact, possess normal bond energy. The fact is that the chemical group attached to squiggle (~) is readily transferable to an acceptor molecule. Therefore, the high energy molecules are said to have high group transfer potential. When chemical groups are transferred, new linkages between molecules (*e.g.*, ester linkages, amide linkages, glycosidic linkages, and ether linkages) are made. This results in the synthesis of small molecules in the cell as well as polymers such as nucleic acids, proteins, polysaccharides and fatty acids.

Group transfer potential

A common chemical group that is transferred between molecules is the phosphoryl group. Various molecules such as ATP, acyl phosphates and phosphoenol pyruvate have a high phosphoryl group transfer potential and undergo similar nucleophilic displacement reactions. The group transfer potential is defined as the negative of the standard free energy of hydrolysis at pH 7. It is the quantitative assessment of the tendency of a molecule to donate the chemical group to a nucleophile. For example, the standard free energy of the phosphate ester bond in particular compound is – 29,000 J/mol. Then its phosphoryl group transfer potential is the negative of this number i.e +29,000 J/mol. Bonds that have a standard free energy of hydrolysis at pH 7.0 equal to or greater than –29,000 J are usually called high energy bonds.

Group transfer potentials are a convenient way of estimating the direction in which a reaction will proceed (Table 4.3). Let us consider the example of ATP and glucose-6-phosphate. The phosphoryl group transfer potential of ATP at pH 7.0 is 35 kJ/mole, and that of glucose-6-

phosphate is only 14 kJ/mol. From this, it is evident that ATP is a more energetic donor of the phosphoryl group than glucose-6-phosphate. It also means that ATP will transfer the phosphoryl group to glucose to form glucose-6-phsophate with the release of 35 – 14 i.e., 21 kJ/mol under standard conditions

$$\text{ATP + glucose} \longrightarrow \text{glucose – 6 – phosphate + ADP; } \Delta G^{\circ\prime} = -21 \text{ kJ/mol}$$

The release of 21 kJ/mol means that the equilibrium lies far in the direction of glucose-6-phosphate.

Table 4.3. Group transfer potentials of some compounds

Compound	**G°' hyd (kJ/mol)**	**Phosphoryl group transfer potential kJ/mol**
PEP + H_2O ⟶ Pyruvate + Pi	–62.0	+ 62.0
1,3 BPGA + H_2O ⟶ 3 – PGA + Pi	– 49.0	+ 49.0
Acetyl – P + H_2O ⟶ Acetate + Pi	– 47.7	+ 47.7
ATP + H_2O ⟶ ADP + Pi	– 36.8	+ 36.8
ADP + H_2O ⟶ AMP + Pi	– 36.0	+ 36.0
ATP + H_2O ⟶ AMP + Pi	– 40.6	+ 40.6
AMP + H_2O ⟶ A + Pi	– 12.6	+ 12.6
PPi + H_2O ⟶ 2Pi	– 31.8	+ 31.8
Glucose – 6 – P + H_2O ⟶ Glucose + Pi	– 14.0	+ 14.0

4.11 ENERGY RICH COMPOUNDS

There are several classes of energy rich compounds which are used in biological systems. Let us now consider some of these molecules and the factors contributing to the large negative free energy changes.

1. Adenosine triphosphate (ATP)

ATP is a nucleoside triphosphate belonging to the family of energy rich compounds (Fig. 4.9). It is highly charged and gets completely oxidized at pH 7.0. The phosphate groups can be removed by hydrolysis in several ways with liberation of free energy. However, the free energy of hydrolysis of ATP is not constant in a cell, it varies from cell to cell depending upon pH and Mg^{2+} ion concentration.

Standard free energy change associated with the hydrolysis of terminal phosphate is –7.3 k cal/mol. When the second phosphate is hydrolysed, another –7.3 k cal of energy is released. These phosphate groups are linked through anhydride bonds which trap a large amount of energy. When the last phosphate group is hydrolysed, the energy released has a smaller value *i.e.,* about –2.0 k cal. The smaller values are associated with ester bonds. Therefore, the terminal and second phosphate bonds in ATP are high energy bonds denoted by ~ P symbol (- P ~ P ~ P)

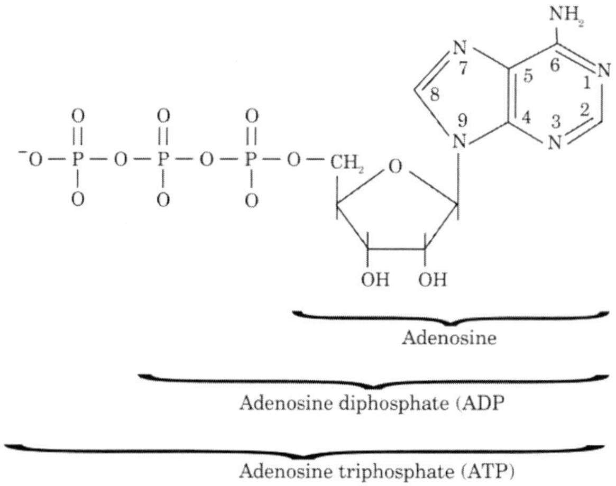

Fig. 4.9. Adenosine triphosphate and Adenosine diphosphate.

2. Enol phosphates

Phosphoenol pyruvate (PEP) is another energy rich compound which has a large free energy of hydrolysis (-14.8 k cals) at pH 7.0. It is an intermediate in carbohydrate metablism, used for regenerating ATP from ADP by phosphorylation (substrate level phosphorylation)

$$CH_2{=}C(OPO_3^{2-}){-}CO_2^- + H_2O \longrightarrow HPO_4^{2-} + CH_2{=}C(CO_2^-){-}OH \longrightarrow CH_3{-}C(CO_2^-){=}O$$

$\Delta G = -6.8$ k cal — unstable enol form

$\Delta G = -8.0$ k cal — pyruvate stable form

PEP is a phosphorylating agent that can be used to synthesise organic phosphates. Hydrolysis of enol phosphate results in the formation of unstable enolic form of pyruvate, which undergoes isomerization to generate stable keto form.

3. Thiol esters

Thiol esters come under the category of energy rich compounds which are largely used to generate ATP. Acetyl coenzyme A is one such compound which yields approximately – 8.0 k cal/mol upon hydrolysis.

$$CH_3{-}\overset{O}{\overset{\|}{C}}{-}S{-}CoA + H_2O \longrightarrow CH_3{-}\overset{O}{\overset{\|}{C}}{-}O^- + CoA{-}SH + H; \quad \Delta G^o = -8.0\,Kcal$$

4. Acyl phosphates

Acyl phosphates are energy rich compounds having a large negative free energy. One such compound is 1,3 biphosphoglyceric acid. When acyl phosphates are hydrolysed, the energy is released.

5. Phosphagens

Phosphagens or phosphoguanidines are important energy–rich compounds which serve as energy storage molecules in muscles of vertebrates and invertebrates. Some important examples of naturally occurring phophagens are phosphocreatine, phosphoarginine and phosphot-aurocyanine. The standard free energy of hydrolysis of such compounds is about 10.0 k cal/mol at normal physiological pH.

$$\text{Phosphocreatine} + \text{ADP} \underset{G^{\circ} = -\,10.3 \text{ k cal/mol at pH } 7.0}{\overset{\text{Creatine kinase}}{\rightleftharpoons}} \text{Creatine} + \text{ATP.}$$

6. Cyclic AMP (cAMP)

Cyclic adenosine monophosphate (c AMP) is also an energy rich compound which not only functions as a second messenger, but also serves to activate a specific enzyme, protein kinase, that transfers a phosphate group from a donor to acceptor. The standard free energy of hydrolysis of this compound is – 12.0 k cals/mol.

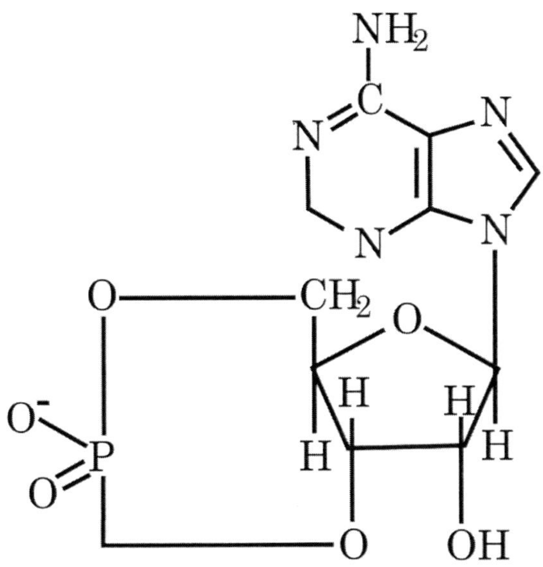

Adenosine 3', 5' cyclic monophosphate

4.12. ATP AND CELL YIELDS

In the earlier discussion, it was made clear that the ATP is the most important energy rich molecule in living system. Bioenergetics of this molecule has also been discussed. At this juncture, it is pertinent to understand the amount of ATP required to synthesise a known amount of the given molecule. It is possible to calculate if the chemical composition of a microbe is known. The maximum possible amount of ATP may be determined on a molar basis from consideration of established synthetic pathways. Calculations of this nature are useful for three reasons: (i) they allow us to compare observed cell yields with what is theoretically possible, (ii) they allow us to determine the extent to which in a given situation, biosynthetic energy is used for the addition of small or larger monomers and (iii) they allow us to determine the extent to which polymerization involves expenditure of large amount of energy for addition of few monomers.

The amount of energy required for synthesis of a constant amount of cell dry weight reflects its chemical composition. Chemical composition and the requirements of ATP of two categories of cells are presented in Table 4.4 and 4.5

Table 4.4. The micromoles of ATP required to synthesize 100 mg of cells composed of 40% DNA, 30% RNA, 20% protein and 5% carbohydrates

Component	Percentage	Monomer weight	Micromoles monomer (100 mg)	ATP per monomer	Micromoles ATP (100 mg cells)
DNA	40	300	133	5	665
RNA	30	300	100	5	500
Protein	20	150	133	3	399
Lipid	5	60	84	3	252
Carbohydrate	5	180	28	2	56
Total	100		478		1872

Total potential micrograms per ATP micromole = 53.4

Table 4.5. The micromoles of ATP required to synthesize 100mg of cells composed of 5% DNA, 5% RNA, 20% protein, 40% lipid and 30% carbohydrate

Component	Percentage	Monomer weight	Micromoles monomer (100 mg)	ATP per monomer	Micromoles ATP (100mg cells)
DNA	5	300	17	5	85
RNA	5	300	17	5	85
Protein	20	150	133	3	399
Lipid	40	60	667	3	2,001
Carbohydrates	30	180	167	2	334
Total	100		1001		2,904

Total potential micrograms per ATP micromole = 34.4

The data presented in Table 4.4 provides some interesting information. The total energy requirement for 100 mg of cells is 1,872 micromoles of ATP. Out of this, 62.2 % is utilized for nucleic acid synthesis, 21.3% is used for protein synthesis, 13.5% is used for lipid synthesis and only about 3% is used for carbohydrate synthesis. Further, as is evident from the table the organisms require 1872 micromoles of ATP to synthesize 100 mg of cells. Each micromole of ATP, therefore, could potentially give rise to 1,00,000/1872=53.4 micrograms/micromole. Interestingly, this theoretical yield is about five times more the yield actually observed for many microbes. Thus, it appears that the cells are only about 20% efficient in the use of ATP for synthesis. Exceptionally, some organisms are about 40% efficient in energy utilization for biosynthesis.

A comparison of table 4.4 and 4.5 also reveals another interesting information. For the cells of table 4.5, synthesis of 100 mg of cells requires 2,904 micromoles of ATP, an approximate 50% increase in the total energy required compared to the energy required for the cells described in Table 4.5. In addition, for the cells of second category (Table 4.5), 68% the total energy is used for lipid synthesis, while the first category organism, 62% of the total biosynthetic energy is used for nucleic acid formation. The amount of energy required for protein synthesis are identical in both cases, however, the fraction of total biosynthetic energy used for protein synthesis in the cells of second category is smaller than that of cells in the first category. The majority of the synthetic energy used for the cells in second category is involved in the addition of relatively small building blocks to monomers, while for the first category, most of the biosynthetic energy is used to incorporate relatively large monomers to nucleic acids. Consequently, the increased energy for lipid synthesis for the cells of second category is compensated for by decrease in the amount of energy required for nucleic acid formation. The amount of energy required to form the carbohydrates in the cells of second category is approximately six times that of the first category cells, but in both cases, the total energy expended for carbohydrate synthesis is a rather small fraction of the total biosynthetic energy.

4.13 FURTHER READING

Anraku, Y. (1988), Bacterial electron transport chains, pp. 101-132. In *Annual Review of Biochemistry*, Vol. 57. Palo AHO.

Baker, R.D.J., Cartledge, T.G., Dewhurst, F. and Jenkins (1992), Principles of cell energetics. Butterworth-Heinemann Ltd., Oxford OX2 8DP.

Bakker, E.P. (1990), The role of alkali-cation transport in energy coupling of neutrophilic and acidophilic bacteria : An assessment of methods and concepts. *FEMS Microbiol Rev.*, **75 :** 319-334.

Boyer, P.D. (1997), The ATP synthase – A splendid molecular machine. *Annu. Rev. Biochem.*, **66 :** 717-749.

Cross, R.L. (1992), The reaction mechanism of F_0-F_1-ATP synthases, pp 317-330. In : Molecular Mechanisms in Bioenergetics, L. Ernster (ed.) Elsevier Science Publishers, Amsterdam.

Dawes, I. (1986), Microbial Energetics : Oxford : Blackwell, pp. 125-158.

Dimroth, P. (1990), Energy transductions by an electrochemical gradient of sodium ions, pp. 114-127. In : The Molecular basis of bacterial metabolism. G. Hauser and R. Thauer (eds.), Springer-Verlag, Berlin.

Driessen, A.J.M. and Konings, W.N. (1990), Energetic problems of bacterial fermentations : extrusion of metabolic end products, pp. 449-478. In : Bacterial Energetics (T.A. Krulwich ed.), Academic Press, New York.

Finel, M. (1993), The proton translocating NADH : Ubiquinone oxidoreductase. A discussion on selected topics. *J. Bioenerg. Biomemb.,* **25 :** 357-366.

Harold, F.M. (1986), The vital force : study of bioenergetics, W.H. Freeman & Co., New York.

Kashket, E.R. (1985), The proton motive force in bacteria : A critical assessment of methods. *Ann. Rev. Microbiol.,* **39 :** 219-242.

Lehninger, A.L., Nelson, D.L. and Cox, M.M. (1993), Principles of Biochemistry, 2nd edn. New York, Scientific American Books.

Mitchell, P. (1966), Chemiosmotic coupling in oxidative and photosynthetic phosphorylation. *Biol. Rev. Cambridge Philos Soc.,* **41 :** 445-502.

Mitchell, P. (1979), Compartmentation and communication in living systems. Ligand conduction : A general catalytic principle in chemical, osmotic and chemiosmotic reaction system. *Eur. J. Biochem.,* **95 :** 1-20.

Nichols, D.G. and Ferguson, S.J. (1992), Bioenergetics 2, Academic Press, London.

Ramasarma, T. (1998), A profile of adenosine triphosphate. *Curr. Sci.,* **74(11) :** 953-966.

Ramasarma, T. and Susheela, L. (1994), In Biomembranes : Architecture, Biogenesis, Bioenergetics and Differentiation (ed. L. Packer), Academic Press, New York, pp. 261-277.

Richardson, D.J. (2000), Bacterial respiration : A flexible process for a changing environment. *Microbiology,* **146 :** 551-571.

Saier, M.H. Jr. (1997), Peter Mitchell and his chemi-osmotic theories. ASM News, **63 :** 13-21.

Senior, A.E. (1990), The proton translocating ATPase of *Escherichia coli. Ann. Rev. Biophys. Chem.,* **19** : 7-41.

Trumpower, B.L. (1990), Cytochrome bc I complexes of microorganisms. *Microbiol. Rev.,* **54 :** 101-129.

Voet, D. and Voet, J.G. (1995), Biochemistry, 2nd edn., New York, John Wiley and Sons.

White, D. (1995), Physiology and Biochemistry of Prokaryotes, Oxford University Press, New York.

■■■■

Chapter - 5

Enzymes, Coenzymes and Cofactors

5.1. INTRODUCTION

Enzymes (en-in, zyme-feast) are macromolecules of biological origin. They are produced within the living cells and catalyse biochemical reactions either in the same cell or sometimes outside the cell. As these catalyse biological reactions, they are also called biocatalysts or biological middlemen. They are made of proteins either partly or wholly and are specific in their function. They are required in very minute quantities and affected by many factors which in turn affect protein configuration. They do not get interfered with several of other biochemical reactions going on simultaneously in a cell. The enzymes produced by the cell, act within the cell are endoenzymes (or intracellular enzymes), where as enzymes active outside the cell after excretion into the environment are called exoenzymes (or extracellular enzymes). Many of the enzymes in bacteria and fungi are extracellular. These are generally hydrolytic and help in nourishment. However, in some bacteria they take part in the synthesis of mucilage of capsule.

The subject of enzymology has a long history. Recorded history of enzymes dates back to 1752 when Rene Reamur demonstrated the gastric juice of kite could dissolve foods. Louis Pasteur attributed the conversion of sugar into alcohol and CO_2 to the catalytic action of some ferment present in the yeast cell. Buchner (1897) was first person to extract the ferment present in the yeast cell and coined the term **zymase**. Willy Kuhne coined the term enzyme to this ferment. James Sumner (1926), for the first time, purified and crystallized the enzyme, urease, from jack bean and demonstrated its proteinaceous nature. It was only in 1963 the first amino acid sequence of bovine pancreatic ribonuclease A was determined. The enzymes which catalyse biochemical reactions show many similarities with catalysts of chemical reactions (Table 5.1).

Table 5.1 : Comparison between catalysts and enzymes

	Catalyst	Enzyme
1.	Remain unchanged qualitatively and quantitatively	Remain unchanged qualitatively and quantitatively
2.	Increase the rate of reaction but not initiator.	Increase the rate of reaction but not initiator
3.	Catalysts are not biological products	Enzymes are biological products
4.	Catalysts lower the activation energy of substratum, but less than that of enzyme	Enzymes greatly lower the activation energy of the substratum

	Catalyst	Enzyme
5.	Catalysts do not catalyse reversible reaction	Enzymes catalyze reversible reaction
6.	Chemical nature of different catalysts vary	Enzymes are partly or totally made of proteins
7.	They arc not affected by pH and temperature	They are affected by pH and temperature
8.	They have small molecular weight	They are large molecular weight compounds
9.	They are specific in their action	They are specific in their action
10.	They are not colloidal.	They are colloidal in nature.

5.2. DISTRIBUTION

Majority of bacteria and fungi being heterotrophs excrete enzymes outside and most of them are hydrolytic in nature. They are extracellular. On the other hand, majority of enzymes are produced and perform their functions inside the cell and are the intracellular or endoenzymes. They regulate the essential metabolic processes. Inside the cell, may be they distributed in cytosol, different cell organelles like nucleus, mitochondria, chloroplasts, peroxisomes, lysosomes etc.

5.3. CHEMISTRY

All the enzymes are partly or wholly proteins of high molecular weight. Depending on the composition they can be divided into two types :

1. Simple enzymes
2. Conjugated enzymes

5.3.1 Simple Enzymes

These are exclusively made of proteins e.g., urease, amylase, papain, pepsin and trypsin.

5.3.2 Conjugated Enzymes

These enzymes consist of atleast two parts (Fig. 5.1). One is protein part called **apoenzyme**, while the other is non-protein and is called **co-factor**. If the cofactor is organic substance, it is called **coenzyme**. Coenzymes are mostly derivatives of vitamins (*e.g.,* biotin, cobalamine, nicotinamide, pyridoxal phosphate). The two components (protein and non-protein) collectively called **holoenzyme**. The organic cofactors generally are loosely bound, while inorganic components are tightly bound. Such tightly bound cofactors are called **prosthetic groups** (*e.g.*, heme). Holoenzymes are relatively stable to heat (thermostable). All enzymes are proteins, however, ribozyme is an exception as it contain only RNA. This was discovered by Thomas Cech and Sydney Altman (1989) in a protozoan, *Tetrahymena thermophila* for which they were awarded Nobel prize in 1989.

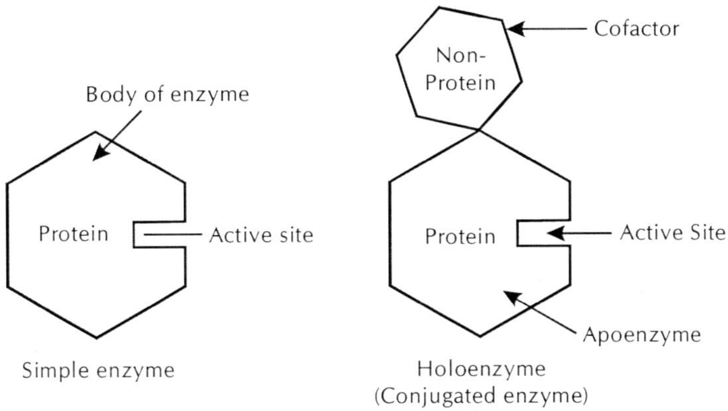

Fig 5.1 Morphology of simple and holoenzymes

Conjugated enzymes are active only when apoenzyme and cofactor are united. Specificity of enzyme is attributed to apoenzyme, while cofactors (coenzyme) are non-specific. A given coenzyme may participate in several types of enzymatic reactions. The specificity of enzyme is due to its definite 3-dimensional conformation of protein and to its active site. Active site is created by the 3-D proteins which provide the correct dimension and topology to accommodate and bind with specific substratum. An enzyme may have only one active site, while others may have more than one active sites. For example, nitrogenase has five active sites. Some enzymes consist of regulatory sites in addition to active sites where regulatory, molecules may bind to activate or inactivate the enzyme.

5.4. TYPES OF ENZYMES

Based on the catalytic functions, enzymes may be divided into the following three types.

1. Isoenzymes
2. Allosteric enzymes
3. Multienzymes

5.4.1. Isoenzymes

Isoenzymes are different physical forms of the same enzyme, performing same function within the same tissues or organism, but at different rates. When the variants of an enzyme are within the same species of an organism, they are called **intraspecific,** but when they are from different species they are called **interspacific**. They are products of different genes and differ in their physico-chemical properties. The isoenzymes differ in their electrophoretic mobility, isoelectric pH and binding with some specific ligands. Isoenzymes also differ in their Km or Michaelis Menton constant for their substratum. More than 100 enzymes are known to have isoenzymes. For example, Lactate dehydrogenase (LDH) in man is reported to have five isoenzymes which catalyse the same biological reactions in most animal tissues.

5.4.2. Allosteric Enzymes

Allosteric enzymes contain two or more receptor sites which are geometrically different and non-overlapping. One is an **active site** and others are **allosteric site(s)** or regulatory sites (Fig. 5.2). Allosteric site is restricted to only allosteric enzymes. The substrates do not bind to allosteric site and they bind only to active site. Substances which bind to allosteric site and cause reversible change in enzyme structure are called allosteric substances or modulators and the phenomenon is called allosteric transition.

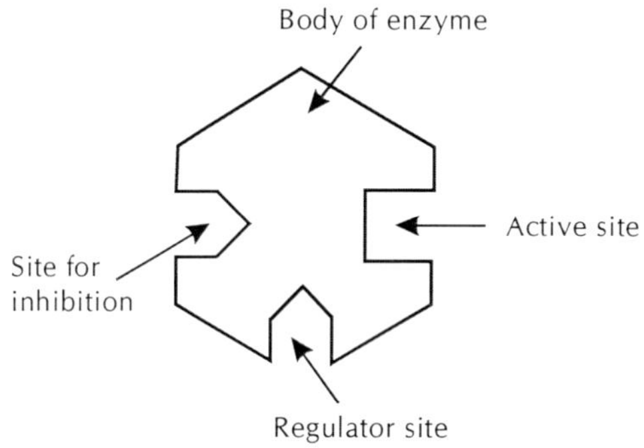

Fig 5.2 Allosteric enzyme showing different sites

Allosteric substances which activate the reaction are called **positive modulators**, which inhibit enzyme activity are called **negative modulators** (allosteric inhibition). Allosteric activator may be same substrate molecule which binds to allosteric site to make the enzyme to function. Such kind of enzymes are called **homotrophic**. But when it is different they are called **heterotrophic allosteric** enzymes, e.g., phosphofructokinase, biphosphofructose phosphatase.

$$\text{Fructose - 6- phosphate} \underset{\text{Biphosphofructose phosphatase}}{\overset{\text{Phosphofructokinase}}{\longleftrightarrow}} \text{Fructose -1,6 biphosphate}$$

5.4.3. Multienzyme or Enzyme Complex

When a substratum is acted upon by more **than one enzyme** in a sequence in such a way that product of one enzyme is the substratum for other enzyme and so on is called **enzyme complex** or **multienzyme system**. Pyruvic acid dehydrogenase of *E.coli* is a good example. In this, there are three independent enzymes.

1. Pyruvate decarboxylase
2. Lipoyl reductase transacetylase and
3. Dihydrolipoyl dehydrogenase.

Fatty acid synthetase enzyme is another example of multienzyme system (Fig. 5.3).

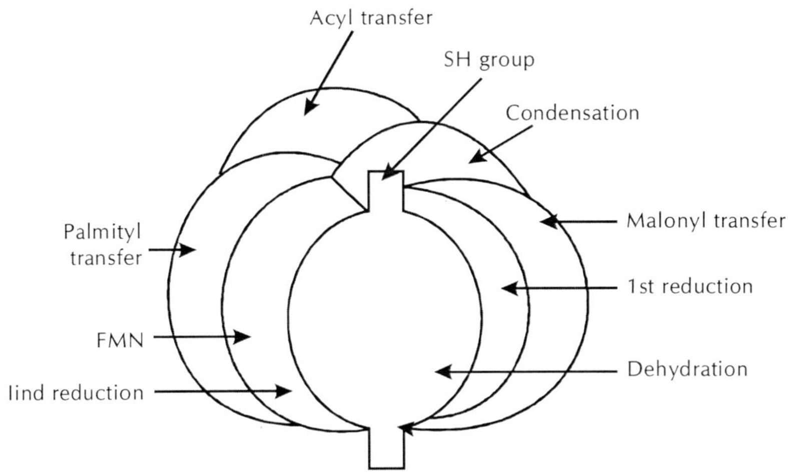

Fig 5.3. Multienzyme (fatty acid synthase)

5.5. NOMENCLATURE

Enzymes are named in different ways. Some enzymes are named after their substratum. For example, sucrase acting on sucrose, lipase acting on lipid, maltase acting on a maltose. Enzymes are also named on the type of reaction they catalyse e.g., Isomerase, dehydrogenase, transaminase, phosphatase, carboxylase etc. Still some enzymes are named arbitrarily such as pepsin and renin. In others its product forms the basis for enzyme name e.g., rhodanase enzyme .

5.6. CLASSIFICATION

Enzymes were previously classified into two types based on the type of reaction they catalyse.

i. **Hydrolases:** Break the substratum in presence of water molecule, e.g., proteases, lipases, amidases.

ii. **Desmolases:** Take part in reactions other than synthesis e.g., aldolases, hydralases, transaminases, oxidases, isomerase.

In an effort to eliminate the confusion and to provide rule for rationality in naming the enzymes, a scheme for the systematic functional classification and nomenclature of enzymes was adopted by the International Union of Biochemistry (IUB) in 1961. According to this system, enzymes are classified and named according to the nature of chemical reactions they catalyze. According to this classification each enzyme name consists of four parts, one after the substratum and other after the type of reaction catalysed with suffix-ase. Further, the enzymes are given a four digit number. For example alcohol dehydrogenase (1111). First digit denotes reaction type (oxidoreductase), second digit denotes the subclass and third digit denotes sub-subclass and the fourth digit denotes particular enzyme named. Each class has been a alloted number, *e.g.,*

1. Oxidoreductases   2. Transferases   3. Hydrolases

4. Lyases   5. Isomerases and   6. Ligases or synthetases

Each class is further classified into several sub-classes on the basis of type of bond split or formed, chemical group transferred or removed. Main sub classes and sub-subclasses are indicated by index number at 2nd and 3rd digit place and several number of specific enzyme with in its own sub-class is indicated by 4th digit, e.g., alcohol dehydrogenase (1111). In this, the first digit denotes type of reaction (oxidoreductase) 2nd digit denotes subclass (acting on CH-OH group of donor) and 3rd sub-sub-class NAD or NADP act as acceptor, 4th digit indicates particular enzyme name. The details of enzyme classification are precised in table 5.2.

Table 5.2. Classification of enzymes (according to IUB 1961)

	Main class and sub-class	Example
I.	**Oxido reductases:** Transfer of group from one kind of molecule to another or from one position to another in the same molecule. Mediate oxidation and reduction reactions which include dehydrogenases, oxidases, oxygenases, hydroxylases, hydroxidases and reductases	
1.1	Acting on CH-OH group of donors	
1.1.1	With NAD or NADP as acceptor	Alcohol dehydrogenase
1.1.3	With O_2 as acceptor	Glucose oxidase
1.2	Acting on CHO or CO-group of donor	
1.2.1	With NAD or NADP as acceptor	Glyceraldehyde 3 P dehydrogenase
1.2.3	With O_2 as acceptor	Xanthine oxidase
1.3	Acting on CH-CH group of donor	
1.3.1	With NAD or NADP as acceptor	Dihydrouracil dehydrogenase
1.3.2	With a cytochrome	Acetyl CoA dehydrogenase
1.4	Acting on CH-NH_2, group of donor	
1.4.1	With NAD or NADP as acceptor	Glutamate dehydrogenase
II.	**Transferases:** Transfer a group from one kind of molecule to another or from one position to another in the same molecule	
2.1	Transferring C_1, groups	
2.1.1	Methyl transferases	Guanidoacetate methyl transferase
2.4	Glycosyl transferases	Phosphorylase
2.6	Transfering amino group	Transaminases
2.7	Transfer of P-containing groups	Hexokinases
III.	**Hydrolases:** Cause cleavage of the substratum. Variety of bonds (ester, glycosidic bonds, peptide bonds) are broken with the help of water.	
3.1	Acting on ester bonds	
3.11	Carboxylic ester hydrolases	Esterases, lipases
3.1.3	Phosphoric monoester hydrolases	Phosphatases

	Main class and sub-class	Example
3.2.1	Glycosidases	Amylase, β-glycosidase
3.4.4	Peptidyl peptide hydrolases	Pepsin, trypsin chymotrypsin
IV.	**Lyases :** Remove groups from substratum by mechanisms other than hydrolysis, leaving double bond	
4.1	C-C lyases	
4.1.1	Carboxyl lyases	Pyruvate decarboxylase
4.1.2	Aldehyde lyases	Aldolases
4.2.1	Aldehyde lyases	Fumarase (Fumarate hydratase)
V.	**Isomerases:** Catalyse rearrangement of atoms or atomic groups inside a molecule to form geometrical, optical or positional isomers (cis, trans isomer)	
5.1	Racemases and epimerases	
5.1.1	Acting on amino acids	Alanine racemase
5.1.3	Acting on carbohydrates	Ribulose 5P epimerase
VI.	**Ligases :** Catalyse the linking of carbon compounds. Utilizes the energy made available due to simultaneous breaking of pyrophosphate bond in ATP or similar compounds (GTP, CTP)	
6.1	Forming C-O bonds	
6.1.1	Amino acid - RNA ligase	Amino acid activating enzyme
6.2	Forming C-S bonds	
6.2.1	CoA ligase	Succinic thiokinase
6.3	forming C-N bonds	
6.3.1	Acid ammonia ligases	Glutamine synthetase
6.3.2	Acid-amino acid ligases	Peptide synthetase

5.7. MECHANISM OF ACTION

Enzyme is active in catalytic action of biochemical reaction. The place where the substratum binds with the enzyme is called **active centre**. The enzyme and substratum form a complex at the active centre.

$$\underset{\text{(Enzyme)}}{E} + \underset{\text{(Substratum)}}{S} \longleftrightarrow \text{E.S. complex}$$

This binding action makes both enzyme and substratum stable. The contact between substratum and enzyme may be due to either ionic bonds, hydrogen bonds or vandervaal forces. The active sites of enzyme have some special groups such as NH_2 , COOH, - SH which bind the substratum through above bonds to form a transitional (intermediate) compound called enzyme-substratum complex (ES). This reaction is a exothermic and releases some energy which raises the energy level of the substratum molecule, thus bringing it to active state. This phenomenon is known as activation energy or energy of activation (Fig. 5.4).

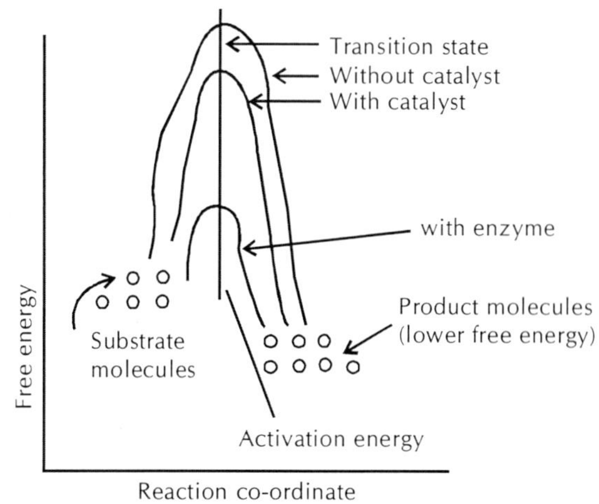

Fig 5.4 Mode of enzyme action

Two mechanisms have been proposed to explain substratum - enzyme complex formation.

1. Lock-key theory (template model)
2. Induced - fit theory.

5.7.1. Lock-key Theory

This theory was proposed by E. Fischer (1898). According to this theory, reaction of substratum and enzyme is analogous to lock and key. Enzyme is analogous to key. The geometrical configuration of socket is fixed. Similarly, substratum has also got fixed geometrical configuration like that of key. A particular lock can be opened or closed by, a particular key. Likewise, a particular substratum can be found at active site of a particular enzyme forming substratum-enzyme complex.

Enzyme–substratum complexing is a tight fitting and active sites of enzymes is complementary to substratum molecule (Fig. 5.5). Subsequently enzyme-substratum complexes, results in the transformation of substratum resulting in the product formation due to activity of reaction site. Since product has lower free energy, it is released. Now the enzymes are free to receive another molecule of substratum and thus enzyme activity continues.

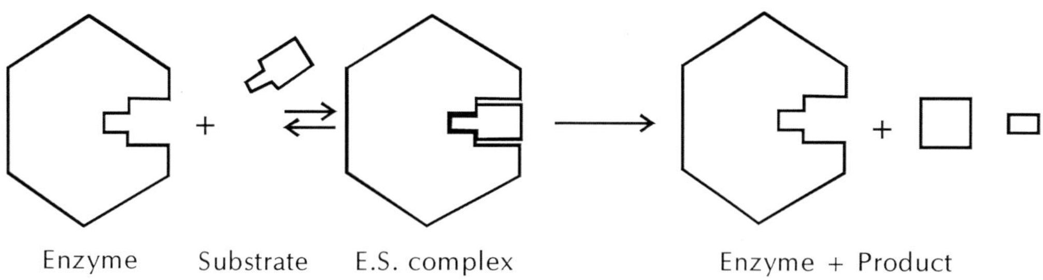

Fig 5.5. Diagrammatic representation of mechanism of enzyme action according to Lock-Key theory

5.7.2. Induced Fit Theory

This theory was proposed by Koshland (1959). In fact this is an improved version of lock-key theory. This theory assumes that active sites of the enzymes are flexible and not pre-fixed.

The geometrical configuration of active sites of enzymes is inducible. Active site changes its shape according to geometrical configuration of substratum to fit completely (Fig. 5.6). According to this theory active site of enzyme comprises of two groups: buttressing and catalytic. The buttresting part supports the substratum, while catalytic part transforms the substratum into product. Substratum binds with buttressing group, while catalytic group lies opposite the substratum bonds to be broken. Electro Static binding forces exert strain to form product.

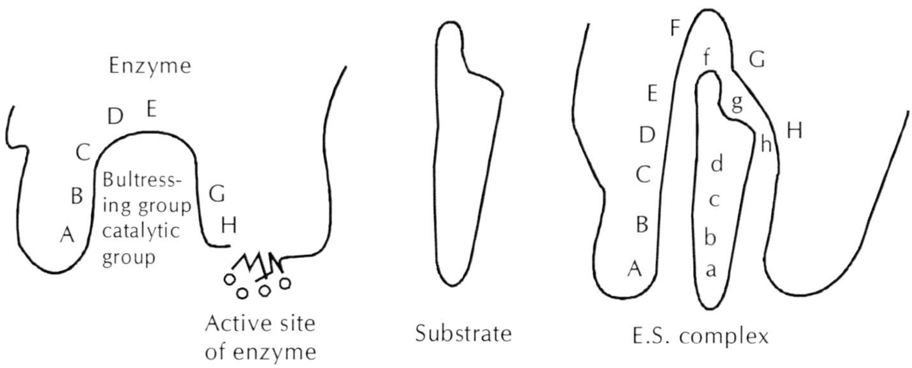

Fig. 5.6. Diagramitc representation of mechanism of enzyme action according to Induced fit theory

5.8. PROPERTIES OF ENZYMES

All enzymes are proteins besides having either a small amount of organic compound (co enzyme) or inorganic substance (prosthetic group) and exhibit several properties which are specific to enzymes. Some of the properties are discussed below.

5.8.1. Catalytic Action

Enzymes mediate biochemical reactions even when they are in small quantity without changing themselves. Small amount of enzyme may produce large number of products. The catalytic power of an enzyme is measured in terms of **turn over number** which is a number of substratum molecules converted into product per unit time. The activity of enzyme is expressed in units. *One unit of enzyme activity is the amount of enzyme which catalyses the transformation of 1μ mol of substratum per minute under defined conditions and it is expressed '1U'.*

5.8.2. Reversibility of Action

Enzymes carry both forward and backward reactions. For example, enzyme lipase brings out synthesis of lipids from fatty acids and glycerol, and causes hydrolysis of lipids into fatty acids and glycerol. However, enzyme urease is an exception as it can hydrolyse urea into carbondioxide and ammonia, but cannot synthesize urea.

5.8.3. Colloidal Nature

As enzymes are either totally or partly made up of proteins, they are colloidal in nature. They have large molecular weight and have low diffusion rate and form colloidal system in water.

5.8.4. Heat Sensitivity

The enzymes are proteins and exhibit sensitivity to temperature. At comparatively higher temperature they get denatured and become non-functional. Their properties are not restored even after bringing back to ideal temperature. At low temperatures enzymes become temporarily inactivated, but are not denatured. The rate of enzyme activity increases with rise in temperature within a limited range of temperature. The optimum temperature for most of the enzymes is between 30 and 40°C. The optimum temperature of an enzyme may be defined as the temperature at which its enzyme activity is maximum.

5.8.5. Sensitivity to pH

Each enzyme acts best at a particular pH which is called optimum pH. Any increase or decrease in pH leads to slow down or inhibit the enzyme activity. Some enzymes such as phosphorylase is active at alkaline pH, while other involved in conversion of glucose phosphate to starch is active at acidic pH.

5.8.6. Specificity

Enzymes are highly specific in their action. A particular enzyme may act on specific substrate or group or structurally related compounds or only one or two stereo isomers. In general, stereo or substrate specificity of an enzyme may be attributed to protein part of enzyme, while reaction specificity is more attributed to the coenzyme/prosthetic group/metal activators of holoenzyme. The specificity of an enzyme may be absolute group specific, stereospecificity and reaction specificity.

(i) **Absolute specificity:** Some enzymes work on only one substratum. Urease acts only on urea

$$NH_2CONH_2 \xrightarrow{\text{Urease}} CO_2 + NH_3$$

(ii) **Group specificity:** These enzymes catalyse reactions which are structurally related group of compounds.

$$\text{Pyruvic acid} \xleftrightarrow{\text{Lactic acid dehydrogenase}} \text{Lactic acid}$$

(iii) **Optical specificity:** Catalyse only one of the isomers (L or D). It acts on only one of the particular isomers. Arginase acts only on L-arginine but not on D-arginine

$$\text{L-alanine} \xleftrightarrow{\text{Alanine racemase}} \text{D-alanine}$$

(iv) **Geometrical specificity:** It exhibits specificity towards cis-isomer or trans-isomer. Enzyme fumarase adds water to trans-isomer (fumaric acid) but does not react with the cis-isomer (maleic acid).

(v) **Reaction specificity:** One enzyme catalyses only one of the reactions. One enzyme does not interfere in the working of another enzyme.

5.8.7 Inhibition

Enzymes get inhibited by the presence of other incompatible substances or negative modulations. In such cases active form of enzyme is converted into inactive form which adversely affects the biochemical reactions.

The inhibition of enzyme activity may be

1. Competitive 2. Non-competitive 3. Feedback

(i) **Competitive inhibition:** Competitive inhibition occurs at the active site of enzyme due to binding of inhibitor molecule in lieu of substrate molecule. Such kind of inhibitors are analogous (similar) in shape to the substratum. Competitive inhibition can be overcome by addition of substrate molecule.

$$E + S \leftrightarrow \text{ES complex} \longleftrightarrow \text{EP complex} \leftrightarrow E + P$$

$$E + 1 \longleftrightarrow \text{E1 complex}$$

E1 complex does not yield product and enzyme is not freed (Fig. 5.7).

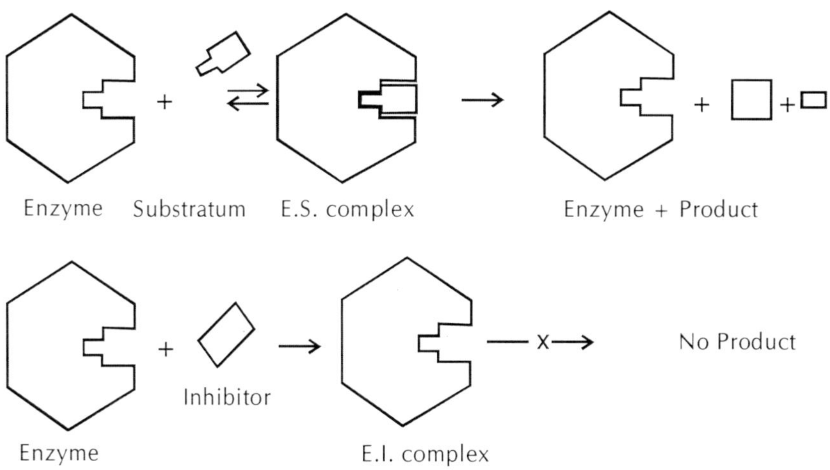

Fig 5.7 (a) Mechanism of enzyme action, (b) Mechanism of enzyme inhibition

Inhibitor competes with the substratum molecule to combine with the enzyme. The degree of inhibition depends upon the relative concentration of the substratum and the inhibitor. Increasing the concentration of substratum competitive inhibition can be decreased. Examples of competitive inhibition: malonic acid, fluorocitric acid, glutaric acid, oxalic acid. Malonic acid inhibits the activity of enzyme succinic acid dehydrogenase because of their structured similarity to enzyme substratum, succinic acid.

Microorganisms like bacteria synthesize folic acid (vitamin B9) from para amino benzoic acid (PABA). Sulfanilamide (a sulfa drug) and other sulfa drugs are structurally analogous to PABA. Hence, sulfa drugs act as enzyme inhibitors and occupy the active site of some bacterial enzymes catalyzing the reaction (*i.e.*, formation of folic acid from PABA). Hence, folic acid synthesis is inhibited and bacteria get killed due to the deficiency of folic acid.

(ii) Non-competitive inhibition: This type of inhibition is restricted to allosteric enzymes. Non-competitive inhibition occurs at allosteric site and not on active live site of enzyme. Non-competitive inhibitors are not analogous to substrate they do not compete with substrates. Non-competitive inhibitions cannot be overcome by adding substratum. Majority of the non-competitive inhibitors are heavy metals (Hg, Pb) or cellular poisons (CO, H_2, CN). The heavy metals combine with -SH group of enzyme while, cyanide and H_2S form metal complexes with metal activators or specific amino acids at or near active sites, Since inhibitor and substrate combine different sites, formation of ES and ES complexes occurs.

(iii) Feedback inhibition: If a reaction proceeds in several steps and the product of first reaction is a substratum for other reaction, then each step is catalyzed by separate enzyme. The end product of reaction if accumulates binds with the enzyme catalyzing first step leading to the inhibition of the process. Such a kind of inhibition is called Feed back inhibition'.

$$\text{Aspartate} \longrightarrow \text{lysine} \longrightarrow \text{aspartyl phosphate} \longrightarrow \text{aspartyl semi-aldehyde}$$

As shown above, in the synthesis of lysine from aspartate, the reaction is inhibited by accumulated lysine by binding it with the enzyme of first reaction.

5.8.8. Kinetics of Action

Enzyme kinetics deals with the rate of enzyme catalytic reactions. It gives information about the specificity of enzyme and mechanism of enzyme action. In general, enzyme catalyzed reactions comprise three steps:

$$E + S \xrightarrow[\text{I step}]{\text{binding}} ES \xrightarrow[\text{II step}]{\text{catalysis}} EP \xrightarrow[\text{III step}]{\text{release}} E + P$$

The first step is binding of enzyme to substratum while, second step is catalysis. In the third step, product is released from the enzyme. Catalysis is faster than the other two steps. The third step is slowest and limits the process.

$$E + S \xleftrightarrow{\text{Fast}} \text{ES complex} \xleftrightarrow{\text{Slow}} P + E$$

If the rate of constant for formation of ES is K_1 and its dissociation into E and S is K_2 then

$$E + S \underset{K_2}{\overset{K_1}{\longleftrightarrow}} ES$$

The conversion of substratum to product from ES complex proceeds with rate constant K_3

$$ES \overset{K_3}{\longleftrightarrow} P + E$$

Functions of the rate constants of three steps are called Michaelis constant, that is

$$K_m = \frac{K_2 + K_3}{K_1}$$

$$K_m = \frac{\text{constant for disappearance of ES}}{\text{constant for formation of ES}}$$

Michaelis-Menton constant

The enzyme catalyzed mechanism was quantified by L. Michaelis and M.L Menton (1913), through the application of law of mass action to the formation of complex (Fig. 5.8).

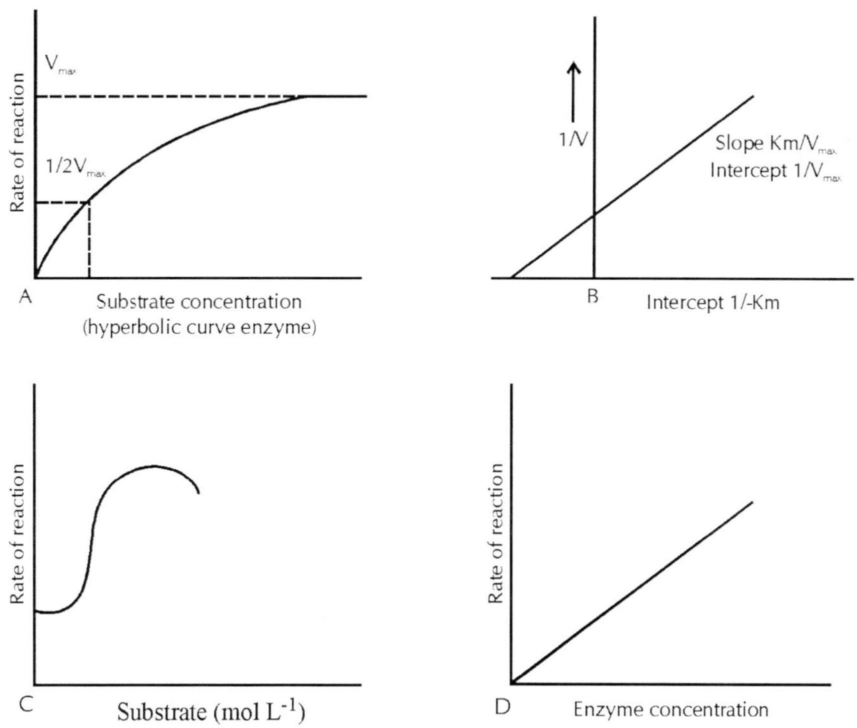

Fig 5.8 Mechanism of enzyme action, (a) Michaelis constant, (b) Line Weaver - Burk plot (c) Sigmoid curve (allosteric enzyme), (d) Michaelis-Menton type curve

They introduced two particularly useful mathematical expressions that for any enzyme relate (S) to V (velocity) and permit quick comparisons of various enzyme catalyzed reactions. The two expressions are called the Michaelis-Menton equation and Michaelis-Menton constant. The equation is

$$V = \frac{V_{max}[S]}{K_m + [S]}$$

where, V = Velocity of the reaction

V_{max} = the maximal velocity

S = substrate concentration and

K_m = Michaelis - Menton constant in moles per litre solution

K_m represents the substrate concentration at which the velocity of the reaction is half of V_{max}.

The Michaelis-Menton equation is sufficient to describe most of the enzyme catalyzed reactions. It can be utilized to determine K_m at various substrate concentrations and the values can be used in predicting rate limiting steps. Therefore V_{max} and K_m should be carefully determined.

5.9. FACTORS AFFECTING ENZYME FUNCTIONS

i. **Concentration of enzyme:** At a specific enzyme concentration, if substratum molecules are in a relatively larger concentration, the enzyme catalysed reaction attains the maximum speed. Further increase of substratum concentration, does not increase the rate of enzyme activity because all the active sites of enzymes are saturated. Now, if the enzyme concentration is increased the rate of reaction increases provided the concentration of substratum is relatively larger.

ii. **Concentration of substratum:** The rate of enzyme activity increases with the increase in substratum. The increase will continue till all active sites of enzyme are occupied by substrate molecules. Further increase in substratum concentration will not increase the rate of enzyme activity because of limiting factor of enzyme concentration.

iii. **Temperature:** The rate of enzyme catalysed reaction is minimum at low temperature. Enzyme activity increases with the increase in incubation temperature upto 40°C. Further increase will inhibit enzyme activity due to protein denaturation.

iv. **pH:** Each enzyme works in a limited pH range. Most of the enzymes are active at neutral pH. Some are active at acidic pH (sucrase), while others are active at alkaline pH (amylase). Every enzyme has optimum pH at which the enzyme exhibits maximum activity. Pepsin has optimum pH 2.0, trypsin 8.0, urease 7.0, arginase 9.9 and catalase 6.8.

v. **Hydration:** For maximum enzyme activity proper hydration is required. Enzyme activity ceases when hydration is low.

vi. **Reaction products:** Accumulation of reaction products may cause feed back inhibition.

vii. **Radiant energy:** High energy radiations (x-ray, ultraviolet rays, γ-rays) are harmful to enzymes since they are able to break various types of bonds and form new bonds (-S-S instead of S-H).

viii. **Industrial applications:** Enzyme renin or rennet obtained from stomach of unweaned calves is used in the cheese making. Now calf rennets are replaced by microbial rennet. Amylases and isomerases are used in the manufacture of soft drinks for hydrolysing starch. In the manufacture of soap and detergents, proteolytic enzymes of microbial origin are used. Similarly, in the leather industry dehairing is done by the use of pancreatic enzymes which hydrolyse the proteins of the hair follicle. Pectic enzymes like pectinlyase and pectin esterase are used in wine industry.

ix **Abatement of pollution:** Hydroxylases from different strains of **Pseudomonas** spp. degrade number of highly toxic detergents, plastics and chemicals (benzene, toluene and xylene).

x. **Genetic engineering:** Restriction enzymes like EcoRI. Hind III, Bam H1 are used in genetic engineering as biological tools.

5.10. FURTHER READING

Berg, J., Tymocko, J. and Stryer, I. (2003), Biochemistry. 5th edn. W.H. Freeman & Co. Ltd.

Bugg, T. (1997), Introduction to Enzyme and Coenzyme Chemistry. Blackwell, Science.

David, Fell (1999), Fundamentals of Enzymology. Oxford University Press.

Fersht, A. (1984), Enzyme structure and Mechanism. W.H. Freeman & Co. Ltd.

Lehninger, A.L., Nelson, D.L. and Cox, M.M. (1993), Principles of Biochemistry. 2nd edn. New York, Scientific American Books.

Palmer, T. and Bonner, P.L. (2007), Enzymes : Biochemistry, Biotechnology, Clinical Chemistry. Woodhead Publishing.

Taylor, K.B. (ed.), Enzyme Kinetics and Mechanisms. Kluwer Academic Publishers.

Voet, D. and Voet, J.G. (1995), Biochemistry, 2nd edn., New York, John Wiley and Sons.

■■■■

Chapter - 6

Microbial Nutrition

6.1. INTRODUCTION

Microbial cells take up nutrients from the environment to achieve the ultimate goal of growth and reproduction. The nutrients serve two purposes : first as a source of building blocks or precursors for the synthesis of new cellular constituents, second to generate energy to drive different metabolic functions. Thus based on physiological purpose, compounds essential for microbial growth can be divided into two major groups. (i) compounds that are either fully or partly incorporated into components of the biomass ("nutrients"), and (ii) compounds that are not incorporated into biomass, but are essential for generation of energy (electron donors or acceptors). However, this distinction cannot always be made in many cases, because there are nutritional categories of organisms in which some compounds can fulfil both functions at the same time. For instance, reduced carbon sources are employed by chemoheterotrophs to obtain carbon precursors for biosynthesis as well as to generate energy (Fig. 6.1).

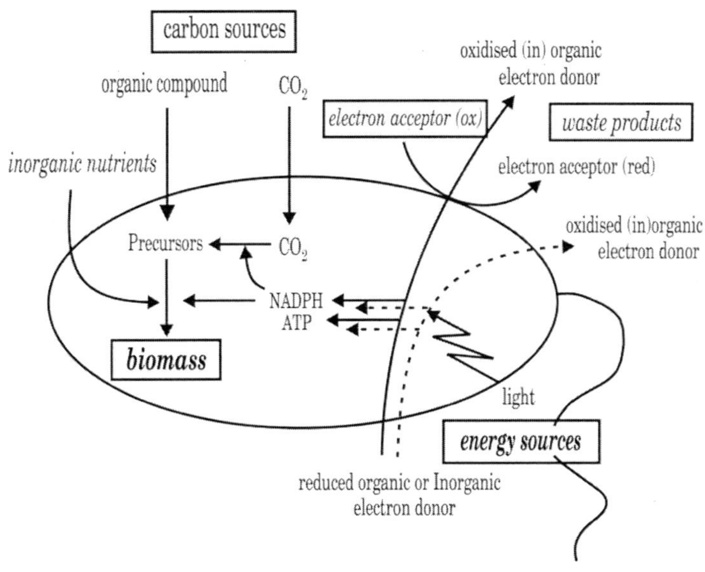

Fig. 6.1. Simplified diagram of the physiological functions of nutrients for the growth of microorganisms.

6.2. COMPOSITION OF MICROBIAL CELL

Chemical composition of a cell either at molecular or elemental level reflects the nutritional requirement of the cell. The composition of a microbial cell is highly dependent on the environment and cultivation conditions. The type of cellular constituents (*e.g.,* ribosomes, particular enzymes, membrane and cell wall components, and compounds of the metabolic pool) and their amount can vary enormously. Despite this variation, the macromolecular and the elemental composition of microbial biomass, including cell material from archaea, eubacteria, and eukaryotes varies in narrow range (Tables 6.1 and 6.2).

Table 6.1. Chemical composition of a prokaryotic cell

Molecule	Dry weight (%)	Molecules/cell	Different kinds
Total macro molecules	96	24,610,000	~2500
Protein	55	2,350,000	~1850
Polysaccharide	05	4,300	2
Lipid	09.1	22,000,000	4
DNA	03.1	2.1	1
RNA	20.5	255,500	~660
Total monomers :	3.5		~350
Amino acids and their precursors	0.5		~100
Sugars and their precursors	2		~50
Nucleotides and their precursors	0.5		~200
Inorganic ions	1		18

Table 6.2. Elemental composition of microbial cell

Element	% of dry weight
Carbon	50
Oxygen	21
Nitrogen	12
Hydrogen	8
Phosphorus	3
Sulphur	1
Potassium	1
Magnesium	0.5
Calcium	1
Chlorine	0.5
Iron	0.5
Sodium	1
Other elements (Mo, Ni, Co, Mn, Zn, W, Se, etc.)	0.5

6.3. CLASSIFICATION OF NUTRIENTS

The chemicals present in the nutrient molecules and incorporated into new cell material can be divided into five different classes (Table 6.3). The division is mainly based on the quantities of these elements required for growth and their occurrence in dry biomass. Class one elements are indispensable for microbial growth. Class two elements are required in significant amounts, whereas those in class three and four are usually referred to as trace elements. For elements categorized in classes two and three, it can be demonstrated experimentally that they are essential, whereas it is difficult to prove that the elements of class four are essential for growth.

Table 6.3. Classes of nutrients used for microbial growth based on their incorporation and occurrence in dry cell mass

Class	Category	Nutrients
Class 1	Always essential	Major elements : C, H, O, N
		Minor elements : P, S, K, Mg, Fe, Ca, Na, Cl^-
Class 2	Mostly essential	Microelements : Mn, Co, Cu, Mo, Zn, Ni, Se, W
Class 3	Essential in special cases	B, Al, Si, V, Cr, As, Sn, I
Class 4	Very rarely essential	Be, F, Sc, Ti, Ga, Ge, Br, Zr
Class 5	Growth factors (essential for special strains)	Amino acids, purines and pyrimidines, vitamins, etc.

We, in this book, through traditional way treat class I elements as macroelements, class II elements as micro or trace elements.

6.3.1. Macroelements

Macroelements are those which are required in large quantities. For instance grams per litre of the medium. The requirements of macronutrients are common to all organisms including the microorganisms. They carry almost the same physiological functions in all organisms. Twelve elements are identified as macronutrients. The source and main physiological role of these elements are presented in Table 6.4.

Table 6.4. Macroelements, their source and some of functions in microorganisms

Element	Source	Functions in metabolism
Carbon	Organic compounds, CO_2	Main constituent of cellular materials.
Oxygen	O_2, H_2O, organic compounds, CO_2	Main constituent of cellular materials.
Hydrogen	H_2, H_2O, organic compounds	Main constituent of cellular materials.
Nitrogen	NH_4^+, NO_3^-, N_2, organic compounds	Main constituent of cellular materials.
Sulphur	SO_4^{2-}, HS^-, S^0, $S_2O_3^{-2}$	Constituent of cysteine, methioine, thiamine, pyrophosphate, coenzyme A, biotin and alpha lipoic acid

Element	Source	Functions in metabolism
Phosphorous	HPO_4^{-2}	Constituent of nucleic acids, teichoic acid, phospholipids and nucleotides.
Potassium	K^+	Principal inorganic cation in the cell, cofactor of pyruvate kinase
Magnesium	Mg^{+2}	Cofactor of many coenzymes; present in cell walls, membrane, ribosomes and phosphate esters.
Calcium	Ca^{+2}	Cofactor of amylases, proteases; cell walls; calcium dipicolinate is an important component of endospores.
Iron	Fe^{+2}, Fe^{+3}	Present in cytochromes, ferredoxins and other iron sulphur proteins; cofactors of enzymes.
Sodium	Na^+	Involved in various transport processes, osmoregulation.
Chlorine	Cl^-	Important inorganic anion in the cell.

6.3.2. Microelements

Apart from the above twelve major elements, organisms require some nutrients in small quantities *i.e.,* micrograms per litre of the medium. Although required in just tiny amounts, micronutrients are, nevertheless, just as critical to cell function as are macronutrients. Microelements are metals, many of which play a structural role in various enzymes. Table 6.5 summarizes the major micronutrients of living systems and gives example of enzymes in which each plays a role. Since the requirements for trace elements is so small, for the laboratory culture of microorganisms it is frequently unnecessary to add trace elements to the culture medium. However, if the culture medium is made out of high grade pure chemicals and high purity distilled water, a trace element deficiency may occur. In such cases, a small amount of a solution of trace elements is added to the medium to make available the necessary metals.

Table 6.5. Elements, source and their functions in microorganisms

Element	Source	Function in metabolism
Zinc	Zn^{+2}	Present in alcohol dehydrogenase, alkaline phosphate, aldolase, RNA and DNA polymerase.
Manganese	Mn^{+2}	Present in bacterial and mitochondrial superoxide dismutase and in photo system II, cofactor in some enzymes (PEP carboxykinase, citrate synthase).
Molybdenum	MoO_4^{-2}	Present in nitrate reductase, nitrogenase, xanthine dehydrogenase and formate dehydrogenase.
Selenium	SeO_3^{-2}	Present in glycine reductase and formate dehydrogenase
Cobalt	Co^{2+}	Present in coenzymes B_{12} containing enzymes (glutamate mutase; methyl malonyl CoA mutase)

Element	Source	Function in metabolism
Copper	Cu^{+2}	Present in cytochrome oxidase, in nitrate reductase of denitrifying bacteria and oxygenases.
Nickel	Ni^{2+}	Present in urease, hydrogenase and in factor F_{430}.
Tungsten	Wo_4^{-2}	Present in formate dehydgenase.

6.3.3. Growth Factors

Growth factors are organic compounds and like micronutrients required in small quantities. They include vitamins, amino acids and purines and pyrimidines.

Of all the growth factors, vitamins are the most commonly needed. Although many microorganisms are able to synthesise the growth factors required, while certain other require one or more of them preformed from the environment. Lactic acid bacteria, which include genera *Streptococcus, Lactobacillus, Leuconostoc* and others, are renowned for their complex vitamin requirements, which are even more extensive than those of humans. A list of various vitamins and their functions in microorganisms are presented in Table 6.6

Table 6.6. Important vitamins and their metabolic functions

Vitamin	Function
Para-amino benzoic acid	Precursor of folic acid
Folic acid	one carbon metabolism; methyl group transfer
Biotin	Fatty acid biosynthesis; beta decarboxylations; carbondioxide fixation
Cobalamine (B_{12})	Reduction of and transfer of single carbon fragments; synthesis of deoxyribose
Lipoic acid	Transfer of acyl group in decarboxylation of pyruvate and alpha keto glutarate
Nicotinic acid	Precursor of NAD^+; electron transfer in oxidation-reduction reactions.
Pantothenic acid	Precursor of co-enzyme A; activation of acetyl and other acyl derivatives.
Riboflavin	Precursor of FMN, FAD in flavoproteins involved in electron transport.
Thiamine (B_1)	Alpha of amino acid and decarboxylations.
Vitamins B_6	Keto acid transformations.
Vitamin K group; quinones	Electron transport; synthesis of sphingo lipids.
Hydroxamates	Iron binding compounds; solubilisation of iron and transport into the cell.
Coenzyme-M	Required by certain methanogens; plays a role in methanogenesis.

Essential nutrients, both macro- and microelements are indispensable for growth of microorganisms. Obviously, they must be getting these nutrients from the substratum while growing naturally. Similarly, while culturing these organisms, these essential elements need to be incorporated into the medium in available form. Table 6.7 provides an information as to how they get the different essential elements in nature as well as growing in culture medium.

Table 6.7. Essential elements and their source in nature and culture media

Element	Usual form of nutrient found in the environment	Chemical form supplied in culture media.
Carbon	CO_2, organic compounds	Glucose, malate, acetate, pyruvate, hundreds of other compounds or complex mixtures (yeast extract, peptone)
Hydrogen	H_2O	H_2O, organic compounds
Oxygen	H_2O, O_2, organic compounds	H_2O, O_2, organic compounds.
Nitrogen	NH_3, NO_3^-, N_2, organic compounds	Inorganic : NH_4Cl, NH_4SO_4, KNO_3 Organic : Amino acids, nitrogen bases, many other nitrogen containing compounds, etc.
Phosphorus	PO_4^{3-}	KH_2PO_4,
Sulphur	H_2S, SO_4^{2-}, organic compounds, metal sulphides, FeS, CuS, ZnS	Na_2SO_4, $Na_2S_2O_3$, Na_2S, cysteine, organic sulphur compounds.
Potassium	K^+ in solution or K salts	KCl, KH_2PO_4
Magnesium	Mg^+ in solution or Mg salts	MgCl, $MgSO_4$
Sodium	Na^+ in solution or NaCl salts	NaCl
Calcium	Ca^+ in solution, $CaSO_4$, Ca salts	$CaCl_2$
Iron	Fe^{++}, Fe^{+++} in solution, FeS, Fe salt	$FeCl_2$, $FeSO_4$, various chelated iron solutions, EDTA etc.

Table 6.8. Characteristics of some complex materials used as ingradients of media

Raw material	Characteristics	Nutritional value
Beef extract	An aqueous extract of lean beef tissue concentrated to a paste	Contains the water soluble substances of animal tissue, which include carbohydrates, organic nitrogen compounds, water soluble vitamins and salts.
Peptone	The product resulting from the digestion of proteinaceous materials *e.g.*, meat, casein and gelatin	Principal source of organic nitrogen; may also contain some vitamins and some times carbohydrates
Yeast extract	An aqueous extract of yeast cells, commercially available as powder	A very rich source of B vitamins, also contains organic nitrogen and carbon compounds.
Agar	A complex carbohydrate obtained from certain marine algae; processed to remove extraneous substances	Used as a solidification agent for media; gels below 45°C; not a nutrient source.

6.3.4. Complex Nutrients

Chemically defined media are needed for the cultivation of autotrophs and are also useful for defining the nutritional requirements of heterotrophs. However, for the routine cultivation of heterotrophs, chemically defined media are not generally used. Instead, certain complex raw materials, such as peptones, meat extract and yeast extract are used, and the resulting media support the growth of wide variety of heterotrophic bacteria. Agar is included as a non-nutritive solidifying agent when a solid medium is desired. The details of these raw materials is presented in Table 6.8.

6.4. CULTURE MEDIA

With the knowledge of information on microbial nutrition and specific habitat of microorganisms, various culture media can be formulated to culture different microorganisms in the laboratory. Sometimes, it become and necessary to devise specialized media for isolation and identification of microorganisms, such as the testing of antibiotic sensitivities, water and food analysis, industrial microbiology and other activities. Media that are commonly used in laboratories can be categorised into:

i. Synthetic or defined media
ii. Complex or undefined media
iii. General purpose media
iv. Enriched media
v. Differential media
vi. Selective media
vii. Characteristic media.

i. Synthetic or defined media: A medium that contains ingredients of known chemical composition of both qualitatively and quantitatively is called a defined medium. *e.g.,*

Asthana and Hawker's medium A

KNO_3	—	3.75 g
KH_2PO_4	—	1.75 g
$MgSO_4$	—	0.75 g
D-Glucose	—	5.0 g
Agar-agar	—	16 g
Distilled water	—	1000 ml

Czapeks dox agar medium

Sodium nitrate	—	2 g.
Dipotassium hydrogen phosphate	—	1 g.
Magnesium sulphate	—	0.5 g.
Potassium chloride	—	0.5 g.
Ferrous sulphate	—	0.01 g.
Sucrose	—	30 g.
Agar-agar	—	15 g
Distilled water	—	1000 ml

ii. Complex or undefined media: A medium that contains some ingredients of unknown chemical composition is called undefined medium. *e.g.,*

Nutrient agar

Peptone	-	5 g
Beef extract	-	3 g
Sodium chloride	-	5 g
Distilled water	-	1000 ml

iii. General purpose media: Media like tryptic soy broth, tryptic soy agar and nutrient agar are called general purpose media because they support the growth of a variety microorganisms, and routinely used in the laboratory. *e.g.,*

Tryptic soy agar

Trypticase (animal peptone)	-	15 g
Phytone (soy peptone)	-	5 g
Sodium chloride	-	5 g
Agar - agar	-	15 g
Distilled water	-	1000 ml

iv. Enriched media: These media support the growth of fastidious microorganisms. These are fortified with specialized nutrients like blood, serum. Enriched media are used primarily to isolate bacteria from cerebrospinal fluids, sputum and abscesses. *e.g.,*

Chocolate agar medium

Protease peptone	-	20 g
Sodium chloride	-	5 g
Disodium phosphate	-	5 g
Dextrose	-	0.5 g
Agar-agar	-	15 g
Distilled water	-	1000 ml

v. Differential media: These media help in distinguishing different groups of bacteria and even permit tentative identification of microorganisms based on their biological characteristics. *e.g.,*

Blood agar medium

Infusion from beef heart	-	5 g
Tryptose	-	10 g
Sodium chloride	-	5 g
Agar-agar	-	15 g
Distilled water	-	1000 ml

vi. Selective media: These media are prepared by the addition of a specific substance to a culture medium that will permit the growth of desired organism, while inhibiting the growth of unwanted organism. *e.g.,*

Mac Conkey agar medium

Peptone	—	20 g
Sodium chloride	—	5 g
Bile salt	—	1.5 g
Lactose	—	10 g
Neutral red solution	—	10 ml
Crystal violet	—	0.001 g
Agar-agar	—	13.5 g
Distilled water	—	1000 ml

Eosin methylene blue (EMB) agar medium

Peptone	—	10 g
Lactose	—	5 g
Dipotassium hydrogen phosphate	—	2 g
Eosin Y	—	0.4 g
Methylene blue	—	0.065 g
Agar-agar	—	15 g
Distilled water	—	1000 ml

(vii) Characteristic media: These media are used to test bacteria for particular metabolic activities, products or requirements. *e.g.,*

Triple sugar iron agar medium

Peptone	-	15 g
Protease peptone	-	5 g
Beef extract	-	3 g
Yeast extract	-	3 g
Lactose	-	10 g
Sucrose	-	10 g
Dextrose	-	1 g
Sodium chloride	-	5 g
$FeSO_4$	-	0.2 g
Sodium thiosulphate	-	0.3 g
Phenol red	-	0.024 g
Agar-agar	-	12 g
Distilled water	-	1000 ml

6.5. NUTRITIONAL CLASSIFICATION OF MICROORGANISMS

All organisms require sources of carbon, energy, hydrogen and electrons for their growth. Microorganisms are classified based on the source of these basic nutrients.

1. Based on carbon source, microorganisms are divided into three types:
   - **i. Autotrophs**: The microorganisms which acquire and assimilate carbon from atmospheric CO_2 are called autotrophs.
   - **ii. Heterotrophs**: The microorganisms which assimilate carbon from reduced and preformed organic molecules are called heterotrophs.

iii. **Mixotrophs**: Some bacteria combine both autotrophic and heterotrophic metabolic processes. These are referred to as mixotrophs.

2. Based on energy source microorganisms are identified into two types :

   i. **Phototrophs**: The microorganisms which acquire energy from sunlight are called phototrophs

   ii. **Chemotrophs**: The microorganisms which acquire energy from the oxidation of chemical compounds (either organic or inorganic) are called chemotrophs.

3. Based on hydrogen or electron sources microorganisms are divided into two types:

   i. **Lithotrophs**: The microorganisms which acquire hydrogen or electrons from reduced inorganic molecules are called lithotrophs.

   ii. **Organotrophs**: The microorganisms which acquire hydrogen or electrons from organic molecules are called organotrophs.

Microorganisms exhibit a great flexibility and versatility in their nutrition and metabolism. Despite the great metabolic diversity seen in microorganisms, most may be placed in one of the following four classes (Table 6.9) based on their primary sources of energy, hydrogen or electron and carbon.

i. **Photolithotrophic autotrophs:** The microorganisms which acquire energy from sunlight, hydrogen or electron from inorganic compounds and carbon from atmospheric CO_2 are called photolithotrophic autotrophs.

ii. **Photolithotrophic heterotrophs:** The microorganisms which acquire energy from sunlight, hydrogen or electron from inorganic compounds and carbon from reduced organic molecules, are called as photolithotrophic heterotrophs.

iii. **Chemolithotrophic autotrophs:** The microorganisms which acquire energy from chemicals, hydrogen or electron from reduced inorganic compounds and carbon from atmospheric CO_2 are called as chemolithotrophic autotrophs.

iv. **Chemoorganotrophic heterotrophs:** Microorganisms which acquire energy from chemicals, hydrogen or electron from organic molecules and carbon from reduced organic molecules are called as chemoorganotrophic heterotrophs.

Table 6.9. Major nutritional types of microorganisms

Nutritional type	Source of energy, H^+/e^-, carbon	Example
Photolithotrophic autotrophy	Light, inorganic compound, CO_2	Algae, purple and green sulfur bacteria, cyanobacteria.
Photolithotrophic heterotrophy	Light, inorganic compound, reduced organic molecule	Purple non-sulfur bacteria, green sulfur bacteria.
Chemolithotrophic autotrophy	Chemicals, inorganic compounds, CO_2	Sulfur oxidising bacteria, hydrogen bacteria, iron bacteria
Chemoorganotrophic heterotrophy	Chemicals, organic compounds, reduced organic molecules	Saprophytic, parasitic, symbiotic bacteria, protozoa, fungi, animals

6.6. UPTAKE AND TRANSPORT OF NUTRIENTS

Obviously, the first step in nutrient utilisation is uptake of the required nutrients by the microbial cell from the environment. Nutrients and other substances needed by the cell must penetrate the cell boundary layers before they can be metabolised. The cell wall does not present much as a barrier to small molecules and ions, but it excludes large molecules with a relative molecular mass above 600. The cell boundary component that really regulates the uptake of more substances into the cell is the cytoplasmic membrane.

Uptake of nutrients is usually specific. Only those nutrients are taken up by the cell for which transport systems are available. Transport is dependent, with few exceptions, on specific permeases and translocases. These are membrane proteins, and the names indicate that they exhibit some of the properties of enzymes. They are substrate specific, may be substrate inducible, and are produced only under conditions that permit protein synthesis.

The term transport gives several meanings in cell biology. In considering only those transport processes, which occur through the cytoplasmic membrane, two main types can be distinguished.

1. Primary transport
2. Secondary transport

6.6.1. Primary Transport

Primary transport consists of those processes that lead to transfer of ions like H^+, Na^+, K^+ and hence to alternations in the electrochemical potential. The forces driving primary transport are either respiration or photosynthetic electron transport, ATP-driven ion pump (ATPases) or ion pumps driven by decarboxylation of metabolites (oxaloacetate, methylmalonyl-CoA).

6.6.2. Secondary Transport

It is applied to all processes resulting in the influx or efflux of cellular metabolites, which are driven by electro chemical potential gradients. Several types of mechanisms for the uptake of substances into the cell can be distinguished (Fig. 6.2), such as :

i. Passive diffusion
ii. Facilitated diffusion
iii. Active transport
iv. Group translocation

6.6.2.1. Passive Diffusion

The non-specific penetration of substances into the cell is called simple or passive diffusion. In this process, solute molecules cross the membrane as a result of difference in concentration which governs the rate of influx of the solute molecule with time. The concentration gradient diminishes until equilibrium is reached. In passive diffusion no substance in the membrane interacts specifically with solute molecules. The rate of passive diffusion is dependent upon the size of concentration gradient between cell's exterior and it's interior. A fairly large concentration gradient is required for adequate nutrient uptake by passive diffusion.

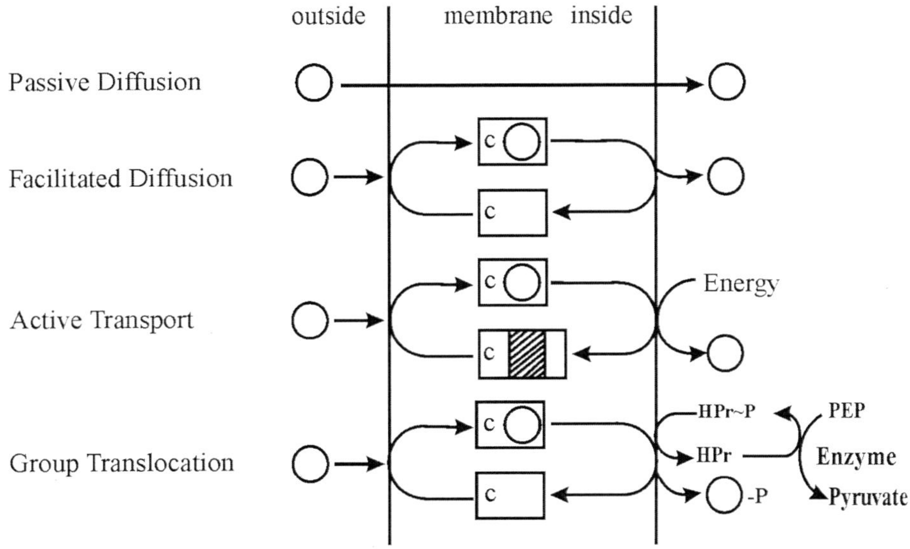

Fig. 6.2. Schematic representation of the four mechanisms for transport of materials into the cell

6.6.2.2. Facilitated diffusion

The rate of diffusion across selectively permeable membrane is greatly increased by using carrier proteins sometimes called permeases, which are embedded in the plasma membrane (Fig. 6.3). As the carrier aids the diffusion process, it is termed as facilitated diffusion.

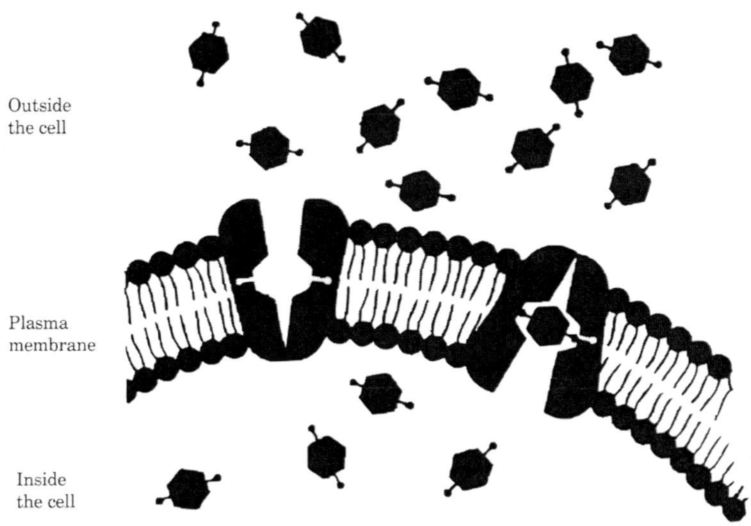

Fig. 6.3. A model of facilitated diffusion.

The rate of facilitated diffusion increases with the concentration gradient much more rapidly. The carrier protein binds and transports as many solute molecules as possible. The curve of concentration gradient against rate of transport reaches a plateau at higher gradient levels and it resembles as enzyme substrate curve (Fig. 6.4). Each carrier is selected and will transport only closely related solutes. A concentration gradient spanning the membrane drives the movement of molecules and no extra energy input is required. If the concentration gradient disappears, net inward movement ceases. This process is much more prominent in eucaryotic cells where it is used to transport a variety of amino acids and sugars.

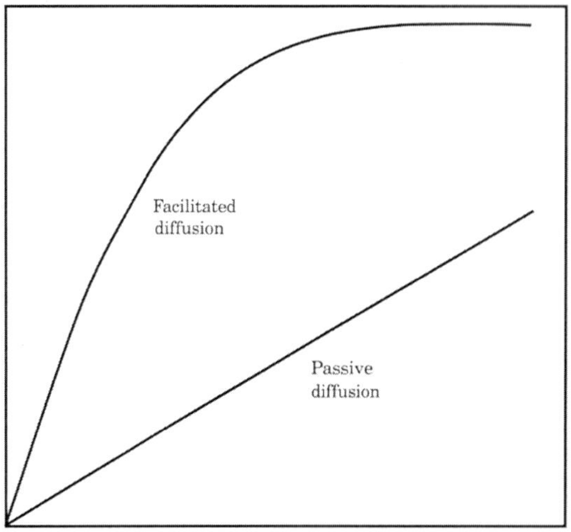

Fig. 6.4. Passive and facilitated diffusion

6.6.2.3. Active Transport

It is the transport of solute molecules to higher concentrations against a concentration gradient with the use of metabolic energy. Because active transport involves protein carrier activity, it resembles facilitated diffusion in some ways. The carrier protein binds particular solutes with grade specificity. Similar solute molecules can compete for the same carrier protein in both facilitated diffusion and active transport. Active transport differs from facilitated diffusion in its use of metabolic energy and its ability to concentrate substances (Fig. 6.5). Metabolic inhibitors that block energy production will inhibit active transport, but will not effect facilitated diffusion. Active transport includes three steps :

i. Binding of a solute to a receptor site on a membrane bound carrier protein.

ii. Translocation of solute carrier complex across the membrane.

iii. Coupling of translocation to an energy yielding reactions to lower the affinity of the carrier protein to solute of the inner membrane surface so that the carrier protein will release solute to the cell interior.

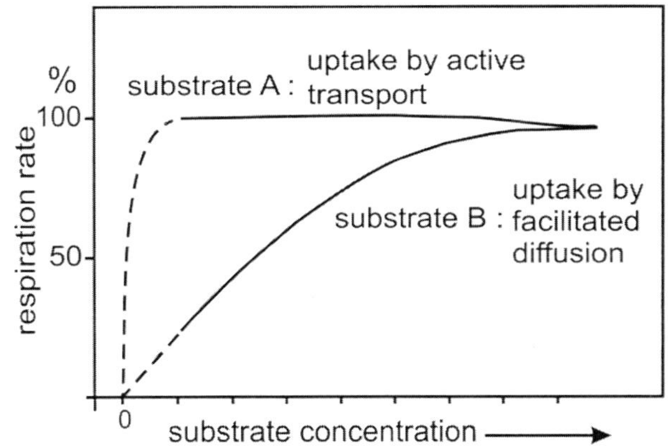

Fig. 6.5. Substrate saturation curves for the uptake of two substrates by intact bacterial cells measured by oxygen consumption (respiration rate).

Various models proposed for active transport include specific transport proteins in the membrane (Fig. 6.6). These have been given names that indicate their presumed functions: permeases, translocases, translocator proteins and carriers. Different transport processes are distinguished mainly by the way in which the energy necessary to drive the transport process is made available : by proton potential, Δp, by ATP or by phosphoenol-pyruvate. The transport of many substances, including inorganic and organic ions, as well as sugars, is driven by the proton potential Δp. The bacterial cell maintains a proton potential by constantly pumping out protons and other ions. This is mediated by transport proteins located on the membrane. Each of these proteins has a specific function. One protein is known, for example, to catalyse the simultaneous transport of sugar molecule and the proton in the same direction. And in the same way permease transport lactose molecule inward as a proton, simultaneously enters the cell. Such linked transport of two substances in the same direction is called **symport**.

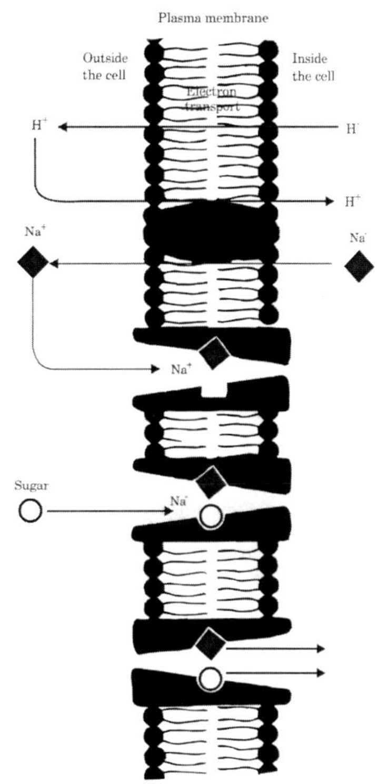

Fig. 6.6. Active transport mechanism

Proton motive force can power active transport indirectly often through the formation of sodium gradient. For example, in *E. coli* sodium transport system pumps sodium outward in response to the inward movement of protons. Such linked transport in which the transported substances move in opposite direction is termed as **antiport** (Fig. 6.7).

6.6.2.4. Group translocation

In group translocation the solute is altered chemically during transport. The best-known group translocation system is the **phosphoenolypruvate dependent phosphotr-ansferase system [PTS]**. It transports a variety of substances into the prokaryotic cells while phosphorylating. They use PEP [phosphoenolpyruvate] as phosphate donor. PTS sugar uptake and phosphorylation requires the participation of several soluble and membrane bound enzymes (Fig. 6.8). The products formed are therefore sugar, sugar phosphate, pyruvate. The overall rections require magnesium ions.

$$\text{PEP} + \text{HPr} \xrightarrow{\text{Enzyme I}} \text{pyruvate} + \text{phosphate} + \text{HPr}$$

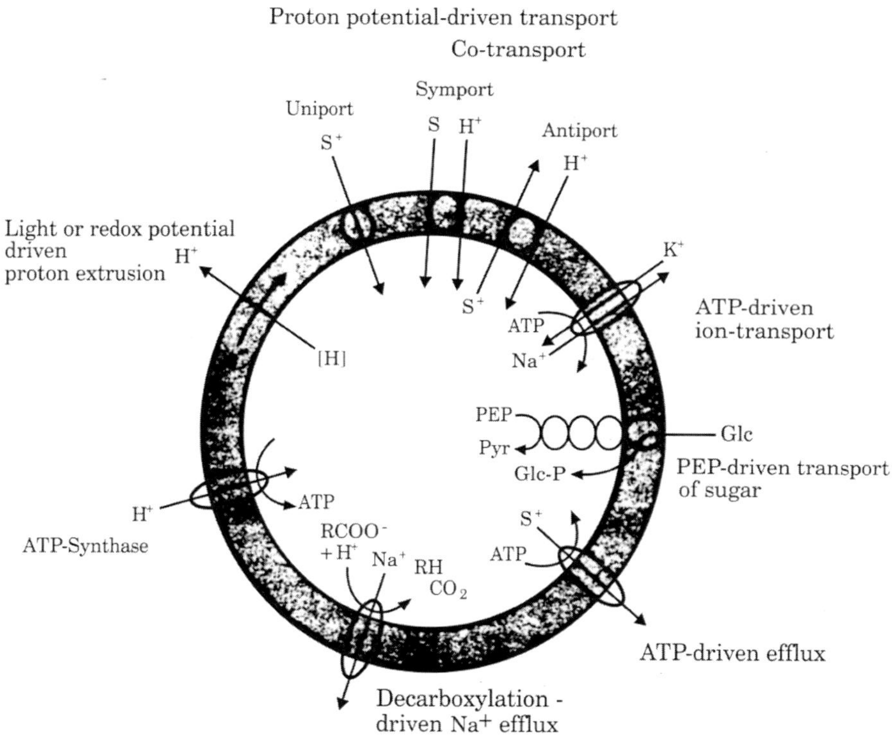

Fig. 6.7. Summary of systems that transport ions and other metabolites through the cytoplasmic membrane.

Enzyme I and HPr (heat stable protein) are soluble proteins and are non-specific components of the process. At the same time the sugar combines with enzyme II at the outer membrane surface and transported to the inner membrane surface. It is then combined with the phosphate group carried by the activated HPr. The sugar phosphate is released by enzyme II.

$$\underset{\text{(outside the cell)}}{\text{Sugar + phospho HPr}} \xrightarrow{\text{Enzyme II}} \underset{\text{(inside the cell)}}{\text{sugar phosphate + HPr}}$$

Enzyme II is the integral membrane protein and catalyses the phosphorylation of sugars. Enzyme I has been partially purified from several bacteria including *E. coli* and *Salmonella typhimurium*. HPr is purified from several bacteria. Enzyme II has been purified from *E. coli.*

$$\underset{\text{(outside the cell)}}{\text{PEP + sugar}} \xrightarrow[\text{Enzyme II}]{\text{Enzyme I, HPr}} \underset{\text{(inside the cell)}}{\text{sugar phosphate + pyruvate}}$$

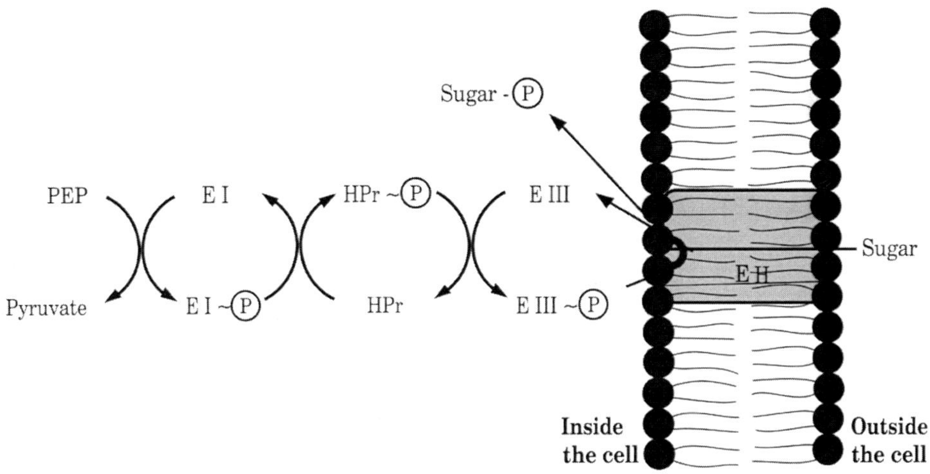

Fig. 6.8. Bacterial PTS transport.

6.7. TRANSPORT OF IRON

Several microorganisms possess specific mechanisms for the uptake of iron. Under anaerobic conditions iron is present as the ferrous ion and its concentration can be as high as 10^{-1} moles/l. It is, therefore, not growth limiting factor. In an aerobic environment at pH 7 it occurs as ferric hydroxide. Its concentration is 10^{-18} moles/l. It was found that microorganisms could excrete substances that render the iron soluble by binding the ferric ions in a complex, which can be transported. Such substances are collectively designated as **siderophores**.

Siderophores are low molecular weight water-soluble compounds. They belong to phenolic group or hydroxamate group. The first of these groups includes enterobactin, which has six phenolic hydroxy groups and is excreted by enterobacteria. The iron free form of enterobactin is excreted into the medium and binds iron, and the cell then takes up the ferri-enterobactin. The yellow-green fluorescent pigments that are excreted by *Pseudomonas putida* and *P. aeruginosa* also function as siderophores. Many fungi produce ferrichromes for the same purpose. These belong to the hydroxamate siderophores. They are cyclical hexapeptides that bind ferric iron through three hydroxamate groups. These are excreted into the medium in the iron free form, combine with iron, and are transported into the cell as ferrichromes (Fig. 6.9).

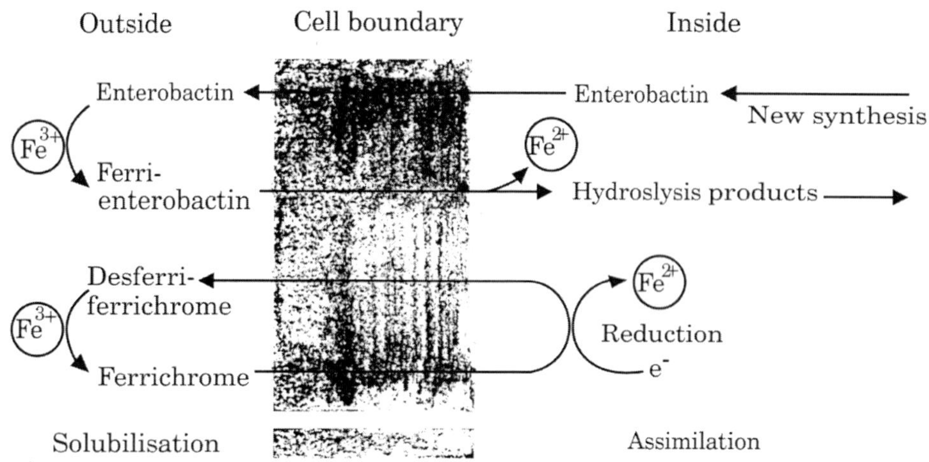

Fig. 6.9. Mechanism of the transport of iron into the cells by siderophores

Microorganisms generally excrete siderophores only when the supply of iron is growth limiting. The excretion is a result of derepession of siderophore synthesis. In the presence of soluble complexed iron, siderophores are synthesized in only small amounts and are retained in the cell wall where they function to transport iron into the cell.

6.8. EFFLUX OF SUBSTANCES

Excretion of substances takes place when the metabolites are overproduced and accumulate in the cell in high concentrations. Such accumulation can be due to incomplete oxidations, defective regulation of metabolism or fermentative process. The mechanism of efflux of substances is not yet clearly known.

6.9. FURTHER READING

Atlas, R.M. (1997), Hand Book of microbiological media. 2nd ed. Boca Raton, Fla. : CRC Press.

Balows, A. *et al.* (1992), "The Prokaryotes. A Hand Book of the Biology of Bacteria" 2nd edn. Springer Verlag, New York.

Cote, R.J. and Gherna, R.L. (1994), Nutrition and media. In Methods for general and molecular bacteriology, 2nd edn. P. Gerhardt (ed.) 155-178, Washington, DC. American Society for Microbiology.

Brawn, V. (1985), The unusual features of the iron transport systems of *E. coli. Trends in Biochem. Sci.*, **10(2) :** 75-76.

Dassa, E. (2000), ABC transport. In Encyclopedia of Microbiology, 2nd edn. Vol. I, J. Lederberg (Ed.), 1-12 San Diego, Academic Press.

Maloney, P.C., Ambudkar, S.V., Anantharam, V., Sonna, L.A. and Varadhachary, A. (1990), Anion-exchange mechanism in bacteria. *Microbiol. Rev.*, **54(1) :** 1-17.

Neidhardt, F.C., Ingraham, J.L. and Schaechter, M. (1990), "Physiology of the Bacterial Cell : A Molecular Approach". Sinaeur, Sunderland, M.S.

Postma, P.W., Lengeler, J.W. and Jacobson, G.R. (1993), Phosphoenolpyruvate: carbohydrate phosphotransferase system of bacteria. *Microbiol. Rev.*, **57(3)** : 543-596.

Whittenburg, R. (1978), Bacterial nutrition. In essays in Microbiology, J.R. Norris and M.H. Richmond, (eds.) John Wiley and Sons, New York.

Chapter - 7

Microbial Growth

7.1. INTRODUCTION

The study of growth of microbial population is most important for understanding the microbial cell physiology because population growth characteristics reflect underlying physiological events in the individual cells. Infact, the cellular events frequently manifest the changes in the growth of population. Therefore, it is important to understand the changes in growth that microbial population undergoes and to measure population growth accurately. Besides, one must be able to manipulate population growth as desired in order to investigate certain aspects of cell physiology. This chapter explains the important aspects of growth physiology, analysis of growth kinetics, methods to measure the growth and various factors influencing the growth.

7.2. DEFINITION

Microbial growth can be defined as an orderly increase in cellular components, resulting in cell enlargement and eventually leading to cell division. This definition is not strictly accurate as it implies that a consequence of growth is always an increase in cell number. Microbiologists usually consider the phenomenon of growth from the view point of population increase. This results from the fact that most current techniques do not allow the detailed study of individual cells. A study of the increase in population implies that each cell is capable of producing new progeny. However, under certain conditions growth can occur without cell division, for example, when cells are synthesizing storage compounds like glycogen and polyhydroxybutyrate. In this situation, the cell number remains constant, but the concentration of the biomass continues to increase. This is also true for coenocytic organisms, such as some fungi, that are not divided into separate cells. Their growth results only in increased size. Growth of a cell is the culmination of an ordered interplay between all of the physiological activities of the cell. It is a complex process involving: (i) entrance of essential nutrients into the cell, (ii) conversion of these compounds into energy and vital cell constituents, (iii) replication of the chromosome, (iv) increase in size and mass of the cell and (v) division of the cell into two daughter cells, each containing a copy of the genome and other vital components. Physiological aspects of the growth include the regulation of rates of synthesis of macromolecules, the regulation of timing of DNA synthesis and cell division, adaptive physiological responses to nutrient availability including changes in gene expression, homeostasis, adaptations to the external environment, and the coupling of rates of biosynthetic pathways to the rates of utilization of products for growth. Thus growth physiology is an extensive and intricate topic.

7.3. GROWTH CURVE

The most common method of bacterial reproduction is binary fission, one cell divides, producing two cells. Thus, if we start with a single bacterium the increase in population is by geometric progression.

$$1 \longrightarrow 2 \longrightarrow 2^2 \longrightarrow 2^3 \longrightarrow 2^4 \longrightarrow 2^5 \text{ ----- } 2^n$$

where 'n' is the number of generations. Population growth is studied by analyzing the growth curve of a microbial culture. When microorganisms are cultivated in liquid medium in an enclosed vessel for a definite period, referred to as 'batch culture', a typical growth curve for a population of cells is obtained as illustrated in Fig 7.1. Thus growth curve, obtained by plotting logarithm of cell number versus the incubation time, can be divided into four distinct phases called the **lag phase, log or exponential phase, stationary phase and death phase**.

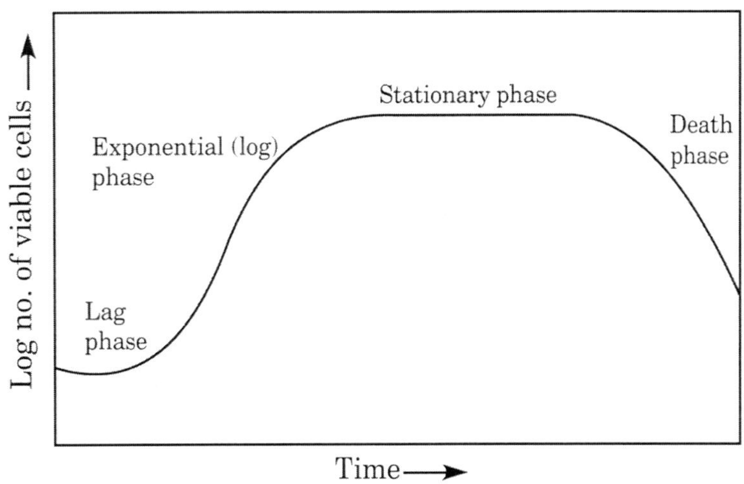

Fig. 7.1. Microbial growth curve in a closed system.

7.3.1. Lag Phase

When a microbial population is inoculated into a fresh medium, growth usually does not begin immediately but only after a period of time called the lag phase. Although cell division does not take place right away, the cell prepares for division by synthesizing new cell components. A lag phase prior to the start of cell division can be necessary for a variety of reasons. The cells may be old and depleted of ATP, essential cofactors, and ribosomes, hence these must be synthesized before growth can begin. The medium may be different from the one the microorganism was growing earlier. Here, new enzymes would be needed to use different nutrients. The other possibility is that the microorganisms have been injured and require time to recover. Whatever the causes, the cells recupe, replicate their DNA, begin to increase in mass and finally divide.

Duration of the lag phase varies, which may be brief or extended depending upon the history of the culture and growth conditions. If an exponentially growing culture is inoculated into the same medium under the same conditions of growth, a lag phase is not seen and exponential growth begins immediately. Inoculation of culture into a chemically different medium results in a long lag phase. A long lag phase also ensues when the inoculum consists of cells that have been damaged by various reasons.

7.3.2. Exponential Phase

During the exponential or log phase the bacteria will be growing and dividing at the maximal rate possible, with in the limits of genetic potential, the nature of the medium and the conditions under which they are growing (Table 7.1). Their rate of growth is constant during the exponential phase. Because, each individual divides at a slightly different moment, the growth curve rises smoothly rather than in discrete jumps. As long as the substrate concentration is not a limiting factor the logarithmic linear growth curve is maintained. It is only when the substrate concentration becomes quite low that the growth rate begins to decrease significantly and enters into stationary phase. In exponential phase the population is most uniform in terms of chemical and physiological properties. Therefore, the exponential phase cultures are usually used in biochemical and physiological studies.

Table 7.1. Exponential growth of *E.coli*

Time[a]	Division Number	2^n	Population (N_0 x 2^n)	$\log_{10} N_t$
0	0	$2^0 = 1$	1	0.000
20	1	$2^1 = 2$	2	0.301
40	2	$2^2 = 4$	4	0.602
60	3	$2^3 = 8$	8	0.903
80	4	$2^4 = 16$	16	1.204
100	5	$2^5 = 32$	32	1.505
120	6	$2^6 = 64$	64	1.806

[a] The hypothetical culture begins with one cell having a 20-minute generation time

7.3.3. Stationary Phase

In batch culture, the exponential growth cannot occur indefinitely. Some thing happens to limit the growth of the population. What generally happens is that either an essential nutrient of the culture medium is used up or some waste product of the organism accumulates in the medium to an inhibitory level, and exponential growth ceases. The population, now, is said to have reached the stationary phase.

The stationary phase is usually attained by bacteria at a population level of around 10^9 cells per ml. Other organisms normally do not reach such high population densities, protozoan and algal cultures often having maximum concentration of about 10^6 cells per ml. In this phase the total number of viable cells remains constant. This may be due to the balance between cell division and cell death (cryptic growth) or the population may simply cease to divide. However, many cell functions may still continue, including energy metabolism and biosynthetic processes. Studies with *E. coli* have identified several genes that are necessary for the survival of cells that have entered the stationary phase. Some of these genes, called *sur* (for survival) genes whose true functions are yet to be identified, but mutations in *sur* genes lead to rapid cell death.

Microbial populations enter the stationary phase for several reasons. One obvious reason is nutrient limitation, if an essential nutrient is severely depleted, population growth will slow. Aerobic organaisms are often limited by O_2 availability. Population growth also may cease due to accumulation of toxic waste products. This factor seems to limit the growth of fermentative bacteria in broth cultures. For example, streptococci can produce so much lactic acid and other organic acids from sugar fermentation that their medium becomes acidic and growth is inhibited. Streptococcal cultures can also enter the stationary phase due to depletion of their sugar supply. Thus, entry into the stationary phase may result from several factors operating simultaneously.

7.3.4. Death Phase

Incubation, if continued after a population reaches the stationary phase, the cells may remain alive and continue to metabolise, but they may also die. If the latter occurs, the population is said to be in the death phase. In some cases the death is actually accompanied by lysis. Detrimental environmental changes like nutrient deprivation and the build up of the toxic wastes lead to the decline in the number of viable cells characteristic of death phase. The death of microbial population, like its growth during the exponential phase, is usually logarithmic (*i.e.* constant proportion of the cells die every hour). This pattern holds even when the total number of cells remains constant because they simply fail to lyse after death. During the initial stage of death phase the number of viable cells decreases exponentially. However, the death rate may decrease after the population has been drastically reduced. This is due to the extended survial of particularly resistant cells.

Certain points may be kept in mind while trying to understand the growth curve. Firstly, a culture proceeds gradually from one phase of the growth to the next. This means that not all the cells are in an identical physiological condition towards the end of the given phase of growth. Time is required for some to catch up with the others. Secondly, the phases of the bacterial growth described so far and shown in Fig. 7.1 are applicable to population of cells. They donot apply to individual cells.

7.4. GROWTH KINETICS

A knowledge of microbial growth rates during the exponential phase is indispensable for the students of microbiology. Growth rate studies contribute to basic physiological and ecological research and the solution of applied problems in industry. Therefore, the quantitative aspects of exponential growth phase will be discussed.

During the exponential phase, each microorganism divides at constant intervals. Thus, the population will double in number during a specific length of time called the **generation time**, doubling time. If we start with a single bacterium, the increase in population is by geometric progression:

$$1 \longrightarrow 2 \longrightarrow 2^2 \longrightarrow 2^3 \longrightarrow 2^4 \longrightarrow 2^5 \longrightarrow 2^n$$

where n = the number of generations. Each succeeding generation, assuming no cell death, the population doubles. The total population at the end of a given time N_t period would be expressed

$$N_t = 1 \times 2^n \tag{1}$$

However, under practical conditions, the number of bacteria N_o inoculated at time zero is not 1 but more likely to be several thousands, so the formula becomes

$$N_t = N_0 \times 2^n \tag{2}$$

Solving for n, the number of generations, where all logarithms are to the base 10, we have

$$\text{Log } N_t = \log N_0 + n.\ \log 2, \text{ and}$$

$$n = \frac{\log N_t - \log N_0}{\text{Log}^2} = \frac{\log N_t - \log N_0}{0.301}$$

$$= 3.3\ (\log N_t - \log N_0) \tag{3}$$

Thus, by use of equation 3, we can calculate the number of generations, that have taken place, provided we know the initial population and the population after growth has occurred (Fig. 7.2).

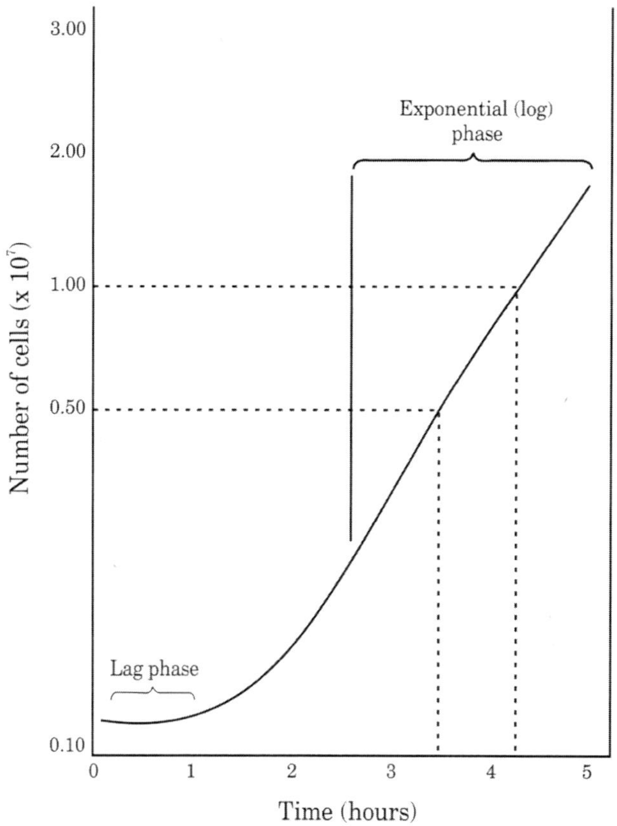

Fig. 7.2. Generation time determination

7.4.1. Growth Rate Constant (k)

The growth rate constant (k) is a measure of the instantaneous growth rate and has the units of reciprocal time. It is also a widely used parameter of growth. The growth rate constant (k) is the number of generations per unit time, often expressed as the generations per hour.

$$k = \frac{n}{t} = \frac{\log N_t - \log N_0}{0.301\,t}$$

The time it takes a population to double in size, that is, the mean generation time or mean doubling time (g), can now be calculated. If the population doubles (t = g), then

$$N_t = 2\,N_o$$

Substituting 2 N_o into the mean growth rate equation, we get

$$k = \frac{\log (2\,No) - \log N_O}{0.301\,g} = \frac{\log 2 + \log N_O - \log N_0}{0.301\,g}$$

$$k = \frac{1}{g} \quad \text{(or)} \quad g = \frac{1}{k}$$

Table 7.2. Generation time for some microorganisms

Microorganism	Temperature (°C)	Generation Time (Hours)
Bacteria		
Beneckea natriegens	37	0.16
Escherichia coli	40	0.35
Bacillus subtilis	40	0.43
Clostridium botulinum	37	0.58
Mycobacterium tuberculosis	37	≈ 12
Anacystis nidulans	41	2.0
Anabaena cylindrica	25	10.6
Rhodospirillum rubrum	25	4.6 – 7.3
Algae		
Chlorella pyrenoidosa	25	7.75
Scenedesmus quadricauda	25	7.9
Asterionella formosa	20	9.6
Skeletonema costatum	18	13.1
Ceratium tripos	20	82.8
Euglena gracilis	25	10.9
Protozoa		
Acanthamoeba castellanii	30	11 – 12
Paramecium caudatum	26	10.4

Microorganism	Temperature (°C)	Generation Time (Hours)
Tetrahymena geleii	24	2.2 – 4.2
Leishmania donovani	26	10 – 12
Giardia lamblia	37	18
Fungi		
Saccharomyces cerevisiae	30	2
Monilinia fructicola	25	30

The mean generation time can be determined directly from a semilogarithmic plot of the growth data and the growth rate constant calculated from the value. The generation time may also be calculated directly from the earlier equation. For example, suppose that a bacterial population increases from 10^3 cells to 10^9 cells in 10 hours.

$$k = \frac{\log 10^9 - \log 10^3}{0.301 \text{ x } (10\,\text{hrs})} = \frac{9-3}{3.01\,\text{hr}} = 2.0 \text{ generations/hr}$$

$$g = \frac{1}{2.0\,\text{gen./hrs}} = 0.5 \text{ hr/gen. or } 30 \text{ min. / gen.}$$

Generation times vary markedly with the species of the microorganism and environmental conditions. They range from less than 10 minutes for a few bacteria to several days with some eukaryotic organisms (Table 7.2). Generation time in nature is usually much longer than in culture.

7.4.2. Use of Growth Equations

The growth equations can be used for finding g and k. Many a times we face situation as to choose proper inoculum size. A convenient equation to remember for this calculation is

$$N = N_o \text{ x } 2^n$$

$$\text{or } \log N_t = \log N_o + (0.301/g)\,(t)$$

Suppose, we have a culture growing exponentially at a density of 10^8 cells/ml and we would like to subculture it, so that 16 hours later the density of the new culture will also be 10^8 cell/ml. The generation time (g) is 2 hours. What should N_o be? One way to do this problem is to estimate the number of generations (y) and then to use above equations. For example, since n = t/g, the number of generations is 16/2 or 8. Using the above first equation $10^8 = N_o \text{ X } 2^8$ or $N_o = 10^8/2^8 = 3.9 \text{ x } 10^5$ cells/ml. This means that the initial cell density in the growth flask should be $3.9 \text{ x } 10^5$ cells/ml. But the cell density of the inoculum is 10^8 cells/ml. This means that the inoculum must be diluted $10^8/3.9 \text{ X } 10^5$ or $2.6 \text{ X } 10^2$ times. If we want to grow 1 litre of cells, the inoculum size would have to be 3.8 ml. Thus, most growth problems that one might have to do routinely can be done with the following equations.

$$N = N_o 2^n$$

$$n = t/g$$

$$g(k) = 0.693$$

7.4.3. Macromolecular Composition – Growth Rate

Many interesting things have been understood regarding the growth physiology by observing changes in macromolecular composition of cells whose growth rates have been altered by changing the nutrient composition of the medium, or by manipulation of the dilution rates of cells growing in continuous culture.

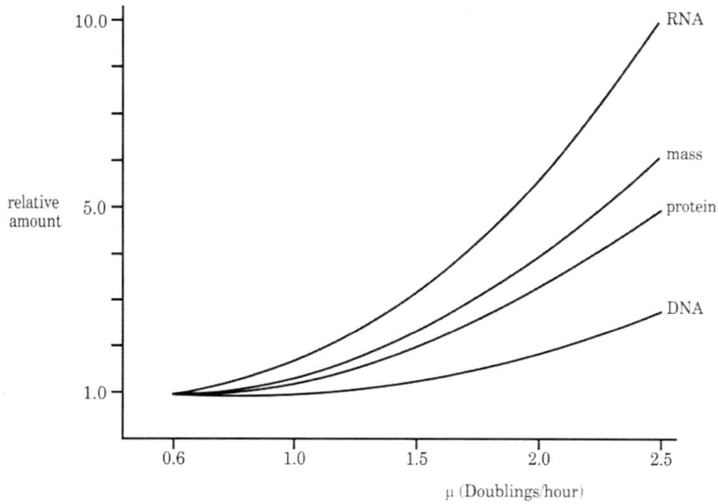

Fig. 7.3. Macromolecular composition of *E. coli* grown at different growth rates. The values are amount per cell and have been normalized to values at a doubling time of 0.6 doublings/hs.

Let us try to understand the Fig. 7.3. It illustrates the changes in the ratios of cell mass, RNA, protein and DNA of *E. coli* grown at increasingly faster growth rates due to difference in the nutrient composition of the growth media. On the y–axis are plotted relative amounts of macromolecule per cell normalized to cell undergoing 0.6 doublings per hours. On the x–axis are plotted the growth rate in units of doubling per hour. It is evident from the figure that the mass of cells as well as the relative concentration of the macromolecules increase as exponential functions of the growth rate. Nevertheless, the relative rates of increase are not equal so that the macromolecular composition of the cells is different at different growth rates. For example, faster growing cells are enriched with RNA when compared to other cell components. It shows a high proportion of ribosomes in faster growing cells. The reason for this is that over a wide range of growth rates, a ribosome polymerises amino acids at approximately a constant rate. When a cell increases or decreases its growth rate, the amount of proteins it must make per unit time, it adjusts the number of ribosomes rather than make each ribosome work substantially slower or faster. Fig 7.3 also shows that faster growing cells have more DNA per cell. The reasons for increased DNA per cell are explained in the following way: In *E. coli*, the minimum time required between the initiation of DNA replication and the on set of cell division for rapidly growing cells is about 60 min. If the generation time is shorter than 60 min, then the DNA replication must begin in a previous cell cycle, which increases the average amount of DNA per cell. It is also evident from the Fig. 7.3 that cells that are grown at a faster growth rate have a greater mass (*i.e.* they are larger). For example, *E. coli* when growing at a doubling rate of 2.5 times an hour is approximately six times larger in mass than when growing at 0.6 doublings per hour.

7.5. SYNCHRONOUS GROWTH

In microbiological research, very often, it is desirable to relate the various aspects of bacterial growth to a particular stage of the cell division cycle. It is not feasible to analyse a single cell because of its small size. However, if all the cells in a culture are at the same stage of division cycle, the results from analysis of the cell crop can be interpreted as that for a single cell. There are several laboratory techniques by which we can manipulate the growth of cultures, so that all the cells will be in the same stage of growth *i.e.,* growing synchronously.

A population can be synchronized by manipulating the physical environment or the chemical composition of the medium. For example, the cells may be inoculated into a medium at a sub–optimal temperature, if they are kept in this condition for some time, they will metabolize slowly but will not divide. When the temperature is subsequently raised, the cells will undergo a synchronized division. The most common method of synchronization takes advantage of the fact that the smallest cells in a lag–phase culture are those which have just divided. When these cells are separated out by filteration or by differential centrifugation, they are reasonably well synchronized with each other. Fig 7.4 shows the growth pattern of a population of synchronous cells.

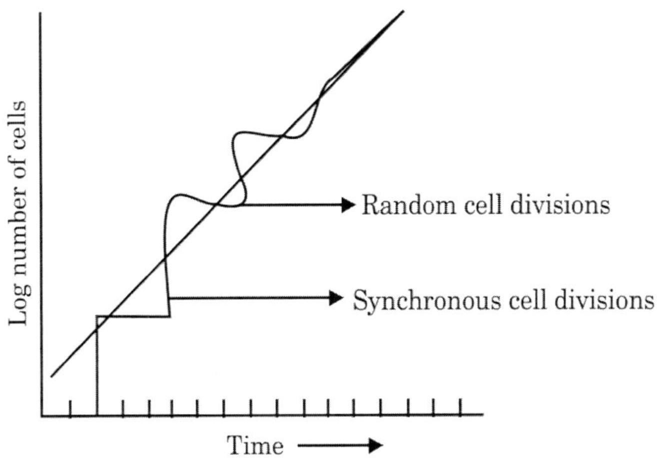

Fig. 7.4. Synchronous growth of bacteria

7.6. DIAUXIC GROWTH

Many bacteria grow preferentially on glucose when presented with mixtures of glucose and other carbon sources. Preferential growth on one carbon source before growth on a second carbon source is called '**diauxic growth**'. The bacteria first grow exponentially on glucose, then enter a lag period and finally grow exponentially on the second carbon source. An example of diauxic growth with glucose and lactose is shown in Fig 7.5.

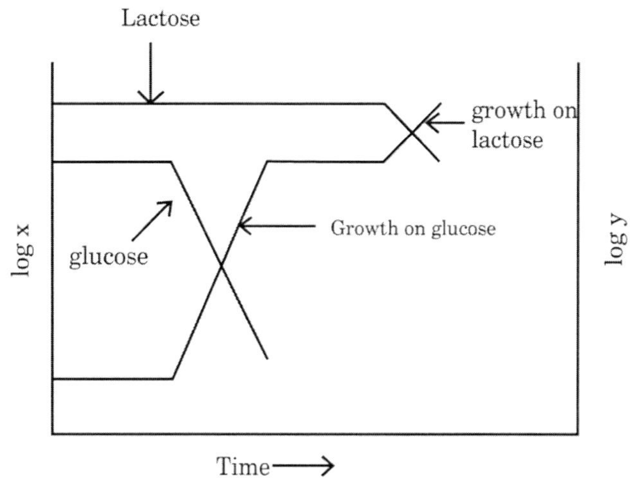

Fig. 7.5 : Diauxic growth of *E. coli* on glucose and lactose X, bacterial mass. Y, glucose or lactose concentration.

However, not all bacteria preferentially grow on glucose. Several obligate aerobic bacteria grow first on organic acids when given a mixture of glucose and the organic acid. For example, *Rhizobium* grows most readily in laboratory cultures on C_4 carboxylic acids such as succinate, malate and fumarate. When presented with a mixture of succinate and glucose, *Rhizobium* shows diauxic growth by growing on the succinate first and the glucose second. This is because, in these bacteria succinate represses key enzymes of the Entner – Doudoroff and EMP pathways that degrade glucose.

7.7. GROWTH YIELDS

When microbial growth is limited by low concentration of required nutrient, the final net growth or yield of cells increases with the initial amount of the limiting nutrient present. The rate of growth also increases with nutrient concentration, but in a hyperbolic manner. The shape of the curve reflects the rate of nutrient uptake by microbial transport proteins.

When the nutrient quantity limits the production of bacteria, it is possible to define a growth yield constant (Y). The growth yield constant is the amount of dry weight of cells produced per weight of nutrient used

$$Y = \frac{\text{mass of microorganism formed}}{\text{mass of substrate consumed}}$$

The molar growth yield constant is Y_m, which is the dry weight of the cells produced (in grams) per mole of substrate used. For example, the $Y_{glucose}$ for aerobically growing cells is about 0.5, which means about 50% of the sugar is converted to cell material and 50% oxidised to CO_2. For certain sugars and bacteria, the efficiency of conversion of cell material can be much lower (*e.g.*, 20%). In dilute media, some bacteria are able to increase their efficiency and assimilate upto 80% of the sugar acquired.

7.8. ENERGETICS OF GROWTH

The difference in the growth yield constant of different nutrient sources can be explained in terms of the amount of ATP generated from unit weight of the nutrient. The more ATP that is made, the greater the growth yields. This is not difficult to understand if we consider the fact that growth is, after all, an increase in dry weight, and it requires a certain number of ATP molecules to synthesize each cell component. A value of 10.5 g of cells per mol of ATP called the Y_{ATP} (or molar growth yield) has been determined for fermenting bacterial cultures growing on glucose as the energy source. Knowing the amount of ATP produced in the fermentation pathways, it is possible to calculate the Y_{ATP} from the $Y_{glucose}$. For example, 22 g *Streptococcus faecalis* cells are produced per mole of glucose fermented. This organism ferments a mole of glucose using the EMP pathway which produces 2 mol of ATP. Thus the Y_{ATP} is 22/2 = 11. On the other hand, *Zymomonas mobilis* produces only 8.6 g of cells per mole of glucose fermented. This organism utilizes ED pathway, which yields only 1 ATP per mole of glucose fermented. Thus the Y_{ATP} is 8.6. A comparison of several different fermentations by bacteria and yeasts has produced an average Y_{ATP} of 10.7. Although this value can vary, knowledge of the growth yields and the Y_{ATP} has allowed some deductions as to which fermentation pathway might be operating. Also, an unexpectedly high growth yield in fermenting bacteria can point to unrecognized sources of metabolic energy.

7.9. THE CONTINUOUS CULTURE

So far, we have discussed the dynamics of microbial growth in a closed system called batch culture in which nutrient supplies are neither renewed nor wastes removed. In a batch culture, the growth passes through all the four phases *viz.,* lag, exponential, stationary and death phases. Exponential growth lasts for only a few generations and soon the stationary phase is reached. For many studies, it is often desirable to keep cultures in constant environments for long periods, and this is done by employing continuous culture. A continuous culture is essentially a flow system of constant volume to which medium is added continuously and from which continuous removal of any overflow can occur. Once such a system is in equilibrium, the cell number and nutrient status remain constant, and the system is said to be in a steady state. A microbial population can be maintained in the exponential growth phase and at a constant biomass concentration for extended period in a continuous culture system. Two types of continuous culture systems are commonly in use : (i) **chemostat** and (ii) **turbidostat**.

7.9.1. The Chemostat

This is the most common type of continuous culture device. Chemostat permits control of both the population density and growth and growth rate of the culture. It is constructed so that sterile medium is fed into the culture vessel at the same rate as medium containing microorganism is removed. Two elements are used in the control of a chemostat : the dilution rate and the concentration of limiting nutrient (e.g., an amino acid). Because of the presence of a limited nutrient, the growth rate is determined by dilution rate in the growth chamber and the final cell density depends on the concentration of the limiting nutrient. The rate of nutrient exchange is expressed as the dilution rate (D), the rate at which medium flows the culture vessel relative to the vessel volume, where f is the flow rate (ml/hr) and V is volume (ml) of the vessel

$$D = f / V$$

For example, if **f** is 30 ml/hr and **V** is 100ml the dilution rate is 0.30 hr^{-1}

The net change in cell concentration over a time period may be expressed as

$$dx/dt = \text{growth} - \text{out put} \quad \text{(or)} \quad dx/dt = \mu x - dx$$

under steady state conditions the cell concentration remains constant, thus $dx/dt = 0$ and

$$\mu x = Dx$$

$$\mu = D$$

Thus, under steady–state conditions the specific growth rate is controlled by the dilution rate. Both the microbial population level and the generation time are related to the dilution rate. The microbial population density remains unchanged over a wide range of dilution rates. The generation time shortens as the dilution rate increases. The limiting nutrient will be almost completely depleted under these balanced conditions. If the dilution rate rises too high, the microorganisms can actually be washed out of the culture vessel before reproducing because the dilution rate is greater than maximum growth rate. The limiting nutrient concentration rises to higher dilution rates because there are fewer microorganisms present to use it (Fig. 7.6). At very low dilution rates, an increase in D causes a rise in both cell density and the growth rate.

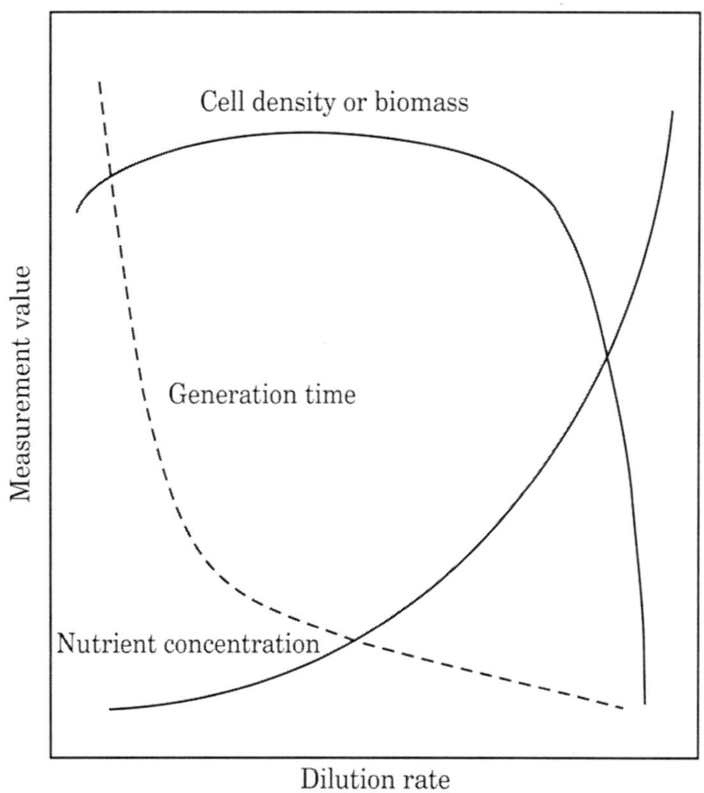

Fig. 7.6 : Chemostat dilution rate and the effects on microbial growth

7.9.2. The Turbidostat

The second type of continuous culture system is the turbidostat. It is provided with a photocell to measure the absorbance or turbidity of the culture in the growth vessel. The flow rate of media through the vessel is automatically regulated to maintin a predetermined turbidity or cell density. This is achieved by monitoring the biomass with photoelectric cell and feeding the signal to a pump supplying the medium to the culture such that the pump is switched on if the biomass exceeds the set point and is switched off if the biomass falls below the set point.

Table 7.3. The differences between the chemostat and turbidostat

	Chemostat	Turbidostat
1.	Regulating factor is a nutrient present in limiting quantities	Regulating factor is turbidity of the medium monitored by photoelectric cell
2.	Dilution rate remains constant	Dilution rate varies
3.	Limiting nutrient present	Limiting nutrient absent
4.	Operates best at lower dilution rates	Operates best at high dilution rates

7.10. BALANCED AND UNBALANCED GROWTH

In a growth culture, if all the cellular constituents are manufactured at a constant rate relative to each other, the culture is said to be in a balanced state. Exponential growth, whether in batch or continuous cultures is a balanced growth. If nutrient levels or other environmental conditions change balanced growth shifts to unbalanced growth because the rates of synthesis of cell components vary relative to one another. However, very soon the culture reaches to a new balanced growth. This response is readily observed in a shift up experiment in which bacteria are transferred from a nutritionally poor medium to richer one. The cells first construct new ribosomes to enhance their capacity for protein synthesis. This is followed by increase in protein and DNA synthesis. Finally, the expected rise in reproductive rate takes place.

Unbalanced growth also sets in when a bacterial population is shifted down from nutritionally rich medium to poor one. The organisms may previously have been able to obtain many cell components directly from the medium. When shifted to nutritionally poor medium, the cells need time to make the enzymes required for the biosynthesis of unavailable nutrients. Consequently, cell division and DNA replication continue after the shift – down, but net protein and RNA synthesis is slow. The cells become smaller and reorganize themselves metabolically until they are able to grow again then balanced growth is resumed.

Thus, shift–up and shift–down experiments demonstrate that microbial growth is under precise coordinated control and responds quickly to changes in environmental conditions.

7.11. MEASUREMENT OF MICROBIAL GROWTH

For various purposes, it is often required for the routine determination of the microbial population, cell number and/or biomass, in order to monitor its progress. There are many ways to measure microbial growth to determine growth rate and generation times. These methods can be broadly divided into two types: (i) direct methods and (ii) indirect methods. Direct procedures involve dry weight determination, cell counting by microscopy and plate counting methods. Indirect methods include turbidimetry, spectrophotometry, estimation of cell components (protein, DNA, RNA or ATP), and online monitoring of carbon dioxide production or oxygen utilization. The method adopted in any given situation depends on the organism involved, conditions under which it is grown and any specific requirement. Several factors must be considered, such as the degree of accuracy and sensitivity needed, and the duration of the analysis. Estimation of unicellular organisms is relatively straight forward, but filamentous organisms like fungi and actinomycetes present additional problems. The speed of analysis is also critical as an instant indication of biomass or cell concentration is often required. Nevertheless, faster methods are frequently less reliable and longer procedures are generally more accurate and reproducible.

7.11.1. Direct Microscopic Counting Methods

In unicellular organisms cell numbers in suspension can be measured by direct microscopic counts, using Petroff-Hauser or Neubauer-type counting chambers. The former is more suitable for counting bacteria. The counting grids, when the chamber is covered with a glass coverslip, hold a known volume of culture. By counting the number of cells within a proportion of the grid, the number of cells per milliliter can be determined. Direct microscope methods are rapid and give correct number. However, there are some disadvantages to the technique. The microbial population must be fairly large (5×10^6 cells/ml) for accuracy because such a small volume is sampled. It is also difficult to distinguish between living and dead cells, unless differentiated by use of vital staining technique.

7.11.2. Electronic Cell Counters

Electronic cell counting method is also rapid, but most methods are more suitable for larger unicellular microbes, such as yeasts, protozoa, some algae and less useful for estimating bacteria. Coutler counter, an instrument used for this purpose is based on the principle the changes in electrical resistance produced by non-conductive particles suspended in an electro type. This method involves drawing suspension of a cell through a small hole or orifice across which an electrical current is maintained. As a cell passes through the orifice it displaces its own volume of electrolyte and changes the electrical resistance. These changes are detected and converted to a countable pulse. However, it essentially counts particles, and is consequently prone to errors due to cell clumping and the presence of particulate debris.

7.11.3. Plate Count

Plate counting methods detect viable cells *i.e.,* those able to form colonies on an appropriate solid nutrient medium. The two methods routinely used are spread plating and pour plating. Prior to plating, it is usually necessary to prepare a serial dilution series in a sterile dilute for concentrated cell suspensions. Alternately, for samples with low cell numbers, as in water analysis, the concentration step is required.

For statistical reliability, the results are recorded only for plates containing 30–300 colonies. Calculation of the cell concentration in the original sample is then carried out, taking into account the dilution and volume plated. Both methods measure **colony forming units (CFU).** This may or may not be the same as number of cells depending upon the extent of clumping and cellular morphology of the organisms. These techniques are accurate, but a minimum of 1–2 days incubation is usually necessary before the colonies are countable. However, more rapid microcolony counting techniques are also available.

Microbial numbers are very often determined from counts of colonies growing on special membrane filters having pores small enough to trap bacteria. In this technique, a sample is passed through a special membrane filter. The filter is then placed on an agar medium or on a pad soaked with liquid medium and incubated until each cell forms a separate colony. A colony count gives the number of microorganisms in the filtered sample, and special media can be used to select for specific microorganisms. Membrane filters are also used to count bacteria directly. The bacterial sample is mixed with a fluorescent dye such as acridine orange and then filtered through a polycarbonate nucleopore filter. The filter is then stained black to provide a good background for counting the fluorescent bacteria. Acridine orange stained microorganisms glow orange or green and are easily counted with an epifluorescence microscope.

7.11.4. Turbidometry and Spectrophotometry

These methods provide a simple, rapid and convenient means of total biomass estimation. They are usually performed at a specific optimum wave length for each microorganism. Turbidometric methods measure the light scattered by a suspension of cells, which is proportional to the cell concentration. Alternatively, the spectroscopy may be employed using absorbance or transmittance of a cell suspension. Some modern fermentation monitoring systems now employ methods based on near – infared spectroscopy.

Turbidimetric and spectrophotometric methods require the construction of appropriate calibration curves, prepared using standard cell suspensions containing known concentration of cells. Care must be taken when interpreting the results of the fermentation broth containing particulate matter or highly coloured.

7.11.7. Dry Weight Estimation

This method determines the weight of the total cells, both living and dead, in liquid culture samples. It involves separating the biomass from a known volume of a homogenous cell suspension. This is usually achieved by filtration under vacuum, through pre-weighed membrane filter with a pore size of 0.2 μm or 4 μm. The filter with collected cells is washed with water to remove any residual growth medium and dried to a constant weight in an oven at 105^0C. Results are normally expressed as milligrams of cells per milliliter of culture. Obviously, any other suspended non-cellular materials above the size of the filter pores is collected and can lead to errors. Further, limitations are the time needed to obtain the results and the relatively large volume of sample required to obtain sufficient biomass for accurate weighing, as an individual bacterium weighs only about 10^{-12}g.

7.11.6. ATP Bioluminimetry

ATP is rapidly lost from dead cells, consequently, it is very useful for determining the concentration of viable microorganisms. The amount of ATP in a sample can be quantified using

ATP bioluminimetry. This technique utilizes an enzyme substrate complex, luciferase – luciferin, obtained from the firefly *Photinus pyralis* which generates a photon of light for each molecule of ATP. When an aliquot of luciferse – luciferin is added to ATP extracted from sample of cell suspension, the light generated can be detected in a bioluminimeter. The resulting signal is amplified and then expressed as a digital or analogue data output.

$$\text{ATP} + \text{O}_2 + \text{luciferin} \xrightarrow{\text{luciferase}} \text{Oxyluciferin} + \text{AMP} + \text{CO}_2 + \text{ppi} + \text{photon of light (562 nm)}$$

This procedure is very rapid and sensitive. Under optimum conditions as little as 10 femtomoles (1 femtomole $=10^{-15}$ mol) of ATP can be detected, which is approximately equivalent to 1 yeast cell or about 10 bacteria. ATP bioluminimetry is most suitable for direct measurement of samples that are not coloured, as quenching of light may be a problem. However, as the method is very sensitive, samples may require considerable dilution, which often overcomes any colour quenching problem.

7.11.7. Determination of Nitrogen Content

The major constituent of cell material is protein. Since the nitrogen is a characteristic part of proteins, one can measure a bacterial population or cell crop in terms of bacterial nitrogen. Nitrogen constitutes about 14 percent of bacteria on dry weight basis. However, this figure is subject to some variation. To measure the growth by this technique, we must first harvest the cells and wash them free of medium and then estimate the nitrogen content by chemical analysis usually by Kjeldahl method. Bacterial nitrogen determinations are somewhat laborious and can be performed only on specimens free of all other sources of nitrogen. Further, the method is applicable only for concentrated populations.

7.11.8. Measurement of a Metabolite

Microorganisms produce and release a variety of metabolites into the environment. In many cases, the amount of metabolite produced is proportional to the size of the bacterial population. For example, we may take a species that produces an organic acid from glucose fermentation. The amount of organic acid produced, provides an indication of growth achieved by that organism. Obviously, the measure of acid or any other end product is a very indirect approach to the measurement of growth and is applicable only in special circumstances.

7.11.9. On-line Estimation

On-line estimation of microbial growth is now possible in automated closed or open fermentor. It can provide real time estimation of biomass and minimizes the requirement for repeated sampling and off-line analysis. Monitoring systems may involve optical density or capacitance – based probes. Also, the microorganisms involved in most fermentation processes will either have a requirement for oxygen and/or will produce CO_2. In such cases, it is theoretically possible to establish a mathematical relationship between factors, such as carbon dioxide evolution or oxygen utilization. Therefore, estimation of biomass concentration and even product formation can be made by measuring these parameters on-line using carbon dioxide or oxygen detectors and biosensors attached to a computer. This can give an accurate estimate of the biomass concentration, provided that the mathematical algorithms developed are reliable.

Different methods of microbial growth measurement are summarized in Table 7.4.

Table 7.4. Methods of measuring bacterial growth

Method	Applications	Manner in which growth is expressed
Microscopic count	Enumeration of bacteria in vaccines and cultures	Number of cells per ml
Electronic enumeration	- do -	- do -
Plate count	Enumeration of bacteria in milk,water, foods, soil, cultures etc	Colony forming units per ml
Membrane filter	- do -	- do -
Turbidimetric measurement	Microbiological assay, estimation of cells in broth cultures or aqueous suspension	Optical density (absorbance)
Dry weight determination	Measurement of cell mass	Dry weight of cells per ml
ATP bioluminimetry	Measurement of cell mass	Femtomoles of ATP equated to a number per cell
Nitrogen determination	Measurement of cell mass	Mg of nitrogen per cell
Measurement of metabolite	Microbiological assays	Milliequivalents of metabolite per ml or per culture

7.11.10. The Choice of Method to Measure Growth

As discussed above, each method has its particular advantages and limitations. As a result, no one method can be recommended universally. One has to select a method according to the requirement. Nevertheless, the colony count is most widely used procedure for general microbiological work, and complete familiarity with this technique, both in principle and practice, is essential. It is important to remember that the colony count is theoretically the only technique that reflects the viable population. Furthermore, it is possible that discrepancies may occur in results of growth of a bacterial population when measured by two different methods. For example, a microscopic count of a culture in the stationary phase would include all cells, the viable and nonviable, whereas the colony count would reveal only the viable populations.

7.12. THE INFLUENCE OF ENVIRONMENTAL FACTORS ON GROWTH

The growth of microorganims is greatly affected by the physical and chemical conditions of their environment. Understanding of environmental influence helps us to explain the distribution of microorganisms in nature and makes it possible for us to devise methods for controlling microbial activities and destroying undesirable organisms. Not all organisms respond equally to a given environmental factor. The fact is that an environmental condition harmful to one organism may be beneficial to another. Further, organisms can tolerate some adverse conditions under which they cannot grow. Hence, it is essential to distinguish the effects of environmental conditions on the viability of an organism, growth, differentiation and reproduction.

Many environmental factors affect the microbial growth. However, six of these factors have been identified that clearly play a critical role in controlling the microbial growth. They are temperature, pH, water activity (a_w), oxygen, radiation and pressure.

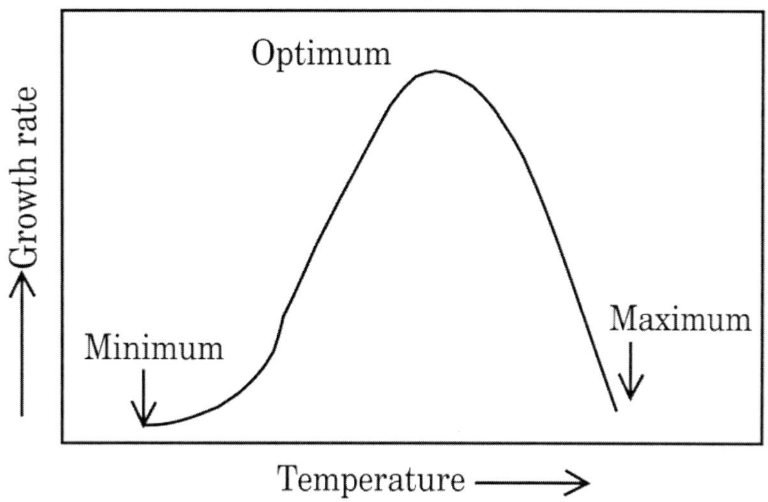

Fig. 7.7. The effect of temperature on microbial growth

7.12.1. Temperature

Temperature is one of the most important environmental factors influencing the growth and survival of the microorganisms. Indeed, microorganisms are particularly susceptible because they are usually unicellular and also poikilothermic *i.e.,* their temperature varies with that of the external environment. Most important factor influencing the effect of temperature upon growth is the temperature sensitivity of enzyme catalyzed reactions and also the stability of the membranes.

As the temperature rises, the rate of chemical reactions increases. This rate doubles for every 10°C rise in temperature. Thus, cell should grow faster as the temperature is raised. However, there are maximum limits beyond which some temperature sensitive macromolecules (proteins, nucleic acids and lipids) will become denatured and hence nonfunctional. There is also a minimum temperature for growth, below which the lipid membrane is not fluid enough to function properly. Due to these opposing temperature influences, microbial growth has a fairly characteristic temperature dependence with distinct cardinal temperatures minimum, optimum and maximum growth temperatures (Fig. 7.7). In most of the cases, the temperature optimum is closer to maximum than to the minimum. The cardinal temperatures for a particular species are not rigidly fixed, but often depend to some extent on other environmental factors such as pH and the available nutrients.

Table 7.5. Temperature ranges for microbial growth

Microorganism	Cardinal Temperature (°C)		
	Minimum	Optimum	Maximum
Nonphotosynthetic Bacteria			
Bacillus psychrophilus	-10	23-24	28-30
Micrococcus cryophilus	-4	10	24
Pseudomonas fluorescens	4	25-30	40
Staphylococcus aureus	6.5	30-37	46
Enterococcus faecalis	0	37	44
Escherichia coli	10	37	45
Neisseria gonorrhoeae	30	35-36	38
Thermoplasma acidophilum	45	59	62
Bacillus stearothermophilus	30	60-65	75
Thermus aquaticus	40	70-72	79
Sulfolobus acidocaldarius	60	80	85
Pyrodictium occultum	82	105	110
Photosynthetic Bacteria			
Rhodospirillum rubrum		30-35	
Nostoc muscorum		32.5	
Anabaena variabilis		35	
Anacystis nidulans		41	
Oscillatoria tenuis			45-47
Synechococcus lividus			74
Synechococcus eximius	70	79	84
Eucaryotic Algae			
Chlamydomonas nivalis	-36	0	4
Fragilaria sublinearis	-2	5-6	8-9
Chlorella pyrenoidosa		25-26	29
Euglena gracilis		23	
Chilomonas paramecium	10	28	35
Skeletonema costatum	6	16-26	>28
Cyanidium caldarium	30-34	45-50	56

Microorganism	Cardinal Temperature (°C)		
	Minimum	Optimum	Maximum
Fungi			
Candida sctottii	0	4-15	15
Saccharomyces cerevisiae	1-3	28	40
Mucor pusillus	21-23	45-50	50-58
Protozoa			
Amoeba proteus	4-6	22	35
Naegleria fowleri	20-25	35	40
Trichomonas vaginalis	25	32-39	42
Paramecium caudatum		25	28-30
Tetrahymena pyriformis	6-7	20-25	33
Blepharisma intermedium	24-25	28	38
Cyclidium citrullus	18	43	47

The cardinal temperatures vary greatly among the microorganisms (Table 7.5). Optima normally range from 0°C to as high as 75°C, while microbial growth occurs at temperatures extending from –20°C to over 100°C. The growth temperature range for a particular microorganism usually span about 30°C. Some species (e.g., *Neisseria gonorrhoeae)* have small range and are called **stenothermal** and other like *Enterococcus faecalis* will grow over a wide range of temperatures and are called **eurythermal.**

Based on temperature range for growth, microorganisms can be placed in one of the following five classes:

i. **Psychrophiles** : The microorganisms which grow well at 0°C and have an optimum growth temperature of 15°C or lower and maximum around 20°C are called psychrophiles, *e.g., Chlamydomonas nivalis* (alga), members of the genera *Pseudomonas, Flavobacterium, Achromobacter* and *Alcaligenes.* These are often killed when exposed to room temperature. They are able to function at low temperature because their membranes contain a high proportion of unsaturated fatty acids. These molecules remain fluid at the lower temperatures, whereas the membranes containing saturated fatty acids become non-functional.

ii. **Psychrotrophs**: These are some times referred to as facultative psychro-philes. They grow most rapidly at temperatures above 20°C, but are capable of slow growth at temperatures down to around 0°C. Psychrotrophs are responsible for spoilage of refrigerated products.

iii. **Mesophiles**: Mesophiles are with growth optima around 20 to 45°C and a temperature minimum of 15 to 20°C. Almost all human pathogens are mesophiles, as might be expected since their environment is fairly constant (37°C)

iv. **Thermophiles**: The microorganisms with optima above 50ºC are termed thermophiles. Several algae, protozoa and fungal species have temperature maxima upto 55ºC. However, it is only certain prokaryotes that are truly thermophilic, as no eukaryotic microbes grow at such high temperatures.

v. **Hyperthermophiles or extreme thermophiles**: Some bacteria, especially archaea, can grow at 100ºC or higher. These are called hyperthermophiles. For example *Pyrolobus fumarii* has an optimum of 106ºC and continues to grow up to 113ºC (Table 7.6). These are mainly anaerobic sulphate reducers or have other metabolism with lower requirements for thermolabile cofactors, such as NADH and NADPH.

Thermophiles and hyperthermophiles flourish in many habitats including compost, self-heating haystocks, hot water lines and hot springs. They have potential industrial use in coal desulphurization, metal leaching, methane generation and the production of commercial enzymes, especially DNA polymerases and restriction endonucleases. Thermophiles differ from mesophiles in having much more heat stable enzymes and protein synthesis systems able to function at high temperature. Their membrane lipids are also more saturated than those of mesophiles and have higher melting points. Therefore, thermophile membranes remain intact at higher temperatures.

Table 7.6. Thermophilic prokaryotes and their temperature optima

Microorganism	Optimum temp (ºC)
Archaea	
Methanobacterium thermoautotropicum	65
Methanococcus jannaschii	85
Pyrococcus furiosis	100
Pyrolobus fumarii	106
Sulfolobus acidocaldarius	75
Eubacteria	
Bacillus stearothermophilus	60 - 65
Clostridium thermocellum	60
Synecococcus lividans	67
Thermoanaerobacter ethanolicus	70

Exposure of non- thermophiles to high temperatures normally causes damage to cytoplasmic membranes, breakdown of ribosomes, irreversible enzyme denaturation and DNA strand breakage. It is difficult to determine which of these heat 'injuries' is the most deleterious to microorganisms. The actual mechanism of thermotolerance is not fully elucidated. However, ribosomes of thermophiles work effectively at higher temperatures and their membranes also remain intact. Enzymes from thermophiles, such as *Bacillus stearothermophilus* are generally more heat stable, which may be attributed to a few amino acid changes within a protein that can markedly increase thermostability. Maintenance of DNA integrity under these conditions is obviously crucial and some extreme thermophiles have DNA with a high guanine – cytosine ratio, giving increased interchain hydrogen bond that provide greater thermostability. They also maintain higher intracellular potassium ion levels, often with the unusual counter-ion, 2,3–diphosphoglycerate which helps to stabilize proteins.

7.12.2. Acidity and Alkalinity (pH)

Like temperature, every microorganism has a pH optimum and a pH range over which it grows. **Acidophiles** have their growth optimum between pH 1.0 and 7.5; **neutrophiles**, between pH 7.5 and 8.0; and **alkalophiles** prefer the pH range of 8.5 to 11.7. Extreme alkalophiles have growth optima at pH 10.0 or more. In general, different microbial groups have characteristic pH preferences. Most bacteria and protozoa are neutrophiles. Most fungi prefer slightly acidic pH, about 4 to 6. Similarly, algae favour slight acidity. Though, microorganisms grow over wide range of pH, there are limits to their tolerance. Drastic variation in pH can harm microorganisms by disrupting the plasma membrane or inhibiting the activities of the enzymes and membrane transport proteins. It has been suggested that whose optimal pH values for growth are other than pH 7.0 have unusual mechanisms for maintaining many of their proteins in the singly ionized form. The ability of some alkalophilic and acidophilic organisms to withstand pH extremes undoubtedly also reflects the intrinsic stability of some or all of their proteins in unusual pH environments. Changes in the external pH also might affect the ionization of nutrient molecules and thus reduce their availability to the organism. Microorganisms frequently change the pH of their own habitat by releasing acidic or basic metabolic waste products. Fermentative microorganisms form organic acids from carbohydrates, while chemolithotrophs like *Thiobacillus* oxidize/reduce sulfur components to sulfuric acid. Similarly, other microorganisms make their environment more alkaline by generating ammonia through amino acid degradation. Buffers are often included in media to prevent growth inhibition by large pH changes. Phosphate is a commonly used buffer. Peptides and amino acids in complex media also have a strong buffering effect.

7.12.3. Solute and Water Activity

Most microorganisms contain approximately 70-80% water and require a certain amount of free water for the performance of specific metabolic activities. When microbial cells with rigid cell walls are placed in hypotonic solutions, they take up water to become turgid, whereas those without walls swell and ultimately burst. In dry or hypertonic environments, water uptake and retention is a problem. Hypertonic environments may contain considerable amount of water, but it is not necessarily available. Infact, the high external levels of solutes, such as salt or sugars, causes water to pass out of cells and the growth is halted.

The actual water availability to a microorganism is indicated by the term water activity (a_w). This is the ratio of the vapour pressure of water in the solution surrounding the microorganism, $P_{soln,}$ to the vapour pressure of pure water, P_{water}

$$a_w = \frac{P_{soln}}{P_{water}}$$

i.e., pure water has an A_w of 1.0

Microbial species vary in their tolerance to dry conditions and environments of high osmotic strength (Table 7.7). Most bacteria, with a few exceptions, cannot grow below an a_w of 0.9, which can provide a valuable means of food preservation. Many fungi that cause biodeterioration of stored grain (a_w = 0.7) including *Aspergillus restrictus*, are xerotolerant and can grow at low moisture levels. However, truly xerophilic filamentous fungi and osmophilic yeasts, *e.g.*, *Zygosaccharomyces rouxii*, have evolved mechanism to colonize environments where the a_w can be as low as 0.6. Some of these species may not grow under conditions of high water availability.

Table 7.7. Water activity and growth of some microorganisms

Water activity (a_w)	Material	Microorganism
1.0	Pure water	*Caulobacter, Spirillum*
0.99	Human blood	*Streptococcus, Escherichia*
0.98	Sea water	*Pseudomonas, Vibrio*
0.95	Bread	Most Gram – positive rods
0.90	Maple syrup, ham	Gram – positive cocci
0.85	Salami	*Saccharomyces rouxii*
0.80	Fruit cake, jams	*Saccharomyces bailii, Penicillium*
0.75	Salt lake, salt fish	*Halobacterium, Halococcus*
0.70	Cereals candy, dried fruits	*Xeromyces bisporus*, xerophilic fungi

Marine bacteria are optimally adapted to the salt concentration in sea water (a_w = 0.98) and possess enzymes that have specific requirements for sodium ions. True halo-philes, e.g., *Halobacterium halobium* found in areas such as Dead Sea and Salt Lakes, will not grow in salt concentration less than 3 mol/l. These are archeans that have highly modified cell walls and unusual membrane lipids.

Osmophilic, xerophilic and halophilic oragnaisms accumulate compensating solutes, which balance the osmotic strength of the external solute. Examples include polyols; arabitol is accumulated by several yeasts and filamentous fungi, and high levels of glycerol are found in the halophilic alga *Dunaliella salina*. Many halophiles also store large amount of potassium ions and some bacteria accumulate amino acids, notably proline and glutamic acid. Certain strains of *E.coli* synthesize betaine (N, N, N – trimethyl glycine).

7.12.4. Oxygen

Oxygen is a critical substance for the vitality of many microbes, Interestingly many bacteria thrive in its absence. Because of the centrality of the oxygen to the survival of many microbes, considerable attention has been devoted to the effect of oxygen upon them. The ability to grow or tolerate in the presence or absence of oxygen is a major taxonomic character, which was initially used to classify the microbes (Table 7.8). Microorganisms are classified into five main groups on the basis of their requirements for oxygen.

i. **Obligate aerobes** grow only in the presence of oxygen, as it is required as the terminal electron acceptor for electron transport in aerobic respiration

ii. **Facultative anaerobes** function with or without oxygen, but grow more efficiently when oxygen is available.

iii. **Microaerophiles** require some oxygen for the synthesis of certain compounds like sterols and unsaturated fatty acids, but cannot grow at normal atmospheric oxygen concentration 21% (V/V). They must have lower oxygen levels of 2 – 10 % (V/V)

iv. **Aerotolerant anaerobes** essentially ignore the presence of oxygen and grow equally well in its presence.

v. **Obligate anaerobes** cannot tolerate oxygen. Exposure to oxygen results in their death. The toxicity of the oxygen to obligate anaerobes is discussed elsewhere in this book (see chapter 18.6)

Table 7.8. Oxygen relationship of microorganism

Group	Relationship to oxygen	Type of metabolism	Example
Aerobes			
Obligate	Required	Aerobic respiration	*Micrococcus luteus*
Facultative	Not required but growth better with O_2	Aerobic / anaerobic respiration and fermentation	*Escherichia coli*
Microaerophilic	Required at lower levels	Aerobic respiration	*Spirillum volutans*
Anaerobes			
Aerotolerant	Not required no better growth when present	Fermentation	*Streptococcus pyogenes*
Obligate	Harmful or lethal	Fermentation or anaerobic respiration	*Methanobacterium formicicum*

The above different relationships with O_2 appear due to several factors, including the inactivation of proteins and the effect of toxic oxygen derivatives (superoxide, hydrogen peroxide and hydroxyl radicals). Enzymes can be inactivated when sensitive groups like sulfahydrayls are oxidized. A notable example is the nitrogen-fixation enzyme nitrogenase which is very oxygen sensitive.

Anaerobic habitats are rather more common than expected. They may arise from the utilization of oxygen by facultative anaerobes to create an anaerobic microenvironment where obligate anaerobes can grow. In general, wherever organic matter accumulates, microorganisms will use oxygen faster than it can be replaced, creating an anaerobic environment.

In order to grow obligate anaerobes in the laboratory, procedures to exclude oxygen from the culture must be used and all manipulations must be performed in an anaerobic atmosphere. Common procedures adapted in the laboratories include:

1. The use of media containing reducing agents, such as thioglycolate, which chemically combine with oxygen

2. Physical removal of oxygen from a growth chamber, by pumping out the air with a vaccum pump, and then flushing the vessel with nitrogen gas and carbon dioxide
3. The use of gaspak jars, which involves sealing the microbial cultures inside the jar, along with a palladium catalyst and a hydrogen gas generating system. The oxygen is removed through its reaction with the hydrogen to form water.
4. Anaerobic cabinets for the growth and manipulation of the organisms, which are flushed with a gas mixture of nitrogen : carbon dioxide: hydrogen (80 : 10 : 10) and have a scrubber system to remove residual oxygen.

7.12.5. Radiation

Radiation exerts a variety of effects on microorganisms. Its effects are largely reflective of two factors, its energy and the extent to which particular microbe can with- stand radiation. Radiation resistance, in turn, is a function of the genetics of the organism.

The ways in which radiation exposure affect microbes are diverse. Radiowaves and infrared rays generally have little or no deleterious effects since they are insufficiently energetic to affect the physiology of the organism. In some cases deleterious effects have been found as a function of visible light exposure at wavelengths greater than 400 nm. However, in many cases, no harmful effects result from visible light exposure. Pigmented organisms may absorb particular wavelengths and both infrared and visible light may be used in energy generation in phototrophic species. Carotenoids and similar pigments may protect these species from excessive radiation exposure. Ultraviolet (UV) and X-ray radiation, because of their relatively high energy, exert substantially more deleterious effects on microbes. The primary effects of UV radiation are four fold : (i) damage to DNA by the formation of pyrimidine dimers: (ii) excitation of orbital electrons in pigments, (iii) fluorescence, in which energy originally absorbed in the UV region is emitted as visible light; and (iv) photosensitization in which energy absorbed by one molecule is transferred to a second molecule activating it.

Due to their intense energy and penetrating power, X–rays and cosmic rays exert more disastrous effect on microbes than any other radiation. In addition to DNA damage, these radiations induce formation of a set of detrimental entities that range from singlet oxygen to free radicals, superoxide and peroxide anions, and other highly reactive molecular and ionic species. Collectively, these elements are highly detrimental to cells through destruction and alteration of DNA and generalised oxidative damage to essential cell components.

Microbes display difference in their ability to resist radiation exposure. For example, certain organisms like *Dienococcus radiodurans* are exceptionally radiation resistant, yet other species are highly radiation affected. Many microorganisms use carotenoids for protection against photooxidation. Carotenoids effectively quench singlet oxygen. Both photosynthetic and nonphotosynthetic microorganisms employ pigments in this way.

7.12.6. Pressure

Terrestrial microorganisms are always subjected to pressure of one atmosphere (atm) and these are never significantly affected by pressure. However, many marine microorganisms are either barotolerant or truly barophilic as they live at depths in oceans where they may be subjected to pressures upto several hundred atmospheres. Barotolerants fail to grow at above 200-600 atm, whereas barophilics are capable of living under 700-1000 atm. Barophiles recovered from great depths of oceans actually donot grow at normal atmospheric pressures. The physiology and biochemistry of the barophiles are not clearly understood.

7.13. MICROBIAL GROWTH IN NATURAL ENVIRONMENT

So far the discussion was centered around the effect on microbial growth of individual environmental factors as visualized under laboratory conditions. However, under natural conditions microorganisms grow in a complex and constantly changing environment. They are exposed to many overlapping gradients of nutrients and various other environmental factors.

7.13.1. Growth Limitation by Environmental Factors

Growth of microorganisms in a natural environment is governed by two laws: One law pertaining to nutrient limitations and another with respect to factors influencing the growth. **Leibig's law of the minimum** states that the total biomass of an organism will be determined by the nutrient present in the lowest concentration relative to the organisms requirements. This law applies in both the laboratory and in natural environments. An increase in a limiting essential nutrient such as phosphate will result in an increase in the microbial population until some other nutrient becomes limiting. If a specific nutrient is limiting, changes in other nutrient will have no effect. **Shelford's law of tolerance**, related to environmental factors, such as temperature, pH, light, salinity etc., states that there are limits to environmental factors below and above which a microorganism cannot survive and grow, regardless of nutrient supply. Each microorganism has its own optimal conditions for growth. Thus the growth of a microorganism depends on both the nutrient supply and its tolerance to the environmental conditions. Another interesting situation wherein growth limitation occurs is when the organisms grow in a eutrophic environments, they release toxic waste products which will limit further growth. The growth and survival of microorganisms living in oligotrophic environments depends upon its ability to sequester critical limiting nutrients, such as iron, making them less available to competitors. Natural substances can also directly inhibit microbial growth and reproduction. The substances include phenolics, tannins, ammonia, ethylene and volatile sulfur compounds.

7.13.2. Quorum Sensing and Microbial Populations

Over the years, microbiologists were under the strong belief that bacterial population as collection of individuals growing and behaving independently. However, more recently it has become clear that many bacteria can communicate with one another and behave cooperatively. This cooperation is mainly achieved by a process known as quorum sensing or autoinduction. Through this phenomenon, the bacteria monitor their own population density through sensing the levels of signal molecules, some times called autoinducers. When the population of the bacteria rises to a specific level, signals are sent to the bacteria that the population density has reached a critical level or quorum. The bacteria then begin expressing sets of quorum-dependant genes. Though quorum sensing has been found among both Gram-positive and Gram-negative bacteria, it is well understood in Gram-negative bacteria. This phenomenon is thoroughly discussed in chapter 26.

7.14. FURTHER READING

Abe, F., Kato. C. and Horikoshi, K. (1999), Pressure regulated metabolism in microorganisms. *Trends Microbiol.,* **7 :** 447-453.

Blomberg, A. and Adler, L. (1992), Physiology of osmotolerance in fungi. In : Advances in Microbial Physiology, vol. 33, A.H. Rose (ed.) 145-212, San Diego, Academic Press.

Brown, A.D. (1976), Microbial water stress. *Bacteriol. Rev.,* **40(4) :** 803-846.

Dunny, G.M. and Leonard, B.A.B. (1997), Cell-cell communication in gram-positive bacteria. *Ann. Rev. Microbiol.,* **51 :** 527-564.

Friedman, S.M. (1992), Thermophilic microorganisms. In Encyclopedia of microbiology, Vol. 4, J. Lederberg (ed.) 217-229. San Diego, Academic Press.

Gray, K.M. (1997), Intercellular communication and group behaviour in bacteria. *Trends Microbiol.,* **5 :** 184-188.

Greenberg, E.P. (1999), Quorum sensing in gram-negative bacteria. An important signaling mechanisms in symbiosis and disease. In Microbial ecology and infectious disease. E. Rosenberg (ed.) 112-122, Washington, DC : ASM Press.

Koch, A.L. (1995), Bacterial growth and form. New York, Chapman & Hall.

Kolter, R., Siegele, D.A. and Tormo, A. (1993), The Stationary Phase of the Bacteria Life Cycle. *Ann. Rev. Microbiol.,* **47 :** 855-874.

Krieg, N.R. and Hoffman, P.S. (1986), Microaerophily and oxygen toxicity. *Ann. Rev. Microbiol.,* **40 :** 107-130.

Madigan, M.T. and Marrs, B.L. (1997), Extremophiles. *Sci. Am.,* 276 (4) : 82-87.

Marr, A.G. (2000), Growth kinetics, bacterial. In : Encyclopedia of microbiology, 2nd edn., Vol. 2, J. Lederberg (ed.), 584-589, San Diego, Academic Press.

Prosser, J.I. and Tough, A.J. (1991), Growth mechanisms and growth kinetics of filamentous microorganisms. *Crit. Rev. Biotech.,* **10 :** 253-274.

Schlegel, H.G. and Jannasch, H.W. (1992), Prokaryotes and their habitats. In : The Prokaryotes, 2nd (edn.) A. Balows *et al.,* (eds.), 75-125, New York, Springer-Verlag.

Siegele, D.A. and Kolter (1992), Life after Rag. *J. Bacteriol.,* **174 :** 345-348.

Tempest, D.W. (1978), Dynamics of microbial growth. In Essays in Microbiology, J.R. Norris and M.H. Richmond, (eds.) 7/1 - 7/32, New York, John Wiley and Sons.

■■■■

Chapter - 8

Homeostasis and Survival

8.1. DEFINITION

Homeostasis is defined as the ability of living organisms to maintain a constant internal environment despite changes in the external environmental conditions. Homeostasis, when applied to bacterial cell means the ability of the bacterium to maintain a steady intracellular pH and a constant osmotic differential across the cell membrane, despite fluctuations in external pH and osmolarity.

8.2. MAINTENANCE OF CONSTANCY OF pH

In nature, bacteria can be found growing in a wide range of habitats that vary in pH from approximately pH 1–2 in acid springs to as high as pH 11 in soda lakes and alkaline soils. Nevertheless, irrespective of the external pH, the internal pH is usually maintained within 1–2 units of neutrality, which is necessary to maintain the physiological functions (Table 8.1). Obviously, bacteria maintain a pH gradient across the cell membrane. In case of acidophiles, which grow optimally between pH 1 and 4 maintain an internal pH of about 8.5–7.0. In other words, they maintain a pH gradient greater than 2.5 units. Neutrophiles maintain a pH gradient of 0.5 – 1.5 units. On the other hand, alkaliphiles, which grow optimally often in the range of pH 10 – 12, have an internal pH of 8.4 – 9.0. Therefore, alkaliphiles maintain a negative pH gradient of about 1.5 – 2.0 units.

Table 8.1. pH homeostasis in bacteria

Bacteria	**pH out**	**pH in**	**ΔpH* (in – out)**
Neutrophiles	8.0 – 8.0	7.5 – 8.0	+
Acidophiles	1.0 – 4.0	8.5 – 7.0	+
Alkaliphiles	9.0 – 12.0	8.4 – 9.0	—

*ΔpH is the gradient of pH.

Many factors influence intracellular pH. These include buffering capacity of the cytoplasm and metabolic reactions that produce acids and bases. However, it has been demonstrated that the regulation of intracellular pH is mainly due to the flow of protons across the cell membrane.

Evidence for this comes from the observations that pH gradient is dissipated when the proton pumps are inhibited or when proton ionophores are added to the medium. Homeostasis with reference to pH can be explained as follows. For example, in the acidic environment, when the cell cytoplasm becomes too acid, protons are pumped out. This is done electroneutrally by bringing K^+ into the cell. Similarly, when the external environment is alkaline, the cytoplasm becomes too basic. Under these conditions, the protons are brought in via exchange with outgoing K^+ or Na^+ (Fig. 8.1). It shows that there exists a feedback mechanism where the intracellular pH can signal proton pumps and antiporters. The regulatory mechanisms at this level is not understood, though this is an area of active research.

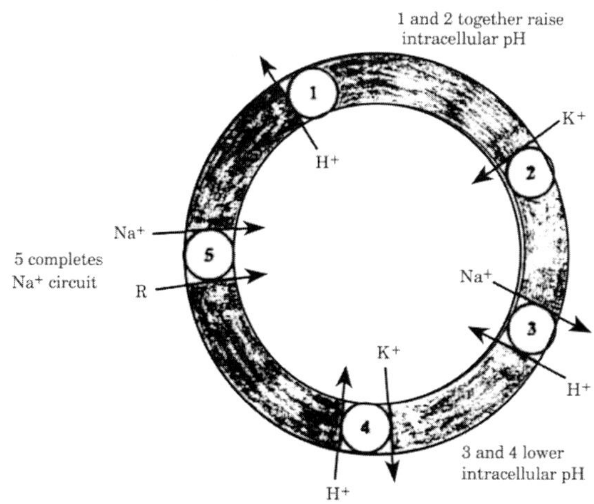

Fig. 8.1. Mechanism of pH homeostasis

The main problem of alkaliphiles is to keep cytoplasmic pH more acidic than the external pH. They solve this problem by bringing in protons and extruding Na^+ by Na^+/H^+ antiporters. When alkaliphiles are placed in a medium without Na^+ at pH 10 or 10.5, the internal pH quickly rises to the value of external pH. However, when Na^+ is present in the external medium, the internal pH does not raise upon shifting to the more basic medium. In acidophiles, the external pH is several units lower than the cytoplasmic pH. The maintenance of the larger pH gradient requires an inverted membrane potential. For example, the membrane potential of *Thiobacillus ferroxidans* is 10 mv at pH 2 and that of *Bacillus acidocaldarius* is +20 to +30 mv at pH 2.5. It has been suggested that the maintenance of inverted membrane potential in acidophiles is due to an inward flux of K^+ greater than an outward flux of protons. This may be due to the electrogenic influx of K^+ catalyzed by an ATP–dependant K^+ pump. Thus, the method of maintaining a pH gradient in acidophiles and neutrophiles may be similar in relying on K^+ influx to depolarize the membrane.

8.3. MAINTENANCE OF CONSTANCY OF OSMOLARITY

The physiological significance of osmotic homeostasis is that it maintains an internal turgor pressure necessary for growth. Bacteria have the capability of adjusting their internal osmolarity to changes in external osmolarity in order to maintain cell turgor. However, how bacteria sense changes in the external osmolarity and transfer the appropriate signals to the adaptive cellular machinery is largely unknown.

8.3.1. Adaptation to High Osmolarity

When cells are placed in a medium of high osmolarity, they spontaneously increase the intracellular concentration of certain solutes called **osmolytes**, thus ensure that internal osmolarity is always higher than the external and thus the cell turgor is maintained. The osmolytes utilized by bacteria for this purpose are called 'compatible solutes' to denote their relative nontoxicity. Some compatible solutes are synthesized by the cell and others are accumulated intracellularly from the medium via transport. The compatible solutes taken from external medium are called **'osmoprotectants'**. Though there are many compatible solutes, the most important compatible solute in bacteria is K^+, whose intracellular concentration is sufficient high to maintain the internal osmolarity. Bacteria that live in high osmolarity media have proportionally higher intracellular concentrations of K^+ because of uptake. Thus, K^+ is important for two major aspects of homeostasis: (i) the maintenance of cell turgor, and (ii) the maintenance of a pH gradient. Bacteria employ several compatible solutes other than K^+ which include amino acids, glutamate, glutamine and proline, the quarternary aminobetaine (also called trimethyl glycine or glycine betaine) and certain sugars like trehalose.

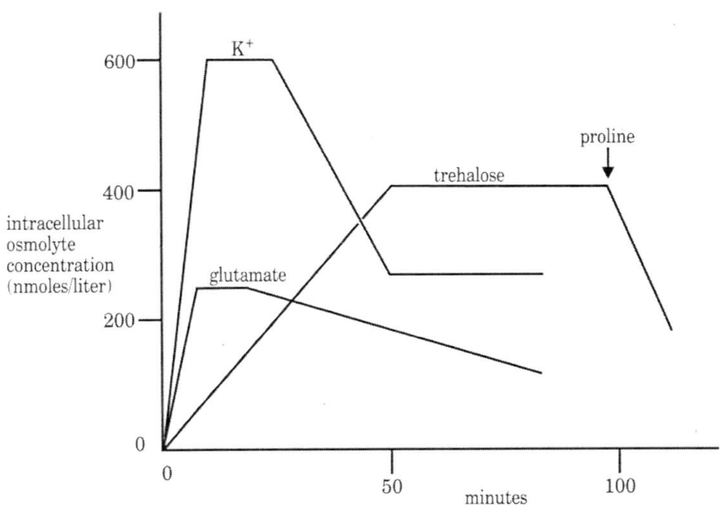

Fig. 8.2. Changes in intracellular osmolyte concentrations when *E. coli* is shifted to a medium of higher osmolarity.

Now let us try to understand the osmotic homeostasis taking in *E. coli* as a model. When *E. coli* cell is transferred to a medium of high osmolarity, a series of internal adjustments take place (Fig. 8.2). First an influx of K^+ occurs via the potassium uptake systems. At about the same time the cells synthesize glutamic acid which ionizes providing counter ions to the increased levels of K^+, and the protons are pumped out of the cell to counteract the membrane depolarization caused by the influx of K^+. Thus, the initial major compatible solute is potassium glutamate. However, after sometime the potassium glutamate is replaced by newly synthesized sugar, trehalose, which then becomes the major compatible solute. If proline or betaine is present in the medium, the *E.coli* transports these into the cell and replaces the trehalose or the excess K^+, indicating that *E. coli* preferentially use proline and betaine as osmotic stabilizers rather than trehalose or K^+. The preferential use of some osmoprotectants over others is common among bacteria. Many bacteria prefer betaine as osmoprotectant. It is known to suppress the uptake of other osmoprotectants, facilitate the excretion of K^+ and the catabolism of trehalose, thus replacing these osmolytes.

8.3.2 Adaptation to Low Osmolarity

Many bacteria possess the ability to adjust to low osmolarity of media, thus limiting cell turgor pressure. However, many details are not available as in the case of adjustment to high osmolarity. One response of bacteria is to decrease the concentration of cytoplasmic osmolytes. This may occur by specific excretion of osmolytes or their catabolism. For example, *E. coli* excretes K^+ via special transporters which may respond to cell turgor.

8.3.3 Osmotic Homeostasis in the Periplasm

In Gram-negative bacteria, periplasm remains as a separate compartment under all conditions of external molarity, with an osmolarity apparently isoosmotic with that of the cytoplasm. It has been suggested that Gram-negative bacteria adapt to low osmolarity media by raising the osmolarity of the periplasm so that cytoplasm never actually experiences the low osmolarity of external medium. In this way shocking of the cell membrane with concomitant compression of the peripalsm are minimized and also the turgor pressure across the cell membrane. These bacteria maintain an osmolarity in the periplasm higher than that of the external medium by synthesis and/or accumulation of periplasmic osmolytes. Indeed many Gram-negative bacteria like *E. coli*, *S. typhimurium*, *Pseudomonas*, *Agrobacterium* and *Rhizobium* synthesize negatively charged periplasmic oligosaccharides (*e.g.* membrane derived oligosaccharides (MDO) or β-glucans) when grown in media of low osmolarity. The increase in the synthesis of the periplasmic oligosaccharides when the bacteria are grown in media of low osmolarity has led to the suggestion that the role of oligosaccharides is to raise the osmolarity in the periplasm.

8.4 THE HEAT SHOCK RESPONSE

Bacteria constantly sense and adapt to their environment in order to optimize their ability to survive and multiply. When an organism is exposed to elevated temperature it quickly responds and adapts to the new conditions. For example, when *E. coli* is shifted transiently from 32ºC to 42ºC it synthesizes a group of proteins, called **heat shock proteins**, that are not present during the growth under normal growth temperature. Many of these heat shock proteins are required for cell growth or survival at more elevated temperatures (thermotolerance). Some of the induced proteins are *dnaJ* and *dnaK*, the RNA polymerase sigma – 70 subunit (rpoD), groES, gro El, lon and lysU. Total of 17 heat shock proteins are identified in *E. coli*. Production of heat shock proteins is a widespread phenomenon in nature and occurs in both microbial and non microbial forms. The precise function of heat shock proteins at present is not clearly known. They may differ from organism to organism. The manner of control of their formation has been, however, established. Formation of heat shock proteins is mediated by formation of a particular sigma factor (δ 32) that allows DNA–dependant RNA polymerase to recognize particular promoters and to allow formation of heat shock proteins.

Heat shock response is also triggered by a variety of environmental agents, such as ethanol, UV radiation and agents that inhibit DNA gyrase. Induction of all these stimuli occur through δ 32.

8.5 STARVATION AND SURVIVAL

The survival of microorganisms in their natural environments depends upon their ability to utilize the nutrients that are available to them. These substrates are often of transient appearance and present at very low concentrations. Organisms are, therefore, likely to experience starvation conditions punctuated by short periods of nutrient supply when relatively rapid growth can occur. Consequently, it is vital that microorganism be able to withstand the periods of famine–starvation and preserve essential cellular functions which will permit growth to occur when a nutrient supply is restored. Lately, increased attention is being given to physiological changes that occur in bacteria that enter the stationary phase when they are experimentally subjected to starvation. These bacteria undergo physiological changes that result in metabolically less active and that are more resistant to environmental hazards. Microbes vary considerably in their survival characteristics. Some are endowed with specialized survival mechanisms such as spore or cyst formation which are resistant to adverse conditions including starvation. More recently, it has been discovered that even bacteria such as *E. coli*, *Salmonella* and *Vibrio* that do not form spores or cysts, undergo adaptive changes when faced with nutrient deprivation. These changes also result in resistant and metabolically active cells. Certain bacteria exist for only a few hours in the absence of nutrients, whereas some soil and aquatic organisms can survive for weeks, months or even years (Table 8.2). The cells under starvation exhibit many adaptations in order to survive. Some of these are discussed below.

(i) **Changes in cell size:** The cells under starvation generally become smaller due to the fact that they keep dividing after growth has ceased. This results in the production of more cells that may be advantageous for dispersing the population. The additional advantage is that the smaller cells have a high surface area to volume ratio which enables them to accumulate nutrients efficiently from dilute solutions. In some cases, there may be several cell divisions in the absence of growth so that the cell size differs radically from the growing cells.

(ii) **Changes in cell walls and surface properties:** When certain marine bacteria are starved, the cell surface becomes hydrophilic and the cells are more adhesive. *Vibrio* synthesises surface fibrils and forms cell aggregates when starved for a long time.

(iii) **Changes in the metabolic activity and chemical composition:** In many bacteria, when they are starved a significant increase in the turnover (metabolic breakdown and resynthesis) of proteins and RNA has consistently been observed. Presumably, in starved cells, the proteins and ribosomal RNA can serve as an energy source to maintain viability and crucial cell functions. The latter would include solute transport systems, an energized membrane and ATP pool levels. There are also specific biochemical changes that occur when cells face starvation. For example, in *E. coli* all the unsaturated fatty acids in the membrane phospholipids become converted to the cyclopropyl derivatives by the methylation of the double bonds. However, the advantage of the cyclopropane fatty acids is not known. Bacteria may also synthesize as many as 30–50 new proteins under conditions of starvation. Many of the proteins serve specific functions related to the nutrients that is low in concentration for example, phosphate starvation induces the synthesis of PhoE porin, which is an outer membrane channel for anions, including phosphate. Similarly, nitrogen fixation genes are induced when the cells are starved for nitrogen. In addition to the proteins of known function, there are also many proteins induced under different conditions of starvation for which there is presently no known function.

Table 8.2. Survival times (in hours) of various copiotrophic bacterial species in non-nutrient buffers

Organism	**ST^{*}_{50} (in hours at 30 – 37°C)**
Arthrobacter crystallopoites	2400
Bdellovibrio bacteriovorus	10
Chromatium vinosum	120 (dark)
Escherichia coli	36
Megasphaera elsdenii	3 - 5
Nocardia corallina	480
Peptococcus prevotti	10 - 12
Rhodospirillum rubram	
Light	344
Dark	12
Sarcina lutea	65
Selenomonas ruminantium	
N - limited	3.6 – 8.1
C - limited	0.5 – 2.5
Staphylococcus epidermidis	
Aerobic	80
Anaerobic	8.0 – 8.5

* ST is the time required for half the original population to die

8.5.1 Global Regulatory System

Many bacteria possess the ability to survive numerous starvations such as nitrogen, phosphate and carbon sources. Through all these nutritional stresses, the cell maintains the required balance between the various physiological components of DNA replication, cell growth and division. The ability to integrate regulatory circuits involved with single aspect of cellular physiology into a coordinated response to an environmental stress is referred to as **global control**. Global regulatory networks include sets of operons and regulators with seemingly unrelated functions with positions and scattered around the chromosome, yet they all are coordinately controlled in response to a particular stress.

8.5.1.1 The Kat F gene

The response to carbon starvation in *E.coli* is mediated by *Kat F* gene product which is required for the synthesis of many proteins. *Kat F* mutants do not have the resistance of the wild type including the stationary phase resistance to heat, osmotic stress or to H_2O_2. They also loose viability faster than wild type when starved under certain conditions. The predicted amino acid sequence of the *Kat F* gene, based upon nucleotide sequence data, suggests that it may be sigma factor. Proteins whose synthesis depends upon *Kat F* include a catalase made during stationary phase (HPII), and which presumably accounts for resistance to H_2O_2 an exonuclease III, and an acid phosphatase. *E. coli* synthesizes about 30 novel proteins after starvation for carbon source.

8.5.1.2. Stringent response

Under amino acid starvation conditions, some bacteria have a unique mechanism for regulating the transcription of specific operons and DNA sequences that code for rRNA and tRNA called as the **stringent response**. This response greatly reduces the rates of protein and other macromolecular synthesis by decreasing the synthesis of ribosomal RNA. Through this mechanism the cell has the ability to shut down a number of energy–draining activities as a survival mechanism under conditions of starvation. The stringent response involves the production of unusual guanosine phosphates, guanosine pentaphosphate (pppGpp) and guanosine tetraphosphate (ppGpp) (earlier referred to as Magic Spot I and Magic Spot II).

The bacteria that exhibit the stringent response produce a protein called stringent factor that is a product of the relA gene. Stringent factor is normally associated with the bacterial ribosome at ratio of about 1 molecule/200 ribosomes. Under conditions of amino acid starvation, uncharged tRNAs can enter the aminoacyl site on the ribosome. When this occurs, stringent factor catalyses the transfer of a pyrophosphate group from ATP to either GTP or GDP, which are involved in protein synthesis. The pyrophosphorylation of GTP and GDP produce pppGpp and ppGpp, respectively. ppGpp is an effector molecule which works in several ways inhibiting the RNA,DNA and protein synthesis. It may specifically bind to promoters of rRNA and tRNA sequences and inhibit their transcription by RNA polymerase. Alternatively, ppGpp causes increased idling or pausing of the translation process and therefore, premature termination of specific polypeptides.

Within seconds, after exposure to starvation conditions, cells shut down RNA, protein and peptidoglycan biosynthesis (Fig. 8.3). At the same time, the rate of proteolysis or protein turn over and degradation of RNA increases. Perhaps protein and RNA may serve as energy sources to drive critical cell functions. Cells start depleting storage polymers such as poly-β-hydroxybutyric acid or glycogen. Half life of mRNA greately increases. As a result, the cells continue to survive for a long period at a low metabolic rate. Newly synthesized proteins efficiently scavenge the deficient nutrient and also confer a more stress tolerant phenotype.

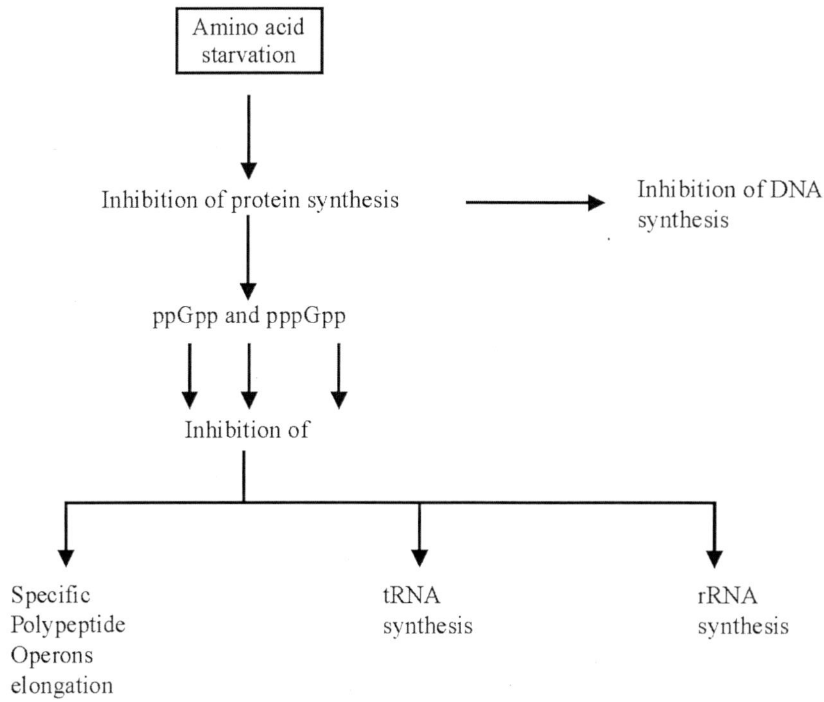

Fig. 8.3 : The stringent response – generalized response to amino acid starvation

8.5.1.3. The Ntr system

Under nitrogen starvation conditions, the Ntr system of the cell becomes activated. When the concentration of ammonia becomes a growth limiting factor, this system enables the cell to scavenge the last traces of ammonia. It also activates additional operons that are involved in the utilization of organic nitrogen sources such as the *nac* operon. This enables the cell to turn to alternate sources of nitrogen to meet its biosynthetic needs.

When ammonia concentration becomes limited, bacteria are able to turn on transcription of genes whose products act on other nitrogen sources to produce ammonia. Similarly, transcription of genes for glutamine synthetase is also induced. The gene for synthesis of glutamine synthetase is regulated as a part of a multigene system called the Ntr system. Other genes regulated by Ntr system are NRII (a histidine kinase) and NRI (a response regulator). The operon controlling the biosynthesis of glutamine synthetase in *Salmonella typhimurium* is *gln-ntrB-ntrC*. The gluA codes for glutamine synthetase, glnL codes for NRII and glnG codes for NRI. Ntr system genes transcription is inhibited by high concentration of ammonia but transcription occurs at high rate under nitrogen limiting conditions. The enzymes of the Ntr system enable bacteria to assimilate very low levels of ammonia by catalyzing the assimilation into glutamine in an ATP dependant reaction. Subsequently, the amino nitrogen group of glutamine can be transferred to β-ketoglutarate to form glutamate, which supplies approximately 85% of the nitrogen to the nitrogen – containing molecules of the cell.

As stated earlier, amino acid starvation conditions also lead to activation of *nif* (nitrogen fixing) genes. It has been estimated that the level of NR_I in ammonia repressed cells is about five molecules per cell, which when phosphorylated under condition of ammonia starvation will activate transcription of gln ALG. When levels of NRI-P becomes sufficiently high, it activates transcription of the other operons in the Ntr regulon system. Infact, NR_I-P activates *nif* AL operon. This leads to increased intracellular concentration of Nif A. Nif A protein can then activate expression of the other nif operons. The activity of Nif A is regulated by the Nif L product in response to a oxygen and probably ammonia. However, the mechanism is not known. The *nif* system is active only under anaerobic and low ammonia concentration

8.5.1.4. The Pho system

Bacteria have evolved a signaling system to induce the phosphate assimilation pathways when the supply of phosphate becomes limiting. Normally, cells obtain phosphate for incorporation into nucleotides from inorganic phosphates. However, under conditions of inorganic phosphate starvation, the phosphate utilization network (pho system) is activated which permits utilization of alternate sources of phosphate. Over 100 proteins are synthesized at elevated rates when Pho system is activated. This system results in the production of high concentration of alkaline phosphatase so that the phosphate can be obtained from organic source. In *E. coli*, alkaline phosphatase can make upto 6% of the cell biomass under conditions of inorganic phosphate starvation.

8.5.1.5. Specific strategies for coping with starvation

Bacteria preferentially growing at low nutrient concentration are called **oligotrophs,** or low nutrient bacteria. Most oligotrophs have slow growth rate. In contrast to oligotrophs, bacteria that grow at high nutrient condition are called **copiotrophs,** such as those growing in most culture media. These bacteria exhibit high rate of reproduction.

(i) **Oligotrophic bacteria:** Oligotrophic bacteria possess physiological properties that permit them to use efficiently the limited nutrient resources available to them. They conserve all the available resources to uptake growth substrates against steep concentration gradient between the cell and its surroundings. Many oligotrophic bacteria have appendages or form very small cells so that they have a high surface area to volume ratio. This enables them to accumulate nutrients efficiently from dilute solutions. Bacteria like *Caulobacter* form appendages called prosthecae which increase the surface area to volume ratio. Besides, the prostheca–forming bacteria, some appendaged bacteria such as *Planctomyces* form stalks which are structures lacking cytoplasm.

(ii) **Endospore formation:** A few bacterial genera, such as *Bacillus* and *Clostridium,* form endospores when an essential growth nutrient is exhausted and exponential growth ceases. Formation of endospore and normal reproduction are mutually exclusive processes. Endospore formation represents a cellular differentiation to a non- reproducing form. The details of physiology of endospore formation are discussed in Chapter 27.

(iii) **Myxobacteria:** Myxobacteria have developed a complex life cycle to cope with alternating conditions of sufficient nutritional resources and starvation. Under conditions of sufficient nutrients, they reproduce by binary fission. As the starvation conditions commence, the binary fission ceases and upto a million cells fuse to form a fruiting body. The formation of fruiting body is initiated by a chemical signal generated from a starving cell. The myxospore produced in the fruit body are resistant to desiccation. They can survive for prolonged periods. These myxospores under favourable conditions germinate and produce vegetative cells. This life cycle strategy permits survival and movement between discrete and widely dispersed sources of nutrients.

8.5.1.6. Energetics of Starvation

Preservation of cellular integrity during starvation depends ultimately upon the availability of an energy source and in the absence of external energy nutrient source or light, it must be derived from endogenous sources. Some microbes possess specialized reserves of carbon and energy that are utilized during starvation. However, others in the absence of such reserves catabolize basal nutrients such as RNA and proteins and obtain the energy. Recently, attempts have been made to identify factors critical for survival including: (i) utilization of specific cellular components, (ii) possession of reserve materials, (iii) endogenous metabolic rate and (iv) adenylate energy. More recently, realising the importance of the energized membranes in transport process and motility, the role of the membrane potential in survival has also been investigated.

(i) Endogenous Metabolism

The endogenous metabolism of a microorganism is defined as the total metabolic reactions that occur when it is deprived of exogenous substrates. Consequently, this metabolism must provide all the energy needed by the bacterium for survival and starvation conditions (Fig. 8.4). Compounds which have been identified as substrates for endogenous metabolism, in addition to recognized reserve materials, include RNA, proteins and in the case of micrococci, staphylococci and streptococci certain amino acids of the intracellular free amino acid pool. It has been suggested that rapid metabolism of endogenous substrates accelerates the death of starved microorganims, whereas low rate prolongs the survival.

(ii) Adenylate Energy Charge and Survival

Bacteria growing under normal conditions maintain an adenylate energy charge of 0.8 to 0.9. Under carbon and energy source starvation conditions, the adenylate energy charge decreases. However, the viability is lost only when the adenylate energy falls below 0.5. Thus, it has been pointed out that energy charge may be useful in the prediction of survival of certain organisms, but not, in all cases.

(iii) Storage and Energy Reserve Compounds

Possession by an organism of storage materials like glycogen and poly–β–hydroxy butyrate generally retard the utilization of cellular components such as RNA and protein during starvation. Glycogen has been shown to prolong the viability of a number of bacteria including *E. coli* and *Klebsiella aerogenes.* Similarly, poly-β–hydroxy butyrate enhances the survival of many microorganisms. It has been implicated as a carbon and energy source for the encystment of

Azotobacter. Recently, it has been reported that some *Streptococcus* species when harvested from certain growth media, contained an elevated levels of phosphoenol puruvate (PEP) and 2-phosphoglycerate which can serve as source of energy during starvation, supporting sugar transport via the PEP dependant phospho-transferase system presumably supplying ATP for purposes of maintenance and survival. Interestingly, these streptococci neither possess reserve polymers nor can they use products of RNA degradation as energy sources. Further, they lack the functional cytochromes for ATP generation. Obviously, at least in the early stages of starvation PEP is a valuable source of energy for driving sugar transport.

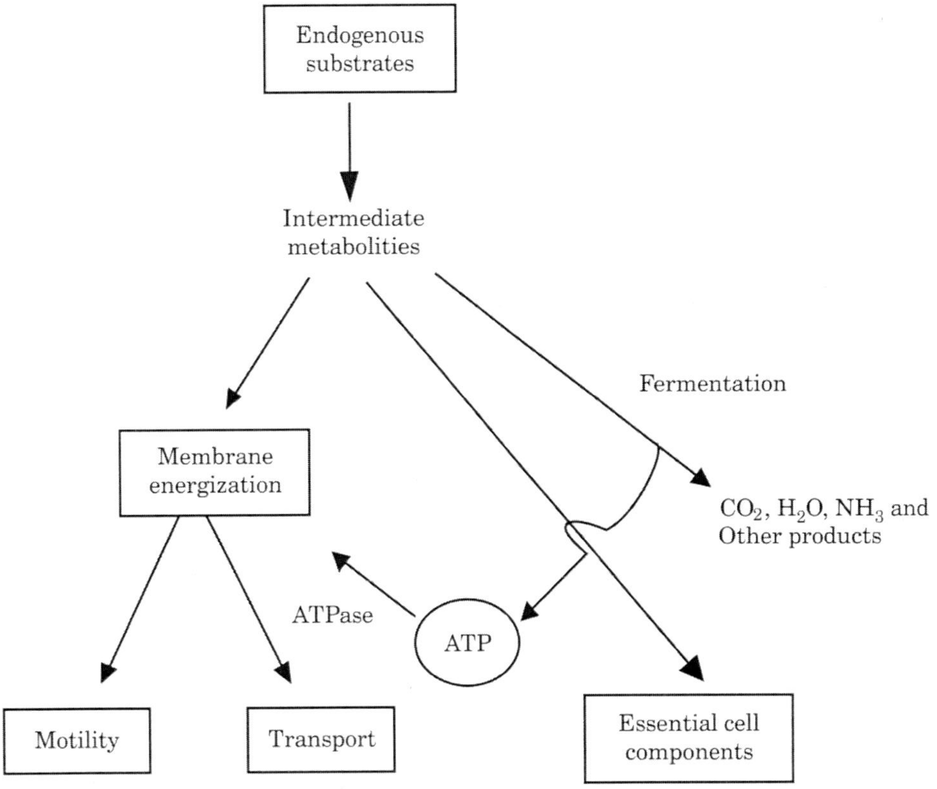

Fig. 8.4. Relationship between endogenous metabolism and maintenance of energy requirements

(iv) Role of Membrane Potential

Preservation of functional transport systems is essential for survival under starvation conditions. These transport systems, more often, are energy dependant and the maintenance of an energized membrane is crucial for their operation. Obligate anaerobes or facultative organisms under anaerobic condition energize their membranes with ATP (derived from substrate level phosphorylation) directly by membrane bound ATPases. ATP depletion with consequent loss of transport activity, would effectively prevent the uptake of nutrients.

(v) Proton–solute Stoichiometry and Gating

Some bacteria possess mechanisms which enable them to retard the rate of energy or metabolite pool dissipation. These comprise symport systems with variable proton solute stoichiometry *i.e.,* the number of protons they translocate per solute molecule varies in response to environmental conditions. This device enables an organism to maintain a high rate of solute uptake and a high intracellular concentration of metabolites and ions by means of increasing the proton solute stoichiometry as the proton motive force decreases. It is obviously a mechanism of importance to the organism during starvation for it favours the uptake of available nutrients, while simultaneously preventing the loss of metabolites from the cell.

Another interesting mechanism is that when the proton motive force falls below a critical value, inhibition of certain transport and energy – transducing systems occurs. This phenomenon is referred to as gating. The consequence of gating is that ATP is not hydrolyzed by the ATPase, ΔP is unable to effect carrier mediated solute uptake, and intracellular metabolites cannot exit via carriers. This retardation of the loss of energy sources enhances the survival under starvation conditions.

8.6. FURTHER READING

Audia, J.P., Webb, C.C., Foster, J.W. (2001), Breaking through acid barrier : an orchastrated response to proton stress by enteric bacteria. *Int. J. Med. Microbiol.*, **291** : 97-108.

Berlett, B.S. and Stadtman, E.R. (1997), Protein oxidation in aging, disease and oxidative stress. *J. Biol. Chem.*, **272** : 313-318.

Booth, I.R. and Louis, P. (1999), Managing hypoosmotic stress : aquaporins and mechanosensitive channels in *Escherichia coli*. *Curr. Opin. Microbiol.*, **2** :166-169.

Clements, M.O., Watson, S.P. and Foster, S.J. (1999), Characterization of the major superoxide dismustase of *Staphylococcus aureus* and its role in starvation survival, stress resistance and pathogenicity. *J. Bacteriol.*, **181** : 3898-3903.

Chatterji, D. and Ojha, A.K. (2001), Revising the stringent response, ppGpp and starvation signaling. *Curr. Opin. Microbiol.*, **4** : 160-165.

Krulwich, T.A., Ito, M., Hicks, D.B., Gilmour, R. and Guffanti, A.A. (1998), pH homeostasis and ATP synthesis : studies of two processes that necessitate inward proton translocation in extremely alkaliphilic *Bacillus* species. *Extremophiles* **2** : 217-222.

Matin, A. (1999), pH homeostasis in acidophiles. In J.C. Derek and G. Gardew (eds.) Bacterial responses to pH. John Wiley & Sons Ltd., Chichester, England, pp. 152-165.

Storz, G. and Hengge-Aronis, R. (eds) (2002), Bacterial stress responses. American Society for Microbiology Press, Washington, DC, pp. 161-178.

Spector, M.P. (1998), The starvation stress response (SSR) of *Salmonella*. *Adv. Microbiol Physiol.*, **40** : 233-279.

White, D. (1995), The Physiology and Biochemistry of Prokaryotes, Oxford University Press, New York.

■■■■

Chapter - 9

Phototrophy

9.1. PHOTOTROPHY

Microorganisms exhibit a great diversity in nutrition. They are able to utilize a variety of energy, carbon and proton and/or electron sources to synthesize their cell materials. Indeed, classifications of microorganisms are based upon mode of nutrition (Chapter 4). Microorganisms can utilize light as well as energy rich organic or inorganic molecules to obtain the energy and drive the biochemical reactions. Organisms using light as energy source are called **phototrophs** (pho = light, trophe = nutrition). Evidently, phototrophs must possess a photosynthetic apparatus that enables them to convert light energy into the high energy phosphate bonds of ATP.

$$ADP + Pi \xrightarrow{\text{Light}} ATP + H_2O$$

Phototrophic metabolism can be differentiated into two types: (i) photolithoautotrophy (ii) photoorganoheterotrophy. The differences between these two types are given in Table 9.1

Table 9.1. Types of phototrophic metabolism

Type	Electron donor	Carbon source	Examples
Photolithoautotrophy	H_2O	CO_2	Plants, cyanobacteria
	H_2S, S^o,H_2	CO_2	Chromatiaceae
			Chlorobiaceae
Photoorganoheterotrophy	Organic substrates	Organic substrates	Rhodospirillaceae

The carbon source commonly used by phototrophs is CO_2 and organisms that derive most of their cell carbon from CO_2 are called **autotrophs**. However, these bacteria when growing in the presence of CO_2 require an electron donor for the reduction of CO_2 to the level of cell material. As in plants and cyanobacteria, the electron donors used are frequently inorganic compounds. Thus, phototrophic bacteria that employ molecular hydrogen or reduced sulfur compounds are called '**photolithoautotrophs**'. A number of phototrophic bacteria can also grow in the light with organic substrates such as acetate or succinate. These organic substrates are used by the microorganism both as source of electron and carbon. Such organisms are referred to as '**photoorganoheterotrophs**'.

9.2. PHOTOSYNTHESIS

Phototrophs can capture light energy and use it to synthesize ATP and $NADH_2$ or $NADPH_2$. This process in which light energy is trapped and converted to chemical energy is called **photosynthesis**. Usually, a photosynthetic organism reduces and incorporates CO_2 in to cell. These reactions are also considered as a part of photosynthesis. Photosynthesis is one of the most significant metabolic processes on earth because almost all our energy is ultimately derived from solar energy. It provides photosynthetic organisms with the ATP and NAD(P)H necessary to manufacture the organic material required for growth. In turn, these organisms serve as the base of most food chains in the biosphere. Photosynthesis is also responsible for replenishing our supply of O_2, a remarkable process carried out by a variety of organisms, both eukaryotic and prokaryotic. Although most people believe that most of the photosynthesis is carried out by plants, but in reality half of the photosynthesis on the earth is carried out by microorganisms.

The unique reaction of photosynthesis is mediated by a class of molecules termed as chlorophylls. Depending upon the photosynthetic system, light energy can do one or both of the following, functions.

1. In all photosynthetic systems light energy can drive the phosphorylation of ADP into ATP (photophosphorylation).
2. In some photosynthic systems, light can also drive the transfer of electrons from H_2O or H_2S to $NADP^+$. This is referred to as photoreduction of $NADP^+$ and in the process water becomes oxidized to oxygen gas or hydrogen sulphide to sulphur and the NADP+ is reduced to NADPH. During both ATP and NADPH2 production, electromagnetic energy is absorbed by photopigments and get converted into chemical energy. In a precise way, this is a light driven oxidation of chlorophyll or bacteriochlorophyll. Martin Kamen defines photosynthesis as a "process in which electromagnetic energy is converted to chemical energy that can be used for various biosynthetic processes". Thus, photosynthesis is the conversion of light energy into chemical energy, where this energy is utilized for reducing carbon dioxide to carbohydrates and related organic products with the help of water.

9.3. DIVERSITY OF PHOTOTROPHS

The photosynthetic prokaryotes (bacteria and cyanobacteria) are a diverse assemblage of organisms that share the common features of being able to use light as a source of energy for growth. Their classification is based upon physiological differences, including whether they produce oxygen, and what they use as a source of electrons for biosynthesis. Outline classification and physiological characteristics of different groups of phototrophic bacteria are summarised in Tables 9.2 and 9.3, respectively. Thus, phototrophic microorganisms can be broadly divided into two groups:

1. Oxygenic phototrophs
2. Anoxygenic phototrophs.

9.3.1. Oxygenic Phototrophs

The oxygenic phototrophic prokaryotes, as the name itself suggests, release oxygen during their photosynthesis especially during light phase reactions. This oxygen is released from the photolysis of water which is used as an electron donor. Most of the species belong to the cyanobacteria which are widely distributed in nature in close association with oxygen, occurring in fresh and marine waters and terrestrial habitats. They possess only one type of chlorophyll (chlorophyll a) and light harvesting pigments called **phycobilins**. Another kind of oxygenic microbial phototrophs are the **prochlorophytes**, which are a diverse group of photosynthetic prokaryotes. They differ from cyanobacteria in having chlorophyll b rather than phycobilins as the light harvesting pigments. There are three genera among the prochlorophytes: *Prochloron, Prochlorothrix* and *Prochlorococcus.* The species of the *Prochloron* are obligate symbionts of certain ascidians (sea squirts). *Prochlorothrix* is a filamentous, free living microorganism that lives in freshwater lakes. *Prochlorococcus* is a free living marine microorganism.

9.3.2. Anoxygenic Phototrophs

Anoxygenic prokaryotes do not produce oxygen during the light phase reactions (*i.e.,* H_2O is not a source of electrons) of photosynthesis. Instead, they use organic compounds, inorganic sulfur compounds, or hydrogen gas as a source of electrons. These organisms will grow phototrophically, only under anaerobic conditions or when oxygen tensions are low.

Table 2. Classification of photosynthetic organisms

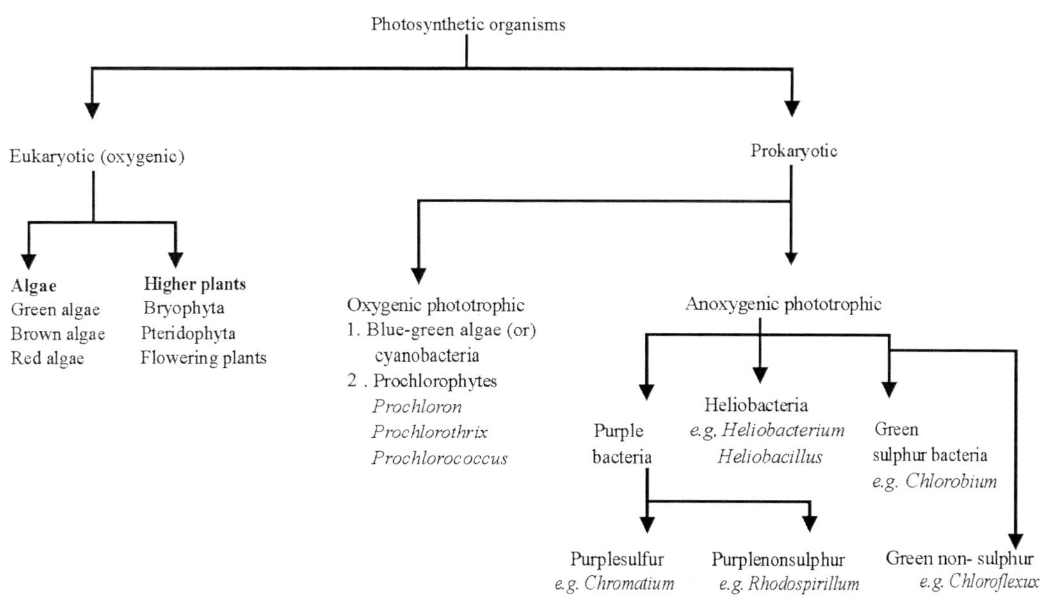

Table 9.3. Characteristics of different groups of phototrophic bacteria

Type	Pigments	Electron donor	Carbon source	Aerobic dark growth
1. Oxygenic :				
Cyanobacteria	Chl a, Phycobilins	H_2O	CO_2	No
Prochloron	Chl a, Chl b	H_2O	CO_2	No
Prochlorothrix	Chl a, Chl b	H_2O	CO_2	No
Prochlorococcus	Chl a, Chl b	H_2O	CO_2	No
2. Anoxygenic:				
Purple sulphur	Bchl a(or) Bchl.b	H_2S, S^o, S_2O_3, H_2	CO_2, Organic	Yes
Purple non-sulphur	Bchl.a, Bchl.b	H_2, Organic	CO_2, Organic	Yes
Green sulphur	Bchl a,c,d,e	H_2, Organic, H_2S	CO_2, Organic	No
Green gliding	Bchl. a,c,d	H_2S, S^o, S_2O_3, H_2, Organic	CO_2, Organic	Yes
Heliobacteria	Bchl. g	H_2, H_2S, Organic	Organic	No

*chl-chlorophyll ; Bchl-bacteriochlorophyll

Anoxygenic phototrophic bacteria have their habitats in the anaerobic zones of many aquatic locations; in shallow ponds, slowly flowing waters, lakes and estuaries. The purple sulfur bacteria often occur as salmon–coloured to dark-burgundy coverings on mud and decomposing plant materials. The purple bacteria also grow in shallow ponds when the surface is covered by a thick layer of duckweed (*Lemna*) or by the leaves of water-lilies. This biological filter apparently absorbs the spectral components that can be utilized by green algae and cyanobacteria, but it transmits the spectral components that are absorbed by the bacteriochlorophylls and the dark red carotenoids. The high carotenoids content of purple bacteria and green sulfur bacteria, which actually determines their colour, enable these bacteria to live at greater depths of water and carry out photometabolism.

There are five major groups of anoxygenic phototrophs.

1. Purple sulfur phototrophs
2. Purple non-sulfur phototrophs
3. Green sulfur phototrophs
4. Green non-sulfur phototrophs (or green gliding bacteria)
5. Heliobacteria

Taxonomically, the first four groups respectively belong to Chromatiaceae, Rhodospirillaceae, Chlorobiaceae and Chloroflexaceae

9.3.2.1. Purple Sulfur Phototrophs

The purple sulfur photosynthetic bacteria grow photoautotrophically in anaerobic environments using hydrogen sulfide as an electron donor and CO_2 as the carbon source. Usually they occur in consortia with sulfate reducers at the bottom of ponds and lakes. They oxidise sulfide to elemental sulfur which accumulates as intracellular granules. *Ectothiorhodospira* is an exception which deposits sulfur extracellularly. Quite interestingly, some can grow photoheterotrophically and still some have been grown chemo-autotrophically under low partial pressures of oxygen with reduced inorganic sulfur as an electron donor and energy source. *e.g. Chromatium.*

9.3.2.2. Purple Non-sulfur Phototrophs

The purple non-sulfur photosynthetic bacteria are extremely versatile with regard to source of both energy and carbon. At one time, it was thought that these bacteria are unable to use sulfide as a source of electrons and hence the name non-sulfur. However, recently it has been found that these bacteria can not tolerate high concentration of sulfide as do the purple-sulfur bacteria. They grow at low concentration of hydrogen sulfide and even utilize it as a source of electrons during the photoautotrophic growth. The purple non-sulfur bacteria can grow photoautotrophically or photoheterotrophically in an anaerobic environment. As shown in Table 9.3, photoautotrophs use H_2 as electron donor and CO_2 as the source of carbon. On the other hand, photoheterotrophs use simple organic acids such as malate or succinate as the electron donor and source of carbon, and light as the source of energy. Interestingly, if these organisms are incubated in darkness in the presence of oxygen, they will carry out an ordinary aerobic respiration and grow as chemoheterotrophs. A few can even grow very slowly and fermentatively in the dark. Thus the purple non-sulfur bacteria, because of their versatility, have received much attention in research. *e.g. Rhodobacter, Rhodospirillum.*

9.3.2.3. Green Sulfur Phototrophs

The green-sulfur phototrophic bacteria are obligate anaerobic photoautotrophs. They use H_2S, S^o, $S_2O_3^{2-}$ (thiosulfate), or H_2 as an electron donor and CO_2 as the source of carbon. The green-sulfur bacteria grow in sulfide–rich anaerobic aquatic habitats, although they are found in separate layers. Their light harvesting pigments are located in special inclusion bodies called **chlorosomes**. Their reaction centres are different from purple-sulfur bacteria. e.g., *Chlorobium.*

9.3.2.4. Green Non-sulfur Phototrophs (green gliding bacteria)

The genera included in this group exhibit an ecological and physiological diversity. For example, the species of *Chloroflexus* are thermophilic, alkalophilic, filamentous and gliding. They grow as photo-heterotrophs and photoautotrophs with either H_2 or H_2S as an electron donor. They can even grow chemoheterotrophically in the dark in the presence of air. As in green sulfur phototrophs, pigments are located in chlorosomes. However, the reaction centre appears to be different from those in the green sulfur bacteria and more closely resembles the reaction centres of purple bacteria.

9.3.2.5. Heliobacteria

Heliobacteria are strict anoxygenic photoheterotrophs. They differ from the other phototrophs in possessing *bacteriochlorophyll* g as their reaction center chlorophyll. The tetrapyrrole portion of bacteriochlorophyll g is similar to chlorophyll a (Fig. 9.1) and infact isomerises into chlorophyll a in the presence of air. They do not possess either internal membranes or chlorosomes. Phylogenetically, based upon 16 S RNA and lack of LPS in cell wall they are closely related to gram–positive bacteria especially clostridia. Like clostridia they form endospores also. The two known genera are *Heliobacterium* and *Heliobacillus*.

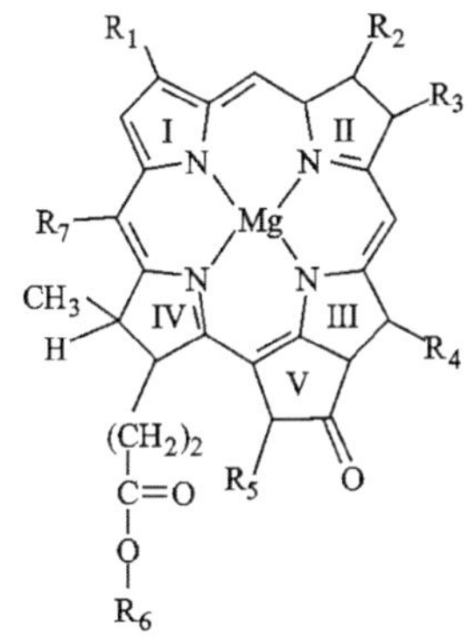

R	bchl a	bchl b	bchl c	bchl d	bchl e	bchl g	chl a
1	$COCH_3$	$COCH_3$	$CHCHCH_3$	$CHCHCH_3$	$CHCHCH_3$	$CH{=}CH_2$	$CH{=}CH_2$
2	CH_3	CH_3	CH_3	CH_3	CHO	CH_3	CH_3
3	CH_2CH_3	$=CHCH_3$	CH_2CH_3	CH_2CH_3	CH_2CH_3	$=CHCH_3$	CH_2CH_3
4	CH_3	CH_3	CH_2CH_3	CH_2CH_3	CH_2CH_3	CH_3	CH_3
5	$COOCH_3$	$COOCH_3$	H	H	H	$COOCH_3$	$COOCH_3$
6	phytyl	phytyl	farmesyl	farmesyl	farmesyl	farmesyl	phytyl
7	H	H	CH_3	H	CH_3	H	H

Fig. 9.1. Structure of bacteriochlorophyll. There is no double bond in ring II of bacteriochlorophyll a,b, and g between C3 and C4 (Nitrogen is in position 1).

9.3.2.6. Isolation of Anoxygenic Phototrophs Byenrichment Culture

Enrichment culture of anaerobic phototropic bacteria is based on their natural stratification in a shallow pond. Selective culture conditions can be established by the use of filters that absorb light in the short wave range, but transmit only in the infra-red region that is appropriate for various groups. In addition to light, availability of hydrogen donor and the concentration of hydrogen sulfide are important in enrichment cultures. By setting up the classical **Winogradsky column,** enrichment culture can be established for different groups of phototropic bacteria Fig. 9.2. In this, a layer of reduced mud is mixed with sodium sulfate, sodium carbonate, a cellulose source in the form of shredded news paper and additional mud, and water are placed in the column, which is then incubated in the light. A series of reactions occur as the column begins to mature, with particular member of microbial community developing in specific microenvironments in response to chemical gradients.

At the bottom of the column, cellulose is degraded to fermentation products by the genus *Clostridium*. With these fermentation products available as reductants and using sulfate as an oxidant, *Desulfovibrio* produces hydrogen sulfide. The hydrogen sulfide diffuses upward towards the oxygenated zone, creating a stable hydrogen sulfide gradient. In this gradient, the photoautotrophs *Chlorobium* and *Chromatium* develop as visible olive green and purple zones.

These microorganisms use hydrogen sulfide as an electron source, and CO_2 from sodium carbonate, as a carbon source. Above this region, the purple non-sulfur bacteria (*Rhodospirillum* and *Rhodopseudomonas)* can grow. These photoheterotrophs use organic matter as an electron donor under anaerobic conditions and function in a zone where the sulfide level is lower. Higher in the column both O_2 and hydrogen sulfide are available, where specially adapted micro-organisms to function. These include *Beggiatoa* and *Thiothrix,* which use reduced sulfur compounds as reductants and oxygen as oxidant. In the upper portion of the column, diatoms and cyanobacteria grow.

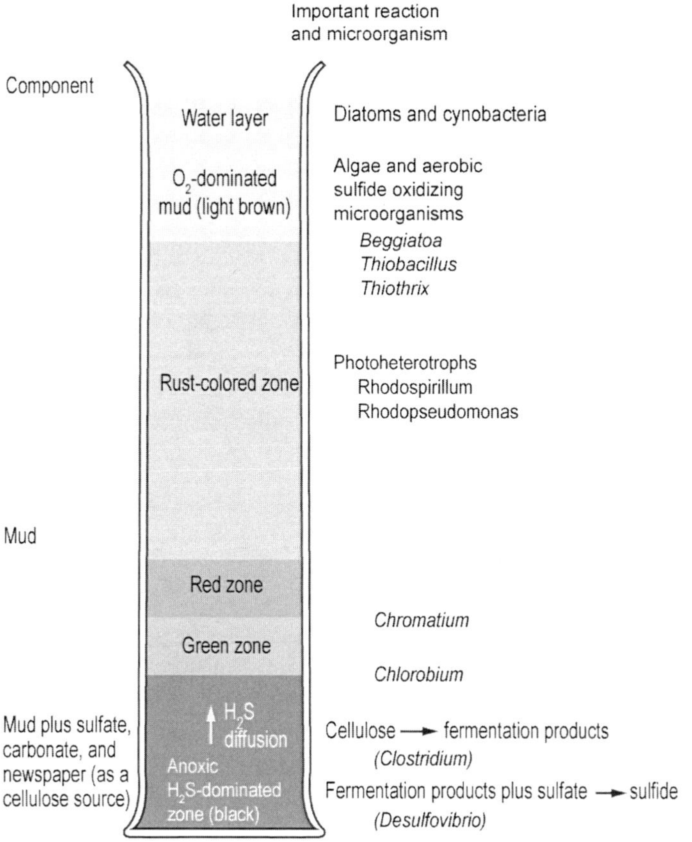

Fig. 9.2. The Winogradsky Column

9.4. PHOTOSYNTHETIC PIGMENTS

9.4.1. Location of Pigments

The photosynthetic pigments, both reaction center and light harvesting pigments, are associated with special membrane systems, the photosynthetic membranes. The location of photosynthetic membranes within the cell differs between eukaryotic and prokaryotic phototrophs. In eukaryotes, photosynthesis is associated with special intracellular organelles, the chloroplasts. In prokaryotes, chloroplasts are not present and photosynthetic pigments are integrated into internal membrane systems. In oxygenic phototrophs (cyanobacteria) photosynthetic membranes are arranged in stacks (thylakoids) with in the cytoplasm. The two forms of chlorophyll a (P_{680} and P_{700}) are attached to specific proteins in the membrane.

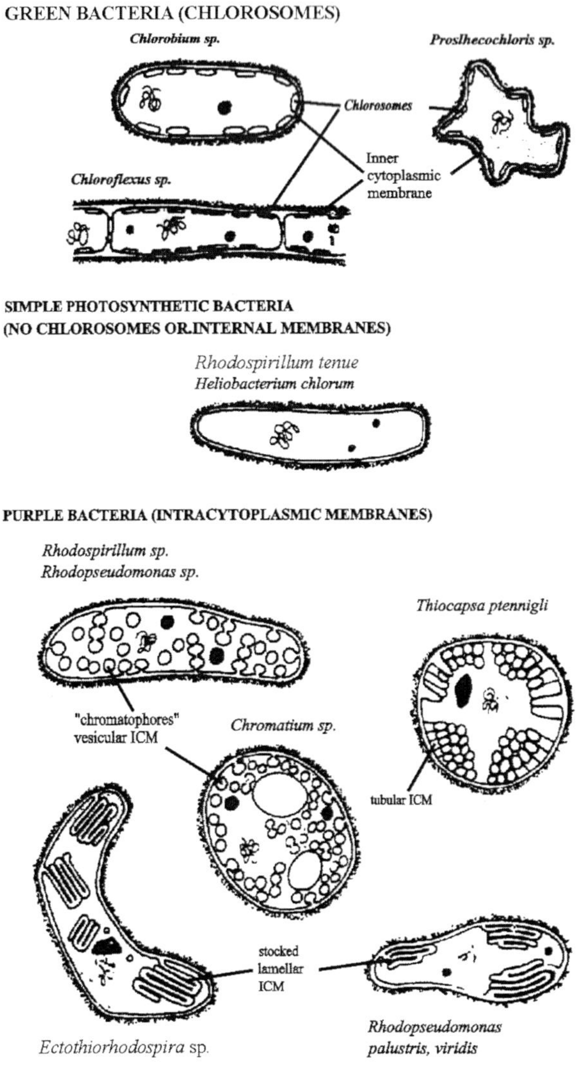

Fig. 9.3. Photosynthetic membranes in anoxygenic phototrophic bacteria.

The distribution and structure of pigment containing membranes vary widely in different groups of anoxygenic phototrophs (Fig 9.2). All the purple-sulfur bacteria possess vesicular thylakoids (chromatophores) which almost fill the cell. Only two exceptions are known so far:

Ectothiorhodospira species and *Thiocapsa pfennigi* which have tubular thylakoids. In purple non-sulfur bacteria, the photosynthetic membrane is probably identical with the cytoplasmic membrane, which shows only a few small invaginations. The members of green sulfur and green non-sulfur bacteria are characterized by the possession of pigmented organelles immediately adjacent to the cytoplasmic membranes, the so–called chlorosomes (earlier called chlorobium vesicles). Like purple non-sulfur phototrophs, in heliobacteria photosynthetic membranes are not distinct, probably the pigments are located in the cell membrane itself.

9.4.2. Types of Pigments

The photosynthetic pigments are broadly divided into two categories. They are

1. The reaction center pigments
2. Light harvesting pigments

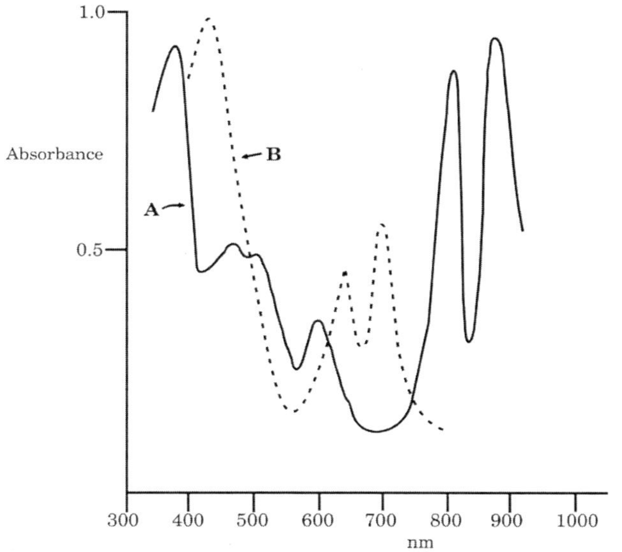

Fig. 9.4. Absorption spectra of photosynthetic bacteria. A *Rhodopseudomonas palustris*, B. Cyanobacterium

Reaction center pigments

The reaction center pigments are mainly chlorophylls. Chlorophylls of the bacteria are differentiated from others by designation as bacteriochlorophylls. They are bacterio-chlorophyll a, b, c, d, e and g. The light harvesting pigments are also called accessory pigments or antennae pigments, dominate over the reaction center pigments. These are critical to the photosynthesis because they absorb light of different wave lengths and tunnel the energy to the reaction centre. The distribution of reaction centre pigments in different groups of phototrophic bacteria is given in Table 9.4. As an example, the whole cell absorption spectra of a purple photosynthetic bacterium and a cyanobacterium are shown in Fig 9.4.

Light hervesting pigments

The nature of light harvesting pigments vary with the type of organism, but some important generalizations about them can be made.

1. They absorb light at wavelengths different from reaction center chlorophyll and therefore extend the range of wave lengths over which photosynthesis is possible.
2. In purple photosynthetic bacteria, they are embedded in the membrane as pigment – protein complexes that are in close physical association with the reaction center.
3. In the green sulfur bacteria and *Chloroflexus*, they exist in chlorosomes, which are separate organelles also called inclusion bodies, attached to the inner surface of the cell membrane.
4. In cyanobacteria and eukaryotic red algae, the light harvesting pigments that transfer energy to reaction center II are located in granules called phycobilisomes that are attached to the photosynthetic membrane. The phycobilisomes consist of proteins called phycobiliproteins that absorb light in the range of 450–660 nm due to open chain tetrapyrroles. There are three classes of phycobiliproteins, *viz.,* phycoerythrin, phycocyanin and allophycocyanin. Cyanobacteria possess phycoerythrin (red) and phycocyanin (blue). However, phycocyanin predominates. Pigment components can be readily recognized by the absorption spectra of intact cells. Absorption maxima is in the blue (<450 nm) and the red as well as infrared spectra (650–1000 nm) are due to chlorophylls. Absorption in the range of 400–550 nm is mainly due to carotenoids and that of cyanobacteria in the range of 550–650 nm is due to phycobiliproteins.

Table 9.4. Photosynthetic pigments and their absorption maxima

Pigment	Wavelength (nm)
Chlorophyll	
a	680
b	675
Bacteriochlorophylls	
a	800-810, 850-910
b	835-850, 1020-1035
c	745-760
d	725-745
e	715-725
g	670-788
Carotenoides	
chlorobactene	458
isorenieratene	517

Pigment	Wavelength (nm)
lycopene, rhodopin	463,490,524
okenone	521
spheroidene	450,482,514
spirilloxanthin	486,515,552
β-carotene	433, 483
γ-carotene	433, 483
Phycobilins	
phycocyannin	620-650
phycoerythrin	560-566

9.4.3. Structure of Pigment Molecules

Chlorophylls: The chlorophylls are substituted tetrapyrroles related to heme. The differences are that chlorophylls have Mg^{2+} in the centre of the tetrapyrrole rather than $Fe^{+2(+3)}$, and a fifth ring is present. The various chlorophyll types differ from each other chiefly by the presence or absence of a double bond between carbon atoms 3 and 4 and by the substituents on the porphyrin skeleton (Fig. 9.1). These differences are responsible for the absorption maxima in the near infra-red.

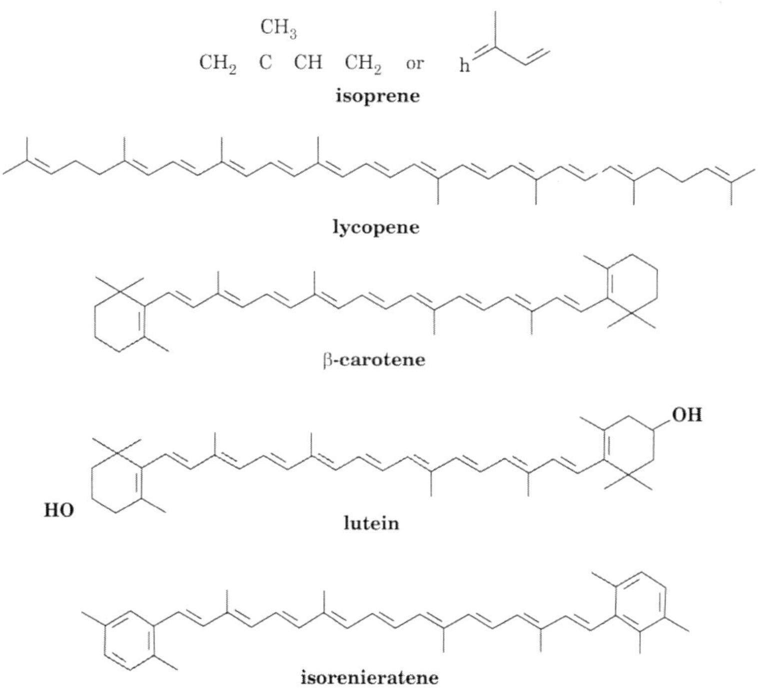

Fig. 9.5. Structure of carotenoids.

Carotenoids: Carotenoids are long isoprenoids, usually C_{40} tetraterpenoids made of eight isoprene units (Fig. 9.5). Carotenoids have a system of conjugated double bonds which account for their absorption spectrum in the visible range at wave lengths poorly absorbed by chlorophylls and makes them valuable as light harvesting pigments. Carotenoid pigments with six-membered rings at both ends of the molecule are called carotenes (*e.g.*, β–carotene). Modification of the carotenes, for example hydroxylation, produces xanthophylls, which comprise major portion of the light harvesting pigments of algae. Other isoprenoids are the side chains of quinones, the phytol chain in chlorophyll, and the retinol chromophole in bacteriorhodopsin, halorhodopsin and rhodopsin. Carotenoids serve two functions, firstly they harvest the light energy and channel it to reaction center. Secondly, they shield the chlorophylls from photo – oxidative damage.

Phycobiliproteins: There are three classes of phycobiliproteins: Phycoerythrin, phycocyanin and allophycocyanin. Allophycocyanin is usually at much lower concentration than phycoerythrin and phycocyanin. They are considered to be an energy funnel to the reaction center. Although, most species of cyanobacteria contain both phycoerythrin (red) and phycocyanin (blue), the phycocyanin generally predominates. The phycobiliproteins are arranged in layers around each other, with phycoerythrin on the outside, phycocyanin in the middle, and allophycocyanin in the inside, closest to the reaction center. A simplified drawing is shown in Fig. 9.6a & b.

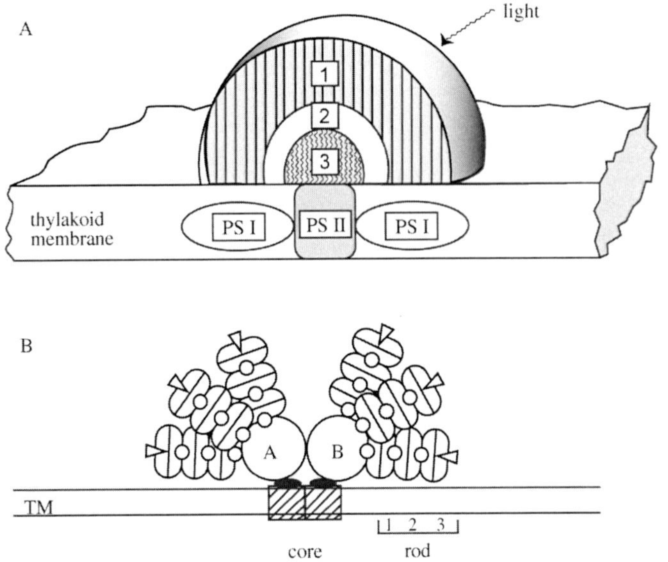

Fig. 9.6a & b: Structure of phycobilisome. (A) Hemispherical phycobilisome (red algae). The major pigment is phycoerythrin. (1) phycoerythrin; (2) phycocyanin; (3) allophycocyanin. (B) Fan shaped phycobilisome (cyanobacteria) is phycocyanin.

Synechococcus, which does not contain phycoerythrin. These are fan-shaped (hemidiscoidal) phycobilisomes that consist of a dicylindrical core (A and B) containing allophycocyanin and six cylindrical phycocyanin rods radiating from the core (in other cyanobaceria the core may consist of three cylinders). They are approximately 48 x 32 nm high x 32 nm. If phycoerythrin cylinders are present, they would be an extension of the phycocyanin rods. The shorter wavelengths of light are absorbed by the phycoerythrin, the longer wavelengths by the phycocyanin, and still longer wave-lengths (650 nm) by the allophycocyanin. These pigments are thus well positioned to transfer energy via inductive resonance to the reaction center.

9.5. PHYSIOLOGY OF PHOTOSYNTHESIS

All the biochemical reactions involved in photosynthesis can be broadly divided into two phases:

1. Light reaction
2. Dark reaction

9.5.1. Light Reaction

In the light reaction, first phase of photosynthesis, light energy is converted to chemical energy in the form of ATP. Another important event of the light reaction is reduction of $NADP^+$ to NADPH. This is accomplished by electrons originating from various electron donors, such as H_2S, organic compounds and water. The ATP and NADPH, collectively known as **assimilatory power**, are required in dark phase reactions for the reduction of CO_2 to carbohydrate.

Phototrophic bacteria including oxygenic cyanobacteria developed different mechanisms to produce the assimilatory power. This largely depends upon the type and number of photosystems, oxygen availability and source of donors of electrons.

9.5.1.1. Light reactions in oxygenic phototrophs

Oxygenic phototrophic bacteria, like higher plants, use light to generate both ATP and NADPH. The electrons for the latter come from the splitting of water into oxygen and electrons. In these organisms, there are two kinds of antennas associated with two different photosystems. Photo-system I (sometimes called pigment system I or PSI) absorbs longer wave length light ($\geq$ 680 nm) and channel the energy to special chlorophyll a molecule called P700. The term P700 denotes that this molecule most effectively absorbs light at a wavelength of 700 nm. Photosystem II (PSII) traps light at shorter wave lengths ($\leq$ 680 nm) and transfers its energy to the special chlorophyll P680.

The path of electron flow in oxygenic phototrophic bacteria roughly resembles the letter 'Z' turned on its side and hence it is usually referred to as Z scheme. When the photosystem I antenna transfers light energy to the reaction center P700 chlorophyll molecule, P700 absorbs the energy and is excited, its reduction potential becomes very negative. It then donates its excited or high energied electron to a specific electron acceptor, probably a special chlorophyll a molecule (A) or an iron – sulfur protein. The electron is eventually transferred to ferredoxin and can then travel in either of two directions (Fig. 9.7). In the cyclic pathway, the electron moves in a cyclic route through a series of electron carriers and back to the oxidised P700. The pathway is termed cyclic because the electron from P700 returns to same P700 after traveling through the photosynthetic electron transport chain. The proton motive force generated during this process is utilized for the generation of ATP. Thus, the process is called cyclic photophosphorylation. However, it should be noted here that only one photosystem participates in this process.

Electrons can also travel in a non-cyclic pathway involving both photosystems. P700 is excited and donates its electrons to ferredoxin as before. In the non-cyclic route, however, reduced ferredoxin reduces $NADP^+$ to NADPH. Since the electrons contributed to $NADP^+$ cannot be used to reduce oxidised P700, photosystem II participation is required. It donates electrons to oxidised P700 and generates ATP in the process. The PSII antenna absorbs light energy and excites 680 nm chlorophyll, which then reduces phaeophytin A. The electrons subsequently travel to Q (a plastoquinone) and donate the electron to P700. Oxidised P680 then obtains an electron from the oxidation of water to O_2. Thus electrons flow from water all the way to $NADP^+$ with the aid of energy from two photosystems. At the same time the proton motive force generated is used for the synthesis of ATP. As such it is called non-cyclic photophosphorylation.

As will be seen later in this chapter, the dark reactions require three ATP and two NADPHs to reduce one CO_2 to synthesize carbohydrate. Technically it is called as assimilatory power.

$$CO_2 + 3ATP + 2\ NADP^+ + 2H^+ + H_2O \longrightarrow (CH_2O) + 3ADP + 3Pi + 2NADP^+$$

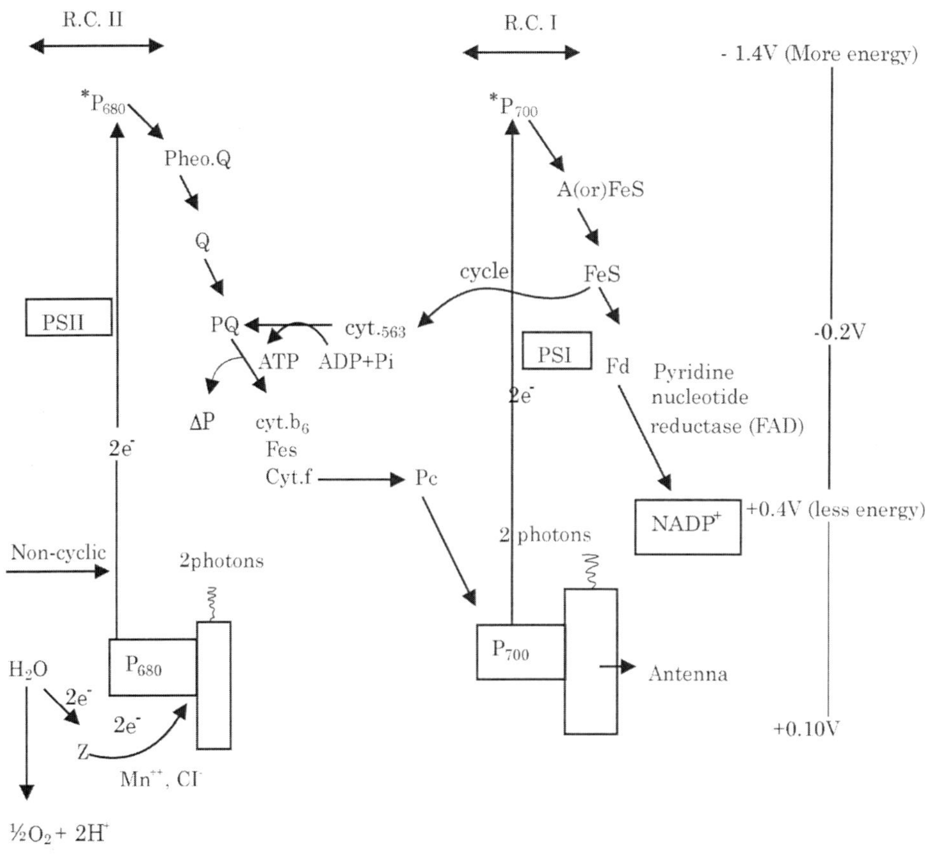

Fig. 9.7. Light reactions in green plants and cyanobacteria (Z-scheme)

The non-cyclic system generates one NADPH and one ATP per pair of electrons. Therefore, four electrons passing through the system will produce two NADPHs and two ATPs. A total of eight quanta of energy (4 quanta for each photosystem) is needed to propel the four electrons from water to $NADP^+$. Because the ratio of ATP and NADPH required for CO_2 fixation is 3 : 2, at least one more ATP must be supplied. Cyclic photophosphorylation probably operates independently to generate the extra ATP. This requires absorption of another 2 to 4 quanta. Around 10 – 12 quanta of light energy is needed to reduce and incorporate one molecule of CO_2 during photosynthesis.

9.5.1.2. Light reactions in anoxygenic phototrophs

Anoxygenic phototrophic bacteria differ from oxygenic phototrophic bacteria and eukaryotic counterparts in several respects (Table 9.5)

Table 9.5. Differences between anoxygenic phototrophic bacteria and oxygenic phototrophs

Character	**Oxygenic phototrophs**		**Anoxygenic phototrophs**
	Eukaryotes	Cyanobacteria	Green and purple bacteria
Photosynthetic pigments	Chlorophyll a	Chlorophyll a	Bacteriochlorophyll
Photosystem II	Present	Present	Absent
Photosynthetic electron donors	H_2O	H_2O	H_2, H_2S, S, organic compounds
O_2 production	Oxygenic	Oxygenic	Anoxygenic
Primary products	ATP+NADPH	ATP+NADPH	ATP
Carbon source	CO_2	CO_2	Organic and/or CO_2

Further, the anoxygenic phototrophic bacteria are characterized by the following characters:

1. Anoxygenic phototrophic bacteria do not use water as an electron source or produce O_2 photosynthetically.
2. The reducing power, NADPH is not directly produced from the light reactions. However, green bacteria can reduce NAD^+ directly during the light reactions. The reduced NAD^+ is utilized for the conversion of $NADP^+$ to NADPH.
3. Anoxygenic phototrophs use electron donors like hydrogen, H_2S, elemental sulfur and organic compounds instead of water.
4. These bacteria possess only one photosystem *i.e.,* photosystem I
5. The photosynthetic pigments are bacteriochlorophylls, many with absorption maxima at longer wavelengths (Table 9.4). The shift of absorption maxima into the infrared region better adapts these bacteria to their ecological niches.

9.5.1.3. Light reactions in purple-bacteria

A tentative scheme of light reactions in purple bacteria is shown in Fig 9.8 (a&b). The electron flow in the above figure can be summarized as follows. Incident light energies bacteriochlorophyll (P870) in the reaction center (It may be remembered here that some purple photosynthetic bacteria have bacteriochlorophyll b which absorbs at 1020-1035 nm). The energied bacteriochlorophyll reduces bacteriopheophytin (B pheo) which is a bacteriochlorophyll without Mg^{2+}. The electrons then travel through two ubiquinones, UQA and UQB in the reaction center. After a second light reaction, UQB becomes reduced to UQH_2 and leaves the reaction center. The electrons return to the reaction center via a proton-translocating bc_1 complex and cytochrome c_2. ATP is synthesized via a membrane ATP synthase driven by the proton motive force (ΔP). Thus the flow of electrons in this group of organisms is cyclic, and the synthesis of ATP is by cyclic photophosphorylation.

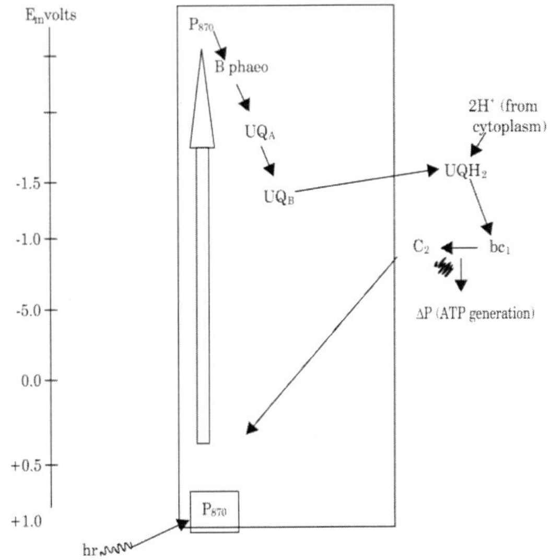

Fig. 9.8a. Light reactions in purple photosynthetic bacteria

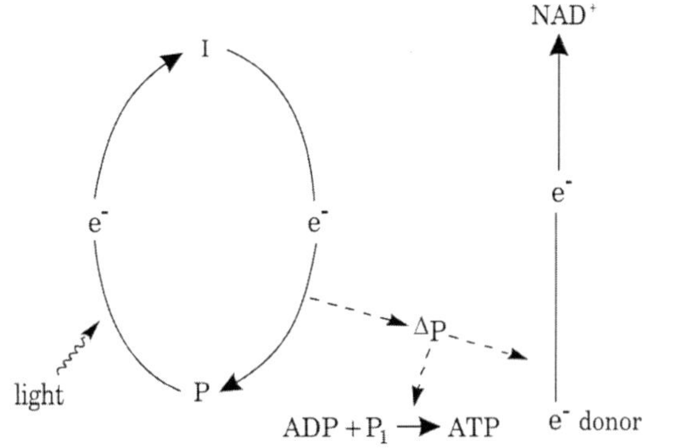

Fig. 9.8b. Relationship between cyclic electron flow and reversed electron transport in the purple photosynthetic bacteria.

The purple bacteria face a further problem because they also require NADH or NADPH for CO_2 incorporation. They may synthesize NADH in at least three ways. If they are growing in the presence of hydrogen gas, which has a reduction potential more than that of NAD^+, the hydrogen can be used directly to produce NADH. Like chemolithotrophs, many photosynthetic purple bacteria use ATP or proton motive force to reverse flow of electrons in an electron transport chain and move them from inorganic or organic donors to NAD^+

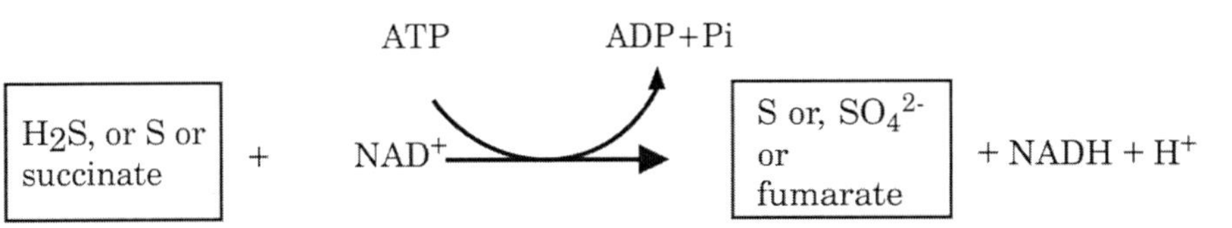

9.5.1.4. Light reactions in green sulfur bacteria

The green sulfur-photosynthetic bacteria have reaction centers that are distinguished from the reaction centers of the purple photosynthetic bacteria by:

1. Producing reducing $NAD(P)^+$ directly instead of NAD^+ or quinones
2. Possessing iron – sulfur centers as reaction center intermediate electron carrriers.

Nevertheless, the principles underlying the transformation of electrochemical energy into a ΔP and a membrane potential are similar to those for purple photosynthetic bacteria. Light energies a *bacteriochlorophyll a* molecule (P840) which reduces a primary acceptor (AO), establishing a redox potential difference greater than volt. The primary electron acceptor (AO) in the green-sulfur bacteria has recently been reported to be an isomer of chlorophyll a called bacteriochlorophyll 663. It may be added here that the primary acceptor of heliobacteria is also chlorophyll derivative, hydroxychlorophyll a, reflecting the similarities known to exist between the reaction centers of the green sulfur bacteria, heliobacteria and photosystem I of chloroplasts and cyanobacteria. The electron flows from AO to quinone like acceptor called A1. The electron is then transferred from A1 through two or three iron – sulfur centers to ferredoxin. There can be both cyclic and non-cyclic electron flow. In cyclic flow, the electron returns to the reaction center via menaquinone (MQ) and a bc_1 complex, creating ΔP. In non-cyclic electron flow, the electrons from ferradoxin are used to reduce NAD^+ to $NADH_2$ and the inorganic sulfur (H_2S) donates the electrons to reaction center According to the scheme in Fig 9.9, the electrons from inorganic sulfur enter at the level of cytochrome c and the bc_1 complex is bypassed. However, from thermodynamics point, it is difficult to justify this entry at this point. It is still an unresolved issue. Production of assimilatory power in green sulfur bacteria can be summarised as follows

$$H_2S + NAD^+ + ADP + Pi \xrightarrow{\text{Light and Bchl}} S^o + NADH + H^+ + ATP$$

9.5.2. Dark Reactions

The second phase of photosynthetic reactions are called dark reactions. The dark reactions are so named because unlike the light reactions, where light is essential, the dark reactions themselves do not require light in order to be carried out. However, the dark reactions do require products of the light reactions, the ATP and reducing power. Thus the dark reactions can be considered 'light dependant' in the sense that without ATP and reducing power generated by the light reactions, these reactions cannot occur.

The net result of dark reaction is the reduction of CO_2 to carbohydrates using the assimilatory power of light reactions. This is known as CO_2 assimilation.

$$CO_2 \xrightarrow{\text{ATP, NADPH} + H^+} C_6H_{12}O_6$$

The pathway of carbon assimilation from CO_2 is also called autotrophic CO_2 assimilation. There are four pathways of autotrophic CO_2 fixation in prokaryotes

1. Calvin cycle
2. Acetyl CoA pathway

3. Reductive tricarboxylic acid cycle
4. Hydroxyl propionate pathway

Phototrophic bacteria can fix the CO_2 through calvin cycle, reductive TCA cycle and hydroxyl propionate pathway. Acetyl CoA pathway is found in some chemoautotrophs like methanogens, acetogens and sulfate reducers (Table 9.6).

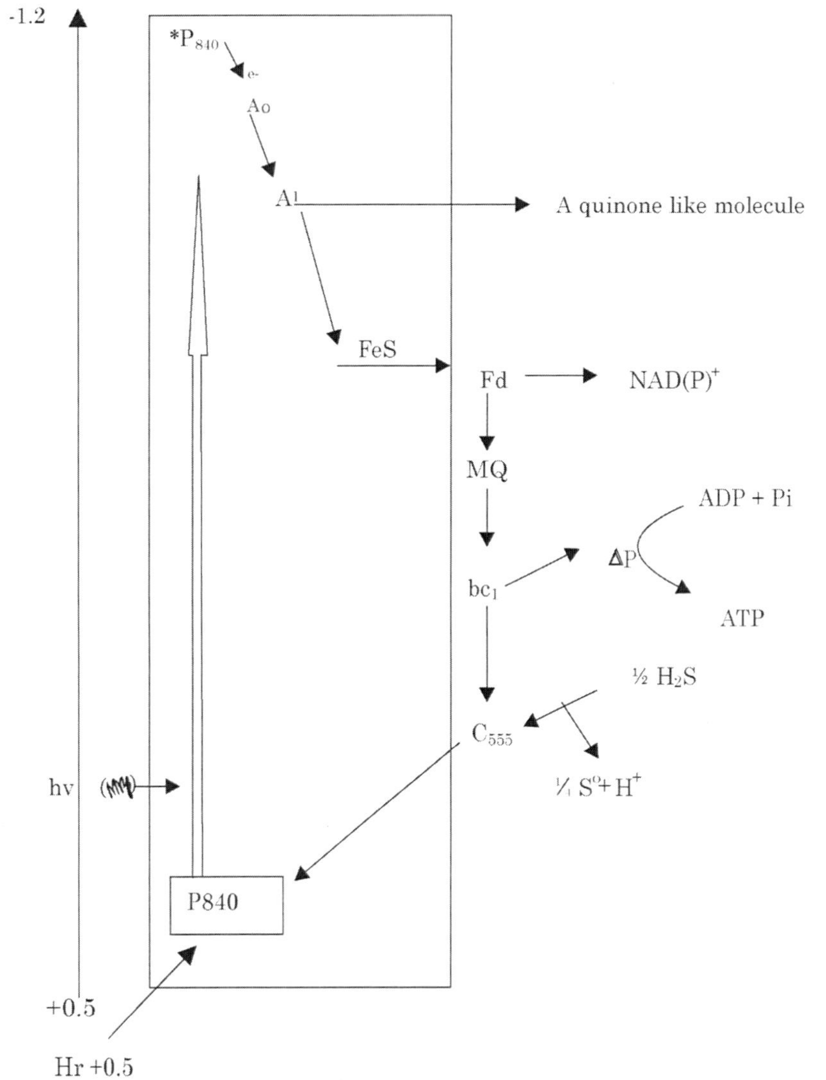

Net reaction :- $H_2S + NAD^+ + ADP + Pi \xrightarrow[\text{Bacteriochlorophyll}]{\text{light}} S^0 + NADH + H^+ + ATP$

Fig. 9.9. Light reactions in green-sulphur and green-non sulphur bacteria.

Table 9.6. Carbondioxide fixation pathways in different groups of organisms

Group	Energy source	CO_2 fixation pathway*
Eukarya		
Green algae	Light, organic carbon	CBB
Red/brown/yellow algae	Light, Organic carbon	CBB, C_4 / CBB
Plants	Light	CBB, C_4 / CBB
Bacteria		
Chemoautotrophic	Metal or H_2 oxidation	CBB, RTCA, Ac CoA
Phototrophic	Light, organic carbon H_2 oxidation	CBB, RTCA, HPP
Cyanobacteria	Light	CBB
Prochlorophytes	Light	CBB
Archaea	H_2 oxidation, organic carbon, metal oxidation	AcCoA, RTCA, CBB

*CBB – Calvin-Benson-Bassham reductive pentose phosphate pathway; RTCA-reductive tricarboxylic acid cycle; C_4/CBB combination of CBB and C_4 pathway; HPP-3-hydroxy propionate pathway, AcCoA – Acetyl CoA pathway.

9.5.2.1. Calvin cycle

The calvin cycle is also known by several names: Calvin – Benson Bassham (CBB) cycle, photosynthetic carbon reduction (PCR) cycle, or reductive pentose cycle (Fig. 9.10). Reactions of this cycle occur in the chloroplast stroma of eukaryotic microbial autotrophs. However, in bacteria where there are no similar organelles, the calvin cycle reactions occur in the cytosol. Cyanobacteria, green-sulfur and non-sulfur bacteria possess chlorosomes. These are polyhedral inclusion bodies that contain the enzyme ribulose –1, 5 biphosphate carboxylase. They may be the site of CO_2 fixation or may store the carboxylase and other proteins.

During Calvin cycle, three CO_2 molecules are reduced to phosphoglyceraldehyde which is at the oxidation level of glyceraldehydes ($C_3H_6O_3$). Twelve electrons are required and these are provided by six NAD(P)H. However, NAD(P)H is not a sufficiently strong reductant to reduce CO_2 without an additional source of energy. Each reduction thus requires an ATP. Therefore, six ATPs are required for the six reductions. However, the overall reaction, shown below, indicates that nine ATP are required. The three extra ATPs are used to recycle an intermediate, ribulose – 1,5–biphosphate(RUBP) that is catalytically required for the pathway.

$$3CO_2 + 9ATP + 6\ NAD(P)H + 6H^+ + 5\ H_2O \longrightarrow \text{Phosphoglyceraldehyde} + 9ADP + 8Pi + 6\ NAD(P)+$$

Calvin cycle is a complicated pathway. It resembles the PP pathway in some specific reactions. It is easy to understand the cycle if it is divided into three stages:

1. The carboxylation phase
2. The reduction phase
3. The regeneration phase.

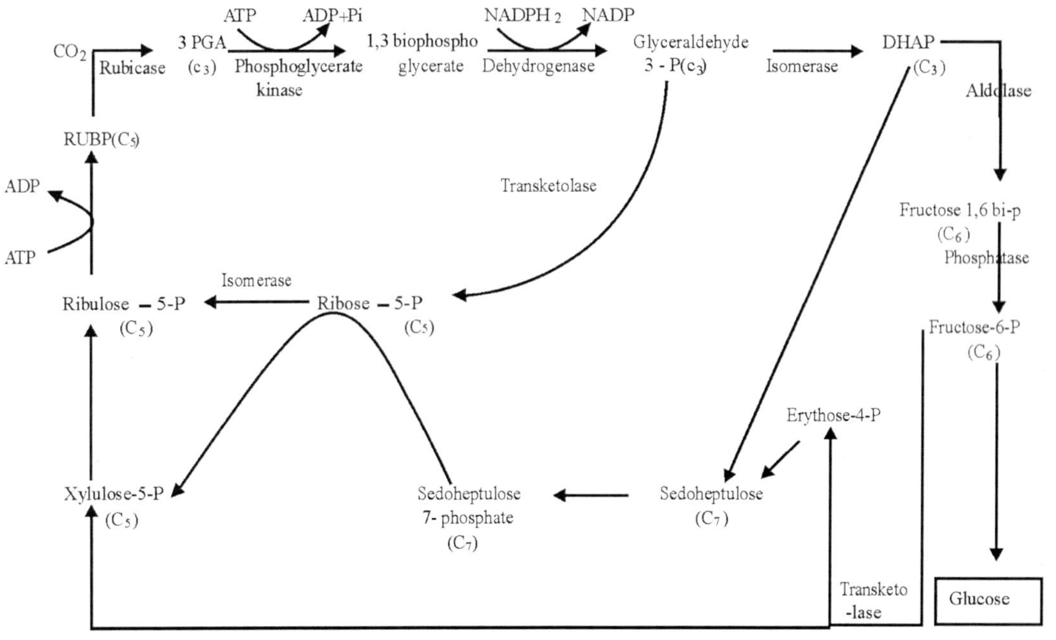

Fig. 9.10. Calvin cycle (or) photosynthetic carbon reduction cycle.

1. The carboxylation phase

In this phase, reductive carboxylation of RUBP occurs to form phosphoglyceraldehyde (PGAL). Carbon dioxide fixation is accomplished by the enzyme ribulose–1,5-biphopshate carboxylase (rubisco) that catalyzes the addition of CO_2 to RUBP, forming two molecules of 3–phosphoglycerate (PGA).

2. The reduction phase

After PGA is formed by carboxylation, it is reduced to glyceraldehyde 3–phosphate. The reduction, carried out by two enzymes, is essentially a reversal of a portion of the glycolytic pathway. However, glyceraldehyde 3–phosphate dehydrogenase differs from the glycolytic enzyme in using $NADP^+$ rather than NAD^+. The reactions of the carboxylation phase and reduction phase can be summarized as follows:

$$3\ CO_2 + 3\ RUBP + 3H_2O \longrightarrow 6\ PGA$$

$$6\ PGA + 6\ ATP \longrightarrow 6\ BPGA + 6\ ADP$$

$$6\ BGPA + 6\ NAD(P)H + 6\ H^+ \longrightarrow 6\ PGAL + 6\ NAD(P)^+ + Pi$$

$$3\ CO_2 + 3\ RUBP + 3\ H_2O + 6\ ATP + 6\ NAD(P)H + 6H^+ \longrightarrow 6\ PGAL + 6\ ADP + 6\ Pi + 6\ NAD(P)^+$$

3. The regeneration phase

The third phase of the Calvin cycle regenerates RUBP and produces carbohydrates such as fructose and glucose. This portion of cycle is similar to the PP pathway, and involves the transketolase and transaldolase reactions. The cycle is completed when phosphoribulokinase regenerates RUBP. The regeneration phase reactions can be summarized as follows.

2 PGAL $\longrightarrow$ 2 DHAP

PGAL + DHAP $\longrightarrow$ FBP

FBP + H_2O $\longrightarrow$ F6P + Pi

F6P + PGAL $\longrightarrow$ Erythrose–4–P + Xylulose–5-P

Erythrose - 4-p + DHAP $\longrightarrow$ Sedoheptulose-1, 7-biphosphate

Sedoheptulose-1,7 biphosphate + H_2O $\longrightarrow$ Sedoheptulose-7-P + Pi

Sedoheptulose-7-P + PGAL $\longrightarrow$ Ribose – 5- P + Xylulose – 5 –P

2 Xylulose – 5 –P $\longrightarrow$ 2 Ribulose – 5- P

Ribose– 5- P $\longrightarrow$ Ribulose – 5- phosphate

3 Ribulose – 5- P + 3 ATP $\longrightarrow$ 3 Ribulose – 1, 5- biphosphate + 3 ADP

5 PGAL + $2H_2O$ + 3 ATP $\longrightarrow$ 3 Ribulose –1,5 biphosphate + 2 Pi + 3 ADP

To synthesize fructose 6 – phosphate or glucose 6 – phosphate from CO_2, the cycle must operate six times to yield the desired hexose and reform the six RUBP molecules. The formation of glucose from CO_2 may be summarized by the following equation

$$6\,CO_2 + 18\,ATP + 12\,NADPH + 12H^+ + 12H_2O \longrightarrow Glucose + 18\,ADP + 18Pi + 12\,NADP^+$$

Relationship between Calvin cycle, glycolysis and pentose phosphate pathway

It is interesting to find that certain reactions are common to Calvin cycle, glycolysis and pentose phosphate pathway (Fig 9.10). In figure reactions 1 – 12 are the Calvin cycle reactions. Reaction 1, i.e the carboxylation of RUBP to PGA, is unique to Calvin cycle. Reactions 2 – 5 are also found in glycolysis. Reaction 7 is the transketolase reaction also found in PP pathway. Reaction 10 is transketolase reaction also found in PP pathway. Similarly reactions 11 are the pentose epimerase and isomerase reactions also found in pentose phosphate pathway. The pentose phosphate pathway differs from the Calvin cycle only in that the pentose phosphate pathway synthesizes sedoheptulose-7-phosphate via a reversible transaldolase reaction, whereas the Calvin cycle synthesizes sedoheptulose-7-phosphate via an aldolase and phosphatase reaction. The Calvin cycle irreversibly converts PGAL to pentose phosphates.

What does this relationship signify? There are atleast two facilities for the organism. Firstly, the intermediates of one pathway can be utilized for other pathways according to the requirement of the organism. Secondly some enzymes, without any multiplicity, can be utilized for different pathways.

9.5.2.2. Reductive tricarboxylic acid cycle

Carbon dioxide fixation in green bacteria like *Chlorobium* occurs through a different pathway known as reductive tricarboxylic acid cycle. TCA cycle, infact, is one of the phases in aerobic respiration and the reactions are oxidative in nature *i.e.,* pyruvic acid is oxidized and decarboxylated successively to CO_2. In contrast, reverse TCA cycle reactions occur in quite opposite manner to normal TCA cycle (Fig. 9.12). It implies that the reductive TCA cycle reactions are reductive in nature. In this pathway, PEP is carboxylated to form oxaloacetate (OAA) using enzyme PEP carboxylase. The oxaloacetate is reduced to succinate which is derivatized to succinyl-CoA. The succinyl CoA is carboxylated to α–ketoglutarate. This is a reductive carboxylation requiring ferredoxin, and carried out by α–ketoglutarate synthase. The α– ketoglutarate is carboxylated to isocitrate which is converted to citrate. Thus, this is really a reversal of the citric acid cycle, substituting fumarate reductase for succinate dehydrogenase and α–ketoglutarate synthase for α–ketoglutarate dehydrogenase. Thus, one mole of PEP is converted to one mole of citrate. The citrate is split by a special enzyme, called ATP – dependent citrate lyase, to acetyl CoA and OAA. The OAA can be used for growth. The acetyl – CoA is used to regenerate the PEP. During this reaction reduced ferredoxin is used.

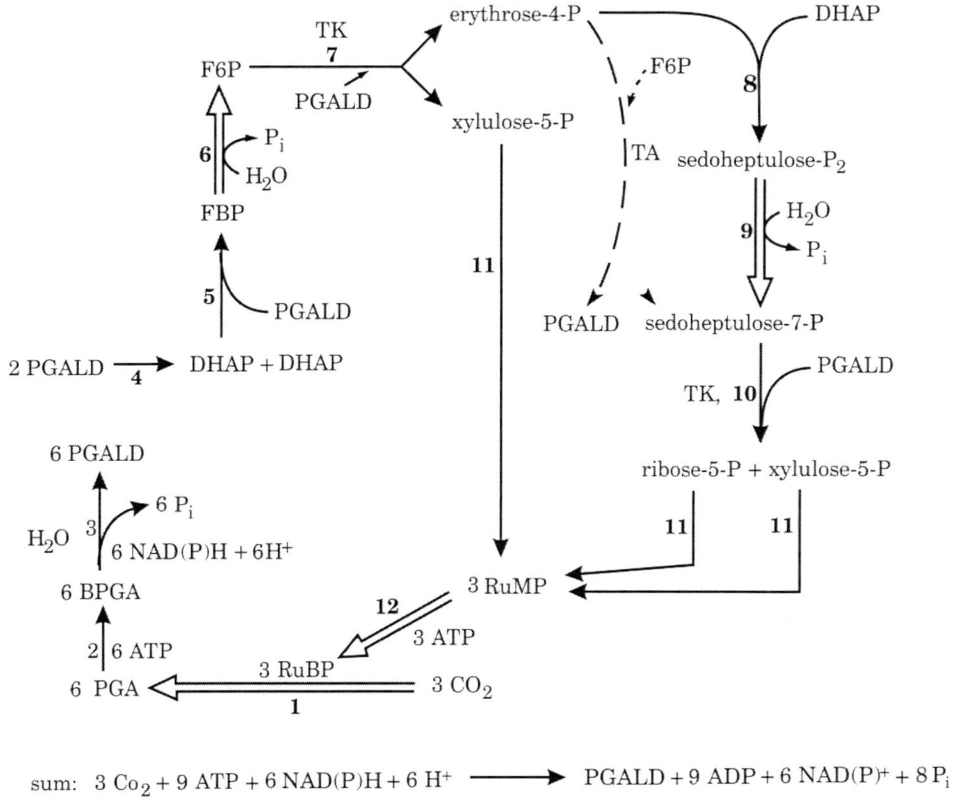

Fig. 9.11. Relationship between the Calvin cycle, glycolysis, and the pentose phosphate pathway.

It may be noted here that reverse TCA cycle requires three new enzymes that are not involved in TCA cycle. They are: fumarate reductase, α–ketoglutarate synthase, ATP – dependant citrate lyase. In addition, pyruvate synthase replaces puruvate dehydrogenase, and PEP synthetase replaces pyruvate kinase.

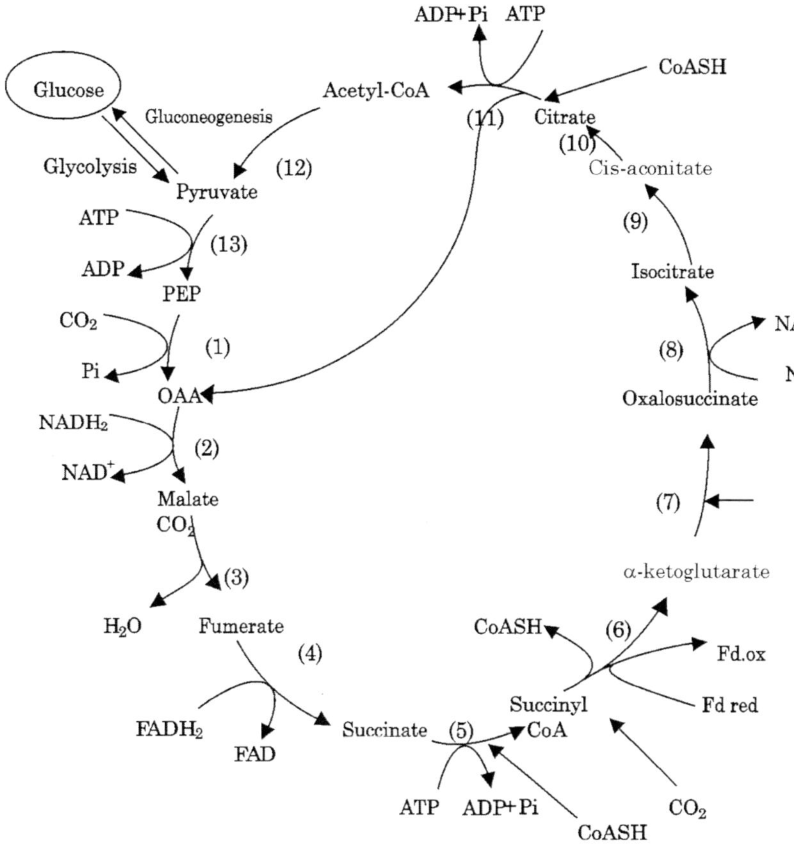

Fig. 9.12. The reductive TCA cycle.

Summary of the reactions of reductive TCA pathway is given below:

CO_2 + PEP ⟶ OAA + Pi

OAA + NADH + H^+ + $FADH_2$ + Fd(red) + ATP + CO_2 ⟶ α-ketoglutarate+ NAD^++FAD^+ + Fd(ox) + ADP + Pi

α-ketoglutarate + NADH + H^+ + CO_2 ⟶ Citrate

Citrate + ATP + CoASH ⟶ OAA + acetyl – CoA

Acetyl – CoA + Fd (red) + CO_2 ⟶ Pyruvate + Fd (ox)

Pyruvate + ATP ⟶ PEP + AMP +Pi

$4CO_2$ + 2NAD(P)H + $2H^+$+2Fd(red)+$FADH_2$+ATP ⟶ OAA+2NAD$(P)^+$ +FAD^+ 2 Fd(ox) + ADP + AMP + Pi

9.5.2.3. Hydroxy propionate pathway (HPP cycle)

This pathway, so far, is reported in *Chloroflexus aurantiacus* only. This organism uses neither Calvin cycle nor reductive TCA cycle for CO_2 assimilation. Two molecules of CO_2 are converted to glyoxylate by this unique autotrophic pathway, the hydroxy-propionate pathway. This pathway is named so because hydroxypropionate is a key intermediate.

In this pathway (Fig. 9.13), acetyl –CoA is carboxylated twice to yield methylmalonyl CoA. This intermediate is rearranged to yield acetyl–CoA and glyoxylate. The latter is converted to cell material through serine or glycine intermediates. Summary of the reactions:

$$2CO_2 + 4\ NADH + H^+ + 3\ ATP \longrightarrow Glyoxylate + 3\ NAD^+ + 3\ ADP + 3Pi$$

The pathway starts at acetyl-CoA, which is carboxylated via acetyl-CoA carboxylase, yielding malonyl-CoA. This compound is then reduced to HPP and eventually secreted by *C. aurantiacus*, particularly when the organism is grown photoheterotrophically in the presence of propionate. The next steps involve the thioesterification of HPP to CoA followed by the reduction of thioester in two steps to propionyl-CoA, which is then carboxylated to yield methylmalonyl-CoA, catalyzed by propionyl-CoA carboxylase. Methylmalonyl-CoA is converted to succinyl-CoA. Succinate is produced by succinyl-CoA : malate-CoA transferase and converted to malate, which can subsequently be utilized in the same reaction yielding malyl-CoA. Malyl-CoA is finally cleaved via a malyl-CoA lyase into glyoxylate (the net product) and acetyl-CoA (the initial CO_2 acceptor), thus completing the cycle. Both carboxylases require biotin and ATP, and are also known in other systems. The unique features of this pathway are the reduction of malonyl-CoA to propionyl-CoA via HPP and the shuttling of the CoA group from succinyl-CoA to malate.

As the genus *Chloroflexus* is considered phylogenetically ancient member, this pathway gains importance. It was suggested that hydroxypropionate pathway may have been one of the earliest attempts to autotrophy in bacteria.

Before we end this chapter on photosynthesis, it is relevant and also worthwhile to know the unique phenomenon occurring in halophilic archaeans by which light energy is converted to chemical energy (ATP) without the involvement of the chlorophyll.

9.6. BACTERIORHODOPSIN AND LIGHT MEDIATED ATP SYNTHESIS

Certain archaea, specially extremely halophilic archaea, have evolved a way to create a ΔP using light energy directly *i.e.,* without the intervention of oxidation-reduction reactions and without chlorophyll. Halophilic archaea (*e.g., Halobacterium salinarium*) are heterotrophic organisms that carry out an ordinary aerobic respiration creating a ΔP which is used to drive ATP synthesis via a membrane ATP synthase. In the presence of light and low oxygen levels the halophiles adapt by making photopigments (rhodopsins), one of which (bacteriorhodopsin) functions as a proton pump that is energized directly by light energy ultimately creating ΔP. Thus, bacteriorhodopsin illustrates the direct transformation of light energy into an electrochemical potential.

This unique pathway *i.e.,* direct conversion of light energy to a chemical energy is based on a purple membrane portion of the cytoplasmic membrane that contains bacteriorhodopsin. Walter

Stoeckinius and his co-workers established the role of purple membrane in light coupled ATP synthesis.

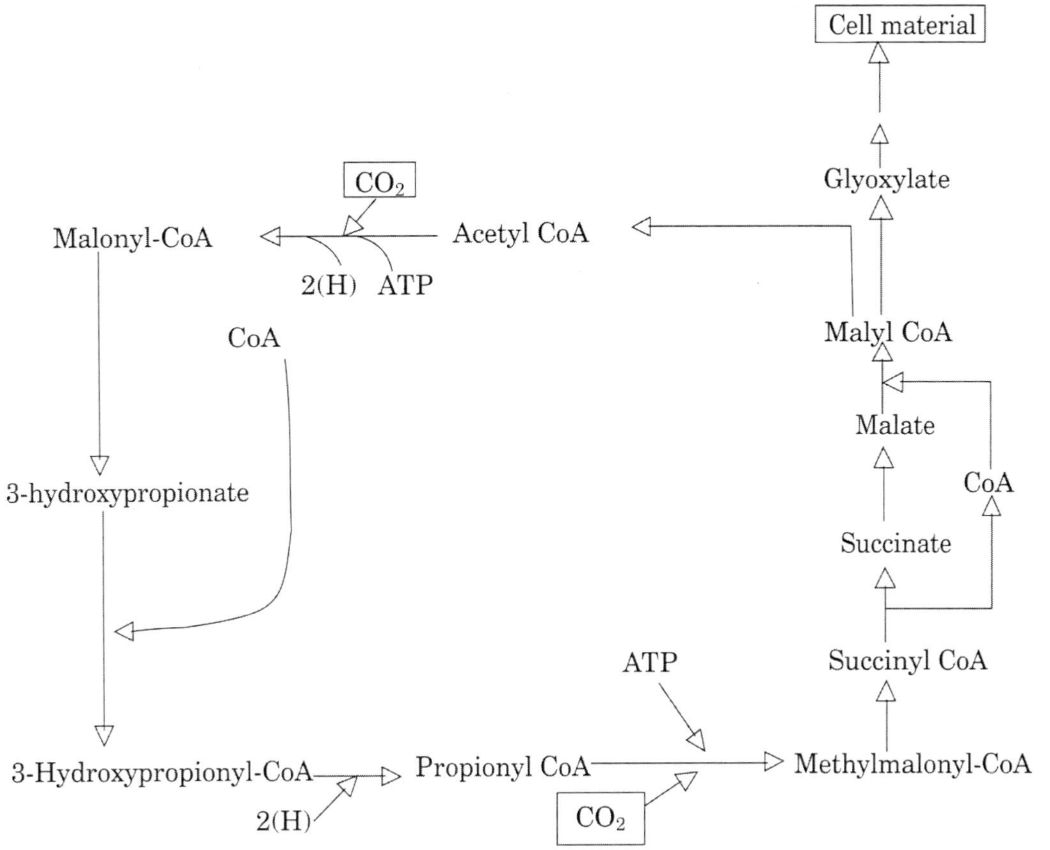

Fig. 9.13: The hydroxypropionate pathway of CO_2 assimilation

Bacteriorhodopsin is a light induced pigment protein inserted into the red membrane. The region of insertion is called purple membrane. Bacteriorhodopsin was named because of its structural and functional similarity to the visual pigment of the eye, rhodopsin. It consists of one large polypeptide folded into seven α–helices that form a transmembrane channel. Located in the middle of the channel (Fig. 9.14) and attached to the bacteriorhodopsin, is a pigment called **retinal (a C_{20} – carotenoid)** which is attached via a Schiff base to a lysine residue on protein. Due to its retinal content, the bacteriorhodopsin is purple in colour.

The photoevents occurring in halophiles can be followed spectroscopically. Bacteriorhodopsin absorbs light in the green region of the spectrum at about 570 nm. The retinal chromatophore of the bacteriorhodopsin, which normally exists in an all – trans-configuration is excited and temporarily converted to the *cis* form following the absorption of light. This transformation results in the transfer of proteins to the out side surface of the membrane. The retinal molecule then relaxes and returns to its more stable all – trans isomer in the dark following the uptake of a proton from the cytoplasm, thus completing the cycle. As protons accumulate on the outer surface of the membrane, the proton motive force increases until the membrane is sufficiently 'charged' to drive ATP synthesis through action of the membrane–bound ATPase (Fig. 9.14).

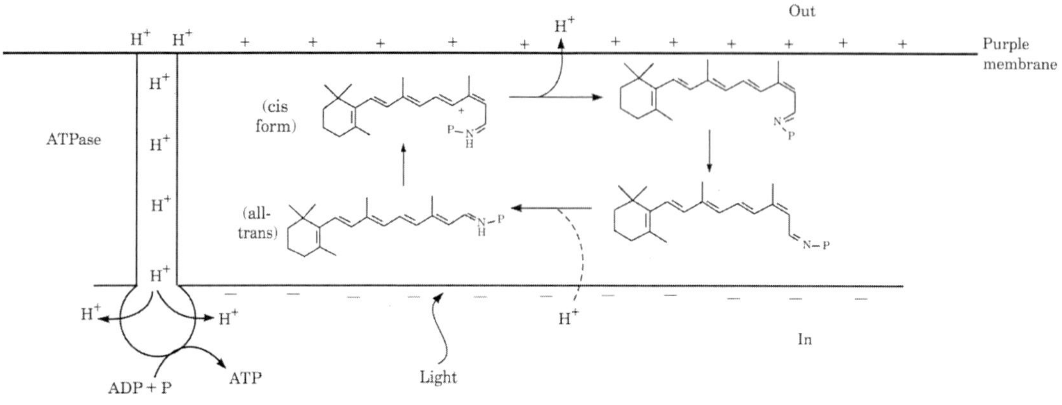

Fig. 9.14. Model of the light-mediated bacteriorhodopsin proton pump in the purple membrane of *Halobacterium.* The P stands for the protein to which the chromophore retinal is attached.

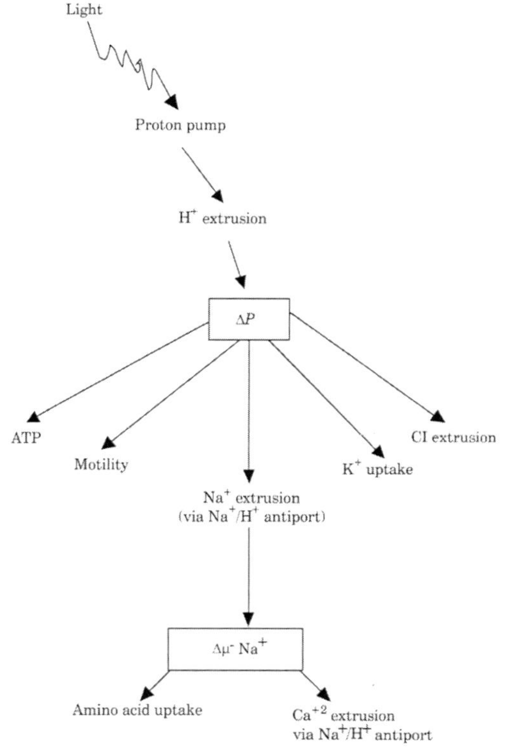

Fig. 9.15. Light dependent reactions in *Halobacterium halobium.*

Light mediated ATP production in *Halobacterium salinarium* has been shown to support slow growth of this organism anaerobically under nutritional conditions, and light has been shown to maintain viability of anoxic cultures of *Halobacterium* incubated in the absence of organic energy sources. The light stimulated proton pump of *H. salinarum* also functions to pump Na^+ out of the cell by action of a Na^+/H^+ antiport system and to drive the uptake of a variety of nutrients, including the K^+ needed for osmotic balance. The uptake of amino acids by *H. salinarium* has been shown to be indirectly driven by light because the transport of amino acids occurs with Na^+ uptake by an amino acid Na^+ symporter. The continued uptake depends on the removal of Na^+ via the light driven Na^+/H^+ antiporter.

9.7. HALORHODOPSIN, A LIGHT – DRIVEN CHLORIDE PUMP

The halophiles have a second light-driven electrogenic ion pump, but one which does not energize the membrane. The second pump, called **halorhodopsin** is structurally similar to bacteriorhodopsin and is used to accumulate Cl^- intracellularly in order to maintain osmotic stability (Fig. 9.15). It may be recalled here that halophiles live in salt water where the extracellular concentration of NaCl can be 3–5 M. The osmotic balance is preserved by intracellular concentrations of KCl that match the extracellular ion concentrations. Since the membrane is negatively charged on the inside, energy must be used to bring it into the cell and halorhodopsin accomplishes this purpose. The dark halophiles use another energy source, probably ATP to accumulate chloride ions actively.

9.8. FURTHER READING

Bullerjahn, G.S and Post, A.F. (1993), The prochlorophytes : Are they more than just chlorophyll a/b containing cyanobacteria ? *Rev. Microbiol.,* **19 :** 43-59.

Codgell, R.J. (1978), Carotenoids in photosynthesis. *Phil.. Trans. R. Soc. Lond. B,* **284 :** 569-579.

Deisenhofer, J., Michel, H. and Huber, R. (1985), The structural basis of photosynthetic light reactions in bacteria. *Trends Biochem. Sci.,* **10:** 243-249.

Diener, B.A. (1986), Photosystems I & II : Structure, proteins and cofactors, pp 422-436. In photosynthesis III. L.A. stachelin and C.J. Arntzen (eds) Springer Verlag, New york.

Dutton, P.L. (1986), Energy transductions in anoxygenic photosynthesis, p 197-239. In L.A. staehelin and C.J. Arntzen (eds). Photosynthesis III : Photosynthetic membranes and light harvesting systems. Springer-Verlag, Berlin.

Grossman, A.R., Schaefer, M.R., Chiang, G.G. and Coler J.L. (1993), The phycobilisome, a light harvesting complex resonsive to environmental conditions. *Microbiol. Rev.,* **57:** 725-749.

Knaff, D.B. (1993), Cytochrome bc, complexes of photosynthetic purple bacteria. *Photosynthesis Res.,* **35 :** 117-13.

Mathis, P. (1990), Compared structure of plant and bacterial photosynthetic reaction centers. Evolutionary implications. *Biochem. Biophys. Acta.,* **1018 :** 163-169.

Meent, van de, E.J., Kobayashi, M Erkelens, C, Van veelan, P.A., Ohe S.C.M. Inoue, K., Watanabe, T. and Amess (1992), The nature of primary electron acceptor in green sulfur bacteia. *Biochem. Bio-phys. Acta.,* **1102 :** 371-379.

Nitschke, W., Feiler, U and Rutherford A.W. (1990), Photosynthetic reaction center of green sulfur bacteria studies by EPR. *Biochemistry,* **29 :** 3834-3842.

Partenskey, F., Hess, W.R and Vaulot, D. (1999), *Prochlorococcus*, a marine photosynthetic prokaryote of global singificance. *Microbiol. Mol. Biol. Rev.,* **63 :** 106-129.

Schlegel, H.G. and Bowien, B. (ed.) (1989), Autotrophic bacteria. Madiron, Wis.: Science Tech Publishers.

Tabita, F.R. (1994), The biochemistry and molecular regulation of CO_2 metabolism in cyanobacteria. In "The Molecular Biology of Cyanobacteria" (D.A. Bryant, ed), pp.437-467, Kluwer, Dordrecht.

Tabita, F.R. (1995), The biochemistry and metabolic regulation of carbon metabolism and CO_2 fixation in purple bacteria. In anoxygenic photosynthetic Bacteria". (R.E. Blankenship, M.T. Madigan and C.E. Bauer, eds.), pp. 885-914. Kluwer, Doudrecht.

White, D. (1995), The Physiology and Biochemistry of Procaryotes, New York; Oxford University Press.

Xiong, J., Fisher, WM., Inoue, K., Nalahara, M. and Bauer, C.E. (2000), Molecular evidence for the early evolution of photosynthesis. *Science,* **289 :** 1724-1730.

Yoon, K., Hanson, T.E, Gibson, J.L. and Tabita, F.R. (2000), Autotrophic CO2. In Encyclopedia of Microbiology" Vol. I, L. Lederberg Ed., San Diego, Academic Press, pp 349-359

Chapter - 10

Chemolithoautotrophy

10.1. DEFINITION

Several groups of soil and aquatic bacteria are able to use inorganic compounds or ions (ammonium, nitrite, thiosulphite, sulphite and ferrous ions) as well as elemental sulphur, hydrogen or carbon monoxide as electron or hydrogen donors, to gain energy and reducing equivalents for synthetic processes by their oxidation. This mode of life with inorganic hydrogen donors is called **chemolithotrophy**. In Chemolithotrophs ATP synthesis is coupled to oxidation of electron donor. Reducing power is obtained either directly from the inorganic compound or by the reverse electron transport reactions. Most of these bacteria grow with carbon dioxide as their sole, or major source of carbon for cell synthesis. They are therefore, also called as autotrophs [chemolithoautotrophs]. It must be remembered here that photolithotrophs and chemolithotrophs mainly differ with respect to synthesis of ATP and reducing power. In the former these are derived from the light driven electron transport, whereas in latter they are procured from oxidation of inorganic compounds (Fig. 10.1). In other respects they are similar. All chemolithotrophic bacteria assimilate their carbon from CO_2 by fixation via the ribulose – biphosphate cycle. They may be aerobic (Table 10.1) or anaerobic (Table 10.2).

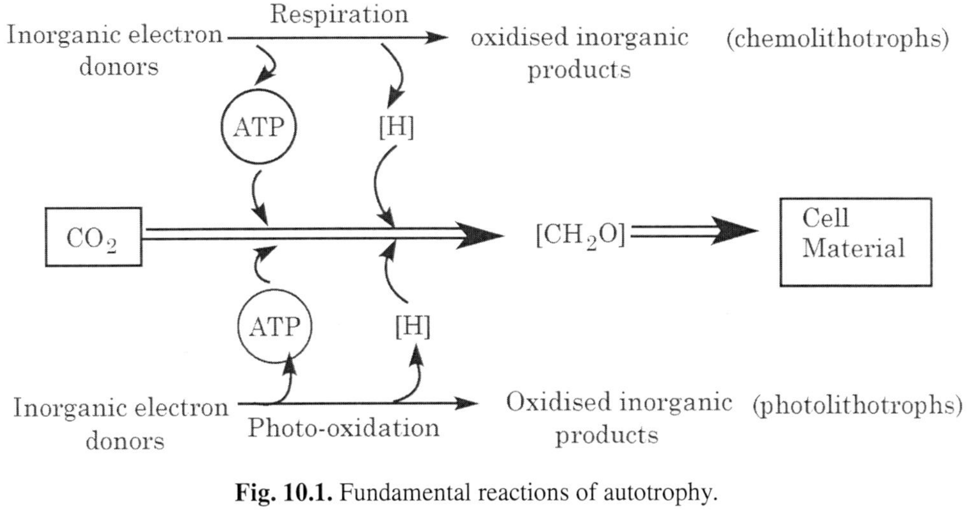

Fig. 10.1. Fundamental reactions of autotrophy.

Table 10.1. Aerobic chemolithotrophic bacteria

Inorganic electron donor / oxidation prodcut	Group	Representative species
$NH_3 \xrightarrow{O_2} NO_2^-$	Ammonia oxidisers	*Nitrosomonas europeaea*
$NO_2 \xrightarrow{O_2} NO_3^-$	Nitrate oxidisers	*Nitrobacter winogradskyi*
$S^o\ S^{-2}\ S_2O_3^{-2} \xrightarrow{O_2} SO_4^{-2}$	Sulphur oxidisers	*Thiobacillus thiooxidans*
$Fe^{+2} \xrightarrow{O_2} Fe^{+3}$	Iron bacteria	*Thiobacillus ferrooxidans*
$H_2 \xrightarrow{O_2} H_2O$	Hydrogen oxidisers	*Alcaligenes eutrophus*
$CO \xrightarrow{O_2} CO_2$	Carboxidobacteria	*Pseudomonas carboxidovorans.*

Table 10.2. Anaerobic chemolithotrophic bacteria

Electron donor / reduction product	Group designation	Representative species
$H_2 \xrightarrow{CO_2} CH_4$	Methanogens (carbonate respiring)	*Methanobacterium thermoautotrophicum*
$H_2 \xrightarrow{CO_2} CH_3\text{-}COOH$	Acetogens (carbonate respiring)	*Acetobacterium woodii*
$H_2 \xrightarrow{SO_4^{-2}} H_2S$	Sulphidogens (sulphate respiring)	*Desulfovibrio desulfuricans*
$H_2 \xrightarrow{SO_4^{-2}} H_2S$	Sulphidogens (sulphate respiring)	*Desulfomonas acetooxidans*
$H_2 \xrightarrow{NO_3^-} N_2$	Denitrifiers (nitrate respiration)	*Paracoccus denitrificans*
$CO \xrightarrow{NO_3^-} N_2$	Denitrifiers (nitrate respiration)	*Pseudomonas carboxydovorans*
$CO \xrightarrow{SO_4^{-2}} H_2S$	Sulphidogens (sulphate respiring)	*Desulfobacterium autotrophicum*

10.2. ENERGETICS OF CHEMOLITHOTROPHY

A review of reduction potentials listed in Table 10.3 reveals that a number of inorganic compounds can provide sufficient energy for ATP synthesis when oxygen is used as an electron acceptor.

Table 10.3. Mid-point potentials of inorganic reductants used by chemoautotrophs

Bacteria	Redox couple	E'_o (V_o)
Carboxydobacteria	CO_2 / CO	-0.54
H-bacteria	H^+ / H_2	-0.41
S-bacteria	SO_4^{-2} / SO_3^{-2}	-0.28
	S / H_2S	-0.25
	SO_3^- / S	+0.05
Nitrite oxidisers	NO_3^- / NO_2^-	+0.42
NH_3 oxidisers	NO_2^- /NH_4^+	+ 0.44
Iron bacteria	Fe^{+3} / Fe^{+2}	+ 0.78

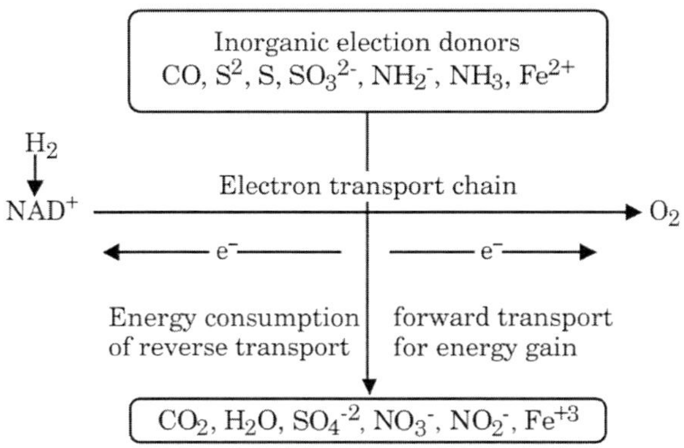

Fig. 10.2 : Reverse electron transport

Energy generation in chemolithotrophs usually occurs by oxidative phosphorylation involving electron transfer chain of conventional type. A few *Thiobacillus* species additionally employ substrate level phosphorylation during chemolithotrophic growth on inorganic sulfur compounds. Besides ATP, the reducing power needed for CO_2 assimilation is provided by NADH generated by the electron transfer chain. However, except molecular hydrogen all the inorganic reductants channel electrons into the respiratory chain at potentials significantly higher than that of the NAD^+/NADH couple, and they are, therefore, unable to reduce NAD^+ by forward electron transfer. The reduction of NAD^+ can only be achieved by energy-dependent process called **reverse electron transport**. The precise energy requirement depends upon the point of electron entry to the chain. Reverse electron flow entails successive reversal of oxidative phosphorylation involving the expenditure of ATP energy. Thus with the exception of molecular hydrogen, a substantial amount of the energy derived from the oxidation of inorganic substrates is employed to drive NAD^+ reduction through reverse electron transport. Thus, a study of the table 10.3 reveals that sulfur bacteria, nitrite oxidisers, NH_3 oxidizers and iron bacteria synthesize their reducing power by reverse electron transport (Fig. 10.2).

The yield of biomass in chemolithotrophic organisms is very less when compared to organotrophs. This is mainly because very small amount of energy is obtained by the oxidation of inorganic substances. In addition several of these bacteria also carry out uncoupled oxidation reactions *i.e.,* they oxidize substrates without concomitant synthesis of cell material (metabolic idling). Thus, synthesis of one gram of cellular dry weight requires the consumption of far greater amount of substrates than is known for other organisms (Table 10.4 and Fig. 10.3).

Table 10.4. Comparison of cell yields with the primary energy source to be converted in different autotrophic and organotrophic microorganisms.

Name of the organisms	Substratum	Amount of substratum (in grams)
Thiobacillus ferrooxidans	Fe^{+2}	156
T. neapolitanus	$S_2O_3^{2-}$	30
Nitrosomonas	NH_3	30
Alcaligenes eutrophus	H_2	0.5
E. coli	glucose	2
yeast	petroleum	1

*The amount required for the synthesis of 1 g cells (dry weight)

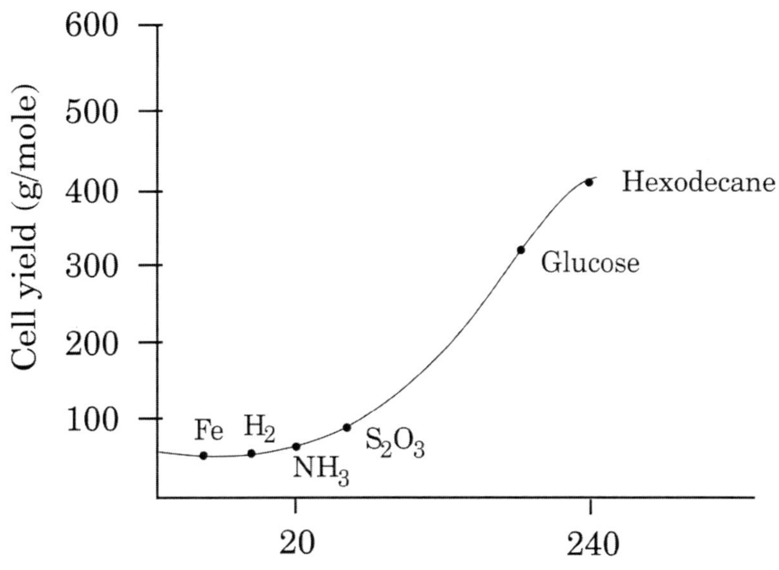

Fig. 10.3: Cell yields versus available energy from inorganic and organic electron sources

10.3. NITRIFICATION – NITRIFIERS

Nitrification is the microbial oxidation of NH_4^+ and organic N into NO_2^- and NO_3. Biological nature of nitrification and the organisms involved were first identified in 1889-1890 by Serge Winogradsky.

Table 10.5. Characteristics of chemoautotrophic nitrifying bacteria

Genus	Morphology	Growth range (pure culture)	Habitat
Ammonium oxidisers			
Nitrosomonas	Straight rod motile or nonmotile; one or two subpolar flagella	5° – 40°C pH 5.8 – 9.5	Soil marine; sewage; fresh water
Nitrosospira	Spiral shaped; motile or nonmotile; peritrichous flagella	25° – 30°C pH 7.5 – 10.0	Soil
Nitrosococcus	Cocci in pairs or tetrads; motile or nonmotile tuft of peritrichous flagella	2° – 30°C pH 6.0 – 10.0	Soil marine; fresh water
Nitrosolobus	Lobular, pleomorphic; motile; peritrichous flagella	15° – 30°C pH 6.0 – 10.2	Soil
Nitrite oxidisers			
Nitrobacter	Short rods; motile or nonmotile; single polar flagellum	5° – 40°C pH 5.7 – 10.2	Soil marine; fresh water
Nitrospira	Long, slender rod; nonmotile	20° – 30°C pH 7.0 – 10.0	Marine only
Nitrococcus	Spherical cell	20° – 30°C pH 7.0 – 10.0	Marine only

Chemoautotrophic nitrification is exclusively bacterial and is performed by a select group of lithotrophic bacteria. Some of the bacteria involved in oxidation of ammonia and nitrite along with their characteristics are given in Table 10.5. These nitrifying bacteria are Gram-negative, and belong to the family Nitrobacteriaceae. They exist by oxidizing NH_4^+ or NO_2^- to NO_2 and NO_3, respectively.

$$NH_4^+ \longrightarrow NH_2OH \longrightarrow NO_2^- \longrightarrow NO_3^-$$

However, no known single bacterium carries both the steps. The first step i.e., conversion of NH_4^+ to NO_2 is carried out by a group of bacteria known as '**ammonia oxidizers**' *e.g. Nitrosomonas, Nitrosolobus, Nitrosospira, Nitrosovibrio, Nitrosococcus* (prefix-nitroso). The second reaction is carried out by another physiological group of organisms known as **'nitrite oxidizers'** e.g. *Nitrobacter, Nitrococcus, Nitrospira* (prefix nitro). The complete oxidation of ammonia to nitrate, an eight electron transfer, is thus carried out by members of these two groups of organisms acting in a sequence. Both these physiological groups co-exist abundantly in aerobic soils.

Nitrification is a multi-step process. The most reduced form of inorganic N serves as the starting point for nitrification, and the first product is hydroxylamine (NH_2OH) (Fig. 10.4).

$$NH_4^+ + H^+ + O_2 \longrightarrow NH_2OH + H_2O$$

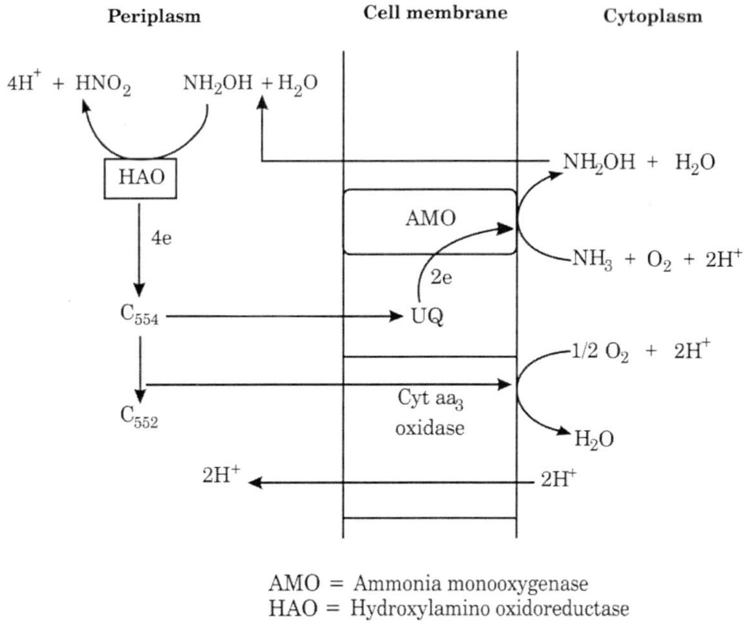

Fig. 10.4 : Electron transport scheme in *Nitrosomonas*

The first step in NO_2 formation is catalyzed by the enzyme ammonia monooxygenase (AMO). The oxygen in NO_2 comes from molecular oxygen, so oxygen is not only required by these bacteria for respiration, but it is also a substrate. Nitrifiers deprotonate ammonium (NH_4^+) to ammonia NH_3 before oxidizing it to hydroxylamine (NH_2OH). The next step in NO_2^- formation is oxidation of NH_2OH by hydroxylamine oxido-reductase.

$$NH_2OH + H_2O \longrightarrow [HNO] \longrightarrow NO_2^- + 5H^+ + 4e^-$$

The last step in nitrification is the oxidation of NO_2^- to NO_3^- (Fig. 10.5).

$$2NO_2^- + H_2O + O_2 \longrightarrow 2NO_3^- + H_2O$$

Interestingly, the oxygen incorporated into NO_2^- actually comes from H_2O, not O_2.

10.3.1. Energy-Yielding Reactions

The electrons from nitrogen compounds enter an electron transport chain and electron flow establishes a membrane potential and proton motive force linked to ATP synthesis. The E_o' of the NH_2OH/NH_3 couple, the first step in the oxidation of NH_3, is about 0 V, E_o' of the NO_3^- / NO_2^-

couple is very high, about +0.43 V. These relatively high reduction potentials mean that nitrifying bacteria must donate electrons to their electron transport chain at rather later steps in the overall process. This effectively limits the amount of ATP that can be produced from each pair of electrons introduced. Several key enzymes are involved in oxidising reduced nitrogen compounds.

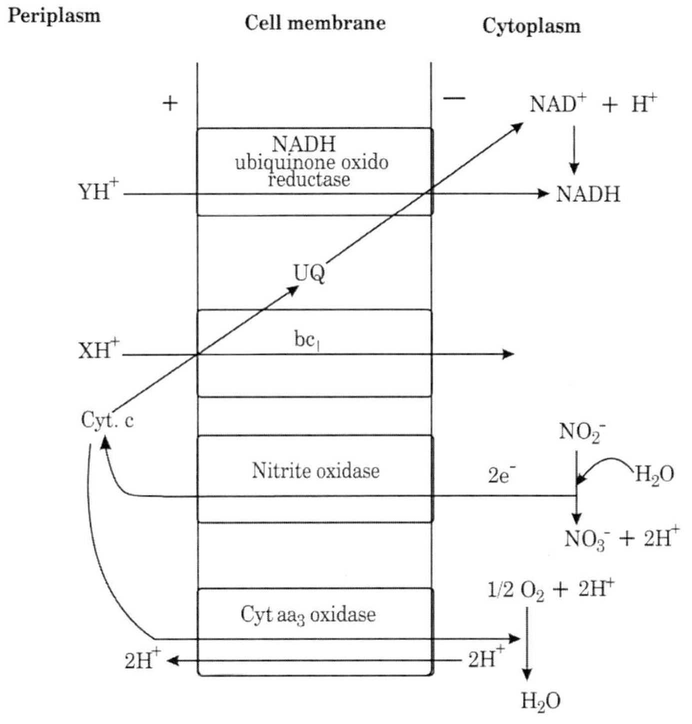

Fig. 10.5 : Electron transport scheme in *Nitrobacter*

In ammonia – oxidising bacteria, NH_4^+ is oxidised by 'ammonia monooxygenase' that produces NH_2OH and H_2O. 'Hydroxylamine oxidoreductase' then oxidises NH_2OH to NO_2^-, removing four electrons in the process.

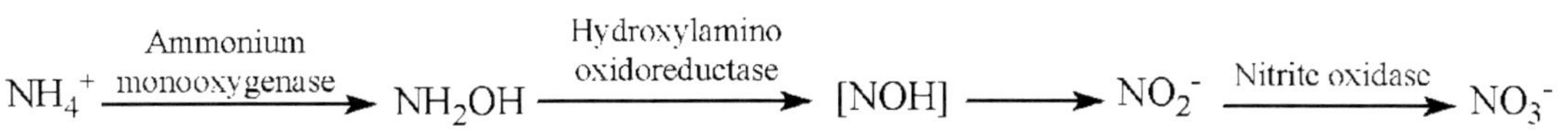

Ammonia monooxygenase is an integral membrane protein, whereas hydroxylamine oxidoreductase is periplasmic.

In the reaction carried out by ammonia monooxygenase, there is a need for two exogenously supplied electrons to reduce one atom of dioxygen to water.

$$NH_3 + O_2 + 2H^+ + 2\,e^- \longrightarrow NH_2OH + H_2O$$

These electrons originate from the oxidation of hydroxylamine and are supplied to ammonia monooxygenase from hydroxylamine oxidoreductase via cytochrome c and ubiquinone. Thus, for every four electrons generated from the oxidation of NH_3 to NO_2, only two actually reach the terminal oxidase (cytochrome aa_3). Nitrite-oxidizing bacteria employ the enzyme nitrite oxidase to oxidise nitrite to nitrate, with electons traveling a very short electron transport chain (because of the high potential of the NO^{-}_{3}/NO_2^{-} couple) to the terminal oxidase.

Cytochromes 'a' and 'c' types are present in the electron transport chain of nitrate oxidiser, and generation of a proton motive force (which ultimately drives ATP synthesis) occurs through the action of cytochrome aa_3. Like the situation with iron oxidation, only small amounts of energy are available. Thus, the growth yields of nitrifying bacteria are relatively low.

10.3.2. Importance of the Nitrifiers

1. The ammonium salts released by the decomposition of the organic matter by ammonifiers accumulate in the soil and render it alkaline. Alkaline soils are not favourable for plant growth due to the non-availability of nutrients like K^+, Ca^{2+}, Mg^{2+} and phosphate. Nitrifiers by their physiological action release the nitrate (anion) into the soil and render the soil slightly acidic which is favourable for plant growth. However, this concept has undergone radical change recently. The nitrates formed by the nitrifiers are very easily leached out of the soil due to its high solubility in water. However, NH_3 sparsely soluble and tenaciously bound to the soil particles and hence it is not easily leached out.
2. Nitrifiers are implicated in the biodeterioration and destruction of lime stone and cement buildings roads and other constructions.
3. Nitrifiers exhibit high oxidising potential. Therefore, they oxidise methane, methanol, CO, ethylene, propylene, cyclohexane, benzyl alcohol, phenol etc.

10.4. SULPHUR OXIDISERS

In nature, the sulphur exists in different forms like $S^{0,}$ SO_2,$S_2O_3^{-2}$, SO_3^{-2}, S^{-2}. These sulphur compounds according to their redox potentials can be arranged as S^{-2} (+6), S^0, $S_2O_3^-$, SO_3^{2-}, SO_4^{2-} (-2). It is evident that sulphides are in highly reduced state where as sulphate is in highly oxidised state. Not surprisingly, some of the microorganisms have developed a mechanism to oxidise the reduced sulphur compounds and to utilize the energy released there upon. This group of bacteria is known as sulphur oxidisers.

Sulphur oxidisers are broadly categorized into : (i) Photosynthetic sulphur oxidisers, and (ii) Non-photosynthetic sulphur oxidisers. The second group of sulphur oxidisers will be discussed in this section.

Most sulphur oxidisers grow near the source of sulphur [*e.g.,* sulphur deposits, sulphide ores, hot sulphur springs, sulphur mines, coal mines that are sites of FeS_2 ferrous sulphide (or) pyrite]. These members are also found in the layers between aerobic and anaerobic environments. Some are extremely acidophilic and grow at pH 1 to 2 and still some are thermophilic.

Sulphur oxidisers are taxonomically heterogenous group of bacteria (Table 10.6). Some belong to eubacteria and other like *Sulpholobus caulderiella* belong to archaebacteria; Some of the common eubacterial genera are *Thiobacillus, Thiosphaera, Thiomicroscopica, Thermothrix, Beggiatoa, Thioplaca, Achromatium, Thiovulum* etc. These bacteria are Gram-negative with polar flagella. However, some are with biflagella and even non-motile. Most of the thiobacilli are obligate chemolithoautotrophs, but some species like *Thiobacillus novellas* and *T. intermedius* can also utilise organic compounds as carbon source. Hence, they are chemoheterotrophs. *Thiobacillus thiooxidans* produces large amount of H_2SO_4 and grows at very low pH. *Thiobacillus denitrificans* can utilize nitrate as electron and proton acceptor and thus it carries anaerobic respiration.

Sulpholobus acidocaldarius is a thermophilic species grows in hot springs [pH 2–3] at temperature 75°C and it is facultatively chemolithotrophic. *Beggiatoa* and *Thiothrix* are filamentous forms where as *Achromatium, Thiovulum* are unicellular forms. It was S.N. Winogradsky while working on *Beggiatoa* established the nutritional class of chemolitho-autotrophy.

Table 10.6. Sulphur oxidising bacteria

Species	pH range for growth	Electron Donor	Type*
Thiobacillus thiooxidans	2 – 5	S^{2-}, $S_2O_3^{-2}$, S^0	O
T. ferrooxidans	2 – 6	Fe^{+2}, $S_2O_3^{-2}$, S^0	F
T. thioparus	6 – 8	CNS^-, $S_2O_3^{-2}$, S^0	O
T. denitrificans	6 – 8	CNS^-, $S_2O_3^{-2}$, S^0	O
T. intermedius	2 – 6	$S_2O_3^{-2}$, S^0 glutamate	F
T. novellas	6 – 8	$S_2O_3^{-2}$, S^0,	F
T. microspira	6 – 8	S^{2-}, $S_2O_3^{-2}$, S^0	O
Sulpholobus acidocaldarius	2 – 8	S^0, glutamate peptone	F

* O – obligate autotrophy; F = facultative autotrophy

10.4.1. Physiology and Biochemistry of Sulphur Oxidisers

Sulphur oxidisers oxidise sulphides, elemental sulphur and sulphite and ultimately convert them to sulphate (Fig. 10.6). In all these cases the end product is sulphate.

$$S^{-2} + 2O_2 \longrightarrow SO_4^{-2}$$

$$S + H_2O + \frac{1}{2}O_2 \longrightarrow SO_4^{-2} + 2H^+$$

$$S_2O_3^{-2} + H_2O + 2O_2 \longrightarrow 2SO_4^{-2} + 2H^+$$

Thiobacilli usually grow very well on thiosulfate. Two enzyme systems have been found to catalyse the oxidation of sulfite to sulfate. Sulfite oxidase is a membrane – bound molybdo-protein and transfers electron to cytochrome **c**. In addition, the enzyme APS reductase is present in many sulphur bacteria. Here also cytochrome functions as electron acceptor. The APS formed is further converted to sulfate and ADP. The oxidation of sulfite to sulfate via APS is accompanied by substrate level phosphorylation.

$$SO_3^{2-} + AMP \longrightarrow APS + 2e^-$$

$$APS + Pi \longrightarrow ADP + SO_4^{-2}$$

$$2ADP \longrightarrow ATP + AMP$$

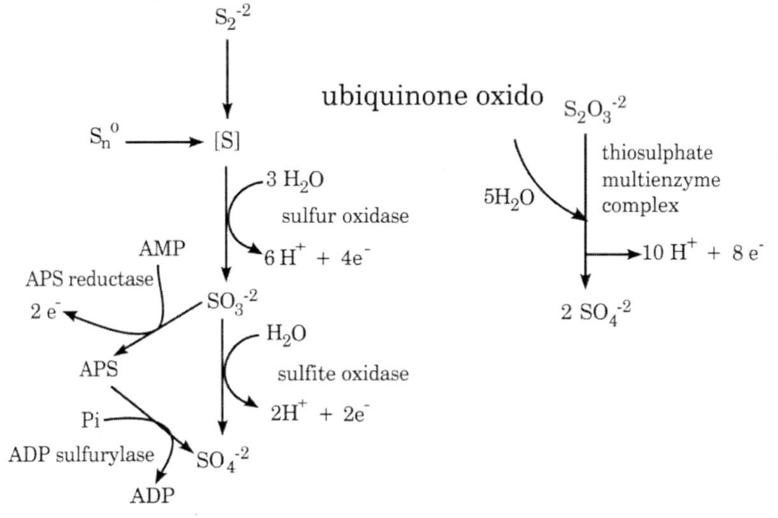

Fig. 10.6: Summary of sulphur oxidising pathways; APS-Adenosine phosphosulfate

In *Thiobacillus versutus*, thiosulfate oxidation takes place in the periplasm by a multienzyme complex consisting of enzyme A and B and cytochromic protons are released in the periplasm during the oxidation and the electrons are transferred electrogenically from the periplasmic side of the membrane to the cytoplasmic side through cytochrome $\mathbf{c}_{552}$ and then cytochrome $\mathbf{aa}_3$ oxidase to oxygen (Fig. 10.7). A proton motive force (ΔP) is created by the release of protons in the periplasm via the oxidations. The consumption of protons in the cytoplasm during oxygen reduction, and electrogenic flow of electrons across the membrane to oxygen, ATP is synthesized by a proton – translocating ATP synthase driven by the ΔP.

In *Thiobacillus tepidarius*, the pathway of oxidation of thiosulfate to sulfate (called the polythionate pathway) begins in the periplasm, but the bulk of the oxidations take place in the cytoplasm. Two molecules of thiosulfate are oxidised to tetrathionate in the perioplasm by a thiosulfate – cytochrome **c** oxidoreductase (Fig. 10.8). The tetrathionate is believed to be transported into the cell where it is oxidised to 4 sulfates via the intermediate sulfite generating 14 electrons. The 14 electrons are transferred to a proton – translocating ubiquinone – cytochrome b complex in the cell membrane, which reduces membrane bound cytochrome **c**. Cytochrome c transfers all of the electrons to cytochrome oxidase which reduces oxygen. Since the oxidations of tetrathionate and sulfite are cytoplasmic rather than periplasmic, the proton current is due to proton – translocation by the ubiquinone cytochrome **b** complex (coupled to the cytoplasmic oxidations), rather than to periplasmic oxidations, as is the case for *T. versutus*.

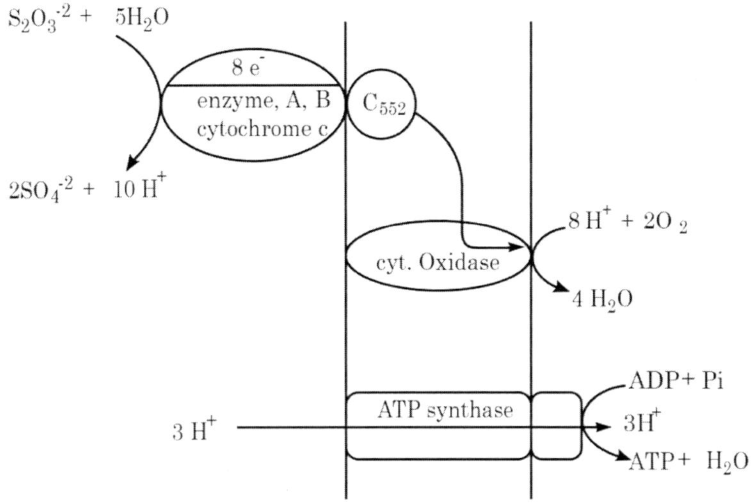

Fig. 10.7: Thiosulphate oxidation in *Thiobacillus versutus*

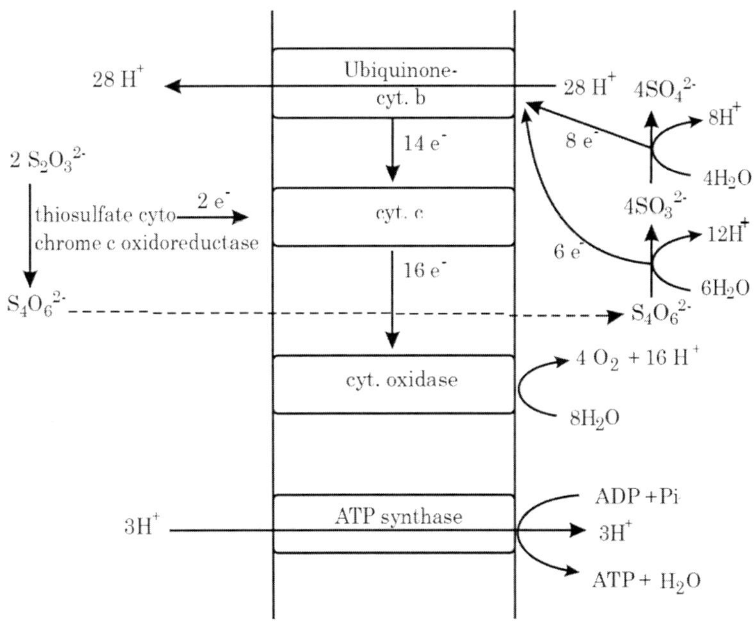

Fig. 10.8 : Thiosulphate oxidation in *T. tepidarius*

10.4.2. Importance of the Sulphur Oxidisers

1. *T. thiooxidans* produces large amounts of sulphuric acid and is specifically adapted to grow at low pH. It can tolerate 1N sulphuric acid. This acidification can be exploited in

a number of ways. Addition of elemental sulphur can be used for de-alkalisation of chalky soils. The sulphuric acid produced by thiobacilli converts calcium carbonate to more soluble calcium sulphate, which is readily leached out.

2. Control of potato scab – caused by *Streptomyces scabies*. *S. scabies* is a acidophobic organism. In problematic soils the disease can be effectively controlled by applying the sulphur to the soil. The principle of control is very simple *i.e.,* sulphur oxidisers oxidise the sulphur to sulphuric acid which then renders the soil acidic. Hence, *S. scabies* is eliminated.

10.5. IRON BACTERIA

In nature, iron exists in ferrous and ferric forms. In between the ferrous and ferric, ferrous is in reduced state and ferric in oxidised state. When ferrous forms are oxidised to ferric forms, energy is released. The redox potential of ferrous and ferric couple is –0.41 v/mol. Some bacteria oxidise the ferrous forms to ferric forms and utilize the energy released there upon *i.e.,* the ferrous salts function as electron donors for these bacteria and such bacteria are called iron bacteria. The acidophilic iron oxidisers can be found growing at the sites of geological deposits of iron sulfide minerals [*e.g.* pyrite (FeS_2) and chalcopyrite ($CuFeS_2$)] where water and oxygen are also present.

Iron bacteria are chemolithoautotrophic. Prominent among these bacteria are *Thiobacillus ferrooxidans, Sulfolobus acidocaldarius, Gallionella ferruginea, Leptothrix ochracea.*

The oxidation of Fe^{+2} to Fe^{+3} by oxygen is carried out at neutral pH, but only slowly at acidic pH. The acidic environment is conducive to growth of the iron oxidisers. An example is, *Thiobacillus ferrooxidans* oxidises both reduced iron and sulfur compounds. It grows on ferrous ion (e.g. $FeSO_4$) at optimal external pH values between 1.8 and 2.4. It maintains an internal pH of around 6.5. The natural habitat of this bacterium is acid waters of iron ore mines which contain metal sulphides and pyrite. *G. ferruginea* and *L. ochracea* grow in drainage pipes and mountain streams where thick covering of iron oxide is present.

10.5.1. Physiology and Biochemistry of Iron Bacteria

Iron bacteria oxidise pyrite [FeS_2] to ferrous sulphate; ferrous sulphate is further oxidised to ferric sulphate.

$$FeS_2 + 3\tfrac{1}{2} O_2 + H_2O \longrightarrow FeSO_4 + H_2SO_4$$

$$S^o + \tfrac{1}{2} O_2 + H_2O \longrightarrow H_2SO_4$$

$$2FeSO_4 + \tfrac{1}{2} O_2 + H_2SO_4 \longrightarrow Fe_2(SO_4)_3 + H_2O$$

$$MoS + 2Fe^{+3} \longrightarrow Mo^{+2} + 2Fe^{+2} + S$$

A model for electron transport and energy production with reference to *T. ferrooxidans* is presented in Fig. 10.9.

The model proposes that extracellular Fe^{+2} is oxidised to Fe^{+3} by a Fe^{+3} complex in the outer membrane. The electrons travel to a periplasmic cytochrome **c**. There is also a copper protein, called **rusticyanin (RC)** in the periplasm, which is thought to be part of the electron transport scheme. From cytochrome **c** the electrons flow across the membrane via cytochrome oxidase of the a_1 type. The ΔP is maintained by the inward flow of electrons and the consumption of cytoplasmic protons during oxygen reduction. The consumption of two protons per two electrons in the cytoplasm and the uptake of three protons through the ATP synthase, indicate that a maximum of 2/3 of an ATP can be made per oxidation of two ferrous ions.

When the bacteria are oxidising iron pyrite (FeS_2), the Fe^{+3} that is produced outside of the cell envelope is chemically reduced to Fe^{+2} by S^{-2}. The resultant S^0 is oxidised to SO^2_4 and the electrons pass through a proton translocating bc_1 complex to periplasmic cytochrome **c** and there to cytochrome oxidase. The figure also illustrates ΔP driven reversed electron transport from S^0 and from Fe^{+2} to NAD^+ through coupling sites that bring protons into the cell.

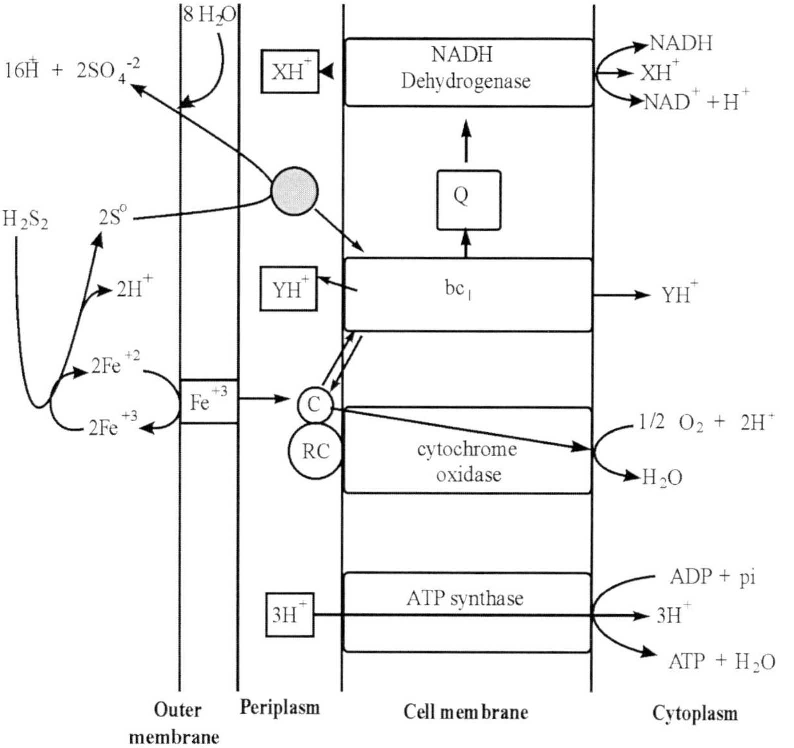

Fig. 10.9 : Electron transport pathway in *Thiobacillus ferrooxidans*

Iron oxidation is not always associated with energy production. Many bacteria such as *Crenothrix, Gallionella, Hyphomicrobium, Leptothrix, Metallogenium, Pedomicrobium, Planctomyces, Seliberia* and *Sphaerotilus* oxidize Fe^{2+} without gaining any apparent energy. The cell walls act as catalytic surfaces for metal precipitation and ferrous iron oxidation leads to precipitated Fe^{3+} around filamentous sheath-forming bacteria. *Leptothrix* species, for example, include heterotrophic fresh water bacteria that oxidize Fe and Mn. *L. discophora* is found in well waters and iron pipes. It forms sheath-covered cells, strings and empty sheaths encrusted with ferromanganous oxides. However, it gains no energy from oxidation.

10.5.2. Importance of the Iron Bacteria

Iron bacteria especially *T. ferrooxidans* is involved in the bioleaching of ores. This organism is employed for biomining of various metals like iron, copper, zinc, molybdenum and chromium. The ability of some acidophilic iron and sulphur oxidising bacteria to convert sulphide ores to water soluble heavy metal sulphates, is utilized for the leaching of low grade ores and to obtain Cu,Zn,Ni,Mo and chromium.

Leaching processes are in large scale use for reclaiming ores from slag–heaps and might also be useful in deep–mining operations. In its simplest form, water is allowed to percolate through columns of broken up ore containing rocks for *i.e.*, iron pyrites (FeS_2) and associated metal sulphides such as Cu_2S, CuS, ZnS, NiS, MoS_2, Sb_2S_3, CoS and PbS and the sulphate containing solutions are collected. The various metals are then obtained by concentration and precipitation. Bringing the heavy metals into solutions is the result of several associated processes.

The bacteria are responsible for the supply of H_2SO_4 and for the regeneration of Fe^{+3}. Both these components are consumed in the solubilisation of ores. These conversions are carried out by *Thiobacillus thiooxidans* and *T. ferrooxidans.* The strains involved are remarkably tolerant to the prevailing concentrations of Cu^{+2}, Co^{+2}, Zn^{+2}, Ni^{+2} and other heavy metal ions. Strains of *Sulfolobus* that oxidize iron and sulphur also take part in the leaching of metal ores.

10.6. HYDROGEN BACTERIA

The hydrogen-oxidising bacteria are usually facultative and can live either autotrophically or heterotrophically. However, some always require organic carbon for growth and are called chemolithoheterotrophs. Hydrogen molecule [H_2] being in reduced state yields energy upon oxidation. It can function as an electron donor. The H_2/H_2O redox potential is about – 0.81 volt/mole. Hydrogen bacteria utilise the energy of hydrogen by the action of hydrogenase enzyme. In nature hydrogen gas is produced both in aerobic and anaerobic environments. As a result, two groups of hydrogen oxidising bacteria have been evolved.

i. Aerobic hydrogen bacteria

ii. Anaerobic hydrogen bacteria

10.6.1. Aerobic Hydrogen Bacteria

Those bacteria that oxidise hydrogen aerobically with oxygen as the terminal electron acceptor are called aerobic hydrogen oxidising bacteria. These are found in aerobic environments where H_2 is available. All of these are able to fix carbon dioxide and are thus autotrophs. They are also able to utilize organic substrates as electron donors. Hence, they are facultative chemolitho-autotrophs. A few aerobic hydrogen – oxidising bacteria can also oxidise carbon monoxide and can grow with this as the only electron donor and carbon source.

Taxonomically, aerobic hydrogen oxidisers are extremely heterogenous e.g., *Alcaligenes eutrophus, Pseudomonas faecalis, Pseudomonas saccharophila, Pseudomonas carboxidovorans, Aquaspirillum autotrophicum, Paracoccus denitrificans, Xanthobacter autotrophicus, Nocardia opacea, Mycobacterium gordonac* and *Bacillus* species.

Utilisation of hydrogen and energy metabolism

Hydrogen is metabolised into the cell by hydrogenases. Two types of hydrogenases are found in aerobic hydrogen – oxidising bacteria. They are : a soluble, NAD^+ reducing hydrogenase [H_2 : NAD^+ oxidoreductase] localized in the cytoplasm, and a membrane – bound hydrogenase. Some bacteria like *Alcaligenes eutrophus* and *A. hydrogenophilus* possess both kinds of enzymes. However, a few contain either of the two only. Both kinds of hydrogenases can catalyse the entry of hydrogen into the respiratory chain. The process of ATP generation is very efficient. The composition of the respiratory chain in *Alcaligenes eutrophus* and *Paracoccus denitrificans* is almost the same as that in mitochondria.

10.6.2. Anaerobic Hydrogen Oxidisers

The bacteria that oxidise hydrogen anaerobically with terminal electron acceptor other than oxygen (*i.e.,* CO_2, sulphate, nitrate, and elemental sulphur) are called anaerobic hydrogen oxidizers (Table 10.2).

As we will come to know in the forthcoming chapter, in nature hydrogen is produced by the activities of different microbial communities as listed below.

(i) Molecular hydrogen is produced during the anaerobic degradation of organic substances in sediments and anaerobic mini-habitats of the soil.

(ii) It is also produced in mixed acid fermentations where pyruvic acid is converted to acetyl CoA and acetyl CoA finally converted to formic acid. Hydrogen is produced by the breakdown of formic acid.

$$\text{Pyruvic acid} \longrightarrow \text{Acetyl CoA} \longrightarrow \text{Formic acid (HCOOH)}$$

$$HCOOH \longrightarrow H_2\uparrow + CO_2$$

(iii) It is produced in butanol-propanol fermentation carried out by clostridia species where hydrogen is produced from pyurvate with acetyl CoA as an intermentate.

$$\text{Pyruvate} \longrightarrow \text{Acetyl CoA} + CO_2 + H_2\uparrow$$

(iv) From ethanol under acidic conditions (acidogenic)

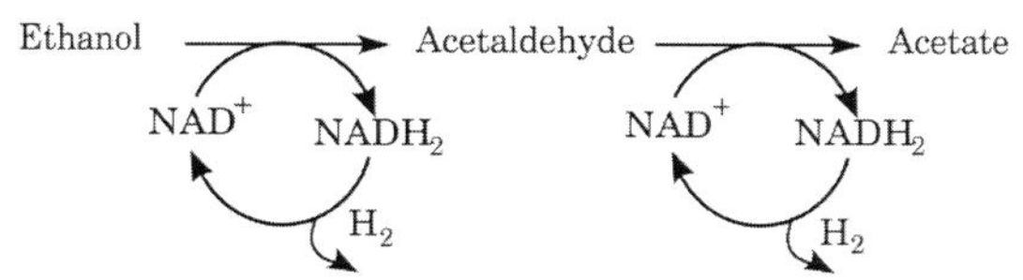

10.7. CARBOXIDOBACTERIA

Carbon monoxide occurs in nature under aerobic and anaerobic conditions. The organisms that are able to grow on carbon monoxide as the sole source of energy and carbon using oxygen as terminal electron acceptor are called **carboxydobacteria**. All carboxydobacteria are aerobic.

Neither the biochemical mechanisms nor the microorganisms involved in these reactions are not clearly known. However, it is known that several bacterial species can oxidise carbon monoxide to carbon dioxide *e.g., Pseudomonas carboxydovorans* metabolises CO during the growth according to the following equation.

$$7CO + 2\frac{1}{2}O_2 + H_2O \longrightarrow (CH_2O) + 6\ CO_2$$

The assimilation of carbon occurs by CO_2 fixation via the ribulose biphosphate cycle. Oxidation of CO is catalyzed by an enzyme containing molybdenum. Therefore, growth on CO requires molybdenum as a trace element, whereas growth on organic substrates does not. Carboxydobacteria, possess a membrane bound hydrogenase and can also grow as hydrogen bacteria.

10.8. FURTHER READING

Ehrlich, H.L., Ingledew, J.W. and Salerno, J.C. (1991), Iron and manganese-oxidizing bacteria, pp. 147-170. In :Variations in autotrophic life. Shively, J.M. and L.L. Barton (eds.), Academic Press, New York.

Hooper, A.B. (1989), Biochemistry of the nitrifying lithoautotrophic bacteria, pp. 239-265. In : Autotrophic bacteria. H.G. Schlegel and B. Bowien (eds.), Springer-Verlag, Berlin.

Ingledew, W.J. (1990), Acidophiles, pp. 33-54. In : Microbiology of extreme environments. C. Edwards (ed.), McGraw-Hill Publishing Company, New York.

Kelly, D.P. (1989), Physiology and Biochemistry of unicellular sulfur bacteria, pp. 193-217. In : Autotrophic bacteria. H.G. Schlegel and B. Bowien (eds.), Springer-Verlag, Berlin.

Kelly, D.P. (1990), Energetics of chemolithotrophs, pp. 479-503. In : The Bacteria, Vol. XII. Krulwich, T.A. (ed.), Academic Press, New York.

Kelly, D.P., Lu, W.P. and Poole, R.K. (1993), Cytochromes in *Thiobacillus tepidarius* and the respiratory chain involved in the oxidation of thiosulfate and tetrathionate. *Arch. Microbiol.,* **160 :** 87-95.

Kuenen, J.G. and Bos, P. (1989), Habitats and ecological niches of chemolitho (auto)trophic bacteria. In : Autotrophic bacteria, H.G. Schlegel and B. Bownien (eds.), 283-288, Springer-Verlag, New York.

Lovely, D.R. (1991), Dissimilatory Fe(III)and Mn(IV) reduction. *Microbiol. Rev.,* **55 :** 259-287.

Nelson, D.C. (1989), Physiology and biochemistry of the filamentous sulfur bacteria. In H.G. Schlegel and B. Bownien (eds.), Autotrophic bacteria, pp. 219-238, Springer-Verlag, New York.

Saier, M.H. (Jr) (2000), Bacterial diversity and the evolution of differentiation. *ASM News,* **66 :** 337-343.

Schlegel, H.G. and B. Bownien (eds.) (1989), Autotrophic bacteria, Springer-Verlag, New York.

Yoon, K.S., Hanson, T.E., Gibson, J.L. and Tabita, F.R. (2000), Autotrophic CO_2 metabolism. In J. Lederberg (ed.) Encyclopedia of Microbiology, 2nd edn. Vol. I, Academic Press, San Diego, pp. 349-3510.

■■■■

Chapter - 11

Chemoheterotrophy - Acetogens

11.1. INTRODUCTION

The term acetogenesis has been used with some reservation by microbiologists. Many microbiologists refer to acetogenesis as a way of production of acetate by any process including fermentation of organic substrates and an autotrophic type of synthesis of acetate from CO_2 and other one-carbon precursors. It is all the more important to distinguish between a fermentative process of acetate formation and an autotrophic synthesis of acetate. Das and Ljungdahl (2000) designated microorganisms producing acetate by linear combination of two molecules of CO_2 and CO as **acetogens**. Microorganisms producing acetate and other acids by fermentation are defined as acid producing or **acidogens**. In this chapter the physiology of only acetogenic bacteria is discussed.

Autotrophic synthesis of acetate from CO_2 was first reported in 1932 in cultures obtained from sewage sludge. However, a true acetogen, *Clostridium thermoaceticum* was isolated in 1942. Acetogens were not investigated much before 1967 partly because of lack of realization of the importance of thcsc bacteria and partly because appropriate techniques to grow and cultivate strictly anaerobic bacteria were not generally available.

11.2. DISTRIBUTION AND TAXONOMY

The acetogens belong to an extremely heterogenous group of anaerobic bacteria. These are obligate anaerobes found virtually in all anoxic environments and produce acetate from a variety of substrates. They occur in digestive tracts of animals and humans, sediments, estuaries, fresh water, salt logoons and soils. Most acetogens are facultative autotrophs, which are capable of growing on a wide range of organic substrates like sugars, alcohols, carboxylic acids, aromatic compounds, halogenated carbon compounds, CO_2 and CO. Recent investigations revealed that in addition to CO_2, they are able to use several other compounds as electron acceptors such as dimethyl sulfoxide (DMSO), sulfite, thiosulfate and nitrate. Some of the acetogens such as *Clostridium mayombei, C. pfennigii* and *Acetonema longum* carry out mixed acid fermentations in which butyrate is produced as one of the final products. Thus, the acetogens have a very complex physiology and they are genetically, phylogenetically and phenotypically very different. Physiological properties of some acetogens are precised in Table 11.1. Interestingly, no archeal member is an acetogen. A total of 64 bacterial species have been isolated, of which most can grow autotrophically on CO_2 and H_2.

Table 11.1. Physiological properties of homoacetogenic bacteria

Genus	No.of species	Morphology	Gram type	Growth temp. (in °C)	Mol%GC
Acetoanaerobium	3	rod	+ or -	37	34 - 40
Acetobacterium	11	rod	+	27 – 30	38 - 48
Acetitomaculum	1	rod	+	38	34
Acetogenium	1	rod	-	66	38
Acetohalobium	1	rod	--	40	34
Acetoneum	1	rod	--	30	52
Clostridium	11	rod	+ or -	30 – 60	22 - 42
Moorella	2	rod	+	58	54
Eubacterium	1	rod	+	37	48
Peptostreptococcus	1	coccus	+	37	45
Ruminococcus	2	coccus	+	37	45
Sporomusa	7	rod	+	30	42 - 48
Syntrophococcus	1	coccus	-	37	52

11.3. CO_2 METABOLISM AND ACETOGENESIS

Three metabolic processes are used by acetogens to produce acetate from CO_2:

i. The acetyl-CoA pathway
ii. Glycine synthetase pathway
iii. The reductive tricarboxylic acid cycle

11.3.1. The acetyl-CoA Pathway

It is also referred to as the **Wood-Ljungdahl pathway**. This pathway is distinguished from the other two pathways in that acetyl-CoA is the first two carbon compound formed from two C_1 precursors.

The acetyl-CoA pathway reactions can be divided into four steps. The first is the series of reactions that result in the reduction of CO_2 to a bound methyl group [CH_3]. The methyl group is then transferred to the enzyme carbon monoxide dehydrogenase

$$CO_2 + 3\,H_2 + H^+ \longrightarrow [CH_3] + 2H_2O$$

The second step reactions involve reduction of second molecule of carbon dioxide to carbonyl group of acetate. For these reactions, during the autotrophic growth, the electrons are provided by hydrogen gas via hydrogenase.

$$CO_2 + H_2 \longrightarrow [CO] + H_2O$$

In the third step, the bound CH_3 and CO are condensed by carbon monoxide dehydrogenase to form bound acetyl [CH_3CO]

$$[CO] + [CH_3] \longrightarrow [CH_3CO]$$

In the fourth and final step the bound acetyl reacts with bound CoASH (CoAS), to form acetyl CoA which is released from the enzyme

$$[CH_3CO] + [COAS] \longrightarrow CH_3COSCOA$$

The acetyl CoA thus produced is converted to acetate via phosphotransacetylase and acetate kinase and excreted, or it is assimilated into cell material. Let us now discuss the acetyl CoA pathway in detail taking *Clostridium thermoaceticum* as a model.

11.3.1.1. The acetyl CoA pathway in Clostridium thermoaceticum

The acetyl CoA pathway was first investigated in *C. thermoaceticum,* an acetogenic bacterium that converts one mole of glucose to three moles of acetate as shown below:

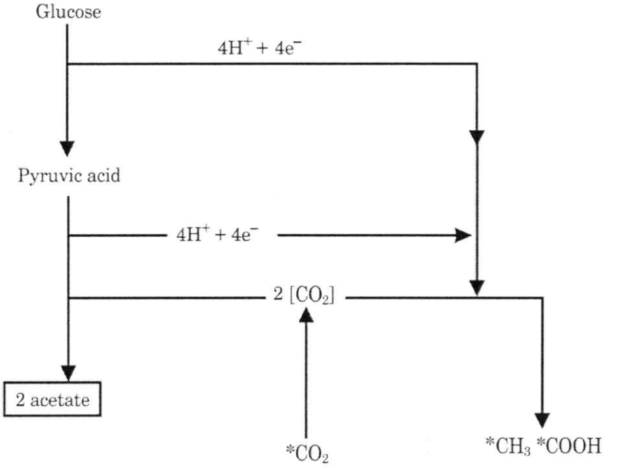

In the reactions of the above diagram, the production of acetate is used as an electron sink during fermentation rather than for acetotrophic growth. The pathway is illustrated in Fig. 11.1. In this pathway, glucose is converted to pyruvate via EMP pathway (reaction 1). Two of the acetates are synthesized from the decarboxylation of pyruvate using pyruvate ferredoxin oxido-reductase, phosphotransacetylase and acetate kinase (reactions 2-4). The third acetate is synthesized from CO_2 using the acetyl CoA pathway.

The acetyl CoA pathway requires two molecules of CO_2. One of these CO_2 is removed from pyruvate is not set free, but instead becomes bound to the enzyme cabonmonoxide dehydrogenase (CODH), where it will be used for acetate synthesis (reaction 2 a). This bound CO_2 will eventually become the carbonyl group in acetyl CoA. The second pyruvate is decarboxylated to release free CO_2 (reaction 2b). In reactions 5 – 9, the free CO_2 is reduced to bound methyl. The free CO_2 first becomes reduced to formate (HCOOH) by formate dehydrogenase (reaction 5). The formate is then attached to tetrahydrofolic acid (THF) in an ATP–dependent reaction to make formyl THF(HCO-THF) (reaction 6). The HCO-THF is then dehydrated to form methenyl – THF (CH-THF) (reaction 7). The methenyl THF is reduced by NADPH to methylene – THF (CH_2-THF) (reaction 8). The methylene – THF is reduced by ferredoxin to methyl – THF (CH_3 – THF) (reaction 9). This reaction is thought to be coupled to ATP formation by a mechanism involving ΔP. The methyl group is then transferred to a corrinoid enzyme [Co]-E, which transfers the methyl group to the CO–CODH (reactions 10 and 11). The CODH makes the acetyl [CH_3CO] moiety, (reaction 12). The CODH also has a binding site for CoASH and synthesizes acetyl CoA from bound intermediates (reaction 13 and 14). The CODH can also bind free CO_2 and reduce it to the level of carbon monoxide (reaction 16).

A perusal of Fig. 11.1. raises certain issues as how to draw a border line between the heterotrophic and autotrophic growth. Nevertheless, the entry of CO_2 at reactions 5 and 15, suggests, that the acetyl CoA pathway allows autotrophic growth on CO_2 and H_2. CO_2 is reduced to [CH_3] and [CO] (reactions 5 to 10 and reaction 15). The [CH_3] and [CO] combine with CoASH to form acetyl CoA (reaction 11 – 14).

Carbon monoxide dehydrogenase (CODH) is a crucial enzyme for acetogenesis for autotrophic growth in anaerobes, and for methanogenesis from acetate. It is a multifunctional enzyme. This enzyme catalyses the breakage and formation of the carbon – carbon bond between the methyl group [CH_3] and the carbonyl group [CO] in the acetyl moiety. It also catalyses the breakage and formation of the thioester bond between CoASH and the carboxyl group of the acetyl moiety. The following exchange has also been demonstrated to be catalysed by CO dehydrogenase

$$[1 - C]\ \text{Acetyl CoA} + \text{CO} \longrightarrow \text{CO} + 1 - \text{C acetyl CoA}$$

The three exchange reactions described above mean that the CODH is capable of disassembling acetyl–CoA into its components [CH_3], [CO] and [CoASH] and resynthesizing the molecule.

11.3.1.2. Importance of acetyl CoA pathway

The primary use of the acetyl – CoA pathway by acetogens is for the fixation of CO_2 and CO, but they also use this pathway for the metabolism of C_1 compounds, including formate, formaldehyde, methanol and methyl chloride, and also for more complex C_1 donors such as methoxylated aromatic compounds, including syringate and vanillate oxalate, glyoxylate and aromatic aldehydes, and various alcohols. The suggested entrances of the C_1 moieties of these compounds into acetyl CoA pathway are shown in Fig. 11.1. The utilization of methoxyl group of methylated aromatic compounds as methyl group precursors of acetate by acetogenic bacteria is of particular interest because these compounds are degradation products of lignin which suggests that acetogens are involved in the degradation of lignin. The involvement of acetyl CoA pathway in the metabolism of the methoxylated aromatic comounds has been established for *A. woodii* and *C. thermoaceticum*. Methanol is oxidised in *C. thermoautotrophicum* by a PQQ – dependent methanol dehydrogenase and it may enter the acetyl–CoA pathway at several levels.

Acetate is a key intermediate in bacterial metabolism in the anaerobic environment. An example is the degradation of cellulose to methane and CO_2. In anaerobic environments, it has been estimated that approximately 10^{10} metric tons of acetate is metabolised per year.

Approximately, 10^{10} or more of the acetate produced is generated from the reduction of CO_2 with hydrogen. Production of acetate from H_2 and CO_2 has been estimated to be approximately 2.3×10^6 and 1.3×10^4 metric tons per day in the hind gut of termites and in the large bowel of humans, respectively. The widespread use of the acetyl CoA pathway by distinctly different groups of microorganisms suggests the global importance of this pathway in the metabolism of anaerobic bacteria. Recently, prebiotic significance of the acetyl CoA pathway in the chemoautotrophic origin of life was suggested.

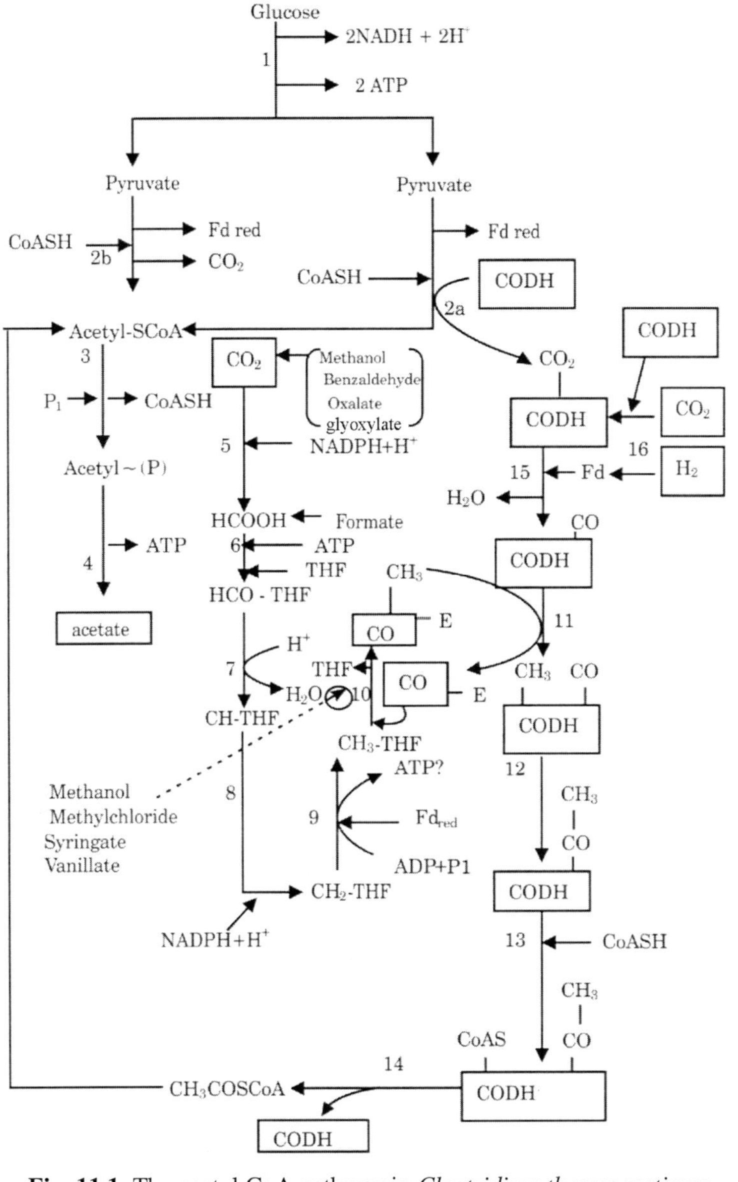

Fig. 11.1. The acetyl CoA pathway in *Clostridium thermoaceticum*

11.3.1.3. Incorporating acetyl–CoA into cell material

For acetogens growing autotrophically, part of the acetyl–CoA that is produced must be incorporated into cell material. The glyoxylate cycle which is generally associated with aerobic metabolism, is not present in these organisms. Therefore, a different pathway must operate. As shown below, the acetyl CoA is reductively carboxylated to pyruvate via pyruvate synthase, a ferredoxin – linked enzyme. The pyruvate can be phosphorylated to phosphoenol pyruvate using PEP synthetase. The phosphoenol pyruvate is used for biosynthesis in the usual pathway (Fig. 11.2). The PEP has two fates. Some of it enters the gluconeogenic pathway and some is carboxylated to oxaloacetate (OAA) via PEP carboxylase. In some acetogens, OAA is reduced to α-ketoglutarate via an incomplete reductive citric acid pathway. Still in some, α-ketoglutarate is synthesized in an incomplete oxidative pathway via citrate and isocitrate.

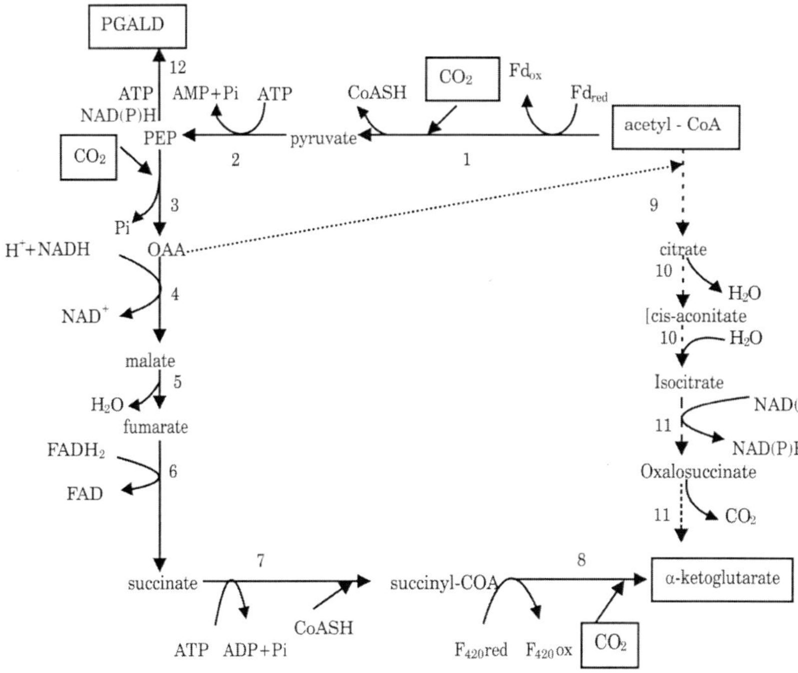

Fig. 11.2. Assimilation of acetyl-CoA into cell material.

11.3.2. The Glycine Synthetase Pathway

Glycine synthetase pathway (Fig. 11.3) has been reported in *Clostridium acidiurici, C. cylindrosporum* and *Streptococcus glycinophilus*. In this pathway, a molecule of carbon dioxide is reduced to formate by the action of formate dehydrogenase and the formate is further converted to reactions analogous to those of the acetyl–COA pathway to methyl-tetrahydrofolate. The last substance combines with CO_2 and NH_3 under the influence of glycine synthetase (decarboxylase), to form glycine. Methylation of glycine, with serine hydroxymethylase can yield serine, and the action of glycine reductase on glycine may yield acetate. In the absence of selenium, serine formation, while in the presence of selenium acetate formation are favoured. As the work progresses, it is assumed that a system of this nature will also be found in other autotrophs. Carbondioxide fixation by acetogens through reductive tricarboxylic acid cycle has already been discussed in photoautotrophy (Chapter 9).

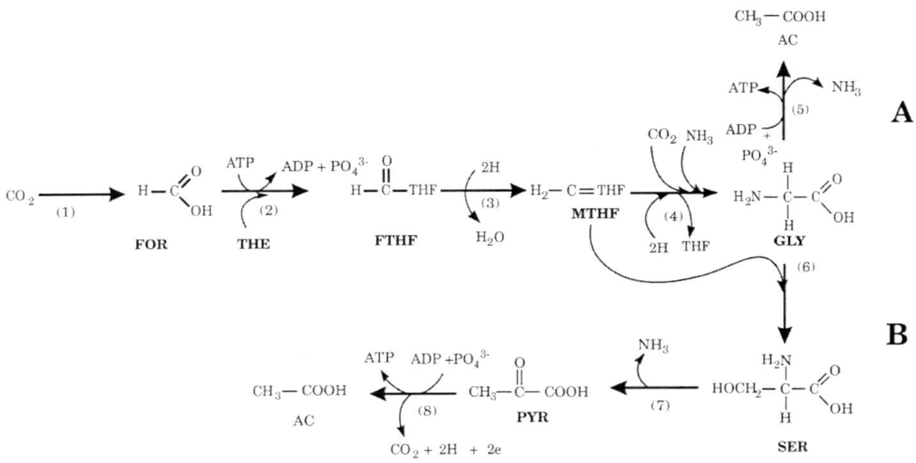

Fig. 11.3. The glycine synthatase pathway for acetate synthesis from carbon dioxide in *Clostridium acidiurici* and similar organisms.

11.3.3. Utilization of Electron Acceptors other than CO_2

Acetogens can use several alternate electron acceptors, such as nitrate, thiosulfate, dimethyl sulfoxide and fumarate instead of CO_2. When these compounds are used as terminal electron acceptors, the acetogens do not produce acetate as a metabolic end product. Instead, reduced forms of terminal electron acceptors (*e.g.* nitrite, sulfide, dimethyl sulfide and succinate) are produced as metabolic end products. However, it is not clearly known how the acetogens conserve energy by reducing these terminal electron acceptors.

11.4. FURTHER READING

Das, A. and L.G Ljungdahl (2000), Acetogenesis and acetogenic bacteria. In : Encyclopedia of Microbiology Vol. I, J. Lederberg (ed), San Diego, Academic Press, pp. 18-27.

Das, A., Ivey, D.M. and Ljungdahl, L.G. (1997), *J. Bacteriol.*, **179 :** 1714-1720.

Drake, H.L (1994), "Acetogenesis" Chapman & Hall, New York.

Fauque,G., LeGall, J. and Barton LL (1991), Variations in autotrophic life (eds. J.M. Shively and LL Barton, pp. 271-337, Academic Press, New York.

Ljungdahl, L.G. (1986), *Ann. Rev. Microbiol.* **40 :** 415-450.

Wood, H.G. and Ljungdahl, L.G. (1991), In variations in autotrophic life (J.M. Shively and LL Barton (eds.)) pp. 210-250, Academic Press, New York.

Chapter - 12

Chemoheterotrophy - Methanogens

12.1. INTRODUCTION

In nature, methane is produced by two methods: abiotic and biotic. Methane is produced abiotically by the degradation of fossil plants and animals in the earth crust and also from petroleum and coal. A large quantities of methane is also produced biotically by bacteria especially archaea.

The biological producers of methane are known as **methanogens** and the process of methane production is known as **methanogenesis**. Methanogens are obligate anaerobes that obtain energy by converting CO_2, H_2, formate, methanol, acetate and other compounds to methane. This is the largest group of archaebacteria. Biological methanogenesis has a significant role in global carbon cycle (Table 12.1). This process is one of the several anaerobic degradative processes, that complement aerobic degradation, by utilizing alternative electron acceptor in habitats, where O_2 is not available. The hierarchy of electron acceptors is O_2 >NO_3^- > Fe^{+3} >Mn^{+4} So_4^{-2} > CO_2, where reduction of CO_2 to form CH_4 is the ultimate terminal process, catalyzed by degradative microbial consortia in the absence of all the electron acceptors.

12.2. GENERAL CHARACTERISTICS

Methanogens, in general, exhibit the following characteristics:

i. All methanogens are obligate anaerobes.

ii. Methanogens are single cell prokaryotic microorganisms and belong to archaebacteria.

iii. They lack two enzymes – catalase and superoxide dismutase

iv. Methanogens occur in highly reduced anaerobic environment and constitute the last trophic level in anaerobic food chain.

v. Nutritionally, they may be chemoautotrophs or chemoheterotrophs.

vi. They possess a collection of unique coenzymes that function as co-catalysts in a number of special enzyme systems.

vii. The energy is conserved by chemiosmotic process with an unusual mechanism of forming a transmembrane ion gradient coupled to an exergonic methyl transfer reaction.

viii. Based upon different substrate utilization, the methanogens are classified into three groups: hydrogenotrophic, acetotrophic and methylotrophic.

Table 12.1. Estimates of CH_4 released into the atmosphere

Source	CH_4 emission (10^{12} g/year)
Biogenic	
Ruminants	80 – 100
Termites	25 – 150
Paddy fields	70 – 120
Natural wetlands	120 – 200
Landfills	5 – 70
Oceans and lakes	1 – 20
Tundra	1 – 5
Abiogenic	
Coal mining	10 – 35
Natural gas	10 – 30
Industrial & pipeline losses	15 – 45
Biomass burning	10 – 40
Methane hydrates	2 – 4
Valconoes	0.5
Automobiles	0.5
Total	350 – 820
Total biogenic	302 – 665 (81-86%)
Total abiogenic	48 – 155 (14-19%)

12.3. HABITAT

Methanogens require highly reduced, anaerobic conditions for their growth. They are ubiquitous in the environment and occur in many habitats, including sewage digesters, waste land fills, marine and fresh water sediments (*M. bryantii, M. bankerii* and *M. voltae*), algal mats, wet wood of living trees (*M. arboriphilus*), mammalian gastrointestinal tracts, ruminants, termite guts (*M. arboriphilus* and *M. bryantii*), geothermal wents and deep surface rocks.

As organic matter accumulates on lake and river basins, the sediment becomes anaerobic. In such eutrophic environment the methanogens inhabit and as a result methanogenesis becomes the ultimate degradative process. The marine sediments receive large amounts of organic matter, in such eutrophic coastal regions and submarine trenches three member methanogenic consortium is involved. In these habitats sulfate reducing bacteria out compete methanogens as the terminal members of the anerobic consortium.

Ruminant animals include both domestic (cows, sheep, camel) as well as wild (deer, bison, giraffes) animals. These animals have a large chamber in their upper alimentary canal called the rumen which is similar to anaerobic bioreactor. This is an ideal habitat for methanogens.

All known termites harbour a dense microbial community of anaerobic bacteria of which methanogens constitute an important component. *Methanobrevibacter arboriphilus* and *M. bryantii* are frequently isolated from the gut of termites. Human colon where undigested polymers, sloughed off intestional epithelium and mucin are dewatered and methane production occurs by bacterial consortia. Methanogenesis occurs in 30-40% of the human population. *Methanosphaera stadmanae* is a common inhabitant of human intestine. Methanogens are present as endosymbionts in many free – living marine and fresh water anaerobic protozoa, where they are often closely associated with hydrogenosomes, organelles that produce H_2, CO_2 and acetate from the fermentation of polymeric substrates. Some more methanogenic habitats include marshes, rice paddies, rotting heart-wood of trees, land fills, terrestrial hot springs and deep-sea hydrothermal vents. Methanogenesis also occurs in high saline environments, such as Great Salt Lake, Mono Lake, California and Solar Salt ponds. Methanogens from these environments generate CH_4 from methylated amines and dimethyl sulfide, which are synthesized by animals and plants as osmolytes.

Table 12.2. Characteristics of methanogenic bacteria

Taxonomic epithet		Morphology	Substrates*
Order – 1	**Methanobacteriales**		
Family:	Methanobacteriaceae		
Genus:	*Methanobacterium*	rod	H,F,2P,2B,CP
	Methanobrevibactes	cocco bacillus	H,F
	Methanosphaera	coccus	H/Me
Family:	Methanothermaceae		
Genus:	*Methanothermus*	rod	H
Order –2	**Methanococcales**		
Family:	Methanococcaceae		
Genus:	*Methanococcus*	irreg,coccus	H,F
Order – 3	**Methanomicrobials**		
Family:	Methanomicrobiaceae		
Genus:	*Methanoculleus*	irreg, coccus	H,F,2B,2P,CP
	Methanomicrobium	curved,rod	H,F
	Methanolacinia	irreg, rod	H,F,2B,2P,CP
	Methanospirillum	sheathed,spiral	H,F
	Methanogenium	irreg, coccus	H,F,E,1P,2P,2B
	Methanofollis	irreg, coccus	H,F,E,1P,2P,2B
	Methanocalculus	irreg, coccus	H,F
Family:	Methanocorpusculaceae		

Taxonomic epithet		Morphology	Substrates*
Genus:	*Methanocorpusculum*	irreg, coccus	H,F,E,1P,2P,2B
Family:	Methanoplanaceae		
Genus:	*Methanoplanus*	irreg, coccus	H,F,1P
Family:	Methanosarcinaceae		
Genus:	*Methanosarcina*	irreg,coccus	AC,NE,MA,DMS,MMP
	Methanolobus	irreg, coccus	MA,ME,DMS
	Methanococcoides	irreg, coccus	MA,ME
	Methanohalophilus	irreg, coccus	MA,ME,DMS
	Methanohalobium	irreg, coccus	ME,MA
	Methanosaeta	sheathed rod	AC
Order-4	**Methanopyrales**		
Family:	Methanopyraceae		
Genus:	*Methanopyrus*	sheathed rod	H

*H, hydrogen/carbon dioxide; F, formate, AC, acetate; ME, methanol; MA, methylamine; H/ME, methanol reduction with hydrogen; E, ethanol; IP, 1 – propanol; 2P, 2-propanol; 2B, 2-butanol; CP, Cyclopentanol; DMS, dimethyl sulfide; MMP, methylmercaptopropionate.

12.4. DIVERSITY AND PHYLOGENY

Methanogens are members of the Archaea, one of three domains of life proposed by C. Woese on the basis of 16 S rRNA sequence. Although the methanogens are a phylogenetically coherent group and have a limited substrate range, they are morphologically and physiologically diverse (Table 12.2).

1. They include psychrophilic species from Antarctic that grow at 1.7°C to extremely thermophilic species from deep submarine vents that grow at 113°C.
2. Acidophiles from marine vents that grow at pH 2.0-5.0 to alkaliphiles from alkaline lake sediments that grow at pH 10.3.
3. Species from fresh water lake sediments that grow at saline concentration below 0.1 M to extreme halophiles from Solar Salt ponds that grow at nearly saturated NaCl concentration.
4. Autotrophs that utilize only CO_2 for cell carbon and heterotrophs that utilize reduced carbon compounds.

Despite the range and diversity of growth habitats, where methanogens are found, methanogens have one common attribute, they all generate methane during their growth. Methane formation is carried out by unique enzyme methyl coenzyme M reductase.

12.5. PHYSIOLOGY

All methanogens generate CH_4 during their growth by three basic catabolic pathways (Fig. 12.1).

1. autotrophic CO_2 reduction with H_2, formate or secondary alcohols,

2. acetotrophic cleavage of acetate,
3. methylotrophic dismutation of methanol, methylated amines or methylthiols.

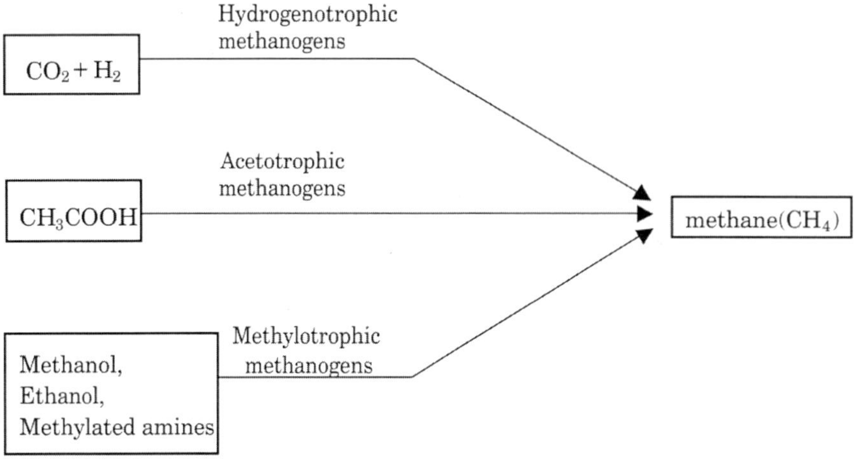

Fig. 12.1. Three different routes of methanogenesis

Table 12.3. Reactions and free energy yields from methanogenic substrates

Substrate	Reaction	$\Delta G^{o'}$ (kJ/mol CH_4)
Hydrogen/CO_2	$4CH_2 + HCO_3^- \longrightarrow CH_4 + 3H_2O$	-135
Formate	$4HCOO^- + 4H^+ \longrightarrow CH_4 + 3CO_2 + 2H_2O$	-145
Carbon monoxide	$4CO + 5H_2O \longrightarrow CH_4 + 3HCO_3^- + 3H^+$	-196
Ethanol	$2CH_3CH_2OH + HCO_3^- \longrightarrow 2CH_3COO^- + H^+ + CH_4 + H_2O$	-116
Hydrogen/methanol	$CH_3OH + H_2 \longrightarrow CH_4 + H_2O$	-113
Methanol	$4CH_3OH \longrightarrow 3CH_4 + HCO_3^- + H_2O + H^+$	-105
Trimethylamine	$4CH_3NH^+ + 9H_2O \longrightarrow 9CH_3 + 3HCO_3^- + 4NH_4 + 3H^+$	-76
Dimethylsulfide	$2(CH_3)_2S + 3H_2O \longrightarrow 3CH_4 + HCO_3^- + 2H_2S + H^+$	-49
Acetate	$CH_3COO^- + H_2O \longrightarrow CH_4 + HCO_3^-$	-31
Pyruvate	$4CH_3COCOOH + 2H_2O \longrightarrow 5CH_4 + 7CO_2$	-31

Depending on the ecosystem, different substrates may be available to support the growth of methanogens. Carbon dioxide and molecular hydrogen are common, and nearly all methanogens are able to obtain energy using H_2 in the reduction of CO_2 to methane. Nonetheless, nearly two thirds of the methane formed in the environment is derived from the cleavage of acetic acid. This methanogenic pathway thus becomes increasingly thermodynamically favourable as H_2 levels in the environment decrease.

12.5.1. Autotrophic Reduction to CH_4

The reduction of CO_2 during growth on H_2 proceeds according to the stoichiometric relationship shown below

$$CO_2 + 4H_2 \longrightarrow 2H_2O + CH_4$$

The different enzyme cofactors utilized in these pathways are categorized, generally, as either electron carriers or one–carbon group carriers and are shown structurally in Fig. 12.2. In this pathway (Fig. 12.3), the CO_2 is first condensed with methanofuran (MFR) and reduced to formyl MFR (reaction-1). The formyl group is then transferred to tetrahydro methanopterin (H_4MPt), to form formyl – H_4MPt (reaction 2). Then a dehydration produces methenyl – H_4MPt (reaction 3). The methenyl – H_4MPt is reduced to methylene – H_4MPt (reaction – 4), which is reduced to methyl – H_4MPt (reaction 5). The pathway diverges at the methyl level. One branch synthesizes acetyl-CoA in reactions quite similar to the synthesis of acetyl-CoA in the bacteria and the other branch synthesizes methane. For the synthesis of acetyl-CoA, the methyl group is transferred to a corrinoid enzyme. A second molecule of CO_2 is reduced by carbon monoxide dehydrogenase (CODH) to bound carbon monoxide (CO) (reaction 14). The CODH then catalyzes the synthesis of acetyl CoA from [CH_3], [CO] and CoASH (reactions 11 – 13).

The reactions described above are meant for incorporating the methyl group [CH_3], derived from CO_2, into acetyl-CoA. Interestingly, methanogens can also reduce methyl group [CH_3] to methane (CH_4) in a series of reactions not found among the bacteria. The production of methane yields ATP and is the only means of ATP formation for the methanogens. For methane synthesis, the methyl group is transferred from CH_3-H_4MPt to CoMSH to form CH_3 – SCoM (reaction 6). The terminal reduction is catalyzed by the methyl reductase system (reaction 7 and 8). The methyl reductase system has two components. One component is methyl reductase that reduces CH_3 – SCoM to CH_4 and CoM-S-S-HTP. The electron donor for the methyl reductase is HTP-SH. A nickel containing tetrapyrrole (F_{430}) is an electron carrier that is part of the methyl reductase. The second component is a FAD – containing heterodisulfide reductase that reduces CoM-S-S-HTP to CoMSH and HTP – SH.

A critical perusal of the acetyl CoA pathway of methanogens and other bacteria such as acetogens reveals certain differences. These differences have to do with the cofactors and the utilization of formate. The differences between the two pathways are:

1. The carriers for the formyl and more reduced C_1 moieties in the methanogens is not tetrahydrofolic acid (THF), but rather tetrahydromethanopterin (H_4MPt), a molecule resembling THF.
2. Methanogens do not reduce free CO_2 to formate nor do they incorporate formate into H_4MPt to form formyl – H_4MPt. Instead, they fix CO_2 into a C_1 carrier called methanofuran (MFR) and reduce it to formyl MFR. The formyl group is then transferred to H_4MPt. It may be recalled here that the acetogenic bacteria reduce free CO_2 to formate and then incorporate the formate into formyl – THF.

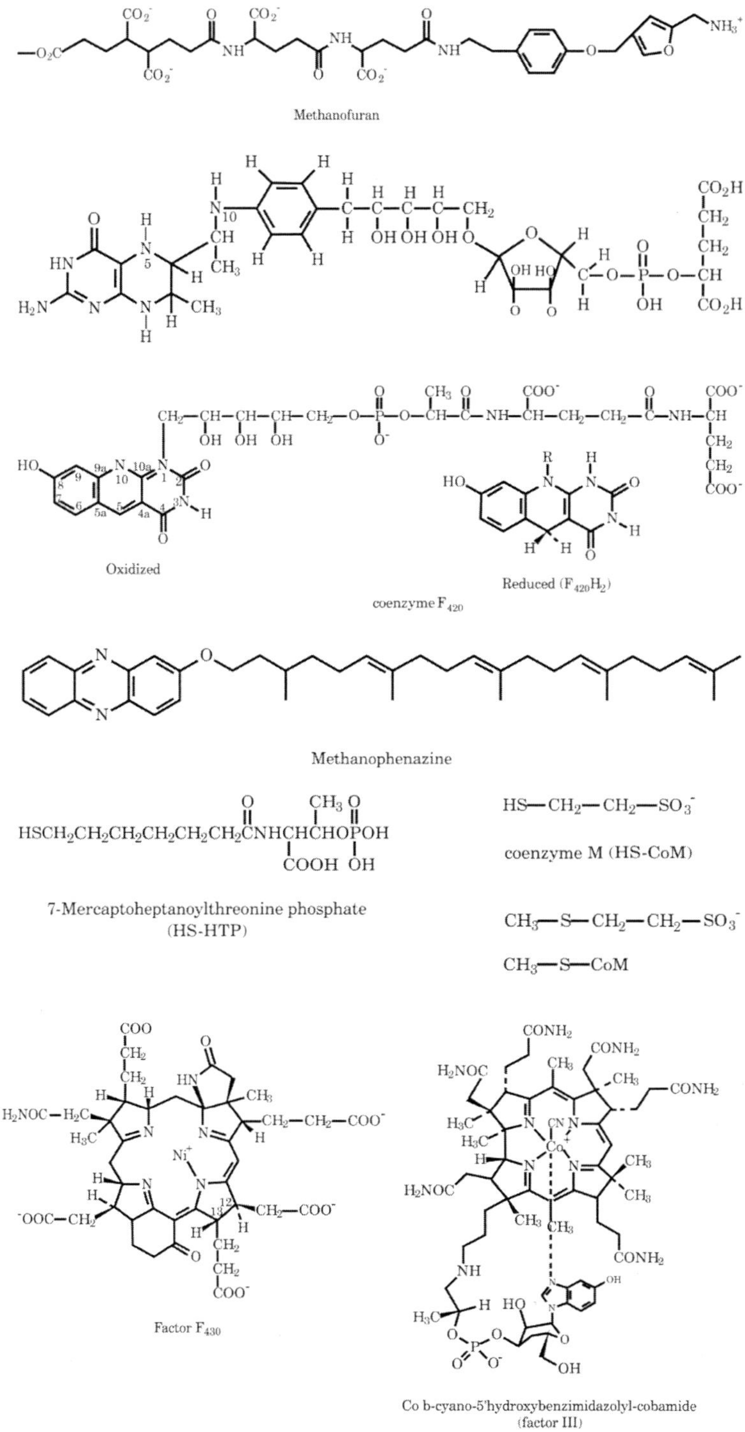

Fig. 12.2. Different enzyme cofactors of methanogens

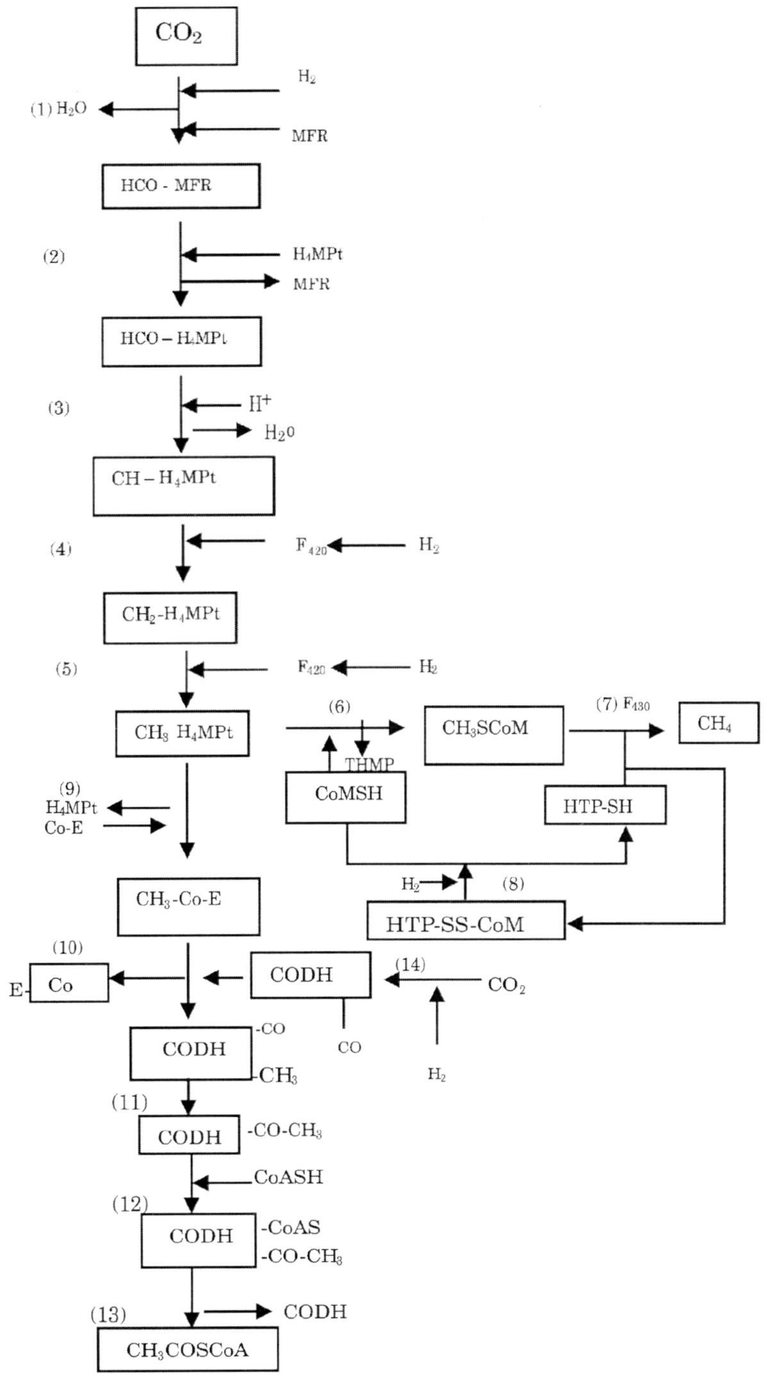

Fig. 12.3. Methanogenesis through autotrophic reduction of CO_2

12.5.2 Acetate Cleavage to CH_4 and CO_2

Acetic acid is a significant product of the decomposition of complex organic materials in anaerobic environments. Further, degradation of acetate accounts for the major portion of the total methane produced biologically. Acetic acid is also oxidized by other anaerobic microorganisms, however, these convert both carbon atoms of acetate to CO_2, which requires a separate oxidizing substrate (*e.g.,* sulfate, nitrate). In contrast, methanogens are capable of disproportionating acetate by a process in which oxidation of the carboxyl group to CO_2 is coupled to reduction of the methyl group to methane (Fig. 12.4).

$$CH_3COOH \longrightarrow CO_2 + CH_4$$

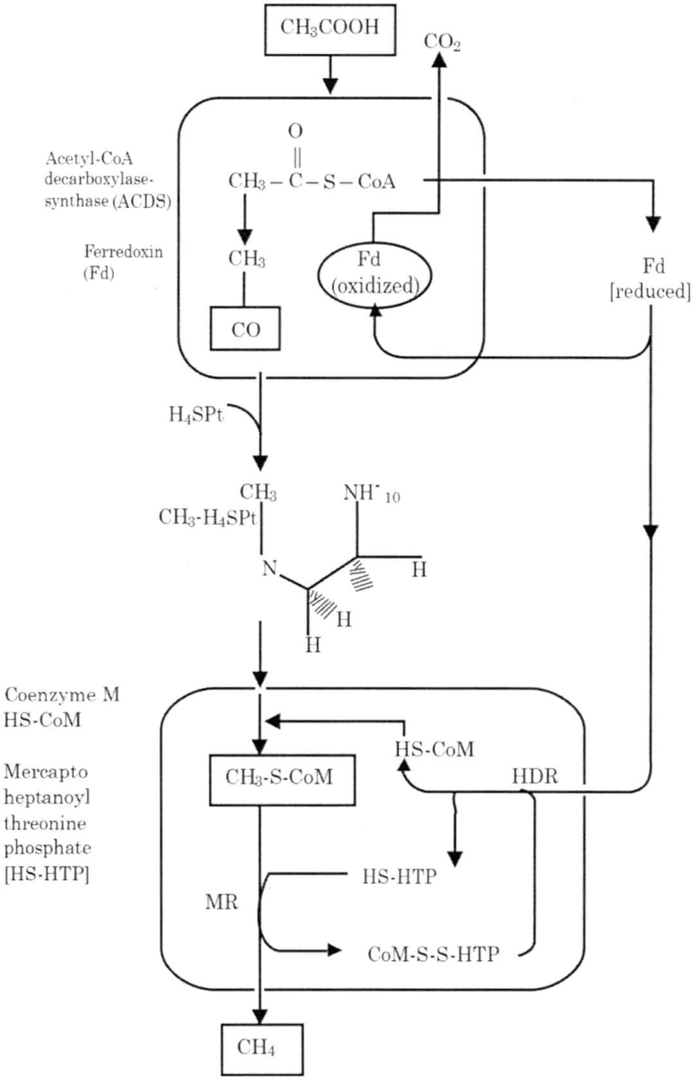

Fig. 12.4. Methanogenesis through acetate pathway.

The central event in the pathway of acetate conversion is the cleavage of the acetate carbon-carbon bond. This reaction occurs following the activation of acetate to form acetyl CoA by the enzymes acetate kinase and phosphotransacetylase. Thereafter, the cleavage of acetyl-CoA takes place in a highly unusual reaction, in which two types of bonds are broken, the thioester C – S and the acetate C – C bond. The reaction is carried out by a multienzyme complex designated as acetyl CoA decarboxylase/ synthase (ACDS) and was formerly called the CO dehydrogenase – corrinoid enzyme complex. In the action of the complex on its substrate acetyl CoA, the cleavage of the carbon – carbon bonds is accompanied by immediate oxidation of carbonyl group to CO_2, producing electrons that are donated directly to reduce ferredoxin. The corrinoid cofactor serves as an intermediate methyl group carrier, acting to remove the methyl group from the site of cleavage and to donate it finally to the substrate tetrahydrosarcinapterin (H_4SPT), forming N^5 – methyl tetrahydrosarcinapterin (CH_3-H_4SPt). The process of carbonyl group excision is not fully understood. The overall reaction is reversible and in environments containing sufficient reducing power, the ACDS complex is used in the reverse direction to synthesize acetyl groups needed for autotrophic growth of methanogens on single-carbon substrates. In this process, carbon fixed in the form of acetyl – CoA is further reduced to generate pyruvate for entry into biosynthetic pathways forming other cellular components and macromolecules, *i.e.,* proteins, polysaccharides, lipids and nucleic acids.

12.5.3 Reduction of Methylated Substrates to CH_4

Methylated compounds, such as methanol and methylamines, are produced in the decay of aquatic animals and plants, and serve as the growth substrates for methylotrophic methanogens, such as those found in the family *Methanosarcinaceae*. Methylotrophic growth generally involves the disproportion of methylated substrate such that, for every three methyl groups reduced to methane (each by two electrons), one methyl group undergoes 6 – electron oxidation to CO_2

$$4CH_3OH \longrightarrow 3\,CH_4 + CO_2 + 2H_2O$$

In the above overall reaction, oxidation reactions remain poorly understood. However, reduction of methanol to methane is more thoroughly understood and illustrated in Fig. 12.5.

Initially, methanol is activated by an enzyme designated methanol : 5-hydroxy benzimidazolyl-transferse (MT1) which transfers the methyl group to a corrinoid protein specific for the methanol pathway. This reaction is important as it results in the breakage of a stable carbon – oxygen bond, with the formation of a much less stable carbon – cobalt bond. The methyl group is then transferred to coenzyme M by the enzyme methylcobomide : coenzyme M methyl transferase (MT2).

A different set of MT1 and corrinoid proteins is employed in the metabolism of each different methylated growth substrate. Several MT2 proteins also exist and exhibit metabolic specificity in that different MT2 isozymes function exclusively in different pathways. Methyl – CoM is the common product formed from the various methylotrophic substrates. Reduction of methyl – CoM to methane proceeds as described under the pathway for CO_2 reduction, but the major electron source most often originates from concurrent methyl group oxidation instead of from hydrogen. The electrons serve to reduce CoM – S – S – HTP, which acts as the penultimate oxidant, is continuously being regenerated by the action of methyl – CoM reductase. Evidence indicates that energy available from electron transport is conserved via a chemiosmotic mechanism. However, individual components that make up the electron transport chain in this pathway and others still remain to be identified and characterized.

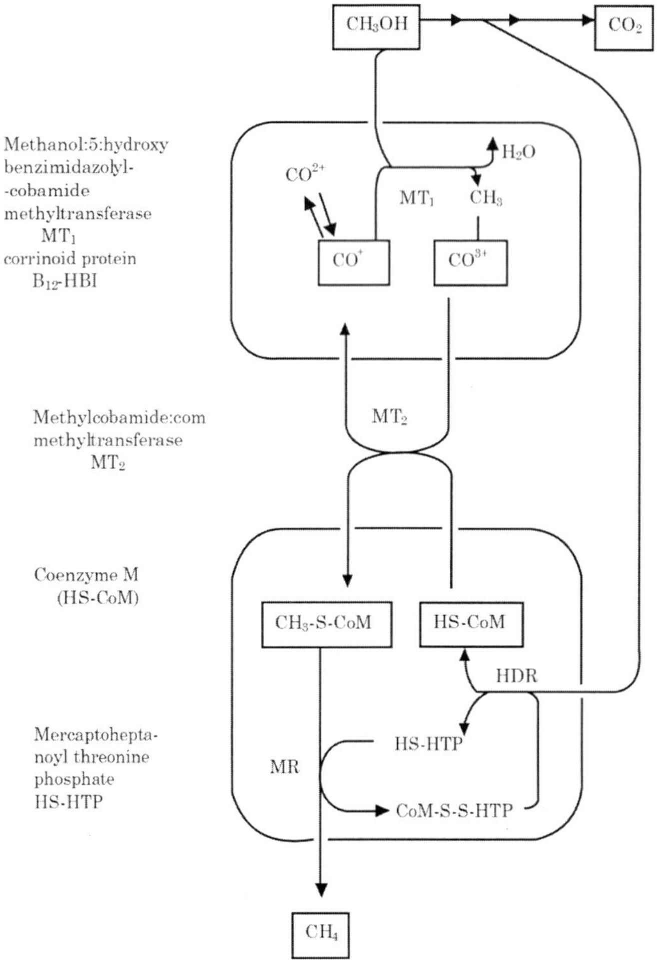

Fig. 12.5. Methanogenesis from methanol in *Methanosphaera stadmanae*

12.6. ECONOMIC IMPORTANCE

The methanogenic archaea have a pivotal role in the global carbon cycle. These bacteria are of great practical importance since methane is a clean burning fuel and an excellent energy source. For many years, sewage treatment plants have been using the methane, they produce as a source of energy for heat and electricity. An anaerobic digester will degrade particulate wastes like sewage sludge to H_2, CO_2 and acetate. CO_2 reducing methanogens form CH_4 from CO_2 and H_2, while aceticlastic (acetotrophic methanogens) methanogens cleave acetate to CO_2 and CH_4. About 2/3 of the methane produced by an anaerobic digester comes from acetate. A kilogram of organic matter can yield up to 600 liters of methane. Rumen methanogens are so active that a cow belches 200 – 400 liters of methane a day.

It is quite likely that future research will greatly increase the efficiency of methane production and make methanogenesis an important source of pollution free energy. The applications of methanogens in biotechnology is the limited production of biogas. It is produced by using cow dung, agricultural wastes, domestic wastes, municipal wastes, industrial wastes, as the substrates by designing suitable biogas plants. Other potential applications of methanogens include the production of novel pharmaceuticals, corrinoids and thermostable enzymes. To date, methanogens have yielded only a few restriction endonucleases as commercial products.

Methanogenesis also can be an ecological problem. Methane absorbs infrared radiation and thus is a green house gas. There is evidence that atmospheric methane concentration has been increasing over the last 200 years. Methane production may significantly promote future global warming. Recently, it has been discovered that methanogens can oxidize Fe^{o} and use it to produce methane and energy. This means that methanogens growing around buried or submerged iron pipes and other objects may contribute significantly to iron corrosion.

However, significant advances in the physiology and biochemistry of methanogens over the past two decades, combined with the recent developments in gene transfer systems and advances in genome sequencing, make the application of methanogens for biotechnology more plausible in the near future.

12.7. FURTHER READING

Chawla, O.P. (1986), Advances in biogas technology. ICAR, New Delhi, P 144.

Daniels, L. (1992), Biotechnological potential of methanogens (review) *Biochem. Soc. Symp* **58:** 181 – 193.

Deppenmeier, U., Muller, V. and Gottschalk, G. (1996), Pathways of energy conservation in methanogenic Archaea. *Arch. Microbiol* **165**: 149 – 163.

DiMarco, A.A., Bobik, T.A., and Wolfe, R.S. (1990), Unusual coenzymes of methanogens . *Annu. Rev. Biochem.* **59:** 355 – 394.

Ferry, J.G. (1992), Biochemistry of methanogenesis. *Crit. Rev. Biochem. Mol. Biol.* **27**: 473 – 503.

Ferry, J.G. (ed.) (1993), Methanogenesis : Ecology, Physiology, Biochemistry & Genetics. Chapman & Hall, New York.

Grahame, D.A. and Gencic, S. (2000), Methane Biochemistry. In J Lederberg (ed). Encyclopedia of Microbiology 2nd edn. Vol. III, Academic Press, San Diego, pp. 188 – 198.

Nagamani, B. and Ramaswamy, K. (1999), Biogas production technology : An Indian perspective *Curr. Sci.* **77**: 44 – 55.

Sowers, K.R. (2000), Methanogenesis. In J. Lederberg (ed). Encyclopedia of Microbiology 2nd edn. Vol. III, Academic Press, San Diego, pp. 204 – 226

Thauer, R.K. (1998), Biochemistry of methanogenesis: A tribute to Marjory Stephenson. *Microbiology* **144 :** 2377 – 2406.

■■■■

Chapter - 13

Chemoheterotrophy - Methylotrophs

13.1. INTRODUCTION

Methylotrophy is the ability of organism to utilise one carbon compounds (but not CO_2) as their sole carbon source. This includes compounds such as methane, methanol, formate, carbon monoxide, cyanide and chloromethane. The definition is usually extended to the utilisation of compounds that contain more than one carbon atom, but no carbon-carbon bonds, such as dimethylamine, trimethylamine and dimethyl sulphite. Methylotrophy is observed in all major groups of microorganisms like bacteria, yeast and filamentous fungi. Methylotrophs are found both in aerobic and anaerobic environments.

Methane is produced by methanogenic bacteria under a number of diverse environments of the biosphere. Methane oxidising bacteria (methylotrophs) are generally found on the fringes of anaerobic environments and probably responsible for oxidising methane derived from methanogens. Methylotrophs appear to be ubiquitous and have been isolated from many different habitats. Both psychrophilic and thermophilic methylotrophs are reported.

13.2. TAXONOMY AND SYSTEMATIC POSITION

Methylotrophs are further differentiated into two categories. (1) Those that utilize only methane are designated as **methanotrophs**. (2) The microorganisms utilizing C_1 compounds other than methane are usually referred to as **methylotrophs**. However, use of methylotrophs for both the groups of organisms has become a common practice. Some of the characteristic features of these groups of organisms are precised in Table 13.1. and 13.2.

Table 13.1. Characteristic features of certain methanotrophs

Genus	Phylogeny	Formaldehyde assimilation pathway	Enzyme type*
Methylobacter	Gamma Proteobacteria	RuMP	pMMO
Methylocaldum	Gamma Proteobacteria	RuMP/Serine	pMMO
Methylococcus	Gamma Proteobacteria	RuMP	pMMO/sMMO
Methylomicrobium	Gamma Proteobacteria	RuMP	pMMO/sMMO
Methylomonas	Gamma Proteobacteria	RuMP	pMMO/sMMO

Genus	Phylogeny	Formaldehyde assimilation pathway	Enzyme type*
Methylosphaera	Gamma Proteobacteria	RuMP	pMMO
Methylocystis	Alpha Proteobacteria	Serine	pMMO/sMMO
Methylosinus	Alpha Proteobacteria	Serine	pMMO/sMMO

* pMMO - particulate methane monooxygenase; sMMO - soluble methane monooxygenase; RUMP - ribulose monophophate pathway.

Table 13.2. Characteristic features of methylotrophs

Group	Organisms	Growth substrates	Major assimilation pathway*
Aerobic methylotrophs			
(a) obligate Gram – negative	*Methylobacillus, Methylophaga, Methylophilus*	CH_3OH, CH_3NH_2	RuMP
(b) facultative Gram – negative	*Acidomonas, Hyphomicrobium, Methylobacterium*	CH_3OH, CH_3NH_2, CH_3Cl	RuMP/Serine
(c) facultative Gram-positive	*Amycolatopsis, Arthrobactor, Bacillus*	CH_3OH, CH_3NH_2, multi-carbon compounds	RuMP
Heterotrophic methylotrophs	**Bacteria**		
	Arthrobacter, Hyphomicrobium, Bacillus PM6	Organic compounds methylated compounds	RuMP/Serine
	Yeast		
	Candida biodinii, Pichia angusta	Organic compounds methylated compounds	XMP
	Filamentous fungi		
	Gliocladium deliquescens, Paecilomyces variotii, Trichoderma lignorum	Organic compounds Methylated compounds	
Obligate methylotrophs	Methylococcus, Methylomonas,	CO_2, CH_3OH	RuMP/Serine Calvin cycle
Autotrophic methylotrophs (CO_2 utilizers)	*Paracoccus denitrificans, Pseudomonas carboxydovorans*	CO_2	

*RuMP = Ribulose monophosphate pathway; Serine – Serine pathway;
XMP = Xylulose monophosphate pathway

13.3. PHYSIOLOGY

To make the understanding of methylotrophy simple, we club the methanotrophs and methylotrophs, and treat them as methylotrophs. The outlines of the physiology of methylotrophy are illustrated in Fig. 13.1. As is evident from the figure, methane, methanol and all other methylated compounds are suitably processed and oxidized to formaldehyde. Methane is oxidized to methanol by an enzyme methane monooxygenase, by introducing one molecule of oxygen. Methanol is further metabolized to formaldehyde by the pyrolloquinoline quinone – (PQQ-) linked enzyme methanol dehydrogenase (MDH), an enzyme found in all Gram-negative methylotrophic bacteria. Thus the formaldehyde is the central key metabolite in methylotrophy.

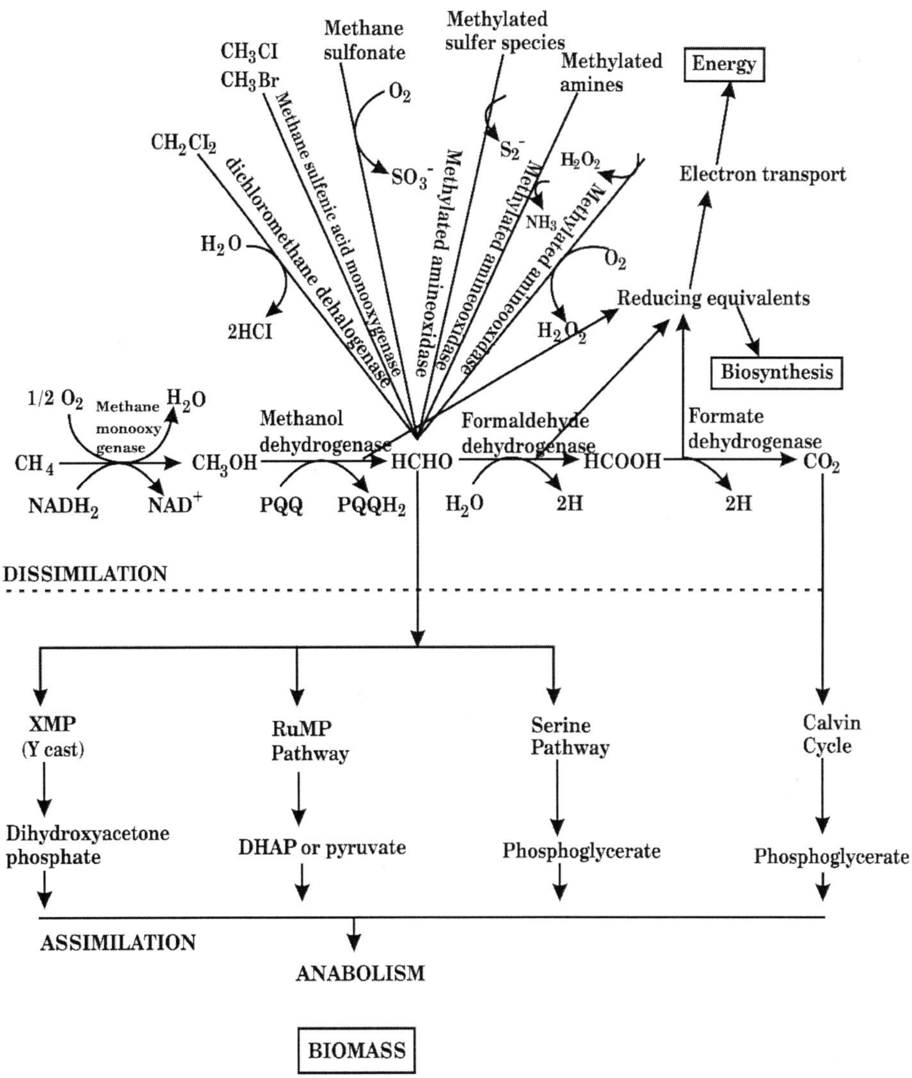

Fig. 13.1. Dissimilation and assimilation pathways in methylotrophs

Approximately, half of the formaldehyde produced is further oxidized to yield CO_2, resulting in generation of reducing power for biosynthesis and the initial oxidation of methane. The CO_2 is not fixed into cell carbon in significant amounts, but is lost to the atmosphere. The remainder of the formaldehyde is assimilated into cell carbon by one of two pathways.

1. Ribulose monophosphate pathway (RuMP pathway)
2. Serine pathway.

13.3.1. Ribulose Monophosphate Pathway (RuMP Pathway)

Ribulose monophosphate pathway has been reported in type I methanotrophs of the genera *Methylomonas, Methylobacter, Methylobacterium, Methylococcus*, and *Methylos-phaera* which possess bundles of intracytoplasmic membranes and are the members of the γ–sub division of the purple bacteria (class : Proteobacteria). Some methanol utilizers also utilize this pathway. The reactions of this pathway are shown in Fig 13.2.

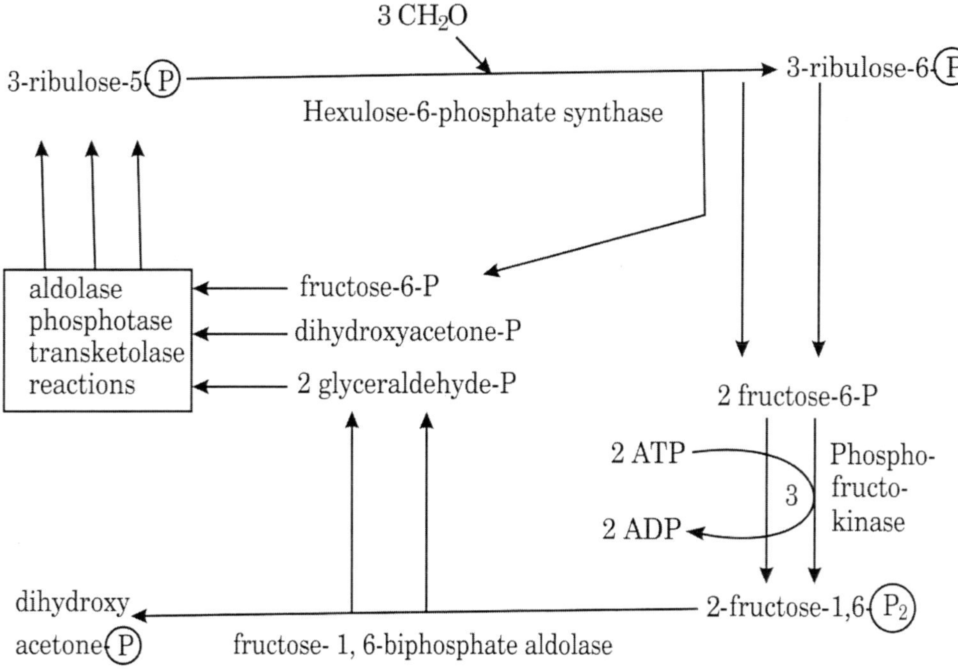

Fig. 13.2. The ribulose monophosphate pathway

The initial reaction in the RuMP pathway involves the addition of 3 mol formal-dehyde to 3 mol ribulose 5- monophosphate to produce 3 mol hexulose 6–phosphate. Hexulose 6–phosphate is converted to fructose–6–phosphate by an isomerase enzyme. Later, two of the three molecules of fructose 6-phosphate are phosphorylated and the resulting fructose–1, 6–biphosphate is cleaved by aldolase into dihydroxyacetone phosphate and glyceraldehyde–3-phosphate. Two molecules of glyceraldehyde-3-phosphate, one dihydroxyacetone phosphate, and one fructose–6–phosphate undergo a series of reactions as in the pentose phosphate cycle (Fig 13.2). The difference is that the reaction sequence here is unidirectional; it proceeds only in the direction of hexose phosphates to pentose phosphates but not vice versa. The reason for this is that sedoheptulose-7-phosphate is not formed from erythrose–4-phosphate in a transaldolase reaction but in a true aldolase reaction with dihydroxyacetone phosphate as the second substrate. Sedoheptulose-1, 7-biphosphate is converted to the Sedoheptulose-7-phosphate by a phosphatase. This reaction is irreversible and determines the direction of entire sequence.

13.3.2. Serine Pathway

Serine pathway has been reported in type II methanotrophs which possess intracytoplasmic membranes arranged around the periphery of the cell, which fall with in γ–subdivision of the proteobacteria. Certain methanol utilizers such as *Methylobacterium* and *Hyphomicrobium* also contain serine cycle for formaldehyde assimilation.

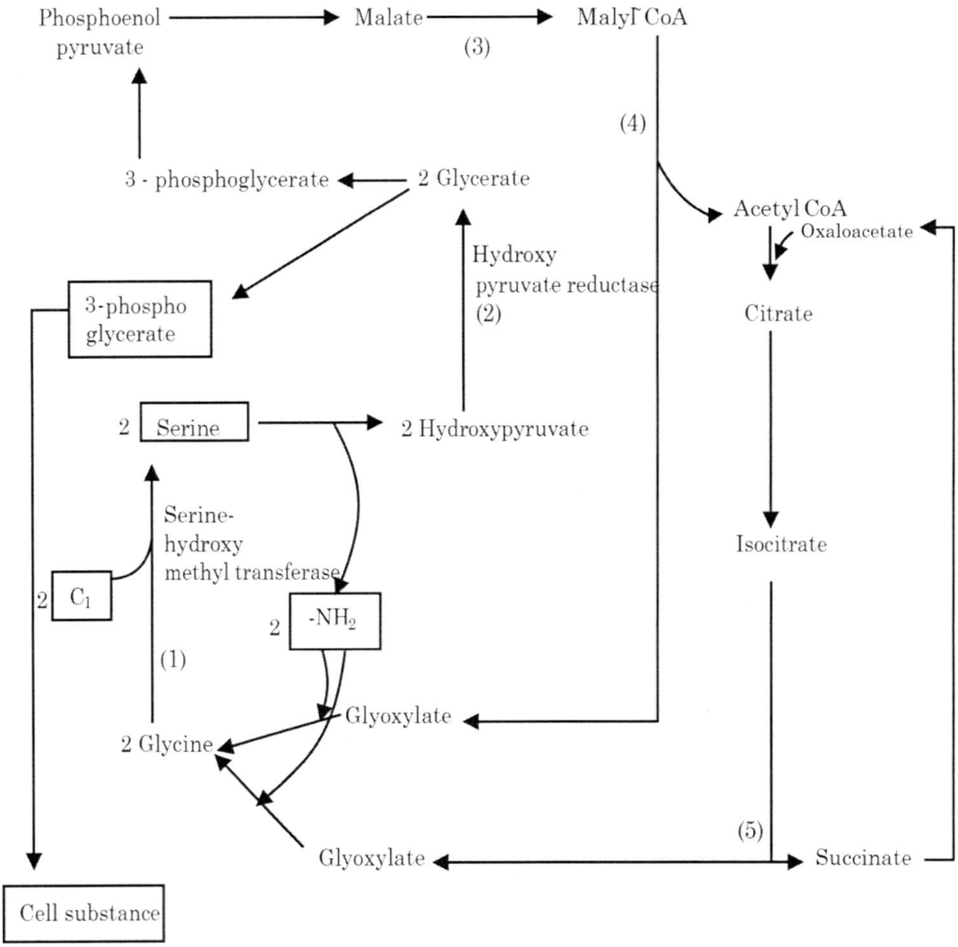

Fig. 13.3. The assimilation of C_1 compounds via the serine pathway

The serine pathway (Fig. 13.3.) is very different from other formaldehyde assimilation pathways. There are no enzymatic reactions in common with the RuMP or methylotrophic yeasts assimilatory pathways. In this pathway, 2 mol of formaldehyde are condensed with 2 mol of glycine to form 2 mol of serine. The serine is converted to 2 mol of phosphoglycerate. One phosphoglycerate is assimilated into cell material, while the other is converted to phosphoenol pyruvate (PEP). The carboxylation of PEP yields oxaloacetate that is subsequently converted to 2 mol glyoxylate. Transmination of glyoxylate with serine as the amino donor regenerates the two molecules of glycine acceptor. The phosphoglycerate, which is assimilated, undergoes transformations in central metabolic routes to provide carbon backbones for the synthesis of all cell materials.

Some methylotrophic bacteria, such as *Paracoccus denitrificans, Pseudomonas carboxydovorans* and species of *Thiosphaera* and *Xanthobacter* can grow on methanol by oxidizing this methylotrophic substrate to carbon dioxide, but then fixing this carbon into cell biomass using the enzyme ribulose biphosphate carboxylase/oxygenase (RuBISCO). These types of organisms are, therefore, considered as autotrophic methylotrophs.

13.3.3. Xylulose Monophosphate Cycle

This cycle was reported in only yeasts. The ability of some yeasts to grow on methanol as a source of carbon and energy has been discovered only recently. They can be isolated from soil, rotting fruits, vegetables, and other plant material, again suggesting that methanol derived from methoxy groups in wood lignin or pectin is an important factor in the ecology of these yeasts. Methylotrophic yeasts belong to the ascomycete yeast genera like *Hansenula, Pichia* and *Candida* and they metabolise methanol via alcohol oxidase in peroxisomes. Assimilation of formaldehyde is accomplished by the xylulose monophosphate cycle (Fig. 13.4).

The key reaction is a specific transketolase in which the first two carbon atoms of xylulose-5-phosphate are transferred to formaldehyde.

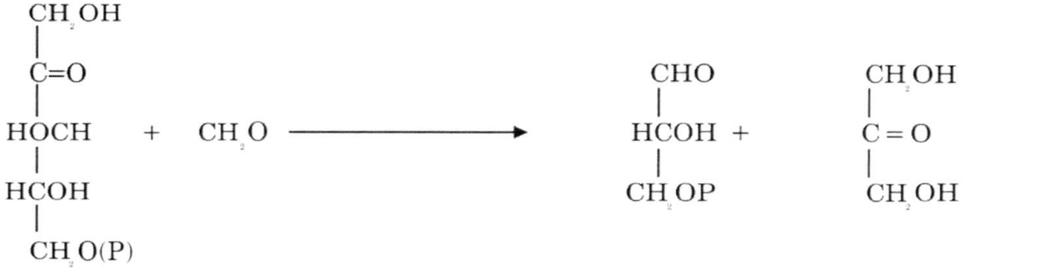

Xylulose-5-P formaldehyde glyceraldehyde-3-p dihydroxyacetone

Dihydroxyacetone is formed which is phosphorylated to dihydroxyacetone phosphate (DHAP) by a specific kinase that can be regarded as the second key enzyme of this cycle. All further reactions are very much similar to those of ribulose monophosphate cycle.

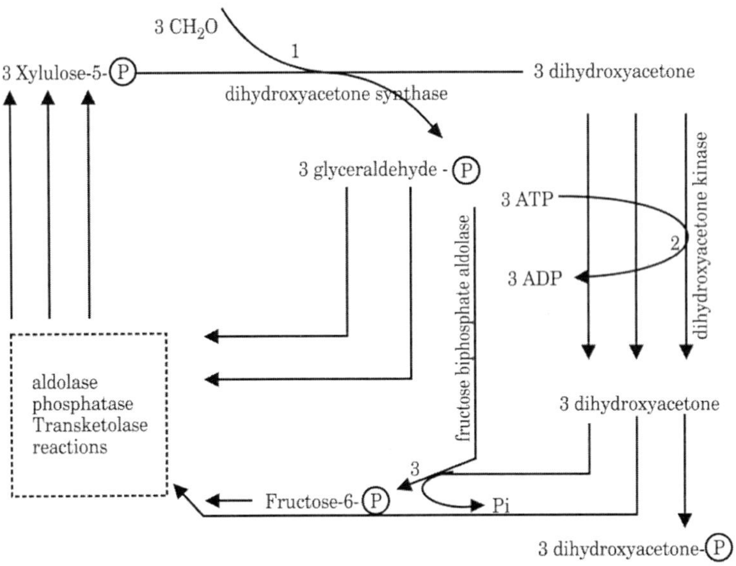

Fig. 13.4. The xylulose monophosphate cycle 1. Dihydroxy acetone synthase 2. dihydroxy acetone kinase 3. Fructose-1,6-bisphosphate aldolase, 4. Fructose-1,6-biphosphatase

13.4. APPLICATIONS OF METHYLOTROPHS

Both methane and methanol are relatively cheap feed stocks for fermentation processes and thus methylotrophs have received considerable attention for a number of biotechnological applications. Initial work on the production of single cell protein (SCP) was carried out with methanotrophs. Methanol utilizing bacteria (*e.g., Methylophilus methylotrophus*) have been successfully used for SCP production (ICI pruteen process). However, SCP from methanol has not been particularly competitive on a commercial scale. Methanol utilizing bacteria have also been exploited for the production of vitamins, polymers, and amino acids and these processes may be more economically viable. The possibilities of genetically engineering methylotrophs for the over production of amino acids is also being explored. Methanotrophs are also being investigated for the production of bulk chemicals such as propylene oxide, a valuable compound in organic synthesis.

Methanotrophs have received a considerable attention for their potential use in bioremediation processes. The enzyme soluble methane monooxygenase (sMMO) not only oxidizes methane but will co-oxidize a wide variety of aliphatic, substituted aliphatic and aromatic compounds. sMMO is able to degrade several pollutants, including vinyl chloride, trichloroethylene, and other halogenated hydrocarbons that contaminate soil and ground water. Research is underway to investigate the metabolism of halogenated methanes and one – carbon compounds containing sulfur, in order to be able to explore the bioremediation potential and to exploit the properties of these novel methylotrophs.

The methylotrophic yeast *Pichia pastoris,* is now becoming one of the best hosts for the production of foreign proteins because of the presence of the strong methanol inducible promoter AOXI. This allows high level expression of a large number of biotechnologically and pharmaceutically important proteins in a controlled fashion during growth of a yeast on relatively cheap substrates.

13.5. FURTHER READING

Anthony, C (1982), The biochemistry of methylotrophs. Academic Press, London, UK.

Hanson, R.S. (1991), The obligate methanotrophic bacteria *Methylococcus, Methylomonas, Methylosinus* and related bacteria. In "The Prokaryotes" (eds. A. Balow *et al.*) Vol. I pp 2350–2365. Springer – Verlag, New York.

Hanson, R.S. and Hanson, T.E. (1996), Methanotrophic bacteria, *Microbial. Rev.* 60: 439–471.

Large, P.J. and Bamforth, C.W. (1988), Methylotrophy and Biotechnology, Longman Scientific and Technical, New York.

Leak, D.J. (1992), Biotechnological and applied aspects of methane and methanol utilizers. In "Methane and Methanol utilizers" (eds. J.C. Murrel and H. Dalton) pp 245 – 279. Plenum press, New York.

Lidstrom, M.E. (1991), Aerobic methylotrophic bacteria. In "The Prokaryotes" (eds. A. Balow *et al.*) Vol. I, pp 431–445. Springer–Verlag, New York.

Lidstrom, M.E. and Tabita, F.R. (1996), Microbial growth on C_1 compounds. 'Kluwer Academic, Dordrecht, The Netherlands.

Murrel, J.C. and Dalton, H (1992), Methane and Methanol utilizers, Plenum Press, New York.

Murrell, J.C. and M.C. Donald, I.R. (2000), Methylotrophy. Encyclopedia of Microbiology, Vol. III, J. Lederberg (ed.), San Diego, Academic Press.

■■■■

Chapter - 14

Chemoheterotrophy - Hydrocarbon Utilizers

14.1. INTRODUCTION

Hydrocarbons are highly reduced water insoluble organic compounds containing carbon and hydrogen atoms in their molecules. Low molecular weight hydrocarbons are gases, whereas those of high molecular weight are liquids and solids at room temperature. These are widely distributed in nature in the form of fossil fuels and also being continuously synthesized by bacteria and plants. Plants synthesize these compounds in large quantities as secondary metabolites (Table 14.1). A large number of hydrocarbons are also produced by the activity of man. Structurally hydrocarbons can be divided into (i) aliphatic hydrocarbons and (ii) aromatic hydrocarbons (Table 14.2).

Table 14.1. Aromatic hydrocarbon compounds of plants

Compound	Examples
Lignin	Coniferyl alcohol and other lignin precursors
Aromatic amino acids	Tyrosine, tryptophan
Quinones	Plastoquinone
Phenolics	Phenols, phenolic acids, coumarins, cinnamic acid etc.
Tannins	Gallotannins, vegetable tannin derivatives
Others	Melanins, flavonoids, hormones, alkaloids and terpenes

Both aliphatic and aromatic hydrocarbons can be either saturated or unsaturated. The saturated hydrocarbons possess single carbon-carbon bonds, where as unsaturated hydrocarbon possess double or triple carbon-carbon bonds.

Due to their carbonaceous nature and high energy content, and wide distribution in nature, many microorganisms have developed a mechanism to utilize the hydrocarbons as source of carbon and energy. These microorganisms are called **hydrocarbon utilizers**. Nutritionally, these are chemoheterotrophs. Almost all major microbial groups are capable of utilizing hydrocarbons (Table 14.3).

Table 14.2. Classification of hydrocarbons

A. Aliphatic hydrocarbons, with a straight or branched carbon open chain but not forming a ring structure. These can be further divided into
i. Simple aliphatic hydrocarbons with straight chain without branching *e.g.,* n-paraffin, Olefinbutene-1
ii. Aliphatic hydrocarbons with branched chains *e.g.,* Iso-paraffin, Olefin isobutene
B. Aromatic hydrocarbons with a closed ring structure which are derivatives of benzene ring
iii. Monocyclic aromatic compounds *e.g.,* Alkyl benzol, mandelate, toluate, alkyl phenol
iv. Polycyclic aromatic compounds *e.g.,* Anthracene, naphthalene, phenanthracene

The susceptibility of hydrocarbons to microbial degradation is determined by the structure and molecular weight of the hydrocarbon molecule. The following are the common observations :

i. Aliphatic hydrocarbons are more susceptible than aromatics to microbial attack

ii. Long chain hydrocarbons are more susceptible than short chain hydrocarbons

iii. Unsaturated hydrocarbons are more susceptible than the corresponding saturated compounds.

iv. Branched chain hydrocarbons are more susceptible than unbranched.

Table 14.3. Hydrocarbon utilizing microorganisms.

Group	Microorganisms
A. Bacteria	
1. Autotrophic bacteria	*Desulfovibrio,* Iron bacteria, *Thiobacillus* spp.
2. Heterotrophic bacteria	*Bacillus megaterium, Azotomonas insalita,*
	Staphylococcus epidermitis, Pseudomonas spp.
	Serratia spp., *Flavobacterium* spp., *Cylindroglea bacterigena, Bacillus mycoides, B. subtilis, Sorangium* spp., *Aerobacter aerogenes, Clostridum* spp., *Pseudomonas fluorescens, Micrococcus* spp., *Brevibacterium* spp., *Escherichia* spp., *Pseudomonas methanica*
3. *Iron Bacteria*	
Sheathed bacteria	*Sphaerotilus*
Stalked bacteria	*Gallionella, Caulobacter, Siderocapsa, Siderobacter, Ferribacterium*
4. *Sulfur bacteria*	
Sulfur oxidizing bacteria	*Thiobacillus thiooxidans, Thiobacillus thioparus, T. denitrificans*
Sulfide oxidizing bacteria	*Beggiatoa, Chlorobium, Chromatium*
Sulfate reducing bacteria	*Desulfo vibrio, Clostridium*

Group	Microorganisms
B. Actinomycetales	
1. Mycobacteriaceae	*Mycobacterium, M. phlei, M. hyalinum, M. lacticola, M. luteum, M. rubram*
2. Actinomycetaceae	*Nocardia, N. opaca, N. parathinae, N. corallina*
3. Streptomycetaceae	*Streptomyces, Micromonospora*
C. Fungi	*Penicillium, Botrytis cinerea, Aspergillus flavus, Pullularia pullulans*

14.2. OXYGENASES

As the hydrocarbons are in a highly reduced state, their oxidation is a prerequisite for their utilization as carbon and energy source. They are suitably modified for their final and complete degradation by a set of unique enzymes called **oxygenases**. Interestingly, these enzymes are distributed only in hydrocarbon utilizers. Oxygenases catalyse the initial reactions of hydrocarbon degradation by incorporating molecular oxygen into the compound. Two classes of oxygenases are identified : **monooxygenases** some times called hydroxylases incorporate a single oxygen atom into the basic molecule, whereas **dioxygenases** incorporate two oxygen atoms.

$$R—CH_2—CH_2—CH_3 \xrightarrow{\text{monooxygenase}} R—CH_2—CH_2—CH_2OH$$

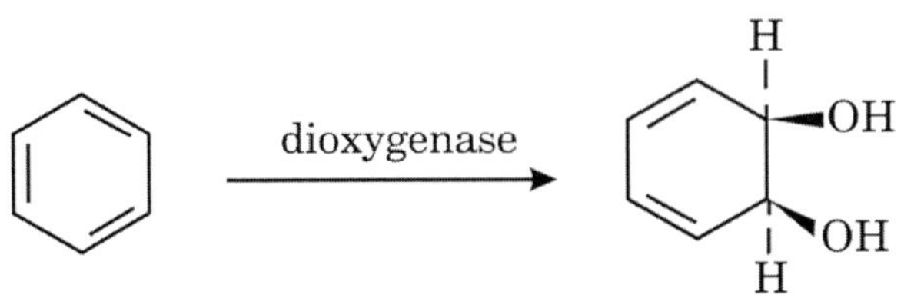

In most monooxygenases, the electron donor is NADH or NADPH, although the direct coupling to O_2 is through a flavin that is reduced by NADH or NADPH donor. Most oxygenases contain Fe(II) in the form of non-heme iron proteins. However, others such as tryptophan oxygenase is a heme protein. These enzymes typically made up of four sub-units with a molecular weight of about 40,000 daltons each, and are active when iron is in ferrous form.

14.3. METABOLISM OF ALIPHATIC HYDROCARBONS

A number of bacteria, moulds and yeasts can use hydrocarbons for growth under aerobic conditions. Oxidation under oxic conditions proceeds via oxygenase system. The reactions involved in the pathway of assimilation of n–octane, a typical aliphatic hydrocarbon, are shown in Fig. 14.1.

The initial oxidation step of saturated aliphatic hydrocarbon involves molecular oxygen (O_2) as a reactant and one of the atoms of the oxygen molecule is incorporated into the hydrocarbon. The reaction is mediated by monooxygenase. The resultant alcohol, in the second phase, is oxidised to aldehyde by an enzyme alcohol dehydrogenase. Aldehyde is further oxidised to acid by aldehyde dehydrogenase. During the latter two reactions reducing power is generated. The fatty acids, thus generated, are oxidized by a process called β–oxidation, in which two carbons of the fatty acid are split off at a time. In eukaryotes, the necessary enzymes are located in the mitochondria, whereas in prokaryotes they are cytoplasmic. The fatty acid is first activated with coenzyme A. After a series of reactions, mostly oxidative in nature, acetyl CoA is released making the original molecule shorter by two carbons. This process of β–oxidation occurs several times and every time an acetyl CoA molecule is released. In this series of reactions, two separate dehydrogenation reactions occur. In the first, electrons are transferred to flavin – adenine dinucleotide (FAD), whereas in the second they are transferred to NAD^+.

Most fatty acids have an even number of carbon atoms, and complete oxidation yields only acetyl CoA. The acetyl –CoA thus formed is oxidized by the citric acid pathway or is converted to hexose and cell constituents via glyoxylate cycle. A perusal of the reactions of Fig. 14.1. reveals that hydrocarbons and their derivative fatty acids are good electron donors. At various levels the electrons are transferred to either NAD^+ or FAD^+ to generate reducing power. The reduced products inturn are oxidized in electron transport to generate ATP or utilized for synthetic purposes.

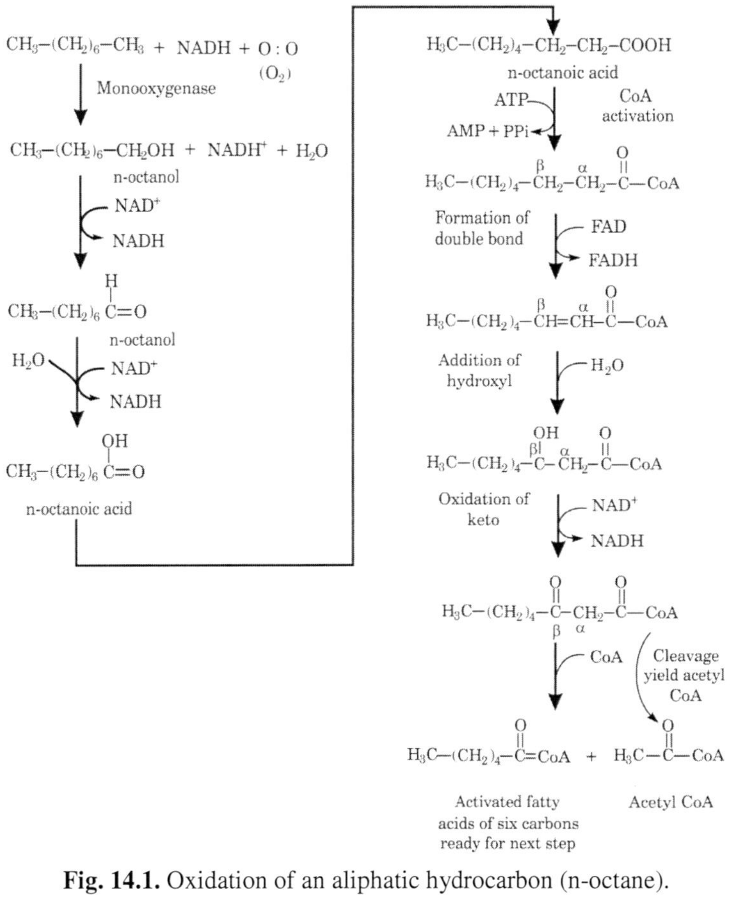

Fig. 14.1. Oxidation of an aliphatic hydrocarbon (n-octane).

14.4. METABOLISM OF AROMATIC HYDROCARBONS

14.4.1. Aerobic Degradation

Aromatic hydrocarbons widely occur in nature. A bulk of these compounds is found in fossil fuels such as petroleum and coal. Plants produce many organic compounds that contain aromatic rings. Quantitatively predominant compound is lignin which accounts for about 20% of the wood. Aromatic amino acids and vitamins are constituents of every organism. Of late, a number of xenobiotic compounds, especially pesticides most of which are aromatic in nature, are added to the natural environments. All these compounds become available to microorganisms and, not surprisingly, many microorganisms have developed the mechanisms to degrade and assimilate these compounds.

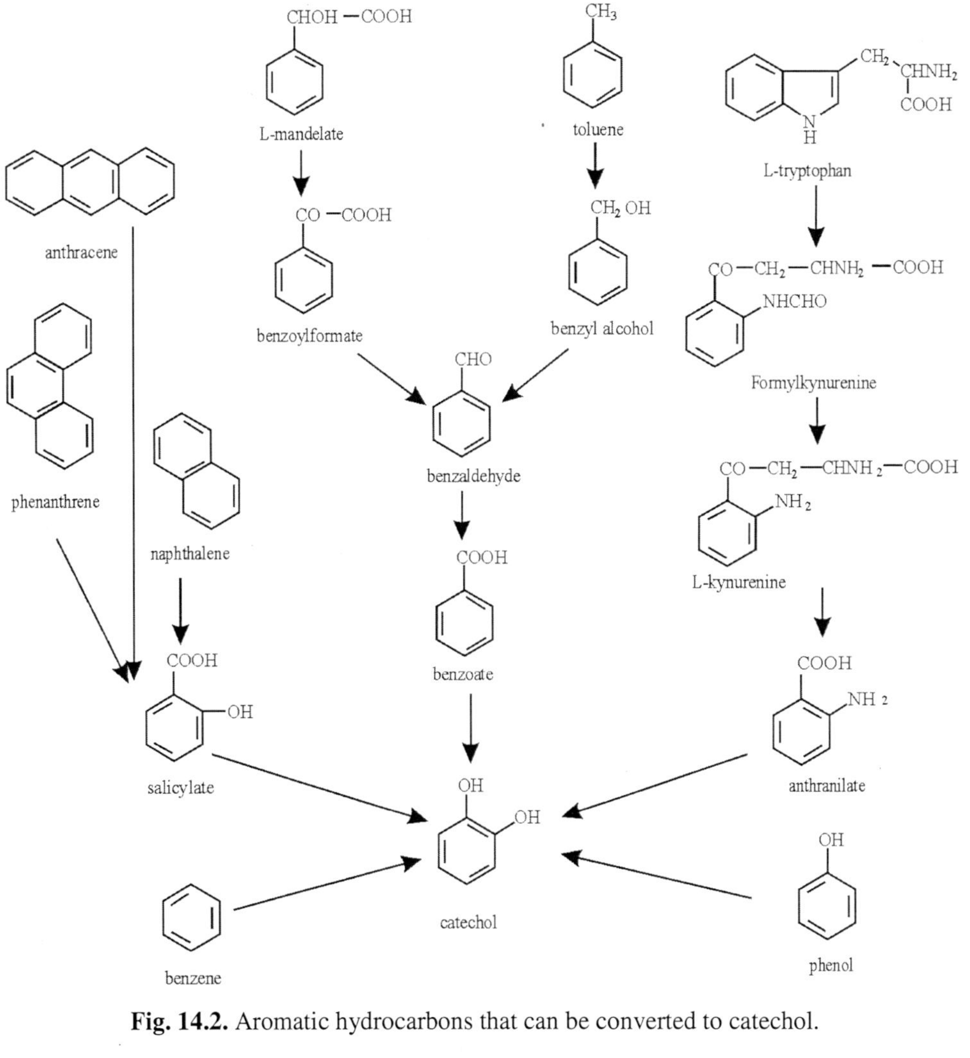

Fig. 14.2. Aromatic hydrocarbons that can be converted to catechol.

A number of bacterial and other microbial species can degrade aromatic compounds. Although many organisms have been studied for their ability to degrade aromatic compounds because of their health hazardous nature and accumulate in nature, much of the work has been focused on members of the genus *Pseudomonas*. The major intermediates recognized in aromatic compound degradation are catechol, protochatechuate, gentisate and homogentisate. Degradation of aromatic compounds depends upon the conversion of the material to one of these compounds and its subsequent oxidation to intermediates of the TCA cycle or to metabolites utilizable by carbohydrate catabolic pathway.

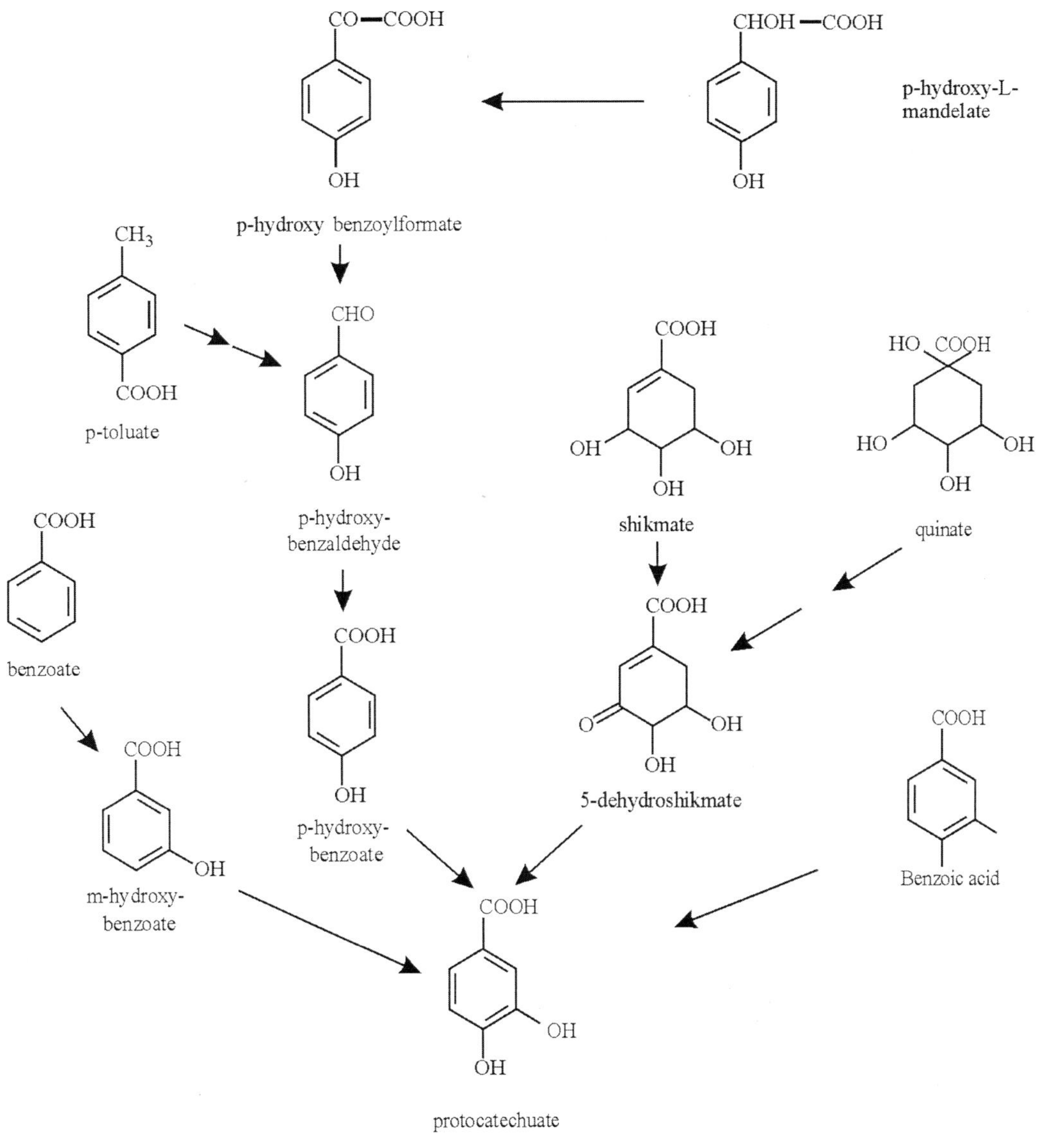

Fig. 14.3. Aromatic hydrocarbons that can be converted to protocatechuate.

It is difficult, if not impossible, to predict from its structure, the precise manner in which a particular aromatic compound may be catabolized. In some situations several pathways may be used for a single compound. To make the understanding of the aromatic hydrocarbon catabolism simple, pathways involving catechol and protocatechuate as intermediates are discussed.

i. Preparation of ring cleavage

Many singly substituted or 1,2-disubstituted aromatic rings, for example, mandelate, phenylalanine, toluol, benzoate, salicylate, phenol and benzene are degraded or modified to catechol (Fig. 14.2). Aromatic rings disubstituted in positions 1,3 and 1,4 and multiple substituted rings, for example 4–hydroxy benzoate, quinate, vanillate and shikimate are degraded to protocatechuate (Fig. 14.3). In all the cases, hydoxy groups are incorporated into the ring. The unsubstituted ring of benzene is hydroxylated by dioxygenase to cis-1,2 dihydro 1,2 dihydroxy benzene which is then dehydrated to protocatechuate.

ii. Ring cleavage

The cleavage of the aromatic rings of catechol and protocatechuate is carried out by dioxygenases through which molecular oxygen is incorporated. The rings are cleaved by two analogous series of reactions viz.,

(i) Ortho cleavage

(ii) Meta cleavage.

i. **Ortho cleavage:** Ortho or intradiol cleavage is the cleavage between two neighbouring hydroxylated carbon atoms and it leads to the formation of dicarboxylic acids (Fig. 14.4). Presumably, there is a primary addition of the O_2 molecule to the carbon atoms next to the hydroxy group with the formation of a cyclic peroxide. Intramolecular rearrangement then dissolves the C-C bond with the formation of cis, cis –muconic acid.

In this cleavage, the ring structures of catechol and protocatechuate are ruptured by 1,2 catechol dioxygenase and protocatechuate 3,4 dioxygenase respectively producing cis- cis muconate (catechol) and β – carboxy cis – cis muconate (protochatechuate). The later two compounds are converted to lactones and subsequently through isomerization (catechol) and decarboxylation (protocatechuate), to 4–oxoadipate enol lactone. Addition of water to 4 – oxoadipate enol lactone by 4 – oxoadipate hydrolase gives β-keto (3-oxo) adipate. Donation of coenzyme A from succinyl CoA, through the action of succinyl CoA transferase, forms β–ketoadipyl CoA, which with the aid of β–ketoadipate CoA thiolase coenzyme A regenerates the succinyl CoA used to coacylate β–ketoadipate and allows the formation of acetyl CoA. Thus, irrespective of structural differences between catechol and protocatechuate, the end products of orthocleavage are acetyl CoA and succinate, which are common intermediates of central metabolism.

ii. **Meta cleavage:** Meta or extra diol cleavage is the cleavage of the ring between the hydroxylated and non-hydroxylated carbon atoms and it is also catalysed by dioxygenases. Meta cleavage of the aromatic ring structure results, once again, in parallel series of reactions with catechol and protocatechuate (Fig. 14.5). Meta cleavage reactions produce 2–hydroxymuconic semialdehyde from catechol and 2–hydroxy-4-carboxy-muconic semialdehyde from protocatechuate. These compounds are acted upon by semialdehyde hydrolase enzymes, yielding 2–oxo pent–4-enoate, and 2-oxo-4-carboxypent-4-enoate respectively. Hydrolysis of the latter compounds by hydrolase enzymes results in the formation of 4–hydroxy-2-oxovalerate and 4-hydroxy-4-carboxy-2-oxovalerate. In the final step of the meta cleavage, these two last compounds are converted by aldolases to pyruvate and acetaldehyde, respectively.

Thus, regardless of whether the ortho or meta mode of metabolism is used, analogous reactions occur with catechol and protocatechuate.

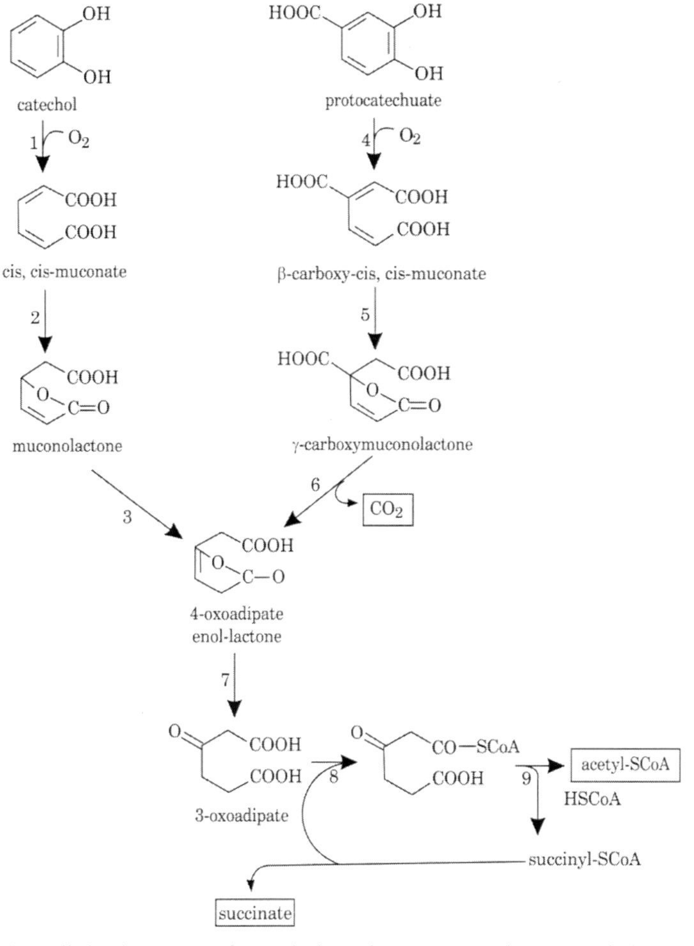

Fig. 14.4. Ortho or intradiol cleavage of catchol and proto-catechuate and 3-oxoadipate pathway. 1. Catechol, 1,2-dioxygenase; 2. muconate-lactonizing enzyme; 3. muconolactone isomerase; 4. protocatechuate 3,4-dioxygenase; 5. β-carboxymuconate-lactonizing enzyme; 6. gamma-carboxymuconolactone decarboxylase; 7, 4-oxoadipate enollactone hydrolase; 8. e-oxoadipate succinyl-CoA transferase; 9. 3-oxoadipate-CoA thiolase.

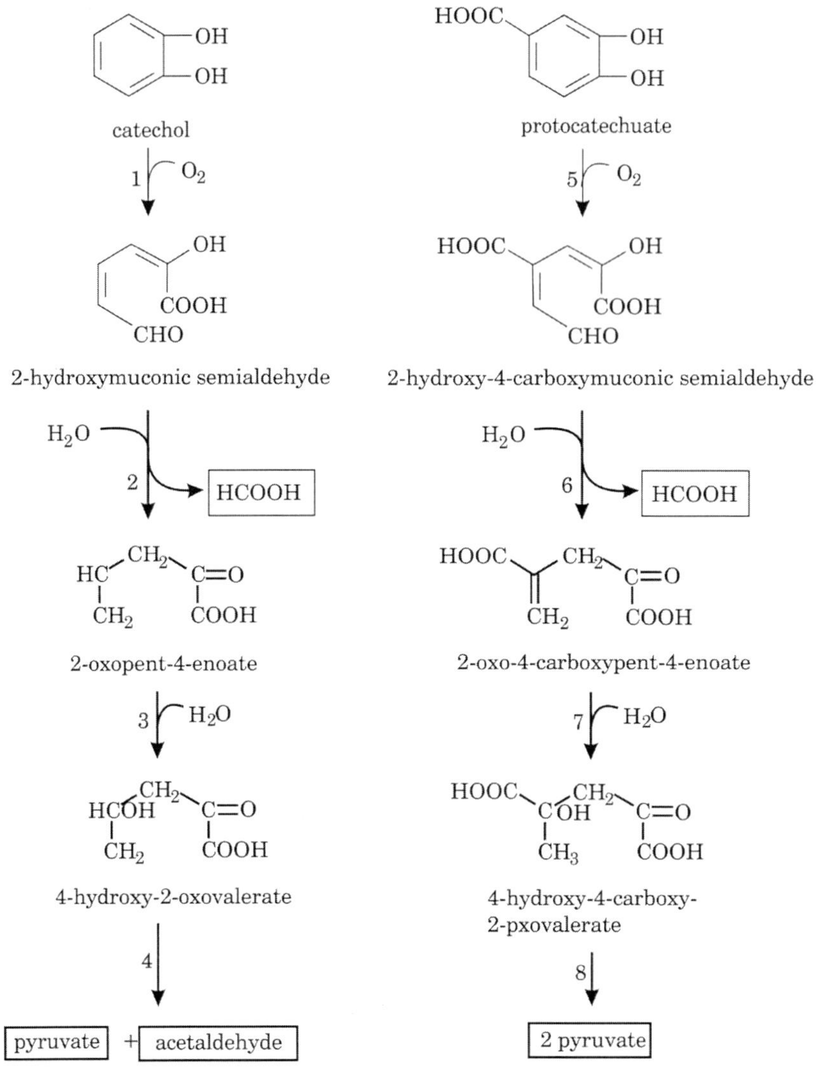

Fig. 14.5. Meta or extradiol cleavage of catechol and protocatechuate. 1. Catechol 2,3-oxygenase; 2. 2-hydroxymuconic semialdehyde hydrolase; 3. 2-oxopent-4-enoic acid hydrolase; 4. 4hydroxy2oxovalerate aldolase; 5. protocatechuate 4,5-oxygenase; 6. 2-hydroxy-4-carboxymuconic semialdehyde hydrolase; 7. 2-oxo-4-carboxypent-4-enoic acid hydrolase; 8. 4-hydroxy-4-carboxy-2-oxovalerate aldolase.

14.4.2. Anaerobic Degradation

Anoxic conditions are generally created when oxygen consumption exceeds supply. Under these conditions, biochemical mechanisms which do not involve molecular oxygen take over during the breakdown of aromatic compounds. As a consequence, some types of aromatic compounds make it difficult to be degraded anaerobically especially the aromatic compounds without the functional groups (benzene, toluene, naphthalene, etc.). However higher chlorinated aromatic compounds (*e.g.*, hexachlorobenzene) or compounds that already contain an atom of oxygen are degraded better under anaerobic conditions. Different types of compounds can act as alternative terminal electron acceptors in the absence of O_2, such as sulphate, nitrate, sulphur, oxidized metal ions (*e.g.* Fe^{3+} and Mn^{4+}), protons and bicarbonate ions. Proton and bicarbonate are employed as primary electron acceptors in methanogenic environment. Phototrophic bacteria can also use aromatic compounds for carbon and electron requirement in the absence of O_2, if light is available. In this case, the degradation product is a biomass rather than CO_2. Major physiological groups involved in anaerobic metabolism of aromatic compounds are presented in Table 14.4. The fundamental differences between aerobic and anaerobic degradation of phenolic compounds are precised in Table 14.5.

Table 14.4. Major physiological groups of bacteria involved in aromatic hydrocarbon metabolism

Energy yielding process	Organisms	Substrates
Photosynthetic	*Rhodopseudomonas palustris*	Benzoate
Phosphorylation	*Rhodopseudomonas gelatinosa*	m, p-Hydroxybenzoate phloroglucinol
Nitrate respiration	*Pseudomonas PN-1*	Benzoate
	Alcaligenes xylosoxidans	Hydroxybenzoates
	Moraxella sp.	Vanillate
	Paracoccus denitrificans	o, m, p-Phthalate
	Bacillus sp.	2-Amino benzoate
		Phenyl, o, m, p – cresol
Sulphate reduction	*Pseudomonas*	Benzoate
	Desulfovibrio sp.	Hydroxybenzoates
	Desulfococcus	Phenol acetate hippurate
	Desulfonema	Phenol, indol
	Desulfosarcina	
Fermentation	*Coprococcus* sp.	Phloroglucinol
	Streptococcus	Resorcinol/acids
	Pelobacter acidigallici	Gallate, pyrogallol
	Eubacterium oxidorensis	Polyphenols, quercetin
Methanogenic	Microbial consortia	Lignin
Fermentation	Fermentative bacteria	Benzoate
	Acetogenic	Tyrosine
	Methanogenic bacteria	Phenol, catechol, hydroquinone, benzoate, benzene, toluene, chlorobenzene, chlorobenzoates

Table 14.5: Aerobic and anaerobic aromatic hydrocarbon metabolism

Step	Aerobic	Anaerobic
Channeling	oxygen	$+H_2O$, $+2(H)$, $-2(H)$, $+(H_2O)$, $+CO_2$, $+CO_4$, +ATP
Central intermediates	catechol, protocatechuate, gentisate	benzoyl CoA, resorcinol, phoroglucinol
(Properties of central intermediates)	easy to oxidize (cleave)	easy to reduce (hydrate)
Attack at the ring	O_2	2 or 4(H) + (H_2O)
Ring cleavage	oxygenolysis of aromatic	hydrolysis of 3-oxo compound
Pathway to central metabolism	3-oxoadipate/TCA cycle	β-oxidation

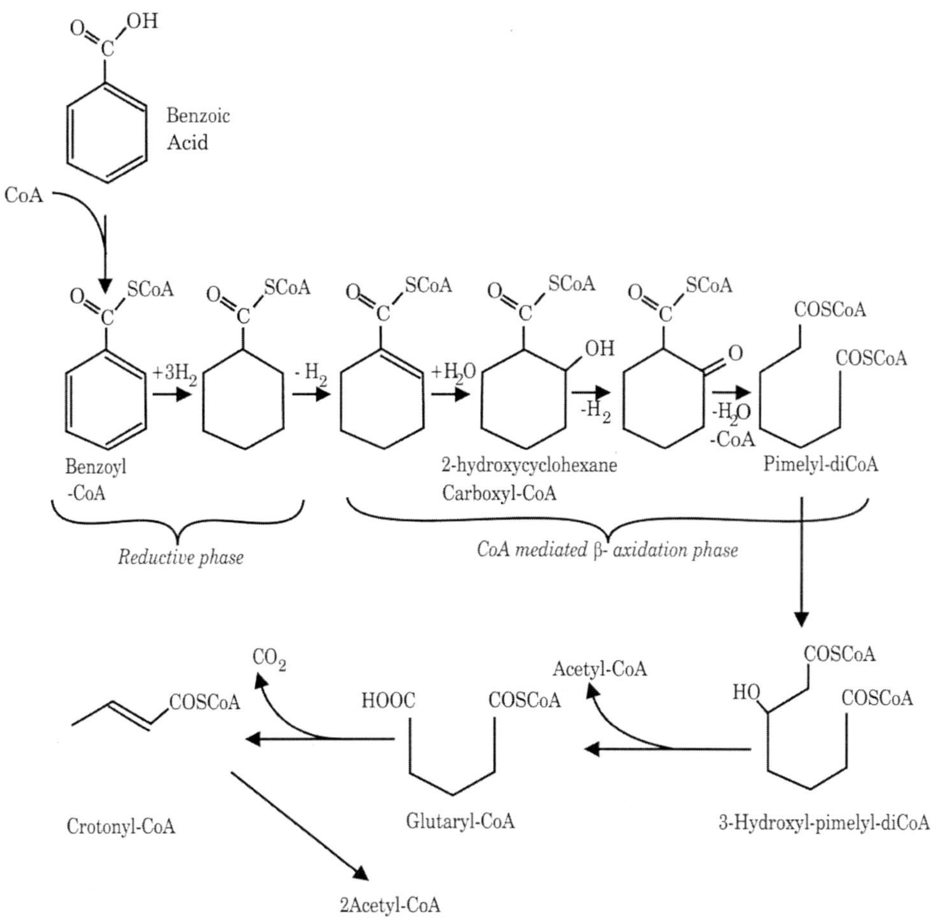

Fig. 14.6. Anaerobic degradation of benzoic acid through pimilic acid pathway.

Metabolic pathways of anaerobic degradation of aromatic compounds : benzoate as a model

Anaerobic degradation of benzoate proceeds by two pathways via pimilic acid and adipic acid, and both involve reduction (saturation) of the aromatic ring and incorporation of oxygen from water leading to an enol prior to ring cleavage. By the reduction of the aromatic ring, the resonant electron distribution is destabilized. The phototrophic purple nonsulfur bacterium, *Rhodopseudomonas,* attacks benzoic acid by the pimilic acid pathway (Fig. 14.6) using coenzyme as a cofactor. Denitrifiers (e.g. *Miraxella*), and *Pseudomonas* PNI degrades by the reduction of the ring followed by decarboxylations i.e., by adipic acid pathway (Fig. 14.7).

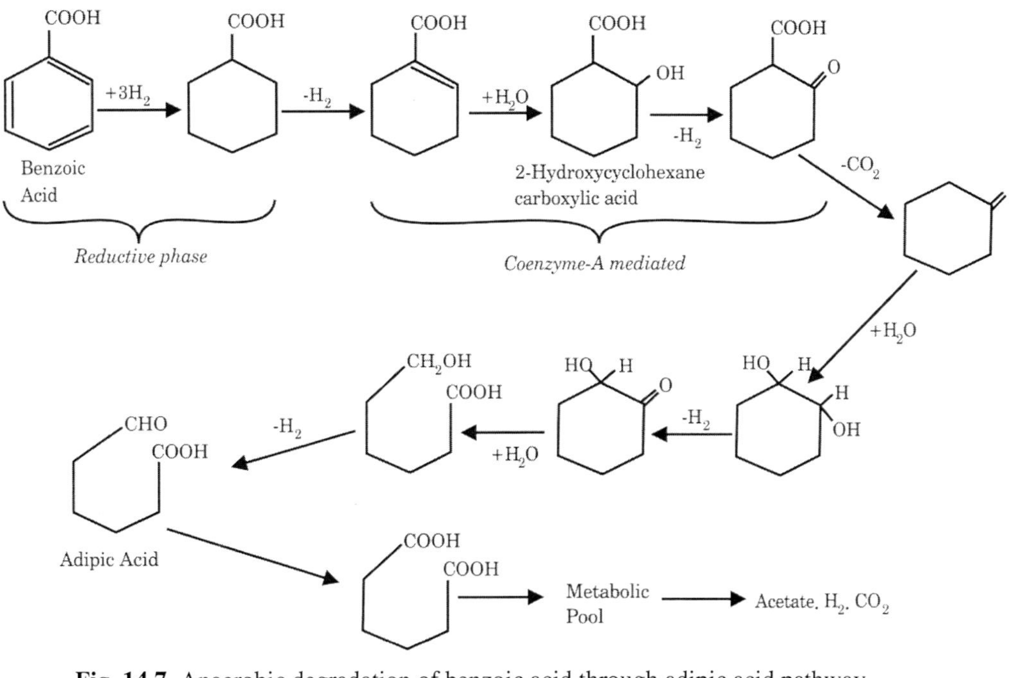

Fig. 14.7. Anaerobic degradation of benzoic acid through adipic acid pathway.

14.5. HYDROCARBON UTILIZERS AND BIOREMEDIATION

The term bioremediation refers to the utilization or transformation of environmental pollutants, such as petroleum hydrocarbons, chlorinated solvents, pesticides that adversely affect human health or environmental quality by microorganisms.

Many microorganisms are capable to transform or degrade the hydrocarbons of both aliphatic and aromatic nature. Such organisms are candidates for use in bioremediation. The use of microorganisms to degrade aromatic pollutants is a low-cost approach for the treatment of industrial effluents, contaminated sediments in soil and ground water. Intensive investigations in this field particularly in the degradation of lignin by white rot fungi resulted in complete understanding of metabolism, enzymes and genes involved. Ideally, bioremediation is practiced by providing a favourable environmental conditions in order to stimulate the degradation of pollutants by naturally occuring microflora. Many attempts are being made to employ microorganisms either singly or in consortia to contaminated sites. Anaerobes play an important role in detoxifying and decomposing many natural and xenobiotic compounds.

14.5.1. Petroleum Biodegradation

Since petroleum is a rich source of organic matter and the hydrocarbons within it are readily attacked aerobically by a variety of microorganisms. Small scale oil pollution of aquatic and terrestrial ecosystems from human as well as natural activities is very common and there exists a diverse microbial community that utilizes hydrocarbons and clean up the polluted sites. In large spills, volatile hydrocarbon fractions evaporate quickly, leaving longer-chain aliphatic and aromatic components for microorganisms to tackle. Hydrocarbon–oxidizing bacteria are able to attach to insoluble oil droplets and their action eventually leads to decomposition of the oil and dispersal of the slick. In oil spills where careful bioremediation studies have been performed, it has been shown that hydrocarbon-oxidizing bacteria increase in number 10^3-10^6 times shortly after the oil spill.

Gasolene and crude oil storage tanks are potential habitats for hydrocarbon-oxidizing microorganisms, which can accumulate and grow at the oil-water interface. If sufficient sulfate is present in the water, sulfate-reducing bacteria can grow, consuming hydrocarbons under totally anoxic conditions. The sulfide (H_2S) that is produced is highly corrosive and can cause pitting of the tanks.

14.5.2. Biodegradation of Xenobiotics

Xenobiotics are man made or chemically synthesized compounds that do not occur in nature. These include a long list of compounds including pesticides, polychlorinated biphenyls (PCBs), munitions, dyes and chlorinated solvents. Many xenobiotics are structurally related to natural compounds and thus can be slowly degraded by enzymes that already exist to degrade these natural compounds. Interestingly many of the xenobiotics especially pesticides possess aromatic structures. As such the hydrocarbon degradation and hydrocarbon utilizers are a point of interest. The organisms that are able to metabolize pesticides and herbicides are fairly diverse genera of both bacteria and fungi. Some pesticides can be both carbon and energy sources and are oxidized completely to CO_2. However, other compounds are much more recalcitrant and are attacked slightly or not at all. Some are degraded either partially or totally provided some other organic material is present as primary energy source, by a phenomenon called co-metabolism or co-oxidation.

Co-metabolism : Co-metabolism is defined as 'a microbial process where an organism tranforms organic compounds without using them as energy for their metabolism resulting in product accumulation with little or no biomass production. The basis for co-metabolism is the supply of energy, cofactors or metabolites at various levels, from the transformation of one substrate to processes, such as substrate transport, enzyme biosynthesis or functioning involved in the transformation of a second substrate. Co-metabolism is affected by a number of bacterial genera that are widespread in natural ecosystems. Halogenated pesticides have been subjected to co-metabolism by a variety of bacteria unable to multiply at the expense of these substances.

14.6. HYDROCABON UTILIZERS AND OIL EXPLORATION

The ability of microorganisms to utilize hydrocarbons for growth is also finding some use in geologic prospecting of new oil deposits. Hydrocarbon gases are liberated from these oil deposits and diffuse upward through the overlying soil. The occurence of a large number of hydrocarbon – oxidizing microorganisms in soil can be taken as an indication of a possible oil reserve below.

14.7. FURTHER READING

Atlas, R.M. (1981), Microbial degradation of petroleum hydrocarbons : An environmental perspective. *Microbiol. Rev.* **45** : 180-209.

Chakravarty, A.M. (1996), Microbial degradation of toxic chemicals : Evolutionary insights and practical considerations, ASM News, **62** : 130.

Deo, P.G., Karanth, N.G. and Karanth, N.K.G. (1994), Microbial degradation of pesticides. *Ind. J. Microbiol.,* **34(4)** : 241-263.

Evans, W.C. and Fuchs, G. (1988), Anaerobic degradation of aromatic compounds. *Ann. Rev. Microbiol.,* **42** : 289-317.

Fuchs, G., Mohamed, M.E.S., Attenschmidt, U., Koch, J., Lack, A., Brackmann, A., Lochmeyer, C. and Oswald, B. (1994), Biochemistry of anaerobic biodegradation of aromatic compounds, Rarledge, C. (ed.) Biochemistry of Microbial degradation, Kluwer Academic Publisher Dordre Cht.

Haggblom, M.M. and Young, L.Y. (1995), Anaerobic degradation of halogenated phenols by sulfate – reducing consortia. *Appl. Environ. Microbiol.,* **61** : 1546.

Jothimani, P., Kalaichelvan, A., Bhaskaran, D., Augustine Selvaseelan and Ramaswamy, K. (2003), Anaerobic biodegradation of aromatic compounds. *Ind. J. Expt. Biol.,* **41** : 1046-1067.

Klug, M.J. and Markovetz, A.J. (1971), Utilization of aliphatic hydrocarbons by microorganisms. *Adv. Microbiol. Physiol.,* **5** : 1-39.

Licht, D., Ahring, B.K. and Arvin, E. (1996), Effects of electron acceptors, reducing agents and toxic metabolites on anaerobic degradation of heterocyclic compounds. *Biodegradation,* **7** : 83.

Salanitro, J.P. (2000), Oil pollution. In Encyclopedia of Microbiology, Vol. 4, 449-455, Academic Press. San Diego

Tiedje, J.M., Boyd, S.A. and Fathepure, B.Z. (1987), Anaerobic degradation of chlorinated aromatic hydrocarbons. *Dev. Ind. Microbiol.,* **27** : 117-127.

■■■■

Chapter - 15

Assimilation of Complex Carbohydrates and Proteins

15.1 INTRODUCTION

Many of the microorganisms are heterotrophic and meet their carbon and energy requirements from different carbon substances available in the nature. Although, green plants have been synthesizing huge quantities of these substances, only a small fraction has been conserved as highly reduced carbon substances (coal, petroleum, natural gas). Thus these substances serve as nutrients to microorganisms. Out of different carbon substances available, simple compounds that too carbohydrates which are water soluble are of primary choice. Subsequently, microorganisms are forced to utilize much more complex carbohydrates and other carbon substances. One or more microorganisms that have the ability to degrade it either completely or partially will be the primary colonizers. The fragments that are produced by the primary colonizers can then be utilized by other microorganisms. In their totality, therefore, the microorganisms appear to be biochemically omnipotent, and this has been referred to as the principle of **microbial infallibility**. However, it is reported that some synthetic low molecular weight compounds and highly polymerised materials are resistant to microbial attack.

Of various substances, complex carbohydrates form the major bulk which include, cellulose, hemicellulose, starch, and pectic substances. Different groups of microorganisms developed various mechanisms to assimilate these compounds more efficiently by an array of enzymes. In this chapter, assimilation of complex carbohydrates and proteins by microorganisms is discussed.

15.2 CELLULOSE

Cellulose is the basic compound of plant cell wall and is produced in greater quantity than any other substances (Fig 15.1 & 15.2). Plant residues in soil consist of 45% cellulose which is, therefore, of central importance next to carbondioxide in the carbon cycle.

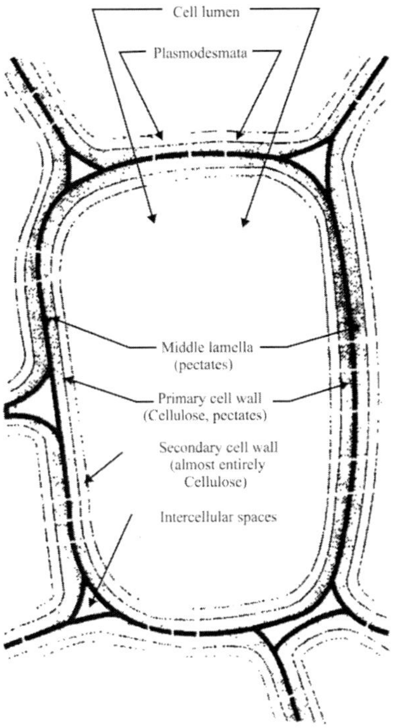

Fig. 15.1. Structure and composition of plant cell wall.

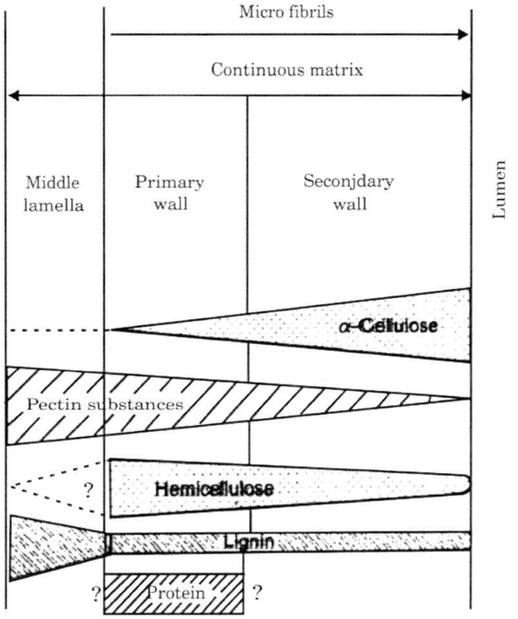

Fig. 15.2. Distribution of major constituents in the plant cell wall.

Cellulose consists of a chain of large number of β-glucopyranose units in 1,4-glucosidic linkage. It has 14000 glucose molecules in plants, 2,500 molecules in *Acetobacter xylinum* and 25,000 in green algae *Valvonia* per molecule. However, in relation to structural conformation of cellulose, the disaccharide cellobiose rather than monosaccharide glucose has to be considered as the elementry unit. In nature cellulose molecules are arranged in fibrils consisting of several parallel cellulose molecules held together by hydrogen bridges. These are microfibrils arranged length wise, oriented randomly in a primary wall and parallely in secondary wall (Muhlather, 1967). Cellulose contains both fibrillar regions (crystalline regions) and amorphous regions with large intercellular spaces filled with amorphous substances (Fig. 15.3). The high mechanical strength and insolubility of cellulose are due to presence of inter- and intramolecular H-bridges.

The cellulose is cleaved by a group of enzymes known as cellulases. The organisms which are capable of producing these enzymes are cellulolytic organisms. Cellulosic substrates are reported to be degraded by atleast three enzymes: (1) Endo-β-1-4 glucanase which attacks β-1,4 bonds randomly at any point of the macromolecule and produces fragments of different chain length with free ends. (2) The exo-β-1,4-glucanase (cellobiohydrolase) releasing the disaccharide (cellobiose) from one end of cellulose chains. (3) The β-glucosidase hydrolyses cellobiose with the liberation of glucose. All these enzymes are hydrolytic and act either in a sequential manner or simultaneously (Fig. 15.4 and Table 15.1). The exo-glucanase can degrade only crystaline cellulose. Cellulose is degraded and utilized in well aerated soil by aerobic microorganisms (fungi, mycobacteria and other eubacteria) and under anaerobic conditions by bacteria, a few anaerobic fungi and protozoa. Cellulases play a key role in the carbon cycle of the biosphere. Cellulases will also be required in biological conversion of plant biomass into fuels and basic chemicals. It has been convincingly demonstrated that different enzymes act in a synergistic manner in the hydrolysis of high ordered cellulose.

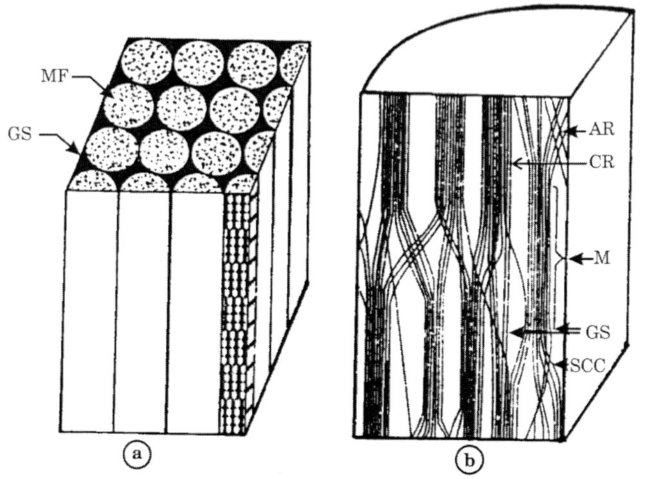

Fig. 15.3. Gross structure of cellulose and microfibrils (a) and of the arrangement of cellulose molecules within a microfibril (b) MF = microfibril, GS = ground substance (pectin, hemicellulose, or lignin), AR = amorphous region of cellulose, CR = crystalline region of cellulose, M = micelle, SCC = single cellulose chain (molecule).

Table 15.1. Classification of cellulolytic enzymes

				Producing organisms
Celluloytic enzymes	β-1,4-glucanase	Endo-β-1,4-Glucanase	(= C_x-cellulase)	Many fungi
		Exo-β-1,4-Glucanase	β-1,4-glucan cellobiohydrolase (= C_1-cellulase)	*Trichoderma, Penicillium, Fusarium*
			β-1,4-glucan glucohydrolase	*Trichoderma*
			(=cellobiase)	Many fungi
	β-1,4-glucosidase			

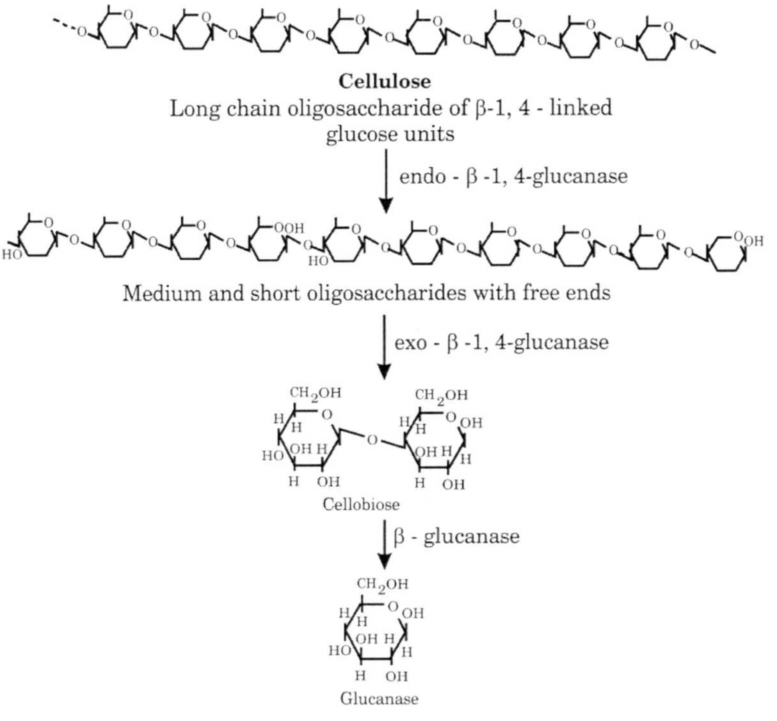

Fig. 15.4. Enzymatic degradation of cellulose.

Montenecourt and Eveleigh (1979) gave a detailed account of stages in the enzymatic hydrolysis of cellulose. According to this, endoglucanase (C_x) acts on amorphous regions in the cellulose fibers. This opens up new chain ends for the attack by cellobiohydrolase (C_1) which then removes cellobiose units from the non-reducing end of the cellulose chains. In this way endoglucanase and cellobiohydrolase act synergistically creating sites of action for each other. β-Glucosidase further enhances the total hydrolysis of cellulose by removing cellobiose, the end product inhibitor of cellobiohydrolase and endoglucanase. In this scheme endoglucanase is believed to be the rate limiting factor. This theory was supported by electron microscopic observations of the enzymatic hydrolysis of cellulose (White and Brown, 1981). Some of the organisms are efficient in secretion of these enzymes and characteristically designated as cellulolytic organisms. These include a variety of aerobes, anaerobes, mesophiles and thermophiles (Table 15.2).

Table 15.2. Cellulose-digesting microorganisms

Eukaryotes		Prokaryotes	
Aspergillus fumigatus	a	*Cytophaga*	a
Aspergillus nidulans	a	*Sporocytophaga*	a
Botrytis cinerea	a	*Archangium, Sorangium*	a

Eukaryotes		Prokaryotes	
Chaetomium globosum	a	*Polyangium*	a
Fusarium	a	*Pseudomonas fluorescens* var. *cellulosae*	a
Myrothecium verrucaria	a	*Cellulomonas fimi*	a
Trichoderma viride	a	*Streptomycetes*	a
Trichoderma reesei	a	*Clostridium thermocellum*	b
Rhizoctonia solani	a	*Clostridium cellobioparum*	b
Neocallimastix frontalis	b	*Ruminococcus albus, R. flavefaciens*	b
Diplodinium	b	*Bacteroides succinogenes*	b
Entodinium	b	*Butyrivibrio fibrisolvens*	b
		Eubacterium cellulosolvens	b
		Cellovibrio flavescens	a

*a-aerobe; b-anaerobe

Cellulose degradation occurs in a variety of ecological niches where plant residues accumulate. As a general rule, cellulolytic microorganisms found in natural habitats constitute mixed populations with non-cellulolytic species which interact synergistically. These interactions lead to efficient cellulose degradation with the formation of carbon dioxide and water under aerobiosis or CO_2, methane and water under anaerobiosis. Rumen of herbivores is one such anaerobic ecological nich where resident flora interact with cellulolytic organisms and they not only degrade cellulose synergistically but also contribute richly to nitrogen metabolism of animal (Fig. 15.5).

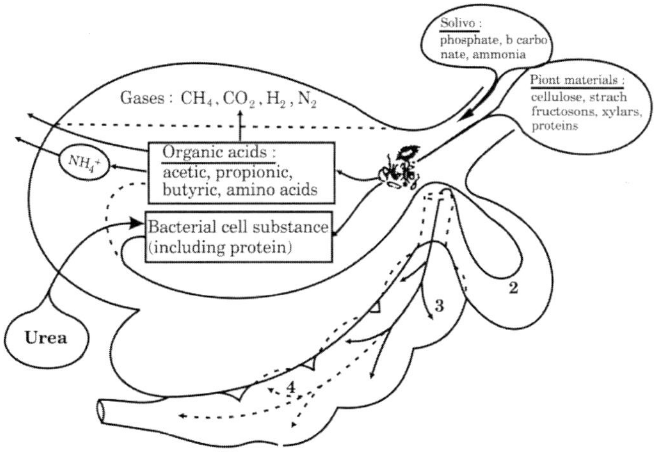

Fig. 15.5. Microbial conversions in the rumen.

In most cellulolytic organisms, cellulolytic enzymes are inducible and are subject to catabolite repression. It has been assumed that cellulolytic enzymes are secreted into the surrounding medium. However, many studies show that these enzymes are not released readily from the cell wall. The concept has developed that the cellulase complex remains in association with the cell and functions in a cell bound multienzyme complex or cellulosomes.

In *Clostridium thermocellum,* cellulosome is a high molecular weight complex that is found both in the culture medium and at the cell surface where it mediates adhesion of the cells to the substratum. Such structures are also reported on the surface of cellulolytic bacteria indicating that the cellulosome concept may be a general feature of these organisms (Fig. 15.6).

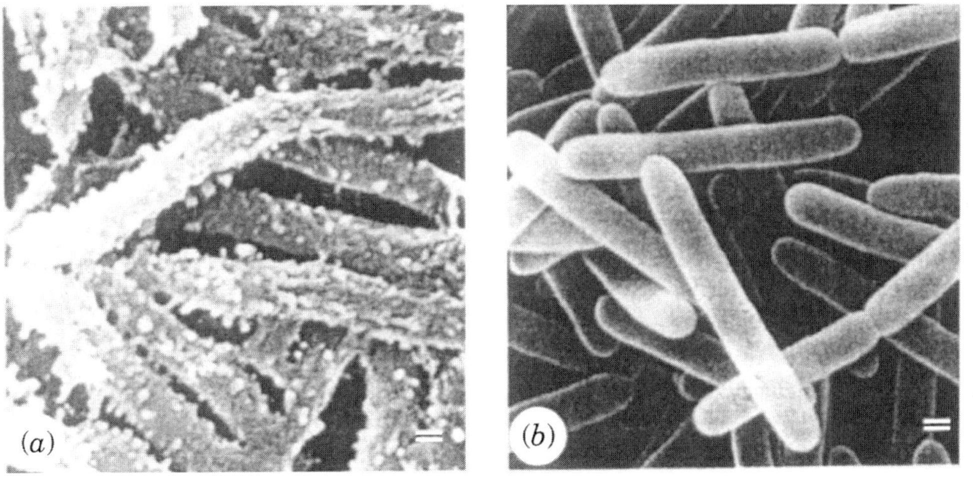

Fig. 15.6. Cellulosomes of *C. thermocellum.* Scanning electron microscopy (SEM) of cationized ferritin-labeled cellobiose-grown cells.

The cellulosome of *C. thermocellum* contains endoglucanase, cellobiohydrolase or hemicellulase activity. A glycoprotein named as CIPA (cellulosome integrating protein) promotes binding of cellulosome to the substratum and acts as scaffolding protein around which the catalytic components are organised as depicted in Fig. 15.7. Two major cellulosome components S_S and S_L act in a co-operative manner to degrade crystalline cellulose. The S_L component appears to serve as an anchor on the cellulose surface for the S_S component.

There is a wide diversity among cellulolytic organisms (Table 15.2). Arthropods associated with the degradation of wood own their ability to digest cellulose to the presence of microorganisms in their digestive system. The woodwasp, *Sirex cyaneus*, acquires cellulases and xylanases from the fungal symbiont, *Amylostereum chailletti*. Termites and wood feeding roaches live on a diet of cellulose as a result of the presence of the cellulose digesting symbiotic flagellated protozoa, *Trichonympha sphaerica*. Cellulolytic microorganisms are important in industrial applications, in certain ecological niches such as rumen and in the digestive systems of arthropods.

In most cases, cellulose fibrils present in plant walls are embedded in a matrix of hemicellulose and lignin. Hemicellulose designates a set of complex carbohydrate polymers, with xylans and mannans as the main components. The xylan backbone carries a variety of side chains including acetyl, arabinofuranosyl and methyl glucouronyl groups. Phenolic components including ferulic and coumaric acid are covalently bound to side chains and some are thought to be involved in cross links between hemicellulose and lignin mediated by ether linkages.

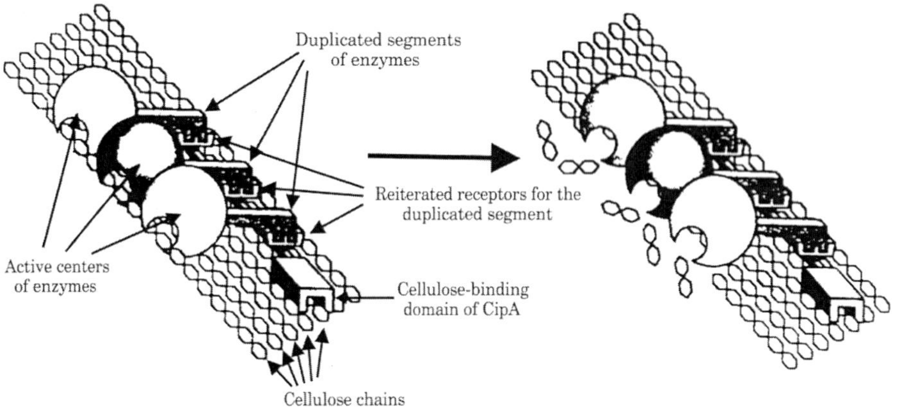

Fig. 15.7. Hypothetical model of the topological organization of subunits within the cellulosome.

Lignin is a phenylpropanoid polymer produced by the free-radical condensation of aromatic alcohols. The resulting compound is an amorphous, highly branched random polymer which is highly resistant to biodegradation and protect cellulose fibers against enzymatic attack by microbes.

The soil surface is the most important aerobic environment in which dead plant material accumulates. The most recalcitrant substratum is wood which is highly lignified. White rot fungus such as *Phaenochaete chrysosporum* a basidiomycetous fungi, can degrade both lignin and cellulose and play a major role in wood decay. Other aerobic fungi such as *Trichoderma reesei* and basidiomycetes can attack cellulose of wood causing brown-rot.

Aerobic cellulolytic bacteria such as *Cellulomonas, Cytophaga, Pseudomonas bacilli*, many actinomycetes like *Thermomonospora* and *Microspora* are thermophilic involved in cellulose degradation. Anaerobic fungi such as *Neocallimastix frontalis*, a resident of rumen of cattle are associated with cellulolytic bacteria *Ruminococcus flavefaciens, Fusobacteria succinogenes* and *Ruminococcus albus*. In other anoxygenic environments, cellulose degradation seems to be mainly by bacteria such as *Acetivibrio cellulolyticus* and various mesophilic clostridia and thermophilic *C. thermocellum* and *C. stercorarium*. The removal of excess sugars derived from cellulose hydrolysis increases the efficiency of various cellulase systems.

True celluloytic organisms are able to grow at the expense of crystalline cellulose, produce a complete cellulase system constituted of several enzymes each with a different specificity. In phytopathogens a restricted cellulolytic action may be sufficient for penetration into the plant tissue. In saprophytic microorganisms, limited cellulolytic activity could be useful for utilization of soluble cellodextrins resulting from the action of true cellulolytic organisms with which they are associated.

Due to extreme insolubility of ligno-cellulose and cellulose, enzymes that can digest these compounds must be either secreted into the medium as in fungi or bound to cell surface as in several cellulolytic bacteria. In addition, the compact structure of wood restricts the accessability of enzymes to their substrates. This may probably be the reason why fungi rather than bacteria are the main degraders of wood.

No organism is known to date which can grow on lignin. The white-rot fungus growing inside the wood first degrades exposed cellulose and hemicellulose sectors and releases sugars and nutrients. Subsequently, the organism shifts its metabolism towards lignin degradation due to prevailing starvation conditions, thereby exposing new cellulose and hemicellulose sectors and process continues. Many bacteria such as *Cellulomonas*, clostridia have been shown to bind to cellulose fibers. This property confers on these bacteria a selective advantage over other organisms being access to soluble sugars released by cellulose hydrolysis.

Cellulases and hemicellulases can be utilized for many applications involving the modification without complete hydrolysis of various lignocellulosic materials. The potential uses of cellulases are, therefore, many and divergent. They are valuable in various specialized nich applications (Table 15.3) even if their use for the bulk commercial conversion of wood depends on further technological advances.

Table 15.3. Applications of cellulases

1. Cellulases are being used to soften fruit pulp and increase the recovery of oil and fruit juice.
2. Pretreatment with cellulase has also been considered for the production of silage.
3. Cellulase and hemicellulase pretreatment are also used to upgrade feed stuffs for chicken by hydrolysing glucans and xylans which increase the viscosity of the bowls content reduce nutrient uptake and lead to the sticky faeces problem.
4. Endoglucanases reduce the viscosity of pulp which is detrimental to the tensile strength of the paper should be avoided allowing benefits accrued pulp processesing.
5. Addition of xylanses significantly reduces the amount of chlorine required for bleaching due to the disruption of lignin-hemicellulose matrix which causes partial delignification
6. Limited action by cellulases removes the microfibrils appeared due to wear and tear of cotton fibers of old clothes and restores the original crisp and texture of the fabric.

15.3 HEMICELLULOSE

Hemicellulose is the most abundant and widely distributed carbohydrate after cellulose. Straw and bark of woods consist of upto 30%, conifer woods about 7-12% and decidious wood 20-25% hemicellulose. This is not structurally related to cellulose nor it contains the same building blocks, but they are atleast partially soluble in water or alkali. It consists of either pentoses (xylose, arabinose) or hexoses (glucose, mannose, galactose) as well as uronic acids. It functions as storage and supporting substance in plants. However, use of term xylan is preferred over hemicellulose as similar polysaccharides were discovered in fungi and bacteria.

Xylan (hemicellulose) consists of β-xylose linked through 1,4-glycosidic bonds. It can be derived from cellulose chain by substituting hydrogen atom for the CH_2-OH groups, but its polymer size (number of units per polymer) is considerably lower (30-100). Some xylans also contain arabinose, glucose, galactose and glucuronate. Thus they have a complex structure in contrast to cellulose and are highly branched.

The xylan is degraded by xylanases. Many microorganisms are capable secreting these enzymes and degrade xylan more rapidly. Many cellulose degrading organisms are also producers of xylanses. Even *Sporocytophaga myxococcoides* which attacks cellulose only when it is in contact with cellulose fibers and *Neocallimastrix* excretes xylanase which attacks xylan depending upon the environmental conditions. In acidic soils fungi predominate, while in neutral and alkaline soils some bacilli and *Sporocytophaga* and other bacteria dominate. Xylan is even an excellent substratum for the cultivation of mushrooms.

Xylanase in some species of clostridia is secreted constitutively, while in other bacteria it is inducible. Its action results in the production of xylobiose and large fragments apart from xylose. Xylanase attacks randomly (endogenously) at many places simultaneously suggesting its complex nature rather than single enzyme.

15.4 STARCH

Starch, one of the most common storage compounds in plants, is found in two forms. α-Amylose consists of long unbranched chains of D-glucose in α (1-4) linkage, while amylopectin is highly branched in which the backbone consists of α(1-4) linkages with α (1-6) linkages at branch point (Fig. 15.8a & b), approximately every 25th glucose molecule. In addition, it contains phosphate residues and magnesium and calcium ions. Although not truly soluble in water, amylose form hydrated micellae in which the polysaccharide chain forms a helical coil. There are about 200-500 units per chain. Amylose is soluble in water without swelling in hot water and is responsible for the typical blue colour with iodine. Amylopectin from different sources differ considerably in their branching, the number of units per chain and other properties. It reacts with iodine to give violet-brown colouration. Starch is usually present as granules, which may be spherical, lens shaped or egg shaped and have a distinctly layered structure. *Aureobasidium pullulans* whose growth resembles that of yeast excretes pullulans (Fig. 15.8c). This glucan consists of maltotriose units in 1,6-α-glycosidic linkage (poly-α-1,6-maltotriose). Pullalan is broken down by *Pullulanase*.

Starch is acted upon by group of enzymes known as amylases. They are classified based on the site of attack, linkage bond they break etc. as detailed in Table 15.5 and Fig. 15.8a and 15.8b. Another form of amylase, β-amylase, also called as glucoamylase or amyloglucosidease cleaves glucose molecule from non-reducing end of amylose.

α-amylase cleaves amylose randomly releasing glucose, maltose and dextrins, while β-amylase acts step wise fashion from non-reducing end of starch molecules to liberate maltose molecule by acting on alternate glucosidic bonds. The β-amylase invert the configuration at the C-1 position where as α-amylases retain the α-configuration at the C_1 position. These enzymes also hydrolyse the α(1-6) glucoside and α-amylase is required to degrade amylopectin completely to glucose and maltose. Microorganisms producing different starch degrading enzymes are precised in Table 15.5.

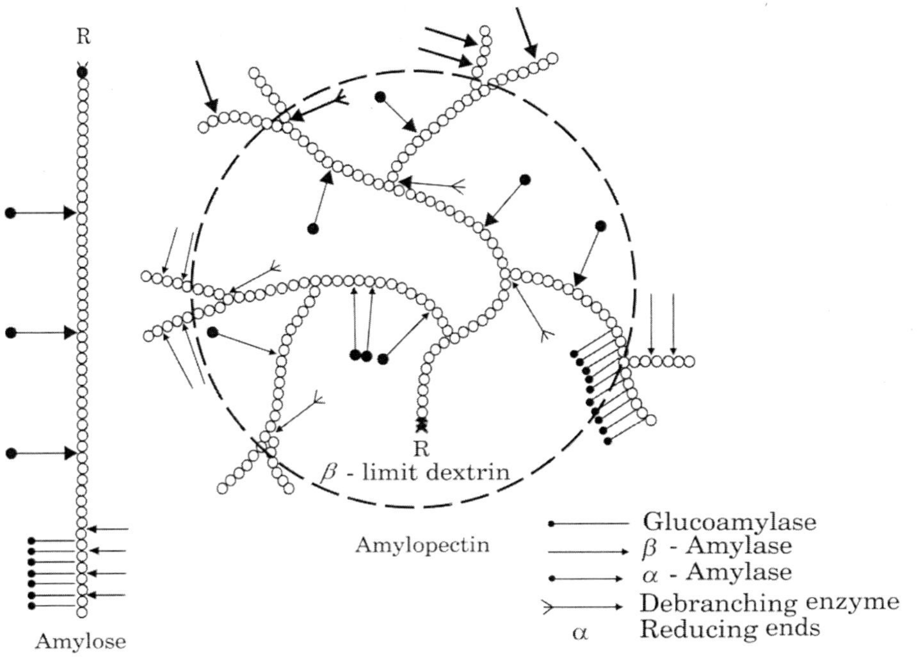

Fig. 15.8a. Target sites of the enzymes degrading amylose and amylopectin

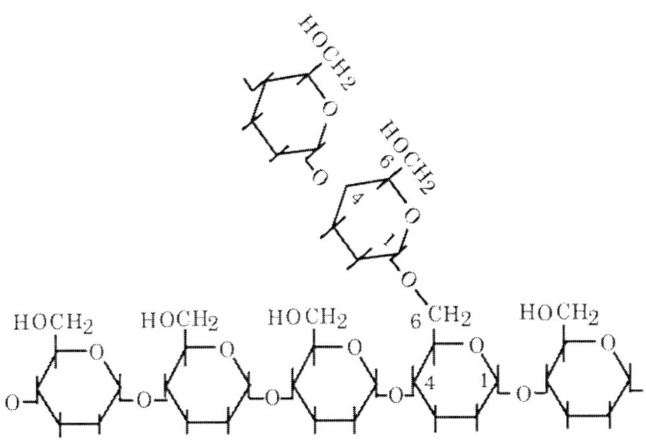

Fig. 15.8b. Branched structure of amylopectin. The structure of glycogen is similar except that there are more frequent a (1 -> 6) branches.

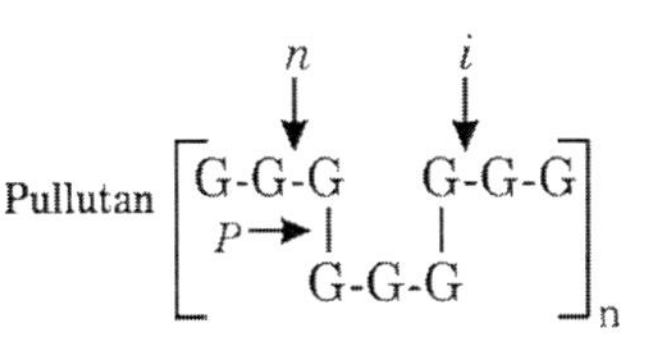

Fig. 15.8c. Structure of pullulan

Table 15.5. Classification of starch degrading enzymes

Enzymes				Producing organisms
Starch-degrading enzyme	α-1,4-glucanase	Endo-α-1,4-glucanase	α-amylase	Bacillus, Aspergillus
		Exo-α-1,4-glucanase	Exo-maltohexohydrolase	Aerobacter
			Exo-maltopentahydrolase	Bacillus
			Exo-maltohexohydrolase	Pseudomonas
			β-amylase	Bacillus, Streptomyces
			Amyloglucosidase	Certain bacteria, Aspergillus, Rhizopus
			Isopullulanase	Aspergillus
	α-1,4-glucosidase	Exo-α-1,6-glucanase	Pullulanase	Aerobacter
			Isoamylase	Pseudomonas
		Exo-α-1,6-glucanase	Exo-pullulanase	Cladosporium

Until recently, it was believed that β-amylase exists only in higher plants but now they are also reported in *Bacillus polymixa*, *B. megaterium* and other *Bacillus* species. The α-amylase from *B. subtilis* is a saccharifying enzyme that produces only glucose and maltose from starch, while α-amylase from *B. licheniformis* and *B. amyloliquifaciens* are liquifying enzymes that produce mostly malto-saccharides.

Starch can be converted to glucose by acid hydrolysis or enzymatically. There are three types of enzymatic degration of starch.

1. Phosphorolysis 2. Hydrolysis and 3. Transglycosylation

Phosphorolysis : Conversion of starch to glucose-6-phosphate is catalysed by a 1,4-glucan phosphorylase (phosphorylase) (Fig. 15.9).

Hydrolysis : Starch is attacked extracellularly by hydrolytic action of amylases.

Transglycosylation : By this process enzymes act on starch and produce closed circular chains of glucose in α 1,4-glycosidic linkage. These α, β, γ cyclodextrins contain 6, 7 or 8 molecules of glucose in the ring. *Bacillus macerans* is known to produce these enzymes.

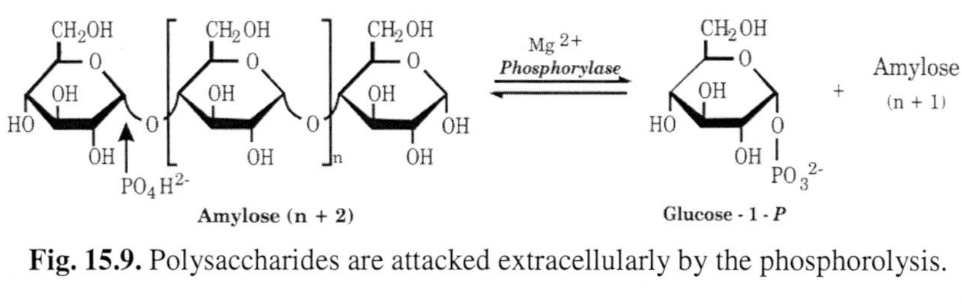

Fig. 15.9. Polysaccharides are attacked extracellularly by the phosphorolysis.

15.5. PECTIN

Pectin polysaccharides have been extensively studied because of their role in softening of fruit and because of wide spread commercial use of these substances as gelling agents. Pectic substances occur in varying amounts in all higher plant tissues and are found in the intercellular spaces (middle lamella) and largely responsible for integrity and coherence of plant tissue (Fig. 15.1). These substances act as cementing material between adjacent cells in the plant tissues. In the primary cell wall of plants, cellulose-microfibrils are found essentially linked through a hemicellulose-pectin network into a more or less rigid matrix (Fig. 15.2).

Inspite of the fact that the pectic substances constitute the major component of cell wall, their chemical composition is yet largely unknown. Talmadge *et al.* (1973) felt that the key pectic component in plant cell wall is rhamnogalacturonone. Pectic substances with about 1000-2000 units of α-1,4-galacturonic acid residues present in the primary cell wall is known as protopectin. Rhamnogalacturonone with 75% uronic acid carboxyl methyl esterified are known as pectinic acid and those with higher degree of esterification are referred as pectin, while unesterified molecule is known as pectic acid (Fig. 15.10).

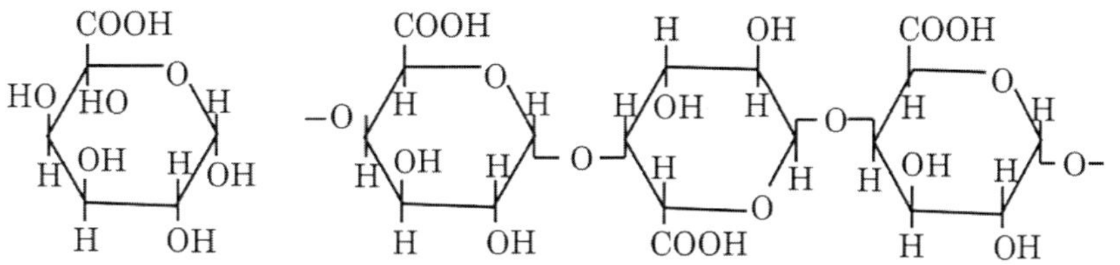

α d-galacturonic acid

Pectic acid

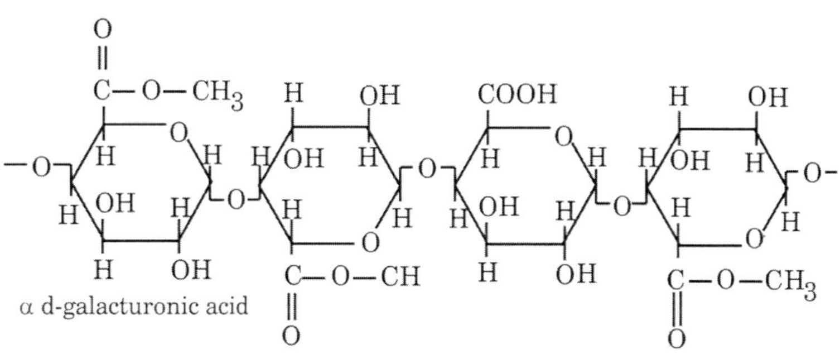

α d-galacturonic acid

Pectinic acid

Fig. 15.10. Structure of α-Dgalacturonic acid, A. Pectic acid B. Pectinic acid

Primary pectin chain consists largely of partially esterified D-galacturonic acid residues linked by α(1-4) linkages with small amount (2-4%) of L-rhamnose linked β(1-2) and β(1-4) to the D-galacturonate units. In addition to the primary chain structure, various side chains composed of arabinose, lactose and xylose are also present. Pectin of apple consists of hairy regions (backbone of rhamnogalacturonone carrying arabinogalactin and xylogalacturonan side chains) and smooth regions consisting of homogalacturonones with 70-80% residues methylation in an apparently random manner.

Protopectin, the parent compound of pectic substances exists as huge insoluble polymers of α-D-galacturonic acid residues. The majority of the COOH groups of the residue are esterified with methyl groups of the residue to form $COOCH_3$. The polymers are cross-linked at unesterified groups with divalent cations, calcium and magnesium.

An array of enzymes is required to degrade these complex substances, which are collectively known as pectinases. Pectic enzymes play an important role in pathogenesis, symbiosis, decomposition of plant debris, digestion of plant foods, retting processes, certain fermentations and fruit and vegetable spoilage. Pectinases are divided conveniently into two main groups.

1. Pectinesterase or saponifying or pectin methyl esterase, which removes methoxyl groups from pectin to yield pectic acid and methyl alcohol.
2. Depolymerases or chain splitting enzymes, which cleave the 1-4 glycosidic bonds of galacturonic acids of pectin. These enzymes are further classified on the basis of mechanism by which 1-4 glycosidic bonds are split.
   a. Glycosidases or hydrolases break the 1-4 glycosidic bonds by introducing a molecule of water.
   b. Transeliminases or lyases cleave the bond by removal of proton of C_5 of one residue, resulting in the formation of unsaturated bond between C_4 and C_5 and unsaturated galacturonyl units.

Newkom (1963) has classified the depolymerases into eight groups based on three criteria.

1. Nature of splitting of the glycosidic bond (hydrolytic or transeliminative).
2. Preference to the substratum (pectin or pectic acid).
3. The site of attack of the bond (exo- or endo-).

Depolymerases

I. Enzymes acting on pectin

A. Polymethyl galacturonase (PMG)
   i. Endo-PMG-hydrolyses pectin in a random fashion
   ii. Exo-PMG-hydrolyses pectin in a sequential fashion
B. Polymethyl galacturonate lyase (PMGL) or pectin lyase (PL)
   i. Endo-PMGL-causes random cleavage of pectin by transelimination process.
   ii. Exo-PMGL-causes sequential cleavage of pectin by transelimination process.

II. Enzymes acting on pectic acid (polygalacturonic acid)

A. Polygalacturonase (PG)
   i. Endo-PG-hydrolyses pectic acid in a random fashion
   ii. Exo-PG-hydrolyses pectic acid in a sequential fashion
B. Polygalacturonate lyase (PGL) or pectic acid lyase (PAL)

i. Endo-PGL-causes random cleavage of pectic acid by a random transelimination process.

ii. Exo-PGL-causes sequential cleavage of pectic acid by a transelimination process.

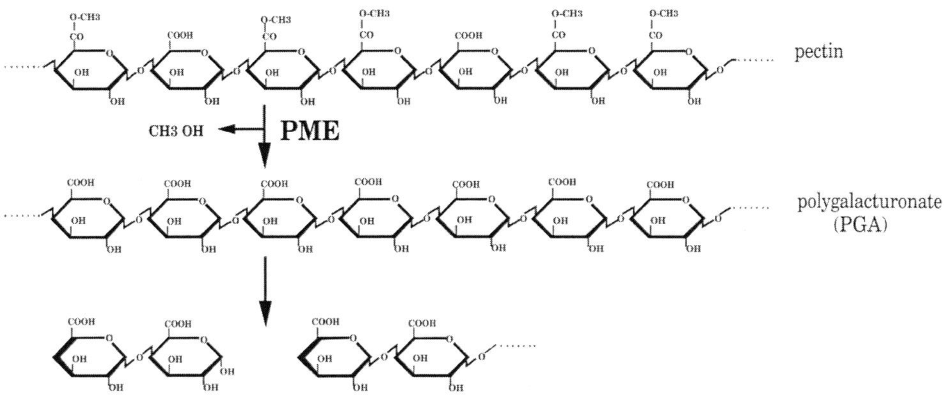

Fig. 15.11. Initial stages in the degradation of pectin. PME-Pectin methyl esterase.

Karr and Albersheim (1970) have reported a wall modified enzyme (WME) from commercial preparation of *Aspergillus niger* which converts the native pectin material to a form of pectic substrate which is more susceptible to the attack of other pectic enzymes.

Table 15.6. Classification of pectolytic enzymes

Enzymes					Producing organisms
Depolymerizing enzyme	Hydrolase	Endo-hydrolase	Endopoly Galacturonase		Bacteria, fungi Coniothyrium, Aspergillus
		Exo-hydrolase	Galacturono-Poly galacturonase	Not cleaving / Cleaving — 4,5-unsaturated galacturonic linkages	
			Digalacturono-Polygalacturonase		Aspergillus Erwinia
	Transeliminase	Endotrans-eliminase	Endo-pectic acid transeliminase		Erwinia, Clostridium Corynebacterium fungi
			Endo-pectin transeliminase		
		Exotrans eliminase	Exopectic acid eliminase		Erwinia, Clostridium
			Oligogalacturonide transminase		Erwinia, Pseudomonas

Many bacteria and fungi have the capacity to degrade pectin (Table 15.6). These enzymes are reported to play significant role in pathogencity and degradation of plant materials. Spore formers like *Bacillus macerans* and *B. polymyxa*, many pseudomonads, rumen bacteria, actinomycetes, thermophilic clostridia and lactic acid bacteria breakdown pectin more efficiently.

Pectin degrading organisms are of technical importance in the retting of flax and hemp under aerobic conditions in which fungi are involved, while under anaerobic conditions bacteria take active part. The degradation of pectin by microorganisms can be depicted as above (Fig. 15.11).

15.6. CHITIN

Chitin is the second most common polysaccharide on earth, after cellulose. Chitin can be derived from cellulose by substitution of the hydroxy group on carbon atom 2 of glucose with an acetyl amino group. These building blocks are linked together by 1,4-glycosidic bonds as per the glucose molecule in the cellulose. Chitin's pronounced stability is based on the hydrogen bonding of the n-acetyl side chain (Fig. 15.12). Chitin acts as a support medium in the animal as well as in the plant kingdom and forms the exo-skeleton of many non-vertebrates such as arthropods and other animal groups. It is also a significant component in the cell walls of large groups of fungi (ascomycetes, basidiomycetes and zygomycetes).

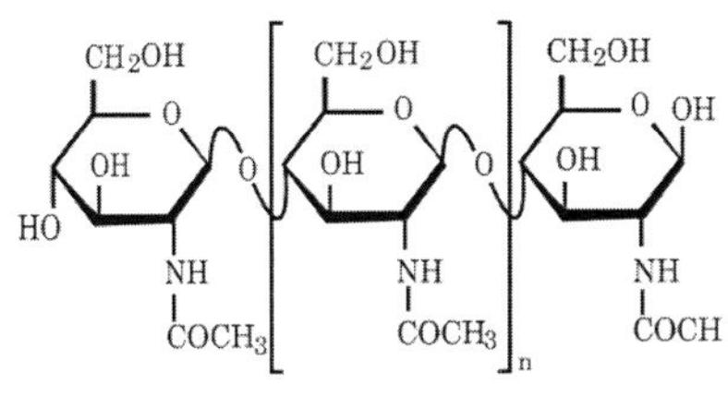

Chilin

Fig. 15.12. Structure of chitin

A large number of soil and aquatic bacteria can utilize chitin. Some of the important genera of bacteria such as *Flavobacterium, Bacillus, Cytophaga, Pseudomonas, Streptomyces, Nocardia* and *Micromonospora* are chitinolytic. Similarly quite a large number of genera of fungi, such as *Aspergillus, Mucor* and *Mortierella* are capable of utilizing chitin in agricultural soils. Presence of such a large number (10^6) of microorganisms in many agricultural soils suggests its common availability. The addition of finely divided chitin to soil provides a rapid response of accelerated growth in actinomycetes. Therefore, media with chitin is employed as a selective medium for the isolation of actinomycetes. Chitin is broken down anaerobically by some highly specialized chitinolytic obligate anaerobic bacteria. However, more detailed information about the chitin breakdown will be a great interest.

Chitinase, secreted extracellularly by *Streptomyces griseus* attacks chitin resulting in the production of N-acetylglucosamine and predominantly chitobiose and chitotriose. The latter are then converted to monomers by chitobiase. Degradation of chitin apparently occurs by simultaneous attack by chitinase at many points along the polymer.

Chitin can also be broken down to chitosan by *Absidia coerula*. Chitosan is a polyglucosamine and is produced by deacetylation of chitin. Chitosan is an important product of biotechnology and is used as an adhesive for wound dressings, as a chelator and as additive for soil and animal feed.

15.7 PROTEINS

The proteins of dead organisms are attacked and broken down by a large number of fungi and bacteria. Like other high molecular weight substances, proteins are first broken down outside the cell by extracellular enzymes, proteases, into oligopeptides and amino acids. The proteolytic activity of microorganisms is often assayed in solid media containing gelatin where proleolytic activity can be recognized by a clear zone formed around the colony in which gelatin is hydrolysed and no longer can be precipitated by hydrochloric acid. Some of the protease producing microorganisms are listed in table 15.7.

Oligopeptides and amino acids are taken into the cell by specific transport systems and peptides are broken down to amino acids by peptidases. The amino acids are then incorporated into cellular proteins or deaminated via specific pathways and then enter intermediatory metabolism and its terminal oxidising system to serve as energy source.

Table 15.7. Classification of proteinases (endo-type)

Enzymes			Producing organisms
Endo-type proteinase	Serine proteinase	Trypsin like serine	Streptomyces
		Alkaline serine	Bacillus, Streptomyces, Aspergillus
		Myxobacter a-lytic	Sorangium
		Staphylococcal	Staphylococcus aureus
		Neutral serine	Bacillus pumilus
		Other serine	Escherichia, Saccharomyces
	Metalloproteinase	Neutral	Bacillus
		Alkaline	Pseudomonas, Proteus, Serratia
	Acid proteinase	Pepsin type	Aspergillus, Mucor
		Rennet-like	Bacillus, Mucor, Endothia
	Thiol proteinase	Clostripain	Clostridium
		Streptococcal	Group A Streptococci

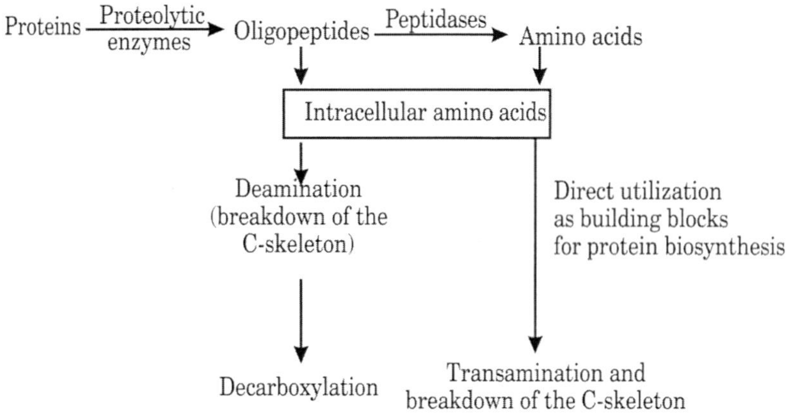

Proteases are enzymes which split peptide bonds of proteins. They may be extracellular which function in protein break down or intracellular which participate in cellular metabolism.

Similarly, peptidases which split peptides into amino acids, may be exo-peptidases which are with carboxy or amino terminal of the polypeptide chain or endo-peptidase (also called proteinases) hydrolyse the peptide bond within the polypeptide chain. There are atleast four types of endopeptidases, serine, cysteine, asparate and zinc proteinases. These intracellular proteinases are highly specific and play an important role in the regulation of cell metabolism by hydrolysis of signal peptides and activating or inactivating different enzymes.

Some of the extracellular proteases which are excreted into the medium act as toxins and virulnce factors (subtilisin, lysostaphine, streptokinase, elastase, haemolysin and other). Large quantities of proteases are produced industrially and are used as additives to detergents in the processing of leather and for other purposes.

Protein degradation in soil is accompained by the formation of ammonia. It is, therefore, referred to as mineralization of nitrogen or ammonification and many bacteria and fungi take part in this process.

Catabolic reactions of amino acid may be either decarboxylation or a deamination. The decarboxylation products are CO_2 and amines (biogenic amines). The best known amine (formerly called protamine) is cadaverine which is the deamination product of lysine. The primary amines occur during normal intestinal digestion and other during anaerobic degradation of proteinaceous materials.

$$H_2N—(CH_2)_4–CHNH_2—COOH \longrightarrow H_2N—(CH_2)_4CH_2NH_2$$

The deamination is the process of liberation of ammonia from amino acid. Deamination can be oxidative, desaturative and hydrolytic, which can be distinguished according to the fate of the carbon skeleton.

Oxidative deamination is the most common type of amino acid catabolism. Glutamic acid is oxidatively deaminated by glutamate dehydrogenase to 2 oxoglutarate. The action is reversible and is the most important in the metabolism of amino acids.

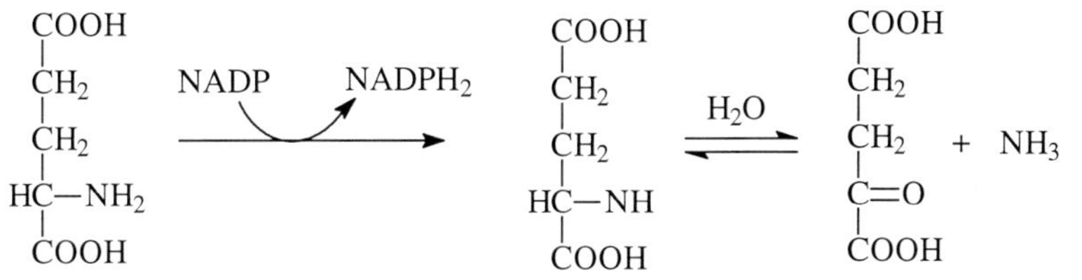

The desaturative deamination of aspartic acid yields fumarate. This reaction is catalysed by asparatase and is also reversible.

$$HOOC{-}CHNH_2{-}CH_2{-}COOH \rightleftharpoons HOOC{-}CH{=}CH{-}COOH + NH_3$$

The hydrolysis of urea is regarded as hydrolytic deamination. A large number of bacteria can utilize urea as a nitrogen source. The urea is hydrolysed by the enzyme urease.

$$H_2N{-}CO{-}NH_2 + H_2O \longrightarrow 2NH_3 + CO_2$$

In most bacteria, the ammonia represses the urease. In few other bacteria, urease is constitutive and are known as urea splitting organisms. In these organisms urease is constitutive and is not repressed by ammonia. Hence these bacteria hydrolyse all urea present to ammonia drifting pH 9 to 0 to which they are adapted.

In transamination, the amino group is transferred to 2-oxoacid which is mediated by transaminase.

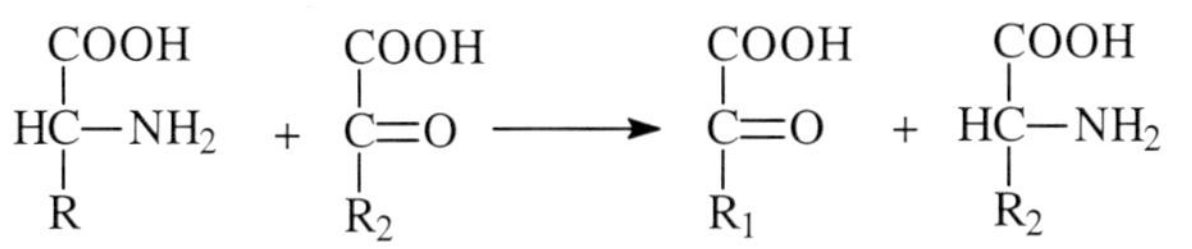

Transamination thus serves for the synthesis of some amino acids that can not be directly aminated by ammonium and at the same time serves for the catabolism of other amino acids. Pyridoxal phosphate is involved in the above reactions of decarboxylation and transamination. This co-enzyme of amino acid metabolism is related to pyridoxal known as vitamin B_6. The reactive group of pyridoxal phosphate is the aldehyde. It participates with the amino group of an amino acid in the formation of Schiff bases. In transamination the amino group remains on the pyridoxal phosphate whilst the carbon skeleton is liberated as oxoacid. The pyridoxal phosphate is then regenerated in the reaction with the (receptor) oxoacid. In the decarboxylation, the schiff base liberates CO_2.

The further fate of the carbon skeleton varies with the amino acid. Only a few of the deamination products are intermediates of the central metabolic pathways (pyruvate, 2-oxoglutarate, oxaloacetate). Other carbon skeletons enter intermediate metabolism in a specific catabolic reactions. The catabolic pathway of amino acid is presented with leucine as an example (Fig. 15.13).

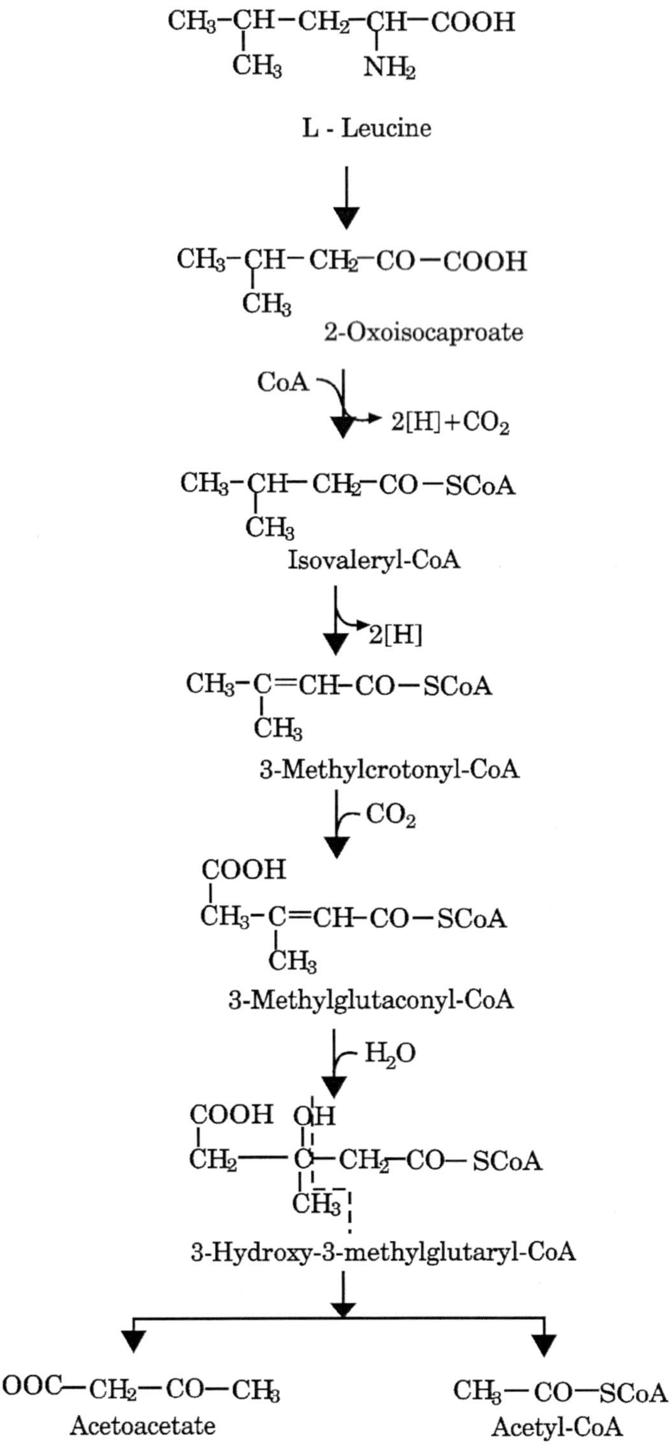

Fig. 15.13. Catabolic pathway of leucine.

15.8. FURTHER READING

Ajay Singh (1999), Engineering enzyme properties. *Indian J. Microbiol.*, **39 :** 65-77.

Bajpai, P. (1997), Bleaching pulp in enzymes in pulp and paper processing, pp. 1-37, Miller Freeman Books, San Francisco, CA.

Beguin, P. and Anbert, J.P. (1994), Biological degradation of cellulose. *FEMS Microbiology Rev.*, **13 :** 25-58.

Biely, P. (1985), Microbial xylanolytic systems. *Trends Biotechnol.*, **311 :** 286-290.

Buswell, J.A. and Odier, E. (1987), Lignin biodegradation CRC. *Rev. Biotechnol.*, **6 :** 1-60.

Eriksson, K.E.L., Blanchette, R.A. and Snder P. (1990), Microbial and enzymatic degradation of wood and wood components, Springer Verlag, Berlin.

Hugouvieu – cotte – Pattat, N.H. Dominguez and J. Robert-Bandouy (1992), Environmental conditions affect transcription of the pectinase gene of *Erwinia chrysanthemi* 3937. *J. Bacteriology*, **174 :** 7807-7818.

Solis, S., Flores, M.E. and Huntron, C. (1997), Improvement of pectinase production by interspecific hybrids of *Aspergillus* strains. *Lett. Appl. Microbiol.*, **24 :** 77-81.

Teeri, T.T. (1997), Crystalline cellulose degradation – New insight into the function of cellobiohydrases. *Trends Biotechnol.*, **15 :** 160-167.

Tomme, P., Jwarsen, R.A. and Giler, N.R. (1995), Cellulose hydrolysis by bacteria and fungi. *Adv. Microbiol. Physiol.*, **37 :** 1-81.

Visser, J. and Voragen, A.G.J. (1996), Pectins and pectinases, In *Progress in Biotechnology*, Elsevier, Amsterdam.

Whitaker, J.R. (1990), Microbial pectolytic enzymes. In Microbial enzymes and Biotechnology (W.M. Fogerty and C.T. Kelly, 2nd ed. Pp. 133-176, Elsevier, London.

Wilson, D.B. (1992), Biochemistry and genetics of actinomycetes cellulases. *Crit. Rev. Biotechnology*, **12 :** 45-63.

■■■■

Chapter - 16

Aerobic Respiration

16.1. INTRODUCTION

The flow of energy and materials through a cell is essential feature of a cell life. As energy flows through living cells, it becomes transformed and redistributed in such a way to allow the cells to grow, multiply, carry out other cellular functions and perform cellular work. The chemical reactions involved in this flow of energy collectively form the process known as **cellular metabolism**. Cellular metabolism consists of a complex network of chemical reactions that capture energy and raw materials from the environment and allows them to be changed into forms that are used to sustain cells. The formation and utilization of ATP and the proton motive force (PMF) are keys to cellular energy transformations and the central focus of the cellular metabolism.

Heterotrophic or chemoorganotrophic metabolism requires a supply of preformed energy rich organic substances for production of cellular constituents and as a source of the chemical energy used to generate ATP. It essentially involves the oxidation of energy-rich reduced organic molecules to oxidized end products via a metabolic pathway that releases sufficient energy to be coupled to the formation of ATP. Obviously, an external molecule in highly oxidized state is required to act as the final electron acceptor (electron sink) whose reduction balances the oxidation of the initial organic substrate molecule (electron donor).

Types of respirations: In nature, a variety of oxidized compounds are available which serve as electron acceptors from the oxidizing electron donors. Depending upon the nature of electron acceptors, three types of energy generation metabolic pathways are identified in microorganisms

1. Aerobic respiration
2. Anaerobic respiration
3. Fermentation.

When molecular oxygen serves as the terminal electron acceptor the metabolic pathway is known as **aerobic respiration** (meaning it requires the presence of air). The pathway of aerobic respiration begins with an organic substrate molecule and typically combines it with oxygen in an oxidation–reduction process that ends with the formation of CO_2 and H_2O. In the process, a substantial amount of ATP is also generated. The respiration in which an inorganic molecule (such as sulfate, nitrate) serves as the terminal electron acceptor is known as **anaerobic respiration** (meaning, it does not require the presence of air). Some bacteria and yeasts, use an organic molecule as the electron acceptor. This type of respiration is **fermentative respiration**. It yields very little amount of energy when compared with aerobic and anaerobic respirations. Important differences of aerobic respiration, anaerobic respiration and fermentation are presented in Table 16.1. Metabolic pathways associated with aerobic respiration are discussed in this chapter. Anaerobic respiration and fermentation are discussed in forthcoming chapters.

Table 16.1. Differences between aerobic, anaerobic respiration and fermentation

Respiration	External electron acceptor	Mode of ATP production	End products
Aerobic respiration	Molecular O_2	electron flow through membrane bound proteins; chemiosmotic mode of ATP generation	CO_2, H_2O
Anaerobic respiration	NO^-_3, SO^-_4, CO_2 etc.	electron flow through membrane bound proteins; chemiosmotic mode of ATP generation	N_2, H_2S, S^0, CH_4 etc.
Fermentation	External electron acceptor not necessary Organic compound; a metabolic intermediate of organic substrate	substrate level phosphorylation; no chemiosmotic mode of ATP generation	Alcohols, acids, CO_2, H_2 etc.

16.2. AEROBIC RESPIRATION

Aerobic respiration is a universal energy generating process in majority of the prokaryotes and in almost all eukaryotes. However, the bacteria exhibit an amazing variation in aerobic respiration. They are able to process and channel a wide range of organic substrates into central metabolic pathway to produce energy and various precursor intermediates. Carbohydrates in general and glucose in particular are taken as a model to explain the oxidation reactions of aerobic respiration. Quite justifiably, these substrates are available in plenty in the environment and further the catabolism of these substrates involves all the steps till the final oxidation and based on which the catabolism of other types of substrates could be very easily understood.

In aerobic respiration glucose molecule is sequentially, nevertheless, completely oxidized to CO_2 and H_2O and releasing energy at different stages. Respiratory metabolism of glucose and other related substrates can be divided into three distinct phases.

i. Glycolysis

ii. Tricarboxylic acid cycle

iii. Electron transport and oxidative phosphorylation

16.3. GLYCOLYSIS

Glycolysis (glyco = sugar or sweet, lysis = break down) is the first phase of the aerobic respiration during which the organic molecule is partially oxidized to smaller molecules, usually with the generation of some ATP and reduced coenzymes. In the case of carbohydrates, a substrate molecule such as glucose is initially broken down to pyruvate. Microorganisms employ several metabolic pathways to catabolize glucose and other sugars. Three important routes of breakdown of glucose to pyruvate have been identified: (i) EMP-pathway (ii) Pentose phosphate pathway and (iii) Entner-Doudroff pathway. The operation of these pathways in various microorganisms is presented in Table 16.2.

Table 16.2. Some alternative pathways of glycolysis in various microorganisms

Microorganism	Cell type	EMP pathway	Phosphorylated ED pathway	Non phosphorylated ED pathway
Pyrococcus sp.	Archeal	--	--	+
Sulfolobus sp.	Archeal	--	--	+
Thermoplasma sp.	Archeal	--	--	+
Arthrobacter sp.	Bacterial	+	--	--
Azotobacter sp.	Bacterial	+	--	--
Alcaligenes chroococcum	Bacterial	+	+	--
Bacillus eutrophus	Bacterial	+	--	--
E. coli	Bacterial	+	--	--
Pseudomonas sp.	Bacterial	--	+	--
Rhizobium sp.	Bacterial	--	+	--
Thiobacillus sp.	Bacterial	--	+	--
Xanthomonas sp.	Bacterial	--	+	--

16.3.1. Embden-Meyerhof-Parnas pathway

Embden-Meyerhof-Parnas pathway also known as glycolysis or glycolytic pathway or Hexose biphosphate pathway is the most common pathway for glucose degradation to pyruvate. It is found in all major groups of microorganisms and functions in the presence or absence of O_2 *i.e.* it is common to all aerobic respiring organisms and to most anaerobic and fermentative bacteria. The exception to this is archaebacteria which donot use this pathway.

In this pathway (Fig. 16.1), glucose is first converted in a series of reactions to fructose 1,6-biphosphate which in turn is cleaved to two interconvertable 3-carbon sugars that enter a common set of catabolic reactions to form two pyruvates. The partial oxidation of glucose to pyruvate by this pathway releases ATP and also accompanied by formation of two reduced coenzyme ($NADH_2$) molecules.

EMP pathway initiates by coupling the conversion of ATP to ADP or, in the case of bacterial cells by using the phosphoenol pyruvate; phosphotransferase system (PEP : PTS) that converts glucose to glucose 6-phosphate during transport across the cytoplasmic membrane. The energy balance of the PEP : PTS is roughly equivalent to that of the direct conversion of glucose to glucose phosphate using ATP. Glucose 6-phosphate is then isomerized to fructose-6-phosphate, which is converted to fructose-1,6-biphosphate in a reaction that requires input of ATP energy.

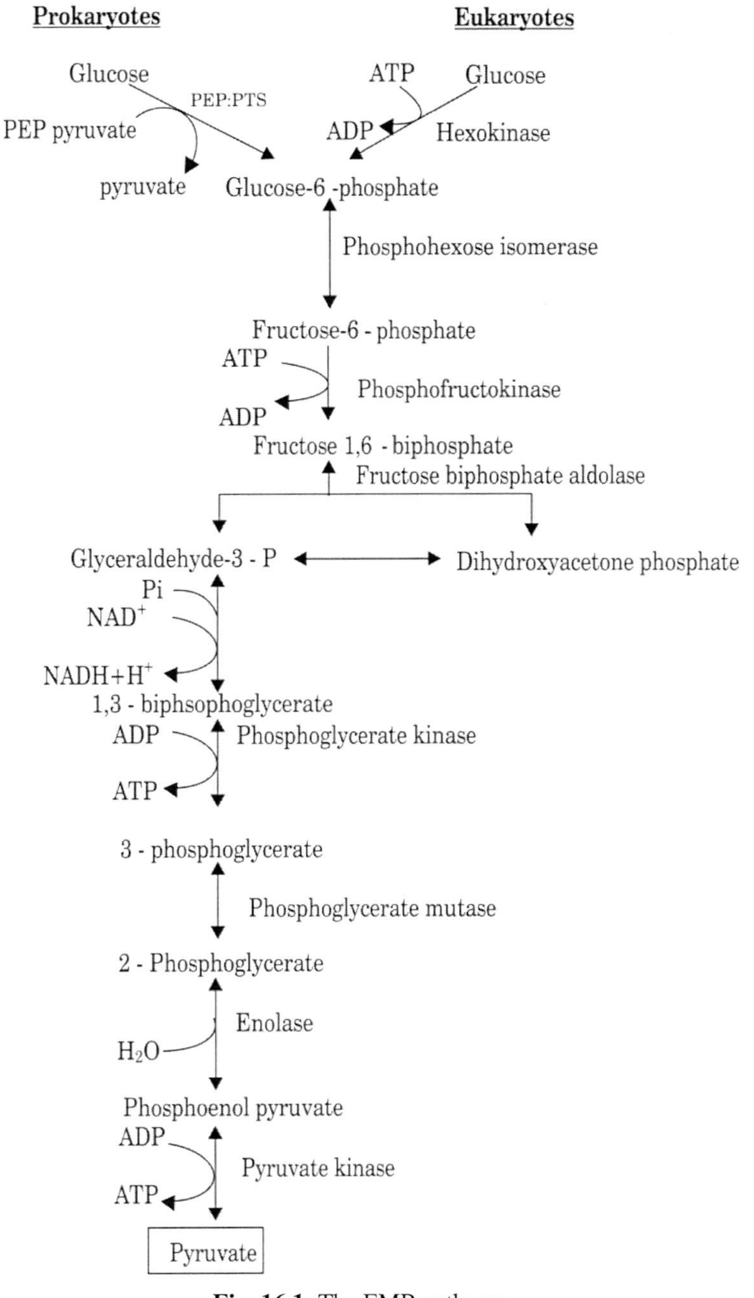

Fig. 16.1. The EMP pathway

The enzyme catalyzing the later reaction is phosphofructokinase. It is a key enzyme in regulating the rate of glycolysis. Thus, the result of first phase of glycolysis is the formation of a C_6–compound that can be broken down into two phosphorylated 3-carbon units without loss of energy. Fructose-1,6-biphosphate is split into two 3-carbon molecules: glyceraldehyde-3-phosphate and dihydroxyacetone phosphate (DHAP) by the action of the enzyme aldolase. DHAP which is not in the direct glycolytic pathway, can be converted to glyceraldehyde-3-phosphate. The equilibrium between DHAP and glyceraldehyde-3-phosphate favours the formation of DHAP. However, the constant removal of glyceraldehyde-3-phosphate shifts the balance of reactant and products, so that the DHAP is converted to glyceraldehyde-3-phosphate. Thus, for each 6-carbon glucose substrate molecule, two molecules of glyceraldehyde-3-phosphate are formed. After the formation of glyceraldehyde-3-phosphate, the next portion of glycolytic pathway is concerned with using the energy stored in this compound to drive the synthesis of ATP through substrate level phosphorylation.

Each glyceraldehyde-3-phosphate molecule is converted to 1,3- biphosphoglycerate by the incorporation of inorganic phosphate (Pi) into the molecule during an exergonic reaction. The oxidative conversion of glyceraldehyde-3-phosphate to form 1,3-biphospho-glycerate is coupled with the conversion of oxidized NAD^+ to the reduced coenzyme NADH. Since there are two molecules of 1,3-biphosphoglycerate generated from each glucose molecule, there is a net production of two NADH molecules per molecule of glucose. The 1, 3-biphosphoglycerate is converted to 3-phosphoglycerate, an exergonic reaction that is coupled with the synthesis of ATP. The formation of ATP in this coupled reaction is a substrate level phosphorylation reaction. As this reaction occurs for each of the two 3-carbon molecules generated from glucose, two molecules of ATP are generated per glucose molecule. Therefore, the synthesis and utilization of ATP are balanced at this point in the metabolic pathway and the net production of ATP is zero.

The 3-phosphoglycerate is then further converted to phosphoenol pyruvate (PEP) and finally to pyruvate. The conversion of PEP to pyruvate is coupled with the synthesis of additional ATP. Thus, glycolytic pathway results in the conversion of the 6-carbon molecule glucose to two molecules of the 3-carbon molecule pyruvate, with the net production of two molecules of reduced coenzyme, NADH and the net synthesis of two ATP molecules.

The overall reaction for glycolysis by the EMP pathway can be written as follows:

$$\text{Glucose} + 2\ \text{ADP} + 2\text{Pi} + 2\ \text{NAD}^+ \longrightarrow 2\ \text{Puruvate} + 2\ \text{NADH} + 2\text{H}^+ + 2\ \text{ATP}$$

The glycolytic pathway serves not only to oxidize carbohydrate to pyruvate and to phosphorylated ADP, but also provides precursor metabolites for many other pathways. **For example, glucose-6-phosphate is a precursor to polysaccharides, pentose-phosphates and aromatic amino acids; fructose-6-phosphate is a precursor to amino sugars; DHAP is a precursor to phospholipids and to the amino acids glycine, serine and cysteine and PEP is a precursor to aromatic amino acids and to the lactyl portion of muramic acid.**

Methyl glyoxal pathway

In some bacteria like *E. coli, Clostridium* and *Pseudomonas* species, the methyl glyoxal pathway operates as an alternate to the EMP pathway when the cell experiences conditions of low phosphate concentration. When phosphate is the rate limiting agent, this pathway converts dihydroxyacetone phosphate to methyl glyoxal and then to pyruvate. This bypasses the phosphorylation step that converts glyceraldehyde 3-phosphate to 1,3-biphosphoglycerate, yet still produces pyruvate, which can be futher metabolized to generate ATP. On the whole, the methyl glyoxal pathway consumes two ATP molecules rather than generating ATP (Fig. 16.2).

16.3.2. Pentose Phosphate Pathway

The second pathway, the pentose phosphate (PP) or hexose monophosphate (HMP) pathway may operate at the same time as the glycolytic pathway or the Entner-Doudoroff pathway operates. It operates both under aerobic and anaerobic conditions and is important in anabolism as well as catabolism.

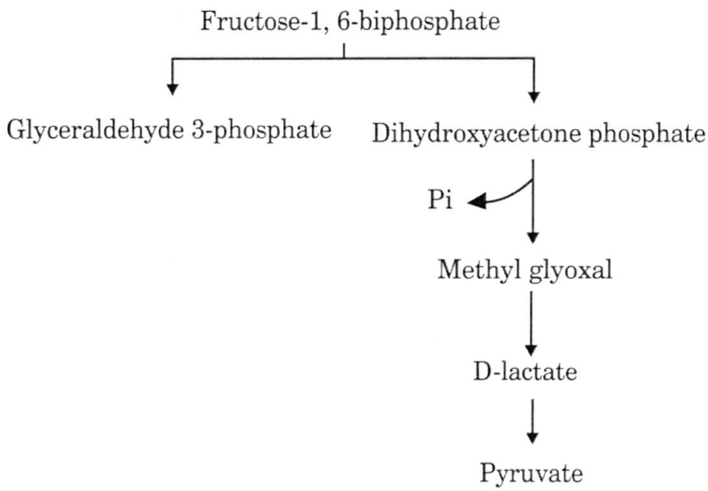

Fig. 16.2 : The methyl glyoxal pathway

The PP pathway (Fig. 16.3) begins with the oxidation of glucose 6-phosphate to 6-phosphogluconate. It is followed by the oxidation of glucose 6-phosphate and CO_2. NADPH is produced during these oxidations. The pentose sugar, ribulose 5-phosphate is then converted to a mixture of three through seven-carbon sugar phosphates. Two enzymes unique to this pathway play a central role in these transformations: (i) transketolase catalyzes the transfer of two-carbon ketol groups, and (ii) transaldolase transfers a three carbon group from sedoheptulose-7-phosphate to glyceraldehyde-3-phosphate. The overall result is that three glucose-6-phosphates are converted to two fructose-6–phosphates, one glyceraldehyde-3-phosphate and three CO_2 molecules as shown in the following equation

$$\text{3 glucose 6-phosphate} + 6\ NADP^+ + 3H_2O \longrightarrow \text{2 fructose-6-phosphate} + \text{glyceraldehyde 3-phosphate} + 3CO_2 + 6\ NADPH + 6H^{+\cdot} \quad \text{– (1)}$$

The intermediates produced in the above pathways are used in two ways. The fructose-6-phosphate can be changed back to glucose-6-phosphate while the glyceraldehyde-3-phosphate is converted to pyruvate by glycolytic enzymes. The glyceraldehyde-3-phosphate also may be returned to the pentose phosphate pathway through glucose-6-phosphate formation. This results in the complete degradation of glucose-6-phosphate to CO_2 and the production of great deal of NADPH.

$$\text{Glucose 6–phosphate} + 12\ NADP^+ + 7\ H_2O + ATP \longrightarrow 6CO_2 + 12\ NADPH + 12H^+ + Pi + ADP \quad - (2)$$

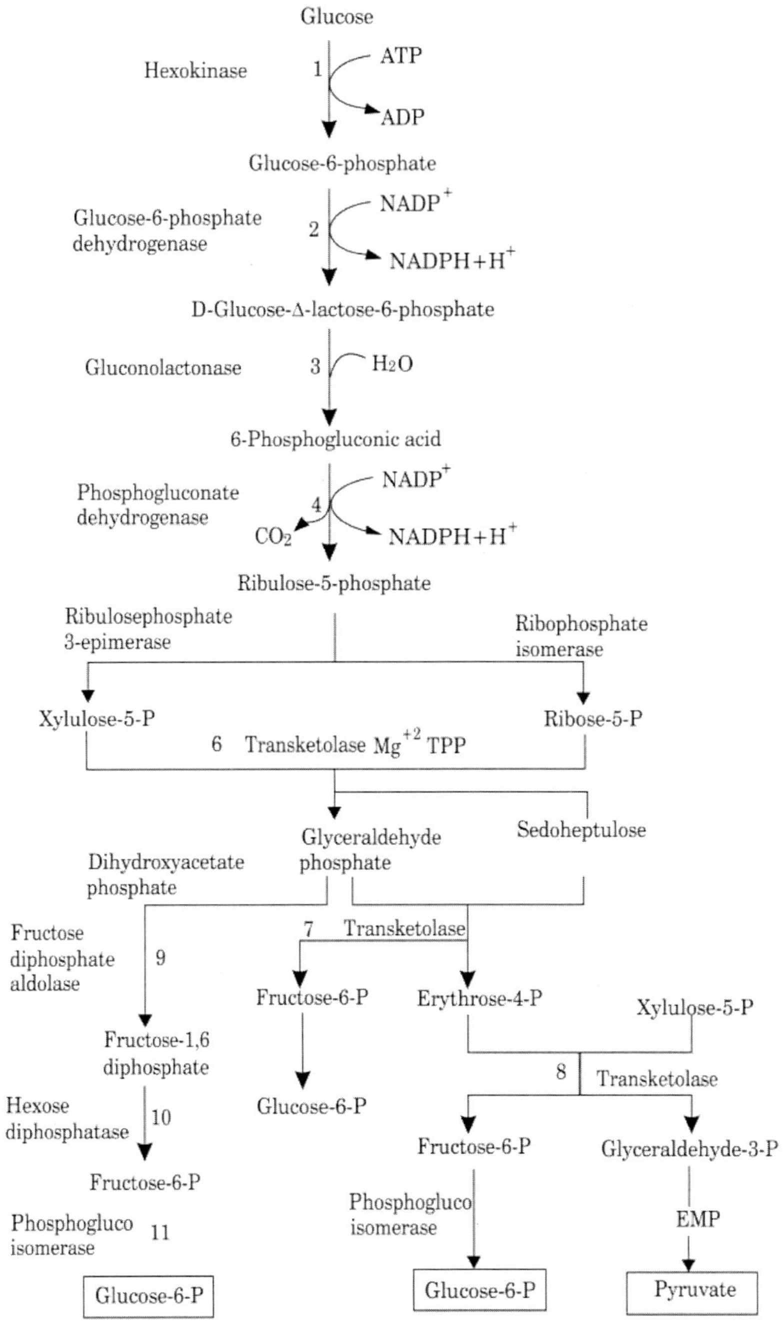

Fig. 16.3. The hexose monophosphate or pentose phosphate pathway.

Reactions 1 and 2 represent partly complete and complete pp pathways, respectively. On comparison of both complete and partly encomplete pathways, it can be concluded that oxidative microrganisms mostly utilize the partly complete pathway to obtain energy via pyruvate and the TCA cycle. However those microorganisms which metabolize glucose mainly though the EMP pathway under anaerobic conditions would utilize the complete HMP pathway only to the purpose of producing the precursors for their purine, pyrimidine, and aromatic amino acids biosynthesis.

As mentioned earlier, the PP pathway has several catabolic and anabolic functions that are summarized as follows:

1. The PP pathway may be used to generate ATP. Glyceraldehyde-3-phosphate from this pathway can enter the three-carbon stage of the glycolytic pathway and be converted to ATP and pyruvate. The later may be oxidized in the tricarboxylic acid cycle to provide more energy. In addition, some NADPH can be converted to NADH, which yields ATP when it is oxidized by the electron transport chain. So also, because five carbon sugars are intermediates in the pathway, the PP pathway can be used to catabolize pentoses as well as hexoses.
2. NADPH from PP pathway serves as a source of electrons for the reduction of molecules during biosynthesis.
3. Four and five-carbon sugars of this pathway serve as precursors for various biosynthetic pathways. The four-carbon sugar, erythrose-4-phosphate is used to synthesize aromatic amino acids. The pentose, ribose-5-phosphate is a major component of nucleic acids and ribulose-1,5-biphosphate is the primary CO_2 acceptor in photosynthesis.

16.3.3. Entner-Doudroff pathway

Some bacteria and archaea use an alternate pathway of glycolysis called Entner-Doudroff (ED) pathway (Fig. 16.4). This pathway is used by many aerobic bacteria, such as *Pseudomonas, Rhizobium, Agrobacterium, Xanthomonas* and *Thiobacillus* species. The net equation for Entner-Doudoroff pathway of glycolysis is

$$\text{Glucose} + NADP^+ + NAD^+ + ADP + Pi \longrightarrow 2\ \text{pyruvate} + NADPH + H^+ + NADH + H^+ + ATP$$

In this pathway, glucose 6-phosphate is oxidized to 6-phosphogluconate and then converted to 2-keto-3-deoxy-6-phosphogluconate (KDPG). KDPG is then cleaved to yield glyceraldehyde-3-phosphate and pyruvate directly. Since pyruvate is formed directly, some of the ATP generating steps are lost. The catabolism of glucose via the ED pathway results in the net production of only one ATP molecule per molecule of glucose, in contrast to the two ATP molecules produced in the EMP pathway. Thus, ED pathway is only 50% as efficient in ATP production as the EMP pathway.

Another difference between the ED pathway and EMP pathway is that in the ED pathway, the reduced coenzyme NADPH is generated from $NADP^+$. Generally, NADH is used in metabolic reactions that generate ATP, whereas $NADP^+$ and NADPH are used in biosynthetic reactions that build up molecules needed by the cell from simpler substrates. **The ED pathway provides an important mechanism for producing NADPH and the 3-carbon building blocks that are used in biosynthetic reactions, where the need for them is greater than the need for ATP.**

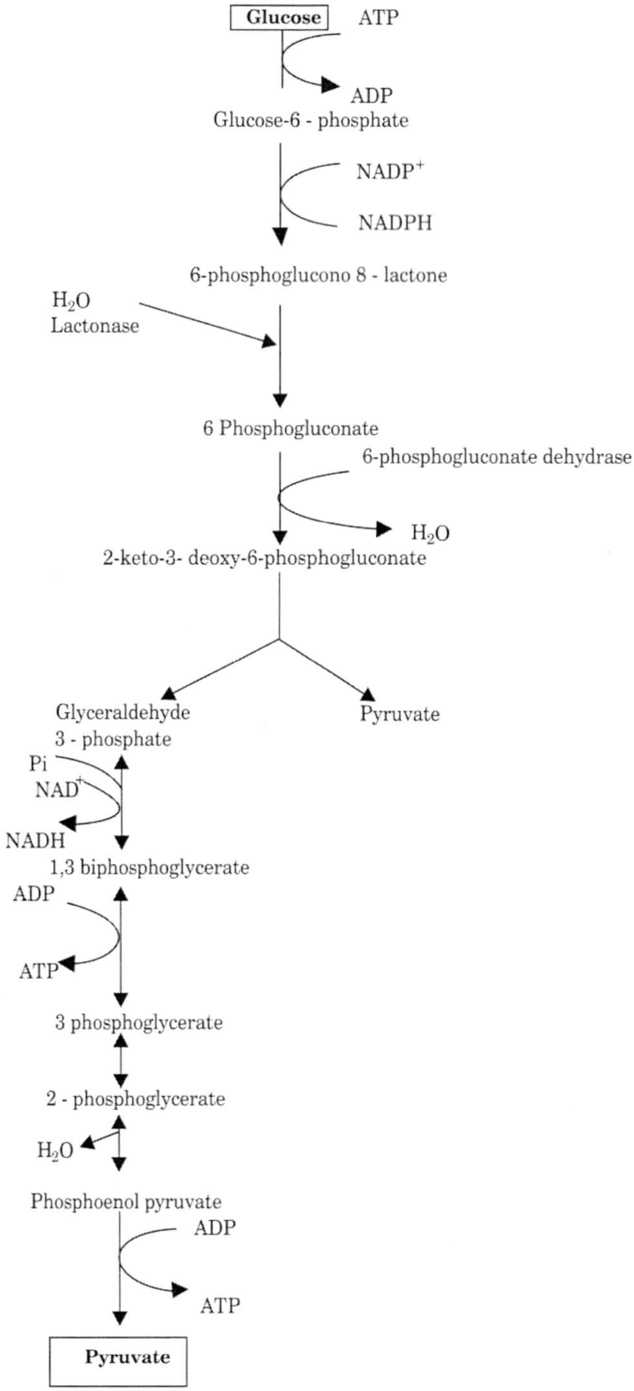

Fig. 16.4. The Entner-Doudoroff pathway.

Some archaea (*Halobacterium saccharovorum)* and bacteria (*Alcaligenes* species, *Achromobacter* species, *Clostridium* species and *Rhodobacter sphaeroides)* have a modified Entner-Doudoroff pathway that is partially nonphosphorylated. In this pathway, the intermediates formed prior to 2-keto-3-deoxygluconate are nonphosphorylated. Glucose is first converted to gluconate, which is subsequently dehydrated to 2-keto-3-deoxygluconate. The 2-keto-3-deoxygluconate is then phosphorylated to KDPG. The remaining steps of this pathway from KDPG to pyruvate are the same as in normal ED pathway.

16.4. TRICARBOXYLIC ACID CYCLE

In the second phase of respiratory metabolism of glucose, acetyl CoA (**also generated from the catabolism of fatty acids and proteins**) is fed into tricarboxylic acid cylce (TCA cycle). This results in the production of CO_2, water, reduced coenzymes and ATP.

TCA cycle is also known as citric acid cycle or the Krebs cycle. (Fig. 16.5) To enter this cycle, pyruvate molecules generated during glycolysis or protein catabolism first react with CoA in a reaction catalyzed by the pyruvate dehydrogenase complex. The carboxylation of pyruvate to acetylCoA and carbon dioxide is coupled with the conversion of the coenzyme NAD^+ to reduced NADH.

$$\text{Pyruvate} + NAD^+ + \text{CoA} \longrightarrow \text{Acetyl CoA} + \text{NADH} + CO_2$$

The resultant acetyl CoA can feed its acetyl group into the TCA cycle (Fig. 16.5). In the first step of TCA cycle, acetyl CoA reacts with oxaloacetate to form citrate and release CoA. Through a series of reactions involving carboxylic acids, the TCA cycle then regenerates oxaloacetate. During the TCA cycle, two reactions liberate carbon dioxide; the conversion of the 6-carbon compound isocitrate to the 5-carbon compound α–ketoglutarate, and the subsequent conversion of α-ketoglutarate to succinyl CoA (a 4-carbon compound). Reduced coenzyme NADH is generated during three reactions of the TCA cycle. The coenzyme flavin adenine dinucleotide (FAD) is also reduced to $FADH_2$ during the conversion of succinate to fumarate.

Conversion of succinyl CoA to succinate is an exergonic reaction. It is coupled with the generation of a high energy phosphate containing compound. In this reaction for some bacteria, ATP is formed in a substrate level phosphorylation by transfer of energy from the high energy succinate-CoA to ADP and Pi. In eukaryotes, in the same reaction, guanosine triphosphate (GTP) is synthesized from guanosine diphosphate and Pi. Some bacteria also form GTP rather than ATP in this metabolic reaction. GTP is the energy source used in some specific cellular reactions, most importantly during the synthesis of protein at the ribosomes. Nevertheless, the energy stored within GTP is equivalent to that stored within ATP. Further, the high energy phosphate group of GTP can be transferred to ADP to form ATP by the reaction:

$$\text{GTP} + \text{ADP} \longrightarrow \text{GDP} + \text{ATP}.$$

For energy accounting purpose, the GTP generated in this reaction will be treated as or equivalent to ATP. The net reactions of the TCA cycle can be summarized as:

$$2\,\text{Pyruvate} + 2\,\text{ADP} + 2\text{Pi} + 2\,\text{FAD} + 8\text{NAD} \longrightarrow 6\,CO_2 + 2\text{ATP} + 2\,FADH_2 + 8\,\text{NADH} + 8H^+$$

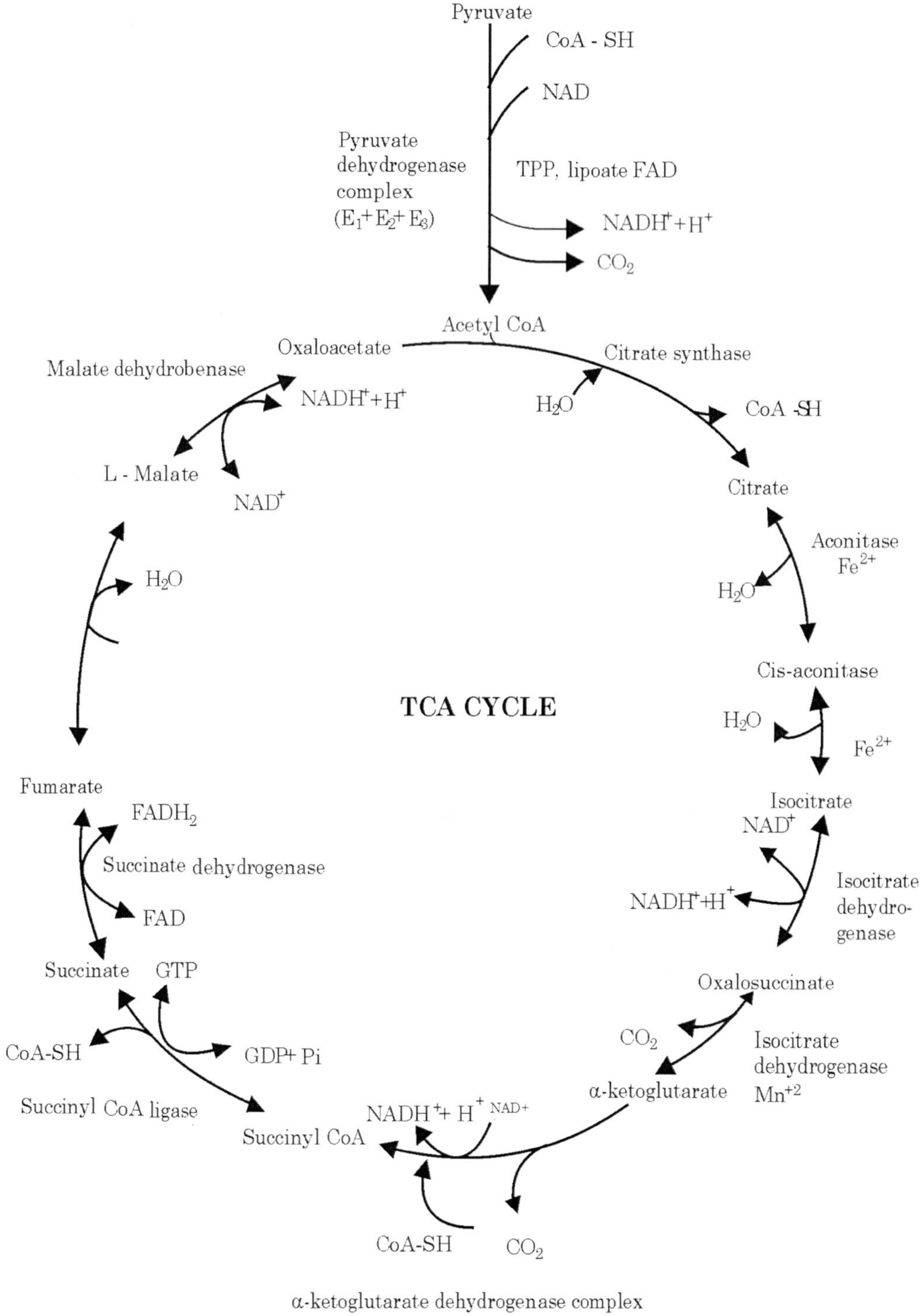

Fig. 16.5. The tricarboxylic acid cycle

At the end of the TCA cycle, all of the carbon from the original glucose molecule has been converted to CO_2. For the catabolism of glucose via EMP pathway, in the TCA cycle there will be a net synthesis of four ATP molecules: two from the EMP pathway and two from the TCA cycle. Ten reduced coenzyme molecules will be generated in the form of NADH (two from the EMP pathway and eight from the TCA cycle). Two reduced coenzyme molecules in the form of $FADH_2$ will come from TCA cycle.

In addition to its role in the overall respiratory generation of ATP, the TCA cycle also plays a central role in the flow of carbon through the cell. In aerobic organisms, the TCA cycle is an **amphibolic pathway**, one that serves in both anabolic and catabolic processes. Besides its role in the oxidative catabolism of carbohydrates, fatty acids and amino acids, the cycle provides precursors for many biosynthetic pathways through reactions that served the same purpose in anaerobic ancestors. α–ketoglutaric acid and oxaloacetate can, for example, serve as precursors of the amino acids, aspartate and glutamate by simple transamination. Through aspartate and glutamate the carbons of oxaloacetate and α–ketoglutarate are then used to build other amino acids as well as purine and pyrimidine nucleotides. Oxaloacetae is converted to glucose in gluconeogenesis. Succinyl CoA is a central intermediate in the synthesis of the porphyrin ring of heme groups, which serve as oxygen carriers (in hemoglobin and myoglobin) and electron carriers (in cytochromes). Citrate is released by some commercially exploited microorganisms.

16.4.1. Anaplerotic Reactions and TCA cycle

A cursory glance at the Fig. 16.6 shows that TCA cycle intermediates are used in the synthesis of pyrimidines and a wide variety of amino acids. In fact, the reactions of this pathway are so essential that most of it must operate anaerobically in order to supply biosynthetic precursors even though NADH is not required for electron transport and oxidative phosphorylation in the absence of oxygen. Thus there will be a heavy demand upon the TCA cycle to supply carbon for biosynthesis, and cyclic intermediates could be depleted if nothing was done to maintain their levels. However, microorganisms have reactions that replenish cycle intermediates so that the TCA cycle can continue to function when active biosynthesis is taking place. Reactions that replace cycle intermediates are called **anaplerotic reactions.**

Anaplerotic reactions vis-à-vis TCA cycle are reported from microorganisms, plant and animals. Usually, CO_2 is added to a an acceptor molecule (pyruvate or phosphoenol pyruvate) to form cyclic intermediate oxaloacetate. Some microorganisms (e.g., *Arthrobacter globiformis,* yeasts) use pyruvate carboxylase in this role.

$$\text{Pyruvate} + CO_2 + \text{ATP} + H_2O \xrightarrow{\text{Biotin}} \text{Oxaloacetate} + \text{ADP} + \text{Pi}$$

Other microorganisms such as *E. coli* and *S. typhimurium*, have the enzyme PEP carboxylase, which catalyzes the following reaction

$$\text{Phosphoenol pyruvate} + CO_2 \longrightarrow \text{Oxaloacetate} + \text{Pi}$$

16.4.2. The Glyoxylate Cycle

In the glyoxylate cycle (Fig. 16.7) acetyl CoA condenses with oxaloacetate to form citrate and citrate is converted to isocitrate, exactly as in the TCA cycle. The next step, however, is not the breakdown of isocitrate by isocitrate dehydrogenase but the cleavage of isocitrate by isocitrate lyase, forming succinate and glyoxylate. The glyoxylate then condenses with a second molecule of acetyl CoA to yield malate in a reation catalyzed by malate synthase. The malate is subsequently oxidized to oxaloacetate, which can condense with another molecule of acetyl CoA to start another turn of the cycle. Each turn of the glyoxylate cycle consumes two molecules of acetyl CoA and produces one molecule of succinate, available for biosynthetic purposes. The succinate may be converted through fumarate and malate into oxaloacetate, which can then be converted to PEP by PEP carboxykinase, and thus to glucose by gluconeogenesis.

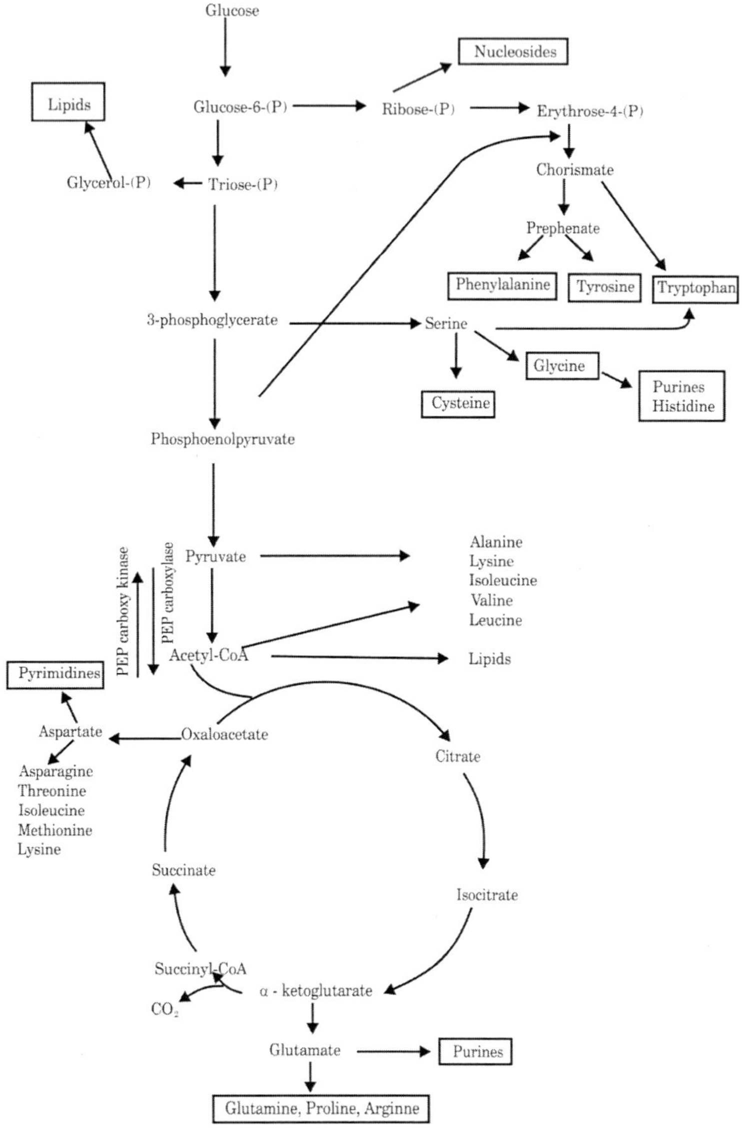

Fig. 16.6. Biosynthetic pathways starting from glycolysis and Krebs cycle.

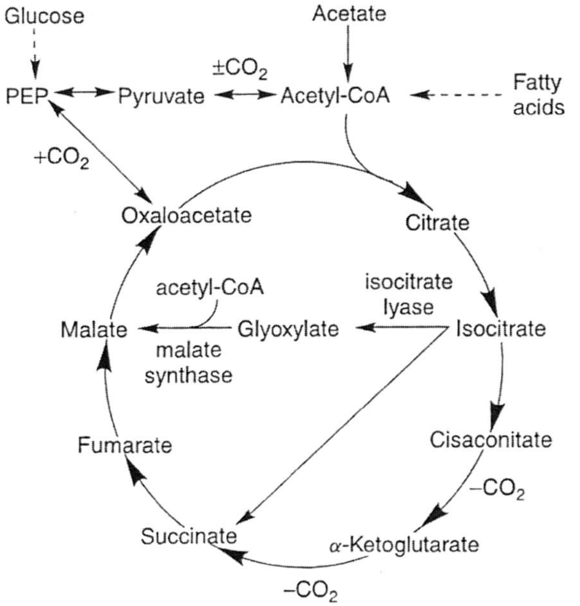

Fig. 16.7. The glyoxylate cycle or glyoxylate shunt.

The glyoxylate cycle cannot be used for the oxidative metabolism of acetate. In addition, since it does not involve formation of succinyl CoA, the sequence does not generate energy by substrate-level phosphorylation as is possible in the TCA cycle, although energy production may result from oxidation of NADH or FADH. The major role of the glyoxylate cycle is replenishment of A-carbon intermediates drained by biosynthesis. Since it tails to synthesize 5-carbon biosynthetic intermediates, the glyoxylate cycle cannot serve as a substitute for the TCA cycle, but only as an adjunct to it. The glyoxylate cycle allows aerobic, but not anaerobic microbes to use acetate for synthetic purposes.

16.5. ELECTRON TRANSPORT AND OXIDATIVE PHOSPHORYLATION

In the third and final phase of the respiratory metabolism of glucose and other substrates, reduced coenzymes generated in glycolysis and the TCA cycle are reoxidized. The process is called **oxidative phosphorylation**. The electrons and protons that are released during this process are transported through a series of membrane-bound carriers to establish proton gradient (proton motive force, ΔP) across a membrane. Electron transport ends with the reduction of a terminal electron acceptor (Fig. 16.8)

During oxidative phosphorylation, electrons from the reduced coenzymes NADH and FADH, are transferred through a series of membrane bound carriers that form an electron transport chain. The electron transport chain carriers reside within the inner membrane of the mitochondrion or in the bacterial plasma membrane. The mitochondrial system is arranged into four complexes of carriers, each capable of transporting electrons part of the way to O_2. Coenzymes Q and cytochrome **c** connect the complexes with each other. The bacterial carriers are also highly organized with in the membrane. The carriers in the electron transport chain participate in a series of reactions with increasing reduction potential difference between that of the primary electron donor and the terminal electron acceptor. Flavoproteins and iron-sulfur proteins transfer hydrogen from NADH to coenzyme Q. Electrons from coenzymes Q reduce a series of cytochromes, usually beginning with cytochrome c or cytochrome b. Cytochromes contain a central iron ion, which can be cycled between oxidized ferric state (Fe^{+3}) and the reduced ferrous state (Fe^{+2}). Ultimately, electrons are passed on to a cytochrome a /cytochrome oxidase (cytochrome o) complex and then to O_2 in aerobic respiration or to an alternate electron acceptor.

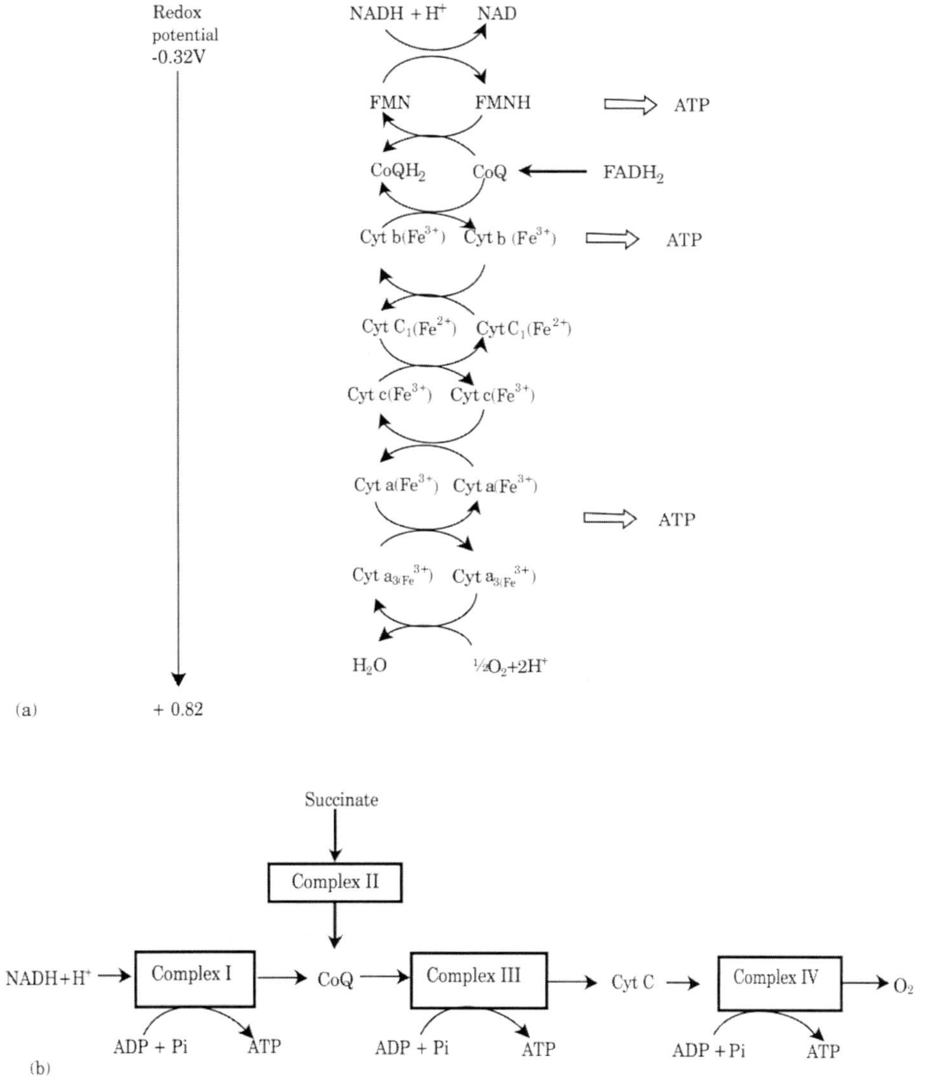

Fig. 16.8. The mitochondrial electron transport chain.

The process by which energy from electron transport is used to make ATP is called **oxidative phosphorylation**. This aspect has been discussed in detail in chapter 3. Thus as many as three ATP molecules may be synthesized from ADP and Pi when a pair of electrons pass from NADH to an atom of O_2. This is the same thing as saying that phosphorus to oxygen (P/O) ratio is equal to 3. Because, electrons from FADH, only pass two oxidative phosphorylation points, the maximum P/O ratio for FADH, is 2.

The interrelationship between glycolysis, TCA cycle, electron transport chain and oxidative phosphorylation is presented in Fig. 16.9.

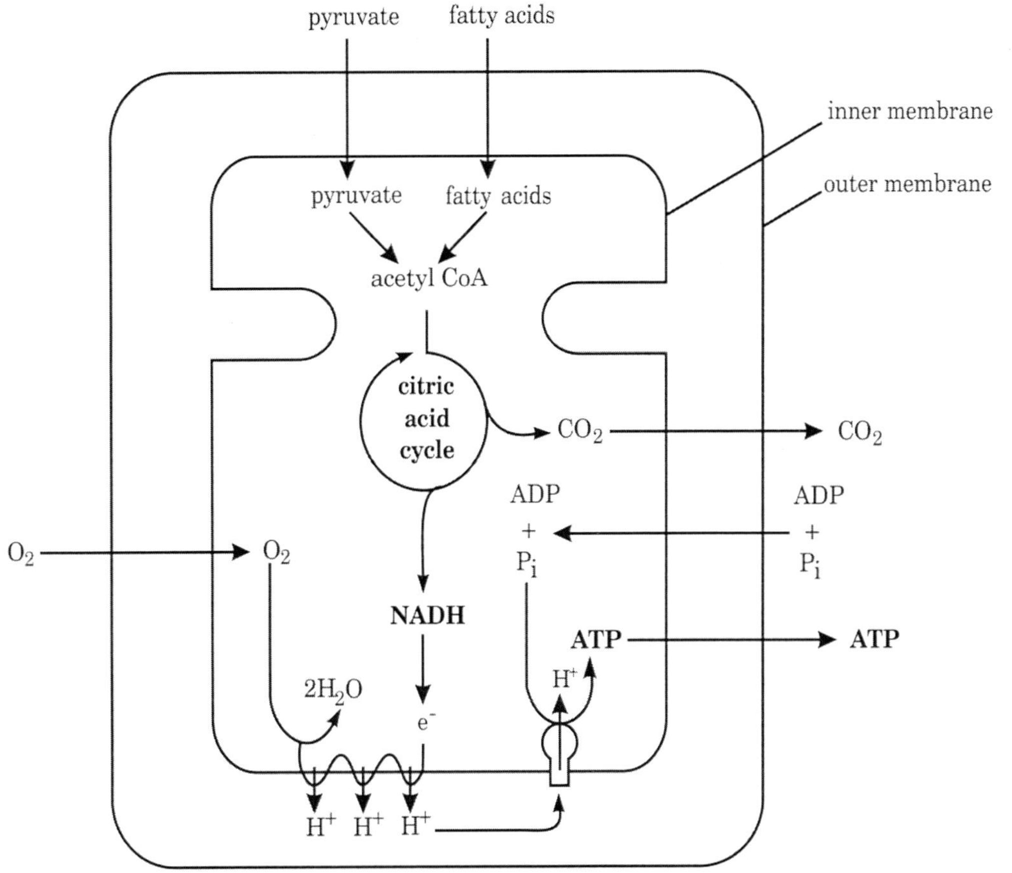

Fig. 16.9. Relationship between glycolysis, Krebs cycle, electron transport and oxidative phosphorylation.

16.6. THE YIELD OF ATP

The yield of ATP by the catabolism of glucose by the route of glycolysis, TCA cycle and electron transport can be readily calculated. Table 16.3 gives the consolidated account of ATP yield in aerobic respiration.

Table 16.3. ATP yield from the aerobic oxidation of glucose

EMP pathway	
Substrate level phosphorylation	2 ATP
Oxidative phosphorylation with 2 NADH	6 ATP
2 Pyruvate to 2 Acetyl-CoA	
Oxidative phosphorylation with 2 NADH	6 ATP
Tricarboxylic acid cycle	
Substrate level phosphorylation (GTP)	2 ATP
Oxidative phosphorylation with 6 NADH	18 ATP
Oxidative phosphorylation with 2 $FADH_2$	4 ATP
Total yield	**38 ATP**

Thus, aerobic oxidation of glucose to 6 CO_2 molecules yields a maximum of 38 ATP. However, only 2 ATP molecules are formed upon the anaerobic conversion of glucose to pyruvate. The fate of pyruvate under anaerobic conditions is the topic of the next two chapters.

16.7. CATABOLISM OF UNUSUAL CARBOHYDRATES

Although glucose is generally considered as the preferred substrate for hexose catabolism for many microbes, many other hexoses and hexitols may be used for energy production and biosynthetic intermediate formation. In addition, microbes exhibit substantial diversity in their ability to use a wide variety of 5-carbon sugars and their alcohol analogues. In many cases, the ability of microbes to use a wide variety of sugars and sugar alcohols reflects their ability to convert these substances into intermediates of central pathways.

Galactose is a sugar that may be catabolized by microbes. It is the hydrolysis product of lactose. Galactose is utilized differently by various species of microorganisms. In members of the genus *Lactobacillus*, lactose uptake is mediated by the phosphotransferase system and the intracellular lactose is hydrolyzed to yield galactose-6-phosphate. The latter is metabolized by the **tagatose pathway**. In the tagatose pathway, galactose-6-phosphate is isomerized to tagatose-6-phosphate which is converted to tagatose-1, 6-diphosphate. The later is then split in an aldolase-like reaction, yielding dihydroxyacetone phosphate and glyceraldehyde-3-phosphate that are converted to pyruvate via EMP pathway reactions.

E. coli exhibits a different mode of lactose utilization. Lactose transport is accomplished by an inducible transport system that converts the lactose to an intracellular mixture of D-glucose and D-galactose. This mixture is converted through a different kinase reaction to galactose-1-phosphate. The galactose-1-phsophate is converted to glucose-1-phosphate by a uridine diphosphate-mediated epimerase. Phosphoglucomutase then converts the glucose-1-phosphate to glucose-6-phosphate that by phosphohexose isomerase becomes fructose-6-phosphate. This is further metabolized by EMP pathway reactions. This pathway of utilization of lactose by *E.coli* is known as **Leloir pathway** (Fig. 16.10).

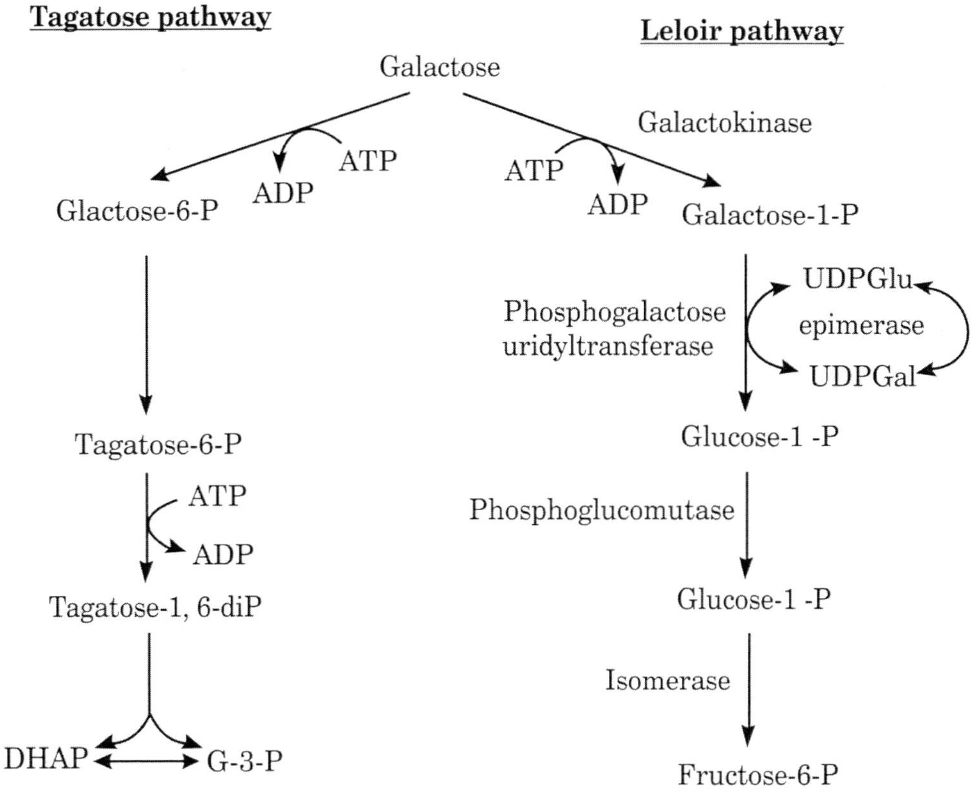

Fig. 16.10. The tagatose and Leloir pathways of galactose metabolism.

16.8. CATABOLISM OF ORGANIC ACIDS

A substantial number of microbes use organic acids as carbon and energy sources. For the most part, they do so by the process of beta-oxidation. The latter is accomplished by removal of acetyl CoA units. Thus with even chain carbon acids, acetyl CoA is the sole degradation product, while with odd-chain acids propionyl CoA is also produced.

16.9. CATABOLISM OF ORGANIC SUBSTRATES OTHER THAN CARBOHYDRATES

Bacteria are ubiquitous and grow in all types of habitats. Thus, they come across a wide variety of substrates which are to be utilized as carbon and energy source. Bacteria are the most versatile of all life forms in their metabolism. Nevertheless, they do not contain separate metabolic pathways for each of the substances they use. In most cases, they channel a diversity of materials into a relatively few number of central pathways, that are essential to life and are found in all cellular forms, not just the bacteria. To a great extent, the ability of bacteria to use a wide diversity of carbon and nitrogen sources for growth reflects their ability to channel unusual substances, or their metabolites into central pathways. In this way, the organisms generate energy from a variety of materials through the use of only a few metabolic sequences. Furthermore, the intermediates of the catabolism are usually the beginning points for synthesis. In a very real sense, synthesis begins where catabolism leaves off. Thus anabolism and catabolism cannot be separated. It is, some times, hard to decide whether a particular metabolic sequence is catabolic or anabolic. Fig. 16.6 shows how a variety of cell molecules use the intermediates of the central metabolic pathway as precursors for their synthesis. It is but natural that when these substances (proteins, amino acids, lipids, purines, pyrimidines, nucleosides etc.) undergo degradation, their products enter the central metabolic pathway through the respective entry points to undergo complete oxidation and yield energy. As such, the various anabolic process shown in Fig. 16.6 should be taken in reverse order for their catabolism.

16.10. FURTHER READING

Caldwell, D.R. (2000), Microbial Physiology and Metabolism, 2nd edn., Belmont, Calif, Star Publishing.

Garret, R.H. and Grisham, C.M. (1999), Biochemistry, 2nd edn., New York, Saunders.

Gottschalk, G. (1986), Bacterial Metabolism, 2nd edn., New York, Springer Verlag.

Jones, C.W. (1982), Bacterial Respiration and Photosynthesis, Washington, DC : American Society for Microbiology.

Lehninger, A.L., Nelson, D.L. and Cox, M.M. (1993), Principles of Biochemistry, 2nd edn., New York : Worth Publishers.

Moat, A.G., Foster, J.W. and Spector, M.P. (2002), Microbial Physiology, 4th edn., John Wiley & Sons, New York.

Peekhaus, N. and Conway, T. (1998), What is for dinner? Entner-Doudroff metabolism in *E. coli*. *J. Bacteriol.,* **180 :** 3495-3502.

Pode, R.K. (2000), Aerobic respiration, Oxidases and globins. In Encyclopedia of Microbiology, 2nd edn. Vol. I, J. Lederberg (ed.) 53-68, San Diego : Academic Press.

Romano, A.H. and Conway, T. (1996), Evolution of carbohydrate metabolic pathways. *Res. Microbiol.,* **147 :** 448-455.

Saraste, M. (1999), Oxidative phosphorylation at the *fin de siecle. Science,* **283 :** 1488-1493.

Vinopal, R.T. and Romano, A.H. (2000), Carbohydrate synthesis and metabolism. In J. Lederberg (ed.) Encyclopedia of Microbiology, 2nd edn., Vol. I, Academic Press, San Diego, pp. 647-668.

White, D. (1995), The Physiology and Biochemistry of Prokaryotes, New York : Oxford University Press.

Zubay, G. (1998), Biochemistry, 4th edn., Dubuque, Iowa : WCB/McGraw Hill

■■■■

Chapter - 17

Anaerobic Respiration

17.1. DEFINITION

In earlier chapter (Chapter 11) we have discussed the aerobic respiration. In aerobic respiration, molecular oxygen (O_2) functions as terminal electron acceptor, accepting electrons from electron carriers such as NADH by way of an electron transport chain. In fermentations, an organic molecule serves as both electron donor and electron acceptor. The electron acceptor is, usually, either pyruvate or derivative of pyruvate. However, some organisms living in environments devoid of oxygen make use a variety of oxidised inorganic or organic compounds to serve as the terminal electron acceptors. These organisms possess the electron transport chains very much similar to aerobes. Thus, the anaerobic energy yielding process in which the electron transport chain acceptor is an oxidised inorganic molecule other than oxygen is called **anaerobic respiration**. In most cases, the energy sources used by organisms carrying out anaerobic respiration are organic compounds, but several chemolithotrophic organisms can also carry out this process. Most of the organisms carrying out anaerobic respiration are prokaryotes and the chemical transformations they carry out during their energy generation processes are of great importance ecological importance for the understanding of the various elementary cycles in nature and the maintenance of equilibria in the biosphere. Certain products of anaerobic respiration are noteworthy: the release of a gas bubbles (N_2, NO_2, CH_4), the odour (H_2S), the formation of flammable gas (methane) and the production of diamagnetic iron oxide from non-magnetic ionic iron.

17.2. DIVERSITY OF MICROORGANISMS

The bacteria carrying out anaerobic respiration generally possess electron transport systems containing cytochromes, quinones, iron-sulfur proteins and other typical electron transport proteins. Their respiratory systems are thus analogous to those of conventional aerobes. In facultative anaerobes, such as denitrifying bacteria, the anaerobic respiration process competes with the aerobic one. In such cases, if O_2 is present, aerobic respiration is favoured, and when O_2 is depleted from the environment, nitrates (or other electron acceptors) are reduced. The organisms living under total anaerobic conditions and totally dependent on electron acceptors other than oxygen are called **obligate anaerobes**. In contrast, certain facultative and microaerophilic microbes can switch to an anaerobic respiratory growth mode. For example, the nonsulfur purple bacteria can respire anaerobically when light is unavailable to support photosynthesis. Thus, many of these microorganisms regulate their use of alternative modes of energy generation in order to maximize growth.

Anaerobic respiration can occur in any habitat devoid of oxygen and if suitable electron donors and acceptors are present to support cellular energy generation by electron transport coupled phosphorylation reactions. Typical habitats include anoxic soils, lake and stream sediments, bogs, waste dumps, composts and intestines of animals. The energy obtained by the cell during anaerobic respiration is then used to support other cellular processes that consume energy. These include the biosynthesis of cell precursors and their assembly into finished cell material, for transport of organic and inorganic nutrients into the cell and for maintenance of cells under stress conditions.

17.3. ELECTRON TRANSPORT CHAIN

Electron transport reactions, under anaerobic conditions, are performed by a variety of membrane bound enzyme systems, commonly termed as electron-transport chains. The electron transport chain couples the oxidation of the reduced electron donor (DH) with the reduction of an oxidized electron acceptor (A).

$$\underset{\text{(Reductant)}}{DH + A} \longrightarrow \underset{\text{(Acceptor)}}{D + AH}$$

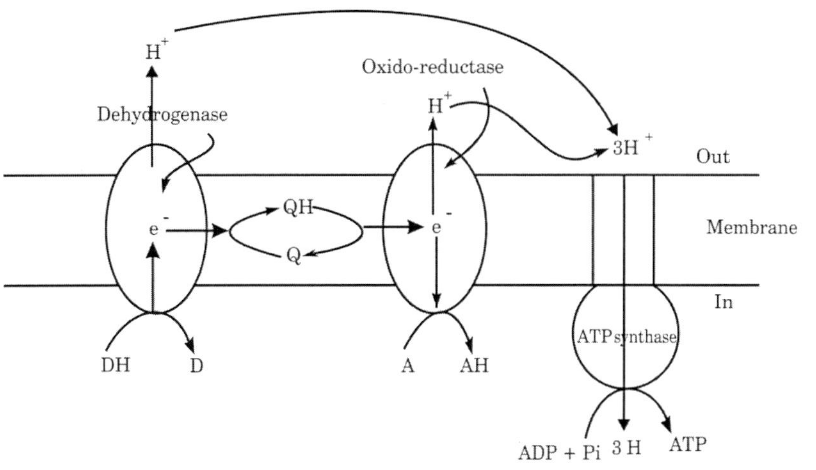

Fig. 17.1: A simple anaerobic respiratory pathway (DH = electron donor; Q = Quinone; A = acceptor)

Electron transport chain of anaerobes consist of atleast two enzyme complexes that are located on the cytoplasmic membrane of the cell. One is dehydrogenase and the second one oxidoreductase. In more complex electron-transport chains, additional enzymes or electron transfer proteins can be present. All these enzymes are freely diffusible in the two dimensional space of the cell membrane. A simple electron transport chain is shown in Fig. 17.1, where only two respiratory enzymes are employed to perform the coupled oxidation-reduction reactions. A quinone (Q) molecule is used to mediate the transfer of electrons from the electron donor reaction to the acceptor reaction. It may be recalled here that in anaerobic bacteria, it is menaquinone

(MK), while in most aerobic bacteria ubiquinone (Q) is used for electron transfer. In most cases, two electrons are transferred per reaction, but some times there is a single electron transfer. In certain complex redox reactions that involve several enzymes, upto 8 electrons can be transferred.

17.4. ELECTRON ACCEPTORS AND ENERGY CONSERVATION

The amount of energy that can potentially be generated by a given respiratory pathway depends on the redox potentials of the electron donor (DH) and electron acceptor (A). The energy released from the oxidation of an electron donor using O_2 as electron acceptor is higher than any other electron acceptor. These energy differences are not difficult to understand if one goes through the redox potentials listed in Table 3.4. It is because the O_2/H_2O couple is the most oxidizing and thus more energy is available. The other electron acceptors that are near to O_2 are Fe^{3+}, NO_3^- and NO_2^-. Still higher up in the scale are S^o, CO_2 and SO_4^{2-}. However, it must be kept in mind that the total energy released in electron transport chain is not utilized for cellular activities. A fraction of the energy generated by electron transport chain is released as heat.

The free energy ($\Delta G^{o\prime}$) released by the coupled oxidation-reduction reactions can be calculated as explained in chapter 3. The $\Delta G^{o\prime}$ values for several anaerobic respiratory pathway reactions are presented in Table 17.1 to illustrate the amount of energy a bacterium may potentially harvest when using different electron acceptors.

Table 17.1. Free energy released during the electron transport.

A	+ DH		AH + D	$\Delta G'^{o} = -nF.\Delta E'_o$
O_2	+ $NADH_2$	$\longrightarrow$	H_2O + NAD	-220 kJ
NO_3^-	+ $NADH_2$	$\longrightarrow$	NO_2^- + NAD + H_2O	- 145 kJ
DMSO	+ $NADH_2$	$\longrightarrow$	DMS + NAD + H_2O	- 92 kJ
TMAO	+ $NADH_2$	$\longrightarrow$	TMA + NAD + H_2O	- 86 kJ
Fumarate	+ $NADH_2$	$\longrightarrow$	Succinate + NAD	- 68 kJ
NO_3^-	+ Formate	$\longrightarrow$	NO_2^- + CO_2 + H_2O	- 167 kJ
NO_3^-	+ G-3-P	$\longrightarrow$	NO_2^- + glyceraldehyde	- 120 kJ

A = Acceptor, D = Donor, DH = Reductant

17.5. DISSIMILATORY AND ASSIMILATORY REDUCTIONS

Inorganic compounds such as NO_3^-, SO_4^{2-} and CO_2 are reduced by microorganisms to achieve two goals. Firstly, these compounds may be reduced as a source of nutrient viz., nitrogen, sulfur and carbon, respectively. The end products of such reductions are primarily amino groups ($-NH_2$), sulfohydryl groups (-SH) and organic carbon compounds. In such cases, the compounds are said to be assimilated and the process is known as assimilation, or **assimilative metabolism**. Secondly, these compounds may serve as terminal acceptors and thus facilitating the electron transport and oxidative phosphorylation releasing the ATP energy. This type of metabolism is referred to as **dissimilative metabolism**.

Table 17.2. Different kinds of anaerobic respirations.

Redox potential		Redox couples	Type of respiration	Representative species
-0.3	ANAEROBIC RESPIRATION	CO_2 → CH_3COO^-	Carbonate respiration homoacctogenic bacteria obligatc anaerobes	*Acetobacterium woodii* *Clostridium aceticum*
-0.27		S^0 → HS^-	Sulfur respiration facultative aerobes & obligate anaerobic bacteria	*Desulfuromonas acetoxida* *pyrodictium occultum*
-0.25		CO_2 → CH_4	Clarbonate respiration Methanogenic archaea; obligate anaerobes	*Methanobacterium thermoautotrophicum* *Methanosarcina barkeri*
		SO_4^{2-} → HS^-	Sulfate respiration (sulfate reduction) obligate anaerobes	*Desulfovibrio desulfuricans* *Desulfotomaculum ruminis* *Desulfonema limicola*
0		Fumarate → Succinate	Fumarate respiration Facultative aerobic bacteria	*Wolinella succinogenes* *Escherichia coli*
+0.4		NO_3^- → NO_2^-, N_2O, N_2	Nitrate respiration (denitrification) Facultative aerobic bacteria	*Paracoccus denitrificans* *Pseudomonas stutzeri*
		Fe^{3+} → Fe^{2+}	Iron respiration Facultative aerobes & obligate anaerobic bacteria	*Alteromonas putrifaciens*
	Aerobic respiration	O_2 → H_2O	Aerobic respiration obligae and facultative aerobes	*Pseudomonas aeruginosa* *Escherichia coli*

Assimilative and dissimilative metabolisms differ substantially. In assimilative metabolism, only the required amount of the compound (NO_3^-, SO_4^{2-} or CO_2) is reduced to satisfy the needs of the nutrients for growth. The reduced atoms are eventually converted to cell material in the form of macromolecules. In dissimilative metabolism comparatively large amount of electron acceptor is reduced and the reduced product is excreted into the environment. Dissimilative metabolism is carried out strictly under anoxic conditions, whereas assimilative metabolism is carried out in both aerobic and anaerobic conditions. Many organisms, such as many bacteria, archaea, fungi, algae and higher plants, are bestowed with the ability to carry out assimilative metabolism whereas dissimilative metabolism is carried out by restricted group of organisms, mostly prokaryotes. In this chapter, only the dissimilatory metabolic pathways are discussed.

17.6. DIVERSITY OF RESPIRATORY PATHWAYS

In anoxic environments, a variety of oxidized inorganic and organic compounds can serve as electron acceptors (A), and can, be used to support anaerobic respiration by different types of microorganisms (Table 17.2). In many cases, electron donor (DH) is NADH. Nevertheless, a variety of other electron donors may be used, depending on the ability of the bacterial species to utilize a particular compound.

17.6.1. Nitrate Respiration (Nitrate Reduction)

Nitrate is one of the major anaerobic electron acceptor in nature. The physiological group that can carry out nitrate respiration can transfer hydrogen and electrons to nitrate which becomes reduced, with the free energy being used for electron transport phosphorylation. This process is called **nitrate respiration** or dissimilatory nitrate reduction. The source of electron donor used for nitrate reduction depends on the microorganism. Many bacteria use NADH as the electron donor, when they are grown on glucose. Other electron donors include hydrogen, formate, glycerol, lactate and certain other substrates. Depending on whether molecular N_2 or NH_4^+ is released as the end product of respiration, two groups of nitrate reducing bacteria can be distinguished. They are :

1. Denitrifiers
2. Nitrate ammonifiers.

17.6.1.1. Denitrifiers

During denitrification NO_3^- is reduced to gaseous nitrogen. Denitrification is the only biological reaction that converts oxidized nitrogen back to molecular N_2. Thus, denitrific-ation plays an important role in the nitrogen cycle. Denitrifiers are aerobic, strictly respiratory bacteria. They cannot grow anaerobically in the absence of nitrate. Under anaerobic conditions, nitrate by serving as electron acceptor gets reduced to gaseous nitrogen dioxide (NO_2) and molecular nitrogen which are released. Thus, denitrification means the conversion of bound nitrogen (nitrate) to free N_2. Denitrifiers are mostly soil bacteria that include *Paracoccus denitrificans, Achromobacter xylosoxidans* (*Pseudomonas denitrificans*), *Thiobacillus denitrificans, Alcaligenes faecalis, Ps. stutzeri, Bacillus licheniformis, Spirillum itersonii, Hyphomicrobium vulgare* and many other bacterial species.

Like aerobic respiration, denitrification allows a complete oxidation of the organic substrate to CO_2

$$\text{Glucose} + 6\,O_2 \longrightarrow 6\,CO_2 + 6\,H_2O;\ \Delta G^{o'} = -2870\text{kJ}\ (-686\ \text{k.cals})$$

$$\text{Glucose} + 4.8\ NO_3^- + 4.8\ H^+ \longrightarrow 6\ CO_2 + 2.4\ N_2 + 8.4\ H_2O;$$

$$\Delta G^{o\prime} = -2669 \text{ kJ } (-638 \text{ k.cals})$$

For instance, if *Bacillus licheniformis* grows with glucose and nitrate under aerobic conditions, the substrate is degraded via EMP pathway; the acetyl CoA formed from pyruvate is oxidised via tricarboxylic acid cycle, and $NADH_2$ and $FADH_2$ thus formed serve as electron donors for the respiratory chain. Apart from $NADH_2$, succinate, lactate, formate, glycerol phosphate and hydrogen can also serve as electron donors. Nitrate, however, does not simply replace oxygen; special types of cytochromes and membrane bound enzyme systems are formed which reduce nitrate to nitrogen. In other words, when the organism switches from aerobic to anaerobic respiration, the organism need to synthesize specific electron carrier enzymes. The electron flow in denitrification, in comparison to aerobic respiration is shown in Fig. 17.2.

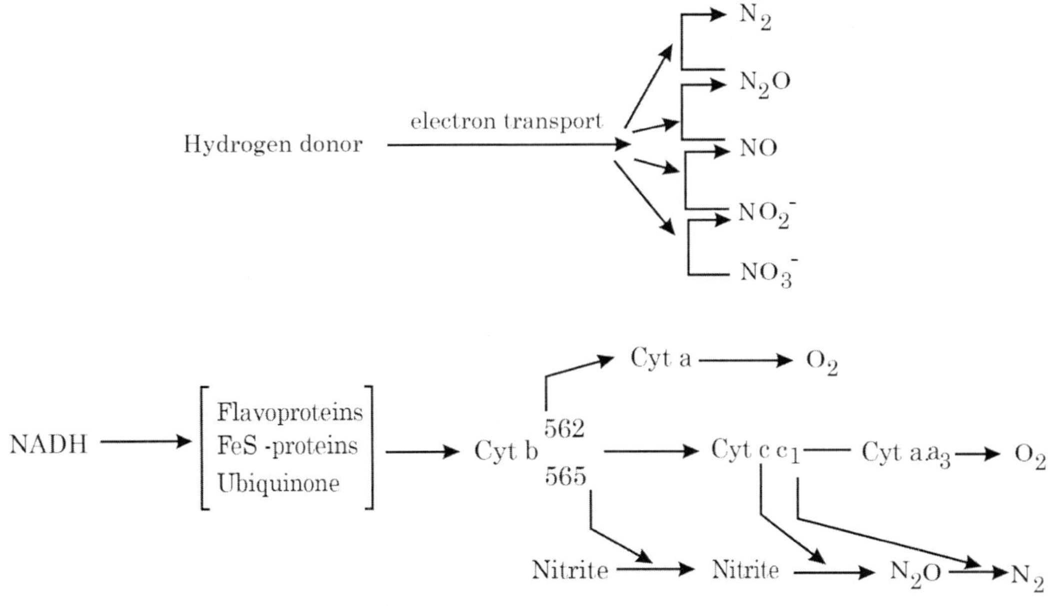

Fig. 17.2 : Electron flow in denitrification (e.g., *Paracoccus denitrificans*)

The enzyme machinery for denitrification is formed only under anaerobic conditions or conditions of low oxygen tension. In most cases nitrate is required as inducer. The enzymes involved in nitrate reduction to N_2 are oxygen sensitive *i.e.,* they are strongly inhibited by the presence of oxygen. Thus, denitrification takes place when oxygen is absent or available in insufficient amounts. All bacteria capable of performing this kind of respiration prefer oxygen respiration, if possible. *Thiomicrospira denitrificans,* an obligate sulfur oxidiser and denitrifer, is a microaerophilic bacterium.

Denitrification proceeds via several steps in which a total of five electrons are transferred from the electron donor substratum to a series of intermediates.

$$NO_3^- \xrightarrow[\text{nitrate reductase}]{2e^-} NO_2^- \xrightarrow[\text{nitrite reductase}]{e^-} NO \xrightarrow[\text{nitric oxide reductase}]{e^-} N_2O \xrightarrow[\text{nitrous oxide reductase}]{e^-} N_2$$

$$\text{Overall reaction : } 2NO_3^- + 10e^- + 12H^+ \longrightarrow N_2 + 6H_2O$$

The reactions are depicted in Fig. 17.3 in greater detail. It is clear from the figure that each step is catalyzed by a specific enzyme. The first step *i.e.,* reduction of nitrate to nitrite is catalyzed by nitrate reductase, a molybdenum containing membrane bound enzyme. It must be noted here that nitrate reductases in denitrifying organisms are very similar to the nitrate reductases employed in nitrate ammonifying organisms.

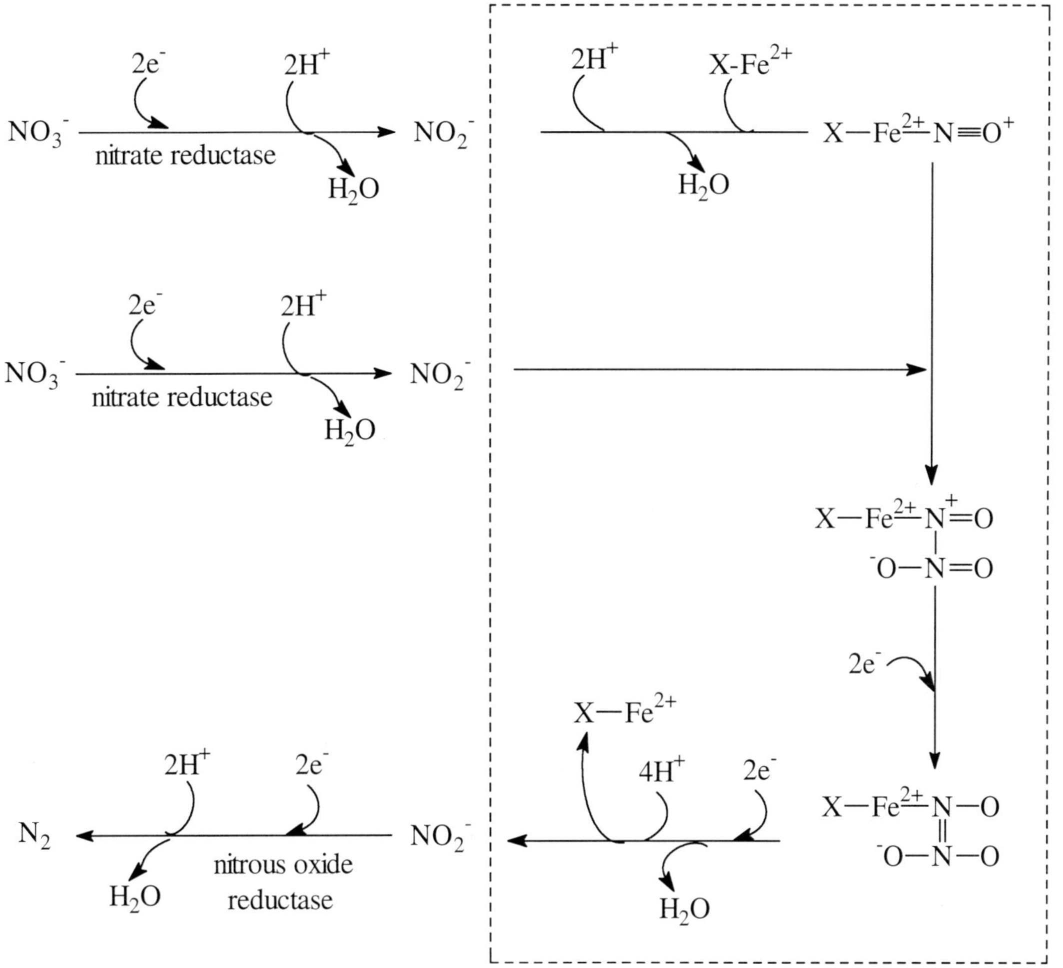

Fig. 17.3 : Scheme of reactions involved in the reduction of nitrate to molecular nitrogen.

Nitrite is reduced to nitric oxide by nitrite reductase. This enzyme contains either c- and d-type haemes or copper atoms as co-factors and it is located in the periplasmic space of the cell. The third step *i.e.,* reduction of nitric oxide to nitrous oxide is catalyzed by nitric oxide reductase which is an integral membrane protein. In fact, the reduction of nitrite to N_2O is very complex. It comprises two 2-electron reduction steps. First an enzyme ferrous-nitrosyl complex is formed, while a second NO_2^- is added to. Reduction yields coordinately bound oxyhyponitrate that is further reduced to N_2O. The last step *i.e.,* reduction of N_2O to N_2 is catalyzed by nitrous oxide reductase, a copper containing enzyme located in periplasmic space of the cell. It is linked to respiration chain via a cytochrome of the c-type.

Denitrifying bacteria, in addition to nitrate, can also use nitrite or N_2O as electron acceptors. Evidences are available to suggest that electron flow to all three compounds is coupled to ATP formation. Under growth conditions where the electron donor is limiting, the intermediates NO_2^-, NO, and N_2O can accumulate in the cell and excreted into the environment. For instance, most enterobacteria and staphylococci reduce the nitrate only to the level of nitrite, which is excreted. Thus, nitrogen oxides can reach the atmosphere not only as a result of oil and coal combustion but also as a product of biological processes in soil and water.

Isolation technique for dentrifiers

Enrichment culture of denitrifiers is straight forward and easy. A nutrient medium containing various hydrogen donors, building blocks for cell synthesis is used in which potassium nitrate is added as an electron source and is inoculated with serial dilutions of soil or mud, and incubated under anoxic conditions. An energy source should be added that is not readily fermentable so fermentative organisms will not be selected. By using suitable energy source different species of denitrifiers can be grown: (a) with a trace of peptone and ethanol and propionate (*Pseudomonas aeruginosa*), (b) with glucose (*P. fluorescens*), (c) with tartarate, succinate or malate (*P. stutzeri*), (d) with organic acids, alcohol, meat extract or with high nitrate (*Bacillus licheniformis*), (e) with a trace of yeast extract and with molecular hydrogen (*Paracoccus denitrificans*), (f) with sulfur or thiosulfate (*Thiobacillus denitrificans*). As most denitrifiers are facultative aerobes, great care to establish anoxic conditions is not required. A simple glass stoppered incubation containers are sufficient. The initial dissolved oxygen present in the culture medium is quickly used up, and then the population switches to anaerobic respiration. Once good growth is obtained, the culture turns to turbid and bubbles under the glass stopper. The bubbles consist of N_2 as well as CO_2 derived from oxidation of the energy source.

Importance of denitrification

Denitrification is common in anaerobic locations. It readily occurs under moist and water-logged conditions, especially when organic fertilizers and nitrates are applied together. In rice fields, for example, fertilization with nitrate causes damage through accumulation of nitrite. Nitrite also accumulates in nitrate-containing drainage systems that may enter the drinking water. Globally, denitrification process is of decisive importance for the maintenance of terrestrial life, because it replenishes the atmospheric nitrogen depleted by nitrogen fixation.

17.6.1.2. Nitrate ammonification

Nitrate ammonification is carried out by several groups of microorganisms, especially the members of Enterobacteriaceae like *Escherichia coli*, *Enterobacter aerogenes*. These are all facultative anaerobes. However, if nitrate is available it is used as electron acceptor which is first reduced to nitrite. The nitrite is further reduced to ammonia (NH_3) via the assimilatory nitrate reduction. Hence, in this process ammonia is released and not molecular nitrogen.

$$NO_3^- \xrightarrow[\text{nitrate reductase}]{2e^-} NO_2^- \xrightarrow[\text{nitrite reductase}]{6e^-} NH_4^+$$

Nitrate ammonification, thus, requires only two enzymes. Eight electrons are transferred to the nitrogen atom of nitrate to produce ammonia. This nitrate reductase is a membrane-bound molybdenum-containing enzyme. It transfers 2-electrons from nitrate to nitrite. Nitrite reductase enzyme is found in cell membrane or periplasm or cytoplasm and contains cytochrome c as a prosthetic group. It catalyzes the 6-electron reduction of nitrite to ammonium.

17.6.2. Sulfate Respiration (Sulfate Reduction)

Under anaerobic conditions, several inorganic sulfur compounds serve as electron acceptors for many microrganisms. Among the sulfur compounds, sulfate is in most oxidised state. The physiological group of sulfate reducing bacteria is characterized by the ability to transfer substrate hydrogen to sulfate as the terminal electron acceptor with the reduction of sulfate to sulfide. This process facilitates electron transport with the participation of cytochrome c, and energy gain comes from electron transport phosphorylation. As this kind of sulfate reduction is formally similar to oxygen respiration, it is also called **sulfate respiration** or **dissimilatory sulfate reduction**. The organisms are referred to as **sulfate reducers** or **desulfuricans**. The end product of this reduction is hydrogen sulfide.

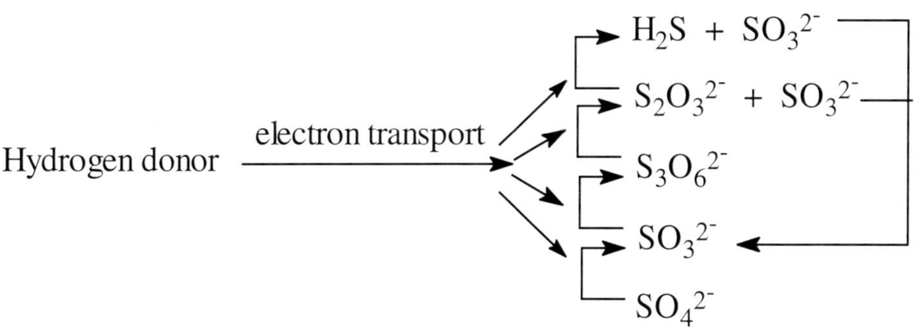

Overall reaction : $8\,(H) + 2H^+ + SO_4^{2-} \longrightarrow H_2S + 4H_2O$

Unlike nitrate reducers, all sulfate reducing bacteria are strict anaerobes. They include flagellated and non-flagellated eubacteria (*Desulfovibrio, Desulfobacterium, Desulfococcus, Desulfomonas, Desulfobacillus, Desulfosarcina, Thermosulfobacterium*), Gram-negative gliding bacteria (Desulfonema), Gram-positive endospore-forming eubacteria (*Desulfatomaculum*) and archaebacteria (*Archaeoglobus*). These are widely distributed in nature and are found in anaerobic environments including water logged soils, streams and lake sediments and brackish, and marine sediments. The smell of hydrogen sulfide gas produced by these organisms provides an indication of their presence. Various low molecular weight compounds can serve as hydrogen donors for different members of this group: lactate, acetate, propionate, butyrate, formate, methanol, ethanol, higher fatty acids, aromatic compounds (*e.g.,* benzoate) and molecular hydrogen. Some species of sulfate reducing bacteria oxidize their hydrogen donors incompletely and excrete acetate. However, some species utilize acetate, short and long chain fatty acids, alcohols, aromatic compounds and oxidise them completely. A few species are able to grow autotrophically with molecular hydrogen and thiosulfate. Thus, this group of sulfate respiring organisms is very heterogenous with regard to their substrate utilization and metabolic pathways.

Sulfate reducing bacteria are loaded with electron and hydrogen carriers. Cytochrome C_3 discovered by Postgate (1984), was the first cytochrome detected in strict ananerobes. All sulfate reducers contain cytochromes, which are of the c-type and/or of the b-type. In addition, menaquinone, several ferredoxins, flavodoxin, rubredoxin, and desulforedoxin are present. Another important constituent of sulfate reducers is a siroheme protein, a protein containing two tetrapyrrol ring systems and iron sulfur centres. This protein is known to exhibit sulfite reductase activity and may catalyze six-electron-transfer reaction.

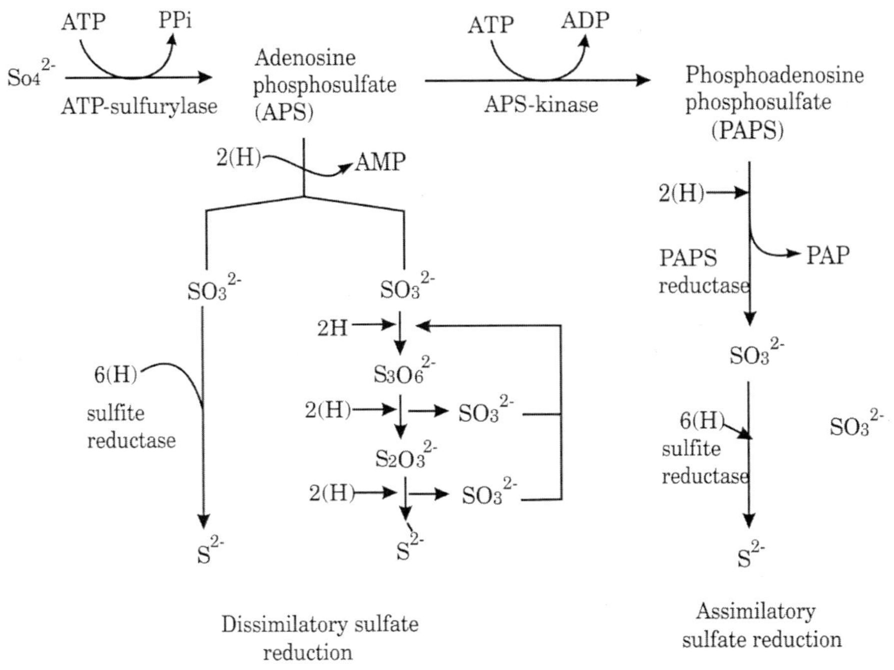

Fig.17.4 : Dissmilatory and assimilatory pathways of sulfate reduction

Almost all bacteria and also other organisms like fungi and green plants can grow with sulfate as their source of sulfur. They produce the sulfide which is necessary for the synthesis of sulfur containing amino acids by assimilatory sulfate reduction. The first step of sulfate reduction is common to assimilatory and dissimilatory pathways. The sulfate ion is stable and cannot be used without first being activated. The sulfate is activated by means of a ATP. In dissimilatory reduction, the reduction of sulfate to sulfide occurs via a sequence of three reactions catalyzed by ATP sulfurylase, adenosine phosphosulfate (APS) reductase and a bisulfite reductase, respectively. These enzymes are common to all sulfate reducers. Futher, all these enzymes are water soluble. Inorganic pyrophosphatase hydrolyzes the pyrophosphate produced and thereby reversible reaction proceeds from left to right.

$$ATP + SO_4^{2-} \rightleftharpoons APS + PPi$$

$$APS + 2e^- \rightleftharpoons AMP + SO_4^{2-}$$

$$HSO_3 + 6H^+ + 6e^- \longrightarrow H_2S + 3H_2O$$

$$PPi + H_2O \longrightarrow 2P$$

Thus the reduction of sulfate effectively expends two ATP equivalents and to account for the net production of ATP during anaerobic growth. The dissimilatory sulfate reduction pathway for APS reduction can proceed via two distinct pathways.

1. Some sulfate-reducing bacteria catalyze the reduction of APS to hydrogen sulfide via a series of free intermediates. The individual enzymes involved in these reduction steps have not been characterized in detail.

$$APS \xrightarrow[\text{APS reductase}]{2e^-} HSO_3^- \xrightarrow[\text{sulfite reductase}]{2e^-} S_3O_6^{2-}$$

$$\xrightarrow[\text{trithionate reductase}]{2e^-} S_2O_3^{2-} \xrightarrow[\text{thiosulfate reductase}]{2e^-} HS^-$$

2. In other bacteria, a distinct sulfite reductase catalyzes the 6-electron reduction of HSO_3^- to form H_2S without the release of free intermediates.

$$APS \xrightarrow[\text{APS reductase}]{2e^-} HSO_3^- \xrightarrow[\text{sulfite reductase}]{6e^-} HS^-$$

A summary of the assimilatory and dissimilatory reductions is presented in Fig. 17.4.

Carbon metabolism and energetics

Reducing equivalents for sulfate reduction and carbon for growth of the chemoheterotrophic organisms are secured from various organic compounds which serve as electron donors. The pathway of dissimilatory sulfate reduction by *Desulfovibrio* species during its growth on lactate is shown in Fig.17.5. However, chemolithoautotrophic sulfate reducing bacteria grow with H_2/SO_4^{2-} and use CO_2 as sole carbon source. For CO_2 fixation they utilize reductive actyl Co-A pathway or reductive tricarboxylic acid cycle, but not the calvin cycle.

Many sulfate reducing bacteria are capable of growth on ethanol, propionate and acetate as sole energy source. The conversion of ethanol to acetate by one group of sulfate reducers proceeds along the usual pathways. Propionate is converted by *Desulfobulbus propionicus* to acetate via propionyl-Co-A, methyl malonyl CoA and succinyl CoA. Organisms like *Desulfobulbus acetoxidans* oxidise acetate completely to CO_2 and reduce sulfate to sulfite.

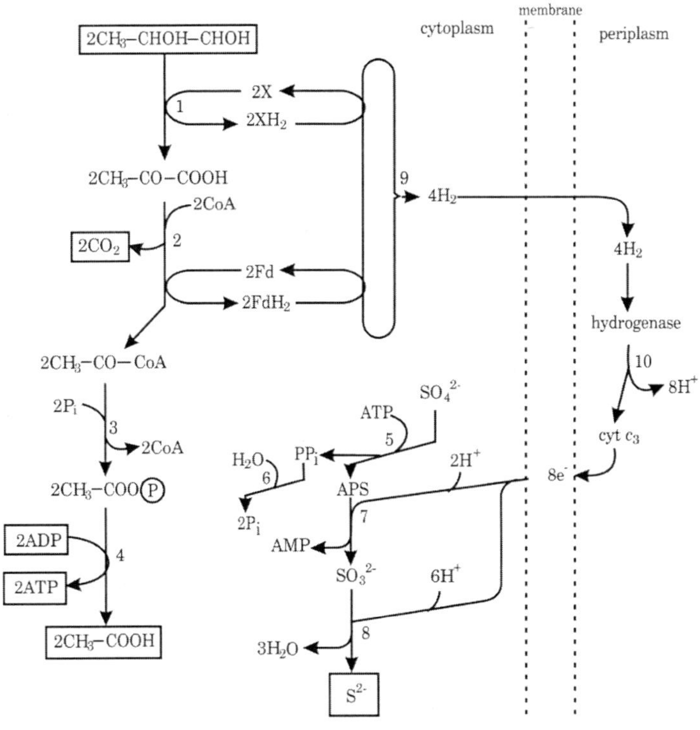

Fig. 17.5. Pathway of dissimilatory sulfate reduction in *Desulfovibrio* species during its growth on lactate 1, Lactate dehydrogenase, membrane-bound, H-acceptor not known; 2, pyruvate-ferredoxin oxidoreductase; 3, phosphotransacetylase; 4, acetate kinase; 5, ATP sulfurylase; 6, pyrophosph-atase; 7, Aps reductase; 8, sulfite reductase; 9, cytoplasmic hydrogenase; 10, periplasmic hydrogenase.

$$\text{Acetate}^- + SO_4^{2-} + 3H^+ \longrightarrow 2CO_2 + H_2S + 2H_2O;\ \Delta G^{0'} = -\ 57.5\ \text{kJ / reaction}$$

Acetate is oxidised through citric acid cycle in a majority of the organisms. However, in acetate oxidising sulfate reducers, two biochemical mechanisms for acetate oxidation are known : a modified citric acid cycle and the acetyl-CoA pathway. Organisms like *Desulfobacter* use the modified citric acid cycle for acetate oxidation. The reactions of this pathway involve most of the enzymes of the citric acid cycle (Fig. 17.6).

Still some acetate-oxidising sulfate-reducing bacteria donot use this modified citric acid cycle, but instead employ the acetyl-CoA pathway for acetic acid oxidation. This pathway allows for the oxidation of acetate to CO_2 via a quite different series of reactions from that of the citric acid cycle, employing the key enzyme carbon monoxide dehydrogenase (see acetyl CoA pathway).

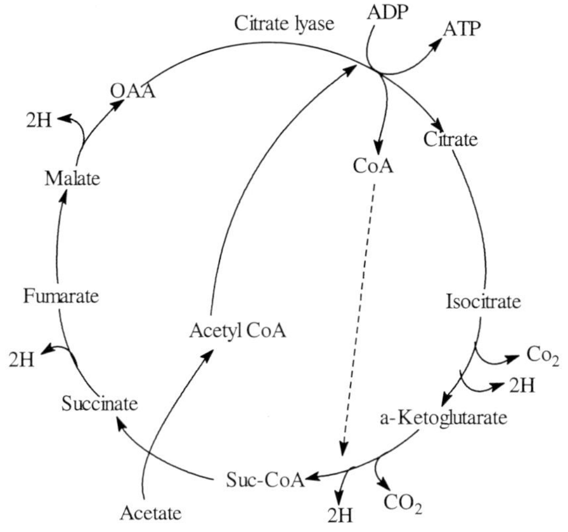

Fig. 17.6. Modified citric acid cycle of acetate oxidizing sulfate-reducers.

Disproportionation of sulfur compounds

Quite interestingly certain sulfate reducing bacteria are capable of carrying out a unique form of energy metabolism called disproportionation. The term disproportionation refers to the splitting of a compound into two new compounds, one of which is more oxidised and another of which is more reduced than the orginal substrate. Some examples are presented here. *Desulfovibrio sulfodismutans* disproportionates sulfur compounds as follows

$$S_2O_3^{2-} + H_2O \longrightarrow SO_4^{2-} + H_2S; \quad \Delta G^{o'} = -21.9 \text{ kJ}$$

In the above reaction one sulfur atom of $S_2O_3^{2-}$ becomes more oxidized forming SO_4^{2-} and other more reduced forming H_2S. Another disproportion involves sulfite.

$$4\,SO_3^{2-} + 2\,H^+ \longrightarrow 3\,SO_4^{2-} + H_2S; \quad \Delta G^{o'} = -235.6 \text{ kJ}$$

In both the above reactions, electrons from either $S_2O_3^{2-}$ or SO_3^{2-} enter the electron transport chain and eventually reduce other molecules of $S_2O_3^{2-}$ or SO_3^{2-}, respectively to H_2S.

Elemental sulfur (S^o) is disproportionated to H_2S and SO_4^{2-}

$$4\,S^o + 4\,H_2O \longrightarrow 3H_2S + SO_4^{2-} + 2H^+; \Delta G^{o'} = +48 \text{ kJ}$$

This reaction, as is evident from free energy, is bioenergetically unfavourable. However, if the H_2S that is formed is oxidized back to S^o by chemical reaction with an electron acceptor such as Mn^{4+} the reaction becomes favourable.

$$3\ S^o + 4H_2O + MnO_2 \longrightarrow 2H_2S + SO_4^{2-} + 2H^+ + Mn^{2+} + 2OH^- ;$$

$$\Delta G^{o'} = -44\ kJ$$

Thus, unlike sulfite or thiosulfide disproportionating bacteria, sulfur dispropor-tionating bacteria require an electron acceptor.

Enrichment and isolation

Sulfate-reducing bacteria can be cultured by setting up enrichment cultures. The nutrient medium must contain a suitable hydrogen donor (like lactate, propionate, ethanol), assimilable substrate (like acetate, SO_2), mineral salts and sulfate. The cultures should be maintained under perfect anaerobic conditions. A low redox potential ($E^{o'}$ = -200 mV) is to be created by way of placing iron nails.

Economic importance

Sulfate reducing bacteria such as the species of *Desulfovibrio* cause an indirect anaerobic corrosion of iron which is of great economic importance. Considerable damage is done to the underground pipe lines in this way. These bacteria are also held responsible for the high hydrogen sulfide content at the bottom of Black Sea and for its colour due to iron corrosion.

17.6.3. Carbonate Respiration : Carbon dioxide as Electron Acceptor : Methanogenesis and Acetogenesis

Carbon dioxide (CO_2) is common in both aerobic and anaerobic environments. It is the major product of the energy metabolism of chemoorganotraphs. Under total anoxic conditions, several procaryotes are able to use CO_2 as terminal electron acceptor in respiration. This type of anaerobic respiration, where CO_2 serves as the terminal electron acceptor is known as **carbonate respiration**. The most important CO_2 reducing procaryotes are the methanogens, a major group of archaebacteria. Many of these organisms utilize H_2 produced in different fermentations as the electron donor.

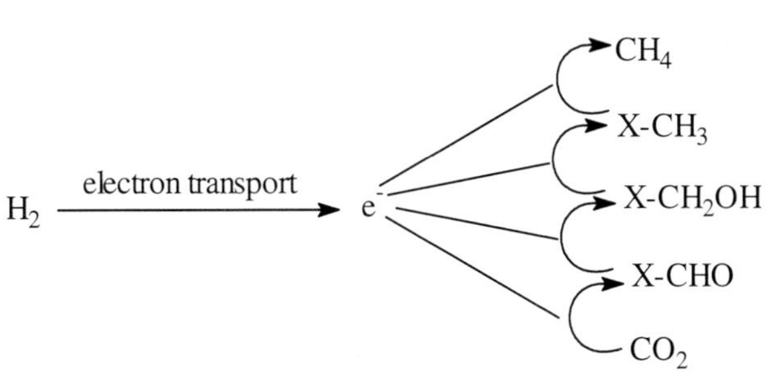

The overall reacton : $4\ H_2 + H^+ + HCO_3^- \longrightarrow CH_4 + 3H_2O$; $\Delta G^{o'} = -135.6$ kJ

Methane bacteria are able to activate hydrogen and to couple its oxidation to the reduction of carbon dioxide. They can fix the carbon of the carbon dioxide and synthesize their cell materials. Their mode of life is evidently regarded as chemolithoautotrophic. However, the process by which CO_2 is fixed is not the calvin cycle, but it is acetyl-CoA pathway. There is another group of methanogens, chemoorganotrophs, which utilize the acetate as the electron donor. Both these groups of methogens are discussed in details in chapter 12.

Another group of anaerobic CO_2^- reducing bacteria is the homoacetogens which reduce the CO_2 to acetate rather than methane. This process is known as homoacetogenesis, the overall reaction is

$$4\,H_2 + H^+ + 2HCO_3^- \longrightarrow CH_3COO^- + 4H_2O;\ \Delta G^{o\prime} = -104.6\ kJ$$

Most acetogenic bacteria can also grow chemoorganotrophically by the fermentation of sugars and a variety of other organic compounds. Acetogenic bacteria, both chemolitho-autotrophs and chemoorganotrophs were discussed in detail in chapter 11.

Bioenergetics of methanogenesis and acetogenesis

Comparison of the overall reactions of the methanogenesis and acetogenesis reveals that some what more energy is released when methane is formed from CO_2 reduction than when acetate is formed. This is not unexpected when redox potentials of the two couples (Table 17.2) are considered.

In methanogens, ATP synthesis is linked to a proton motive force and proton translocation ATPase; the terminal step in the reduction of CO_2 to CH_4 and establishes a proton motive force whose energy is captured by membrane bound ATPases.

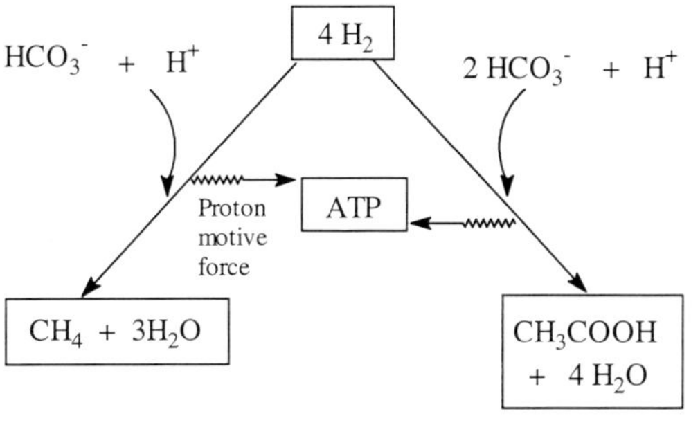

Two bioenergetic patterns have been observed in homoacetogens. Some homoace-togens, such as *Acetobacterium woodii*, require Na^+ for growth. These species link the synthesis of acetate to formation of a sodium motive force that is instrumental in generating ATP. Other homoacetogens, such as *Clostridium thermoaceticum*, do not require Na^+ and generate a proton motive force during the acetogenesis that functions to drive ATP synthesis. In addition to oxidative phosphorylation, substrate-level phosphorylation is possible during acetogenesis. This is evident from the final step of acetate formation

$$\text{Acetyl} \sim P + ADP \longrightarrow \text{acetate} + ATP$$

17.6.4. Fumarate Respiration

Under anaerobiosis, fumarate is formed from a variety of sources such as malate, aspartate, pyruvate and CO_2 and is, therefore, available as an electron acceptor. Due its higher redox potential ($E^{o\prime}$ of fumarate / succinate, -30 mV), it can, therefore, accept electrons that are supplied by hydrogen-transferring coenzymes and that have already passed through portion of the electron transport chain. Fumarate reduction can allow electron-transport phosphorylation, and such electron transport phosphorylation with fumarate as the final electron can, therefore, be referred as anaerobic respiration or **fumarate respiration.**

Fumarate respiration is widely distributed among chemoorganotrophic anaerobic bacteria. An enzyme which reduces fumarate to succinate *i.e.,* fumarate reductase is genetically distinct from succinate dehydrogenease. It is present in many facultative and obligate anaerobes. The experimental evidence is that the addition of fumarate to nutrient media promotes faster growth and higher cell yields in many bacteria. This indicates that fumarate affords effective regeneration of ATP. A good example is the anaerobic *Vibrio succinogenes* which grows with H_2 as the energy source using fumarate as the electron acceptor, but other organisms include *Desulfovibrio gigas, Clostridium formicoaceticum,* Propionibacteria, some bacteroides and the facultative *Escherichia coli* and *Proteus rettgeri.* A few eukaryotes such as anaerobic protozoa and parasitic helminths employ fumarate as the terminal electron acceptor of the mitochondrial respiratory chain in place of oxygen.

The electrons required for the reduction of fumarate reductase can be supplied by the oxidation of formate to CO_2 or by the oxidation of $NADH_2$, H_2 or lactate by respective dehydrogenases. These reactions lead to the formation of proton motive force across the cytoplasmic membrane and thus allow electron transport phosphorylation (Fig. 17.7).

Fumarate reductase is a membrane bound iron-sulfur flavoprotein enzyme with a covalently linked FAD prosthetic group. It is not constitutive in many bacteria, and anaerobic growth on fumarate or fumarate precursors induces the enzyme. In the membrane, the fumarate reductase is associated with electron carriers that link it to electron donors such as H_2, formate, lactate, glycerol phosphate or NADH (Fig. 17.7). Although some variations are found in the carrier systems of different organisms and the electron donors they employ, the lipophilic compound menaquinone and *a b* type cytochromes are characteristic electron carriers of the bacterial fumarate reductase system. The mitochondria of the anaerobic helminths which reduce fumarate contain rhodoquinone rather than ubiquinone.

Experiments with permeant and impermeant electron donors provided evidence that the fumarate reductase of *Vibrio succinogenes* is located on the inner side of the cytoplasmic membrane, while the formate dehydrogenase (FDH) associated with fumarate reduction (FR) is on the outer side, indicating that fumarate reduction is transmembrane redox process.

17.6.5. Iron Respiration : Reduction of Ferric Iron to Ferrous

In nature iron exists in two forms : ferric (Fe^{3+}) and ferrous (Fe^{2+}). Ferric is in more oxidised state than ferrous. Ferric iron is one of the most common metals present in soils and rocks, and its reduction leads to the production of ferrous iron, a more soluble form of iron. Reduction of ferric iron to ferrous is brought about by a variety of microorganisms. Since ferric is abundant in many microbial habitats, its reduction can be a major form of anaerobic respiration often referred to as **'iron respiration'**.

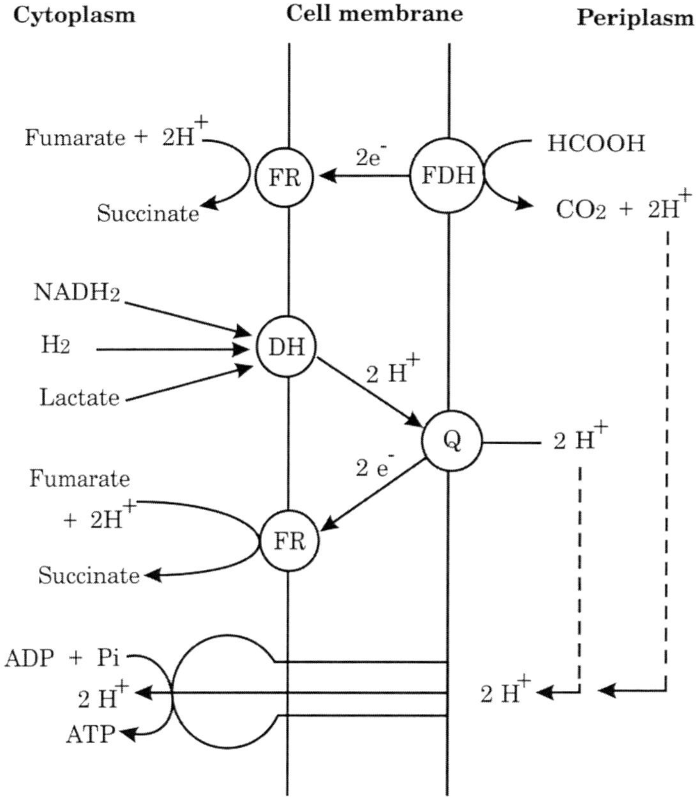

Fig. 17.7. Electron transport phosphorylation in fumarate respiring bacteria, utilizing formate, $NADH_2$, H_2 and lactate as electron donors

Iron respiration has been reported in certain fungi and in a wide variety of both chemoorganotrophic and chemolithotrophic bacteria. The redox potential of Fe^{3+} / Fe^{2+} couple is very electropositive ($E^{o'}$ = +0.77 v) hence thermodynamically Fe^{3+} is well suited for reduction by the oxidation of variety of both organic and inorganic electron donors. Various organic compounds including aromatic compounds can be oxidised anaerobically by ferric iron reducers. The electron transport chain, in these organisms, terminates in a ferric iron reductase system.

Most research on energetics of ferric iron reduction has been done in the Gram-negative bacterium *Shewanella putrefaciens* and a strain of *Alteromonas*, GS-15. The later organism uses acetate as carbon source and H-donor, and Fe^{3+} as H-acceptor. During the reduction of ferric iron by acetate a mixture of Fe^{2+} and Fe^{3+} is formed which is transformed to the oxide Fe_3O_4 (= magnetite) which is strongly ferromagnetic. Thus diamagnetic iron is produced from non-magnetic iron. The metabolic products of these iron reducing organisms can therefore, be detected with a magnet. As ferric oxide is practically insoluble and must be converted to a soluble and cell permeable form, probably by siderophores. The growth under these conditions is very slow and meagre. Bacterial iron reduction leads to solubilization of iron, thus, it is vital in geochemical process.

17.6.6. Reduction of Manganese and other Inorganic Substances

Under anoxic conditions several other inorganic substances can function as electron acceptors. These include manganese (Mn^{4+}), arsenic and selenium. Anoxic reduction of Mn^{4+} to Mn^{2+} is carried out by a variety organisms, mostly chemoorganotrophs. Experiments with *Shewanella putrefaciens* revealed that $Mn^{4\pm}$ supported anaerobic respiration is possible. Arsenic and selenium compounds usually donot occur in large amounts in natural systems. However, they occur in substantial quantities in polluted environments where they serve as electron acceptors for anaerobic bacteria. The reduction of SeO_4^{2-} to SeO_3^{2-} and eventually to Se^o (metallic selenium) is an important method of selenium removal from water and has been used as a means of bioremediation of selenium contaminated soils. Interestingly, most bacteria capable of selenate or arsenate reduction can also use several other electron acceptors, such as Fe^{3+}, Mn^{4+} and organic compounds and in most cases show a facultatively aerobic form of metabolism.

17.6.7. Organic Electron Acceptors in Anaerobic Respiration

Several organic compounds can function as electron acceptors in anaerobic respiration (Table 17.3). One such acceptor is fumarate. One may here, quite justifiably put the question that if the organic substance is electron acceptor, in what way this type of anaerobic respiration differs from fermentation (in both the cases the terminal electron acceptor is an organic compound). However, it should be borne in mind the fact that during fermentation, the organic electron acceptor is produced internally as a part of the metabolic process then reduced internally but in anaerobic respiration organic electron acceptor is externally supplied.

Table 17.3 : Some alternate electron acceptors for anaerobic respiration

Acceptor	Reaction	E_o' of couple (V)	Product
Manganic ion	$Mn^{4+} \xrightarrow{2e^-} Mn^{2+}$	+0.798	Manganous ion
Ferric ion	$Fe^{3+} \xrightarrow{e^-} Fe^{2+}$	+ 0.77	Ferrous ion
Selenate	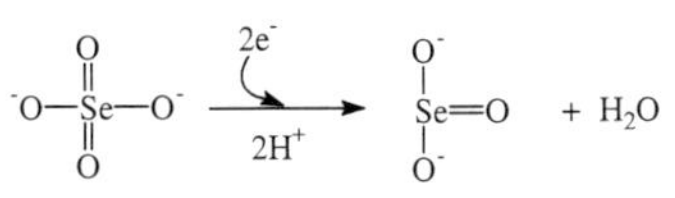	+0.475	Selenite
Dimethyl sulfoxide (DMSO)	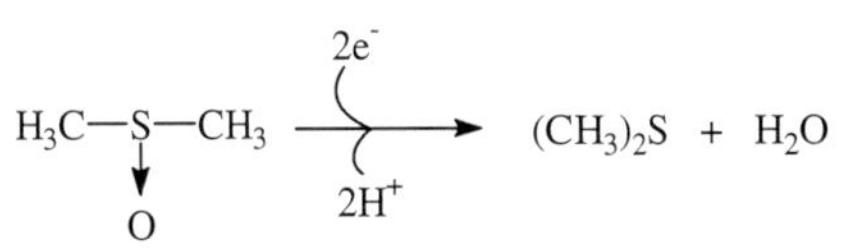	+0.16	Dimethyl sulfide (DMS)

Acceptor	Reaction	E_o' of couple (V)	Product
Arsenate	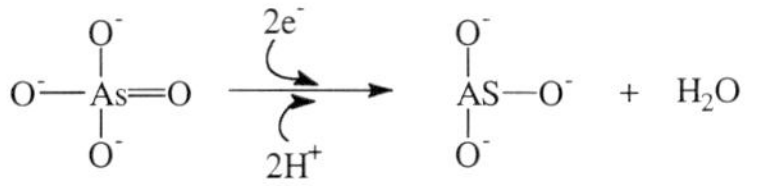	+0.139	Arsenite
Trimethyla-mine-N-oxide (TMAO)	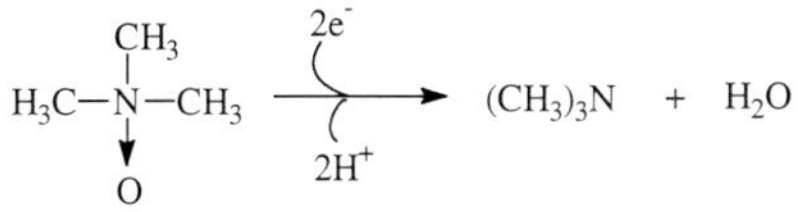	+0.13	Trimethyl amine (TMA)
Fumarate	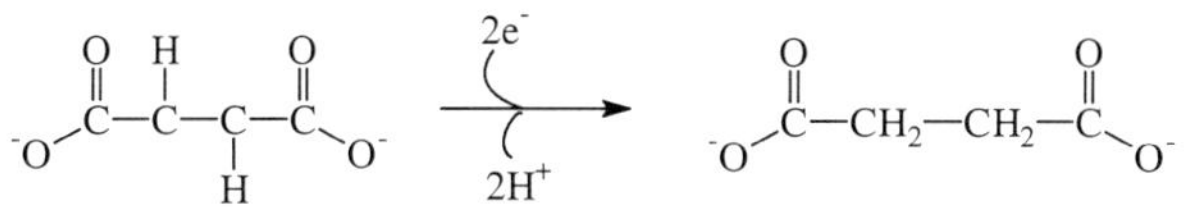	+0.03	succinate
Glycine	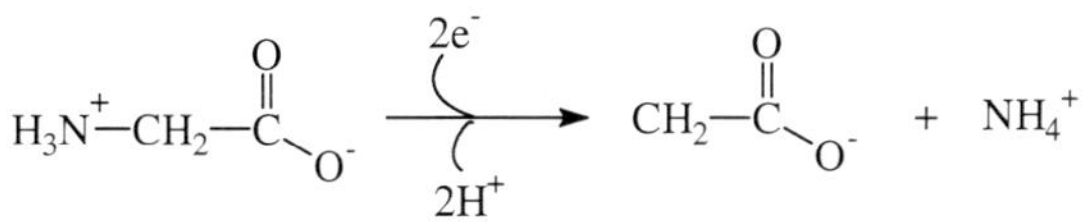	-0.01	Acetate + ammonia

Trimethylamine oxide (TMAO), an important osmotic solute in marine fish is utilized by a variety of facultative aerobic bacteria as an alternate electron acceptor and is reduced to trimethylamine (TMA). In addition, several phototrophic purple bacteria are able to use TMAO as an electron acceptor for anoxic metabolism in darkness. Dimethylsulfoxide (DMSO), a compound analagous to TMAO, is also reduced by a variety of bacteria to dimethyl sulfide (DMS) during anaerobic respiration. These bacteria include *Campylobacter, Escherichia* and many purple bacteria. In most cases of TMAO and DMSO reductions, cytochromes of the b type (with reduction potentials near O V) have been identified as terminal oxidases.

Glycine, an amino acid, can also function as an electron acceptor for certain obligate anaerobic bacteria such as clostridia. During this metabolism, glycine is reduced to acetate and the electron donor is another amino acid. The reader should recollect that when a pair of amino acids is used in metabolism, one as the electron donor and another as the electron acceptor, the metabolism is called **'Stickland reaction'** (refer chapter-18).

17.7. FURTHER READING

Ferguron, S.J. (1987), Denitrification : A question of the control and organization of electron and ion transport. *TIBS*, **12 :** 354-357.

Ferry, J.G. (ed.) (1993),. Methanogenesis : Ecology, Physiology, Biochemistry & Genetics, Chapman & Hall, New York.

Gennis, R. and Stewart, V. (1996), Respiration. In *Escherichia coli* and *Salmonella* (2nd edn.) (F.C. Neidhardt, ed.), pp. 217-261.

Gunsalus, R.P. (2002), Anaerobic respiration, In Encyclopedia of Microbiology, Vol. I, 2nd edn., (ed. J. Lederberg) 180-188, San Diego, Academic Press.

Hamilton, W.A. (1988), Energy transduction in anaerobic bacteria, pp. 83-149. In Bacterial energy transduction (Anthony (ed.), Academic Press, New York.

Hochstein, L.I. and Tomlinson, G.A. (1988), The enzymes associated with denitrification. *Annu. Rev. Microbiol.*, **42 :** 231-261.

Kroger, A., V. Geisler, E. Lemma, F. Theis and R. Lenger (1992), Bacterial fumarate respiration. *Microbiology*, **158** : 311-314.

Lovely, D.R. (1991), Dissimilatory Fe(III) and Mn(IV) reduction. *Microbiol. Rev.*, **5 :** 259-287.

Moodle, A.D. and Ingledeu, W.J. (1990), Microbial anaerobic respiration. *Adv. Microbial Physiol.*, **31 :** 225-239.

Postgate, J.R. (1984), The sulfate reducing bacteria, 2nd edn. New York, Cambridge University Press.

Richardson, D.J. (2000), Bacterial respiration : A flexible process for a changing environment. *Microbiology*, **146** : 551-571.

Speece, R.E. (1996), "Anaerobic Biotechnology : For Industrial Waste Waters" Archae Press, Nashville, TN.

Zumfit, W.G. (1997), Cell biology and molecular basis of denitrification. *Microbiol. Mol. Biol. Rev.*, **61 :** 533-536.

■■■■

Chapter - 18

Fermentaion

18.1. INTRODUCTION

Microbial fermentations has been one of the favourite subjects for mankind. The ancients in every civilization were aware of the importance of fermentations in producing alcoholic beverages, leavening of bread and various types of fermented milk, delicious, flavouring and nutritious food products, though they were unaware of the nature of the actual organisms involved. **Louis Pasteur** and **Buchner brothers** unfolded the facts underlying the fermentation that it is a biochemical process mediated by microbes. It is only during the last few decades we are able to understand the every detail of different fermentations *i.e.,* specific microorganisms involved, biochemical pathways, role of enzymes and the end products formed. Similarly, now it has become possible to manipulate every step of the fermentations so that desired end product could be produced in substantial quantities.

18.2. DEFINITION

The term fermentation was first defined by Pasteur; he described fermentation as life in the absence of oxygen. It can also be defined as those biological processes that occur in the dark and that do not involve respiratory chains with oxygen or nitrate as electron acceptors. In a physiological sense, fermentation is defined as a type of respiration where in both electron donor and acceptor is an organic compound. However, for industrial mycologists and biotechnologists it implies a different meaning.

Table 18.1. Fermentation : A word with many meanings

1. Any process involving the mass culture of microorganisms, either aerobic or anaerobic.
2. Any biological process that occurs in the absence of O_2.
3. Food spoilage
4. The production of alcoholic beverages
5. Use of an organic substrate as the electron donor and acceptor
6. Use of an organic substrate as a reductant and of the same partially degraded organic substrate as an oxidant.
7. Growth dependent substrate level phosphorylation.
8. An alternate means of reducing power disposal.

As given in the Table 18.1, the term can have several meanings. Each of the above annotations of fermentation has merits, but none of them provide the complete understanding of the fermentation process. Nonetheless, keeping in view the objectives of this book, we subscribe to the definition **"Fermentation as a process in which organic compounds serve as both electron donor and electron acceptor.**

18.3. PHYSIOLOGICAL SIGNIFICANCE OF FERMENTATION

In aerobic respiration, NADH produced in different steps of glycolysis and Krebs cycle are oxidized by utilizing the oxygen as terminal electron acceptor. However, in fermentation in the absence of O_2 or any suitable electron acceptor, NADH produced in glycolysis is not oxidized by the electron transport chain. As a result, NADH gets accumulated and the accumulated NADH exerts a feedback inhibition on entire glycolysis process. Consequently, ATP required for the organisms through substrate level phosphorylation is not produced. Hence, there need to be a provision for continuous removal of NADH and also recycling of NAD to drive the conversion of glyceraldehyde-3-phosphate to 1,3-biphosphoglycerate. Many microorganisms solve this problem by slowing or stopping pyruvate dehydrogenase activity and using pyruvate or one of its derivatives as an electron and hydrogen acceptor in the reoxidation of NADH (Fig. 18.1). This will pave the way for the production of more ATP. It is worthwhile to remember here that obligate fermentative microorganisms depend completely on the ATP produced during glycolysis by way of substrate level phosphorylation.

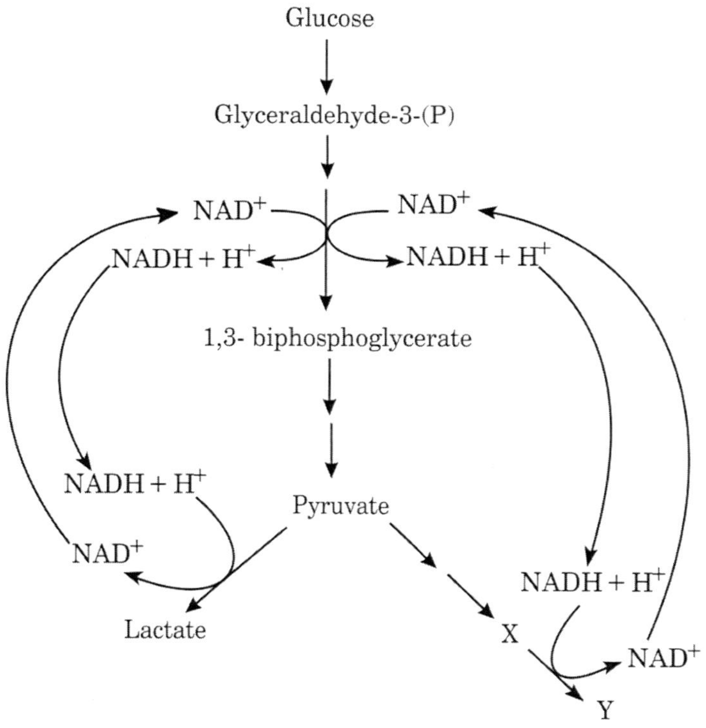

Fig. 18.1 : Reoxidation of NADH during fermentation

Fermentations have ecological significance too. In aerobic respirations, carbohydrates are completely oxidized releasing the ATP, CO_2 and H_2O. Thus, the energy and carbon will be available to only one trophic level. However, if the same carbohydrate like cellulose is degraded in anoxic environments, it is oxidized in a step-wise manner through fermentation. The fermentation products, such as alcohols, organic acids, CO_2 and H_2 are then at the disposal of other bacteria, which can produce methane and in the presence of sulphate, hydrogen sulfide by means of other specific fermentations or anaerobic fermentations. These fermentations thus initiate anaerobic food chain. In the sediments of fresh water lakes and in the rumen of ruminants, hydrogen and acetate are fermented to methane by methanogenic bacteria. This methane produced by methanogens serves as the substrate for methylotrophs. In marine anoxygenic ecosystems hydrogen and sulfate are converted to hydrogen sulfide by sulfidogenic bacteria. Thus, fermentative organisms play a considerable role in the metabolic cycle of nature.

18.4. TYPES OF FERMENTATIONS

There are many kinds of fermentations. They are named according to the exclusive or predominant end product formed. The unifying features of all fermentations are: (i) NADH is oxidized to NAD^+, (ii) electron acceptor is either pyruvate or a derivative of pyruvate, (iii) the products formed by particular organisms are characteristic of them and may be used in their identification and (iv) the end product formation is a means of reducing power disposal.

In some fermentations, a single end product is produced, but in many a number of end products are formed. Some important and common fermentations are outlined in the Fig 18.2 and Table 18.2.

Table 18.2 : Some important fermentations, their main end products and group of microorganisms involved.

Fermentation	Microorganisms	End product/products
Lactic acid fermentation	*Bacillus, Streptococcus* *Lactobacillus*	Lactic acid
Alcoholic fermentation	Yeast, *Zymomonas*	Alcohol, CO_2
Propionic acid fermentation	Propionibacterium	Propionic acid
Mixed acid fermentation	Enteric bacteria *Escherichia* *Enterobacter* *Salmonella, Proteus*	Formic acid, ethanol acetate, lactate, succinate
Butanediol fermentation	*Enterobacter, Serratia, Bacillus* Clostridia	Butanediol
Butyric acid-butanol fermentation		Butyrate, butanol propanol, acetate

Fermentations other than mentioned above also exist. Protozoa and fungi often ferment sugars to lactate, ethanol, glycerol, succinate, formate, acetate, butanediol and additional products. Substances other than sugars are also fermented. For example, many members of the genus *Clostridium* ferment amino acids, oxidizing one amino acid with a second one as the electron acceptor.

18.5. PYRUVATE, A KEY COMPOUND IN FERMENTATIONS

As is evident from the Fig. 18.2, pyruvate is the key compound for all types of fermentations. Pyruvate undergoes cleavage and gives rise to intermediates and end products by condensation. Pyruvate cleavage occurs so that each molecule gives rise to a 2-carbon moiety and a 1-carbon. The 1-carbon arises from the carboxyl group of pyruvate. A number of different systems for pyruvate cleavage are known (Table 18.3).

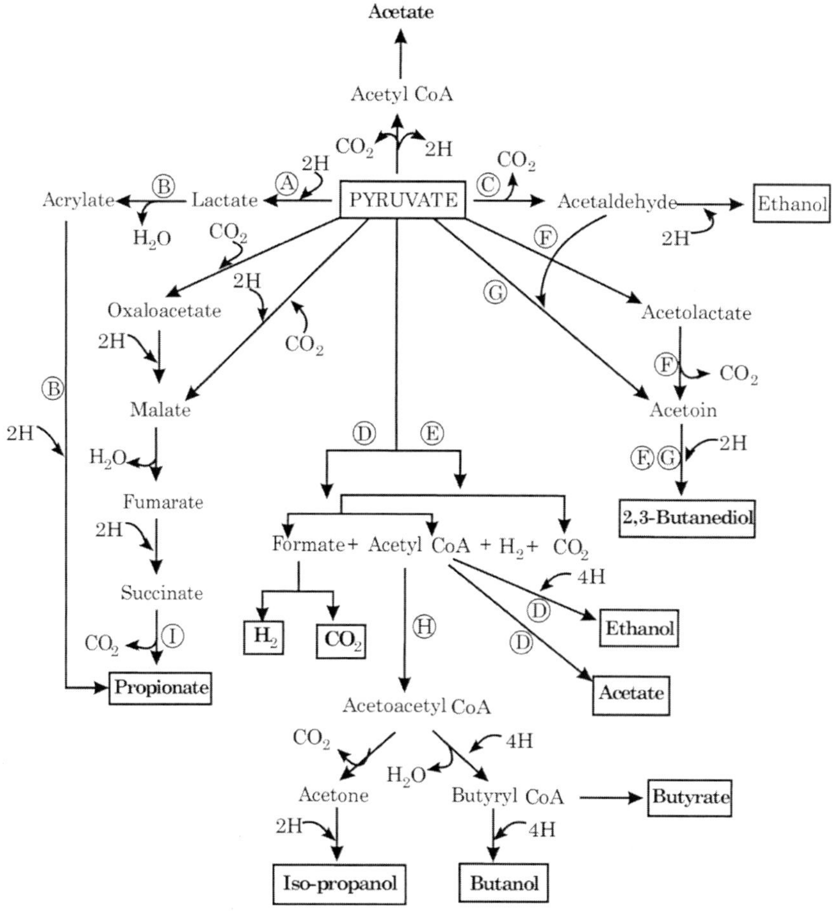

Fig. 18.2 : Bacterial fermentation products of pyruvate. A Lactic acid bacteria (*Streptococcus, Lactobacillus*), B. *Clostridium propionicum,* C. Yeast, *Acetobacter, Zymomonas, Sarcina ventriculi, Erwinia amylovora*, D. Enteric bacteria, E. Clostridia, F. *Klebsiella*, G. Yeast, H. Clostridia (butyric, butylic organisms), I. Propionic acid bacteria.

Table 18.3 : Enzymes and products of pyruvate cleavage

Pyruvate dehydrogenase	Acetyl CoA, CO_2, NADH
Pyruvate oxidase	Acetyl PO_4,CO_2,FADH
Pyruvate decarboxylase	Acetaldehyde, CO_2
Pyruvate formate lyase	Acetyl CoA, formate
Pyruvate ferredoxin oxidoreductase	Acetyl CoA, CO_2,FDH

Many products are formed by metabolic sequences in which pyruvate is cleaved. Fermentation end products are formed from further metabolism of the cleavage products. When pyruvate is cleaved, the cleavage occurs in a manner such that 1-carbon compounds arise from the carboxyl carbon of pyruvate, while 2 carbon compounds arise from pyruvate carbons 2 and 3. Many of the fermentation end products arise from acetyl CoA. Four carbon products, and compounds derived from them, typically arise by condensation of two acetyl CoA molecules or by condensation of acetyl CoA with other two carbon molecules derivated from pyruvate cleavage. The condensations may occur in two ways.

i. Head to head condensation

ii. Head to tail condensation

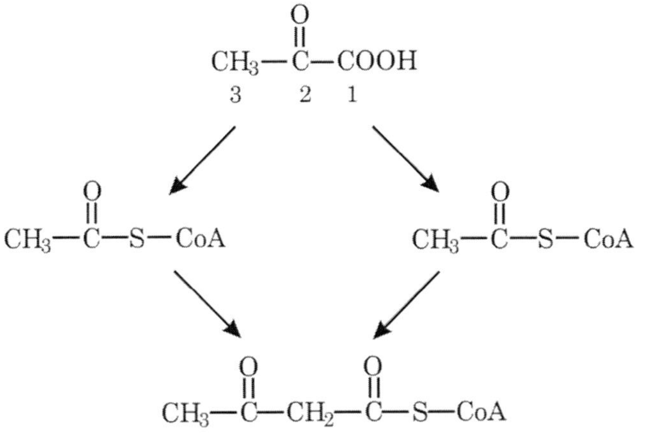

Acetoacetyl CoA

In **head-to-head condensations**, the carbon corresponding to the 2–carbon moieties derived from pyruvate are joined to form a 4–carbon entity whose exterior carbons are equivalent to the methyl carbons of pyruvate or acetate, both of which arise from carbon 3 of pyruvate. The interior of the 4–carbon entities formed by condensation in this manner are equivalent to the carboxyl carbons of the acetate moiety. Diacetyl, acetoin and 2,3-butanediol are the major products formed from head-to-head condensation of two carbon moieties derived from pyruvate cleavage.

In **head-to-tail condensations**, acetyl CoA moiety or equivalent two carbon units, are joined so that the methyl carbon of the acetate molecule is joined to carboxyl carbon of a second acetate molecule, condensations of this type produce 4-carbon molecules in which carbons 1 and 3 arise from the carboxyl carbons of the acetate moiety and carbons 2 and 4 arise from methyl carbons. Butanol and isopropanol are the end products formed from head-to-tail condensation.

In many fermentations, carbon dioxide is a frequent fermentation end product. However, for certain organisms its fixation is required for end product formation (Table 18.4). In some systems, succinate and propionate formation require CO_2 fixation. In these organisms, CO_2 may serve four physiologicl functions.

i. As a fermentation end product

ii. As an essential agent in energy generation through formation of terminal electron acceptor

iii. As an agent in fermentation end product formation and

iv. In the formation of critical biosynthetic intermediates.

Table 18.4: Enzymes of CO_2 fixation in microbes

Enzyme	Acceptor	Product
Pyruvate carboxylase	Pyruvate	Oxaloacetate
Phosphoenol pyruvate carboxylase	PEP	Oxaloacetate
Malic enzyme	Pyruvate	Malic acid
Phosphoenol pyruvate carboxykinase	PEP	Oxaloacetate, ATP

18.6. FERMENTATION VERSUS OXYGEN

Fermentation is carried out by either facultative or obligate anaerobes. Facultative anaerobes, such as enterobacteria, grow as aerobes in the presence of oxygen; under anaerobic conditions they carry out fermentative metabolism. On the other hand, obligate anaerobes cannot grow as aerobes because they do not possess the linked oxygen respiratory chain. Moreover, many of them do not tolerate oxygen and are killed in air for the reasons explained below.

Many strict anaerobes such as clostridia and methanogens die when they are exposed to air. In these organisms, when reduced flavoproteins or reduced iron-sulfur proteins come together with oxygen, two toxic compounds are formed *i.e.,* hydrogen peroxide (H_2O_2) and superoxide radical.

$$FADH_2 \rightleftharpoons FAD + 2H^+ + 2e^-$$

$$O_2 + 2e^- + 2H^+ \xrightarrow{\text{Oxidase}} H_2O_2$$

$$FADH_2 \rightleftharpoons FAD + 2H^+ + 2e^-$$

$$2O_2 + 2e^- + 2H^+ \xrightarrow{\text{Oxidase}} 2O_2^- + 2H^+$$

It has now been proved that superoxide radicals are more toxic than H_2O_2. Aerobes as well as aerotolerants, therefore, possess catalase and superoxide dismutase which convert the H_2O_2 and superoxide radicals into non-toxic form.

$$2H_2O_2 \xrightarrow{\text{catalase}} 2H_2O + O_2$$

$$O_2^- + O_2^- + 2H^+ \xrightarrow{\text{sup eroxide dismutase}} H_2O_2 + O_2$$

E. coli produces two types of superoxide dismutases, one is a iron containing protein and the second one is mangano-protein. Mammalian superoxide dismutse contains copper and zinc. Distribution of these two enzymes in aerobes, aerotolerants and strict anaerobes is shown in Table 18.5.

Table 18.5 : Activity of superoxide dismutase and catalase in microorganisms

Organism	**Specific activity (μg/mg)**	
	Superoxide dismutase	Catalase
Aerobes		
Escherichia coli	1.8	6.1
Salmonella typhimurium	1.4	2.4
Micrococcus radiodurans	7.0	289.0
Pseudomonas species	2.0	22.5
Aerotolerant anaerobes		
Eubacterium limosum	1.6	0
Streptococcus faecalis	0.8	0
Streptococcus lactis	1.4	0
Clostridium oroticum	0.6	0
Strict anaerobes		
Veillonella alcalescens	0	0
Clostridium pasteurianum	0	0
C. butyricum	0	0
Butyrovibrio fibrisolvens	0	0.1

It is evident from the Table 18.5 that aerobes contain both catalase and superoxide dismutase, aerotolerants are without catalase, but contain superoxide dismutase. Obviously, strictly anaerobes lack both the enzymes.

18.7. ALCOHOL FERMENTATION

Production of ethanol by the fermentation of sugars is widely distributed among microorganisms. Accumulation of ethanol has been detected in plants and fungi under anaerobic conditions. However, yeasts notably *Saccharomyces* species are the principal producers of alcohol. A number of anaerobic and facultatively aerobic bacteria can also produce ethanol as the main or subsidiary product during fermentation.

18.7.1. Alcohol Fermentation by Yeasts

It was Gay-Lussac (1815) who first established that yeast produce the alcohol with utilization of glucose and other hexose sugars as per the following equation.

$$C_6H_{12}O_6 \longrightarrow 2CO_2 + 2C_2H_5OH$$

Yeasts, like fungi, are aerobic, but in the absence of oxygen in air they ferment carbohydrates to ethanol and carbon dioxide. Alcoholic fermentation reactions carried out by yeasts are summarized in Fig. 18.3.

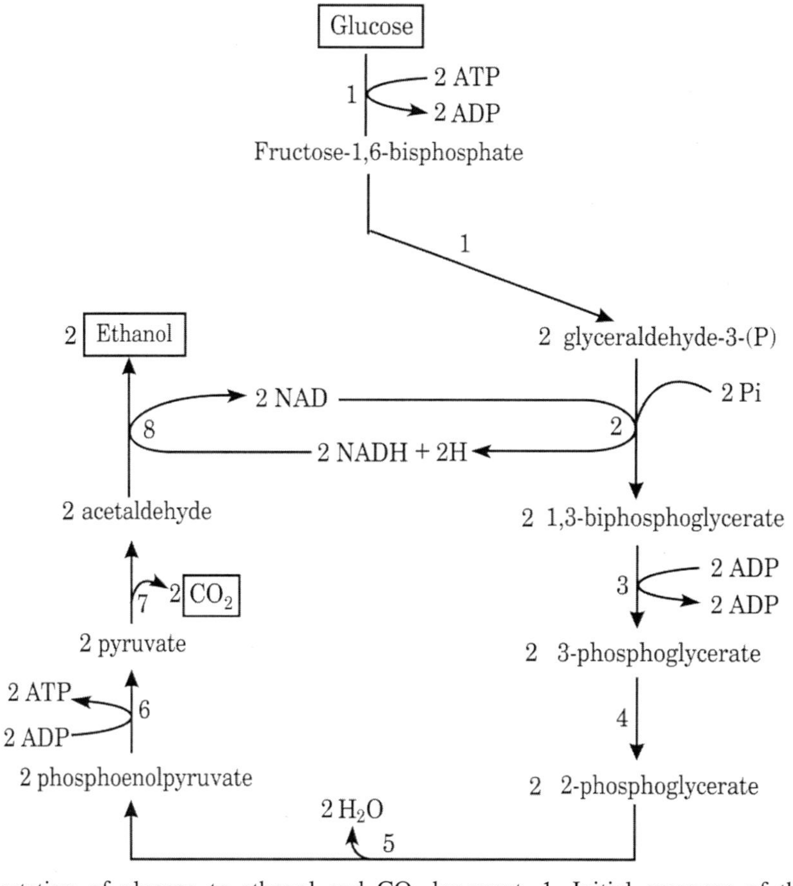

Fig. 18.3: Fermentation of glucose to ethanol and CO_2 by yeast. 1. Initial enzymes of the Embden-Meyerhof-Parnas pathway; 2. Glyceraldehyde-3-phosphate dehydrogenase; 3. 3-phosphoglycerate kinase; 4. phosphoglycerote mutase; 5. enolase; 6. pyruvate kinase; 7. pyruvate decarboxylase; 8. alcohol dehydrogenase.

Yeasts employ the Embden-Meyerhof-Parnas pathway for glucose breakdown. The pyruvate generated through this pathway is not converted to acetyl-CoA, but it is decarboxylated to acetaldehyde. This reaction is catalyzed by an enzyme pyruvate decarboxylase, which is a key enzyme in the alcohol fermentation. Pyruvate decarboxylase is a complex enzyme with thiamine pyrophosphate as cofactor. Acetadehyde is reduced to ethanol by alcohol dehydrogenase.

Pasteur effect: Yeast, while growing on glucose anaerobically obtain net yield of 2 moles of ATP for every one mole of glucose, which is much lesser than the ATP production under aerobic conditions. In order to cope up with this situation large amount of glucose is broken down to obtain the necessary energy. When the yeasts are transferred from anaerobic to aerobic condition, the utilization of glucose is drastically reduced by a factor of 3 to 4 with concomitant stoppage of alcohol production. This phenomenon was first observed by Pasteur and in his honour it is called as 'Pasteur effect'. Apparrently the Pasteur effect is result of difference in the cell energy charge under aerobic and anaerobic conditions.

Neuberg fermentations: C. Neuberg while working on fermentation of glucose by yeasts and employing trapping method discovered a second type of fermentation known by his name, **Neuberg second fermentation**. The intermediary fermentation of acetadehyde during fermentation was demonstrated by adding hydrogen sulfite ($NaHSO_3$), which is practically nontoxic to yeasts. On addition of hydrogen sulfite to fermenting broth, acetaldehyde was precipitated.

$$CH_3CHO + NaHSO_3 \longrightarrow CH_2\text{-}CHOH\text{-}SO_3Na$$

Under these conditions, glycerol was formed as a new fermentation product and at the same time production of ethanol and carbon dioxide diminished. This shift in metabolism be explained as follows: when the acetaldehyde is precipitated by hydrogen sulfite it can no more function as the hydrogen acceptor. In order to overcome this bottle neck, dihydroxyacetone phosphate of EMP pathway functions as hydrogen acceptor and thus it is reduced to glycerol-3-phosphate, which is further dephosphorylated to glycerol as shown in the equation.

$$\text{Glucose + hydrogen sulfite} \longrightarrow \text{glycerol + acetaldehyde-sulfite} + CO_2$$

Further, the addition of alkali ($NaHCO_3$, Na_2HPO_4) to fermenting yeast extracts also leads to the formation of glycerol because acetaldehyde undergoes dismutation to ethanol and acetate and hence does not function as hydrogen acceptor. The equation for this fermentation is known as **Neuberg's third fermentation**

$$\text{2 glucose} + H_2O \longrightarrow \text{ethanol + acetate + 2 glycerol} + 2\ CO_2$$

In light of the Neuberg's second and third fermentations, the normal yeast fermentation has been designated as **Neuberg's first fermentation**.

18.7.2. Ethanol Fermentation by Bacteria

Although alcohol fermentation is the domain of yeasts, some bacterial species also carry out this fermentation. However, different bacterial species opt for different pathways and end products also vary both qualitatively and quantitatively.

Zymomonas mobilis isolated from pulque, the fermenting juice of agave, metabolizes glucose via the 2–keto–3 deoxy–6–phophogluconate pathway (ED pathway Fig. 16.4) and splits the pyruvate by pyruvate decarboxylase to acetaldehyde and carbon dioxide. Acetaldehyde is then reduced to ethanol. Small amounts of lactate is also produced in this fermentation.

Sarcina ventriculi, a strict anaerobic acidophilic bacterium and *Erwinia amylovora* a facultative anaerobic enterobacterium ferment the sugars much the same as those of yeasts. During fermentation by Enterobacteriaceae and clostridia, ethanol is produced as subsidiary end product. The acetaldehyde, however, is not produced by pyruvate decarboxylase directly from pyruvate, but arises via reduction of acetyl CoA (Fig. 18.9).

The heterofermentative lactic acid bacteria (*e.g. Leuconostoc mesenteroides)* produce ethanol by utilizing pentose phosphate pathway. Glucose is first metabolized by the initial steps in the pentose phosphate cycle to pentose phoshate. The xylulose-5-phosphate is then acted by phosphoketolase.

$$\text{Xylulose-5-phosphate + Pi} \longrightarrow \text{Acetyl phosphate + glyceraldehyde–3–phosphate}$$

The acetyl phosphate thus formed is reduced to ethanol by acetaldehyde dehydrogenase and alcohol dehydrogenase. Glyceraldehyde-3-phosphate, is converted to pyruvate and later reduced to lactate (Fig. 18.5).

18.8. LACTIC ACID FERMENTATION

Lactic acid is a very common end product of many bacterial fermentations. Some bacterial genera produce large quantities of lactate as end product which are often referred to lactic acid bacteria (LAB) and are collectively assigned to the family Lactobacteriaceae.

Lactic acid bacteria are heterogenous morphologically and include, rods, cocci and variation within these morphological types. However, these bacteria some exhibit stable physiological characteristics. They are all Gram-positive, do not form spores, non-motile, mostly saccharolytic and excrete large quantities of lactic acid as fermentation end product. Further, they lack the most anabolic pathways and thus exhibit very complex nutritional requirements which are met in special natural habitats, such as milk, intestinal tract of animals, human babies and plant materials. As they lost capacity to synthesize a number of metabolites, they are referred to as 'metabolic cripples' and 'fastidious organisms'. They do not contain haemins (cytochromes, catalase). However, streptococci are able to grow in the presence of oxygen, thus they are anaerobic but aerotolerant. Lactic acid bacteria, like the members of Enterobacteriaceae have the ability to utilize lactose, a milk sugar. The utilization of lactose by these organisms can be regarded as an adaptation to the ecological conditions in the mammalian digestive tract.

$$\text{Lactose} + H_2O \xrightarrow{\beta\text{ - galactosidase}} \text{D-glucose + D-galactose}$$

The enzyme β-galactosidase that hydrolyzes the lactose occurs in only a few bacteria. Galactose, a hydrolysis product of lactose, is isomerized to glucose-1-phosphate by an enzyme galactokinase. Table 18.6 summarizes the morphological and fermentative abilities of different species of lactic acid bacteria.

Lactic acid bacteria are classified into three groups according to their ability to ferment lactose to lactate or to additional products. They utilize three different pathways:

1. Homofermentative pathway
2. Heterofermentative pathway
3. Bifidum pathway

18.8.1. Homofermentative Pathway

In this pathway, lactate is the sole end product (about 90%). It yields 2 moles lactate per mole of glucose. The bacteria carrying this pathway (Table 18.6) metabolize glucose via fructose 1-6 biphosphate pathway to pyruvate. The latter is, however, not decarboxylated to acetaldehyde as in the alcohol fermentation, instead, used directly as H-acceptor. A very small portion of the pyruvate is decarboxylated and converted to acetate, ethanol, CO_2 or acetoin. The homofermentative pathway reactions are illustrated in Fig. 18.4.

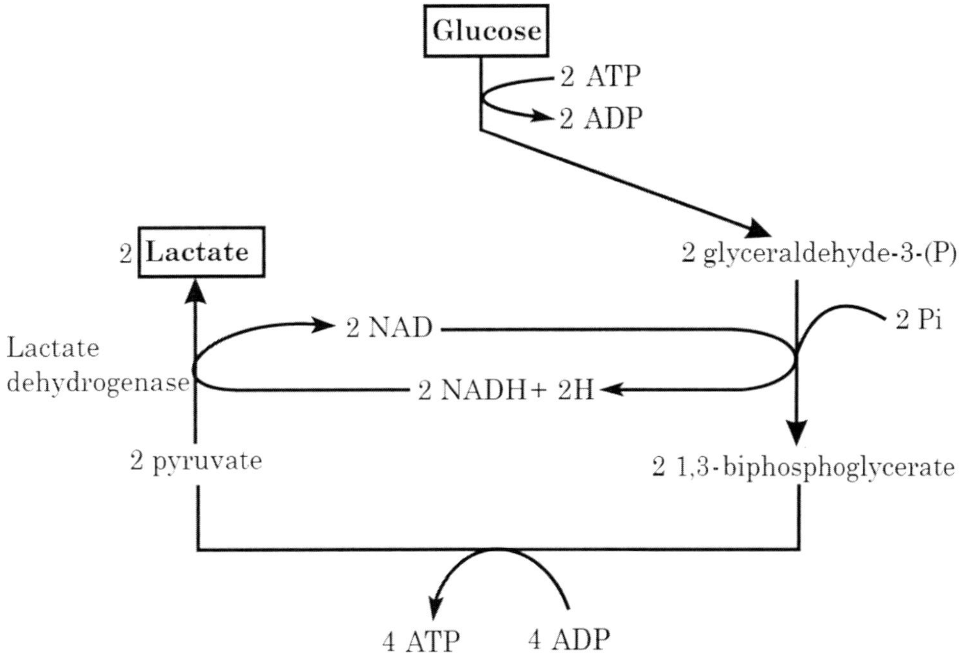

Fig. 18.4. Formation of lactate from glucose by homofermentative pathway

Table 18.6 : Morphology, fermentation types and configuration of lactic acid produced by different lactic acid bacteria

Genera and species	Morphology	Homofer-mentative	Hetero-fermentative	Configuration of lactic acid
Lactobacillus	rod			
L. delbrueckii		+	-	D(-)
L. lactis		+	-	D(-)
L. bulgaricus		+	-	D(-)
L. casei		+	-	L(+)
L. plantarum		+	-	DL
L. curvatus		+	-	DL

Genera and species	Morphology	Homofer-mentative	Hetero-fermentative	Configuration of lactic acid
L. brevis		-	+	DL
L. fermentum		-	+	DL
Sporolactobacillus	rod			
S. inulinus		+	-	D(-)
Streptococcus	coccus			
S. faecalis		+	-	L(+)
S. cremoris		+	-	L(+)
S. lactis		+	-	L(+)
Leuconostoc	coccus			
L. mesenteroides		-	+	D(-)
L. dextranicum		-	+	D(-)
Pediococcus	coccus			
P. damnosus		+	-	DL
Bifidobacterium	rod			
B. bifidum		-	+	L(+)

18.8.2. Heterofermentative Pathway

The heterfermentative lactic acid bacteria are devoid of enzymes of EMP pathway *viz*, aldolase and triose phosphate isomerase. Glucose is degraded through pentose phosphate pathway *i.e.,* via glucose-6-phosphate, 6-phosphate gluconate and ribulose-5-phosphate. The latter is epimerized to xylulose-5-phosphate, which is cleaved into glyceraldehyde-3-phosphate and acetyl phosphate by an enzyme phosphoketolase. Acetyl phosphate formation is facilitated through an enzyme bound α,β dihydroxyethylthiamine pyrophosphate, which is converted to acetyl-TPP-E via the α-hydroxyvinyl derivative.

The heterofermentative pathway yields 1 mol each of lactate, ethanol and CO_2 per mole of glucose

$$\text{Glucose} \xrightarrow{\text{Heterofermentative pathway}} \text{Lactate} + \text{ethanol} + CO_2$$

The detailed reactions of this pathway are illustrated in Fig. 18.5.

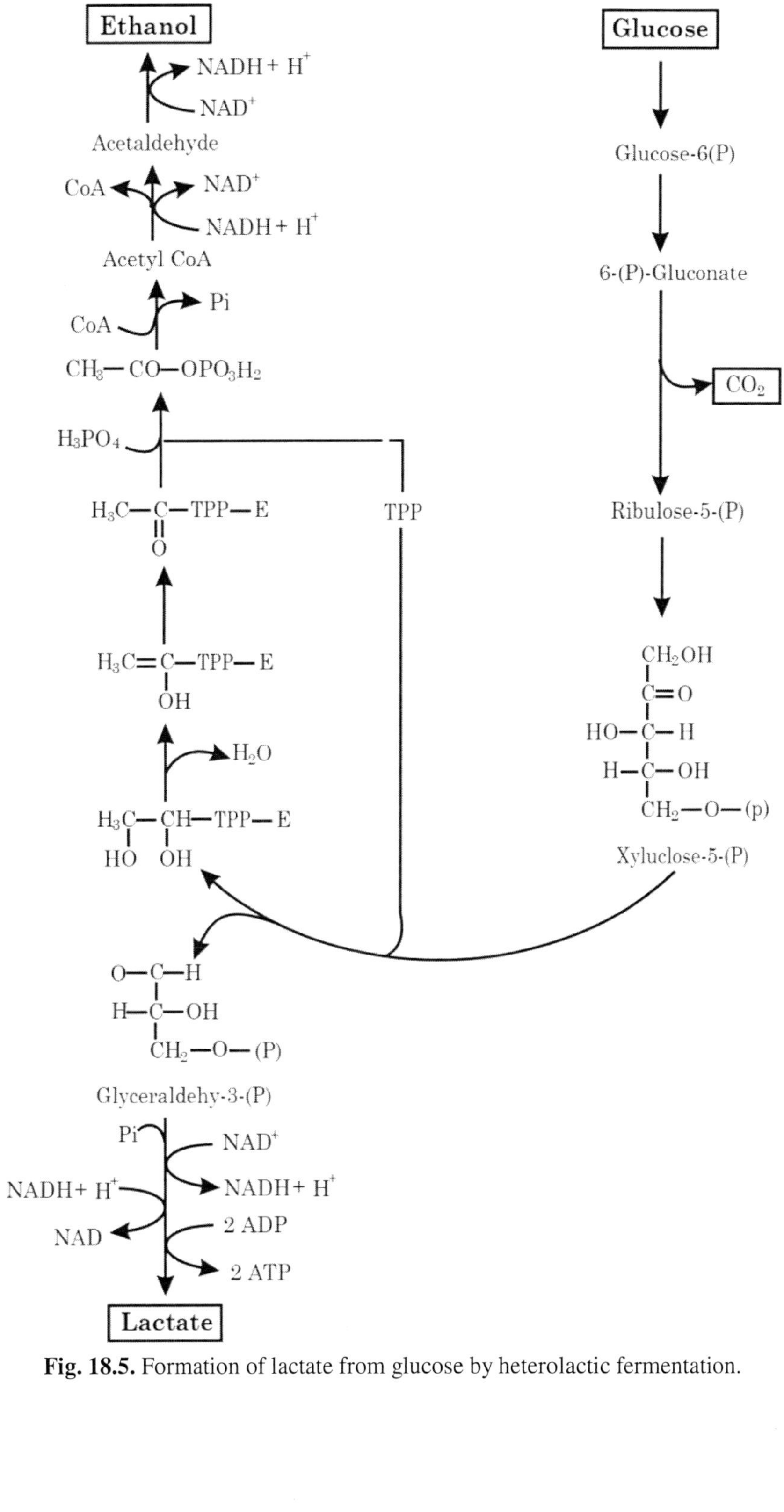

Fig. 18.5. Formation of lactate from glucose by heterolactic fermentation.

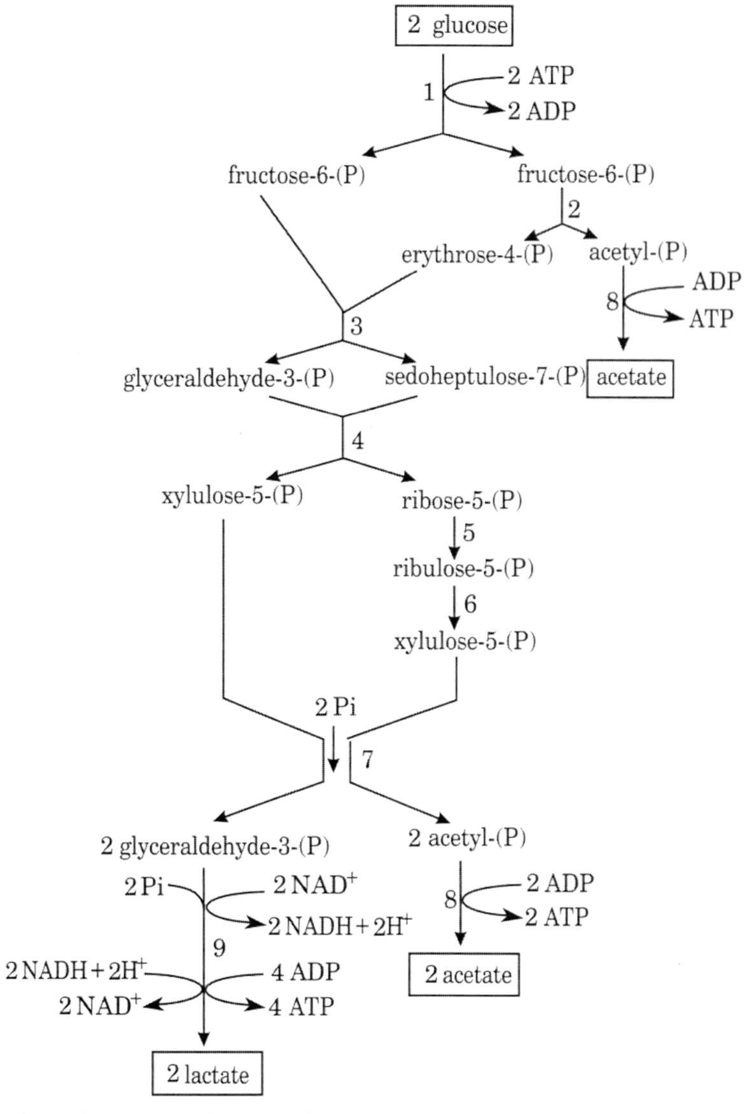

Fig. 18.6. Formation of acetate and lactate from glucose by the bifidum pathway. 1. Hexokinase and glucose-6-phosphate isomerase; 2. fructose-6-phosphate phosphoketolase; 3. transaldolase; 4. transketolase; 5. ribose-5-phosphate isomerase, 6. ribulose-5-phosphate 3-epimerase; 7. xylulose-5-phosphate phosphoketolase; 8. acetate kinase; 9. enzymes as in homofermentative pathway.

18.8.3. Bifidum Pathway

The bacterium *Bifidobacterium bifidum,* a V or Y shaped bacterium is an obligate anaerobe found in intestinal tract of breast fed human babies, where N-acetyl glucosamine is found. This bacterium metabolizes glucose according to the equation.

$$2C_6H_{12}O_6 \longrightarrow 2\ CH_3\text{-CHOH-COOH} + 3\ CH_3COOH$$

This pathway yields acetate and lactate in the ratio of 3:2. The bifidum pathway is illustrated in Fig 18.6. In glucose breakdown through this pathway two phosphoketolases are involved. One specific for fructose-6-phosphate and another specific for xylulose-6-phosphate. These two phosphoketolases that split fructose-6-phosphate and xylulose-5-phosphate to acetyl phosphate and erythrose-4-phosphate or glyceraldehyde-3-phosphate, respectively. The conversion of the later (G-3-P) to lactate involves glyceraldehyde-3-P and lactate dehydrogenase. The formation of acetate from acetyl phosphate is coupled to the formation of ATP from ADP. This reaction is catalyzed by acetate kinase. Thus in bifidum pathway ATP yield is more than that of the homo- and heterofermentative pathways.

Lactic acid bacteria also ferment a large number of sugars other than glucose such as fructose, galactose, mannose, saccharose, lactose, maltose and pentoses. Certain variations in the established pathways have been observed during the utilization of these substrates. For instance, *L. casei*, a homofermentative lactobacillus utilizes the phosphoketolase,a key enzyme of the fermentative pathway. Similarly, *Leuconostoc mesenteroides* ferments fructose producing lactate, acetate, CO_2 and mannitol. In this case, fructose is reduced to mannitol by NADH. Some lactic acid bacteria like *L. plantarum, L. casei* and *L. mesenteroides* which are involved in the wine fermentation and decrease the acidity ferment the malate, in presence of sugars, to lactate by a special malic enzymes.

$$\text{L-malate} \xrightarrow{Mn^{2+},\ NAD^{+}} \text{L-lactate} + CO_2$$

Some lactic acid bacteria like *Streptococcus lactis* subsp. *diacetylactis* and *Leuconostoc cremoris*, in contrast to their designation do not produce lactate as an end product. These bacteria ferment citrate, a milk component, to acetoin and diacetyl formation. Citrate lyase enzyme breaks down citrate under anaerobic conditions. Acetoin and diacetyl produced are responsible for characteristic flavour of butter

$$3\ \text{Citrate} \longrightarrow \text{lactate} + 3\ \text{acetate} + 5CO_2 + \text{diacetyl}$$

18.9. PROPIONATE FERMENTATION

Several bacteria under anaerobic conditions ferment glucose, sucrose, lactose and pentoses as well as lactate, malate, glycerol and other substrates to propionate. These bacteria include propionibacteria (*Propionibacterium freudenreichii* subspecies *shermanii*, *P. acidi-propionici*), *Veillonella alcalescens* (*Micrococcus lactilyticus*), *Clostridium propionicum*, *Bacteroides ruminicola*, *Megasphaera elsdenii*, *Selenomonas ruminantium* and species of *Micromonospora*. The species of *Propionibacterium* are Gram-positive, nonmotile rods and non-spore formers. Earlier, they were thought to be obligate anaerobes, but experimental evidence shows that they are microaerotolerants. These bacteria are natural inhabitants of rumen and intestines of ruminants where they play a part in the formation of fatty acids like propionate and acetate. They are responsible for the conversion of lactate which arises from various fermentations in the rumen, to predominantly propionate. Propionibacteria can be isolated by enrichment culture under anaerobic conditions into a lactate-yeast extract medium inoculated with fresh cheese. Two pathways have been identified for fermentative production of propionate. In both the pathways, catabolism of hexose proceeds via the fructose-biphosphate pathway.

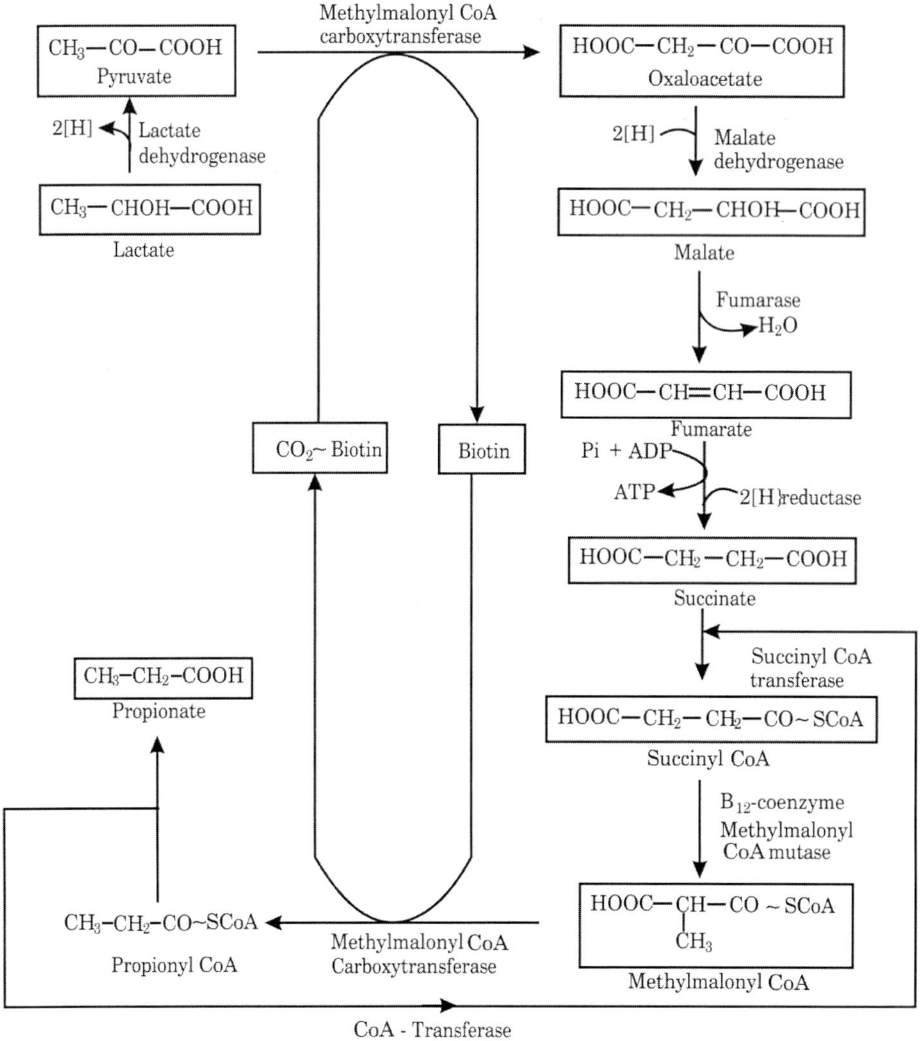

Fig. 18.7. Methylmalonyl-CoA pathway of propionate formation

18.9.1. Methylmalonyl–CoA Pathway (Succinate – Propionate Pathway)

As the methylmalonyl CoA is a characteristic intermediate in this pathway it is known as methylmalonyl CoA pathway. In this pathway (Fig. 18.7) the formation of propionate from lactate occurs according to the overall reaction :

$$3CH_3\text{-}CHOH\text{-}COOH \longrightarrow 2CH_3\text{-}CH_2\text{-}COOH + CH_3\text{-}COOH + CO_2 + H_2O$$

The various steps of this pathway in a sequence are explained below.

1. In the first step, lactate is converted to pyruvate catalysed by lactate dehydrogenase.

2. Pyruvate is carboxylated to oxaloacetate (OAA) by methylmalonyl-CoA carboxytransferase with the participation of biotin-carbon dioxide complex.
3. The OAA is then reduced to succinate via malate and fumarate. Reduction of fumarate to succinate is coupled to an electron transport phosphorylation (see fumarate respiration).
4. The inactive succinate is activated to succinyl CoA by a CoA transferase.
5. Succinyl CoA is then converted to methylmalonyl-CoA by an enzyme methylmalonyl–CoA mutase with the participation of coenzyme B_{12} (cyanocobalamin).
6. Methylmalonyl-CoA loses the carbon dioxide to yield propionyl-CoA. This CO_2 is taken up by methylmalonyl carboxytransferase (step-2) and transfers to pyruvate to give rise to OAA.
7. Propionyl-CoA by with the loss of CoA to succinate, is convered to propionate.

The unique features of this pathway are:

1. Recycling of CO_2 and CoA between the product and precursors without occurring in free form.
2. Participation of three cofactors viz., biotin, coenzyme A and B_{12}.

18.9.2 Acrylyl CoA Pathway

Clostridium propionicum, Prevotella ruminicola and *Megasphaera elsdeni* produce propionate fermentatively by a route known as Acryloyl-CoA pathway. This sequence is commonly referred to as the direct reduction or non-randomizing pathway. In this pathway, acrylyl CoA is the key intermediate. The reactions of this pathway are shown in Fig. 18.8 and important steps are explained below:

1. A portion of the available lactate is converted to its D-isomer by racemase enzyme
2. Propionic acid formation by this sequence begins with the conversion of L-lactate to L-lactyl-CoA by a CoA transferase reaction
3. The L-lactyl-CoA is then converted to acryloyl-CoA by a dehydrogenase reaction.
4. The acryloyl-CoA is reduced to propionyl CoA with the reducing equivalents supplied by a reduced electron transferring flavoprotein.
5. The propionyl-CoA transfers its CoA moiety to a molecule of L-lactate regenerating L-lactyl-CoA and forming free propionic acid.
6. D-lactate is converted to pyruvate with the aid of flavoprotein dependant lactic dehydrogenase
7. Pyruvate is then cleaved by pyruvate ferredoxin oxidoreductase, giving carbon dioxide, acetyl CoA and reduced ferredoxin. The ferredoxin is reoxidised by a transhydrogenase reaction, providing a portion of the reducing equivalents required for conversion of acrylyl-CoA propionate

8. The acetyl-CoA is converted by phosphotransacetylase to acetyl phosphate which through acetyl kinase, becomes free acetic acid. The last reaction also produces a molecule of ATP.

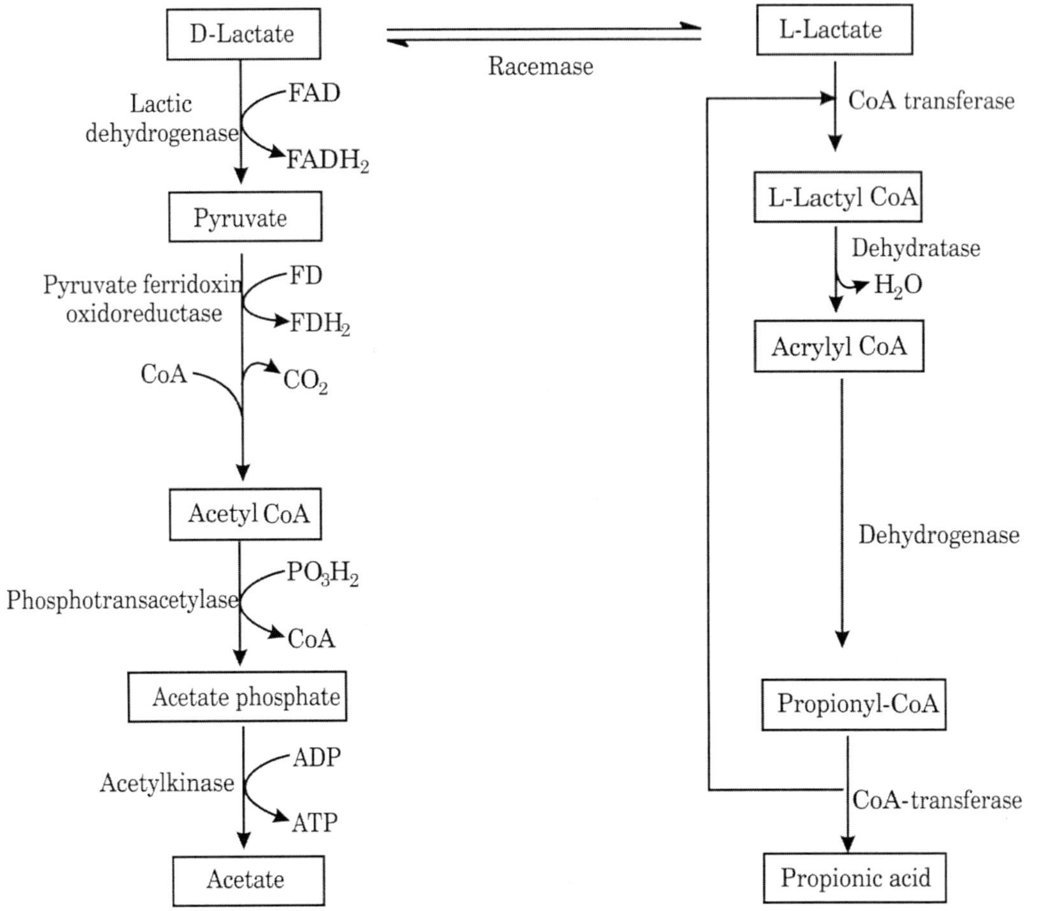

Fig. 18.8. The acrylyl CoA pathway for propionate formation

18.9.3 Wood–Werkman Reaction

Some propionibacteria produce succinate in addition to propionate. Hence, the carboxylase alone cannot be accountable for the formation of C_4–dicarboxylic acid (OAA). It was presumed that an enzyme system must be present in these microorganisms that catalyzes the synthesis of C_4–dicarboxylic acids from a C_3 compound and CO_2. Wood and Werkman (1936) while studying the glycerol fermentation by *P. acidi-propionic* discovered that this involved the net fixation of carbon dioxide. The fixed CO_2 was detected in excreted succinate, thus proving the point that carboxylation of pyruvate with the formation of dicarboxylic acid has taken place. This carboxylation, on the name of discoverers, was called Wood-Werkman reaction. This type of carboxylation is not restricted to propionibacteria but it is also found in animals and plants and in all heterotrophic organisms

$$\text{Pyruvate} + CO_2 + \text{ATP} \longrightarrow \text{OAA} + \text{ADP} + \text{Pi}$$

18.9.4. The PEP Carboxytransphosphorylase of Propionibacteria and its Physiological Significance

The propionibacteria produce succinate as well as propionate as an end product of fermentation when growing on carbon sources such as glucose that enter the glycolytic pathway. Succinate is a 4–carbon compound whereas the end product of glycolysis that is pyruvate is a C_3 compound. It indicates that they must have an enzyme to carboxylate a C_3 intermediate to form the C_4 product. The C_3 intermediate that is carboxylated is phosphoenol pyruvate (PEP). The PEP is carboxylated to oxaloacetate, which is then reduced to succinate via malate and fumarate. The enzyme that catalyzes the carboxylation of PEP is called PEP carboxytransphosphorylase and it catalyzes the following reaction.

$$\text{PEP} + CO_2 + \text{Pi} \longrightarrow \text{oxaloacetate} + \text{ppi}$$

The pyrophosphate formed in the above reaction is a high-energy compound and propionibacteria have enzymes that phosphorylate fructose-6-phosphate to fructose-1, 6-biphosphate and serine to phosphoserine using pyrophosphate as the phosphoryl donor.

The carboxylation of PEP or pyruvate to oxaloacetate and the reduction of the oxaloacetate to succinate is a widespread pathway in many fermenting bacteria. This pathway is extremely important for anaerobes and serves three purposes: (i) The fumarate is an electron sink enabling $NADH_2$ to be reoxidized, (ii) the fumarate reductase is a coupling site (*i.e.*, a ΔP is generated), and (iii) the succinate can be converted to succinyl CoA, which is required for the biosynthesis of tetrapyrroles, lysine, diaminopimelic acid and methionine. From evolution point of view, Gest (1983) has suggested that the carboxylation of the C_3 glycolytic intermediate and reductive pathway from oxaloacetate to succinate may have evolved when the earth's atmosphere was still anaerobic and served the purpose of balancing fermentations, thus sparing one of the two pyruvates derived from glucose for biosynthesis. The reactions between oxaloacetate and succinate may have later became part of the oxidative citric acid cycle during the evolution of aerobic metabolism.

Economic importance: Propionibacteria play an important role in the nutrition of ruminants. In the rumen, they ferment the lactate to propionate which along with other volatile fatty acids is absorbed into the blood stream. They also play an important role in maturation and flavour formation of cheese. Renin, an aqueous extract of calves stomach, contains numerous viable propionibacteria. This was earlier used for the coagulation of milk for cheese production. Their role in other habitats like soil is not clear.

18.10. MIXED ACID FERMENTATION

Fermentative microbes can be broadly divided into two categories: homofermentative and heterofermentative. In homofermentations a single end product is produced, whereas in heterofermentations more than one end products are produced. Mixed acid and butanediol fermentations are good examples for heterofermentations.

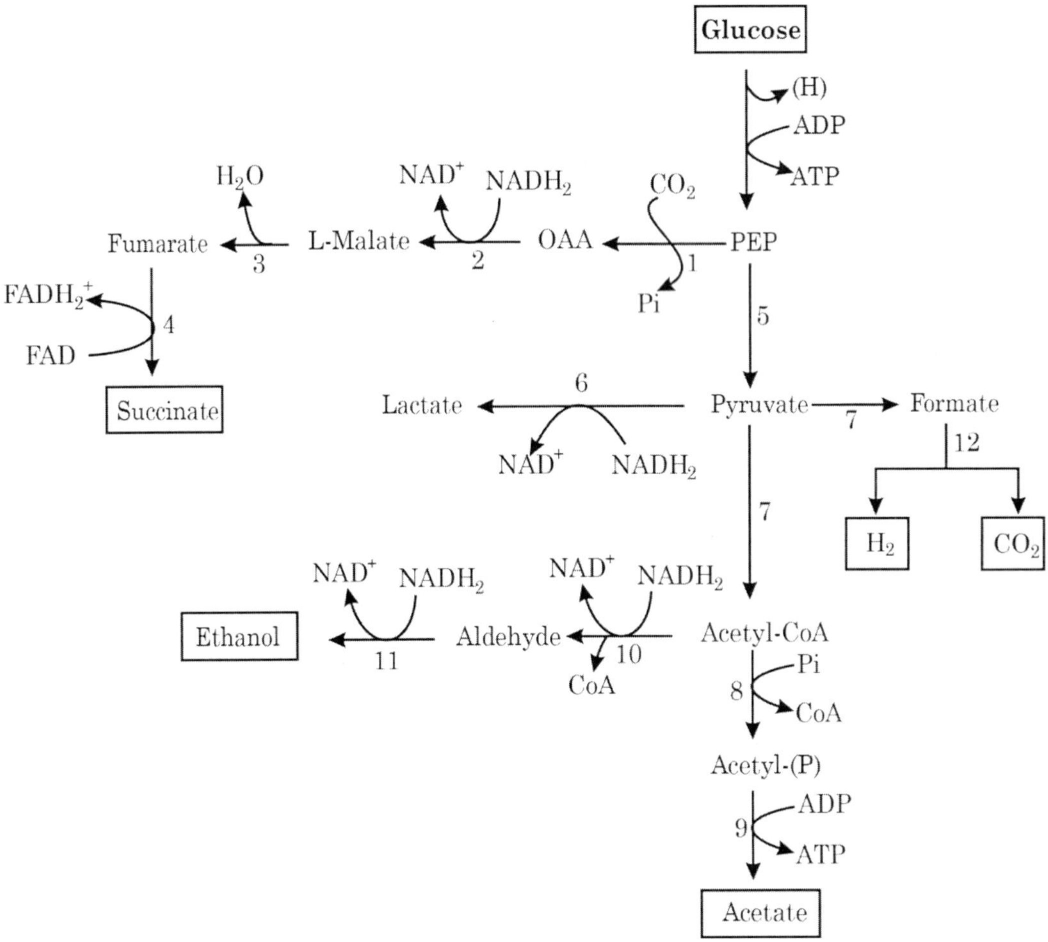

Fig. 18.9. Mixed acid fermentation by *Escherichia coli.*1. PEP carboxylase, 2. Malate dehydrogenase, 3. Fumarase, 4. Fumarate reductase, 5. Enolase, 6. Lactate dehydrogenase, 7. Pyruvate ferredoxin, 8. Phosphotransacetylase, 9. Acetyl kinase, 10. Aldehyde dehydrogenase, 11. Alcohol dehydrogenase, 12. Formic dehydrogenase and hydrogenase.

Mixed acid fermentation is carried out by the enterobacteria organisms belonging to the genera *Escherichia, Salmonella* and *Shigella,* that ferment sugars to lactate, acetate, succinate and formate. In addition CO_2, H_2 and ethanol are also formed. The reactions of this fermentation are shown in Fig 18.9. The salient features of this fermentation are:

1. The enterobacteria ferment glucose by the EMP pathway and use intermediates for the formation of succinate, ethanol, lactate, formate, CO_2 and hydrogen.
2. Succinate is formed by the carboxylation of PEP to produce oxaloacetate (OAA). The OAA is converted to succinate by the combined actions of malate dehydrogenase, fumarase, and fumarase reductase.
3. Lactate is formed by direct pyruvate reduction with lactate dehydrogenase.

4. Formate and acetyl CoA are produced by the action of pyruvate ferredoxin oxidoreductase.
5. Acetyl CoA is converted to free acetic acid, with the intermediate formation of acetyl-P by the action of phosphotransacetylase and acetyl kinase. This sequence allows the generation of ATP.
6. A portion of the acetyl CoA is converted to ethanol by the action of acetadehydedihydrogenase and alcohol dehydrogenase.
7. Formate is converted to CO_2 and hydrogen by the joint action of formic dehydrogenase and hydrogenase.

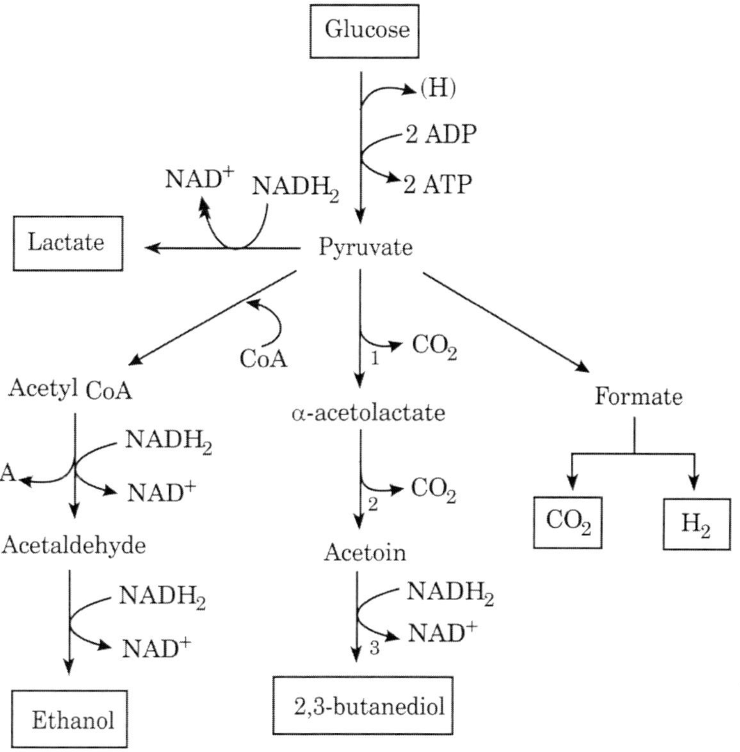

Fig. 18.10. Butanediol fermentation by *Enterobacter aerogenes*.1. Acetolactate synthetase, 2. Acetoin decarboxylase, 3. Butanediol dehydrogenase

18.11. BUTANEDIOL FERMENTATION

Species of *Enterobacter*, *Serratia* and *Erwinia* display this type of fermentation as shown in Fig. 18.10. They produce less acids, but more gas (CO_2), ethanol and above all large amount of 2,3-butanediol (for comparison the two fermentation balances are presented in table 18.7. Salient features of this fermentation are:

1. Pyruvate is formed from glucose by the EMP pathway.

2. Lactate, ethanol, formate, carbon dioxide and hydrogen are formed by mechanisms identical to those of the mixed acid fermentation.
3. A protion of the pyruvate is converted to acetolactate by acetolactate synthetase.
4. The acetolactate is further converted to acetoin and CO_2 by acetoin decarboxylase, and the acetoin is then reduced to butanediol with butanediol dehydrogenase.

Table 18.7. Products formed by mixed acid and butanediol fermentations

Product	***Escherichia coli*** **(mole formed / 100 mol glucose fermented)**	***Enterobacter aerogenes***
Formate	2.4	17.0
Acetate	36.5	0.5
Lactate	79.5	2.9
Succinate	10.7	--
Ethanol	49.8	69.5
2,3-butanediol	0.3	66.4
CO_2	88.0	172.0
Hydrogen	75.0	35.4

Enterobacteria, carrying the mixed acid and butanediol fermentations, are the inhabitants of the intestinal tract, hence the family has been named as Enterobacteriaceae. They exhibit the following common characteristics:

1. Gram-negative, peritrichously flagellated, motile and non-sporing rods.
2. Facultatively aerobic, possess haemins (cytochromes and catalase) and can obtain their energy aerobically (by respiration) and anaerobically (by fermentation).
3. Nutritionally non-demanding and can be grown on simple synthetic media containing mineral salts, carbohydrates and ammonium salts.
4. They all ferment glucose with the formation of acids.

Enterobacteriaceae members are important from sanitation and hygiene point of view. Some are even pathogenic. Hence, their diagnostic characters are presented below Table 18.8.

Table 18.8 : Diagnostic characteristics of genera of Enterobacteriaceae.

Genus	**Motility**	**Urea utilization**	**Glucose fermentation**	**Lactose fermentation**	**H_2 formation**	**Indole formation**	**Acetoin fermentation**	**Proteolysis**
Escherichia	+	-	+	+	+	+	-	-
Klebsiella	-	+	+	+	+	-	+	-
Enterobacter	+	+	+	+	+	-	+	+

Genus	Motility	Urea utilization	Glucose fermentation	Lactose fermentation	H_2 formation	Indole formation	Acetoin fermentation	Proteolysis
Serratia	+	-	+	-	+	-	+	+
Proteus	+	+	+	-	+	+	-	+
Citrobacter	+	+	+	+	+	+	-	-
Salmonella	+	-	+	-	+	-	-	-
Shigella	-	-	+	-	-	+	-	-
Erwinia	+	-	+	+	-	-	+	+

Escherichia coli and *Enterobacter aerogenes* are collectively designated as coliforms and share many common characteristics. Hence, they are often referred as 'twins'. However, they can be differentiated by well known IMViC (Indole, methyl red, Voges-Proskauer and citrate) tests (Table 18.9). *E. aerogenes* is commonly distributed in soils and produces large amount of gas.

Table 18.9. The IMViC test reactions for distinguishing *Escherichia coli* and *Enterobacter aerogenes.*

	Indole Formation	Methyl red test	Acetoin formation	Citrate utilization
Escherichia coli	+	+	-	-
Enterobacter aerogenes	-	-	+	+

18.12. BUTYRATE-BUTANOL FERMENTATIONS

The fermentation of sugars to butyric acid was first discovered by Louis Pasteur in 1896. Soon after, microorganims responsible for the production of butyrate were isolated and it was found that several clostridia species carry out this type of fermentation. Along with butyrate, butanol, acetone, 2-propanol and other organic acids, and alcohols are typical products of carbohydrate fermentation by these bacteria. Different species of clostridia utilize different carbohydrates, amino acids, alcohols, organic acids and produce butyrate and several other compounds as end products. Some of the details are presented in Table 18.10. In addition to clostridia, *Butyrivibrio fibrisolvens, Eubacterium limosum* and *Fusobacterium nucleatum* also produce butyrate as fermentation end product. All these bacterial species are obligate anaerobes.

Clostridia species are gram-positive rods, motile with peritrichous flagella. They form characteristic oval or spherical endospores. Most of these are strict anaerobes, however facultative aerobes and aerotolerant species are also common (*C. pasteurianum, C. acetobutylicum).* They contain no haem derivatives (catalase, cytochromes). Most species are mesophiles, but *C. thermoacetium C. thermohydrosulfuricum* are thermophiles. They grow at either neutral or alkaline pH. Clostridia can utilize a wide range of natural material as substrates. They can metabolize polysaccharides (starch, glycogen, cellulose, hemicellulose and pectins), nucleic acids, proteins, amino acids, purines and pyramidines. Some require complex media. Species like *C. pasteurianum* can utilize gaseous nitrogen as their sole source of nitrogen.

Table 18.10. Fermentative properties of clostridia species

Clostridum species	Substrates	Fermentation products
C. butyricum	Glucose, starch, dextrin	Butyrate, acetate, CO_2, H_2
C. tyrobutyricum	Glucose, lactate, glycerol + acetate	Butyrate, acetate, CO_2, H_2
C. pasteurianum	Glucose, starch, mannose, inulin	Butyrate, acetate, CO_2
C. pectinovorum	Pectin, starch, glycogen, dextrin	Butyrate, acetate
C. butylicum	Glucose	Butyrate, acetate, butanol, 2-propanol, CO_2, H_2
C. acetobutylicum	Glucose, glycerol, pyruvate	Butyrate, acetate, butanol, acetone, acetoin, ethanol, CO_2, H_2
C. kluyveri	Ethanol + acetate + CO_2	Caproate, butyrate, CO_2
C. tetanomorphum	Glutamate, histidine	Butyrate, acetate, NH_3, CO_2, H_2

Biochemistry of fermentation

Clostridia are heterofermentative organism. As end products of fermentation varying amount of acids (butyrate, acetate, lactate) and alcohols (butanol, 2–propanol), as well as acetone and gas (CO_2, H_2) are produced. Their yields are variable and depend on conditions. The fermentation of glucose by *C. butyricum* and *C. acetobutylicum* are often regarded as the prototypes of butyrate fermentation. Important reactions of this pathway are shown in Fig. 18.11 and explained below:

1. Glucose is metabolized via fructose-biphosphate pathway.
2. Pyruvate by the actions of pyruvate-ferredoxin oxidoreductase and hydrogenase, yields acetyl CoA and hydrogen.
3. The acetyl CoA is converted to acetoacetyl CoA by acetyl CoA: acetyl transferase.
4. A portion of the acetoacetyl CoA is then converted to free acetoacetate by acetoacetyl CoA acetate coenzyme A transferase.
5. Decarboxylation of the free acetoacetate yields acetone and carbon dioxide.
6. The remaining portion of the acetoacetyl CoA is converted to β-hydroxybutyric acid by L (+) –β-hydroxybutyryl CoA dehydrogenase.
7. The β–hydroxybutyryl CoA is converted to crotonyl CoA by L-3-hydroxyacyl CoA hydrolase and the crotonyl CoA is reduced to butyryl CoA by butyryl CoA dehydrogenase.
8. Finally, butyryl CoA is reduced to butanol via butyraldehyde by two enzymes: butyraldehyde dehydrogenase and butanol dehydrogenase.
9. Hydrogen derived from the oxidation of pyruvate is liberated in gaseous form.

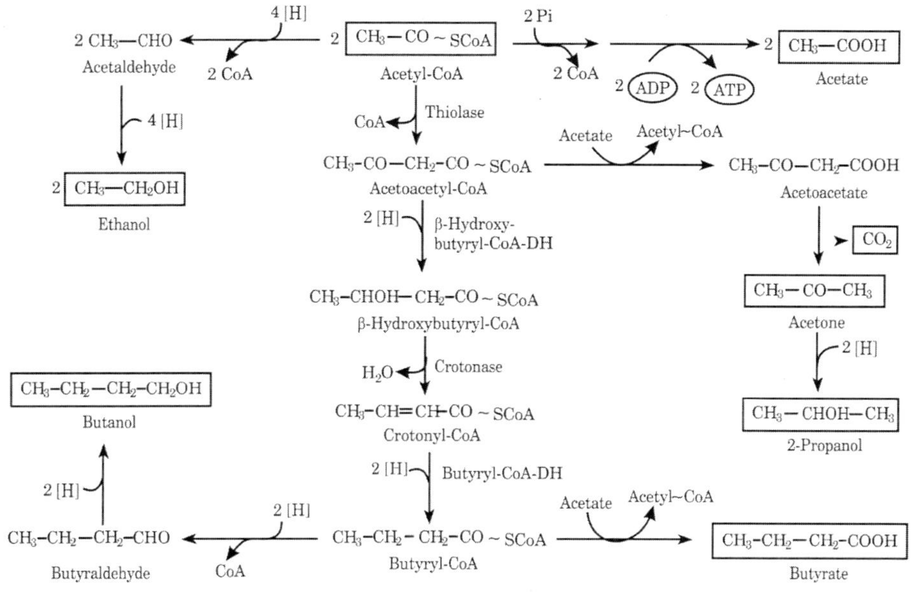

Fig. 18.11. The formation of acetate, ethanol, n-butanol, butyrate, acetone, and 2-propanol during clostridial fermentation.

Thus, the end products of butyric acid fermentation by *C. acetobutylicum* are butyrate, acetate, butanol, ethanol, 2-propanol, acetone, CO_2 and H_2. During the fermentations, butyrate is produced initially. With the increasing acidity, however, enzymes are induced, among them acetoacetate decarboxylase, whose function gives rise to the accumulation of acetone and butanol. There is a close connection between the formation of acetone and butanol. The decarboxylation of the acetoacetate to form acetone results in the loss of a potential hydrogen acceptor, which could have taken up two times 2[H] in the reduction of butyrate. This hydrogen, therefore, must be transferred to other hydrogen acceptors, including the excreted butyrate already formed. However, the butyrate to be converted into butanol, must first be activated by conversion to butyryl CoA. When the fermentation is carried out under alkaline conditions, for example in the presence of $CaCo_3$, *C. acetobutylicum* behaves like *C. butyricum* and produces substantial quantities of butyrate with the inhibition of butanol (Table 18.11).

Table 18.11. Acetone-butanol fermentation by *C. acetobutylicum* in presence and absence of $CaCO_3$

Fermentation products	$CaCO_3$ (Without)	$CaCO_3$ (With)
Butyrate	32.4	630
Butanol	411.5	45.7
Acetate	102.1	230.7
Ethanol	44.5	2.4
Acetone	222.3	18.2

18.13. THE ETHANOL-ACETATE FERMENTATION

Clostridium kluyveri carries out a very interesting fermentation. It produces butyrate, caproate and hydrogen from ethanol and acetate. Ethanol alone is not fermented. It requires either acetate or propionate. Acetate serves as an additional hydrogen acceptor, though it is also formed during fermentation. The ratio in which butyrate and caproate are formed can vary. An increase of the ethanol concentration of the medium favours caproate formation. A typical fermentation balance is

$$6 \text{ ethanol} + 3 \text{ acetate} \longrightarrow 3 \text{ butyrate} + 1 \text{ caproate} + 2H_2 + 4H_2O + H^+$$

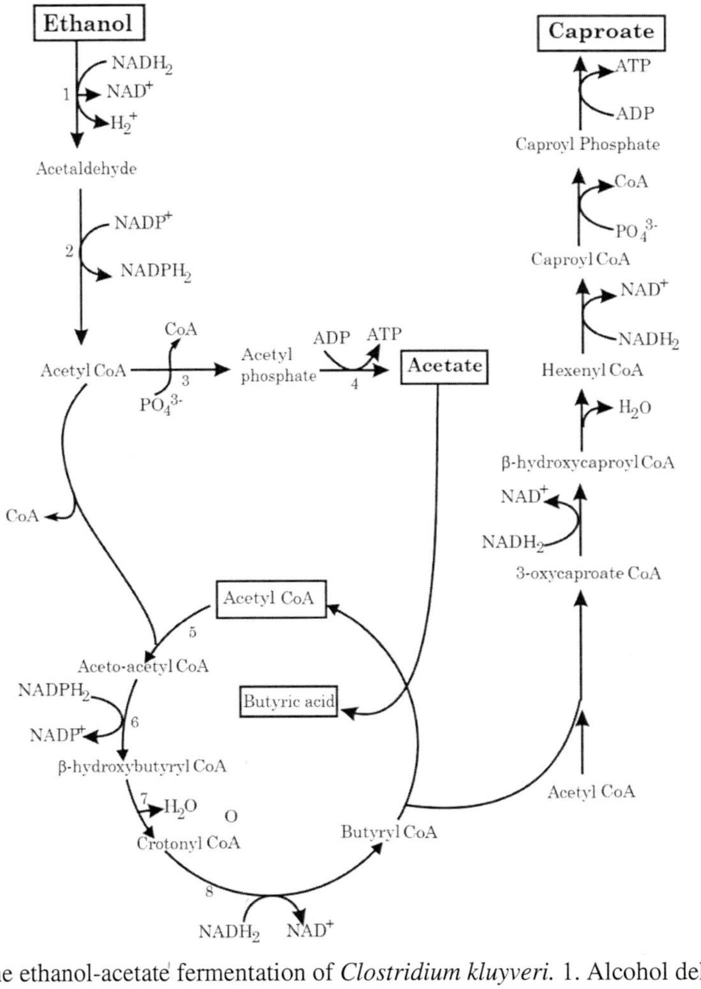

Fig. 18.12. The ethanol-acetate fermentation of *Clostridium kluyveri*. 1. Alcohol dehydrogenase, 2. Aldehyde dehydrogenase, 3. Phosphotransacetylase, 4. Acetyl kinase, 5. Thiolase, 6. B-hydroxybutyryl CoA dehydrogenase, 7. Crotonase, 8. Butyryl CoA dehydrogenase.

The detailed reactions of this fermentation are shown in Fig. 18.12. The salient features of this pathway are :

1. Ethanol is converted to acetaldehyde by the action of alcohol dehydrogenase. The acetaldehyde is further oxidized to acetyl CoA by aldehyde dehydrogenase.
2. Alcohol dehydrogenase is NAD^+ mediated whereas acetaldehyde dehydrogenase is $NADP^+$ mediated. Metabolism of $NADH_2$ and $NADPH_2$ produces hydrogen
3. A portion of the acetyl CoA becomes free acetate and ATP with the intermediate formation of acetyl phosphate by the actions of phosphotransacetylase and acetyl kinase. Two molecules of acetyl CoA condense in a reaction mediated by thiolase to produce acetoacetyl CoA.
4. Acetoacetyl CoA is converted to L (+)-β-hydroxy-butyryl CoA by NADPH-dependant L (+)-β-hydroxy butyryl CoA dehydrogenase
5. Crotonase enzyme removes water from -β-hydroxy butyryl CoA and converts it to crotonyl CoA.
6. Crotonyl CoA is reduced to butyryl CoA by NADH-dependant butyryl CoA dehydrogenase
7. Reaction of acetate with butyryl CoA, mediated by CoA transerase form free butyric acid and regenerates acetyl CoA.
8. Caproic acid is formed by the reaction of butyryl CoA with a molecule of acetyl CoA to yield 3-oxycaproyl CoA.
9. The remaining reactions of caproate formation are analogous to those involved in butyrate *i.e.,* reduction of β-ketocaproyl CoA to β-hydroxy caproyl CoA and the dehydration and reduction of β-hydroxycaproyl CoA to produce caproyl CoA.

18.14. FERMENTATION OF AMINO ACIDS

Most bacteria ferment carbohydrates or the end products of other fermentation products. However, some bacteria ferment aminoacids like glutamate and produce fatty acids, CO_2 and ammonia under anaerobic conditions. For example, *C. tetanomorphum* and *Peptostreptococcus asaccharolyticus* ferment L-glutamate and produce butyrate as an end product. This fermentation excited attention because it led to the discovery of the biochemical functions of vitamin B_{12}, upto then only known as vitamin. The schemes of the fermentation by these two organisms are shown in Figs. 18.13 & 18.14.

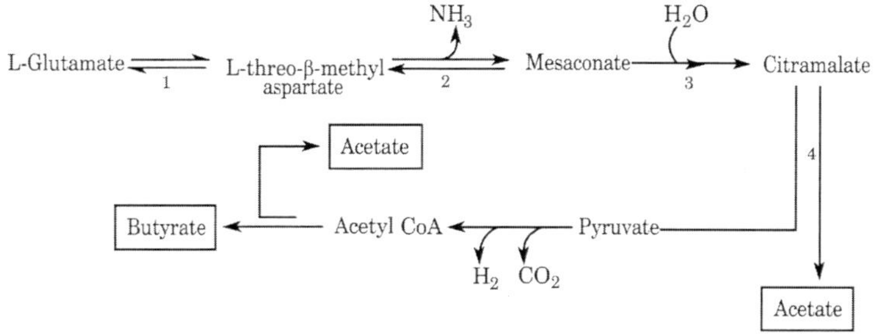

Fig. 18.13. Fermentation of glutamate by *C. tetanomorphum.* 1. L-glutamate mutase, 2. β-methyl aspartase, 3. Citramalate dehydratase, 4. Citramalate lyase.

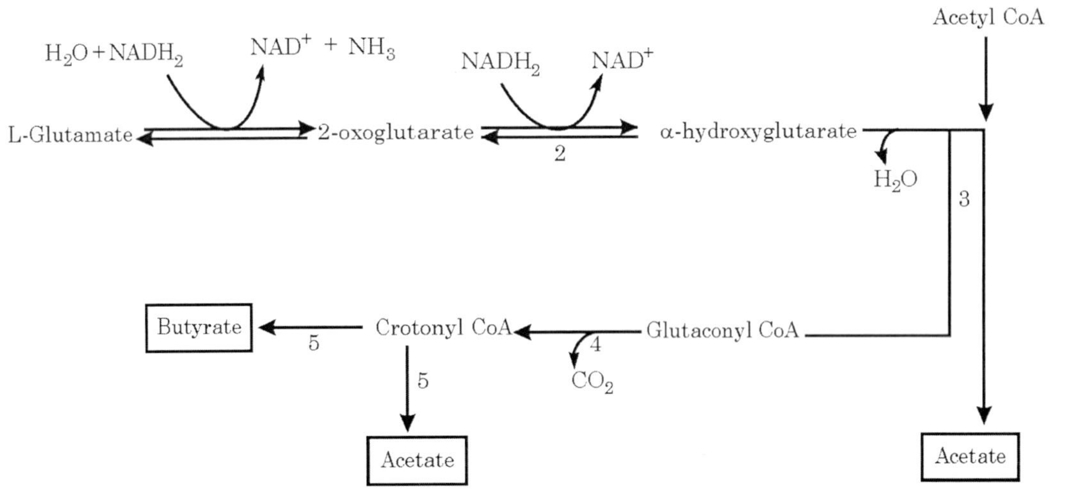

Fig. 18.14. Fermentation of glutamate by *P. asaccharolyticus.* 1. Glutamate dehydrogenase, 2. α-hydroxyglutarate dehydrogenase, 3. Dehydratase and CoA transferase, 4. Glutaconyl CoA decarboxylase, 5. Crotonyl CoA dismutase.

In case of *C. tetanomorphum,* L-glutamate is isomerized to L-threo-β-methyl aspartate by the action of L-glutamate mutase. The L-threo-β-methyl aspartate is deaminated to mesaconatate by β-methyl aspartase. Hydration of mesaconita by citramalate dehydratase, operating in the reverse direction produces citramalate. Citramalate lyase then produces equimolar amount of acetate and pyruvate. Further metabolism of pyruvate via centrol pathways yields additional acetate as well as butyrate, CO_2 and hydrogen.

Fermentation of L-glutamate by *P. asaccharolyticus* occurs by a different set of reactions. In the first step, L-glutamate is oxidatively deaminated by glutamate dehydro-genase to form 2-oxoglutaric acid (2 OG). The 2OG is then reduced to α–hydroxyglutaric acid by a hydroxyglutarate dehydrogenase. The combined action of a dehydratase and a CoA transferase on α-hydroxyglutaric acid produces glutaconyl CoA. Glutaconyl CoA upon decarboxylation is converted to crotonyl CoA. Finally crotonyl CoA dismutase produces acetate and butyrate in the ratio of 2:1.

Fermentation of a pair of aminoacids-Stickland reaction

The peptolytic clostridia bring about hydrolysis of proteins and ferment the resulting amino acids. Interestingly, some clostridia cannot ferment a single amino acid. They require a pair of amino acids, one functioning as hydrogen donor and another as hydrogen acceptor, for the fermentation process. This phenomenon was first discovered by Stickland (1934) in *C. sporogenes,* while it was growing on a mixture of two amino acids, alanine and glycine. The overall reaction in this case is

$$\text{Alanine} + 2\text{ glycine} + 2\text{ }H_2O \longrightarrow 3\text{ acetate} + 3\text{ }NH_3 + CO_2$$

In this fermentation, alanine functions as hydrogen donor and glycine as the hydrogen acceptor. Obviously, the energy is gained from a coupled oxidation-reduction reaction. In general aminoacids like alanine, leucine, isoleucine, valine and methionine can function as hydrogen donors and glycine, proline, arginine, tryptophane etc., can serve as hydrogen acceptors.

18.15. FERMENTATION BALANCES

Any fermentation is said to be balanced when the hydrogens produced during the oxidations equal the hydrogens transferred to the fermentation end products. Only under these conditions all of the $NADH_2$ and reduced ferredoxins are recycled to the oxidized forms. It is often essential to understand the fermentation balances. The joint study of carbon and oxidation–reduction balances is often useful in dispelling confusion. For example, the identification of the missing products in carbon balances may help to balance an otherwise unbalanced oxidation-reduction balance. Sources of confusion regarding fermentations include: (i) the possibility of simultaneous fermentation of more than one substrate, (ii) the loss of volatile products, (iii) carbon dioxide fixation for growth or end product production, (iv) inaccuracies in product measurement, and (v) incomplete product recovery. There are two methods used to balance fermentations *viz.*,

1. The O/R method
2. The available hydrogen method

1. The O/R method: This is a book keeping method to keep track of the hydrogens. An oxidation-reduction (O/R) balance is computed as described below:

i. Assign an oxidation number to each of the products of the fermentation by comparing the number of oxygen and hydrogen atoms in the molecule, assuming that each oxygen is numerically equal to +1 and two hydrogens together are equal to –1.
ii. Determine the number of molecules of products formed, on a molar basis, per unit of substrate catabolized
iii. Multiply the oxidation number of each product by the number of molar units per unit of substrate catabolized that the product represents.
iv. Sum up the resulting positive and negative numbers and divide the total of the positive numbers by the total of negative numbers. In a perfect balance, one should obtain a number numerically equal to 1.0. The sign (+ or -) of the calculated number is not important, but only its obsolute value. A number less than 1.0 indicates that oxidized products are missing, and a number greater than 1.0 indicates that reduced products are missing.
v. Compare the ratio of C_2 products and C_1 products. If the products result from pyruvate cleavage only, the ratio should be 1.0

This method of fermentation balance is determined in complex fermentations such as one carried out by *C. acetobutylicum*, where it is difficult to judge whether the hydrogen balance is even or not (Table 18.12).

Table 18.12: Balancing an acetone butanol fermentation

Substrate and products	Yield (mol/100 mol substrate)	Carbon (mol)	O/R value	O/R value mol/100 mol	Available H	Available H (mol/100 mol)
Glucose	100	600	0	--	24	2400
Butyrate	4	16	-2	-8	20	80
Acetate	14	28	0	--	8	112
CO_2	221	221	+2	+442	0	----
H_2	135	----	-1	-135	2	270
Ethanol	7	14	-2	-14	12	84
Butanol	56	224	-4	-224	24	1344
Acetone	22	66	-2	- 44	16	352
Total		569		-425, +442		2242

2. The available hydrogen method : Like the O/R method, this procedure is merely one of book-keeping. It has nothing to do with the chemistry of the reactions. According to this method, one oxidizes the molecule to CO_2 using water to obtain the 'available hydrogen'. For example.

$$C_6H_{12}O_6 + 6H_2O \longrightarrow 24\ H + 6\ CO_2,$$

thus glucose has 24 available H.

The available H in all of the products must add up to the available H in the starting material. In Table 18.12, the concentration of the products is given per 100 moles of glucose used. The available H in the glucose is 24 X 100 = 2,400. The available H in the products adds upto 2,242. Thus, the balance is 2,400/2,242 = 1.07

18.16. *RUMINOCOCCUS ALBUS*: A MODEL FOR INTERSPECIES HYDROGEN TRANSFER : SYNTROPHY

Some anaerobic bacteria use protons as the major or sole electron sink. They include the obligate proton-reducing acetogens that oxidize butyrate, propionate, ethanol and other compounds to acetate, H_2 and CO_2. The information on the physiology of these organisms is scanty. It is thought that electrons travel from organic substrate to an intermediate electron carrier such as NAD^+ to hydrogenase to H^+. Some of these oxidations are shown below:

$$CH_3CH_2CH_2COO^- + 2H_2 \longrightarrow 2CH_3COO^- + H^+ + 2H_2\ ;\ \Delta G^{o\prime} = +\ 48.1\ kJ$$

$$CH_3CH_2COOH + 3H_2O \longrightarrow CH_3COO^+ + HCO_3^- + H + 3H_2\ ;\ \Delta G^{o\prime} = +\ 76\ kJ$$

$$CH_3CH_2 + H_2O \longrightarrow H_3COO^- + H^+ + H_2\ ;\ \Delta G^{o\prime} = +\ 9.6 kJ$$

A critical analysis of the above reactions reveals that they are thermodynamically unfavourable under standard conditions at pH 7.0. The fact is that proton reducers live symbiotically with hydrogen utilizers that keep the H_2 levels low and pull the reactions towards H_2 production. The hydrogen utilizers are methanogens and sulfate reducers. Sulfate concentrations in the environment determine the dominance of methanogens and sulfate reducers. Under the conditions of sufficient sulfate, the sulfate reducers predominate over methanogens. The symbiotic relationship between the obligate proton reducers and the hydrogen utilizers is called a **'syntrophic association'**. It is also called interspecies hydrogen transfer. Apart from obligate proton reducers, many other fermenting bacteria have hydrogenases that transfer electrons from $NADH_2$ to protons and can use protons as an electron sink when grown in the presence of hydrogen gas utilizers, an example is *Ruminococcus albus*.

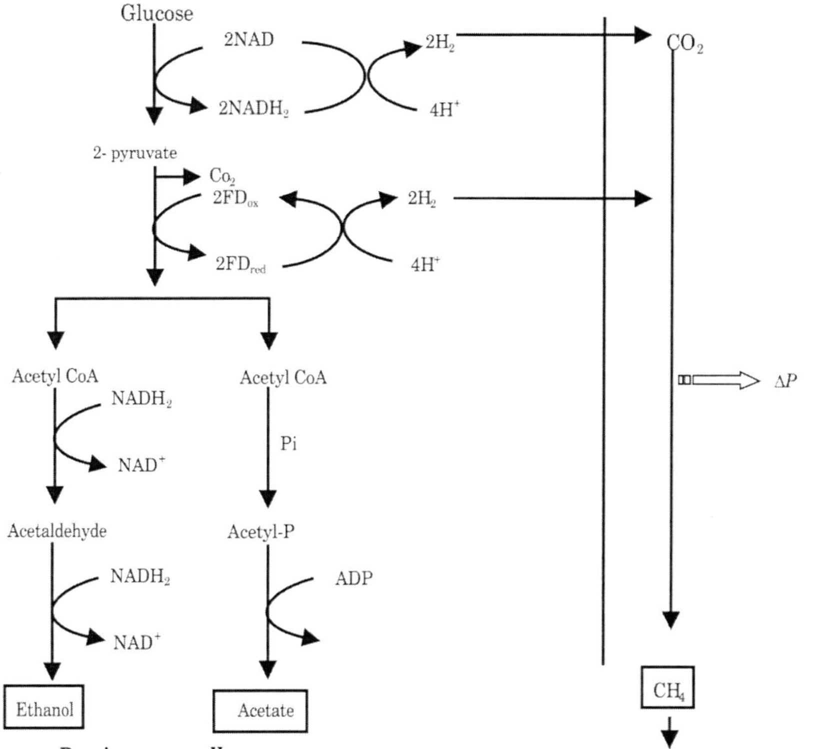

Fig. 18.15. Fermentation of glucose by *R. albus* and utilization of H_2 by methanogen to reduce CO_2

R. albus is a rumen bacterium that can ferment glucose to ethanol, acetate, H_2 and CO_2 using the fructose-1, 6-biphosphate pathway (Fig. 18.15). This fermentation is easy to understand and also explains how the growth of bacteria in mixed populations can influence fermentation and products. In pure culture growth, the glucose is first oxidized to two moles of $NADH_2$. Then each mole of pyruvate is oxidized to acetyl Co-A, CO_2 and H_2 using the pyruvate ferredoxin oxidoreductase and hydrogenase. One of this acetyl CoA molecules is converted to acetate allowing an ATP to be synthesized. The second acetyl CoA molecule is converted to ethanol using the two $NADH_2$. Thus the fermentation is balanced. However, interestingly, something different happens when the bacterium is grown in a co-culture with a methanogen. Growth with a methanogen shifts the fermentation in the direction of acetate with the concomitant production of more ATP. This shift in metabolism is explained in the following way. *R. albus* has a hydrogenase that transfers electrons from $NADH_2$ to produce hydrogen gas. When hydrogen accumulates in the surroundings, the hydrogenase does not oxidize $NADH_2$ because of the feed back mechanism. The $NADH_2$, therefore, reduces acetyl CoA to ethanol. However, methanogen utilizes hydrogen gas as reducing power for methane production and keeps the hydrogen levels very low. Thus, in the presence of methanogen, the $NADH_2$ in *R. albus* is utilized to reduce the protons to hydrogen gas instead of reducing the acetyl CoA. The result is that *R. albus* converts the acetyl CoA to acetate. This is also an advantage to the methanogen since methanogens can also use acetate as a carbon and energy source. The transfer hydrogen gas from one species to another is called interspecies hydrogen transfer and is a good example of nutritional synergism found among mixed populations of bacteria.

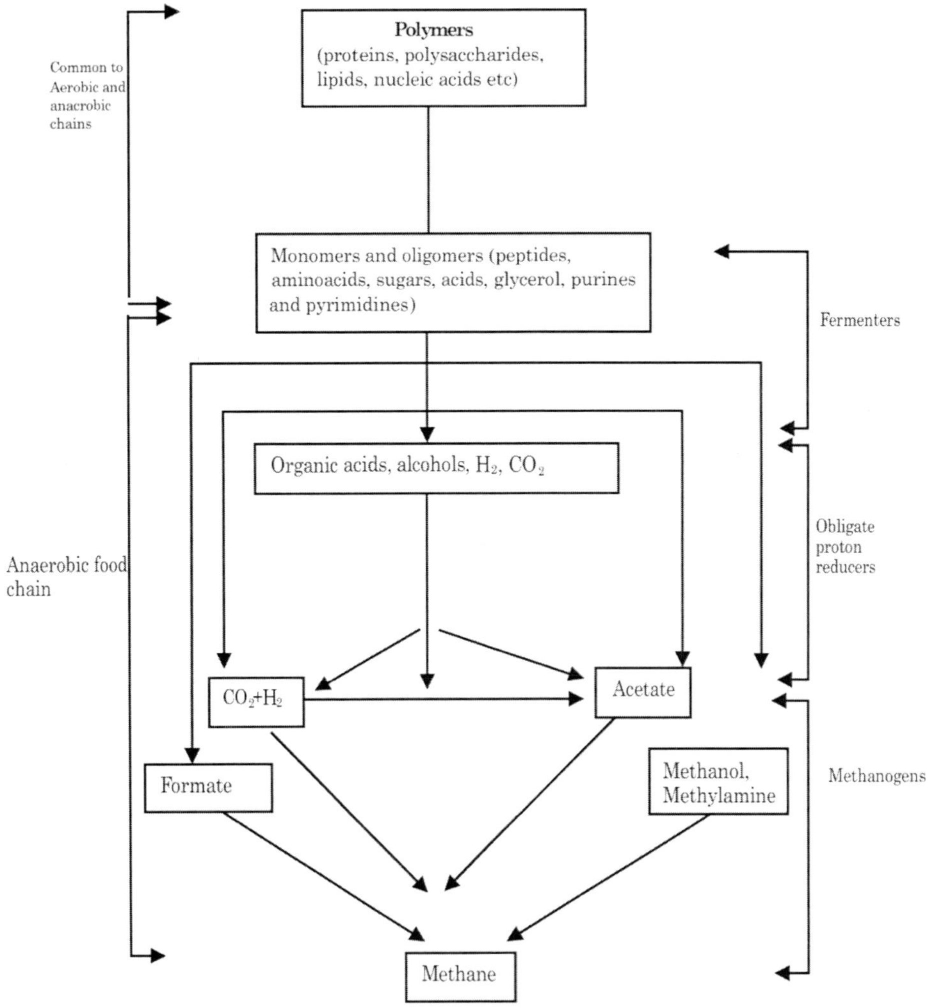

Fig. 18.16. The anaerobic food chain

18.17. ANAEROBIC FOOD CHAIN

So far, the discussions were on various fermentations that are carried out by different groups of microorganisms. Naturally, a question arises as to what is the ecological significance of these fermentations in a broader perspective. These fermentations, definitely play an important role in bio-geochemical cycle especially in anaerobic environments, such as in muds, the botton of lakes and sewage treatment plants. In these environments, various organisms are lined up with their catabolic activities and form a so-called anaerobic food chain. The relationships between fermentative organisms are such that the products of one group of organisms can serve as substrates for another group of organisms. Thus they form the food chain.

The fermentation of carbohydrates, amino acids, purines, pyrimidines etc. to organic acids and alcohols (acetate, ethanol, butanol, propionate, succinate, butyrate etc.) by prokaryotes and the conversion of these fermentation end products to CO_2, CH_4 and H_2 by the combined action of several different groups of bacteria is called the **anaerobic food chain** (Fig. 18.16). This is an important part of the carbon cycle and serves to regenerate gaseous carbon (*i.e.* CO_2 and CH_4) which is reutilized by other microorganisms and plants throughout the biosphere. If sulfate is present in sufficient amounts the terminal process of this food chain is the sulfide fermentation or at limiting sulfate concentrations such as fresh water environments it is the methane fermentation. As shown in Fig. 18.16, the anaerobic food chain can be divided into three stages. In the first stage, fermenters produce organic acids, alcohols, H_2 and CO_2 by utilizing the simple carbohydrates, amino acids, nitrogen bases etc. In the second stage these end products are oxidised to CO_2, H_2 and acetate by obligate proton-reducers such as acetogens. Finally, in the third stage methanogens grow on the acetate, H_2 and CO_2 converting these to CH_4 and CO_2.

18.18. FURTHER READING

Adams, M.W.W. and Stiefel, E.J. (1998), Biological hydrogen production, not so elementary, *Science,* **282 :** 1842-1843.

Bock, A. (2000), Fermentation. In : Encyclopedia of Microbiology, J. Lederberg (ed.), 2nd edn. Vol. 2, Academic Press, San Diego, pp. 343-349.

Caldwell (2000), Microbial metabolism, 2nd edn., Star Publishing,Belemont,California

Cheryan, M. (2000), Acetic acid production. In : Encyclopedia of Microbiology, J. Lederberg (ed.), 2nd edn., Vol. I, Academic Press, San Diego, pp. 13-17.

Fridovich, I. (1997), Superoxide anion radical (O_2^{*-}), superoxide dismutases and related matters. *J. Biol. Chem.,* **272 :** 515-517.

Gottschalk, G. (1986), Bacterial metabolism 2nd edn. Springer-Verlag, New York, Berlin, Heidelberg, Tokyo.

Honegraaf, P.P.F., Stouthamer, A.H. and Kooijman, S.A.L.M. (2000), A mathematical model for yeast respirofermentative physiology. *Yeast,* **16 :** 423-437.

Jones, D.T. and Woods, D.R. (1986), Acetone-butanol fementation revisited. *Microbiol. Rev.* **50 :** 484-524.

Litchfield, J.H. (2000), Lactic acid, microbially produced, In : Encyclopedia of Microbiology, J. Lederberg (ed.), 2nd edn., Academic Press, San Diego, pp. 9-17.

Lowen, P.C., Klotz, M.G. and Hassett, D.J. (2000), Catalase-an "old" enzyme that continues to surprise us. ASM News, **66 :** 76-82.

Mitchell, W.J. (1999), Physiology of carbohydrate to solvent conversion by clostridia. *Adv. Microb. Physiol.*, **39** : 31-130.

Moat, A.G. (1985), Biology of lactic, acetic and propionic acid bacteria. In: A.L. Demain and N.A. Solomon (eds.), Biology of Industrial Microorganisms. Benjamin Cummings, Menlo Park, CA, pp. 143-188.

Walker, G.M. (1998), Yeast Physiology and Biotechnology, John Wiley & Sons, Chichester, U.K.

White, D. (1995), The Physiology and Biochemistry of Prokaryotes. Oxford University Press, New York

Chapter - 19

Nitrogen Metabolism – Biological Nitrogen Fixation

19.1. INTRODUCTION

Nitrogen is an important and essential element for the growth of the plant. It comes next to carbon, hydrogen and oxygen. Nitrogen is one of the main structural components of living organisms. The element is found in the composition of proteins, amino acids, vitamins, nucleic acids (DNA and RNA), plasma membrane, chlorophyll and growth hormones. Generally, leaves contain 1-15% nitrogen of their dry weight.

Though nitrogen is included in the list of essential mineral nutrient elements, it is not a mineral element in the true sense of the word. The air comprises 79% nitrogen in gaseous form. Plants cannot absorb nitrogen in the gaseous form from the atmosphere. They absorb nitrogenous compounds present in the soil. However, some lower organisms like bacteria and cyanobacteria can utilize the gaseous nitrogen directly from the atmosphere.

The biosynthetic pathways leading to amino acids and nucleotides share a requirement for nitrogen. Because of easy solubility, biologically useful nitrogen compounds are generally scarce in natural environmental and most organisms maintain strict economy in their use of ammonia, amino acids and nucleotides. Indeed, as an economy measure, free amino acids, purines and pyrimidines formed during metabolic turnover of proteins and nucleic acids are often salvaged and reused.

19.2. DISTRIBUTION AND PROPERTIES OF NITROGEN AND ITS COMPOUNDS

In nature, reservoirs of nitrogen include the lithosphere, atmosphere, hydrosphere and biosphere. Most of the nitrogen on earth exists in the lithosphere (~64×10^{20} kg), and most of this is locked up in igneous rocks that make up the crust and mantle and is unavailable for use of biosphere. Most of the remainder of the nitrogen exists as N_2 in the atmosphere, with very small amounts of NOx (oxides of nitrogen) and organic-N, totaling ~3.86×10^{18} kg. In fact, the atmosphere above each square meter of the earth contains about 24, 300 kg of N, more than any other element. In the hydrosphere, nitrogen exists as dissolved N_2, NH_4^+, NO_3^-, NO_2^- and organic-N and it accounts for a total of ~2.3×10^{16} kg. The forms of nitrogen found in the biosphere vary from complex organic-N compounds to nitrogen oxides and NH_4^+. Because nitrogen is in a constant state of flux due to cycling through the various nitrogen pools in the biosphere (*i.e.*, N2, NO_x, NH_4 and organic-N), it is difficult to quantify each of the nitrogen species. Nitrogen is introduced into the biosphere at the rate of about 92 million metric tons per year. The total amount that is denitrified and returned to the atmosphere is about 83million metric tons per year. The difference (9 million metric tons) represents the rate at which fixed N is building up in the biosphere.

The total amount of N in a given reservoir at any one time represents a balance between N gains and losses. For example, N-gains to the hydrosphere are from the land (lithosphere) and atmosphere from which fixed N is carried by rivers and rain. In contrast, denitrification represents a loss of N from lithosphere and hydrosphere. These gains to the hydrosphere represent losses from the lithospheres and the atmosphere, and the loss of N from the hydrosphere through denitrification represent a gain of N to the atmosphere (as N or N_2O). The transfer of N among the various reservoirs constitutes the global nitrogen cycle.

Nitrogen can exist in the soil in the form of nitrates (NO_3), nitrites (NO_2), ammonia (NH^+_4) or urea or some other nitrogenous compounds. The most usable source of nitrogen is nitrate ion (NO_3). However, nitrite (NO_2) can also be absorbed by the plants, but nitrites are not a good source of nitrogen for many plants. Ammonia is also directly absorbed by the plants.

The organic form of nitrogen occurs in the soil in the form of proteins, amino acids, urea and other organic nitrogenous compounds. These organic nitrogenous compounds in the soil are formed by death and decay of plants and animals including man. Amino acids, which are formed by the decomposition of proteins, are directly absorbed by the plants or they are converted into ammonia by an enzyme urease, produced by soil microorganisms, before being absorbed.

$$\underset{\text{Urea}}{NH_2-\overset{\overset{\displaystyle O}{\|}}{C}-NH_2} \xrightarrow{\text{Urease}} \underset{\text{Ammonia + carbon dioxide}}{2\,NH_3 + Co_2}$$

14.3 NITROGEN CYCLE

In nature, nitrogen exists in avariety of oxidation states ranging from +5 (nitrate) to -3 (ammonium), and biological systems can transform nitrogen between these states. Thus, assimilatorynitrate reduction converts nitrate (+5) to ammonia (-3), dissimilatory nitrate reduction (denitrification) converts nitrate (+5) to nitrous oxide (+1) and molecular nitrogen (0), while in opposite direction, nitrification convers ammonium (-3) to nitrite (+4) and nitrate (+5). A cycling of nitrogen, therefore, occurs in the biosphere. Apart from other phenomenon, nitrogen fixation *i.e.,* reduction of molecular nitrogen (0) to ammonium (-3) is an integral part of this process. Microorganisms play an important role in the nitrogen cycle. Nitrogen cycle consists of five separate nitrogen transformation reactions that are linked to form the cycle (Fig. 19.1). The transformations are oxidation-reduction reactions involving a change in the valency state of nitrogen. The five nitrogen transformation reactions are:

i. **Nitrogen fixation**: It is the conversion of dinitrogen (N_2or N=N) to fixed nitrogen usually NH_3 with the help of nitrogenase.

ii **Assimilation**: It is the conversion of NH_3 to organic nitrogen and biomass.

iii. **Ammonification:** The complex organic nitrogenous compounds present in dead plants and animals are broken down to simple inorganic substances such as ammonia and the process is known as ammonification. Bacteria, called ammonifying bacteria, play a major role in this process. Actinomycetes and fungi also play a significant role in this process. Since, the ammonifying agents use carbohydrates as a source of energy more readily, the quantity of ammonia is appreciably reduced if large amounts of carbohydrates are present.

iv. **Nitrification:** It is accomplished in two steps. In the first step, the ammonia arising from the decomposition of proteins and other activities is acted upon by nitrifying bacteria like *Nitrosococcus* and *Nitrosomonas* and it is oxidized to form nitrite. In the second step, a different group of organisms such as *Nitrobacter* further oxidize nitrites to nitrates. Thus, this process of conversion of ammonia to nitrite and then to nitrate is known as nitrification. Both these groups of organisms are obligate aerobes.

v. **Denitrification:** In contrast to nitrification, a large number of other organisms reduce nitrates to nitrites and to ammonia. This generally takes place in plants in the process of assimilation of nitrogen (assimilatory denitrification). Similarly, some soil microorganisms are capable of reducing nitrates to molecular nitrogen and such organisms are collectively designated as denitrifying bacteria and the process is known as denitrification. Thus, bound nitrogen in living systems is released into the atmosphere.

Under normal conditions, due to balancing processes of nitrogen fixation and denitrification, the quantity of the nitrogen in the atmosphere is maintained at equilibrium state.

It should be remembered that a N atom moving from one oxidation state to another is a random and neednot necessarily go through the complete nitrogen cycle. Further, all transformations need not be mediated by microorganisms. For instance, some N containing organic compounds can undergo abiotic chemical reactions with other organic compounds.

The different steps of the nitrogen cycle require different environmental conditions. Thedifferences in the conditions of different habitats determine the spatial distribution patterns of different organisms involved in nitrogen cycle. Accordingly, they occur in soils, sediments, fresh water and marine systems that are both aerobic and anaerobic. Thus, a particular process of nitrogen cycle dominates in a given habitat. For example, nitrification occurs primarily in aerobic habitats, whereas denitrification predominates in anaerobic habitats such as sediments and water-logged soils.

14.4. NITROGEN FIXATION

Molecular nitrogen present in the atmosphere cannot be utilized by the higher plants. However, certain higher plants like legumes can utilize the gaseous nitrogen with the help of bacteria making a symbiotic association with them. Higher plants by themselves can utilize the nitrogen when it is present in the form of nitrogenous compounds such as nitrates, nitrites and ammonia etc. But certain prokaryotes like bacteria, cyanobacteria can convert free nitrogen into nitrogenous compounds either in free-living state or making symbiotic association.

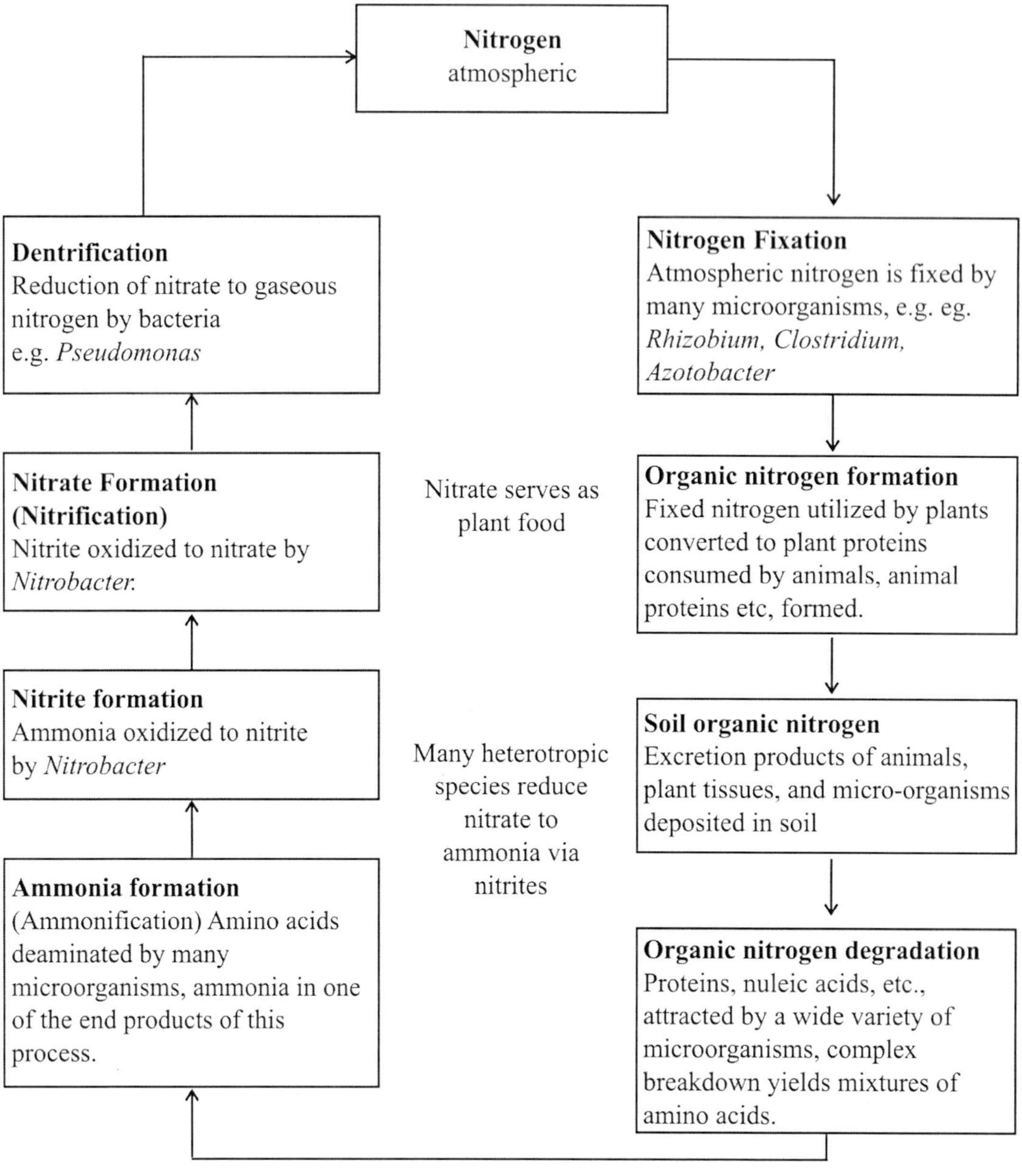

Fig. 19.1. Nitrogen cycle in nature.

N_2 is free in diatomic state in the atmosphere and combined on clute salt petre. The N_2 molecule contains triple bond ($N \equiv N$). The nitrogen bond has the shortest bond length of 1.095 A°, the highest ionization potential (15.58 ev) and highest stretching frequency. Conversion of molecular nitrogen to atomic nitrogen requires at least 160 kcal of energy. The actual formation steps in which two atoms of nitrogen combine with six atoms of N_2 to form two molecules of ammonia releases 13 kcal of energy. Therefore, the whole process of nitrogen fixation requires 147 kcal.

Nitrogen fixation can be defined as a phenomenon of conversion of free molecular nitrogen into nitrogenous salts to make it available for absorption by the plants. The fixation of nitrogen can be both by physico-chemical process and biological process. Thus, nitrogen fixation can be of two types:

1. Physical nitrogen fixation
2. Biological nitrogen fixation

19.4.1 Physical Nitrogen Fixation

About 10% of total nitrogen fixation occurs due to physical processes such as lightning and thunder storms. During the lightning and thunder storms heavy electrical discharges takes place. Under the effect of heavy electrical discharges, the gaseous nitrogen and oxygen present in the atmosphere combine to form nitric oxide as follows:

$$N_2 + O_2 \xrightarrow[\text{discharge}]{\text{Electrical}} 2\,NO \text{ (Nitric oxide)}$$

The nitric oxide is again oxidized with O_2 to form nitrogen peroxide (NO_2).

$$2\,NO + O_2 \xrightarrow{\text{Oxidation}} 2\,NO_2 \text{ (Nitrogen peroxide)}$$

During the rain, the nitrogen peroxide combines with rain water to form nitrous acid (HNO_2) and nitric acids (HNo_3). The acids fall on the soil along with the rain water and react with alkaline radicals to form nitrate and nitrite salts, which are soluble in water and can be absorbed by the plants.

19.4.2. Biological Nitrogen Fixation

The conversion of atmospheric nitrogen into nitrogenous compounds through the agency of the living organisms is called "Biological Nitrogen Fixation". The biological nitrogen fixation can be carried out by either free-living microorganisms or the organisms which live in symbiotic association with the other plants. There are three types of nitrogen fixation.

i. Asymbiotic or Non-symbiotic N_2 fixation

ii. Symbiotic N_2 fixation

iii. Assosiative symbiosis

19.5. NITROGEN FIXING MICROORGANISMS

Although the total number of organisms is relatively small, a wide range of microorganisms are able to fix nitrogen (Table 19.1). Nitrogen fixers, also called **diazotrophs**, are generally divided into free-living and symbiotic forms. However, some organisms (notably certain cyanobacteria) appear to be equally efficient to fix nitrogen independently and in symbiotic association.

The free-living diazotrophs include representatives of both archaebacteria and eubacteria. Among the latter there is an almost random distribution of the ability to fix nitrogen among the various families. Diazotrophy also spans a range of physiological behaviour, covering autotrophy and heterotrophy and also anaerobiasis, facultative anaerobiasis and aerobiasis. Symbiotic diazotrophs include the various genera of rhizobia that form the well documented symbiosis with legumes. *Parasponia* is an exception as it is a non-legume plant infected by a strain of *Rhizobium*. Some rhizobia like *Azorhizobium caulinodans* and *Photorhizobium thompsonianum* form stem nodules on certain plants. *Frankia*, an actinomycete, forms nitrogen fixing nodules (or actinorhiza) on the root of some non-leguminous woody plants. Finally, cyanobacteria participate in a variety of diazotrophic symbiosis. The co-symbiont may be an angiosperm, a gymnosperm, a pteridophyte, a bryophyte, a fungus (lichens), a diatom or an animal like coral reef sponge.

The amount of nitrogen fixed by these bacteria depends on the environmental conditions encountered by the organisms and can range from trace amounts for free-living soil bacteria to more than 300 kg/ha/year for the rhizobia in some legumes.

Table 19.1. Nitrogen fixing microorganisms

Symbiotic association of Rhizobium and Bradyrhizobium spp. with legumes

R. trifolii, clover *(Trifolium, Crotolaria)*

B. japonicum soybean (*Glycine max*)

R. meliloti, alfalfa (*Medicago sativa*)

R. leguminosarum, cowpea (*Vigna unguiculata*)

Symbiotic association of actinomycetes with nonlegumes

Frankia sp., Alder (*Alnus*)

Frankia sp., Bog myrtle or sweet gale (*Myrica*)

Frankia sp., Oleasters (*Shepherdia, Eleagnus, Hippophae*)

Frankia sp., New Jersey tea (*Ceanothus*)

Symbiotic association with leaf nodulating plants

Klebsiella aerogenes

Associative interaction with grasses

Azospirillum brasiliense, tropical grasses

A. lipoferum, tropical grasses

A. paspali, tropical grass (*Paspalum notatum*)

Symbiotic association of marine bacteria with bivalves

Aerobic chemoheterotrophic sp., bivalves (Teredinidae)

Symbiotic association with marine diatoms

Richelia intracellularis (cyanobacterium), *Rhizoselenia*

Free living bacteria and cyanobacteria

Aerobic, heterotrophic

Azotobacteriaceae (Azotobacter, Derxia, Azomonas, Beijerinkia), Nocardia, Pseudomonas

Aerobic phototrophic

Anabaena, Calothrix, Nostoc, Gleotheca, Cylindrospermum, Aphanocapsa

Facultative heterotrophic

Enterobacter cloacae, Klebsiella pneumoniae, Bacillus polymyxa, Desulfovibrio desulfuricans, D. gigas, A chromobacter

Anaerobic, heterotrophic

Clostridium pasteurianum, C. butyricum, Propionispira arboris

Anaerobic, photoautotrophic

Chromatium vinosum, Rhodospirillum rubrum, Rhodopseudomonas sphaeroides, R.capsulata, Rhodomicrobium vernielli, Rhodocyclus, Chlorobium limicola.

Nonphotosynthetic, autotrophic

Methanobacterium, Methylococcus, Methylosinus, Methanococcus, Methanosarcina

---

19.6. OUTLINES OF NITROGEN FIXATION

Nitrogen (N_2) is a very non-reactive molecule. The reason for its chemical stability lies in the electronic structure of the molecule, i.e., presence of triple bonds and requirement of high energy for activation. Thus, in biological nitrogen fixation the challenging job is inert molecular gaseous nitrogen should be reduced to active ammonia. Reduction of single molecule of N_2 yields two molecules of ammonia. The simplest equation for nitrogen fixation is

$$N_2 + 3\,H_2 \longrightarrow 2NH_3;\ \Delta G^{\circ\prime} = +340\ \text{kJ mol}^{-1}\ NH_3$$

The standard free energy ($\Delta G^{\circ\prime}$) of this reaction is positive, which indicates that the reaction requires the input of large amount of energy. This must ultimately be derived from the oxidation of carbohydrates which are produced either directly or indirectly by photosynthesis.

Atmospheric N_2 is also reduced to ammonia industrially by the Haber-Bosch process. The nominal reaction is

$$N_2 + 3H_2 \longrightarrow 2NH_3\ ;\ \Delta G = -16.6\ \text{kJ mol}^{-1}$$

In this case, the negative value for $\Delta G^{\circ\prime}$ indicates that energy is released as N_2 and H_2 react *i.e.*, the process is exergonic. However, inspite of the apparently favourable thermodynamics, this reaction does not occur spontaneously at room temperature and pressure, but requires high temperature (400-600°C) and pressure (100-200 atm). In biological nitrogen fixation, these expensive demands (temperature and pressure) are scuttled away by enzyme catalyst nitrogenase. Biological nitrogen fixation process is accomplished at considerable expense of energy, requiring between 12 to 16 molecules of ATP and 6 to 8 electrons, depending on the manner in which the equation is viewed. In one form, the equation may be written:

$$N_2 + 6H^+ + 6e^- + 12\ ATP \longrightarrow 2NH_3 + 12ADP + 12Pi$$

This form of equation does not take into account the fact that dihydrogen (H_2) is an obligate product of nitrogenase reaction. If H_2 is considered, then equation can be shown as

$$N_2 + 8H^+ + 8\,e^- + 16\,ATP \longrightarrow 2NH_3 + H_2 + 16ADP + 16Pi$$

Most aerobic nitrogen fixers rarely evolve H_2 because a membrane bound, uptake hydrogen system recycles the H_2 produced by nitrogenase and produces ATP needs of the system. An anerobic environment is essential for nitrogenase activity due to the oxygen liability of both proteins in the complex. In addition, oxygen represses the formation of the hydrogen uptake system. A bird's eye view of nitrogenase complex and associated activities required for nitrogen fixation is presented in Fig. 19.2.

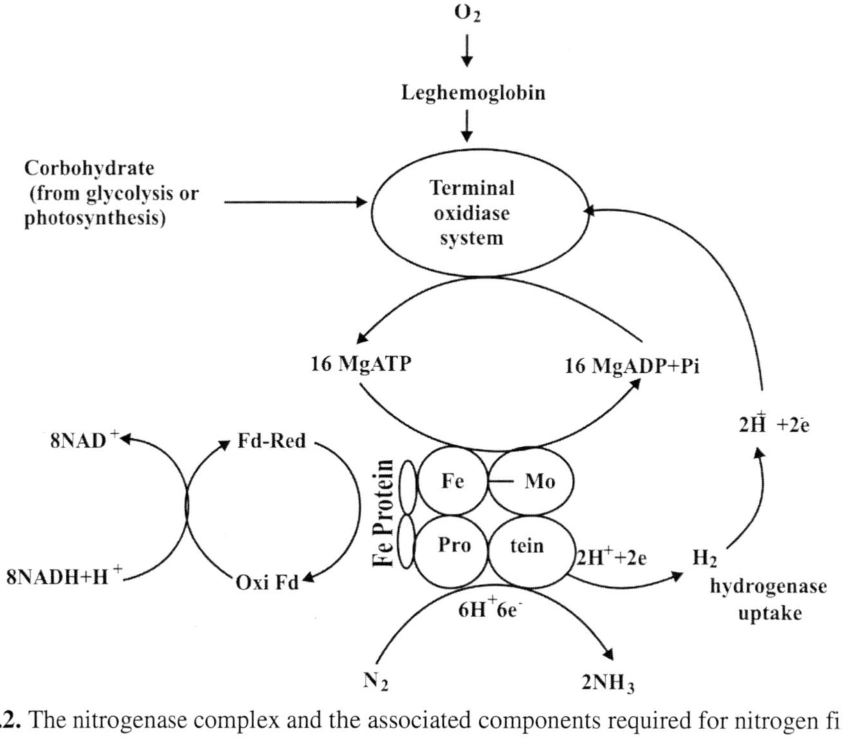

Fig. 19.2. The nitrogenase complex and the associated components required for nitrogen fixation.

19.7. REQUIREMENTS OF NITROGEN FIXATION

The principal requirements for nitrogen fixation in any system are:

i. Nitrogenase
ii. Mechanism to protect nitrogenase from O_2.
iii. Energy requirement
iv. Source of electrons for reduction.

19.7.1. Nitrogenase

Nitrogen fixation is catalyzed by an enzyme complex known as **nitrogenase** (or dinitrogenase). This enzyme is found only in diazotrophic prokaryotes, because in their genome they have the genes coding for this enzyme. Nitrogenase has been purified from virtually all known nitrogen fixing prokaryotes. It is a multimeric protein complex made up of two proteins of different sizes (Fig. 19.3). The smaller protein is a dimer consisting of two identical subunit polypeptides. The molecular mass of each subunit ranges from 24 to 36 kD, depending on the bacterial species. It is called the **Fe protein** because the dimer contains a single cluster of four iron, ions bound to four sulfur groups (Fe_4S_4). The larger protein in the dinitrogenase complex is called the **MoFe protein**. It is a tetramer consisting of two pairs of identical subunits with a total molecular mass of approximately 220 kD. Each MoFe protein contains two molybdenum ions in the form of an iron-molybdenum-sulfur cofactor. The MoFe protein also contains Fe_4S_4 clusters although the exact number is uncertain. It varies as a function of the species or its physiological condition.

Till 1980, only type of enzyme described above was known. However, now three genetically distinct nitrogenase enzymes are recognized: i. the original enzyme, which contains molybdenum and iron (**Mo-nitrogenase**), ii. an enzyme that contains vanadium and iron (**V-nitrogenase**) and, iii. an enzyme that contains iron, but only very few levels of molybdenum and vanadium (**Fe-nitrogenase or alternative nitrogenase**). All the three enzymes catalyze the reduction of nitrogen to ammonium and are related. However, there are some differences in their properties as shown in Table 19.2.

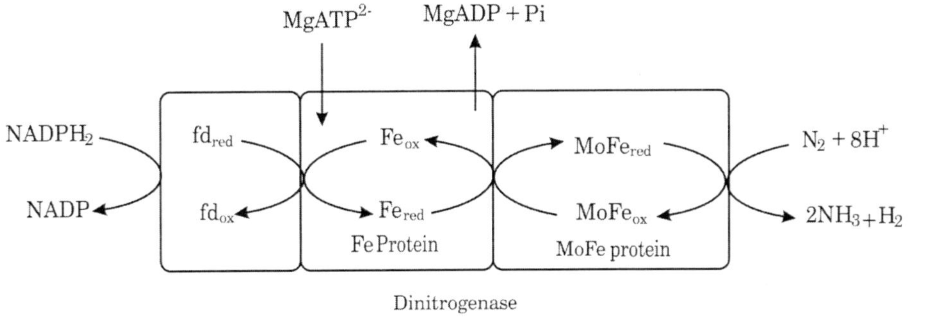

Fig. 19.3. Nitrogenase activity in bacteriods.

Table 19.2. A comparison of molybdenum nitrogenase, vanadium nitrogenase and alternative nitrogenase

Property	Mo-nitrogenase	V. nitrogenase	Alternative nitrogenase
	Fe-protein	**Fe-protein**	**Fe-protein(2)**
Molecular weight	57-72kDa	63kDa	65kDa
Subunit structure	γ_2	γ_2	γ_2
Metal content	4Fe:4S	4Fe:4S	4Fe:4S
	MoFe-protein	**VFe-protein**	**Fe-protein(1)**
Molecular weight	220kDa	220-210kDa	216kDa
Subunit structure	$\alpha_2\beta_2$	$\alpha_2\beta_2\delta_2$	$\alpha_2\beta_2\gamma_2\delta_2$
Metal content	2Mo:24-32Fe:24-30S	2V:17-21Fe:18-20S	24Fe:18S
Product of C_2H_2 reduction	C_2H_4	C_2H_4+ C_2H_6	C_2H_4+ C_2H_6

There is circumstantial evidence that Mo-independent nitrogenases also occur in a variety of diazotrophs, though these enzymes do not appear to be universal. However, it is not clear why there could be three different nitrogen fixing enzymes at all. It may be related to the relative availability of Mo and V in nature or to some other environmental factors.

Nitrogenase synthesis is directed by a set of genes known as *nif* genes. They are under intensive study in a variety of organisms and progess in nitrogen fixation research is particularly active in this area. Organization of the *nif* genes is well understood in *Klebsiella pneumoniae.* In this organism, all the *nif* genes are clustered into a single regulon containing 23,280 base pairs. This entire region has been sequenced. It consists of 20 genes organized most probably into seven operons (Fig. 19.4). The characteristics of the protein products of these genes are shown in Table 19.3.

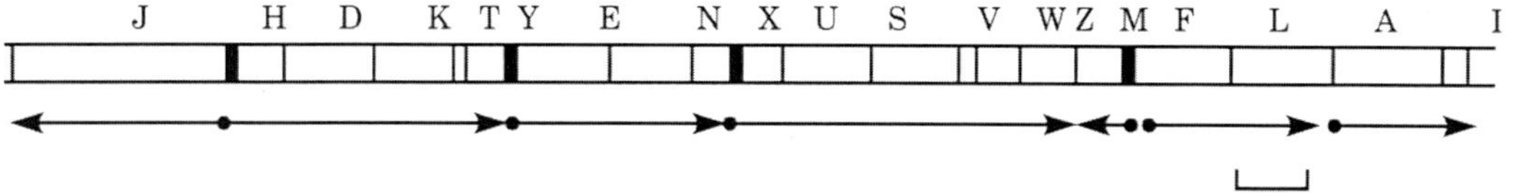

Fig. 19.4. The *nif* gene cluster in *Klebsiella pneumoniae*. The arrows indicate the direction of transcription of individual operons. The bar represents 1000 base pairs.

Table 19.3. *Nif* gene products and their functions in *Klebsiella pneumoniae*

Gene	Function
J	Electron transport (pyruvate: flavodoxin oxidoreductase)
H	Fe-protein of nitogenase
D	α-subunit of MoFe-protein of nitrogenase
K	β-subunit of MoFe-protein of nitrogenase
T	Unknown
Y	Processing of MoFe-protein
E	Synthesis of FeMoco
N	Synthesis of FeMoco
X	Unknown
U	Processing of MoFe-protein
S	Processing of MoFe-protein
V	Synthesis of FeMoco (homocitrate synthase)
W	Unknown
Z	Unknown
M	Processing of Fe-protein
F	Electron transport (flavodoxin)

Gene	Function
L	Negative regulator (inactivator of *nif* A product)
A	Positive regulator (activator of *nif* transcription)
B	Synthesis of FeMoco
Q	Uptake and/or processing of Mo

The genes *nif* H, *nif* D, and *nif* K encode the structural proteins of the Fe^{-} and MoFe-proteins of nitrogenase. These genes are highly conserved among diazotrophs, emphasizing the similar nature of the enzyme in various nitrogen-fixing organisms. Of the other *nif* genes whose function is known, all are involved in the processing of the three structural proteins. *nif* J and *nif* F are involved in the provision of reductant for nitrogen fixation. *nif* L and *nif* A are regulatory genes.

19.7.1.1. Nitrogenase and oxygen

A critical problem faced by the diazotrophs is the sensitivity of nitrogenase to molecular oxygen. Nitrogenase is rapidly and irreversibly inactivated by molecular oxygen. This extreme sensitivity of nitrogenase to oxygen raises a problem for nitrogen fixing organisms. The large amount of energy required (in the form of ATP and reductant) is produced through a cellular respiratory pathway that can operate efficiently only when molecular oxygen present. Then, how come these organisms reconcile with the conflicting demands of the respiratory pathway for oxygen and the sensitivity of nitrogenase to oxygen. This is the most fascinating area of research in nitrogen fixation. An understanding of the different systems that operate to protect nitrogenase from inactivation by oxygen, is essential if we are to succeed in our aim of transferring the ability to fix nitrogen to new crop plants.

Several strategies for minimizing oxygen inactivation of nitrogenase have been identified among diazotrophs and it is also common more than one strategy to be found in a single organism. At the same time, there is no universal method by which diazotrophs protect nitrogenase from the damaging effects of oxygen. Many free-living bacterial nitrogen fixers have retained an anaerobic life style or, if facultative, fix dinitrogen only under anaerobic conditions. Other organisms, such as cyanobacteria, have structurally isolated the nitrogen fixing apparatus. The nitrogen-fixing cells of the cyanobacteria are specialized cells called **herterocyst**. Heterocysts have thickened, multilayered cell walls that restrict the diffusion of oxygen. They are also characterized by a high respiratory activity that maintains a low intracellular oxygen concentration. Finally, although heterocysts are photosynthetic cells, they have eliminated oxygen evolving portion of photosynthesis and reserved only that portion that generates ATP. In legume nodules, the oxygen supply is regulated to a large extent by an oxygen-binding protein called **leghemoglobin**. This is synthesized by the host plant and is located within the bacteroids-infected cell. It binds oxygen and controls the release of oxygen in the region of bacteroids. The equilibrium concentration of oxygen in the bacteroid zone is thus kept at a level (about 10nM) sufficient to support bacteroid respiration and the production of ATP and reducing potential and at the same time preventing oxygen from inactivating nitrogenase.

Some diazotrophs possess electron transport chains that are poorly coupled to ATP synthesis and that probably function more to generate a low intracellular concentration of oxygen than to provide energy. In some diazotrophs, the electron donor to this chain may be hydrogen (generated during nitrogen fixation) in addition to, or instead of NAD(P)H derived from carbon metabolism. Diazotrophs also possess a battery of enzymes associated with the destruction of toxic oxygen derived radicals. Another technique is the conformational protection best exemplified in *Azotobacter*. In this aerobic organism, when grown under aerobic conditions the nitrogenase forms an iron-sulfur protein (FeS_{11}) that is oxygen resistant, but catalytically inactive. After oxygen is removed, FeS_{11} becomes reduced, and the complex dissociates and nitrogenase activity is recovered. Finally, diazotrophs can simply use *de novo* synthesis of nitrogenase in order to replace oxygen-inactivated enzymes. Strategies adapted in different groups of diazotrophs are summarized in Table 19.4.

Table 19.4. Strategies for protecting nitrogenase from inactivation by oxygen

Organism	Strategies
Clostridium pasteurianum	Avoidance
Klebsiella pneumoniae	Avoidance, limited respiratory consumption of oxygen
Azospirillum	Migration
Azotobacter	Respiration (C- and H_2- supported), 'conformational' protection
Photosynthetic bacteria	Avoidance, limited respiratory consumption of oxygen
Legume root nodules	Respiration (C- and H_2- supported), variable diffusion barrier
Actinorhizal nodules	Respiration (C- and H_2- supported), variable diffusion barrier
Actinorhizal nodules	Respiration (C- and H_2- supported), barrier (vesicle formation)
Cyanobacteria :	
Heterocystous strains	Spatial separation, barrier (heterocyst envelope), respiration (C- and H_2-supported)
Non-heterocystous strains	Temporal separation, clumping, respiration, continued synthesis of nitrogenase

19.7.1.2. Nitrogenase and hydrogen production

Yet another problem confronted by diazotrophs is the evolution of hydrogen. Although the mechanism is not well understood, it appears that hydrogen production by nitrogenase is an inescapable process of the nitrogen fixation. As noted earlier, at least one molecule of hydrogen (H_2) is evolved for every molecule of N_2 reduced. It has been estimated that as much as 25 to 30 percent of the ATP and electrons supplied to nitrogenase may be consumed in hydrogen production

Needless to say, H_2 production along with nitrogen fixation is a wasteful energy expenditure that might otherwise be used to reduce dinitrogen. However, although all nitrogen fixers produce hydrogen, not all release hydrogen into the atmosphere. Many nitrogen fixing organisms contain an oxygen-dependent enzyme called **uptake hydrogenase**, which recovers some of the energy lost to hydrogen production. This is accomplished by coupling H_2 oxidation to ATP production.

The electrons are returned to the reductant pool for nitrogenase (Fig. 19.5). In recent years, considerable interest has been focussed on the biochemistry and physiology of hydrogenase and the genes regulating this process are known as ***hup*-genes**.

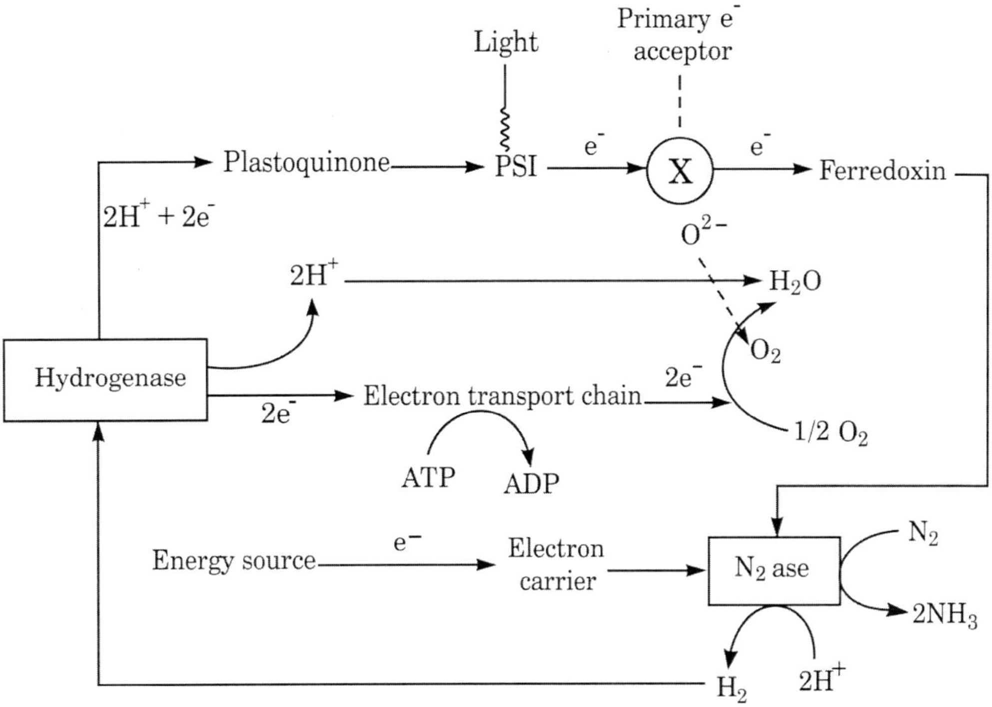

Fig. 19.5. The role of uptake hydrogenase during nitrogen fixation

19.7.2. Energy

Biological reduction of dinitrogen is costly in terms of energy. One measure of energy cost is the number of adenosine triphosphates (ATP) required. At least 16 ATP are required for each molecule of nitrogen reduced; two for each electron transferred, while only 3 ATP are required to fix one molecule of CO_2 in photosynthesis. The total energy cost of biological nitrogen fixation, however, must take into account the requirement for reduced ferredoxin as well. It has been estimated that the reducing potential used in nitrogen fixation is equivalent to atleast a further 9ATP bringing the total investment to a minimum of 25 ATP for each molecule of dinitrogen fixed. A similar calculation for CO_2 fixation brings the total to 9 ATP, about one third of the cost for nitrogen. Such high cost of investment for nitrogen fixation will ultimately reflect on the carbohydrates produced by photosynthesis in the host plant. A portion of that carbohydrate is diverted from the plant to the bacteroid where it is metabolized to produce the required reducing potential and ATP. Thus, the nitrogen fixation represents a considerable drain on the carbon resource of the host plant in symbiotic systems.

The source of generation of ATP for nitrogen fixation varies from diazotroph to diazotroph. In obligate anaerobes the only possible source of ATP are substrate level phosphorylation, and in some cases anaerobic electron transport. Aerobes, however, can generate ATP through respiratory electron transport using oxygen as a terminal electron acceptor. Phototrophs exploit photosynthetic electron flow, either cyclic or non cyclic for ATP generation.

19.7.3. Source of Electrons for Reduction

In order to reduce nitrogenase and eventually dinitrogen, a reductant with a low redox potential (about –0.43V) is needed. Natural electron donors that meet this requirement are flavodoxins and ferridoxins both of which occur in diazotrophs. Which of these proteins actually functions to reduce nitrogenase varies from organism to organism. Similarly, how these proteins are themselves reduced differs from organism to orgnism. For example, in *Clostridium pasteurianum* ferredoxin is reduced directly by a pyruvate ferredoxin oxidoreductase. This enzyme forms part of the system that also generates ATP in this strict anaerobe. In *Klebsiella pneumoniae*, flavodoxin acts as the electron donor to nitrogenase and is itself reduced by the action of a pyruvate flavodoxin oxidoreductase. Photosynthetic diazotrophs can generate reduced ferredoxin as a direct consequence of photosynthetic electron transport. However, in many diazotrophs, including *Azotobacter, Rhizobium*, metabolically generated NADPH (E^0 = -0.32V) is used to reduce ferredoxin or flavodoxin. The thermodynamic unfavourability of this reduction is overcome by the involvement of an energized membrane.

19.8. MECHANISM OF NITROGEN FIXATION

Biological nitrogen fixation can be carried out by either free-living microorganisms or the organisms living in Symbiotic association with other plants. Accordingly, three types of nitrogen fixations are identified.

1. Asymbiotic or non-symbiotic nitrogen fixation
2. Symbiotic nitrogen fixation
3. Associative Symbiosis

19.8.1. Asymbiotic or Non-Symbiotic Nitrogen Fixation

Among free-living diazotrophs, cyanobacteria have a distinct advantage in that, as photoautotrophs the energy and reducing power required for nitrogen fixation is supplied by light. Their simultaneous use of both atmospheric carbon and nitrogen has enabled them to colonize environments that are unfavourable for other organisms. However, diazotrophic cyanobacteria possess an oxygen evolving photosystem II very similar to that found in higher plants. Consequently, their photoautotrophic metabolism is based on two incompatible processes. The extreme oxygen sensitivity of both nitrogenase proteins effectively prevents nitrogen fixation and photosynthesis from operating in the same cell and at the same time. Being prokaryotes, the cyanobacteria do not possess intracellular inclusions apart from the thylakoid membranes. They are unlikely, therefore, to effect an intracellular compartmentalization of the two processes. Nevertheless, cyanobacteria have evolved answers to this paradox. Unicellular forms of cyanobacteria separate the two processes in time, whereas multicellular filamentous forms effect a spatial separation of photosynthesis and nitrogen fixation. To achieve this, the filamentous forms developed a specialized, differentiated cell called **heterocyst**.

19.8.1.1. Heterocystous cyanobacteria

Filamentous cyanobacteria such as *Nostoc, Anabaena* species, when deprived of an external nitrogen source develop specialized, thick-walled cells called heterocysts. These are spaced along the filament, separated by around 10-12 vegetative cells. The heterocyst differs from the vegetative cells in: i. possessing nitrogenase enzyme, ii. having only photosystem I (PSI), iii. being surrounded by a thick cell wall consisting of glycolipids and polysaccharides and iv. not dividing.

The prime function of the heterocyst is to harbour nitrogenase in an anaerobic environment. The thick cell wall prevents the entry of oxygen. In an unknown manner, the cell wall is more impermeable to oxygen than to nitrogen. In addition, the ends of heterocyst, where it joins the adjacent vegetative cells, have plugs that may function to limit the entry of oxygen (Fig. 19.6). Microplasmdesmata connect adjacent vegetative cells with the heterocyst and presumably permit entry of nitrogen and other necessary metabolites while preventing the entry of oxygen.

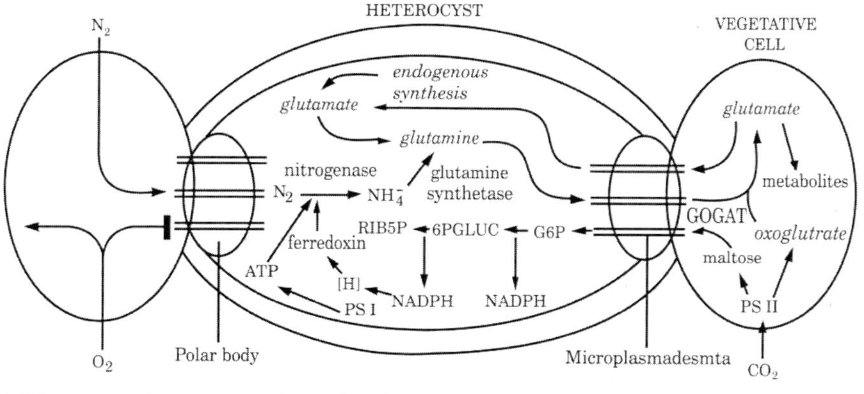

Fig. 19.6. Diagrammatic representation of carbon and nitrogn exchange between the heterocyst and vegetative cells.

The heterocyst of filamentous cyanobacteria is the site of nitrogen fixation. The heterocyst lacks photosystem II activity, which prevents the formation of oxygen in the vicinity of the oxygen-labile nitrogenase, but also prevents carbon fixation. Consequently carbon compounds possibly in the form of disaccharides such as maltose, must be provided from the vegetative cell in order to supply the necessary reductant (NADPH) via the oxidative pentose phosphate pathway enzymes (glucose 6-phosphate and 6-phosphogluconate dehydrogenases). The ammonia produced by nitrogen fixation is used to form glutamine through condensation with glutamate by the enzyme glutamine synthetase (GS). The glutamate is generated either by endogenous synthesis or by transport from the vegetative cell. In return the glutamine is transported back to the vegetative cell, where it forms glutamate via glutamate synthase (GOGAT).

19.8.1.2. Non-heterocystous cyanobacteria

Unlike their heterocystous relatives, the non-heterocystous cyanobacteria do not usually effect spatial separation between the oxygen-sensitive process of nitrogen fixation and oxygen evolving photosynthesis. Most non-heterocystous cyanobacteria, under diurnal cycles of alternating light and dark, fix nitrogen mainly or exclusively in the dark. In these organisms, photosynthesis supports nitrogen fixation only indirectly through accumulation of carbon reserves in the light that are broken down to support nitrogen fixation in the dark.

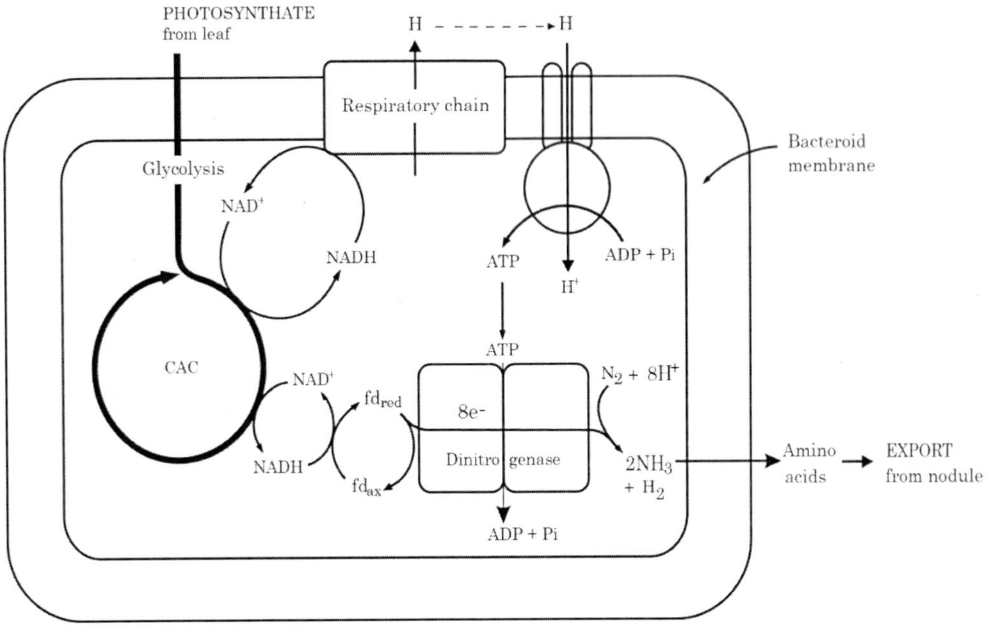

Fig. 19.7. Interactions between photosynthesis, respiration and nitrogen fixation in bacteroids.

19.8.2. Symbiotic Nitrogen Fixation

The symbiotic association between higher plants and the bacterial diazotrophs takes many forms, ranging from the complex, invasive nodulation of legumes by *Rhizobium* species and of woody and non-leguminous plants by *Frankia* to the simpler colonization of leaf cavities of *Azolla* and *Anthoceros* by *Anabaena* and *Nostoc,* respectively. The manner of these associations may vary, but they all show the basic exchange of fixed carbon and nitrogen compounds.

19.8.2.1. The Rhizobium – Legume symbiosis

With the single known exception of *Parasponia*, a non-leguminous tree, rhizobia fix nitrogen symbiotically only with legumes. The most prominent feature of rhizobial symbiosis is the formation of nodules on the root of the host plant. About 90% of all leguminous plant species are capable of becoming nodulated. However, there is a marked specificity between species of legume and strains of *Rhizobium*. A single *Rhizobium* strain is generally able to infect certain species of legumes and not others. A group of *Rhizobium* strains able to infect a group of related legumes is called a cross inoculation group. The major rhizobial cross inoculation groups are listed in Table 19.5. It should be noted that under normal conditions, neither legume nor *Rhizobium* alone can fix nitrogen; but the interaction between the two leads to the development of nitrogen-fixing ability.

Table 19.5. Species of rhizobia in genera I and II, and the cross-inoculation groups of legumes nodulated by these rhizobia

Rhizobia	**Cross-inoculation group**	**Legumes in cross-inoculation group**
Genus I : Rhizobium		
Rhizobium leguminosarum bv. viceae	Pea	Peas (*Pisum* spp.), Vetches; (*Vicia* & *Lathyrus* spp.); lentils (*Lens esculenta*)
R. leguminosarum bv. trifolii	Clover	Clovers (*Trifolium* spp.)
R. leguminosarum bv. phaseoli	Bean	Common beans (*Phaseolus vulgaris);* scarlet runner bean (*Phaseolus coccineus*)
R. meliloti	Alfalfa	Alfalfa/medics (*Medicago* spp.); sweet clovers (*Melilotus* spp.); fenugreek (*Trigonella foenumgraecum*)
R. loti	Lotus	Trefoils (*Lotus corniculatus* and *L. tenuis*); lupine (*Lupinus densiflorus*); serradella (*Ornithopus sativus*); kidney vetch (*Anthyllis vulneraria*).
R. galegae		Goat's rue (*Galega orientalis*)
R. fredii	Soybean	Soybean (*Glycine max*)
Rhizobium spp.		Leucaena (*Leucaena* spp); *Gliricidia sepium, Sesbania grandiflora, Calliandra callothyrsus, Pithocellobium dulce, Prosopis pallida, P. juliflora, Acacia senegal, A. farnesiana, Robinia, Pseudoacacia*
Rhizobium sp.	Chickpea	Chickpea (*Cicer arietinum*)
Genus II : *Bradyrhizobium*		
Bradyrhizobium japonicum	Soybean	Soybean (*Glycine max*)
Bradyrhizobium spp.	Cowpea	Pigeon pea (*Cajanus cajan*); peanut / groundnut (*Arachis hypogaea*); *Acacia mearnsii, A. mangium, A. auriculiformis,* limabean (*Phaseolus lunatus*); winged bean (*Phosphocarpus tetragonoloba*); siratro (*Macroptilium atropurpureum*); guar bean (*Cyamo-psis tetragonolobus*); cowpea, mungbean, black / green gram, rice bean (*Vigna* spp.), *Desmodium* spp., *Stylosanthes* spp., hyacinth bean (*Lablab purpureus*).

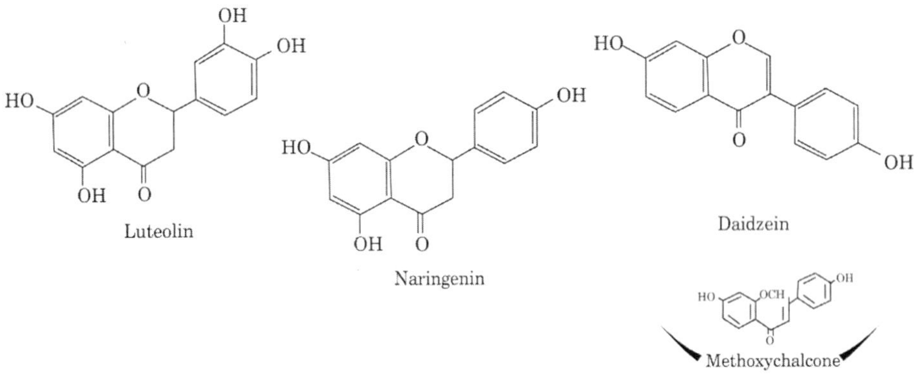

Fig. 19.8. Structures of four common flavonoids implicated in rhizobia-host interactions.

Symbiotic association begins with specific infection of a root hair, which is dependent upon recognition by the *Rhizobium* of the host plant proteins called **lectins**. Lectins function as chemical stimulants for rhizobia which showing positive chemotactic movement and swarm around the infection point. Another group of chemicals that have been implicated in attraction of rhizobia are the **flavonoids**. A wide variety of flavonoids (Fig. 19.8) have now been characterized in root exudates many of which stimulate nodulation. Lectin proteins bind to polysaccharides present on the cell surface of *Rhizobium* species. This binding provides a recognition system and explains why a particular host plant species is infected only by certain species of rhizobia. Once rhizobia colonize the rhizosphere, they begin to synthesize morphogenic signal molecules called **nodulation factors or *nod factors***. Nod factors are coded by *nod* genes, which in turn are located in a circular piece of rhizobial DNA (or plasmid) known as *Sym* (for symbiosis) plasmid. Three nod genes (*nod* A, *nod* B and *nod* C) are basic nodulation genes common to all rhizobia. Nod factors are chemically lipochitooligosaccharides (Fig. 19.9). Nod factors secreted into the soil solution by rhizobia induce several significant changes in the growth and metabolism of host roots as a prelude to rhizobial invasion of the root hair and subsequent nodule development. These changes include increased root hair production and the development of shorter, thicker roots. Stimulated by the nod factors to renew their growth, the root hairs develop branching and curling. Infection proceeds via an infection thread through which rhizobia penetrate to the cells of the inner cortex of the root. The infected cortical cells increase in size and divide to form a ball surrounded by uninfected cells and an outer fibrous layer. Within the infected cells the rhizobia divide and enlarge to form bacteroids which are separated from the cytoplasm of the cell by an envelope membrane. Development of an active nodule requires a number of **nodule-specific proteins** (nodulins) contributed by the host cells. Early nodulins are expressed during the infection process and nodule development. Early nodulins appear to be involved with the infection thread, plasma membrane and in the formation of the nodule primordia. The expression of late nodulin genes coincides almost with the onset of nitrogen fixation and appears to be involved in the nodule function and maintenance. Leghemoglobin is the most abundant late nodulin. Signal interactions between the host and the rhizobia in the initial stages of infection are presented in Figs. 19.10, 19.11 and 19.12.

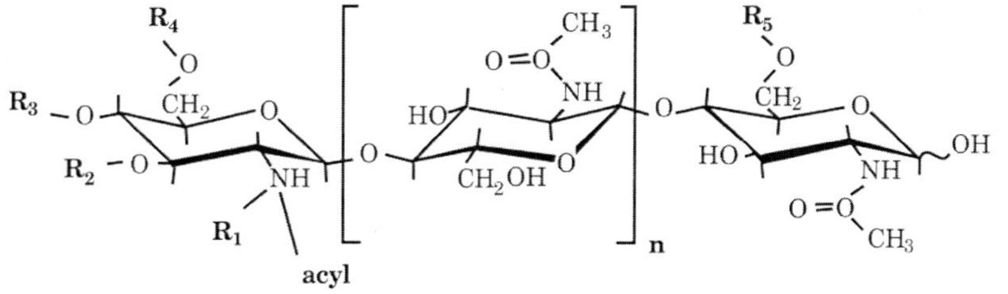

Fig. 19.9. The general structure of Nod factors elicited by rhizobia. In this structure n=2 or 3, and the substitutions possible at each indicated position are R_1 for H or methyl; R_2 and R_3 for H, carbomoyl, or 4-0 methyl carbomoyl; R_4 for H, acetyl, or carbomoyl; R_5 for H, surface, acetyl 2-0 methyl fucose, 4-O-sulfo-2-0 methylfucose, 3-0-acetyl-2-0-methyl-fucose, 4-0-acetyl-fucose, or D-arabinose; and Acyl for C16-C20.

One function of the nodule is to protect nitrogenase with in the bacteroids from oxygen. Shortly before the maturing nodule becomes capable of nitrogen fixation, leghemoglobin is synthesized. This can accumulate to form 25% of the soluble protein in the infected cell and confers a characteristic pink/red colour in an effective nodule. Leghemoglobin regulates the concentration of free oxygen in the nodule and facilitates the controlled diffusion of oxygen to the bacteroids. The outer fibrous layer of the nodule also acts as an effective and variable barrier to diffusion. Channels with in the fibrous layer permit the diffusion of oxygen and nitrogen into the nodule, but limit the rate of entry and so may prevent the saturation of leghemoglobin.

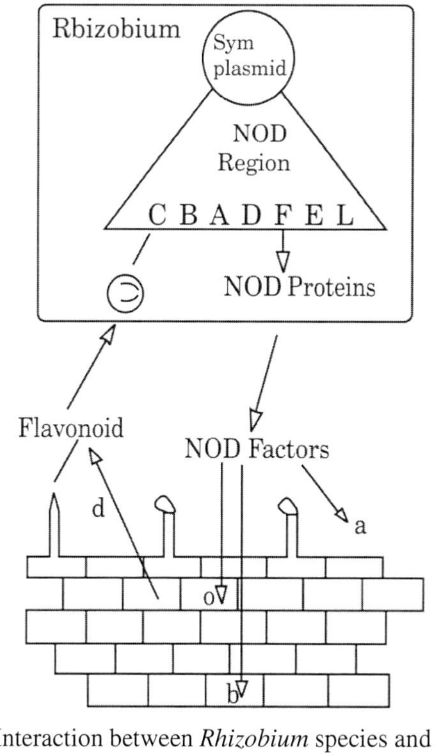

Fig. 19.10. Interaction between *Rhizobium* species and legume roots.

19.8.2.1.1. Physiology of the Rhizobium infected cell

The energy and reducing power for nitrogen fixation in the nodule are thought to be supplied entirely by the plant in the form of photosynthate, mainly sucrose, transported from leaves. Sucrose that enters the infected plant root cell is converted into usable sugars by a catabolic form of sucrose synthetase. This enzyme is also a nodulin. As the first enzyme in the catabolism of carbohydrate in the cell it may control carbon metabolism and therefore, indirectly, the supply of carbohydrate to the bacteroids. The glucose-6-phosphate generated from sucrose may get converted to two relatively inert storage compounds : starch within the amyloplasts of the plant root cell and polyhydroxy butyrate (PHB) granules in the bacteroid. Presumbly, these compounds function as buffers against fluctuations in the supply of photosynthate.

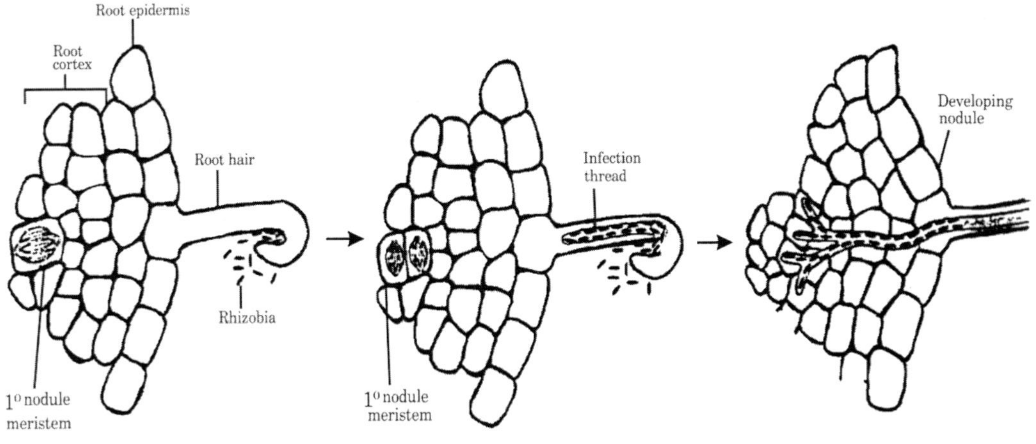

Fig. 19.11. Infection process leading to nodule formation, in lagumes

(a) **Carbon metabolism in host cell and bacteroid:** In infected cells, glucose is catabolized by the Embden-Meyerhoff (glycolysis) pathway to form acetyl CoA. The plant cells of the nodule are rich in PEP carboxylase which has an anaplerotic role in providing oxalocetate that, by condensing with acetyl-CoA produces the 2-oxoglutarate needed for assimilation of ammonium by GS/GOGAT. Carbon metabolism in bacteroid is complicated. However, it has been noted that *Rhizobium* species catalyze glucose using ED pathway, while *Bradyrhizobium* species use the oxidative pentose phosphate pathway.

(b) **Transport of ammonia from bacteroid to host cell:** Both glutamate dehydrogenase (GDH) and GS/GOGAT are present in host cytoplasm. These are involved in the assimilation of ammonia produced by nitrogenase. However, the synthesis of these enzymes is suppressed in the bacteroids, thus due to concentration gradient, ammonia diffuses freely across the bacteroid and peribacteroid membrane into the cytoplasm of the host cell.

(c) **Transport of fixed nitrogen out of the nodule:** Two different pathways are used by legumes to transport fixed nitrogen out of nodule. In the first pathway, aspartate synthesized by the transamination of oxaloacetate is converted to asparagine by glutamine dependent asparagine synthetase. The asparagine produced is transported via surrounding uninfected cells to the xylem for distribution. The second pathway is rather complicated found in ureide-producing nodules. Fixed nitrogen is channeled from glutamine, via serine/glycine and aspartate into purine biosynthesis. The purine is transported to large peroxisomes of surrounding uninfected cells where in the form of xanthine, it is converted to uric acid and then to allantion and other ureides before export via xylem. Ureide transport is a feature of tropical legumes, whereas asparagine transport is more common in temperate zone legumes. However, the significance of the different pathways is not clear.

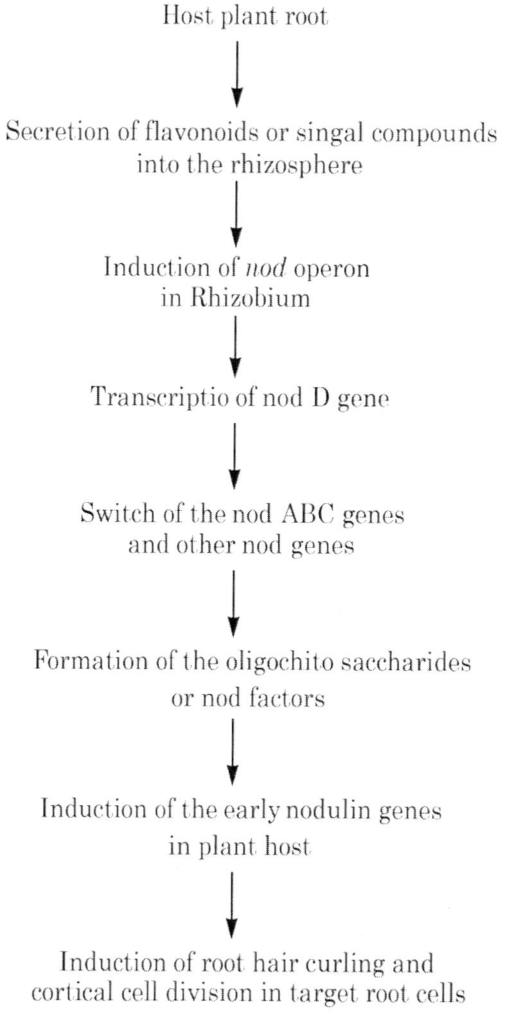

Fig. 19.12. Initial events in the signal interaction between host plant and rhizobia.

(d) Hydrogen production: Hydrogen produced as a byproduct of the nitrogenase catalyzed reduction is consumed through a hydrogen oxidizing system. This system consists of an uptake hydrogenase and an electron transport chain that uses oxygen as its terminal electron acceptor. During transport of electrons derived from hydrogen, ATP can be generated, thereby recovering some of the energy input to the nitrogenase reaction. In addition, oxygen consumption may benefit nitrogen fixation by creating a relatively low molecular concentration of oxygen. There are evidences to show that legumes infected with strains of *Rhizobium* or *Bradyrhizobium* that contain an active uptake hydrogenase (Hup^+) produced a greater crop yield than did legumes infected with strains lacking uptake hydrogenase (Hup^-). The overall picture of physiology of root nodule is presented in Fig. 19.13.

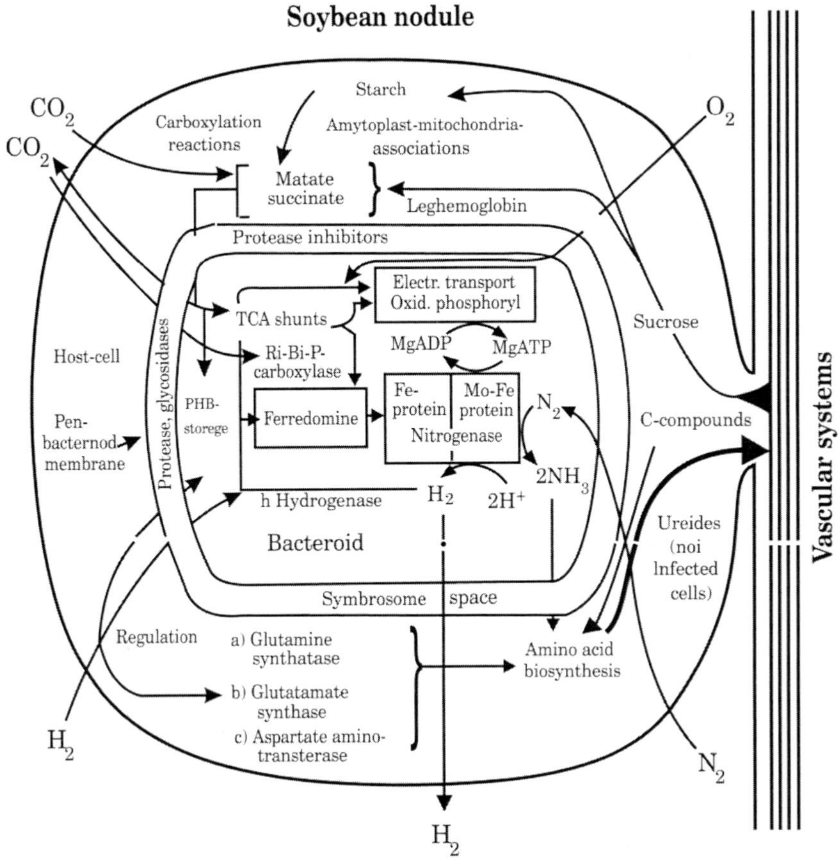

Fig. 19.13. Nitrogen fixation and associated physiological processes in soybean root nodule.

19.8.2.2. Non-legume nitrogen fixing symbiosis

Nitrogen fixing symbiosis involving microorganisms other than rhizobia occur in a variety of non-leguminous plants. *Frankia*, a filamentous nitrogen-fixing actinomycete can develop symbiotic association in atleast 21 genera belonging to eight families of dicotyledons and form nitrogen-fixing nodules (Table 19.6).

Table 19.6. Angiosperm genera nodulated by *Frankia*

Family	Genus	Remarks
Betulaceae	*Alnus*	Nodulation a generic attribute
	Arctostaphlyos	Nodulates only in Alaska
Casuarinaceae	*Casuarina*	Nodulation rather variable
Coriariaceae	*Colletia*	One report on C. *paradoxa*
	Coriaria	Nodulation a generic attribute?
Datiscaceae	*Datisca*	One sp. each from Pakistan and California
Eleagnaceae	*Eleagnus*	Nodulation generic attribute?
	Hippophae	
	Shepherdia	
Myricaceae	*Comptonia*	*C. pergrina* also known as *Myrica asplenifolia*
	Myrica	Nodulation a generic attribute?
Rhamnaceae	*Ceanothus*	
	Discaria	
	Trevoa	One report on *T. trinervis*
Rosaceae	*Cercocarpus*	
	Chaemaebatia	One report on C. *foliolosa*
	Cowania	One report on C. *mexicana*
	Dryas	Nodulates in Alaska and Canada, not in Europe or Japan
	Purshia	
	Rubus	

Many similarities exist between rhizobial and *Frankia* symbiosis. Both lead to the formation of nodules on the roots of host plant in which nitrogen fixation occurs. In both the cases, there is complimentarity with some of the '*nif*' gene sequences of *K. pneumoniae* and a number of associated proteins are related. However, *Frankia* does not invade the host cell and is not, therefore, surrounded by peribacteroid membrane. Instead, it is encapsulated in carbohydrate material derived from the host plant. *Frankia* symbiosis involves a hemoglobin similar to that found in legumes and that the mechanisms regulating expression of nitrogenase in the presence of oxygen and ammonium are very similar in *Rhizobium* and *Frankia. Frankia* protects its nitrogenase by localizing it in terminal swellings on the cells called vesicles. The vesicles contain thick walls that retard O_2 diffusion, thus maintaining the O_2 tension within vesicles at levels compatible with nitrogenase activity. In this regard, *Frankia* vesicles resemble the heterocysts produced by some filamentous cyanobacteria as the sites of N_2 fixation.

Unlike the *Rhizobium*-legume relationship, a single strain of *Frankia* can form nodules on several different species of plants, suggesting that the Frankia/root nodule symbiosis is less specific. A comparison of symbiotic characteristics between rhizobia and *Frankia* in presented in Table 19.7.

Table 19.7. Symbiotic characteristics of rhizobia and *Frankia*

Feature	Rhizobia	Frankia
Host plants	**Many legumes, *Parasponia***	**A few genera (usually woody)**
Reaction to process of microsymbiont in rhizosphere	Root hairs may curl or branch cortical cells may divide	Root hairs may curl or branch
Method of invasion	Through root hairs, wounds, or between epidermal cells	Through root hairs or between epidermal cells
Site of nodule initiation	Root cortex but exceptions possibly occur in woody spp.	Prenodule may form in cortex; nitrogen-fixing nodule originate in pericycle
Location of vascular tissue	Cortical (except in *Parasponia*)	Central
Location of infected tissue	Central (except in *Parasponia*)	Cortical
Division of infected cells	Present in most spp. with wound infection and those with determinate nodules	Rare
Infection threads	Absent in some spp. with wound infection and where rhizobia are surrounded by matrix of unknown origin	Absent as such, although *Frankia* always surrounded by host cell wall material
Nitrogen-fixing structure	Bacteroid, commonly but not always released from infection thread	Usually a vesicle
Oxygen tension in infected region	Low in those few spp. studied	Low or high, depending on host genus and nodule anatomy
Control/transport of oxygen	Variable diffusion resistance in some spp. coupled with hemoglobin.	By nodule structure, structure of vesicle wall, sometimes with hemoglobin
Longevity of nodules	Few weeks to perennial	Perennial

19.8.2.3. Associate symbiosis

Azospirillum lipoferum is a nitrogen fixing bacterium in association with C_4 plants particularly tropical grasses such as *Paspalum notatum* and *Zea mays*. It is found in the rhizosphere without inducing any apparent morphological changes in the host root. Photosynthate released through root exudates attract the *Azospirillum* in to the rhizosphere. These bacteria fix nitrogen and release organic nitrogen which is taken up by the plant. Although there is neither apparent invasion of the plant root cell by *Azospirillum* nor any formation of specialized structures such as nodules, the association appears to be a symbiosis involving mutual benefit to the two participating species.

19.9. FURTHER READING

Dilworth, M.J. and Glenn, A.R. (Eds.) (1991), Biology and biochemistry of nitrogen fixation", Elsevier, Amsterdam.

Downie, J.A. (2010), The roles of extracellular proteins, polysaccharides and signals in the interactions of rhizobia with legume roots. FEMS Microbiol. Rev 34(2): 150–170.

Eaddy, R.R. (1980), Methods for studying nitrogenase. In methods for evaluating biological nitrogen fixation (ed. F.J. Bergersen), pp. 213-264, John Wiley, Chichester.

Elkan, G.H. and Upchurch, R.G. (eds.) (1997), Current issues in symbiotic nitrogen fixation. Kluwer, Dordrecht, The Netherlands.

Gallon, J.R. and Chaplin, A.E. (1987), An introduction to nitrogen fixation. Cassell, London.

Gresshof, P.M. (ed.) (1990), Molecular biology of symbiotic nitrogen fixation. CRC Press, Roca Raton, FL.

Hirsch, A.M. (1992), Development biology of legume nodulation, *New Phytol.,* 122: 211-237.

John, M., Rohrig, H., Schmidt, J., Walden, R. and Schell, J. (1997), Cell signalling by oligosacc-harides. *Trends Plant Sci.,* 2: 111-115.

Long, S.R. (1996), *Rhizobium* symbiosis : *Nod* factors in perspective. *Plant Cell,* 8: 1885-1898.

Pawlowski, K. (1997), Nodule-specific gene expression. *Physiol. Plant,* 99: 617-631.

Rai, A.N. (ed.) (1990), A Handbook of symbiotic cyanobacteria. CRC Press, Boca Raton.

Rhizobium Web site : http://www.rhizobium.umn.edu

Schultze, M. and Kondorosi, A. (1996), The role of lipochitooligosaccharides in root nodule organogenesis and plant cell growth. *Curr. Opin. Genet. Dev.,* 6: 631-638.

Smith, R.J. and Gallon, J.R. (1993), Nitrogen fixation. In 'Plant Biochemistry and Molecular Biology' ed. P.J. Lea and R.C. Leegood, John Wiley & Sons Ltd, New York.

Somasegaran, P. and Hoben, H.J. (eds.) (1994), Hand book of Rhizobia. Springer-Verlag, New York.

Sprent, J.I. and Sprent, P. (1990), Nitrogen fixing organisms: Pure and Applied Aspects, Chapman and Hall, London.

Stacey, G., Burris, R. and Evans, H.J. (eds.) (1992), Biological nitrogen fixation, Chapman & Hall, New York.

Vance, C.P. (1990), Symbiotic nitrogen fixation : Recent genetic advances. In the biochemistry of plants, Vol. 16 (eds. B.J. Miflin and P.J. Lea), pp. 43-88, Academic Press, San Diego.

Wolk, C.P. (1982), Heterocysts. In the biology of cyanobacteria (eds. N.G. Carr and B.A. Whitton), pp. 359-386. Blackwell Scientific Publications, Oxford

Chapter - 20

Amino Acid Metabolism

20.1 INTRODUCTION

Most amino acids are made up of four elements C H O N, however some aminoacids like methionine, cysteine contain S also. Structurally, they have a carboxyl group and an amino group bonded to the same carbon atom (the α-carbon). Different amino acids differ from each other in their side chains or R groups, which vary in structure, size, electric charge which in turn influence the solubility of amino acid in water. Twenty amino acids are required for protein synthesis and all of them are α-amino acids. In addition to 20 amino acids of proteins, there are also less common one. Some amino acids undergo modifications ones they become a constituent part of proteins. Some amino acids also exist in the protoplasm in free state. As the α-carbon of all amino acids (except glycine) is asymmetric, the amino acid and exist in at least two stereo isomeric forms. In soluble state, amino acid exists as the dipolar ion or zwitterions. Substances having this dual nature are amphoteric and often are called ampholytes. Amino acids are classified in to five types based on the polarity and charge (at pH 7.0) of their R groups. The common amino acids of proteins have been assigned three letter abbreviations and one letter symbols which are used as shorthand to indicate composition and sequence of amino acids polymerized in proteins (Table 20.1).

In addition to their role as the building blocks of proteins, amino acids are precursors of many specialized biomolecules, including hormones, coenzymes, antibiotics, pigments and neurotransmitters.

20.2 BIOSYNTHESIS OF AMINO ACIDS

Bacteria and plants synthesize all 20 common amino acids on their own. However, mammals can synthesize about 10 amino acids, the requirement of remaining amino acids are met through the external food.

Twenty amino acids, required for the biosynthesis of proteins are formed from the metabolic precursors of central metabolic pathway (Table 20.2). A critical perusal of the table reveals that six of the 20 amino acids are synthesized from oxaloacetate and four from α-ketoglutarate, both are intermediates of the TCA cycle. In addition, succinyl-CoA donates succinyl group in the formation of intermediates in the biosynthesis of lysine, methionine and diaminopimelic acid. This again emphasizes the fact that TCA cycle is not only important for catabolic processes but also for anabolism. The table also reveals that three glycolytic intermediates (pyruvate, 3-PGA and PEP) are precursors to seven more amino acids. The PP pathway provides erythrose–4-P which is a precursor to aromatic aminoacids. Finally, 5-phosphoribosy-1-pyrophosphate (PRPP), which donates the sugar for nucleotide synthesis, is also the precursor to the amino acid histidine.

Table 20.1. Classification and three letter abbreviations and one letter symbols of amino acids

Amino acid	Abbreviation/symbol	
Nonpolar		
Aliphatic R groups		
Glycine	Gly	G
Alanine	Ala	A
Proline	Pro	P
Valine	Val	V
Leucine	Leu	L
Isoleucine	lle	I
Methionine	Met	M
Aromatic R groups		
Phenylalanine	Phe	F
Tyrosine	Tyr	Y
Tryptophan	Trp	W
Polar		
Uncharged R groups		
Serine	Ser	S
Threonine	Thr	T
Cysteine	Cys	C
Asparagine	Asn	N
Glutamine	Gln	Q
Positively charged R groups		
Lysine	Lys	K
Histidine	His	H
Arginine	Arg	R
Negatively charged R groups		
Aspartate	Asp	D
Glutamate	Glu	E

Before proceeding for the synthesis of individual amino acids, it is pertinent to understand how the amino group, a key functional group of all amino acids, is introduced into the cell materials and how it is transferred to key amino acids that function as the donors of amino group for all other amino acids.

Table 20.2. Precursors for amino acid biosynthesis

Precurosr	Amino acid
pyruvate	L-alanine L-valine L-leucine
oxaloacetate	L-aspartate L-asparagine L-methionine L-lysine L-threonine L-isoleucine
α-oxoglutarate	L-glutamate L-glutamine L-arginine L-proline
3-phosphoglycerate	L-serine glycine L-cysteine
phospho-enol pyruvate erythrose-4-phosphate	L-phenylalanine L-tyrosine L-tryptophan
5-phosphoribosyl-1-pyrophosphate + ATP	L-histidine

The synthesis of ammonia from nitrate and gaseous nitrogen is described in Chapter 19. The ammonia group is then incorporated into key amino acids, glutamate and glutamine. The ability to synthesize glutamate and glutamine is of extreme importance because this is the only route for incorporation of inorganic nitrogen into cell material. Glutamate is the amino donor for most of the amino acids, and glutamine is the amino donor for the synthesis of purines, pyrimidines, amino sugars, histidine, tryptophan, asparagine, NAD^+ and P-amino benzoate.

20.2.1 Glutamate and Glutamine Synthesis

Glutamate is synthesized by two alternate routes. The first one requires the enzyme glutamate dehydrogenase. This enzyme catalyzes the reductive amination of α-ketoglutarate.

$$\alpha\text{-ketoglutarate} + NADPH + H^+ + NH_3 \rightleftharpoons \text{Glutamate} + NADP^+ + H_2O$$

However, under conditions of low ammonia concentration, bacteria use two enzymes in combination for ammonia incorporation. One is L-glutamine synthetase and the second one is glutamate synthase or glutamine: α-oxoglutarate amino transferase (GOGAT). Glutamine synthetase incorporates ammonia into glutamate to form glutamine. This reaction uses ATP which drives the reaction to completion. The second enzyme, glutamate synthase transfers the newly incorporated ammonia from glutamine to α-ketoglutarate to form glutamate (Fig. 20.1). Thus, glutamine is used catalytically.

$$\text{Glutamate} + \text{ATP} + \text{NH}_3 \longrightarrow \text{Glutamine} + \text{ADP} + \text{Pi}$$

$$\text{Glutamine} + \alpha\text{-ketoglutarate} + \text{NADPH} + \text{H}^+ \longrightarrow 2 \text{ glutamate} + \text{NADP}^+$$

$$\alpha\text{-ketoglutarate} + \text{ATP} + \text{NH}_3 + \text{NADPH} + \text{H}^+ \longrightarrow \text{glutamate} + \text{ADP} + \text{Pi} + \text{NADP}^+$$

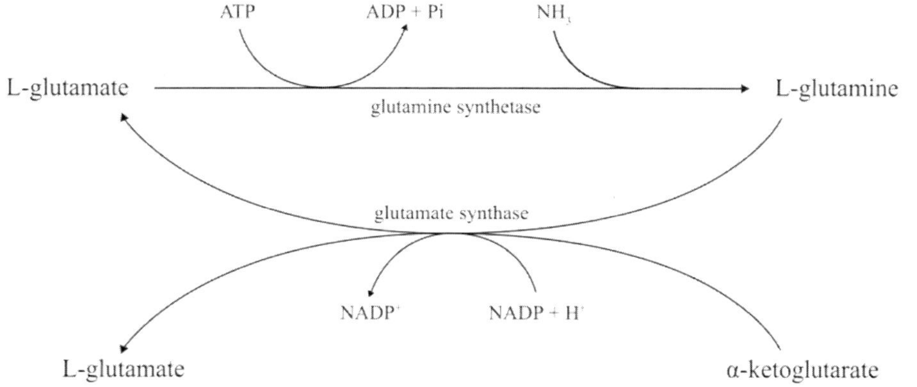

Fig. 20.1 : Synthesis of glutamate by glutamine synthetase and glutamate synthase reactions.

Transamination: Once the glutamate is synthesized, it functions as a principal donor of amino group to an α-keto carboxylic acid to synthesize different other amino acids. The enzyme that catalyzes the amino group transfer is called a **transaminase**. Transaminases use a coenzyme called pyridoxal phosphate to carry the amino group from glutamate to the α-ketoglutaric acid. In transamination reactions, after the glutamate has donated its amino group it becomes α-ketoglutarate again, which then becomes aminated once more to form another molecule of glutamate. Thus, glutamate can be thought of as a conduit through which ammonia passes into other amino acids.

20.2.2 The Oxaloacetate and Pyruvate Families of Amino Acids

Oxaloacetate is the intermediate of TCA cycle and pyruvate is the end product of glycolysis. Utilizing both these, either alone or in combination, nine amino acids (aspartate, asparagine, lysine, threonine, methionine, isoleucine, valine, leucine and alanine) are synthesized. The pathways leading to these nine amino acids are summarized in Fig. 20.2. It is evident from the figure that L-aspartate and L-threonine are both products and intermediates.

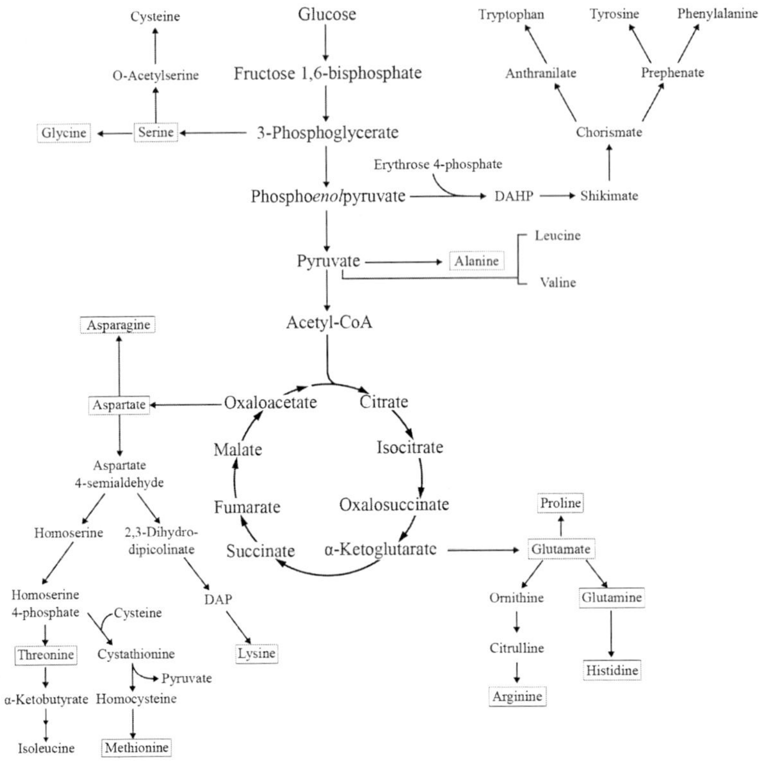

Fig. 20.2 : Outline pathways for synthesis of different amino acids

Synthesis of L-aspartate and L-asparagine

Simple transamination reactions are employed for the formation of L-aspartate from oxaloacetate

$$\text{Oxaloacetate} \xrightarrow[\text{Transaminase}]{\text{L-glutamate} \curvearrowright \alpha\text{-oxoglutarate}} \text{Aspartate}$$

L-aspartate is converted into L-asparagine by L-asparagine synthetase

$$\text{L-aspartate} + Nh_3 + \text{ATP} \xrightarrow[\text{Asparagine synthetase}]{} \text{Asparaging} + ADP_3 + Pi$$

Synthesis of L-threonine and L-methionine

Synthesis of L-threonine and L-methionine makes use of a common pathway starting fom oxaloacetate (Fig. 20.3). Aspartate semialdehyde, which is synthesized in two steps from L-aspartate is reduced to homoserine by homoserine dehydrogenase. Homoserine is phosphorylated to homoserine phosphate which is finally converted to threonine. These two reactions are catalyzed by homoserine kinase and threonine synthase respectively.

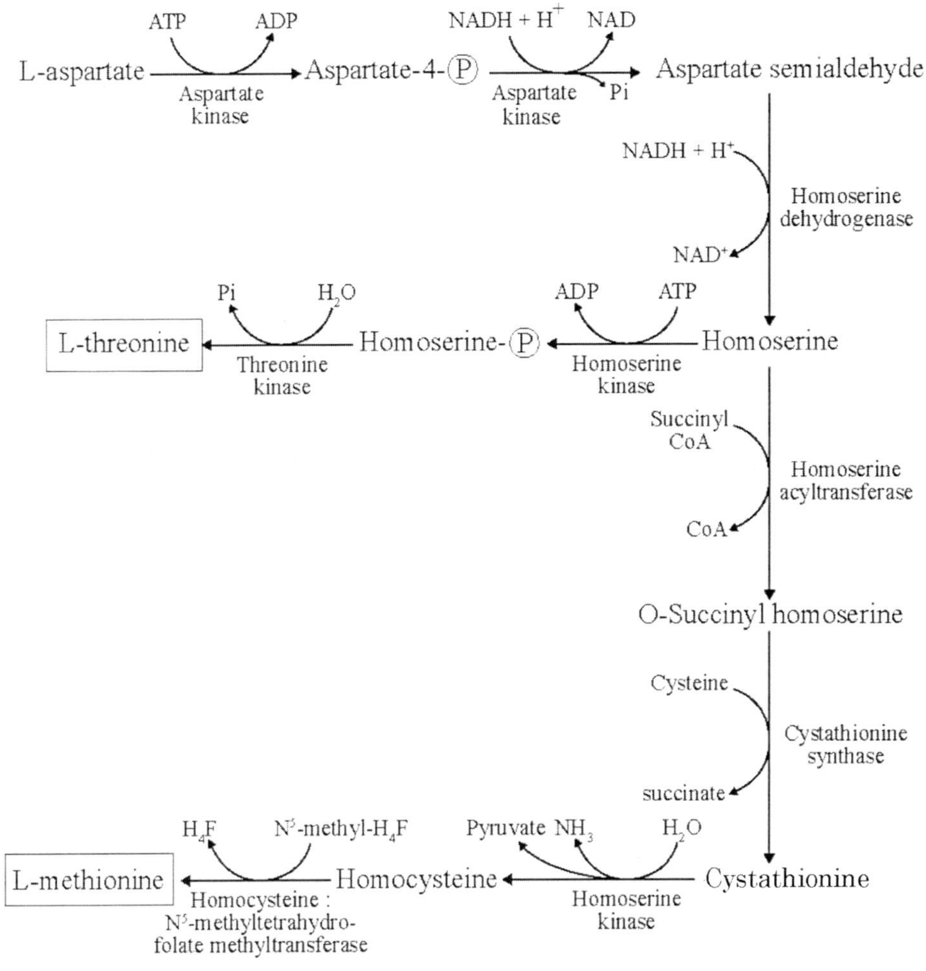

Fig. 20.3 : Synthesis of L-threonine and L-methionine from oxaloacetate

Another enzyme acting on homoserine is homoserine acyltransferase. Its reaction product is succinyl homoserine. The succinyl residue is then exchanged with a cysteinyl residue, and elimination of ammonia and pyruvate yields homocysteine. The final reaction is the methylation with N5 –methyltetrahydrofolate as donor of the methyl group and thus forming L-methionine.

Synthesis of lysine: For the synthesis of lysine, reactions upto the formation of aspartate semialdehyde are similar to threonine and methionine synthesis. However, aspartate semialdehyde is condensed with pyruvate to yield 2,3 dihydrodipicolinate (Fig. 20.4). Reduction and ring opening by succinylation gives N-succinyl-2-amino-6-oxopimelate. One more amino group is transferred at this stage giving rise to N-succinyl-2,6-diaminopimelate. This intermediate is desuccinylated to yield L, L-α, e- diaminopimelate which is finally decarboxylated to L-lysine.

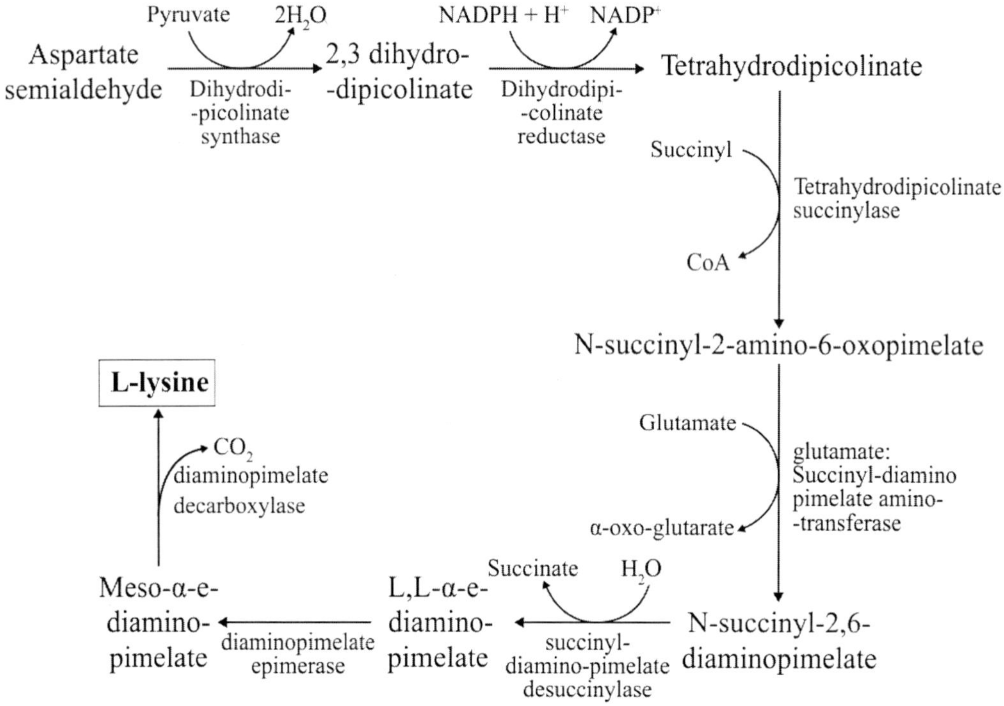

Fig. 20.4 : Synthesis of lysine

Synthesis of L-isoleucine, L-valine and L-leucine

Isoleucine is synthesized from threonine (Fig. 20.5). Threonine is deaminated to 2-oxybutyrate, an intermediate required for isoleucine synthesis by an enzyme threonine deaminase. At this stage, pyruvate condenses with 2-oxybutyrate to produce α-aceto-α-hydroxy butyrate. This reaction is catalyzed by acetohydroxy acid synthase. α-aceto-α-hydroxy butyrate is in turn reduced to α,β-dihydroxy-β-methyl valerate by acetohydroxy acid isomer reductase. α,β-dihydroxy -β- methyl valerate is a substrate for the activity of dihydroxy acid dehydratase which produces the direct precursor of isoleucine α-oxo-β-methyl valerate. Finally, a branched chain amino acid transaminase catalyzes the transfer of an amino group from glutamate onto the oxo acid of isoleucine.

The pathway of valine synthesis is a four-step pathway that shares all of its steps with the parallel pathway of isoleucine biosynthesis (Fig. 20.5). Synthesis of this amino acid starts with the synthesis of α-acetolactate from two molecules of pyruvate. α-acetolactate is reduced by keto acid reductoisomerase to yield α, β-dihydroxy isovalerate. This metabolite is acted upon by dihydroxy acid dehydratase to produce α- oxoisovalerate. As in case of isoleucine, the final step of valine synthesis is catalyzed by transaminase.

α-oxoisovalerate, which is the direct substrate for valine synthesis is also a substrate in the pathways which yields leucine (Fig. 20.5). Four out of subsequent six reactions utilize specific enzymes, followed by one spontaneous reaction and eventually, a transamination as in case of isoleucine and valine. α-oxoisovalerate is converted to α-isopropylmalate catalyzed by an enzyme α- isopropyl malate synthase. 3-isopropylmalate undergoes a NAD^+ dependent oxidation catalyzed by 3-isopropyl malate dehydrogenase. This reaction is followed by a spontaneous decarboxylation producing an oxoacid of leucine, α-oxoisocaproate. The last reaction in this pathway is catalyzed by transaminase.

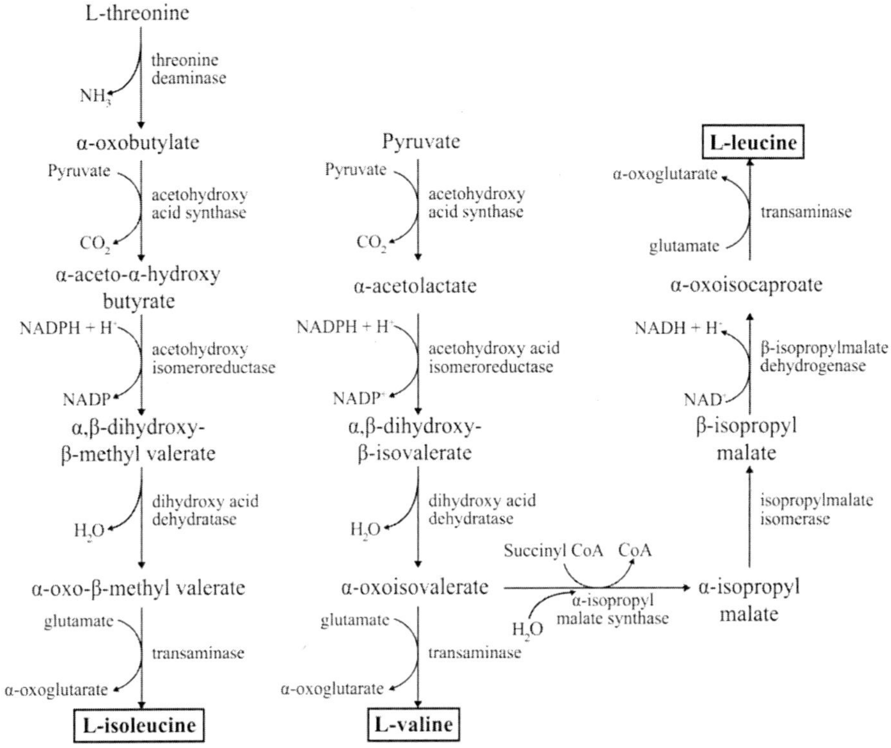

Fig. 20.5 : Synthesis of L-isoleucine, L-leucine and L-valine

Synthesis of alanine

Alanine is directly synthesized from pyruvate by transamination. Glutamate is the donor of amino group.

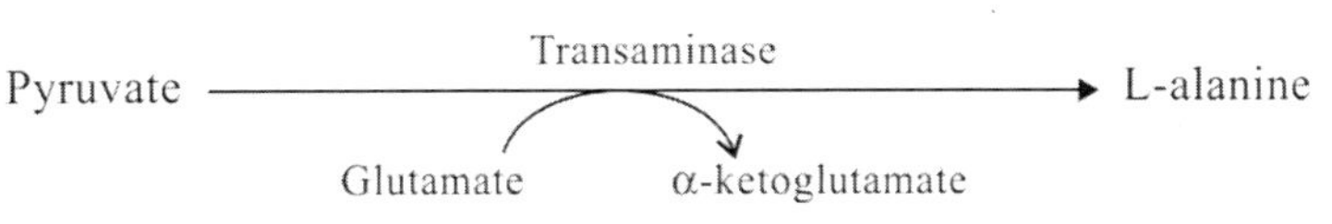

20.2.3 The Phosphoglycerate Family of Amino Acids

Three amino acids *viz.,* serine, glycine and cysteine are synthesized utilizing 3-phosphoglycerate as the precursor (Table 20.2). Serine synthesized from 3-phosphoglycerate is also a precursor for both cysteine and glycine.

In the synthesis of serine, the first step is the oxidation of 3-phosphoglycerate into 3-phosphohydroxy pyruvate (Fig. 20.6). The enzyme that catalyzes the oxidation is 3-

phosphoglycerate dehydrogenase. The oxidation generates an α-keto acid which can now be transmitted by transferring an amino group form glutamate to 3-phosphoglycerate to generate 3-phosphoserine and α-ketoglutarate. The phosphate group is removed by phosphoserine phosphatase to produce serine.

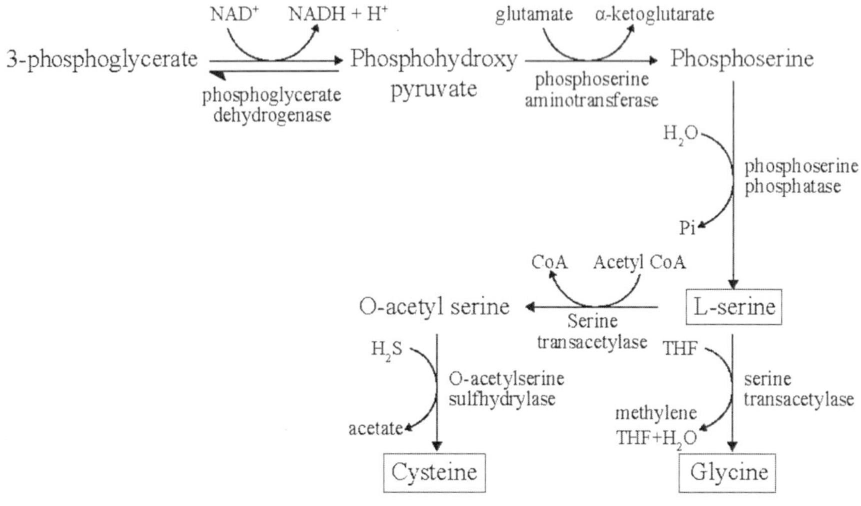

Fig. 20.6 : The synthesis of serine, glycine and cysteine from 3-phosphoglycerate

Serine is converted to glycine by serine hydroxymethyl transferase which contains both pyridoxal phosphate cofactor and tetrahydrofolate co-factor. In the course of this reaction, carbon 3 is transferred to tetrahydrofolic acid, which functions as donor of methyl groups in a number of biosynthetic reactions.

The route to cystine biosynthesis is initiated when serine accepts an acetyl group from acetyl-CoA to become O-acetyl sarine. Sulfide then displaces the acetyl group from O-acetyl serine forming cystine (see sulphate assimilation section).

20.2.4 The α-Ketoglutarate Family of Amino Acids

Glutamate is the common precursor for the synthesis of arginine and proline. L-proline is synthesized along a pathway consisting of only of a few reactions. This pathway utilizes glutamate, which is reduced to glutamate semi-aldehyde by pyrroline carboxylate synthetase enzyme (Fig. 20.7). Subsequently, glutamate-semialdehyde is spontaneously converted to pyrroline-5-carboxylate. Δ^1 pyrroline-5-carboxylate reductase further reduces the Δ^1-pyrroine-5-carboxylate intermediate to proline. Alternatively, proline can be synthesized from ornithine, which is first transaminated by ornithine delta transaminase producing glutamate-semialdehyde.

In comparison to proline, synthesis of arginine is more complicated. In the first step, glutamate is acetylated using acetyl-CoA. The reaction is catalyzed by N-acetyl glutamate synthase The second step is the phosphorylation of N-acetylglutamate by N-acetylglutamate kinase. N-acetyl glutamyl-phosphate then undergoes a NADPH-dependent diphosphorylation reduction catalyzed by N-acetyl-glutamyl reductase to produce N-acetyl glutamyl semialdehyde. A product of this reaction is further converted into N-acetylornithine by acetylornithine transaminase with glutamate as the amino group donor. A subsequent reaction catalyzed by action of amidoacylase produces ornithine and acetate. The conversion of ornithine to arginine proceeds in three steps (which are part of urea cycle). Ornithine is converted to citrulline by reaction with carbamoyl-phosphate catalyzed by ornithine carbamoyl transferase. In a subsequent reaction citrulline reacts with aspartate in a ATP dependent reaction catalyzed by argininosuccinate synthase. The product of this reaction, argininosuccinate is split by argininosuccinate lyase to arginine and fumarate.

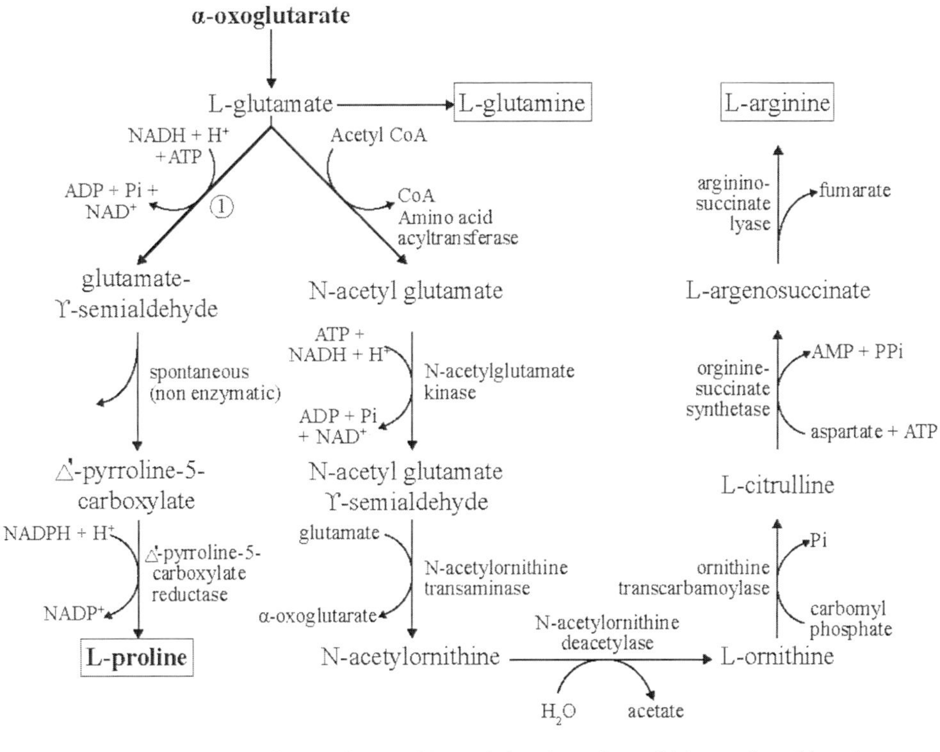

Fig. 20.7 : The Synthesis of L-proline and L-arginine (reaction of 1 is catalyzed by glutamate kinase + glutamate semialdehyde dehydrogenase)

20.2.5 Synthesis of Aromatic Amino Acids

Biosynthesis of aromatic amino acids is very complicated. The three aromatic amino acids, phenylalanine, tyrosine and tryptophan, are derived from a common precursor, chorismate (Fig. 20.8 and 20.9). The synthesis of chorismate begins with condensation of erythrose-4-phosphate and phosphoenol pyruvate in a reaction catalyzed by 3-deoxy-D-arabino-heptulosonate. The reaction product, 3-deoxy-o-arabinoheptulosonate-7-phosphate is used to form 5-dehydroquinate in a reaction of cyclisation catalyzed by 5-dehydroquinate synthase. A third enzyme of chorismate biosynthesis pathway, 5-dehydroquinate dehydrogenase converts 5-dehydroquinate into 5-dehydroshikimate, which is reduced to shikimate in the fourth step by shikimate-NADP oxidoreductase. A subsequent phosphorylation of shikimate to form shikimate-5-phosphate is catalyzed by shikimate kinase. Shikimate 5-phosphate and phosphoenol pyruvate are used to produce 3-enol pyruvyl shikimate-5-phosphate and orthophosphate. This reaction, 3-enolpyruvylshikimate-5-phosphate to form chorismate, is catalyzed by chorismate synthase.

Prephenate is the next intermediate in the synthesis pathway of tyrosine and phenylalanine. It is produced from chorismate by the action of chorismate mutase. Prephenate gives rise to phenyl pyruvate and p-hydroxyphenyl pyruvate by the action of prephenate dehydratase and prephenate dehydrogenase respectively. These products are transaminated to L-phenylalanine and L-tyrosine. Glutamate is the donor of amino group.

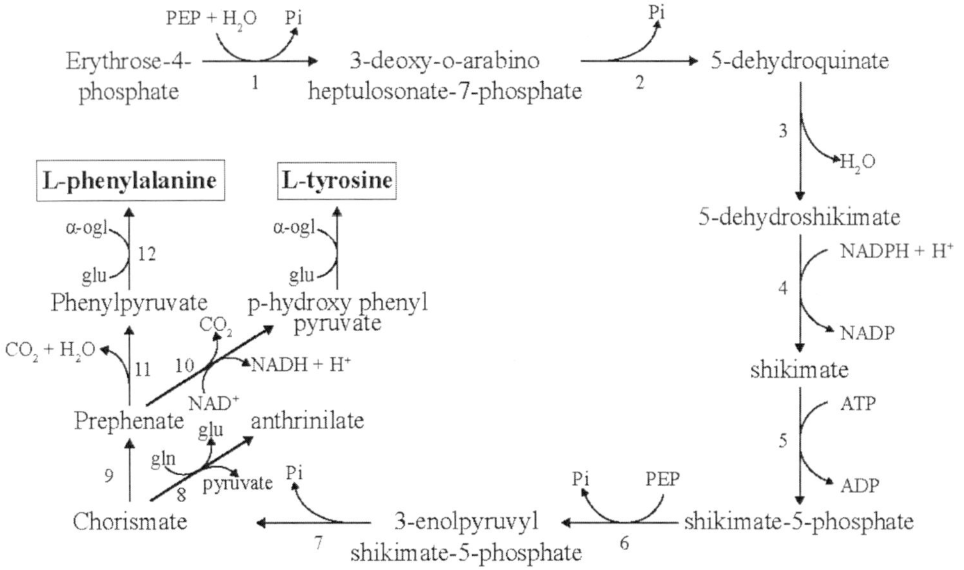

Fig. 20.8 : Synthesis of aromitc amino acid 1, 3-Deoxy-D-arabino-heptulosonate (DAHP) synthase; 2, 5-dehydroquinate synthase; 3; 5-dehydroquinate dehydratase; 4, shikimate dehydrogenase; 5, shikimate kinase; 6, enolpyruvylshikimate-5-phosphate synthase; 7, chorismate synthetase; 8, anthranilate synthase; 9, chorismate mutase; 10, prephenate dehydrogenase; 11, prephenate dehydratase; 12, transaminase B. PEP, Phosphoenolpyruvate; gln, glutamine; glu, glutamate; α-oxyglutarate.

The biosynthesis of tryptophan begins with the conversion of chorismate to anthranilate by the elimination of the enolpyruvyl side chain by transfer of the amino group from glutamine (Fig. 20.9). This reaction is catalyzed by the heteromeric enzyme anthranilate synthase. In the next step, anthranilate phosphoribosyl transferase converts anthranilate to 5^1-phosphoribosylanthra-nilate. 5^1-phosphoribosylanthranilate undergoes isomerization by the action of the phosphoribosylanthranilate isomerase. In the following reaction is indole-3-glycerol-phosphate synthase produces indole-3-glycerolphosphate from. 1-(carboxyphenyl amino)-1-deoxy-D-ribulose-5-phosphate. The least two steps in this pathways are catalyzed by a single enzyme, tryptophan synthase, which is composed of two α and two β subunits. Each of the subunits is responsible for catalyzing an individual step. Thus the activity of tryptophan synthase can be divided into two separate steps in the pathway, each catalyzed by the corresponding subunit. First subunit cleaves glyceraldehyde phosphate from indole-3-glycerolphosphate giving one indole molecule, which in turn is combined with serine giving rise to tryptophan.

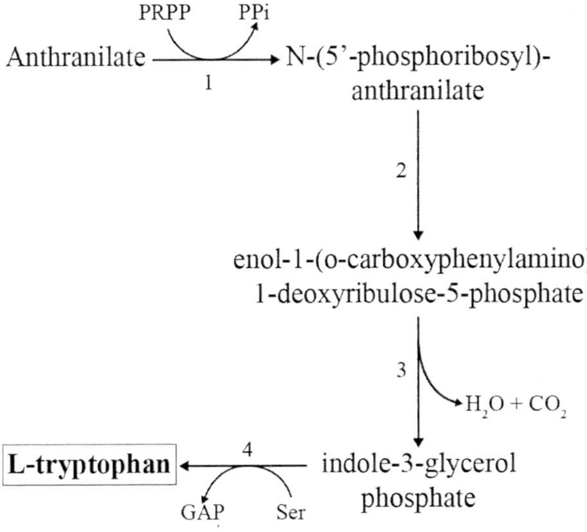

Fig. 20.9 : Synthesis of L-tryptophan from anthranilate (refer Fig. 20.8). 1. Anthranilate phosphoribosyl transferase, 2, phosphoribosyl-anthranilate isomerase; 3, indole-3-glycerol phosphate synthase; 4, tryptophan synthase. PRPP, 5-phosphoribosyl-1-pyrophosphate; serine; GAP, glyceraldehydes-3-phosphate

20.2.6 Biosynthesis Histidine

Histidine biosynthesis pathway is completely an independent of pathways used for all other amino acids. The histidine synthesis begins with the ATP-dependent phosphorylation of ribose-5-phosphate catalyzed by ribose-phosphate diphosphokinase to form 5-phosphoribosyl diphosphate (Fig. 20.10). In the following reaction, 5-phosphoribosyl diphosphate and ATP are bound by ATP phosphoribosyl transferase to phosphoribosyl-ATP. The product is further hydrolyzed by bi-functional enzyme phosphoribosyl-ATP pyrophosphase/phosphoribosyl-AMP cyclohydrolase to phosphoribosyl-AMP. In the next step, the same enzyme opens the aromatic ring of adenine giving rise to N-5-phosphoribosyl-formimino-5-amino-imidozole-4-carboxamide ribonucleotide. It is further converted to phosphoribosyl formimino-5-amino carboxamide ribonucleotide by an isomerase. The following step, a conversion of phosphoribulosylformimino-5-amino-imidazole-4-carboxamide ribonucleotide to erythro-imidazoleglycerol phosphate and 5-amino imidozole-4-carboxamide ribonucleotide, requires two catalytic functions: transmidation (with glutamate as amide donor) and a cyclization reaction. Prokaryotes use two separate enzymes to catalyze this reaction. In the reaction catalyzed by imidazoleglycerol phosphate dehydratase, erythro-imiazoleglycerolphosphate produces imidazole-acetol-phosphate, which is converted to histidinol-phosphate in subsequent reaction by an enzyme histidinol-phosphate transaminase. This reaction also consumes glutamate as the amino group donor. The penultimate and the final reactions of histidine biosynthesis are catalyzed by histidinol dehydrogenase. The same enzyme catalyzes the synthesis of histidine from histidinol in two subsequent oxidation reactions with histidinal as the intermediate.

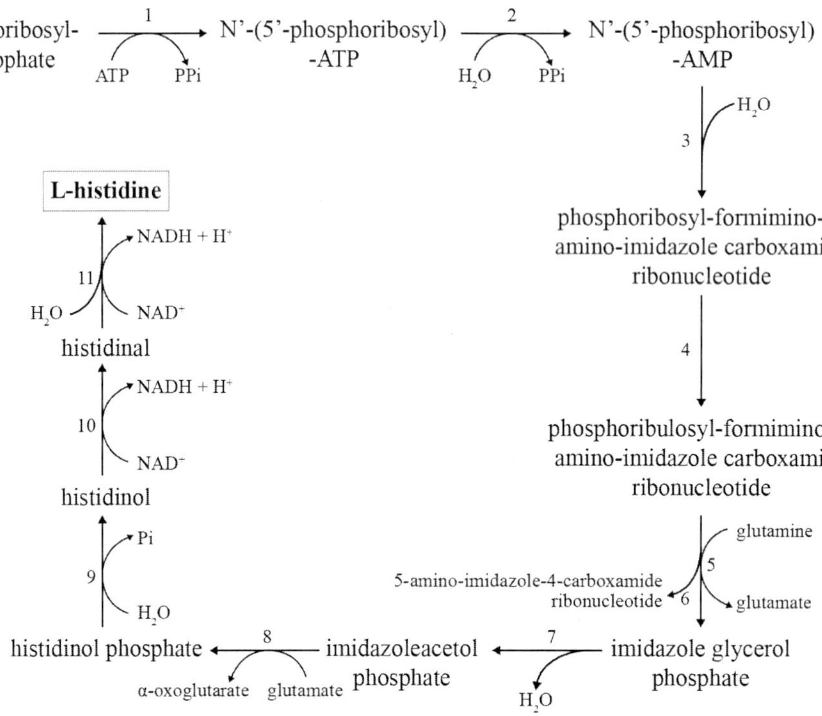

Fig. 20.10 : Synthesis of L-histidine. 1, Phosphoribosyl ATP pyrophosphorylase; 2, Phosphoribosyl-ATP pyrophosphohydrolase; 3, Phosphoribosyl-ATP cyclohydrolase; 4, phosphoribosyl formimino-5-aminoimidazole carboxamide ribonucleotide isomerase; 5, Phosphoribulosyl-formimin05-aminoimidazole carboxamide ribonucleotide glutamine aminotransfersae; 6, cyclase, the name of the compound releasesd is 5-aminoimidazole-4-carboxamide ribonucleotide; 7, imidazole-glycerol phosphate dehydratase, 8, histidinol phosphate transaminase; 9, histidinol phosphatase; 10, 11, histidinol dehydrogenase (enzyme catalyses the two-step dehydrogenation of histidinol to L-histidine). Reactions 7 and 9 are catalyzed by the same enzyme.

20.3 CATABOLISM OF AMINO ACIDS

Amino acids are identified into two types, glucogenic and ketogenic amino acids based on the fate of the degradation end product. Glucogenic amino acids are broken down into pyruvate, α-ketoglutarate, succinyl Co-A, fumarate or oxaloacetate whose carbon skeleton is utilized to synthesize all of the cell components via glucose. Ketogenic amino acids are broken down into acetoacetate or acetyl Co-A whose carbon skeleton is oxidized to CO_2. Some amino acids are both glucogenic and ketogenic.

Glucogenic amino acids: Arginine, glutamate, glutamine, histidine, proline, valine, methionine, aspartate, asparagine, alanine, serine, cysteine and glycine.

Ketogenic amino acids: Lysine and leucine

Glucogenic and ketogenic amino acids: Tryptophan, phenylalanine, tyrosine, isoleucine and threonine.

Each of the 20 amino acids has a separate catabolic pathway, yet all 20 pathways converge into 5 intermediates all of which can enter the TCA cycle. From the TCA cycle the carbon skeletons can be completely oxidized into CO_2 or diverted into gluconeogenesis or ketogenesis. A scheme showing the overall pattern of carbon flow during amino acid catabolism is shown in Fig. 20.11. The first step is always deamination *i.e.,* the removal of the amino group to generate the α-keto acid which eventually enters TCA cycle. As shown in the Fig. 20.11, all twenty amino acids are degraded to seven intermediates that enter the TCA cycle. These seven intermediates are: pyruvate, acetyl Co-A, acetoacetyl Co-A, α-ketoglutarate, succinyl Co-A, fumarate and oxaloacetate. Acetoacetyl Co-A itself is a precursor to acetyl Co-A.

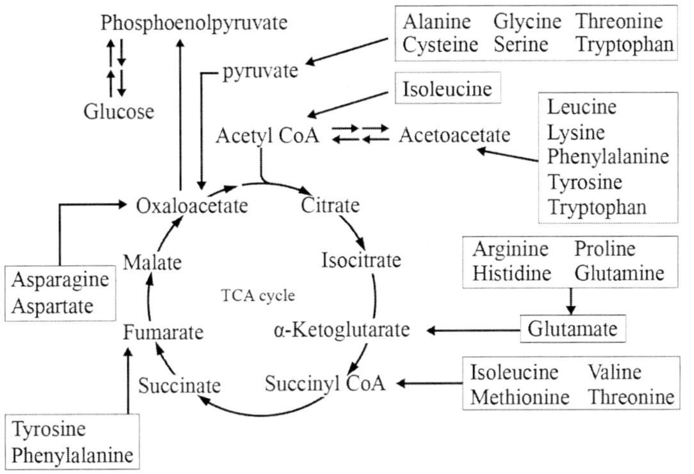

Fig. 20.11. Summary of catabolism of different amino acids.

Removal of the amino group

Usually, amino acids are oxidatively deaminated to their corresponding keto acids by one of the following four processes:

i. Oxidation by nonspecific flavoprotein oxidases: These enzymes oxidize any one of amino acids and feed the electrons directly into electron transport chain. These oxidases can be D-amino oxidases or L-amino oxidases. The D-amino acid oxidases are more useful because several biological molecules like peptidoglycan and some antibiotics have D- amino acids.

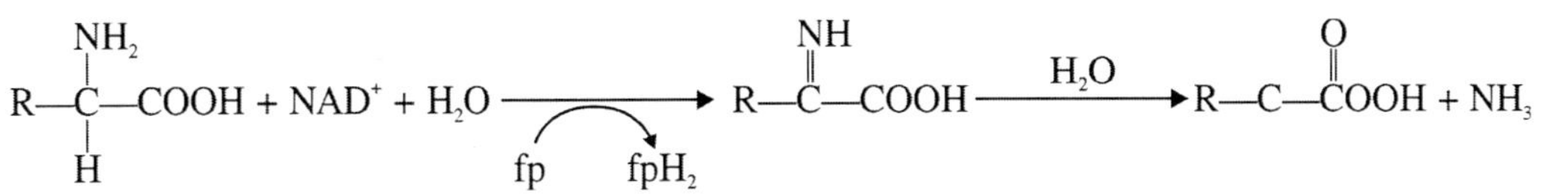

ii. $NADP^+$ linked dehydrogenases: These are more specific than the flavoprotein oxidases. The dehydrogenases catalyze the reversible reductive amination of the keto acids.

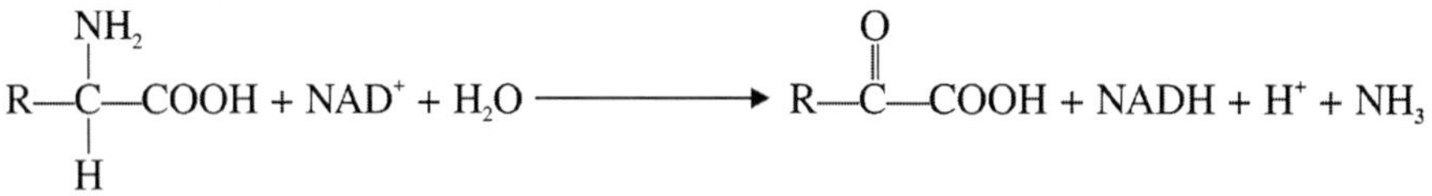

iii. Transamination by nonspecific transaminase: In this reaction, a transaminase is coupled to dehydrogenase. For example, a nonspecific transaminase transfers an amino group to pyruvate forming L- alanine. The amino group is released as ammonia and pyruvate is regenerated by L-alanine dehydrogenase. α-ketoglutarate and its dehydrogenase serve the same function in amino acid catabolism.

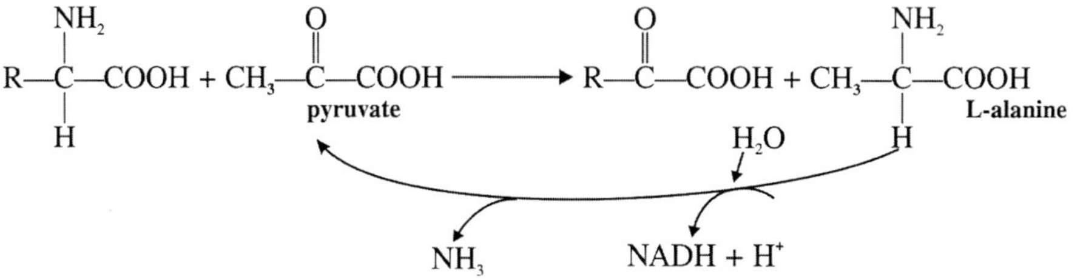

iv. Deamination by specific deaminases: Certain amino acids like serine, threonine, aspartate and histidine are deaminated by amino acid specific deaminases. However, in these reactions, redox reactions are not involved and thus reducing power is not generated. Amino group is released in the form of ammonia. Such deaminations usually, proceed by the elimination of water.

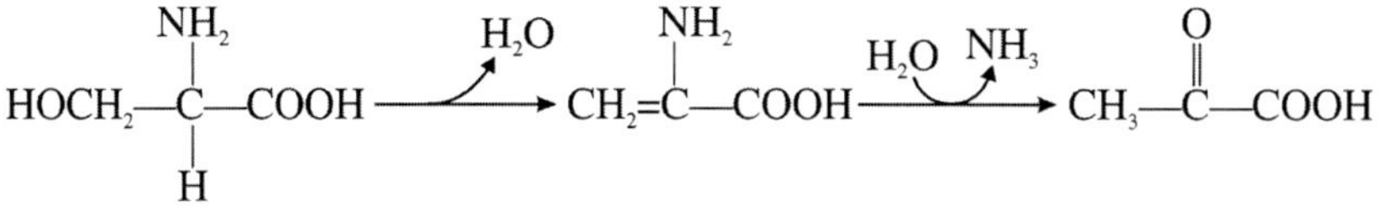

20.3.1 Amino Acids Degraded to Pyruvate

Six amino acids, alanine, tryptophan, cysteine, serine, glycine and threonine are degraded to pyruvate. The carbon skeleton of these six amino acids are converted to in whole or in part to pyruvate. The pyruvate can then be converted to either acetyl Co-A (a ketone body precursor) or oxaloacetate (a precursor for gluconeogenesis). Thus, these six amino acids are both ketogenic and glucogenic.

Alanine transaminase reversibly transfers the amino group from alanine to α-ketoglutarate to form pyruvate and glutamate. This reaction requires a pyridoxal phosphate cofactor. The α-ketoglutarate is regenerated by glutamate dehydrogenase

$$\text{Alanine} + \alpha\text{-ketoglutarate} \xrightarrow[\text{alanine transaminase}]{} \text{pyruvate} + \text{glutamate}$$

$$\text{Glutamate} \xrightarrow{\text{NAD(P)+} \quad \text{NAD(P) H}} \text{Glutarate}$$

Serine dehydratase, which also requires pyridoxal phosphate cofactor, catalyzes the β-elimination of the hydroxyl group of serine to form an amino acrylate intermediate which tautomerizes into the imine which is then hydrolyzed to produce ammonia and pyruvate.

$$\text{Serine} \xrightarrow[\text{Serine dehydratase}]{} \text{amino acrylate} \longrightarrow \text{imine} \xrightarrow{H_2O \;\; NH_4^+} \text{pyruvate}$$

Glycine is converted into pyruvate via serine formation by serine hydroxymethyl transferase. This enzyme is an unusual and interesting one. It contains two cofactors pyridoxal phosphate and a N^5, N^{10}-methylene-tetrahydrofolate. This enzyme is produced by the glycine cleavage system which transfers a methylene group from glycine to THF. The THF cofactor is a one carbon acceptor and donor.

$$\text{Glycine + THF} \xrightarrow[\text{glycine cleavage system}]{NAD^+ \;\; NADH} {}^+NH + CO_2 + N^5, N^{10}\text{–methylene-THF}$$

$$\text{Glycine} + N^5, N^{10}\text{-methylene-THF} \xrightarrow[\text{serine hydroxymethyl transferase}]{THF} \text{serine} \longrightarrow \text{pyruvate}$$

Cysteine is converted to pyruvate in two steps; in the first one sulphur atom is removed and in the second one transamination

Threonine is both glucogenic and ketogenic. There are couple of routes for the degradation of threonine. The major route is as shown below. Threonine is first converted into acetyl Co-A and glycine. The glycine is then converted into serine by serine hydroxymethyl transferase, and serine is then converted into pyruvate by serine dehydratase.

$$\text{Threonine} \xrightarrow{NAD^+ \;\; NADH{+}H^+} \alpha\text{-amino-}\beta\text{-ketobutyrate} \xrightarrow{CoA} \text{glycine + Acetyl Co-A}$$

$$\text{Glycine} \rightarrow\rightarrow\rightarrow \text{Pyruvate}$$

20.3.2 Amino Acids Degraded to Asparagine and Aspartate

Both asparagine and aspartate are degraded into oxaloacetate. Aspargine is hydrolyzed into aspartate and ammonia. This reaction is catalyzed by asparginase. Aspartate is converted into oxaloacetate by aspartate amino transferase which is a PLP enzyme that transfers an amino group from aspartate to α-ketoglutarate to form glutamate and oxaloacetate

$$\text{Asparagine} \xrightarrow[\text{Asparaginase}]{H_2O \;\; NH_4^+} \text{Aspartate}$$

$$\text{Aspartate} + \alpha\text{-ketoglutarate} \xrightarrow[\text{Aspartate aminotransferase}]{} \text{Glutamate + oxaloacetate}$$

20.3.3 Amino Acids Degraded to α-Ketoglutarate

Five amino acids arginine, proline, glutamine, glutamate and histidine are converted to glutamate which is then deaminated by a transaminase to form α-ketoglutarate (Fig. 20.12)

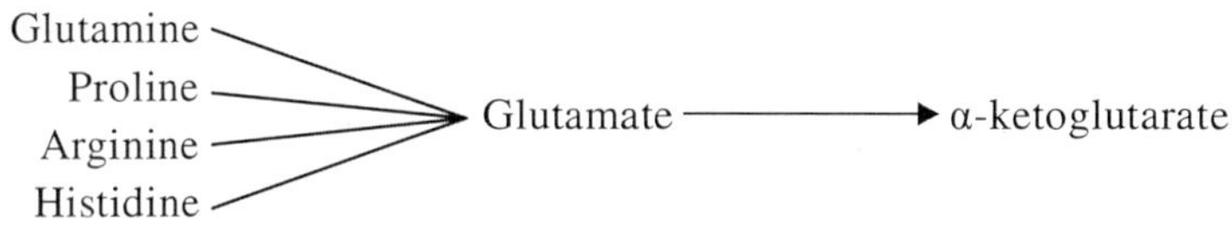

In a single step reaction, glutamine is converted into glutamate by glutaminase. Proline is oxidized by proline oxidase to form pyrroline 5-carboxylate which spontaneously hydrolyzes to form glutamate γ-semialdehyde. Arginase converts the arginine into ornithine and urea. Ornithine δ-amino transferase transfers the δ-amino group of ornithine to α-ketoglutarate to form glutamate γ-semialdehyde and glutamate. The former is oxidized to form glutamate by glutamate 5-semialdehyde dehydrogenase.

Histidine is deaminated by histidine ammonia lyase which forms urocanate. Urocanate hydratase adds water to form 4-imidazolone-5-propionate which is hydrolyzed by imidazalone propionase to form M-formiminoglutamate. Glutamate formimino transferase transfers the formimino group to tetrahydrofolate to generate glutamate and N^5-formimino-THF.

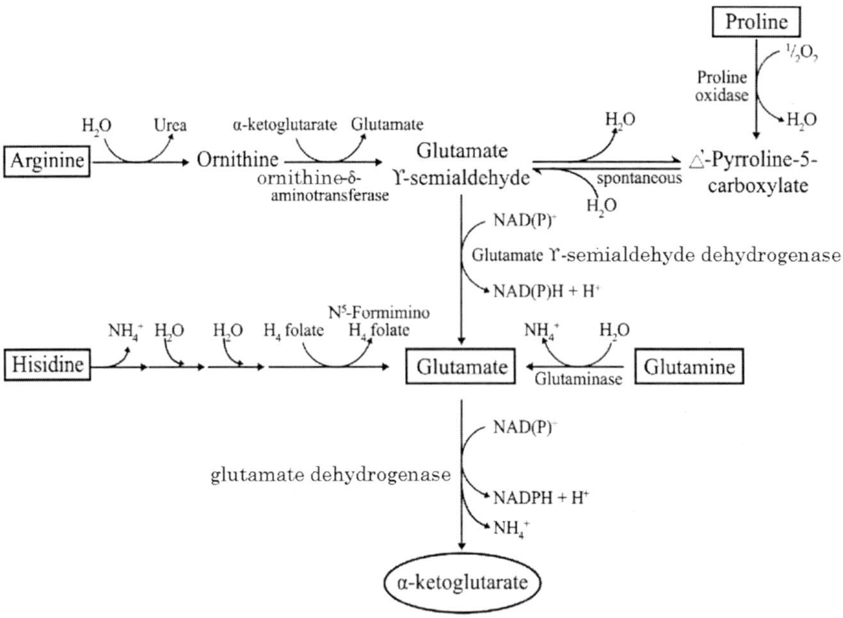

Fig. 20.12 : Degradation of arginine, histidine, glutamate, glutamine and proline

20.3.4 Amino Acids degraded to succinyl-CoA

Three amino acids methionine, valine and isoleucine are broken down into propionyl-CoA. Propionyl-CoA is converted into D-methylmalonyl-CoA by propionyl Co-A carboxylase. D-methylmalonyl-CoA is racemized into L-methylmalonyl-CoA by methylmalonyl-CoA.

The degradation of methionine requires nine steps (Fig. 20.13). The first is catalyzed by methionine adenosyl transferase, which transfers the adenosyl group of ATPs to sulfur of methionine to form S-adenosylmethionine (SAM)

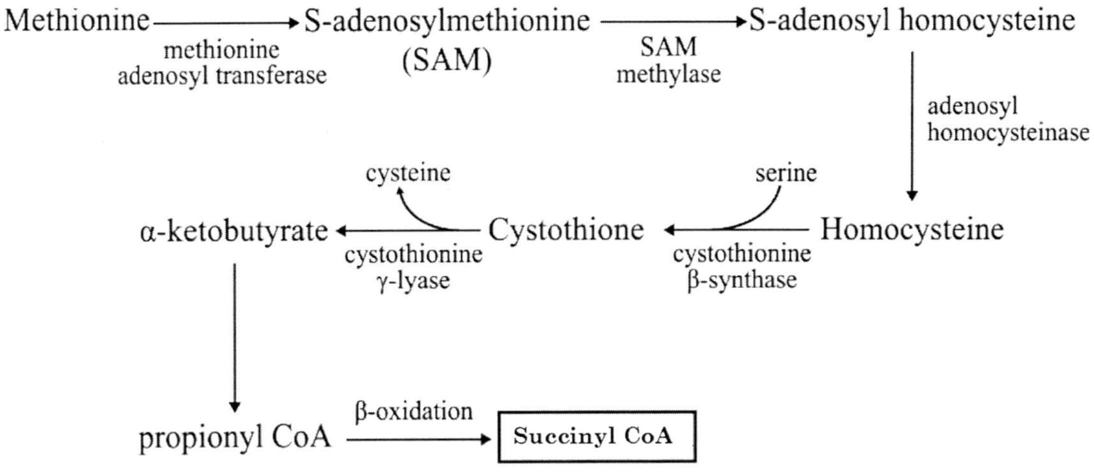

Fig. 20.13 : Degradation of methionine to succinyl CoA

SAM methylase transfers the activated methyl group to an acceptor to form S-adenosylhomo cysteine which is hydrolyzed by adenosyl homocysteinase to form homocysteine. Cystithionine β-synthase is a OPLP dependent enzyme that catalyzes the condensation of a serine residue with homocysteine to form cystothionine. Cystathionine ϒ-lyase cleaves cystathionine into cysteine and α-ketobutyrate. α-ketobutyrate is converted into propionyl-CoA by α-ketobutyrate dehydrogenase which catalyzes a reaction that is analogous to pyruvate dehydrogenase and α-ketoglutarate dehydrogenase.

The degradation of branched chain amino-acids (isoleucine and valine) uses some enzymes of TCA cycle and β-oxidation. Isoleucine undergoes transamination, followed by oxidative decarboxylation resulting, α-keto acid (Fig. 20.14). The remaining five carbon skeleton is further oxidized to acetyl-CoA and propionyl-CoA. Valine undergoes transamination and decarboxylation, then a series of oxidation reactions that convert the remaining four carbons to propionyl-CoA. Some steps in valine and isoleucine degradative pathways closely resemble the steps in fatty acid degradation (Chapter-22).

20.3.5 Amino Acids Degraded to Acetyl Co-A and Acetoacetate

There are only two amino acids leucine and lysine that are purely ketogenic.

Leucine catabolism is similar to the branched amino acids valine and isoleucine (Fig. 20.14). Leucine is, first, transaminated by branched amino acid amino transferase to form α - ketoisocaproate which is then oxidatively decarboxylated to form isovaleryl-CoA by the branched chain α-ketoacid dehydrogenase complex. In the next step, isovaleryl-CoA is dehydrogenated to form β-methylcrotonyl-CoA. The enzyme that catalyzes this dehydrogenation is isovaleryl-CoA dehydrogenase. β-methylcrotonyl-CoA is then carboxylated by a biotin containing enzyme called methylcrotonyl-CoA carboxylase to form β-methylglutaconyl-CoA which is then hydrated by β-methylglutacomyl-CoA hydratase to form β-hydroxy β-methylglutaryl-CoA. β-hydroxy β-methylglutaryl is finally cleaved into acetyl-CoA and acetoacetate by HMG-CoA lyase.

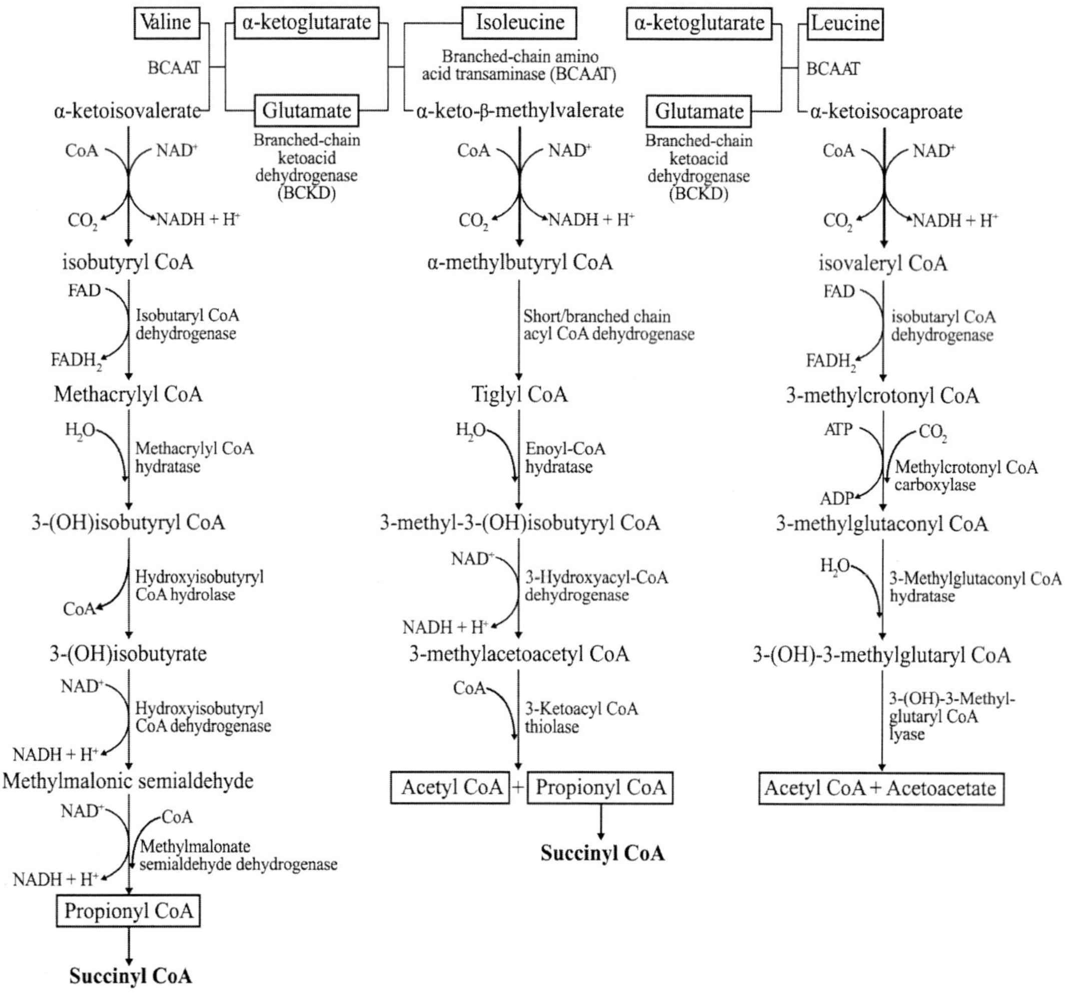

Fig. 20.14 : Catabolism of three branched chain amino acids, valine, isoleucine and leucine.

A complete degradation pathway for lysine has not yet been identified (Fig. 20.15) However, the initial steps are well characterized. Conversion of lysine to saccharopine and subsequently to 2-aminoadipic 6-semialdehyde is catalyzed by a bifunctional cytosolic enzyme lysine ketoglutarate reductase. The conversion of 2-aminoadipic 6-semialdehyde to acetyl-CoA probably requires eight reaction steps. First, cytosolic dehydrogenase 7B4 catalyzes the oxidation to 2-aminoadipate, which is transaminated to 2-ketoadipate and then oxidatively decarboxylated to glutaryl-CoA by the 2-oxoglutarate dehydrogenase complex. Isovaleryl-CoA dehydrogenase catalyzes the oxidation of glutaryl-CoA to glutaconyl-CoA. Glutaconyl-CoA metabolism then proceeds via crotonyl Co-A, 3- hydroxybutyryl-CoA, and acetoacetyl-CoA. However, enzymes catalyzing the decarboxylation of either glutaconyl-CoA or keto adipate have not yet been identified, and experimental evidence proving the functional significance for this part of the pathway is still missing.

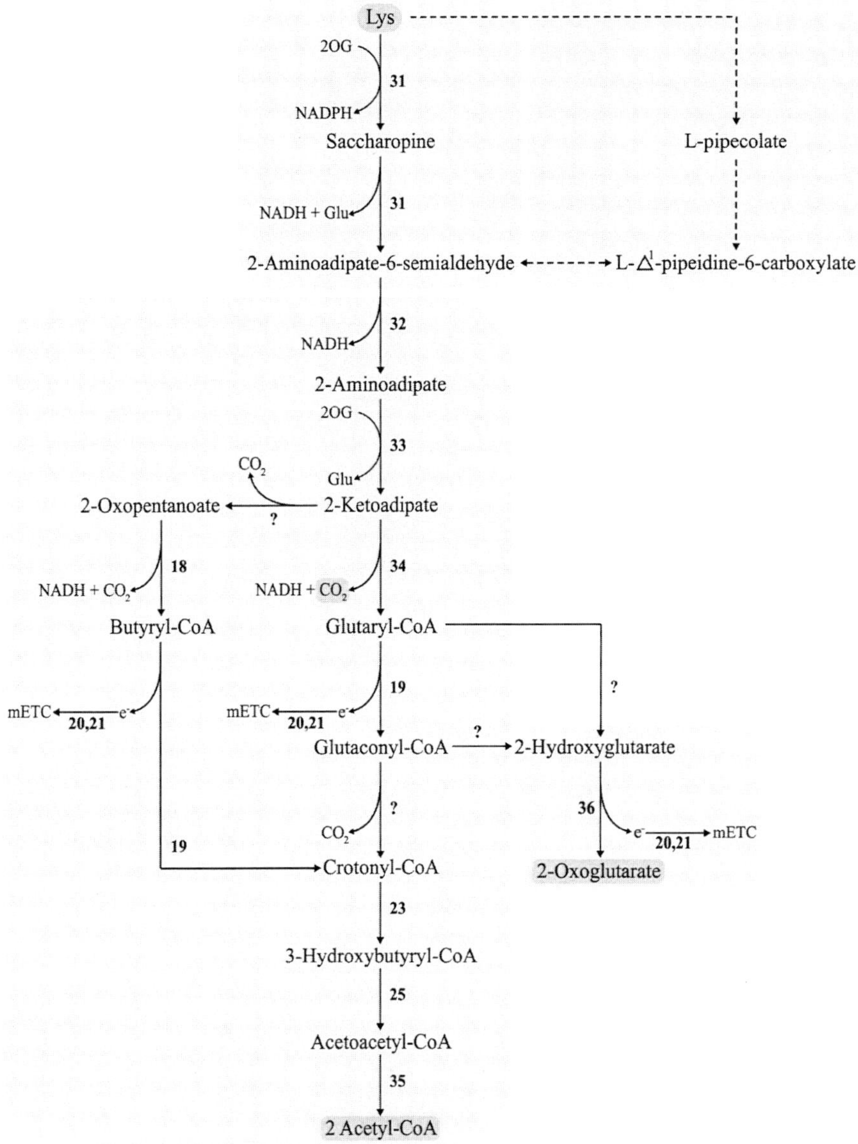

Fig. 20.15: Catabolism of lysine

20.3.6 Catabolism of Aromatic Amino Acids

The three aromatic amino acids are phenylalanine, tyrosine and tryptophan. Degradation of aromatic amino acids requires molecular oxygen to breakdown the aromatic rings.

Degradation of phenylalanine begins with a monooxygenase phenylalanine hydroxylase which adds a hydroxyl group to phenylalanine to form tyrosine (Fig.20.16). From these the remaining five catabolic pathways for phenylalanine and tyrosine are the same. Tyrosine amino transferase deaminates tyrosine to form β-hydroxyphenyl pyruvate which is converted into homogentisate by the enzyme B-hydroxy phenylpyruvate dioxygenase. The final three steps in the catabolism of tyrosine and phenylalanine are carried out by the enzymes homogentisate 1,2-dioxygenase, maleylacetoacetate hydrolase resulting in the production of fumarate (glucogenic) and acetoacetate (ketogenic).

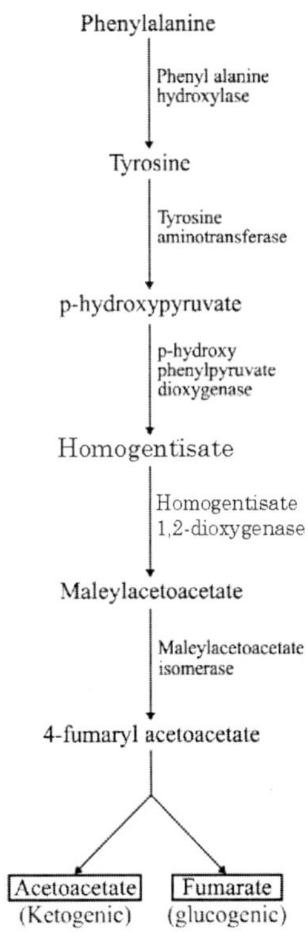

Fig. 20.16: Catabolism of phenylalanine and tyrosine.

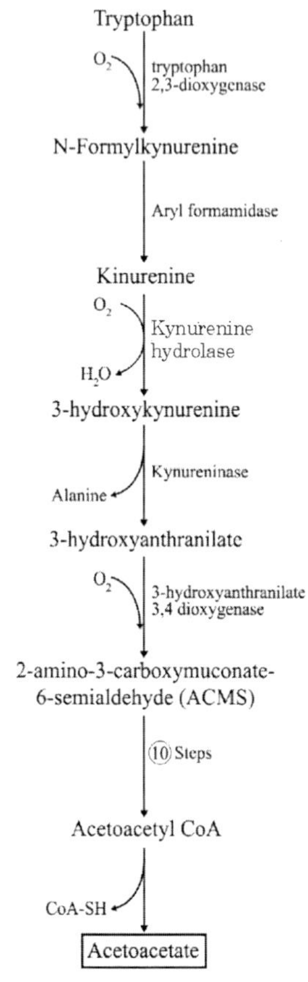

Fig. 20.17: Catabolism of tryptophan

The catabolism of tryptophan includes the generation of the glucogenic amino acid, alanine, which allows tryptophan to be classified as glucogenic and ketogenic amino acid. The tryptophan catabolic pathway (Fig. 20.17) starts with either tryptophan 2,3 –dioxygenase, or indoleamine 2,3 dioxygenase 1 which open the indole ring to form N-formyl-kynurenine. N-formyl-kynurenine is then converted to kynurenine by the enzyme, arylformamidase. Kynurenine then undergoes a series of catabolic reactions resulting in complete oxidation to acetoacetyl CoA. The first is catalyzed by kynurenine hydrolase forming 3-hydroxykynurenine. The next step in the pathway is catalyzed by kynureninase and produces 3- hydorxyxanthranilic acid and alanine. The next step in the pathway is the conversion of 3-hydroxyanthranic acid to 2-amino 3-carboxymuconate 6-semialdehyde (ACMS) and quinolinic acid by the enzyme 3-hydroxyanthranilate 3,4-dioxygenase. ACMS can be metabolized via the action of the enzyme ACMS decarboxylase followed by ten other enzymes to acetoacetyl CoA the end product of complete tryptophan catabolism.

Finally, it may be noted that some amino acids are degraded through more than one metabolic pathway.

A summary of the amino acid metabolism showing start point and end point is presented in the Table 20.3.

Table 20.3. Precursors and degradation end products of amino acids

Amino acid	Made from	Degraded to	Glyco / keto
Alanine	pyruvate	pyruvate	glycogenic
Arginine	glutamate	glutamate	glycogenic
Asparagine	aspartate	aspartate	glycogenic
Aspartate	oxaloacetate	oxaloacetate	glycogenic
Cysteine	(methionine)*	pyruvate	glycogenic
Glutamate	oxoglutarate	oxoglutarate	glycogenic
Glutamine	glutamate	glutamate	glycogenic
Glycine	serine	one-carbon pool***	glycogenic
Histidine	5-phosphoribosyl-1 - pyrophosphate+ATP	glutamate	glycogenic
Isoleucine	oxoglutarate	acetyl-CoA + propionyl-CoA	mixed
Leucine	pyruvate	acetyl-CoA	ketogenic
Lysine	oxaloacetate	not known	mixed
Methionine	oxaloacetate	propionyl-CoA	glycogenic
Phenylalanine	PEP, erythrose-4 phosphate	tyrosine	mixed
Proline	glutamate	glutamate	glycogenic
Serine	phosphoglycerate	pyruvate	glycogenic
Threonine	oxaloacetate	disputed****	Glycogenic
Tryptophan	PEP, erythrose-4 phosphate	not known	Mixed
Tyrosine	PEP, erythrose-4 phosphate	fumarate + acetoacetate	mixed
Valine	pyruvate	propionyl-CoA	glycogenic

20.4 FURTHER READING

Berg J., Tymocko, J. and Stryer I. (2003), Biochemistry 5th edn. W.H. Freeman & Co. Ltd.

Caldwell, D.R. (2000), Microbial Physiology and metabolism 2nd edn. Bilmont, California, Star Publishing.

Doelle, H.W. (1975), Bacterial Metabolism 2nd edn. Academic Press, New York

David A. Beuder (2012), Amino acid metabolism, John Wiley & Ltd.

Gottschalk, G. (1986), Bacterial Metabolism 2nd edn. Springer-Verlag, New York.

Harveg R.A. and Ferrier, D.R. (2011), Lippincott's Illustrated Reviews : Biochemistry. 5th edn. Wolters Kluwer Health, Philadelphia.

Nelson D.L. and Cox M.M. (2007), Lehpinger Principles of Biochemistry 4th edn. W.H. Freeman and Co. New York.

Voet D.J., Voet I.G and Pratt, C.W. (2008), Principles of Biochemistry 3rd edn., John Wiley & Sons. Inc.

Victory W.R. *et al.* (2015), Harpis Illustrated Biochemistry (30th edn.) The McGraw-Hill Education.

White, D. (1995), The Physiology and Biochemistry of prokaryotes. Oxford University Press, New York.

■■■■

Chapter - 21

Nucleotide Metabolism

21.1 INTRODUCTION

Nucleotides, nitrogenous compounds like amino acids, are the building blocks of two crucial macromolecules DNA and RNA, the genetic material of all the living organisms. Apart from their crucial importance in genetic material, they are important to control and maintain important cellular functions. For example, ATP and GTP are important for energy storage. NADP and FAD are nucleotides, which act as cofactors. Nucleotides like CAM (cyclic adenosine mononucleotide) are essential for cell signaling pathways.

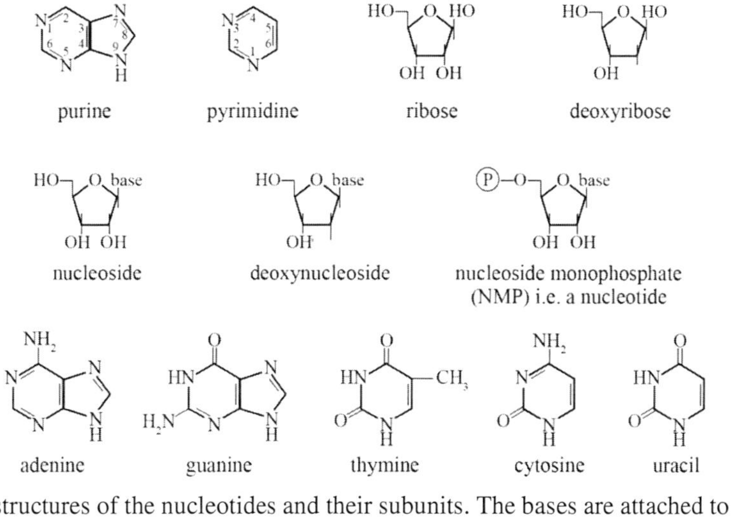

Fig. 21.1 The structures of the nucleotides and their subunits. The bases are attached to the C1 of the sugars and the phosphates to the C5.

A nucleotide is a complex molecule composed of three units: nitrogenous base (purine or pyrimidine), a sugar (ribose or deoxyribose) and a phosphate if there is no phosphate, it would be called nucleoside (nitrogen base + sugar) thus nucleotides are nucleoside phosphates. If one phosphate is present, then the molecule is called a nucleoside monophosphate. If two phosphates are present, then it is called nucleoside diphosphate and so on. There are mainly two groups of nitrogenous bases as pyrimidines and purines. Pyrimidines are smaller heterocyclic, aromatic, six member rings containing nitrogens at 1 and 3 positions (Fig 21.1). The three major pyrimidines are cytosine, thymine and uracil. The corresponding nucleosides are called cytidine, thymidine, and uridine, respectively. The cytidine nucleotides are called cytidine monophosphate (CMPP), and cytidine diphosphate (CDP), and cytidine triphosphate (CTP). A similar nomenclature is used for uridine nucleotides (UMP, UDP, UTP) and the thymine nucleotides (TMP, TDP, TTP). The two major purines are adenine and guanine which are much larger than pyrimidines. Other than heterocyclic aromatic ring, they have an imidazole fused to that. In this case,` the nucleosides are adenosine and guanosine and similarly the nucleotides are adenosine mono-, di-and triphosphate and guanosine mono-, di-, and triphosphate. All the five monophosphates are also known by alternate names adenylic acid (AMP), guanylic acid (GMP), uridylic acid (UMP), cytidylic acid(CMP) and thymidylic acid (TMP).

In nucleosides or nucleotides if the sugar present is deoxyribose rather than ribose then they are indicated by a suffix for example dATP, dGTP, dCTP, dTTP (Table 21.1).

Table 21.1. Nomenclature of bases, nucleosides, and nucleotides

		RNA
Base	Ribonucleoside	Ribonucleotide (5'-monophosphate)
Adenine (A)	Adenosine	Adenylate (AMP)
Guanine (G)	Guanosine	Guanylate (GMP)
Uracil (U)	Uridine	Uridylate (UMP)
Cytosine (C)	Cytidine	Cytidylate (CMP)
		DNA
Base	Deoxyribonucleoside	Deoxyribonucleotide (5'-monophosphate)
Adenine (A)	Deoxyadenosine	Deoxyadenylate (dAMP)
Guanine (G)	Deoxyguanosine	Deoxyguanylate (dGMP)
Thymine (T)	Thymidine	Thymidylate (TMP)
Cytosine (C)	Deoxycytidine	Deoxycytidylate (dCMP)

21.2 OVERVIEW OF NUCLEOTIDE BIOSYNTHESIS

Understanding of biosynthetic pathways of nucleotides is tremendously important from two aspects, first to follow the biosynthesis of two most important molecules of living systems, DNA and RNA, secondly to know the intervention points for therapeutic agents. Many of the most widely used drugs in the treatment of cancer block steps in nucleotide biosynthesis, particularly steps in the synthesis of DNA precursors.

Synthesis of nucleotides takes place through different pathways: (i) *de novo* pathway and (ii) salvage pathway. these two pathways may operate simultaneously depending upon the demand. The salient features of *de novo* synthesis are:

i. This pathway is nearly identical in all living organisms.

ii. In *de novo* (from scratch) pathways, the nucleotide bases are assembled from simpler compounds.

iii. The frame work for a pyrimidine base is assembled first and then it is attached to ribose (Fig. 21.2). In contrast, the frame work for a purine is synthesized piece by piece directly on to a ribose based structure.

iv. These pathways comprise a small number of elementary reactions that are repeated with variation to generate different nucleotides.

In salvage pathways, the bases resulting from the partial degradation of nucleic acids are recovered, reused and reconnected to ribose unit. These pathways, thus economize the cell in terms of both energy and basic cell components.

In any case, both *de novo* and salvage pathways lead to the synthesis of ribonucleotides. However, DNA is built from deoxyribonucleotides. It is interesting to note that all deoxyribonucleotides are synthesized from the corresponding ribonucleotides which provides the strength to the notion that RNA preceded the DNA during evolution. To complete this transformation deoxyribose sugar is generated by the reduction of ribose and the methyl group, that distinguishes the thymine of DNA from the uracil of RNA, is added at the last step.

21.3 DE NOVO SYNTHESIS OF NUCLEOTIDES

As mentioned earlier, the nucleotide bases are synthesized from simpler compounds, as simple as CO_2 and NH_4 and using some elementary reactions.

21.3.1 The Synthesis of Pyrimidine Nucleotides

In de novo synthesis of pyrimidines, the ring synthesized first and then it is connected to ribose to form a pyrimidine nucleotide (Fig. 21.2). Pyrimidine rings are assembled from bicarbonate, aspartic acid and ammonia. Although ammonia can be used directly, it is usually produced from the hydrolysis of the side chain of glutamine.

The first step in *de novo* pyrimidine biosynthesis is the synthesis of carbamoyl phosphate from bicarbonate and ammonia catalyzed by an enzyme carbamoyl phosphate synthetase (CPS) (Fig. 21.3). This enzyme requires two ATPs, one ATP is used to form a carboxylated biotin intermediate, and the second ATP is used to phosphorylate carboxyl group. In step 2, carbamoyl phosphate reacts with aspartate to form carbamoylaspartate in a reaction catalyzed by aspartate transcarbamoylase. In this reaction, phosphate of carbamoyl group is displaced by the nitrogen atom of aspartate. Carbamoylaspartate then cyclizes to form dihydrorotate which is then oxidized by $NADP^+$ to form orate. At this stage, orate displaces ppi from phosphoribosyl pyrophosphate (PRPP) to form the first nucleotide, orotidine-5- phosphate, also called orotidylate. The reaction is driven to completion by the hydrolysis of PPi catalysed by a pyrophosphatase. It may be noted that PRPP is synthesized from ribose-5-phosphate and ATP. Orotidine -5- phosphate is decarboxylated to form uridylate (UMP), a major pyrimidine nucleotide that is a precursor to RNA. The reaction is catalyzed by orotidylate decarboxylase. Phosphorylation of UMP using ATP as the donor yields UDP and UTP.

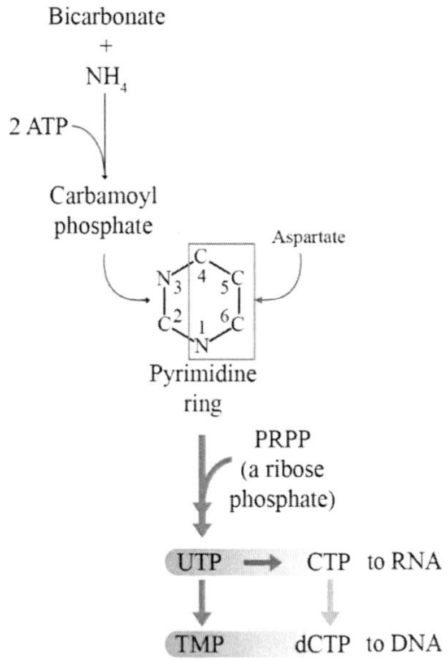

Fig. 21.2: *De novo* pathway for pyrimidine nucleotide synthesis. The C-2 and N-3 atoms in the pyrimidine ring come from carbamoyl phosphate, whereas the other atoms of the ring come from aspartate.

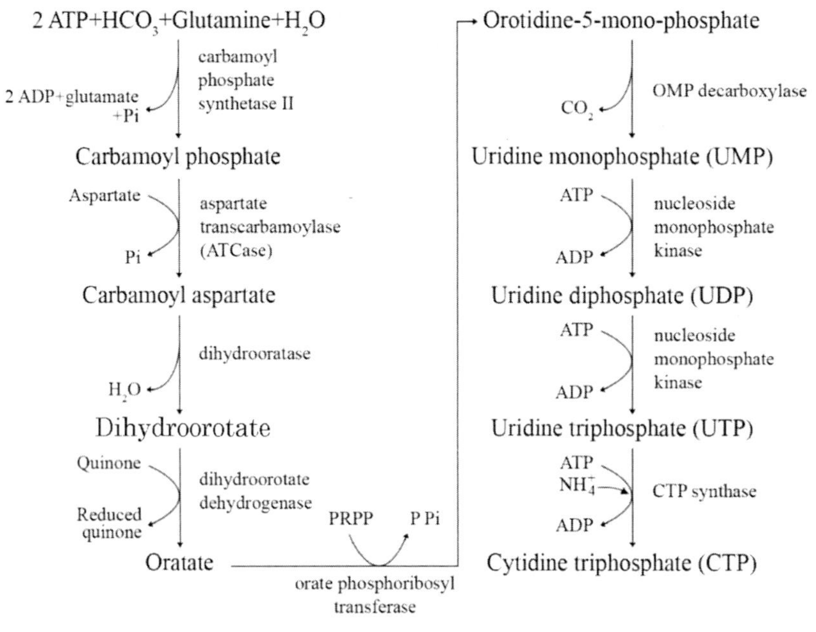

Fig. 21.3. Biosynthesis of pyrimidine nucleotides (UTP, CTP)

Formation CTP from UTP: After UTP formation it can be transformed into cytidine triphosphate by ATP-dependent amination using NH^+_4. The CTP can be dephosphorylated to CDP, which is the precursor to the deoxy ribonucleotides as shown below

CDP ⟶ dCDP —— dCTP ⟶ dUTP ⟶ dUMP ⟶ dTMP ⟶ dTDP ⟶ dTTP

UDP can also be reduced to dUDP which can be converted to dUTP. Thus, as show below there are two routes to dUTP, one from CDP and another from UDP. However, the former is the major route.

CDP ⟶ dCDP —— dCTP ⟶ dUTP

UDP —— dUDP —— dUTP

21.3.2 The Synthesis of Purine Nucleotides

The two common purines, adenine, guanine, possess a heterocyclic aromatic ring fused with imidozole ring (Fig. 21.1). Experiments conducted by John Buchanan demonstrated that N1 of purines arises from the amino group of aspartate; C2 and C8 originate from formate; N3 and N9 are contributed by the amide group of glutamine; C4, C5 and N7 are derived from glycine and C6 comes from HCO_3^- (Fig. 21.2). Purine nucleotides are synthesized in two phases. In the first phase, inosine monophosphate (IMP) is synthesized and in the second phase, by divergent reactions, AMP and GMP are synthesized from AMP (Fig 21.4).

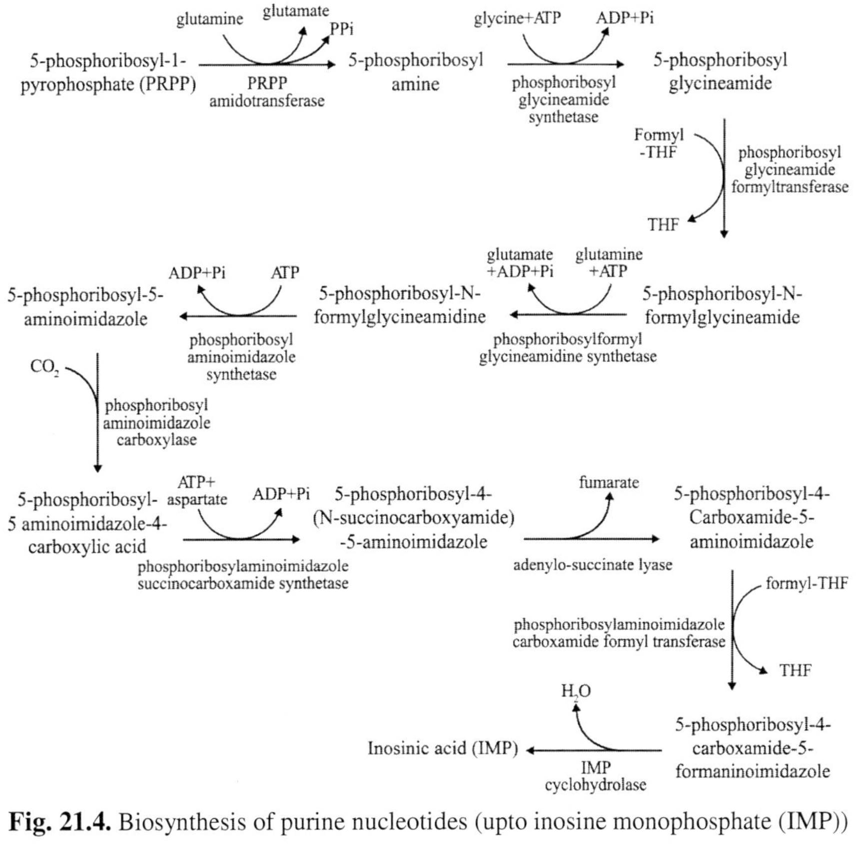

Fig. 21.4. Biosynthesis of purine nucleotides (upto inosine monophosphate (IMP))

i. Synthesis of IMP

As mentioned earlier, unlike in pyrimidine synthesis, purine molecular are synthesized step-by-step while attached to ribose phosphate, which in turn comes from PRPP. In the other words ribose-5-phosphate functions as a scaffolding molecule on which the purine is built step –by–step.

IMP is synthesized in a pathway composed of 10 reactions (excluding the synthesis of ribose-5-phosphate) (Fig. 21.4).

Step 1 : Acquisition of purine atom N9: This reaction is unique to purine biosynthesis, amidophosphoribosyl transferase catalyzes the displacement of PRPP's pyrophosphate group by glutamines amide nitrogen in a nucleophilic displacement. The pyrophosphate that is released has a high free energy of hydrolysis and the amination is driven to completion by the hydrolysis of the pyrophosphate by the pyrophosphatase.

Step 2 : Acquisition of purine atoms C4, C5and N7: In this reaction, glycine is added. Glycine's carboxyl group forms an amide with the amino group of phosphoribosylamine, yielding 5-phosphoribosylglycineamide or glycine amide (GAR). This reaction is reversible, despite its concomitant hydrolysis of ATP to ADP +Pi. It is the only step of purine biosynthetic pathway in which more than one purine ring atom is acquired.

Step 3 : Acquisition of purine atom C8: 5-phosphoribosylglycineamide's free α-amino group is formylated to yield 5-phosphoribosyl-N-formyle glycine amide or formylglycinamideribotide (FGAR).

This requires a molecule with high formyl group transfer potential to donate the formyl group. The formyl donor in the reaction is N10-formyltetrahydrafolate (N10 -Formyl-THF), a coenzyme transfers C1 units.

Step 4 : Acquisition of purine atoms N3: In this step amide group is changed into an amidine, the nitrogen being donated by glutamine. This reaction which yields 5-phosphoribosyl-N-formyleglycine amidine or formylglycinamide ribotide(FGAM) is driven by the coupled hydrolysis of ATP to ADP + Pi.

Step 5 : Formation of purine imidazole ring 5-phosphoribosyl-N-formylglycineamidine cyclizes to 5-phosphoribosyl-5-aminioimidazole or 5-aminoimidazoleribotide (AIR). This cyclization or aromatization of the imidazole ring is facilitated by the tautomeric shift of the reactant from its imine to its enamine form.

Step 6 : Acquisition of C6: In this step 5-phosphoribosyl-5-aminoimidazole is carboxylated to form 5–phosphoribosyl – 5 aminoimidazole – 4 – carboxylic acid or carboxyaminoimidazole ribotide (CAIR). Carbon (C6) is introduced as HCO_3^- (CO_2) by phosphoribiosylaminoimidazole carboxylase.

Step 7 : Acquisition of N1: Purine atom N1 is contributed by aspartate in an amide forming condensation reaction yielding 5-phosphoribosyl-4 (N-succinocarboxyamide)-5-aminoimidazole or 5 - aminoimidazole - 4 - (N-succinylocarboxamide) ribotide (SACAIR). It is an ATP dependent step.

Step 8 : Elimination of fumarate: In this step, fumarate is removed, leaving the nitrogen on the chain, to form 5-phosphoribosyl-4-carboaminoimidazole or 5-aminoimidazole-4-carbixamide ribotide (AICAR). The reaction is catalyzed by adenylosuccinate lyase.

Step 9 : Acquisition of C2: The final purine is finished through formylation by N10-formyl-THF, yielding 5-phosphoribosyl-4-carboxamide-5-foraminoimidazole or 5-formaninoimidazole-4-carboxamide ribotide (FAICAR)

Step 10 : Cyclization to form IMP: The final reaction in purine biosynthetic pathway, ring closure to form IMP, occurs through the release of water. the reaction catalyzed by IMP cyclohydrolase does not require ATP.

The purine synthesized through the above series of reaction is hypoxanthine.

ii. Biosynthesis of AMP and GMP from IMP

IMP does not accumulate in the cell but is rapidly converted to AMP and GMP (Fig. 21.5). AMP which differs from IMP only in the replacement of its 6-ketogroup by an amino group, is synthesized in a two-reaction pathway. In the first reaction, aspartate's amino group is linked to IMP in a reaction powered by the hydrolysis of GTP to GDP + Pi to yield adenylosuccinate. The next reaction adenylosuccinate lyase eliminates fumarate from adenylosuccinate to form AMP. This reaction is similar to reaction 8 of IMP synthesis.

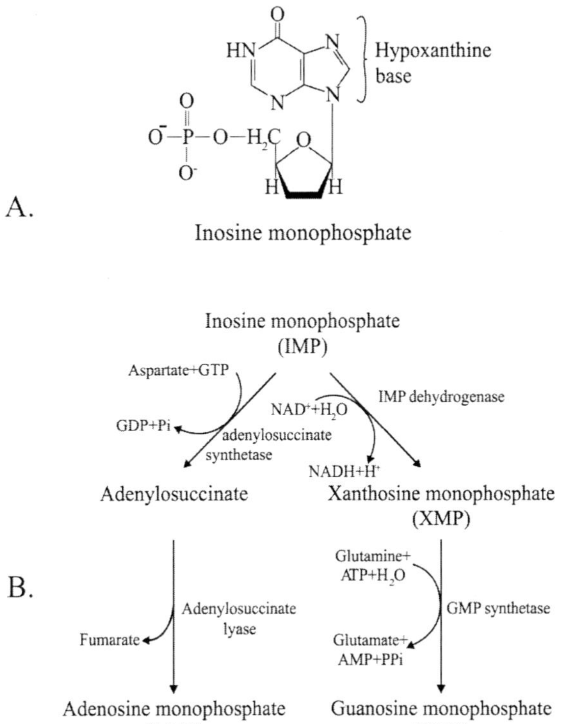

Fig. 21.5: A. Structure of IMP. B. Synthetic pathways of AMP and GMP from AMP

GMP is also synthesized from IMP in a two-reaction pathway. In the first reaction IMP is dehydrogenated to form xanthosine monophosphate (XMP). XMP is then converted to GMP by the transfer of the glutamine amide nitrogen in a reaction driven by the hydrolysis of ATP to AMP + PPi and subsequently to 2pi

iii. Nucleoside monophosphates are converted to nucleoside triphosphates

Nucleoside monophosphates (NMP) to be used in biosynthesis of RNA and DNA are first converted to nucleoside triphosphates (NTP). The conversion pathways are similar in all cells. Phosphorylation of AMP to ADP is brought out by adenylate kinase.

$$\text{ATP} + \text{AMP} \xrightarrow[\text{Adenylate kinase}]{} 2\text{ADP}$$

The ADP so formed is phosphorylated to ATP by substrate level phosphorylation or oxidative phosphorylation

$$\text{ADP} + \text{Pi} \longrightarrow \text{ATP}$$

ATP also brings about the formation of other nucleoside diphosphates by the action of nucleoside monophosphate kinases, which are specific for a particular base, but nonspecific for the sugar nucleoside diphosphates are converted to triphosphates by the action of a ubiquitous enzyme nucleoside diphosphate kinase. This enzyme is nonspecific for the base (purines or pyrimidines) or the sugar (ribose or deoxyribose).

$$\text{ATP} + \text{NMP} \rightleftharpoons \text{ADP} + \text{NDP}$$

$$\text{NTP (donor)} + \text{NDP (acceptor)} \rightleftharpoons \text{NDP (donor)} + \text{NTP (acceptor)}$$

iv. Deoxyribonucleotides are derived from ribonucleotides

Deoxyribonucleotides, the building blocks of DNA are derived from the corresponding ribonucleotides, rather than by their *de novo* synthesis from deoxy ribose-containing precursors, by direct reduction at the 2'-carbon atoms of the D-ribose to form the 2'-deoxy derivatives. For example, adenosine diphosphate (ADP) is reduced to 2' deoxy adenosine diphosphate (dADP) and GDP is reduced to GDP. The reaction is catalyzed by ribonucleotide reductare, the best characterized in *E. coli*, in which its substrates are ribonucleotide diphosphates. The reduction of the D-ribose portion of ribonucleoside diphosphate to 2'deoxyribose requires a pair of hydrogen atoms, which are ultimately donated by NADPH via an intermediate hydrogen carrying protein, thioredoxin.

$$\text{ADP} \longrightarrow \text{dADP}$$

$$\text{GDP} \longrightarrow \text{dGDP}$$

$$\text{CDP} \longrightarrow \text{dCTP}$$

$$\text{UDP} \longrightarrow \text{dUDP}$$

v. Thymine monophosphate is derived from CDP and dUMP

DNA contains thymine rather than uracil. The dTMP component of DNA is synthesized by methylation of dUMP. The dUMP is generated through the hydrolysis of dUTP by dUTP di phosphohydrolase (dUTPase). Conversion of dUMP to dTMP is catalyzed by thymidylate synthetase (Fig. 21.6). dTMP, once it is formed, is phosphorylated to form dTTP.

21.4. SALVAGE PATHWAY FOR SYNTHESIS OF NUCLEOTIDES

During the metabolic degradation of free nucleotides and nucleic acids especially RNA, purine and pyrimidine bases are constantly released in the cells. Most of the free purines are salvaged and reused to make nucleotides, in a pathway much simpler than the *de novo* synthesis of purine nucleotides. This pathway consists of single reaction catalyzed by adenosine phosphoribosyl transferase, in which free adenine reacts with PRPP to yield the corresponding adenine nucleotide.

$$\text{Adenine + PRPP} \xrightarrow[\text{Adenosine phosphoribosyltransferase}]{} \text{AMP + PPi}$$

Free guanine and hypoxanthine are salvaged in the same way by hypoxanthine –guanine phosphoribosyltransferase.

A similar pathway is reported to exist for pyrimidine bases in microorganisms.

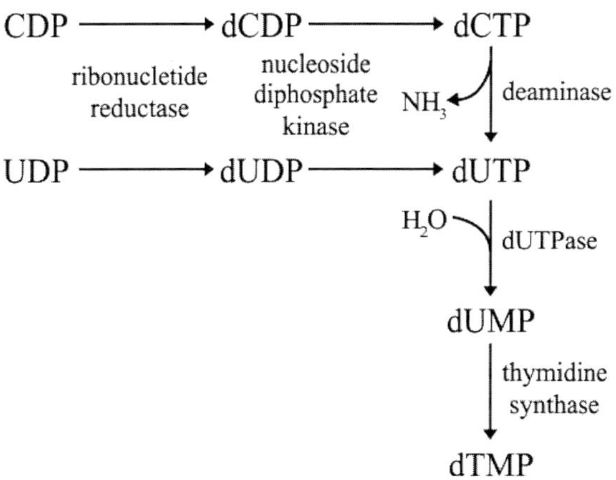

Fig. 21.6 : Formation of thymidine monophosphate from CDP and dUMP

21.5 DEGRADATION OF NUCLEOTIDES

Two major classes of nucleotides (purine and pyrimidine) are degraded through different pathways. The degradation of purines yields uric acid whereas degradation of pyrimidines yields malonyl CoA.

21.5.1 Degradation of Purines

A summary of degradation pathways of four purine nucleotides is presented in Fig. 21.7.

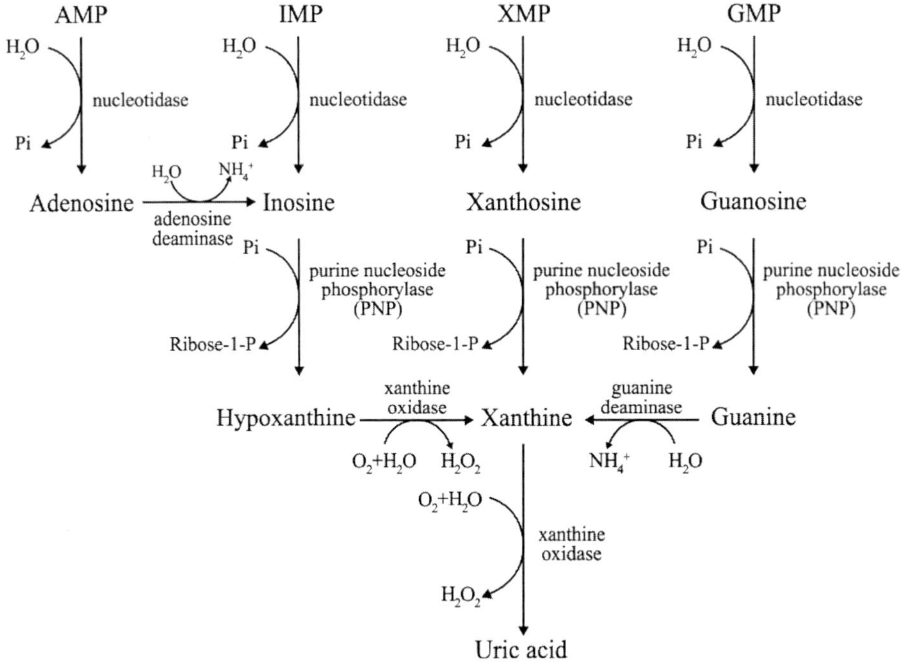

Fig. : 21.7: Degradation of purine nucleotides to uric acid

Purine nucleotides are degraded by a pathway in which they lose their phosphate through the action of 5'nucleotidase. Adenylate upon hydrolysis yields adenosine, which is then deaminated to inosine by adenosine deaminase. Similarly, inosine is hydrolyzed to hypoxanthine and D-ribose. Hypoxanthine is oxidized successively to xanthine and then uric acid by xanthine oxidase.

GMP catabolism also yields uric acid as the end product. GMP is first hydrolysed to guanosine, which is then cleaved to free guanine. Guanine undergoes hydrolytic removal of its amino group to yield xanthine, which is finally converted to uric acid by xanthine oxidase.

As shown in Fig. 21.8 uric acid by a series of reactions is finally degraded to ammonia via allantois, allantoic acid and urea. In animal systems, these compounds are excreted: uric acid (primates, birds, reptiles, insects), allantoin (mammals other than primates), allantoic acid (teleost fish), urea (cartilaginous fish, amphibian) and ammonia (marine invertebrates and fresh water fish). However, in bacteria glyoxylate appears to be a major degradation product of purine bases. In bacteria under aerobic conditions, the allontoic acid is further metabolized by one of the two reactions illustrated in Fig. 21.9. *Pseudomonas aeruginosa*, *P. fluorescense* and *Penicillium* species form first 2 mol of urea and 1 mole of glyoxylate. *P. acidovorans*, *Streptococcus allantoicus*, *Arthobacter allantoicus* and *E. coli* on the other hand, first remove ammonia and CO_2 from allontoic acid, so that finally, 1mol of urea per mole of glyoxylate is formed. These differences are probably not important for microorganisms because they synthesize the enzyme urease if ammonia becomes growth limiting. Urease cleaves urea to ammonia and CO_2.

$$\text{Urea} + H_2O \xrightarrow{\text{Urease}} 2NH_3 + CO_2$$

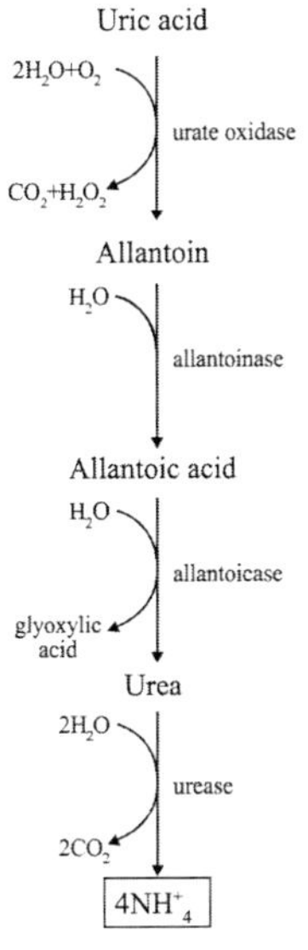

Fig. 21.8 : Degradation of uric acid to ammonia

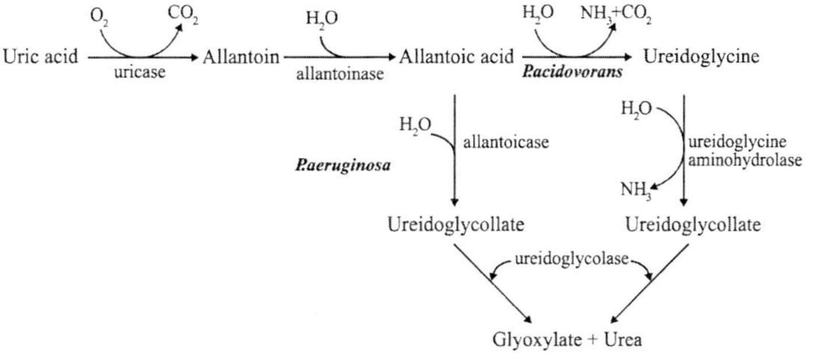

Fig. 21.9 : Degradation of uric acid to glyoxylate by *P. aeruginosa* and *P. acidovorans*

The glyoxalate formed from allontoic acid is converted into an intermediate of the central metabolic pathway. This pathway consists of three enzymes and is known as glycerate pathway (Fig. 21.10). The first enzyme, glyoxylate carboxylase, catalyzes the condensation of two molecules of glyoxylate to CO_2 and tartonate semialdehyde. The latter is reduced to glycerate which is phosphorylated to yield 3-phosphoglycerate, an intermediate of EMP pathway.

Paracoccus denitrificans metabolizes glyoxylate to oxaloacetate through a different pathway (Fig. 21.11). Here glycine is formed from glyoxylate by transamination. It then condenses with - β-hydroxyaspartate, which is converted to oxaloacetate and ammonia.

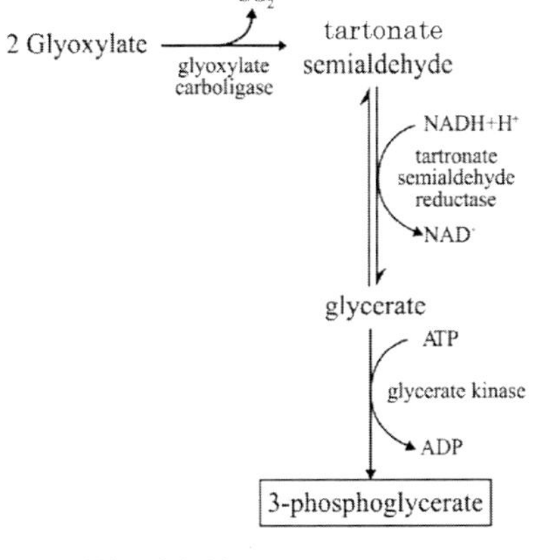

Fig. 21.10 : The glycerate pathway

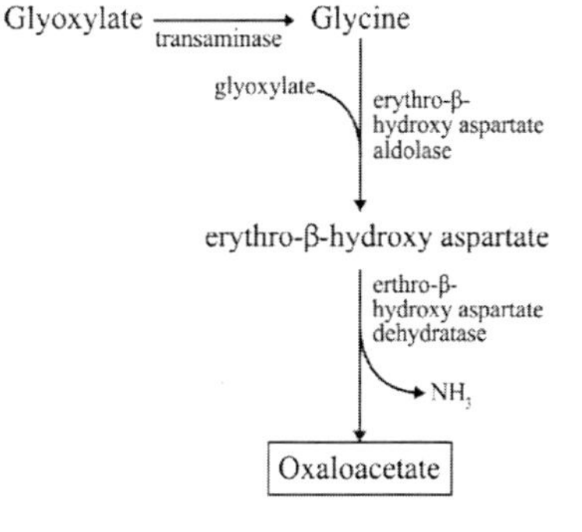

Fig. 21.11 : The β-hydroxyaspartate pathway

21.5.2 Degradation of Pyrimidines

Pyrimidine nucleotides are degraded to their component bases, uracil and thymine (Fig. 21.12). The reactions, like those of purine nucleotides, occur through dephosphorylation, deamination and glycoside bond cleavages. The resulting uracil and thymine are broken down through reduction rather than by oxidation as occurs in purine catabolism. The end products of pyrimidine catabolism, β-alanine and β-aminoisobutyrate, are amino acids and are metabolized as such. They may be converted through transamination and activation reactions, to malonyl-CoA and methylmalonyl-CoA.

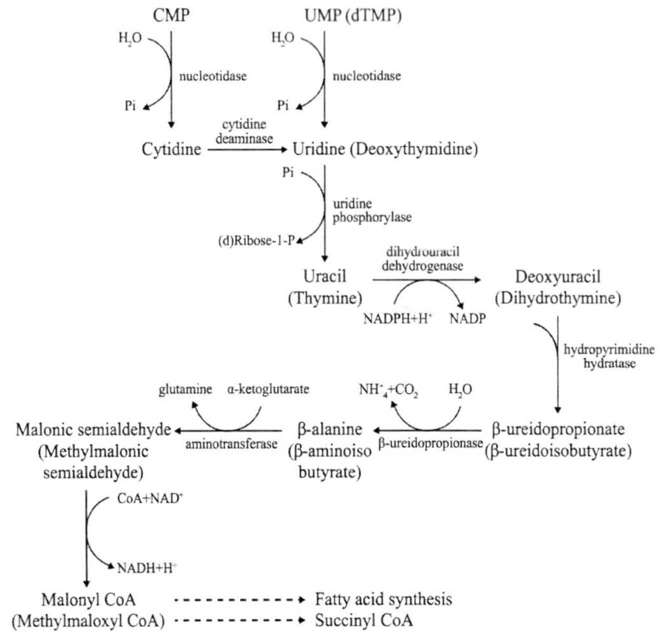

Fig. 21.12. Catabolism of pyrimidine nucleotides (cytidine, uridine and thymine). The pathway for dTMP degradation is shown in brackets

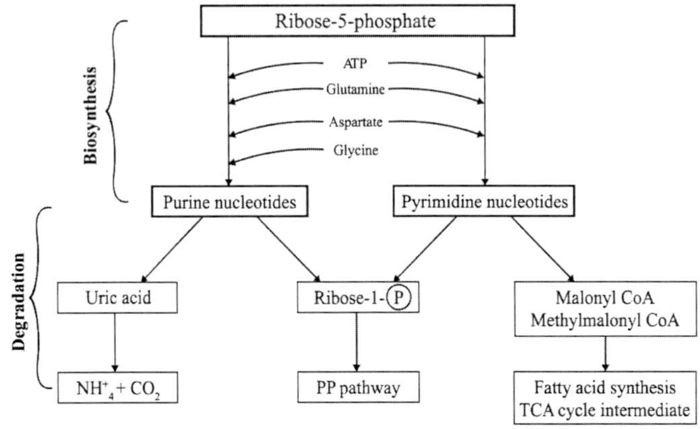

Fig. 21.13 : A summary of nucleotide metabolism

Methylmalonyl-CoA is a precursor of fatty acid synthesis (see section 22.2.1), and methylmalonyl-CoA is converted to the TCA cycle intermediate succinyl CoA. Thus, to a limited extent, catabolism of pyrimidine nucleotides provides the precursors or intermediates of other anabolic pathways.

Finally, a summary of nucleotide metabolism (biosynthesis and degradation) is shown in Fig. 21.13.

21.6 FURTHERS READING

Berg J., Tymocko, J. and Stryer (2003), Biochemistry 5th edn. WH Freeman & Co. Ltd.

Doelle, H.W. (1975), Bacterial Metabolism. 2nd edn. Academic Press, New York.

Gottschalk, G. (1986), Bacterial Metabolism 2nd edn. Springer – Verlag, New York.

Kappock, T.J., Ealick, S.E. and Stubbe J. (2000), Modular evolution of the furine biosynthetic pathway. Curr. Opin. Chem. Biol. 4 : 567-572.

Neidhardt, F.C. (ed.), 1996. *Escherichia coli* and *Salmonella* : Cellular and Molecular Biology 2nd edn. ASM Press, Washington, DC.

Nelson DL and Cox M.M. (2007), Lehninger Principles of Biochemistry 4th edn. W.H. Freeman and Co. New York.

Voet D.J. Voet, I.G. and Pratt, C.W. (2008), Principles of Biochemistry 3rd edn. John Wiley & Sons. Inc.

White, D. (1995), The Physiology and Biochemistry of Prokaryotes. Oxford University Press, New York.

■■■■

Chapter - 22

Lipid Metabolism

22.1 INTRODUCTION

Lipids (Greek: lipos, fat) are the fourth major group of molecules found in all living cells. They are primarily made of C, H, O, however, some complex lipids may contain nitrogen also. Lipids are in highly reduced state as compared to carbohydrates, proteins and nucleic acids, and hence are energy rich thus their biosynthesis requires lot of energy (endergonic) and oxidation releases a large amount of free energy (exergonic).

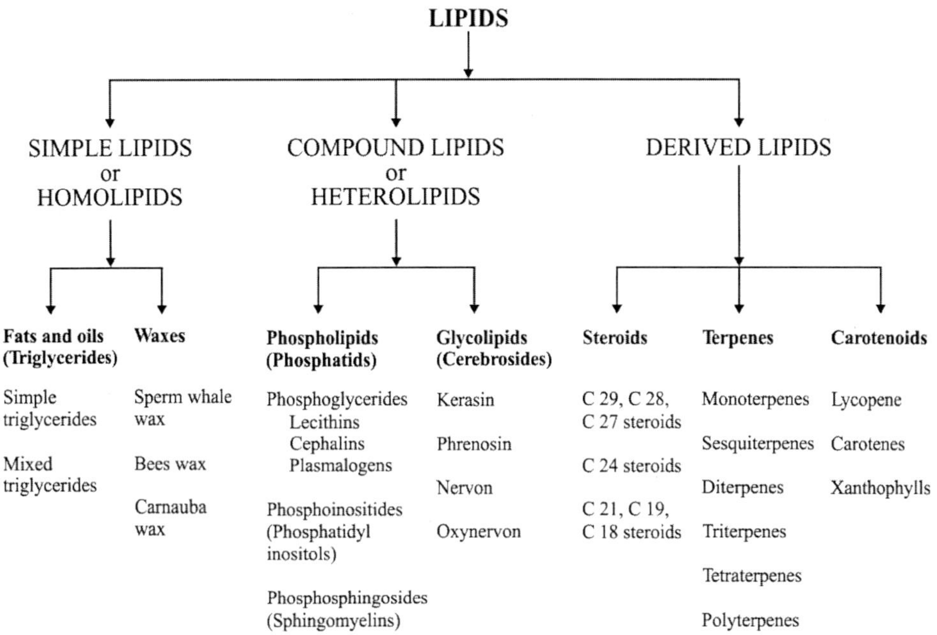

Fig. 22.1 : Classification of lipids

Lipids, though made of only three elements, exhibit greater structural variation than the other classes of biomolecules (Fig. 22.1). Lipids are not polymeric, but form aggregates and, in fact, in aggregate state only they perform most of their functions. Despite their great structural variation, they share the common property of being highly soluble in polar solvents (*e.g.* methanol, chloroform etc.) and relatively insoluble in water. Lipids play a variety of cellular roles which can be broadly categorized into three types:

i. Lipid molecules in the form of lipid bi-layers are essential components of biological membranes.

ii. Lipids are the principal form of stored energy in most organisms.

iii. Specialized lipids serve as pigments (retinal), cofactors (vitamin K), coenzymes (ubiquinone), detergents (bile salts), transporters (dolichols), hormones (vitamin D derivatives, sex hormones), extracellular and intracellular messengers (eicosanoids and derivatives of phosphatidylinositol) and anchors for membrane proteins.

The ability to synthesize a variety of lipids is, therefore, essential to all organisms. With respect to lipids, prokaryotes differ from eukaryotes in that (i) lipids are not the energy store houses, as their reserve food materials do not contain lipids or lipid derivatives, (ii) all lipids in prokaryotes are in the cell membrane. Hence, needless to mention, microbial physiologists and biochemists interest will be on membrane lipids. The major membrane lipids in prokaryotes (as well as eukaryotes) are phospholipids consisting of fatty acids esterified to glycerol phosphate derivatives and are called phosphoglycerides.

Fatty acids are the major components of both triacylglycerols and phospholipids. Triacylglycerols, the fatty acid triesters of glycerol, function as energy reservoirs in animals and plants. However, they are not components of cellular membranes. Phospholipids, on the other hand, are the major membrane lipids in prokaryotes as well as eukaryotes. They consist of glycerol-3-phosphate whose C_1 and C_2 positions are esterified with fatty acids. In addition, the phosphoryl group is linked to another usually polar group, X (Fig. 22.2). In either case, the fatty acids are important components. Hence, the major discussion in this chapter will be on the metabolism of fatty acids with less attention to other forms of lipids and their components.

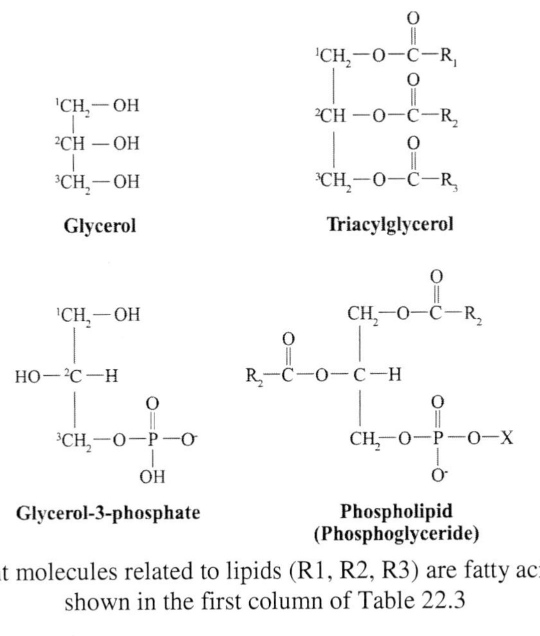

Glycerol **Triacylglycerol**

Glycerol-3-phosphate **Phospholipid (Phosphoglyceride)**

Fig. 22.2 : Some important molecules related to lipids (R1, R2, R3) are fatty acids, X may be any moiety shown in the first column of Table 22.3

22.2 FATTY ACIDS

Fatty acids are carboxylic acids with long chain hydrocarbon side groups. The fatty acids in cells are almost never found in the form of "free" fatty acids. They usually occur in esterified form as major components of the various lipids. The more common biological fatty acids are those of the C_{16} and C_{18} species. Fatty acids with less than 14 or more than 20 carbon atoms are uncommon. Most fatty acids have an even number of carbon atoms, because they are synthesized by the concatenation of C_2 units (acetyl-CoA). They can be branched (methyl), saturated, unsaturated or hydroxylated. Fatty acids differ with each other in the number of carbon atoms, the number of double bonds they contain, where the double bonds are located in the molecule and whether the molecule is branched or whether it is cyclopropane (Fig. 22.3; Table-22.1). Unsaturated fatty acids and branched chain fatty acids make the membrane more fluid, which is a necessary condition for membrane function. Bacterial fatty acids are rarely polyunsaturated but are commonly branched, hydroxylated, or contain cyclopropane rings. Generally, Gram-positive bacteria are richer in branched chain fatty acids than are Gram-negative bacteria.

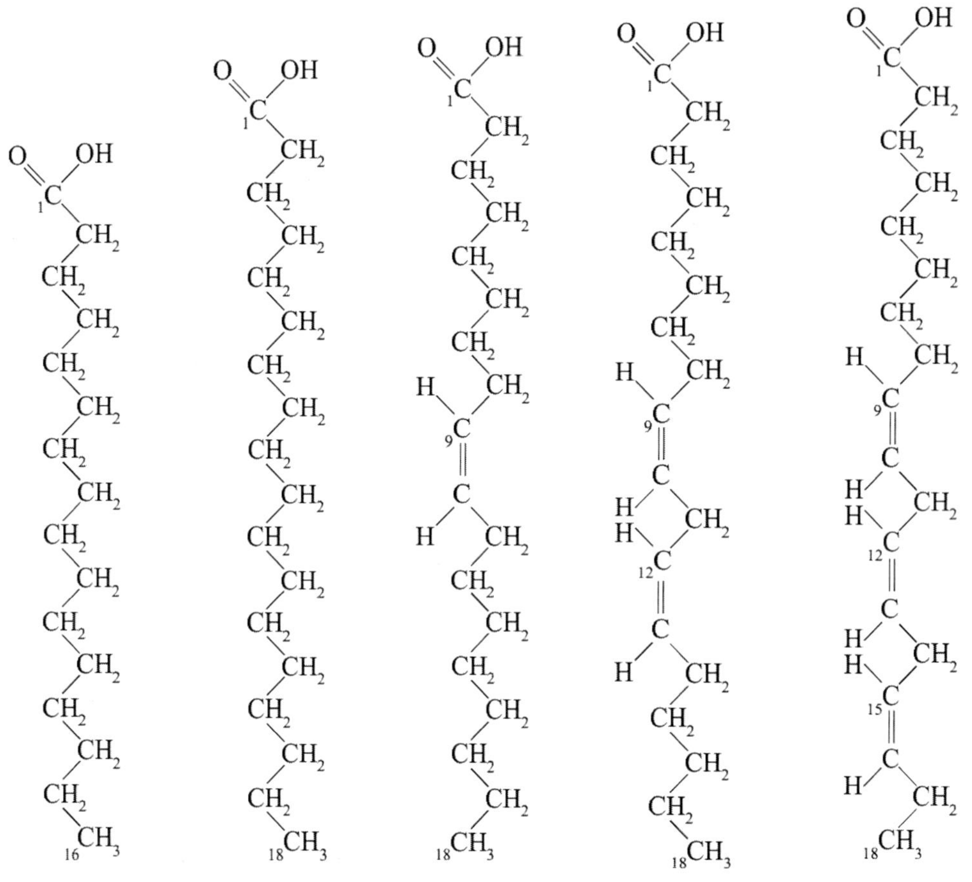

Palmitic acid **Stearic acid** **Oleic acid** **Linoleic acid** **α-Linoleic acid**

Fig. 22.3 : The structural formulas of some fatty acids.

Table 22.1 : The common biological fatty acids

Symbol[a]	Common Name	Systematic Name	Structure	mp (°C)
Saturated fatty acids				
12:0	Lauric acid	Dodecanoic acid	$CH_3(CH_2)_{10}COOH$	44.2
14:0	Myristic acid	Tetradecanoic acid	$CH_3(CH_2)_{12}COOH$	53.9
16:0	Palmitic acid	Hexadecanoic acid	$CH_3(CH_2)_{14}COOH$	63.1
18:0	Stearic acid	Octadecanoic acid	$CH_3(CH_2)_{16}COOH$	69.6
20:0	Arachidic acid	Eicosanoic acid	$CH_3(CH_2)_{18}COOH$	77
22:0	Behenic acid	Docosanoic acid	$CH_3(CH_2)_{20}COOH$	81.5
24:0	Lignoceric acid	Tetracosanoic acid	$CH_3(CH_2)_{22}COOH$	88
Unsaturated fatty acids (all double bonds are cis)				
16:1*n*-7	Palmitoleic acid	P-Hexadecanoic acid	$CH_3(CH_2)_5CH{=}CH(CH_2)_7COOH$	–0.5
18:1*n*-9	Oleic acid	9-Octadecanoic acid	$CH_3(CH_2)_7CH{=}CH(CH_2)_7COOH$	12
18:2*n*-6	Linoleic acid	9,12-Octadecadienoic acid	$CH_3(CH_2)_4(CH = CHCH_2)_2(CH_2)_6COOH$	–5
18:3*n*-3	α-Linolenic acid	9,12,15-Octadecatrienoic acid	$CH_3CH_2(CH{=}CHCH_2)_3(CH_2)_6COOH$	–11
18:3*n*-6	γ-Linolenic acid	6,9,12-Octadecatrienoic acid	$CH_3(CH_2)_4(CH{=}CHCH_2)_3(CH_2)_3COOH$	–11
20:4*n*-6	Arachidonic acid	5,8,11,14-Eicosatetraenoic acid	$CH_3(CH_2)_4(CH{=}CHCH_2)_4(CH_2)_2COOH$	–49.5
20:5*n*-3	EPA	5,8,11,14,17-Eicosapentaenoic acid	$CH_3CH_2(CH{=}CHCH_2)_5(CH_2)_2COOH$	–54
22:6*n*-3	DHA	4,7,10,13,16,19-Docosahexenoic acid	$CH_3CH_2(CH{=}CHCH_2)_6CH_2COOH$	–44
	Nervonic acid	15-Tetracosenoic acid	$CH_3(CH_2)_7CH{=}CH(CH_2)_{13}COOH$	39

[a]Number of carbon atoms: Number of double bonds. For unsaturated fatty acids, the quantity "*n*-*x*" indicates the position of the last double bond in the fatty acid, where *n* is its number of C atoms, and *x* is the position of the last double-bonded C atom counting from the methyl-terminal (ω) end. *Source:* LipidBank (http://www.lipidbank.jp)

As mentioned earlier, fatty acids do not occur free in bacteria, but are covalently attached to other molecules like glycerol, glycerol phosphate, carbohydrates, proteins.

i. Fatty acids esterified to glycerol form triacylglycerols or triglycerides, which occur in plants and animals in the form of fats and oils.

ii. Fatty acids esterified to glycerol phosphate and their derivatives form phosphoglycerides, which mostly occur in cell membranes of all organisms.

iii. Fatty acids may be esterified to carbohydrates to form glycolipids. In bacteria, for example, the lipid A portion of lipopolysaccharides consists of fatty acids esterified to glucosamine.

iv. Fatty acids may be esterified to proteins to form lipoproteins. For example, Gram-negative bacteria have a lipoprotein that is covalently attached to the peptidoglycan and protrudes into the outer membrane. Lipoproteins impart stability to the outer membrane.

22.2.1 Biosynthesis of Saturated Fatty Acids

David Ritternberg and Konrad Bloch, for the first time in 1945, demonstrated that fatty acids are synthesized by the condensation of C_2 units (acetyl-CoA) which are derived from acetic acid. Later, it was shown that both acetyl-CoA and bicarbonate are required, and that a C_3 unit malonyl-CoA, is an intermediate of fatty acid biosynthesis.

In eukaryotes, fatty acid biosynthesis takes place in the cytosol and the degradation occurs in the matrix of mitochondria. However, in prokaryotes as similar organelles are absent, both synthesis and degradation take place in cytosol itself. It is now clear that, with minor differences, the pathways of fatty acid biosynthesis are similar in all living organisms. All the reactions taking part in fatty acid biosynthesis, for convenience, can be broadly divided into two stages.

i. Formation of malonyl-CoA

ii. Synthesis of fatty acid by condensation of C_2 units.

The reactions of these two phases are catalyzed by two separate enzymes, acetyl-CoA carboxylase (ACC) and fatty acid synthase, respectively.

i. Formation of malonyl–CoA

Formation of malonyl–CoA (a C_3 compound) from acetyl-CoA, which is irreversible, is catalyzed by acetyl-CoA carboxylase (ACC). AC carboxylase has three functional regions: biotin carrier protein; biotin carboxylase, which activates CO_2 by attaching it to a nitrogen in the biotin ring in an ATP-dependent reaction; and transcarboxylase, which transfers activated CO_2 from biotin to acetyl-CoA, producing malonyl CoA. The reaction occurs in two steps; CO_2 activation and carboxylation.

$$\text{E-biotin} \xrightarrow[\text{biotin carboxylase}]{HCO_3^- + \text{ATP} \quad \text{ADP+Pi}} \text{E-biotin-}CO_2^- \xrightarrow[\text{transcarboxylase}]{\text{Acetyl CoA}} \text{Malonyl-CoA}$$

+ E-biotin (E- enzyme carrying biotin)

The *E. coli* acetyl-CoA carboxylase, which is a multisubunit protein, is regulated by guanine nucleotides so that fatty acid synthesis is coordinated with cell growth.

ii. Synthesis of fatty acid by condensation of C_2 units

The pathway of biosynthesis of palmitic acid (C_{16}) serves as a model for understanding the synthesis of all other fatty acids.

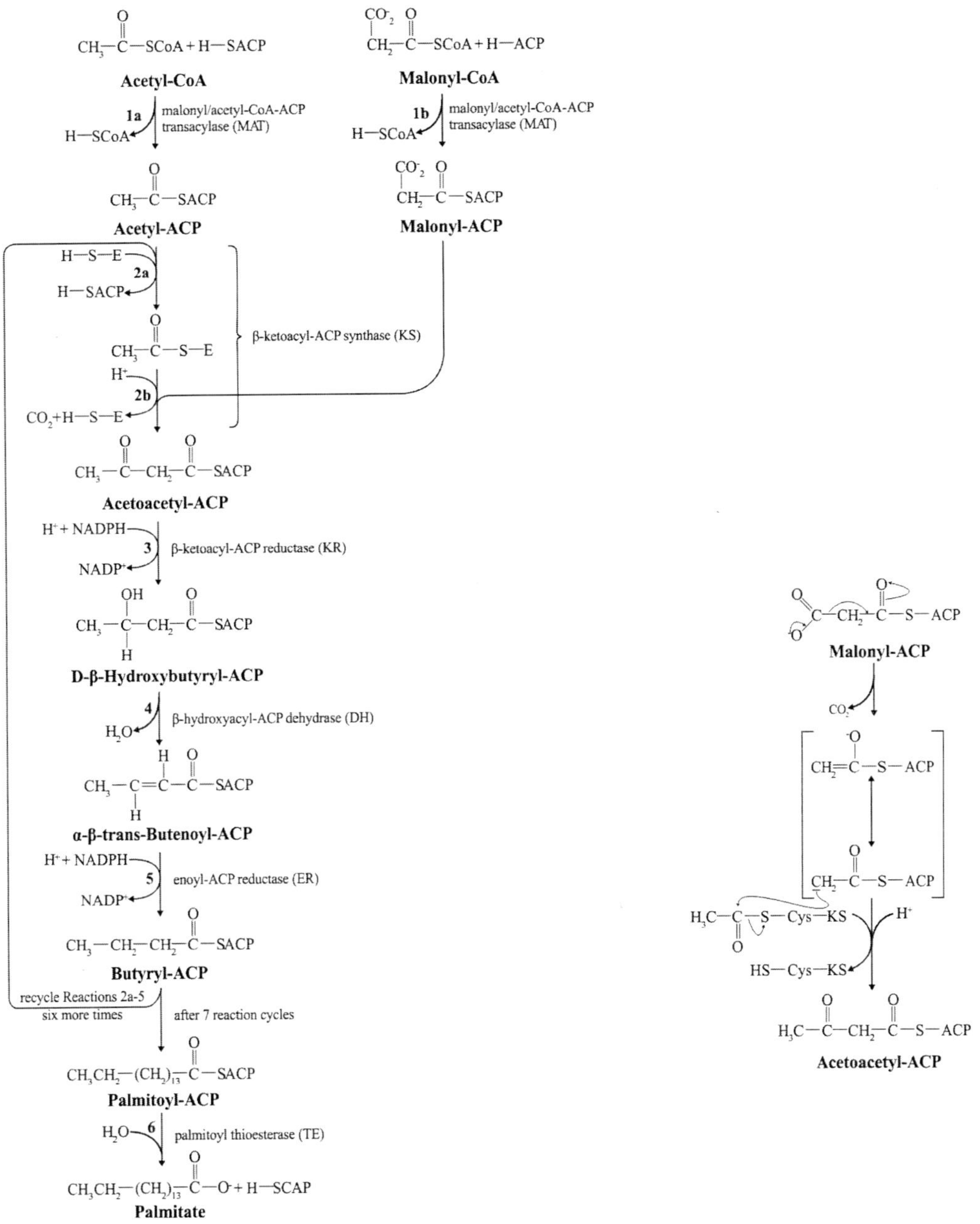

Fig. 22.4 : Reaction sequence for biosynthesis of fatty acids. Synthesis of acetoacetyl-ACP from malonyl-ACP is shown separately (right side)

The synthesis of fatty acids, mainly palmitic acid, from acetyl CoA and malonyl- CoA involves seven enzymatic reactions (Fig. 22.4). These reactions were first studied in cell-free extracts of *E. coli.* All the reactions in the synthetic process are catalyzed by a multi-enzyme complex, the **fatty acid synthase**. The detailed structure of this multi-enzyme complex and its location in the cell differ from one species to another, but the reaction sequence is identical in all organisms.

Fatty acid synthase: The fatty acid synthase system of *E. coli* consists of seven separate polypeptides that are tightly associated in a single organized complex (Fig. 22.5). The proteins act together to catalyze the formation of fatty acids from acetyl-CoA and malonyl-CoA (Table 22.2). Throughout the process, the intermediates remain covalently attached to one of two thiol groups of the complex. Fatty acid synthase complex of plants is located in chloroplasts (plant fatty acid synthesis does not occur in the cytosol). In yeasts the fatty acid synthase is cytosolic, 2500-KD multifunctional enzyme with the composition α_6 β_6, whereas in animals it is a 534-KD multifunctional enzyme consisting of two identical polypeptide chains.

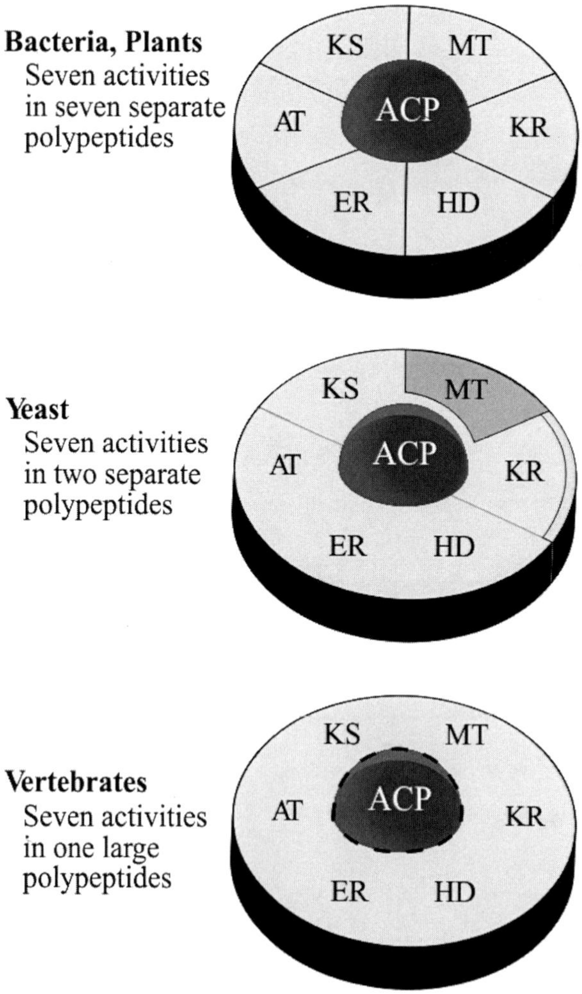

Fig. 22.5: Structures of fatty acid synthases in bacteria, plants yeast and vertebrates

Table 22.2 : Proteins of the fatty acid synthase complex of *E. coli*

Component	Function
Acyl carrier protein (ACP)	Carries acyl groups in thioester linkage
Acetyl-CoA-ACP transacetylase (AT)	Transfers acyl group from CoA to Cys residue of KS
β-Ketoacyl-ACP synthase (KS)	Condenses acyl and malonyl groups (KS has at least three isozymes)
Malonyl-CoA-ACP transferase (MT)	Transfers malonyl group from CoA to ACP
β-Ketoacyl-ACP reductase (KR)	Reduces β-keto group to β-hydroxyl group
β-Hydroxyacyl-ACP dehydratase (HD)	Removes H_2O from β-hydroxyacyl-ACR creating double bond
Enoyl-ACP reductase (ER)	Reduces double bond, forming saturated acyl-ACP

Biochemical reactions

The fundamental reaction sequence by which the long chains of carbon atoms in fatty acids are assembled consists of four steps (Fig. 22.4). The saturated acyl group produced during this set of reactions is recycled to become the substrate in another condensation with an activated malonyl group. With each passage through the cycle, the fatty acyl chain is extended by two carbons. When the chain length reaches 16, the product (palmite) leaves the cycle. The methyl and carboxyl carbon atoms of the acetyl group become C_{16} and C_{15}, respectively of the palmate; the rest of the carbon atoms are derived from malonyl-CoA.

Anchoring of growing chain to acetyl carrier protein (ACP)

Although fatty acid synthesis begins with the synthesis of a CoA ester, malonyl CoA, the growing fatty acid is anchored to **acyl-carrier protein (ACP)**. ACP, like CoA, contains a phosphopantetheine group that forms a thioester with an acyl group. In *E. coli,* ACP is a 10-KD polypeptide, whereas in animals it is a part of the multifunctional fatty acid synthase. The subsequent reactions catalyzed by fatty acid synthase are described here (Fig. 22.4).

i. In the priming reactions, synthase is loaded with the precursors for the condensation reactions. Acetyl group of malonyl-CoA and acetyl group of acetyl- CoA are transferred to ACP by malonyltransacetylase and acetyl transacetylase, respectively (Ia and Ib).

ii. An enzyme β-ketoacyl – ACP synthase (KS; also known as condensing enzyme) first transfers the acetyl group from ACP to an enzyme Cys residue (2a). In the condensation reaction (2b), the malonyl-ACP is decarboxylated and the resulting carbanion attacks the acetyl-thioester to form a four-carbon acetoacety-ACP (ii, iii & iv). Two reductions (3, 5) and a dehydration (4) convert the β-keto group to an alkyl group. The coenzyme in both reductive steps is NADPH.

At this point, the acyl group, originally an acetyl group, has been elongated by a C_2 unit. This butyryl group is then transferred from ACP to the Cys-SH of the enzyme (a repeat of reaction 2a) so that it can be extended by additional rounds of the fatty acid synthase reaction sequence. In each reaction cycle, the ACP is reloaded with malonyl group, and the acyl chain grows by two carbon atoms. It may be noted that each new acetyl unit donated by malonyl-CoA adds to the growing acyl chain at its point of attachment to the enzyme. Thus, the fatty acid grows from its thioester end, not from its methyl end.

Seven cycles of elongation are required to form palmitoyl–ACP. The thioester bond is then hydrolyzed by palmitoylthioesterase, yielding palmitate and regenerating the enzyme for a new round of synthesis. The acyl portion may be immediately transferred to membrane phospholipids or used for lipid A biosynthesis.

The stoichiometry of palmitate synthesis is

$$\text{Acetyl-CoA}+7\text{ malonyl-CoA}+14\text{ NADPH}+7H^+ \longrightarrow \text{palmitate}+7CO_2+14\text{ NADP}^+ + 8\text{ CoA}+6H_2O$$

However, the 7 malonyl CoA are derived from acetyl CoA as follows:

$$7\text{ Acetyl-CoA} + 7\ CO_2 + \text{ATP} \longrightarrow 7\text{ malonyl-CoA} + 7\text{ ADP} + 7\text{ Pi} + 7H^+$$

Therefore, the overall stoichiometry for palmitate biosynthesis is

$$8\text{ Acetyl-CoA} + 14\text{ NADPH} + 7\text{ ATP} \longrightarrow \text{Palmitate} + 14\text{ NADP}^+ + 8\text{ CoA} + 6\ H_2O + 7\text{ ADP} + 7\text{ Pi}$$

22.2.2 Synthesis of Unsaturated Fatty Acids

Eukaryotic fatty acids are mostly polyunsaturated, whereas bacterial fatty acids if occur, are mostly monounsaturated. Unsaturated fatty acids are formed in two different ways depending upon the organism.

i. The anaerobic pathway

ii. The aerobic pathway

i. Anaerobic pathway: This pathway is restricted to prokaryotes. It is widespread in bacteria, found in *Clostridium, Lactobacillus, Escherichia, Pseudomonas,* the cyanobacteria and photosynthetic bacteria. The two-common unsaturated fatty acids in bacteria synthesized in this pathway are palmitoleic acid which has 16 carbons (cis– Δ^9–hexa–decenolic acid) and cis–vaccinic acid which has 18 carbons (cis–Δ^{11}– octadecenoic acid). In palmitoleic acid the double bond is between C_9 and C_{10} in the final product. As vaccinic acid is two carbons longer, the double bond is between C_{11} and C_{12}.

In this pathway, a special dehydratase, 3-hydroxydecenoyl–ACP dehydratase, desaturates the C_{10} hydroxyacyl–ACP intermediate to the trans–α, β–decenoyl–ACP which is then isomerized to cis – β, γ – ACP. In the β-γ derivative the double bond is between C_3 and C_4. This derivative instead of undergoing reduction, elongates leading to fatty acid with a double bond.

ii. Aerobic pathway: The aerobic pathway is found in *Bacillus, Mycobacterium, Corynebacterium, Micrococcus* and in all eukaryotes. In this pathway, a special desaturase (oxidase) introduces a double bond into the completed fatty acyl-CoA or fatty acyl–ACP derivative. The enzyme system requires molecular oxygen and NADPH. Oleic acid (cis – Δ^9 –

octadecenoic acid), which has double bond between C_9 and C_{10}, is formed from stearic acid, a C_{18} saturated progenitor.

22.3. PHOSPHOLIPIDS AND PHOSPHOGLYCERIDES

Phospholipids are abundant in all biological membranes. A phospholipid molecule is constructed from four components: fatty acids, a platform to which the fatty acids are attached, a phosphate and an alcohol attached to the phosphate. The platform on which phospholipids are built may be glycerol, a 3-carbon alcohol, or sphinosine, a more complex alcohol. Phospholipids derived from glycerol are called **phosphoglycerides.** Thus, a phosphoglyceride consists of a glycerol backbone to which two fatty acid chains and a phosphorylated alcohol are attached (Fig. 22.6).

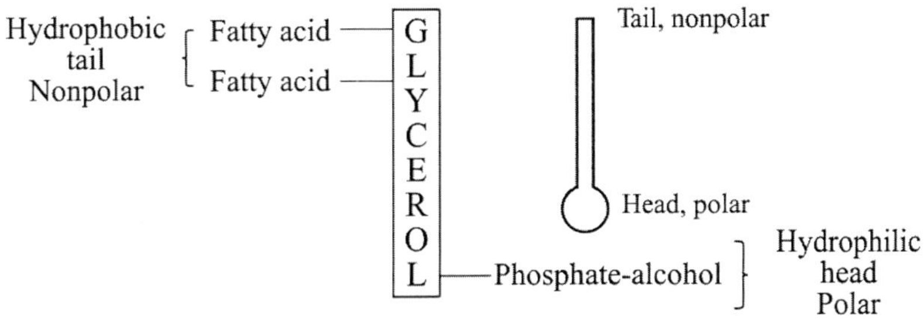

Fig. 22.6 : Outline structure of phosphoglyceride

In phosphoglycerides, the hydroxyl groups at C_1 and C_2 of glycerol are esterified to the carboxyl groups of the two fatty acid chains. The C_3 hydroxyl group of the glycerol backbone is esterified to phosphoric acid. When no further additions are made, the resulting compound is phosphatidate (diacylglycerol 3-phosphate), the simplest phosphoglyceride. Though it is in small amounts in the membranes, it is a key intermediate in the biosynthesis of the other phosphoglycerides. The major phosphoglycerides are derived from phosphatidate by the formation of an ester bond between the phosphate group of phosphatidate and the hydroxyl group of one of several alcohols. The common alcohol moieties of phosphoglycerides are the amino acid serine, ethanolamine, choline, glycerol and the inositol (Table 22.3). Phosphoglycerides are amphiphilic molecules with nonpolar aliphatic (hydrocarbon) "Tails" and polar phosphoryl – X "Heads".

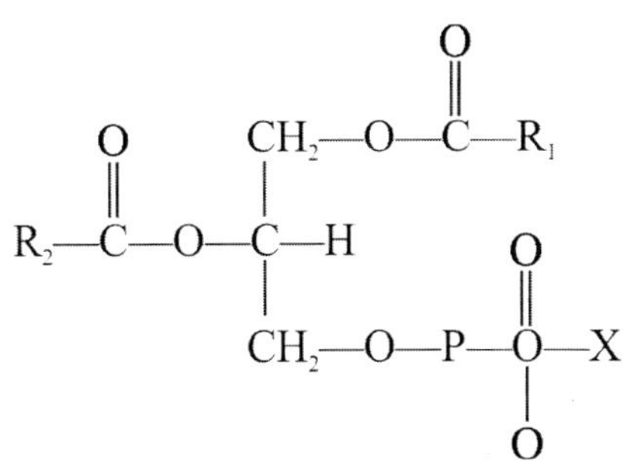

Table 22.3. The common of glylcerophospholipids

Name of X-OH	Formula of —X	Name of phospholipids
Water	—H	Phosphatidic acid
Ethanolamine	$—CH_2\,CH_2\,NH_3^+$	Phosphatidylethanolamine
Choline	$—CH_2CH_2N(CH_3)_3^+$	Phosphatidylcholine (lecithin)
Serine	$—CH_2CH(NH_2^+)COO^-$	Phosphatidylserine
myo-Inositol	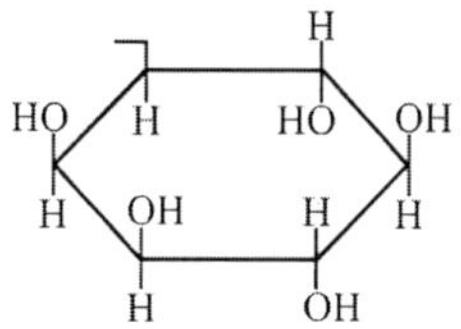	Phosphatidylinositol
Glycerol	$—CH_2CH(OH)CH_2\,OH$	Phosphatidylglycerol
Phosphatidylglycerol	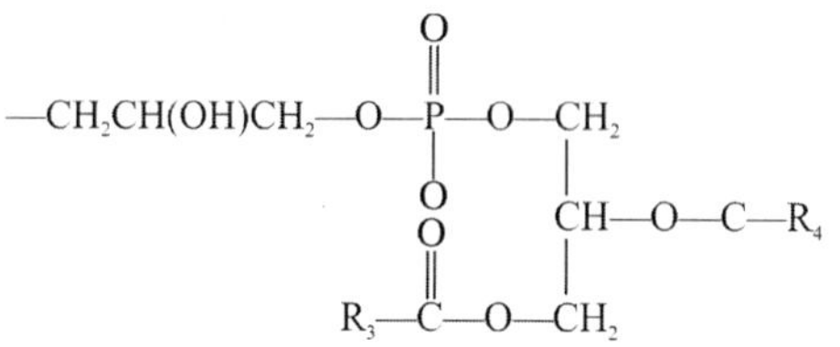	Diphosphatidylglycerol (Cardiolipin)

The tail part of the molecule is hydrophobic and head part is hydrophilic. The amphibolic nature of the phospholipids explains their orientation in membranes. Phospholipids are oriented in membrane as a bilayer in such a way that hydrophilic head regions are pointed towards the aqueous phase and the hydrophobic fatty acid tails interacting with each other in a hydrophobic interior.

Many different phospholipid species can be constructed by combining various fatty acids and polar head groups with the glycerol backbone. Individual phosphoglycerides are named according to the identities of these fatty acid residues.

22.3.1 Synthesis of Phosphoglycerides

All the biosynthetic pathways destined to produce different phosphoglycerides follow a few basic patterns. In general, the assembly of phosphoglycerides from simple precursors requires: i. synthesis of the backbone molecule, ii. attachment of fatty acid(s) to the backbone through an ester or amide linkage, iii. addition of hydrophilic head group to the backbone through a phosphodiester linkage and iv. alteration or exchange of head groups to yield the final phospholipid product.

Triacylglycerols and phosphoglycerides are synthesized from the same precursors

Synthesis of both triacylglycerols and phosphoglycerides starts with dihydroxyacetone phosphate (DHAP), an intermediate in glycolysis. DHAP is reduced to glycerol 3-phosphate by an enzyme glycerol phosphate dehydrogenase

$$\text{DHAP} \xrightarrow[\text{biotin carboxylase}]{\text{NADH+H}^+ \quad \text{NAD}^+} \text{Glycerol 3-phsophate}$$

In the next step, transfer of fatty acids from fatty acyl-S-ACP to glycerol 3-phosphate. The reactions are catalyzed by membrane bound enzymes called G3P acyl transferases. Some bacteria are also capable of using acyl-SCoA derivatives as the acyl donor. The first phospholipid made is phosphatidic acid, which is glycerol phosphate esterified to two fatty acids. The other phospholipids are made from phosphatidic acid.

$$\text{Glycerol 3-phosphate} \xrightarrow[\text{G3P acyl transferase1-acyl-G3P acyltransfease}]{\text{R2-C-SACP}} \text{Phosphatidic acid}$$

Phosphatidic acid (phosphotidate) is the common intermediate in the synthesis of phospholipids and triacylglycerols

As stated earlier, phosphatic acid is the common precursor for synthesis of triacylglycerols and phospholipids. The pathways diverge at phosphatic acid. In the synthesis of triacylglycerols, phosphatic acid is hydrolyzed by a specific phosphatase to give a diacyl glycerol (DAG). This intermediate is acylated to a triacylglycerol in a reaction that is catalyzed by diglyceride acyltransferase. Both enzymes are associated in a triacylglycerol synthetase complex that is bound to membranes.

$$\text{Phsophatidic acid} \xrightarrow[\text{Phosphatase}]{H_2O \quad \text{Pi}} \text{diacylglycerol (DAG)} \xrightarrow[\text{Diglyceride acyl-transferase}]{\text{R2Co-CoA} \quad \text{CoA}} \text{Triacylglycerol}$$

Phospholipid synthesis requires the combination of a diacylglyceride with an alcohol. As in most anabolic reactions, one of the components must be activated. In this case, either of the two components may be activated, depending on the source of the reactants.

i. synthesis from an activated diacylglycerol.

ii. synthesis from an activated alcohol.

In *E. coli*, however, synthesis of phospholipid takes place through activated diacylglycerol. This strategy is illustrated by the synthesis of phosphatidylserine, phosphatidylethanolamine and phosphotidylglycerol (Fig. 22.7).

The diacylglycerol is activated by condensation of phosphatidic acid with cytidine triphosphate (CTP) to form CDP–diacylglycerol, with an elimination of pyrophosphate. Displacement of CMP through nucleophilic attack by the hydroxyl group of serine or by the C_1 hydroxyl of glycerol 3-phosphate yields phosphotidyl serine or phosphatidylglycerol 3-phosphate, respectively. The latter is processed further by cleavage of the phosphate monoester (with release of Pi) to yield phosphatidylglycerol.

Phosphatidylserine and phosphatidylglycerol can serve as precursors for other membrane lipids in bacteria. Decarboxylation of the serine moiety in phosphatidylserine, catalyzed by phosphatidylserine decarboxylase, yields phosphatidylethanolamine.

In *E. coli*, condensation of two molecules of phosphatidylglycerol, with elimination of one glycerol, yields cardiolipin, in which two diacylglycerols are joined through a common head group with release of one molecule of glycerol.

Pathways for synthesis of phosphatidylserine, phosphatidylethanolamine and phosphatidylglycerol, in yeasts, are like those in bacteria.

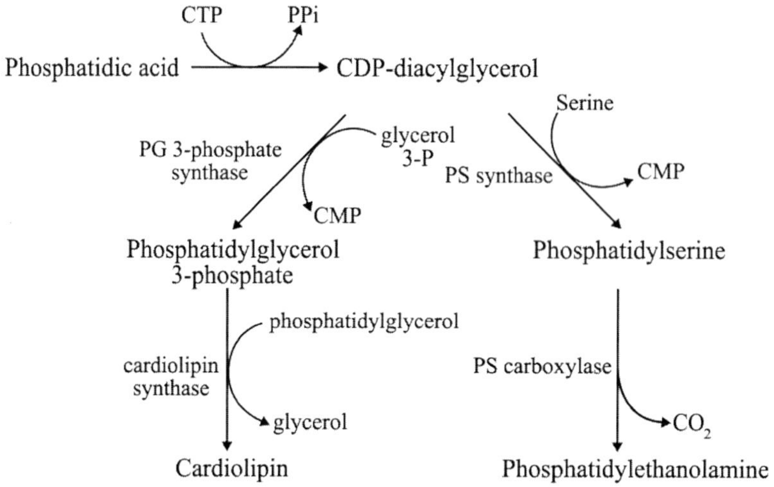

Fig. 22.7 : Synthesis of phosphadylethanolamine and cardiolipin from phosphatidic acid.

22.4 CATABOLISM OF FATTY ACIDS

Fatty acids are energy rich molecules due to their highly reduced state. In animal and plant systems, during the energy utilization stages, triacylglycerols and phosphoglycerides are broken down to fatty acids which in turn are oxidized and decarboxylated to CO_2, releasing ATP energy. Many bacteria like pseudomonads, various bacilli and *E. coli* can grow on long chain fatty acids. Thus, both in prokaryotic and eukaryotic organisms fatty acid oxidation assumes significance. β–oxidation is the most common pathway for degradation in all organisms despite the differences in cell organelles and reactions. Outlines of β–oxidation pathway of fatty acids is presented in Fig. 22.8.

Complete oxidation of fatty acids takes place in three stages. In the first stage, through β-oxidation, fatty acids undergo oxidative removal of successive two-carbon units in the form of acetyl-CoA starting from the carboxyl end of the fatty acyl chain. For example, 16-carbon palmitic acid undergoes seven passes through the oxidative sequence, in each pass losing two carbons as acetyl-CoA. At the end of seven cycles, the last two carbons of palmate (C_{15} and C_{16}) remain as acetyl-CoA. The overall result is the conversion of the 16-carbon chain of palmitate to eight 2-carbon acetyl groups of acetyl-CoA molecules. Formation of each acetyl-CoA requires removal of four hydrogen atoms (two pairs of electrons and four H+) from fatty acids moiety by dehydrogenases.

In the second stage, the acetyl-CoAs generated in the first stage are fed into TCA cycle and oxidized to CO_2. In the third stage, the reduced electron carries NADH and $FADH_2$ produced in first two stages feed the electrons to respiratory chain in which the electrons pass to oxygen with the concomitant phosphorylation of ADP to ATP. The energy released by fatty acid oxidation is thus conserved in ATP.

Though, β-oxidation is the universal pathway for oxidation of fatty acids, two less-general modes of fatty acids catabolism, α-oxidation and ω -oxidation are also known.

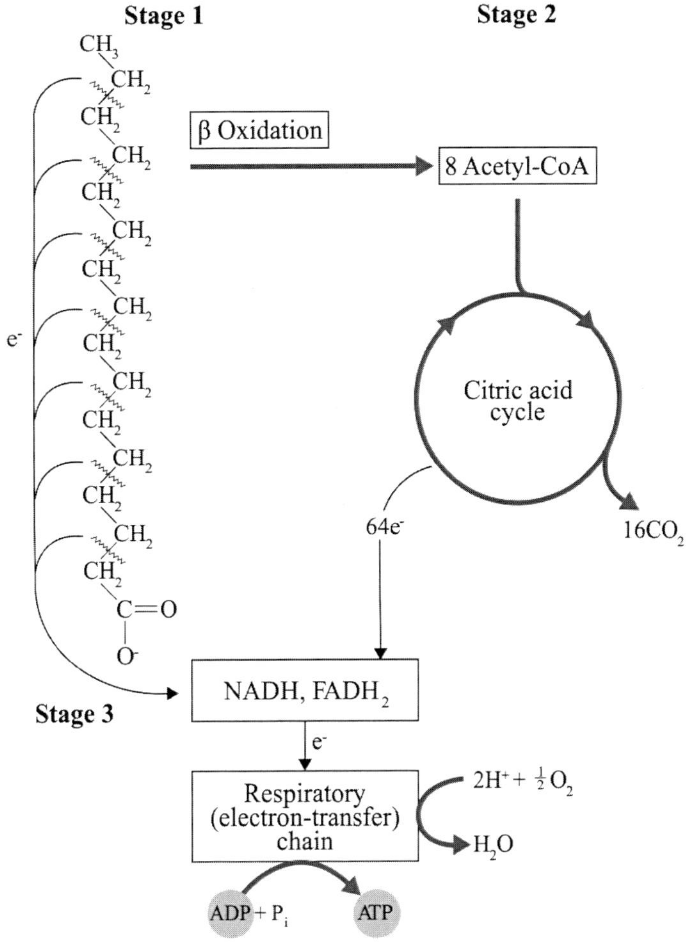

Fig. 22.8 : Stages of fatty acid oxidation.

22.4.1 β-Oxidation of Fatty Acids

In this pathway, the oxidation is on β-carbon, hence this series of reactions is called β-oxidation pathways.

Oxidation of saturated fatty acids

In the initial step, the fatty acids are activated by their attachment to coenzyme A. It is an ATP dependent acylation reaction to form fatty acyl-CoA. The activation process is catalyzed by acyl-CoA synthetase. Once the fatty acyl-CoA is formed its oxidation leading to release of acetyl CoA is completed in four enzyme catalyzed reactions (Fig. 22.9).

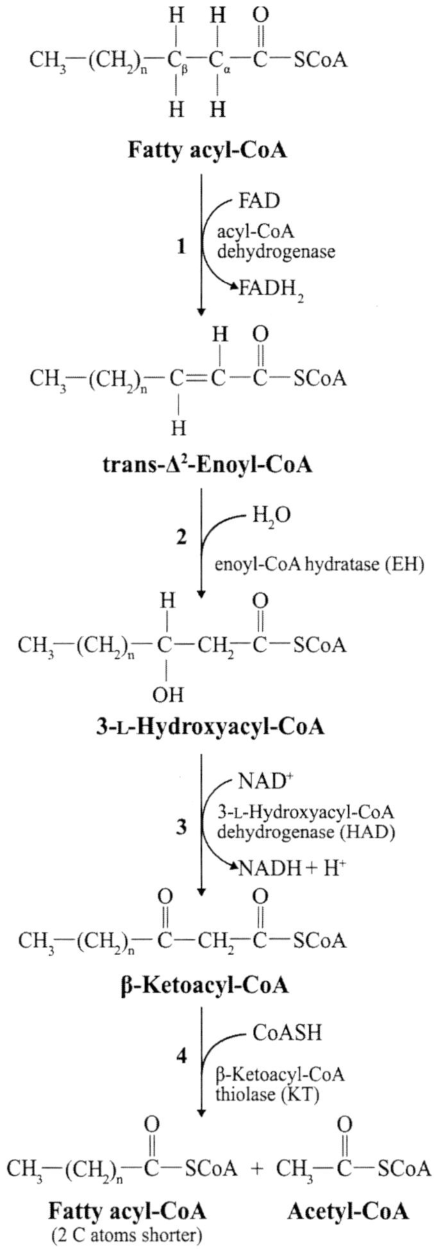

Fig. 22.9 : The β-oxidation pathway of fatty acyl-CoA

i. Formation of a trans-α, β double bond through dehydrogenation by the flavo enzyme fatty acyl-CoA dehydrogenase (AD).

ii. Hydration of the double bond by enoyl CoA hydratase (also called 3-hydroxyacyl CoA hydrolyase) to form a 3-L-hydroxyacyl-CoA.

iii. NAD^+-dependent dehydrogenation of the β-hydroxyacyl-CoA by 3-L-hydroxyacyl-CoA dehydrogenase (HAD) to form corresponding β-ketoacyl-CoA.

iv. C_α-C_β cleavage in a thiolysis reaction with CoA as catalyzed by β-ketoacyl-CoA thiolase to form acetyl-CoA and a new acyl-CoA containing two fewer C atoms than the original one.

The above reactions along with their enzymes are summarized in Table 22.4.

Table 22.4 : Principal reactions in fatty acid oxidation

Step	Reaction	Enzyme
1	Fatty acid + CoA + ATP ⇌ acyl CoA + AMP + PP_i	Acyl CoA synthetase [also called fatty acid thiokinase and fatty acid:CoA ligase (AMP)]
2	Acyl CoA + E-FAD → *trans*-Δ^2 -enoyl CoA + E-$FADH_2$	Acyl CoA dehydrogenases (several isozymes having different chain-length specificity)
3	*trans*-Δ^2-Enoyl CoA +H_2O ⇌ 1-3-hydroxyacyl CoA	Enoyl CoA hydratase (also called crotonase or 3-hydroxyacyl CoA hydrolyase)
4	1 -3-Hydroxyacyl CoA + NAD^+ ⇌ 3-ketoacyl CoA + NADH + H^+	1-3-Hydroxyacyl CoA dehydrogenase
5	3-ketoacyl CoA + CoA ⇌ acetyl CoA + acyl CoA (shortened by C_2)	β-Ketothiolase (also called thiolase)

In one series of reactions of β-oxidation, one molecule of acetyl-CoA, two pairs of electrons and four protons (H^+) are removed from the long chain fatty acyl-CoA, shortening it by two carbon atoms. The equation for one series of reactions, beginning with the coenzyme A ester of palmitate is

$$\text{Palmitoyl-CoA+CoA+FAD+NAD+}H_2O \longrightarrow \text{myristoyl CoA + acetyl CoA+}FADH_2\text{+NADH+}H^+$$

This series of reactions are repeated until the last two carbons (C15 and C16) remain in the form of acetyl CoA. Altogether, 7 series of reactions are required to oxidize one molecule of palmitoyl-CoA to eight molecule of acetyl CoA. The overall reaction is

$$\text{Palmitoyl-CoA+7CoA+7FAD+7NAD+7}H_2O \longrightarrow \text{8 acetyl CoA + 7}FADH_2\text{+ 7NADH + 7}H^+$$

Oxidation of acetyl CoA

The acetyl CoA produced from oxidation of fatty acids are further oxidized to CO_2 and H_2O by the TCA cycle. The following equation represents the balance sheet for second stage in the oxidation of palmitoyl CoA, including the coupled phosphorylations of the third stage.

$$\text{8 Acetyl-CoA+16 }O_2\text{+80 pi + 80 ADP} \longrightarrow \text{8 CoA + 80 ATP + 16 }CO_2\text{+ 16 }H_2O$$

For complete oxidation of palmitoyl CoA the yields of NADH, $FADH_2$ and ATP are summarized in Table 22.5.

Table 22.5. Yield of ATP during oxidation of one molecule of palmitoyl-CoA to CO_2 and H_2O

Enzyme catalyzing the oxidation step	Number of NADH or $FADH_2$ formed	Number of ATP ultimately formed
Acyl-CoA dehydrogenase	7 $FADH_2$	10.5
β-Hydroxyacyl-CoA dehydrogenase	7 NADH	17.5
Isocitrate dehydrogenase	8 NADH	20
Ketoglutarate dehydrogenase	8 NADH	20
Succinyl-CoA synthetase		8
Succinate dehydrogenase	8 $FADH_2$	12
Malate dehydrogenase	8 NADH	20
Total		108

As shown in the Table 22.5, complete oxidation of palmitate yields 108 molecules of ATP. However, the equivalent of 2 molecules of ATP is consumed in the activation of palmitate in which ATP is split into AMP and 2 molecules of Pi, thus, the complete oxidation of a molecule of palmitate yields 106 molecules of ATP.

Odd-chain fatty acid oxidation - End products propionyl CoA and Acetyl CoA

Fatty acids having an odd number of carbon atoms rarely occur. They are oxidized in the same way as fatty acids having an even number, except that propionyl CoA and acetyl CoA, rather than two molecules of acetyl CoA, are produced in the final round of degradation. In *E. coli* propionyl CoA is converted to acrylyl-CoA, which is hydrated to lactyl-CoA. The lactyl-CoA is oxidized to pyruvate, which can be oxidized via TCA cycle.

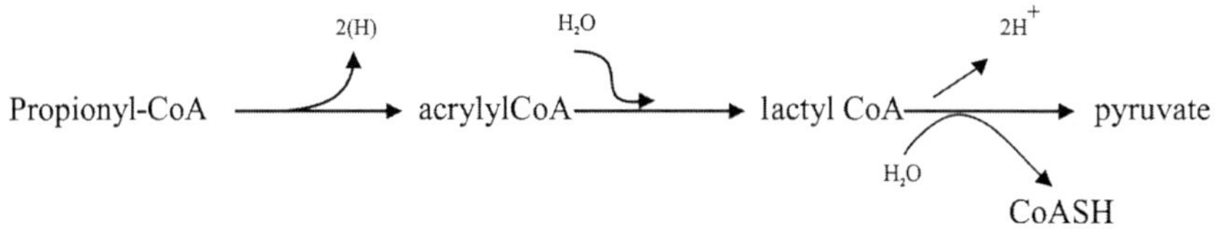

However, in plants and animals the propionoyl-CoA is converted with help of coenzyme B_{12} to succinyl-CoA. The latter is oxidized through TCA cycle.

Oxidation of unsaturated fatty acids

Most of the fatty acids in the triacyglycerols and phospholipids of animals and plants are unsaturated. However, in bacteria fatty acids if unsaturated are mostly monounsaturated. The double bonds are cis configuration and cannot be acted upon by enoyl-CoA hydratase, the enzyme catalyzing the addition of H_2O to the *trans* double bond of the Δ^2 enoyl-CoA generated during β-oxidation. Two more enzymes are required for β-oxidation of the common unsaturated fatty acids; an isomerase and a reductase. Oxidation of oleate, a C_{18} carbon mono unsaturated fatty acid with a double bond between C_9 and C_{10} is explained here.

In the first step, oleate is converted to oleoyl-CoA. Oleoyl-CoA then undergoes three passes through the fatty acid oxidation cycle to yield three molecules of acetyl-CoA and the coenzyme A ester of a Δ^3, 12-carbon unsaturated fatty acid, Cis-Δ^3 dodecenoyl- CoA. The problem is that this product cannot serve as a substrate for enoyl-CoA hydratase, which acts only on *trans* double bonds. The auxiliary enzyme $\Delta^3\Delta^2$ enoyl- CoA isomerase isomerizes the cisΔ^{3-}enoyl-CoA to the trans Δ-enoyl-CoA, which is converted by enoyl-CoA hydratase into corresponding L-β-hydroxyacyl-CoA. This intermediate is now readily acted upon by the remaining enzymes of β-oxidation to yield acetyl-CoA and the coenzyme-A ester of a 10-carbon saturated fatty acids, decanoyl-CoA. The latter undergoes four more series through the pathway to yield five more molecules of acetyl-CoA. Altogether nine acetyl-CoAs are produced from one molecule of the 18-carbon oleate.

The other enzyme, a reductase is required for oxidation of polyunsaturated fatty acids.

22.5 FATTY ACIDS SYNTHESIS Vs DEGRADATION

In most of the metabolic pathways, anabolic and catabolic reactions are almost reverse except at few points where an irreversible step exists, so that both reactions operate independently. For example, glycolysis has two irreversible reactions, the phosphofructokinase and pyruvate kinase reactions and these must be bypassed in order to reverse glycolysis. However, in some cases entirely two different sets of reactions are used for biosynthesis and catabolism of compounds. Fatty acid metabolism is one such example. Here, the biosynthetic pathway consists of reactions completely different from β-oxidation pathway. Some important differences between the pathways are:

1. In eukaryotes, the synthesis takes place in the cytosol, in contrast with degradation, which takes place primarily in the mitochondrial matrix. However, in prokaryotes both synthesis and degradation take place in the same compartment. *i.e.,* cytosol.
2. Intermediates in fatty acid synthesis are covalently linked to the sulfhydryl groups of an acyl carrier protein (ACP), whereas intermediates in fatty acid breakdown are covalently attached to the sulfhydryl group of coenzyme A.
3. The enzymes of fatty acid synthesis in higher organisms are joined in a single polypeptide chain called fatty acid synthase. In contrast, the degradative enzymes do not seem to be associated.
4. The growing fatty acid chain is elongated by the sequential addition of two-carbon units derived acetyl-CoA. The activated donor of two carbon units in the elongation is malonyl ACP. The elongation reaction is driven by the release of CO_2.
5. The reductant in fatty acid synthesis is NADPH, whereas the oxidants in fatty acid degradation are NAD^+ and FAD.
6. Elongation by the fatty acid synthase complex stops on formation of palmitate (C_{16}). Further elongation and the insertion of double bonds are carried out by another enzyme system.

22.6 FURTHER READING

Berg, J., Tymoczko, J. and Stryer, L. (2013), Biochesmistry (5th edn.) W.H. Freeman & Co. Ltd.

Gurr, M.I. Harwood, I.L. and Frayn, K.N. (2002), Lipid Biochemistry. An Introduction (5th edn.) Blackwell Publishing Co.

Harwood, I.L. (1988), Fatty acid metabolism Annu. Rev. Plant Physio1. Plant Mol. Biol 39: 101-138.

Helge, Jans, JanBen and Alexander Stein Luchel (2014), Fatty acid synthesis in *E. coli* and its applications towards the production of fatty acid based biofuels.

Hoekstra, D. (Ed.) (1994), Cell lipids. Current topics in membranes. (Vol. 4) Academic Press Inc. San Diego.

Kunau, W.H., Dommes, V. and Schulz, H. (1995), B-oxidation of fatty acids in mitochondria, peroxisomes and bacteria; a century of continued progress Prog. Lipid Res. 34: 267-342.

Magunson, K.S. et. al. (1993), Regulation of fatty acid synthesis in *E. coli*. Microbial Rev. 57: 522-542.

Nelson, D.L. and Cox, M.M. (2007), Lehringer Principles of Biochemistry (4th edn.) W.H. Freeman and Co., New York.

Raez, R.V. (Ed.) (2013), Lipid Metabolism : Open Access Book, in Tech Publishers DOI 10.5772/2928. Science, Oxford UK.

Vance, J.E. and Vans D. (Eds.) (2008), Biochemisrty of lipids, lipoproteins and Membranes (5th edn.) New Comprehensive Biochemistry Publishing Co. Inc. New York.

Victory, W.R. et al. (2015), Harpis Illustrated Biochemistry (30th edn.) The McGraw-Hill Education.

Voet, D.J., Voet J.G. and Pratt C.W. (2008), Principles of Biochemistry (3rd edn.) John Wiley & Son, Inc.

Zubay, G. (1997), Biochemistry (4th edn.) McGraw-Hill Education, Europe.

■■■■

Chapter - 23

Secondary Metabolism

23.1. INTRODUCTION

During the early years of twentieth century, microbiologists clearly identified that a microbial culture will pass through different phases *viz.,* lag phase, log or exponential phase, stationary phase and death phase. They further understood that in the logarithmic phase of growth there is extensive metabolism with the microbes rapidly replicating their cell components as a prerequisite to growth and cell division. It was also assumed that the stationary phase of growth represented virtually total metabolic inactivity on the part of the microorganisms. During the period 1920-1940, microbial biochemists found that stationary phase cultures of filamentous fungi to be rich and almost ever ending source of complex organic compounds. As the structures of these excreted compounds were described, it became apparent that they were not compounds which play an important role during exponential phase of microorganisms. Poineered by the works of J.D. Bu' Lock and A.L. Demain, several workers attempted to delineate the microbial metabolites produced during the different phases of growth and to link their importance in the growth and reproduction of microorganisms. Vigorous and intensified investigations eventually led to the concept of primary metabolites (the products of primary metabolism) and secondary metabolites (products of secondary metabolism). Though the criteria for distinguishing the primary and secondary metabolites are ambiguous and the delimitation is hazy, the terminology became well established in literature.

23.2. THE CONCEPT OF PRIMARY AND SECONDARY METABOLITES

Apart from kinetics of growth, the behaviour of a culture may also be described according to the products which it produces during the various stages of the growth. During the lag phase of growth the products produced are essential to the growth of cells which include amino acids, nucleotides, proteins, nucleic acids, lipids, carbohydrates etc. These products referred to as the products of **primary metabolism** (or **primary metabolites**) and the phase in which these products are produced is known as the **tropophase**. Primary metabolites are essentially required for growth and reproduction. The synthesis of primary metabolites by wild-type microorganisms is such that their production is sufficient to meet the requirements of the organism. They rarely accumulate since the primary metabolism is finely balanced.

During the deceleration and stationary phase, some microbial cultures synthesize compounds which are not produced during the tropophase and which do not appear to have any obvious function in cell metabolism. These compounds are referred to as the compounds of **secondary metabolism** (or secondary metabolites) and the phase in which they are produced is known as **idiophase** (equivalent to stationary phase (Fig. 23.1).

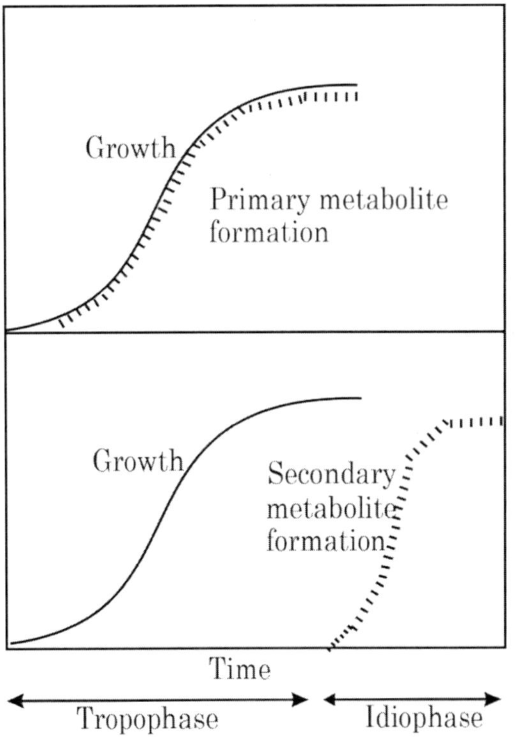

Fig. 23.1 : Production of primary and secondary metabolites in relation to growth.

23.3. CHARACTERS OF SECONDARY METABOLITES

Secondary metabolites are nonessential for growth and reproduction of the producer organism and their formation is highly influenced by the growth conditions, especially by the composition of the culture medium. In principle, they are expected to be produced in a batch culture. However, the secondary metabolites may be produced in continuous cultures at low growth rates. Many secondary metabolites have chemical structures that are infrequent in the biological world. These molecules belong to many classes of organic compounds: aminocyclitols, amino sugars, quinones, coumarins, epoxides, glutarimides, glycosides, non-ribosomal peptides, phenazines, polyacetylenes, polyenes, pyrroles, quinolines, terpenoids and tetracyclines. In addtion, secondary metabolites possess unusual chemical linkages, such as β-lactam rings, cyclic peptides made of normal and modified amino acids, the unsaturated bonds of polyenes, and the large rings of macrolides. As a result of the broad substrate specificity of some biosynthetic enzymes, secondary metabolites are typically produced as members of a particular family of compounds. Secondary metabolites are frequently species specific, and the mechanism of evolution of their biosynthetic genes is intriguing. The biosynthesis of secondary metabolites is closely related to the primary metabolism of the producer cells, since all secondary metabolites derive from the intermediates of primary metabolism (Fig. 23.2). In many cases, the intermediates of primary metabolism are converted to specific modifying reactions. It should also be noted that not all microorganisms carry secondary metabolism. It is common amongst filamentous bacteria and fungi, and the sporing bacteria but it is not found, for example, in the Enterobacteriaceae. Thus, the taxonomic distribution of secondary metabolism is quite different from that of primary metabolism.

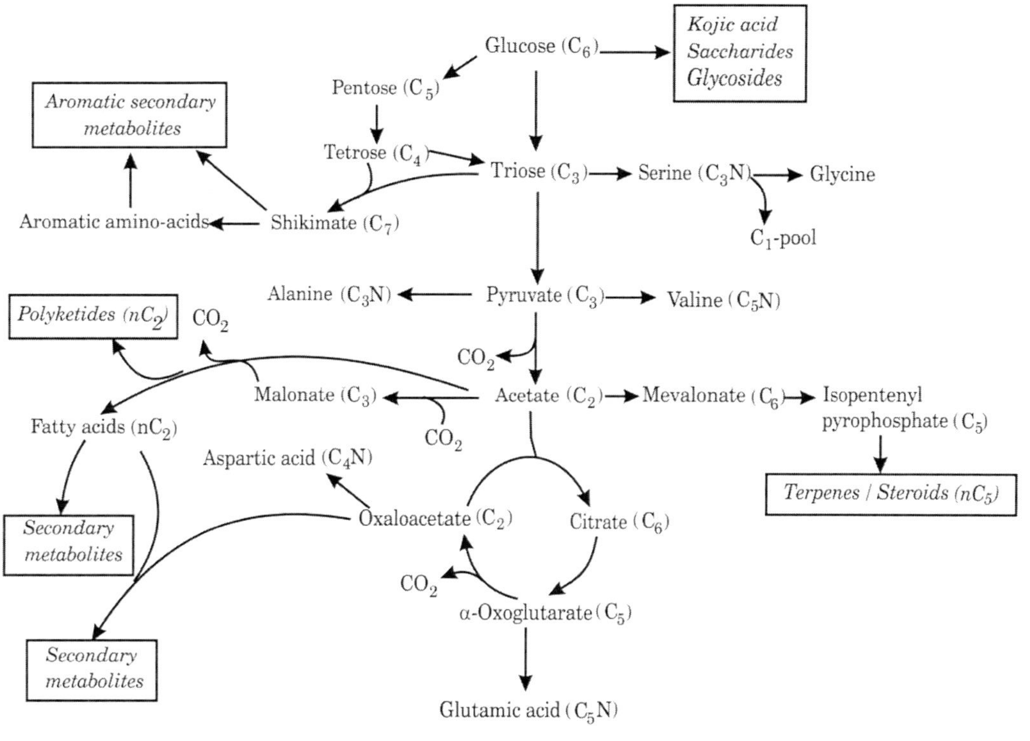

Fig. 23.2 : The interrelationships between primary and secondary metabolism. Secondary metabolites are shown in italics and in boxes

The classic separation of secondary metabolites from primary metabolites is convenient one, but in some cases appears artificial one. The separation has to be reconsidered in the light of new findings about the genetics of the biosynthesis of a variety of secondary metabolites. The biosynthetic pathways of the so-called secondary metabolites follow the same rules as any other biosynthetic process in living cells, except for peculiar aspects related to the formation of precursors and polymerization reactions. Bushall (1988) felt that classification of metabolites into primary and secondary should not be allowed to act as a conceptual straight jacket, forcing the reader to consider all products as primary or secondary metabolites. In many cases, it is difficult to categorize a product as primary or secondary and the kinetics of certain compounds may change depending on the cultural conditions.

23.4. THE SIGNIFICANCE OF SECONDARY METABOLISM

The physiological role of secondary metabolism in the producer cells has been the subject of considerable debate. One early suggestion was that they are reserve food materials accumulated in an environment at times of abundant food supply and utilized when there is a shortage of nutrients. They have been thought to confer upon an organism, for example antibiotic producing, an advantage in its natural environment. In some instances, this explanation may be valid but, it is not compatible with the lack of antibiotic activity in the majority of metabolites. The explanation currently favoured by most microbial physiologists is that it is the process of secondary metabolism that is important to microorganisms rather than the chemical nature of the metabolites. This explanation proposes that microbes which enter the stationary phase of growth, and are unable to shut off production of low molecular-weight intermediates that are required during growth for synthesis of cell constituents, convert these precursors into innocuous compounds that do not repress synthesis of low molecular weight compounds. The advantages to the organism are that it avoids the metabolic problems that would arise from intracellular accumulation of a large amount of low molecular weight compounds, while at the same time allowing the synthetic processes to remain operational during metabolic stress.

23.5. BIOSYNTHETIC PATHWAYS OF SECONDARY METABOLISM

Biosynthetic pathways leading to the formation of secondary metabolites can be broadly divided into four stages.

1. Direct utilization of primary metabolites as precursors of secondary metabolites.
2. Conversion of primary metabolites into specific precursors for secondary metobolite synthesis.
3. Entry of the precursors into specific pathways for secondary metabolite synthesis.
4. Polymerization and condensation reactions giving rise to polyketides, peptides, isoprenoids, terpenes, etc.
5. Late modification of the compounds into secondary metabolites.

23.5.1 Direct utilization of Primary Metabolites as Precursors

Almost all biosynthetic pathways of secondary metabolism originate from the intermediates of central metabolic pathways *i.e.,* glycolysis and Kreb's cycle. Through central metabolic pathway, carbohydrates are converted to smaller five-, four-, three-, and two carbon units (*e.g.,* pentoses, tetroses, trioses, acetate, 2-oxoglutarate, oxaloacetate). Some of these intermediates may be directly used as precursors of secondary metabolites or in many cases specific precursors of secondary metabolites arise by the modification of intermediates of glycolysis or Krebs cycle (Table 23.1).

23.5.1.1 Short Chain Fatty Acids

Many secondary metabolites are derived from acetate, propionate, butyrate and other short-chain organic acids through the polyketide pathway (polyacetate and polypropionate). Acetate and propionate linked to coenzyme A (CoA) in the form of acetyl CoA and propionyl CoA and act as C_2 and C_3 starter units respectively, whereas the normal elongation units are malonyl CoA and methyl malonyl CoA, which act as donors of two and three carbon units respectively. Malonate is formed by carboxylation of acetate via the action of acetyl CoA carboxylase. In actinomycetes propionate is an important precursor of the polyketide derived secondary metabolites. Propionate is used as a precursor in the synthesis of the lactone rings of some macrolide (*e.g.,* erythromycin) and ansamycin (rifamycin) antibiotics.

Table 23.1. Intermediates of primary metabolism that are precursors of secondary metabolites

Type of compound	Precursors
Short-chain fatty acids	Acetate, propionate, malonate, methylmalonate, butyrate
Isoprene units	Isopentenyl pyrophosphate, mevalonate
Amino acids	Normal protein amino acids, unusual amino acids
Sugars and amino sugars	Hexoses, pentoses, tetroses, modified hexoses, modified amino sugars (*e.g.,* N-methyl-L-glucosamin, mycosamine, mycominose), branched sugars (*e.g.,* streptose)
Cyclitols and aminocyclitols	Inositol, inosamine streptidine, 2-deoxystreptamine, actinamine
Amidino group	Arginine (amidino moiety)
Purine and pyrimidine bases	Adenine, guanine, cytosine, dimethyladenine, 3'-deoxypurines
Aromatic intermediates and aromatic amino acids	Shikimic acid, chorismic acid, prehenic acid, p-aminobenzoic acid, p-hydroxybenzoic acid, tyrosine, tryptophan, phenylalanine
Methyl groups (C_1 pool)	S-Adenosylmethionine, formyltetrahydrofolate

23.5.1.2 Isoprene units

A variety of secondary metabolites from plants and fungi are derived from condensation of isoprene units, CH_3-$CH(CH_3)$-$CH=CH_2$. Isoprenoid compounds are formed by polymerization of isopentenyl pyrophosphate, an activated isoprene. This isopentenyl pyrophosphate is synthesized from acetyl CoA through acetoacetyl CoA and mevalonic acid. It can also be formed from leucine by deamination and conversion into the intermediate, 3-hydroxy-3-methyglutaryl-CoA.

Terpenes, a class of compounds formed through these precursors exhibit antibiotic activity. Some isoprene units are also involved in the biosynthesis of complex antibiotics such as novobiocin. Others include ubiquinones, plastoquinones, vitamin K and the ring systems of coumarins and quinolines. Geosmin, a metabolite of *Streptomyces*, is also derived from isoprene units.

23.5.1.3 Amino acids

Amino acids found in proteins or modified non-protein amino acids are used as building units of homopeptide as well as heteropeptide antibiotics. Amino acids are synthesized from intermediates of glycolysis, the pentose phosphate cycle and the citric acid cycle by several biosynthetic pathways. Many unusual amino acids, formed by the modification of normal amino acids, also participate in the biosynthesis of secondary metabolites (Table 23.2).

Table 23.2. Some unusual amino acids and hydroxy acids existing in peptide secondary metabolites[a]

Type	Amino acid	Antibiotic
D-Amino acids	Allo-D-hydroxyproline	Etamycin
	D-Allothreonine	Stendomycin
	D-Serine	Quinomycin
	D-Leucine	Linear gramicidin
	D-Ornithine	Bacitracin
	D-Phenylalanine	Gramicidin S, tyrocidine, polymyxin B
N-Methyl amino acids	N-Methyl-L-threonine	Stendomycin
	N-Methyl-L-valine	Quinomycin
	N-Methyl-L-alloisoleucine	Quinomycin B
	N,β-Dimethyl-L-leucine	Etamycin, triostin C
	Sarcosine	Etamycin
	Phenylsarcosine	Etamycin
C-Methyl amino acids	3-Methyltryptophan	Telomycin
	3-Methylphenylalanine	Bottromycin A
	3-Methylvaline	Bottromycin A
	N, β-Dimethyl-L-leucine	Etamycin, triostin C
	3-Methyl-L-lanthionine	Nisin, subtilin
β-Amino acids	β-Lysine	Streptothricin, viomycin
	L-β-Methylaspartic acid	Aspartocin
	β-Phenylalanine	Edeine
	β-Serine	Edeine
	β-tyrosine	Edeine
Imino acids	cis-3-Hydroxy-L-proline	Telomycin
	4-Oxo-L-pipecolic acid	Vernamycin
S-Amino acids	L-Lanthionine	Nisin
	2-Thiazolyl-L-alanine	Bottromycin
Dehydro amino acids	Dehydrotryptophan	Telomycin
	Dehydroalanine	Nisin
	Dehydroleucine	Albonoursin

Type	Amino acid	Antibiotic
	Dehydrophenylalanine	Albonoursin
	Dehydroproline	Osteogrisin A
Complex amino acids	α,β-Diaminopropionic acid	Edeine
Basic amino acids	L-Ornithine	Gramicidin S
	L-2,4-Diaminobutyric acid	Polymyxin
Hydroxy acids	D-Allo-4-hydroxyproline	Etamycin
	2-Hydroxyisovaleric acid	Enniantin
Rare amino acids	D-2-Aminoadipic acid	Cephalosporin, cephamycin

[a]Modified from Martin (1998)

23.5.1.4 Sugars and aminosugars

Sugars, both normal and unusual sugars, and amino sugars very frequently occur in antibiotics and other secondary metabolites (*e.g.,* macrolides, aminocyclitols, tetracyclines, anthracyclines and nucleoside antibiotics). More frequently used sugar precursors include trioses, tetroses, pentoses, hexoses and heptoses. Sugars in secondary metabolites are frequently found attached to glycosidic bonds to the other moieties referred to as aglycones. Attachment is generally through a hydroxyl group, giving O-glycosides. However, N-glycosides, S-glycosides (involving amines and thiol groups), and C-glycosides also occur.

23.5.1.5 Aromatic intermediates

Most of the aromatic antibiotics of actinomycetes and some aromatic secondary metabolites of fungi and plants utilize the aromatic amino acid synthetic pathway. The secondary metabolites synthesized through this pathway include quinic acid, the quinoline alkaloids, naphthoquinone, quinozoline, ergot alkaloids and the phenazines. In addition to aromatic amino acids, some fungi utilize polyketide pathway and shikimate pathway to synthesize the aromatic secondary metabolites. Examples include chromophores of chloramphenicol, the C, N unit of rifampicin and the phenazine skeleton of pyocyanin. Shikimate, a precursor for aromatic amino acids, is formed by the condensation of erythrose-4-phosphate with phosphoenol pyruvate.

23.5.1.6 Purine and pyrimidine bases

Purines and pyrimidines, both in modified and unmodified forms are used as precursors for the synthesis of many secondary metabolites. Methylated purines (*e.g.,* caffeine, theophylline) and substituted purines having an isopentenyl group are formed as secondary metabolites in plants and microorganisms. Several antibiotics are analogs of nucleosides. They are either modified in the ribose moiety or in the purine or pyramidine bases. Nucleoside antibiotics containing an unmodified purine or pyrimidine base (*e.g.*, Cordycepin, psicofuranine, angustmycin) are derived directly from normal nucleosides.

23.5.1.7 Methyl (one-carbon) groups

In the biosynthesis of many antibiotics, methylation is a common phenomenon. All methylations use methionine as the methyl group donor in a transmethylation reaction catalysed by methyltransferases. Prior to the transfer, the methyl group of methionine is activated in the presence of magnesium and ATP through the formation of S-adenosyl methionine (SAM) which serves as a high energy methyl donor. The transmethylations occuring in primary and secondary metabolism include C-, O- and N-methylations.

23.5.2 Conversion of Primary Metabolites to Specific Precursors

Synthetic processes of many secondary metabolites do not use intermediates of primary metabolism directly as precursors, instead they use specifically modified primary metabolites. These modifications include (i) modification of the carbon skeleton, (ii) change in the oxidation-reduction level the molecule, and (iii) cyclizations giving rise to heterocyclic rings. Peptide of type of secondary metabolites contain unusual amino acids that are different from 23 normal protein amino acids. There are about 200 non-protein amino acids that include D-amino acids, N- and β-methylated amino acids, dibasic amino acids and amino acids that are intermediates of biosynthetic routes such as ornithine and α-aminoadipic acid (Table 23.2). Many of the sugars of secondary metabolites, for example those occuring in aminocyclitol and macrolide antibiotics, are rare sugars. They include deoxy sugars, branched chain sugars and unusual amino sugars. Many antibiotics contain rare chemical groups, such as azide, pyrrolo(1,4)benzodiazepine nucleus, cyano, mitosan which have equivalents in primary metabolism. Such abnormal precursors are synthesized by only a few microorganisms. Similarly, many amino acids and other intermediates of metabolism act as specific antibiotic precursors.

23.5.3 Entry of the Precursors into Specific Pathways

Chaneling of precursors into routes leading to biosynthesis of secondary metabolites is often considered as a topic of great interest. Surprisingly, little information is available on the enzymology of this aspect. Out of many enzymes involved, the first enzyme is considered as of great importance, because it not only recognizes the key precursor but also determines the amount of that precursor and also plays a role in the feedback regulation. Most of these enzymes are inducible. For example, the first enzyme of the biosynthetic pathway of ergot alkaloids is induced by tryptophan and tryprophan analogs. Some of these enzymes have been studied in some detail. They include dimethylallyltryptophan (DMAT) synthetase, which forms the phenoxazinone chromophore of actinomycin, amidinotransferase, a key enzyme in the biosynthesis of the streptidine moiety of streptomycin. Guanosine triphosphate-8-formyl hydrolase, which forms the pyrrole ring of pyrrolopyramidine nucleoside antibiotics; p-aminobenzoate synthetase, the enzyme that converts chorismic acid to p-aminobenzoic acid and lysine-6-aminotransferase which directs lysine to the cephamycin pathway in two different actinomycetes.

23.5.4 Polymerization and Condensation Reactions

In many cases, the monomeric structural units are polymerized to form polymeric secondary metabolites. It is a common process in the formation of polyketides, non-ribosomal peptides and isoprenoid molecules. Polymerization reactions are catalyzed by modular polyketide or polypeptide synthases with unusually high molecular weights. The products of polymerization usually undergo structural modifications. In some antibiotics different moieties synthesized by separate biosynthetic pathways, are assembled together to form complex structures.

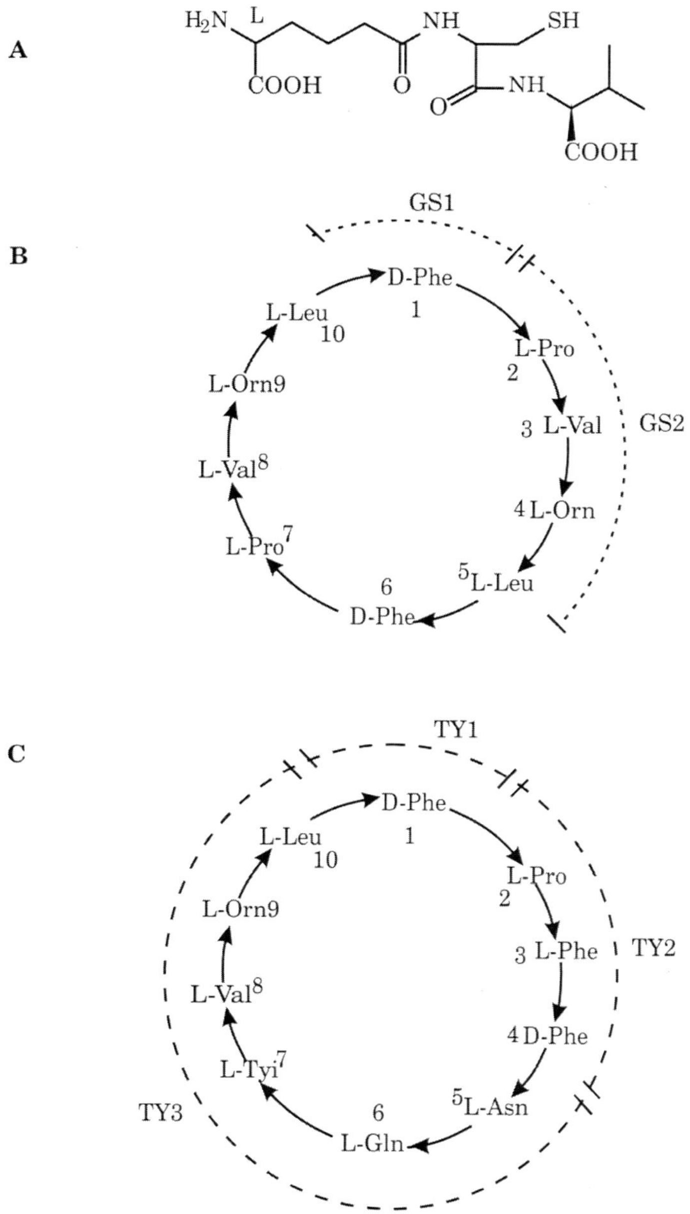

Fig. 23.3. Structures of some peptide secondary metabolites and modules required for their biosynthesis.

23.5.5 Late Modification of the Secondary Metabolites

Once the polymeric secondary metabolites are formed, they undergo further modification to give rise to functionally active molecules. Late modifications to a larger extent depends upon the physiological condition of the cell. The biochemistry of these late-modification reactions is only begining to be understood. These include glycosylation, hydroxylation, oxidation of hydroxyl to ketogroups, amination and transaminations; N- and O-methylations, N- and O-acetylations, dehydrations, epoxidations and chlorination.

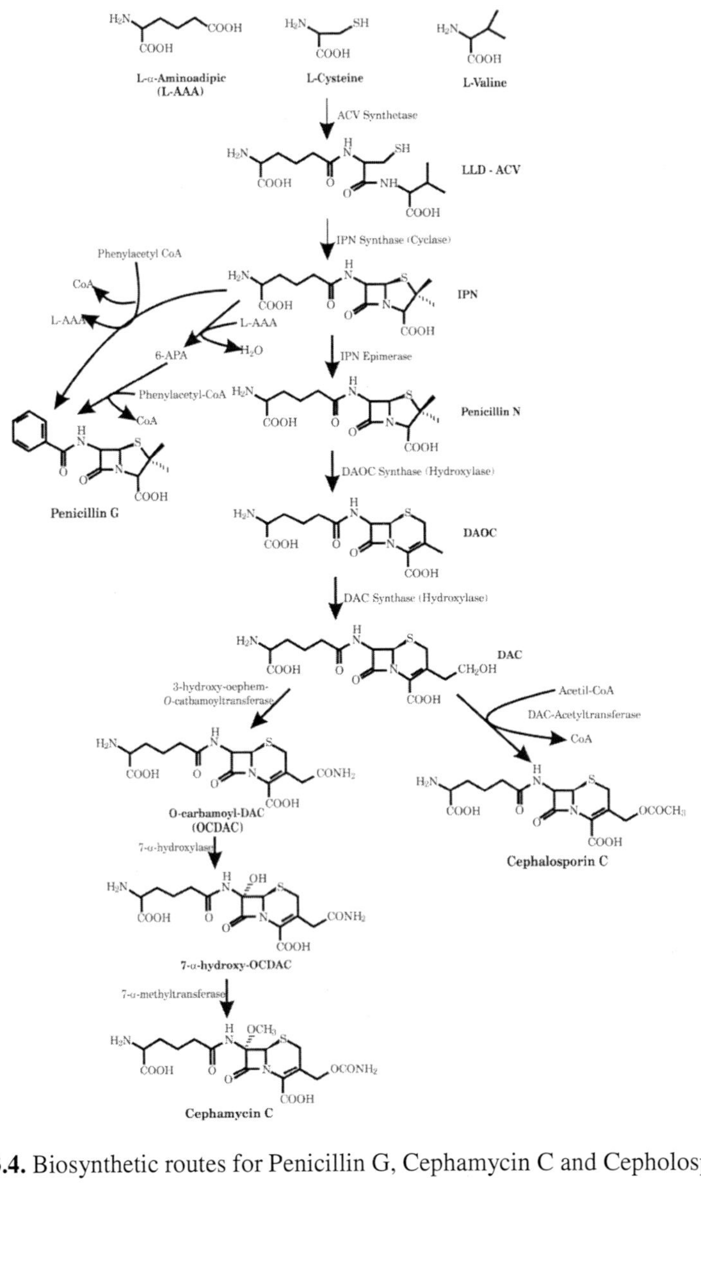

Fig. 23.4. Biosynthetic routes for Penicillin G, Cephamycin C and Cepholosporin

23.6. IMPORTANT MICROBIAL SECONDARY METABOLITES

23.6.1 Non-Ribosomal Peptides

Microbial peptide antibiotics are broadly divided into two categories : (i) Ribosomal peptides (ii) non-ribosomal peptides. Majority of the peptide antibiotics belong to the latter category. Non-ribosomal peptides are synthesized by thiotemplate mechanism without using ribosomes.

The non-ribosomal peptide secondary metabolites exhibit several particular characteristics that enable us to distinguish from ribosomal proteins.

1. Non-ribosomal peptides are generally smaller ($\leq$ 3000 Da) than proteins.
2. Most non-ribosomal peptides are synthesized by bacteria and fungi as families of closely related substances.
3. Non-ribosomal peptides are made up of non-protein amino acids.
4. They often form cyclic structures with no free α-amino or carboxyl ends.
5. These are, to a great extent, resistant to cleavage by proteases. This is due to presence of rare amino acids and cyclic structures.
6. Some non-ribosomal antibiotics are made exclusively of amino acids, whereas others contain fatty acids, hydroxy acids, amines, amino sugars, alcohols and pyrimidines linked to the amino acids (Fig. 23.3).

23.6.2. β-Lactams

These antibiotic secondary metabolites have a β-lactam ring in their chemical structure. They include penicillins, cephalosporins and cephalomycins. A brief account of biosynthesis of these three antibiotics is presented (Fig. 23.4).

The penam nucleus of the penicillins and the cephem nucleus of both cephamycins and cephalosporins are formed from three precursor amino acids, L-α-aminoadipic acid, L-cysteine, and L-valine. These three amino acids are condensed by a mechanism designated as non-ribosomal peptide synthesis. The initial reactions upto the formation of isopenicillin N are common to all the three antibiotics. At this stage, penicillin G synthetic pathway deviates from the other two, due to the exchange of α-aminoadipyl side chain for phenylacetic acid in penicillin producing organism. This reaction is catalyzed by isopenicillin N-acetyltransferase (IAT).

Further down, formation deacetylcephalosporin C (DAC) from isopenicillin N, through penicillin N and deacetoxycephalosporin (DAOC) is common for both cephalo-sporin C and cephalomycin C. In cephalosporin producing fungi, DAC is acetylated to form cephalos-porin C. The reaction is catalyzed by DAC-acetyl transferase. The late reactions in the cephalomycin biosynthesis include the attachment of the C-7 methoxyl group and the C-3' carbamoyl group.

23.6.3 Polyketides

Polyketides are a large class of natural metabolites produced through the successive condensation of simple carboxylic acids. They are found in most groups of organisms, but they are specially abundant in actinomycetes. These compounds have generated much interest in synthetic and biosynthetic chemistry because of their staggering structural complexity and also due to their broad spectrum biological activity (Fig. 23.5). The broad range biological activities include antibacterial, antifungal, anticancer, insecticidal, antiparasite and immunosuppressive activities. Some polyketides have more than one property.

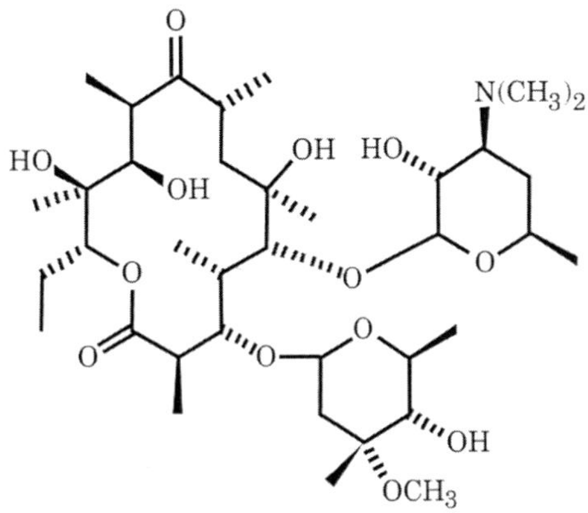

Erythromycin A

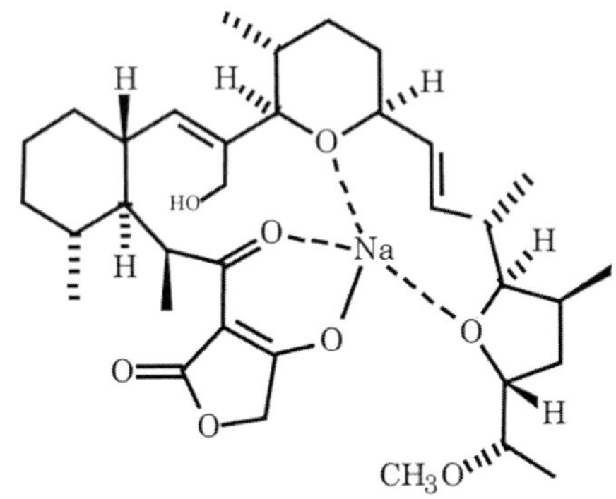

Tetronasin

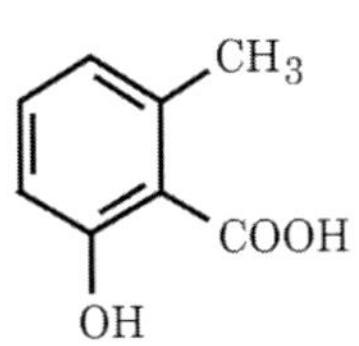

6-methylsalicylic acid

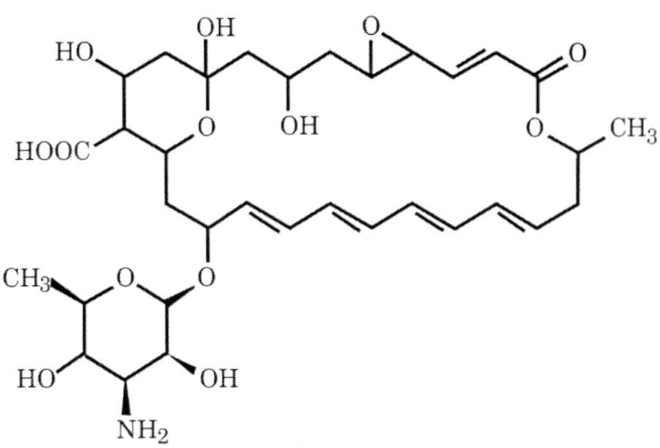

Pimaricin

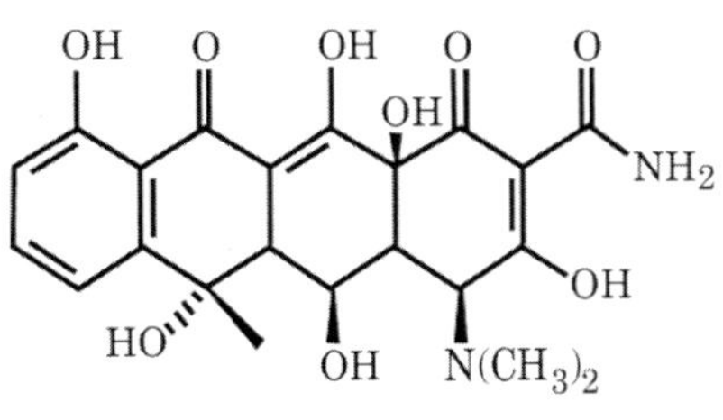

Oxytetracycline

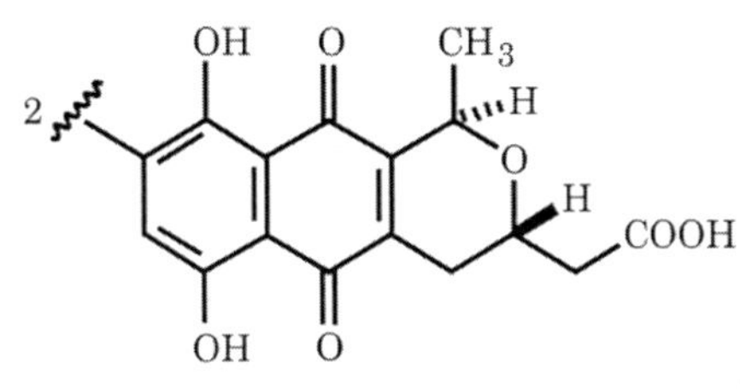

Actinorhodin

Fig. 23.5. Structure of some fungal and bacterial polyketides

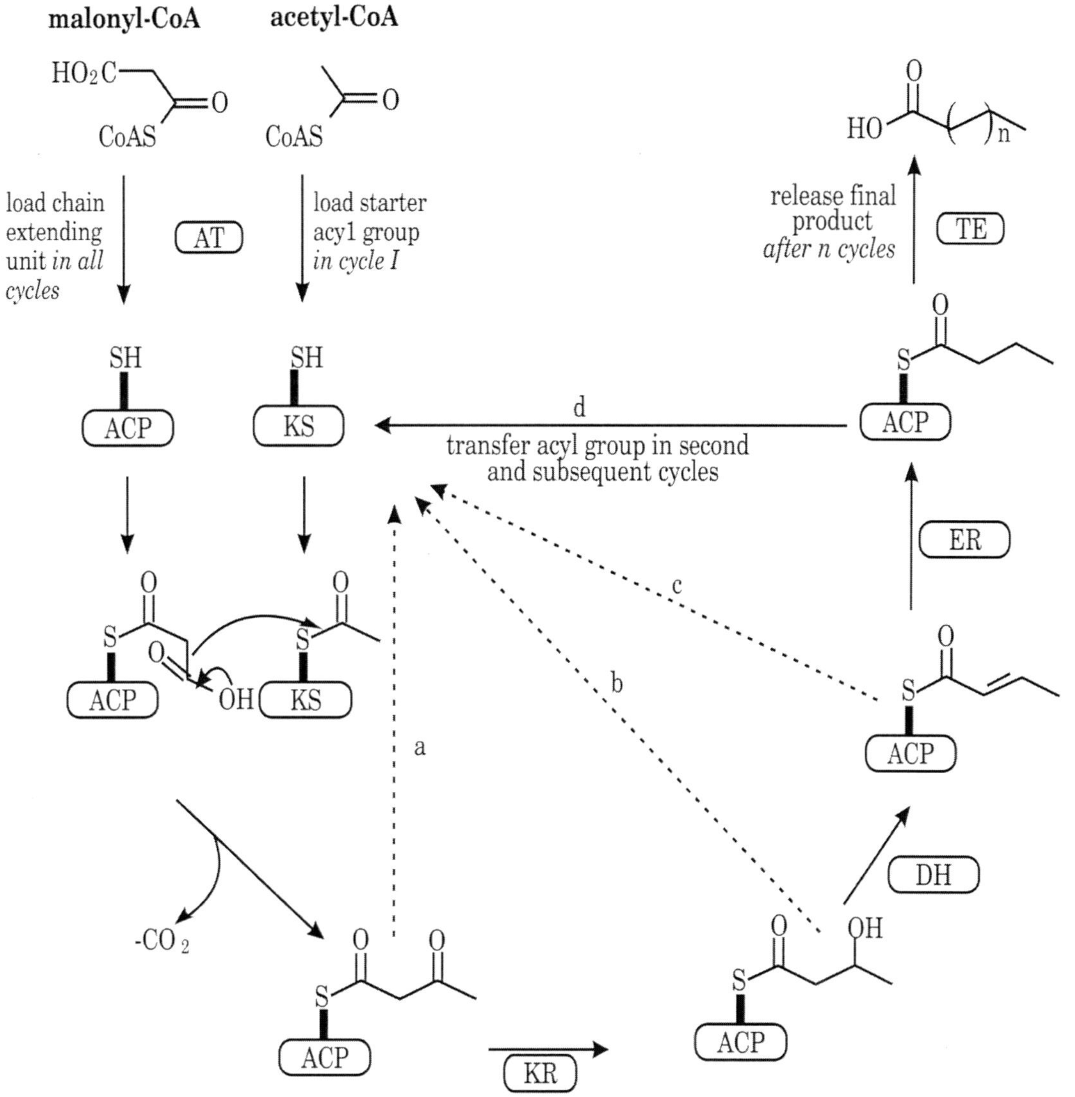

Fig. 23.6: The basic pathway of polyketide biosynthesis. Dashed arrows represent shortcuts of the pathway that may occur in a given cycle to produce a (a) keto, (b) hydroxyl, (c) enoyl, or (d) methylene functionality in the polyketide backbone. The reaction steps are labeled: AT, acyltransferase; KS, ketosynthase, ACP, acyl carrier protein; KR, ketoreductase; DH, dehydratase; ER, enoylreductase; TE, thioesterase.

Polyketides are basically formed from acetate, propionate and more rarely butyrate units (derived from malonyl-, methylmalonyl-, or ethylmalonyl CoA respectively). Inspite of their structural diversity, all polyketides share a common mechanism of biosynthesis (Fig. 23.6). A polyketide synthase (PKS) catalyzes the formation of a polyketide through repeated decarboxylative condensations between enzyme bound acyl carrier protein (ACP) thioesters. Extender units are transferred from coenzyme A (CoA) to the pantetheine arm of the ACP by the acyltransferase (AT). After each condensation, carried out by a β-ketoacyl ACP synthase (KS), the ketogroup of the growing chain remains unchanged or is reduced to a hydroxyl, enoyl, or methylene group by the action of a β-ketoreductase (KR), dehydratase (DH), enoyl reductase (ER), respectively. Structural variations among natural occuring polyketides arise largely from the ways in which PKS control the number and type of starter and extender untis used, and from the extent and stereochemistry of reduction at each cycle. For aromatic polyketide synthesis, the extension units are derived from malonyl CoA. However, the condensation reactions are catalyzed by type II polyketide synthases.

23.6.4 Isoprenoids

Isoprepoids belong to both primary and secondary metabolism. Microbial isoprenoids include carotenoids, steroids, gibberellins, phytoalexins, antibiotics and anticancer agents.

Isoprenoid molecules are derived from the C_5 unit isopentenyl pyrophosphate (IPP), which in turn is synthesized by condensation of three acetate units (activated as acetyl CoA) via the intermediate 3-hydroxy-3-methylglutaryl-CoA (HMG-CoA) that undergoes reduction to mevalonic acid and then activation and decarboxylation to isopentenyl-pyrophosphate. A second biosynthetic route to synthesize IPP also exists that involves neither acetate nor mevalonate. In this pathway, first, pyruvate and glyceraldehyde-3-phosphate undergo a thiamine pyrophosphate-dependant condensation to give 1-deoxy-D-xylulose-5-phosphate, which is later condensed to IPP. This route, also called deoxyxylulose phosphate pathway, is the source of isoprenoids in prokaryotic organisms (except archaebacteria) and occur side by side with the mevalonate pathway in higher plants.

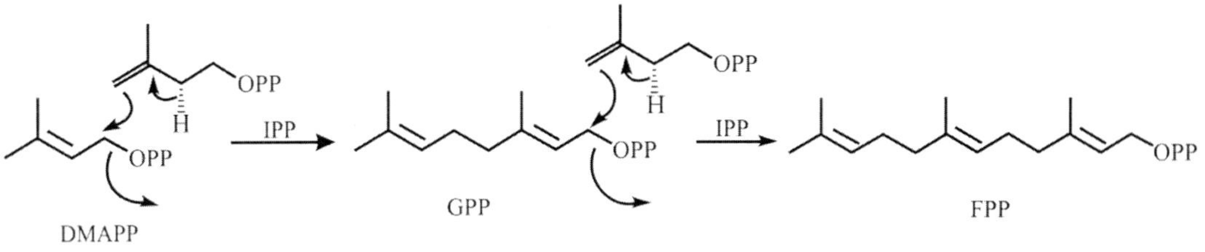

Fig. 23.7. Enzymatic chain elongation of isoprenoids by the prenyltransferases to form geranyl diphosphate (GPP) and farnesyl diphosphate (FPP).

Linear long chain isoprenoid compounds are derived by enzymatic chain elongation using as substrates for these transformations the acylic isoprenoids such as dimethylallyl diphosphate (DMAPP), geranyl diphosphate (GPP), farnesyl diphosphate (FPP) and geranylgeramyl diphosphate (GGPP) (Fig. 23.7). Isoprenoid chain elongation is catalyzed by prenyl transferases that contain a divalent metal cofactor Mg^{2+}, Mn^{+2}. The length of the isoprenoid chain is determined by each biosynthetic prenyltransferase.

Many natural isoprenoids are cyclic monoterpenes, sesquiterpenes and diterpenes that are formed by cyclization of the linear geranyl, farnesyl and geranylgeramyl diphos-phate.

Cyclization requires previous isomerization of all trans molecules. Biosynthesis of the more complex terpene molecules such as the triterpenes and steroids involves the head-to-head condensation of two farnesyl diphosphate molecules which is followed by reduction step.

23.7. FURTHER READING

Aparicio, J.F., Caffrey, P., Marsden, A.F.A., Staunton, J. and Leadlay, P.F. (1994), Limited proteolysis and active site studies of the first multienzyme component of the erythromycin-producing polyketide synthase. *J. Biol. Chem.* 269, 8524-8528.

Cane, D.E. (1999), Isoprenoid biosynthesis : Overview *In* "Comprehensive Natural Products Chemistry" (D. Barton, K. Nakanishi and O. Meth-Cohn, eds.), Vol. 2, pp. 1-13, Pergamon, Oxford.

Drew, S. and Demain, A.L. (1977), Effect of primary metabolites on secondary metabolism. *Ann. Rev. Microbiol.,* 31, 343-356.

Hopwood, D.A. and Sherman, D.H. (1990), Molecular geneties of polyketide and its comparison to fatty acid biosynthesis. *Ann. Rev. Genet.,* 24, 37-66.

Katz, L. and Donadio, S. (1993), Polyketide synthesis: Prospects for hybrid antibiotics. *Ann. Rev. Microbiol.,* 47, 875-912.

Kleinkauf, H. and von Dohren, H. (1990), Non-ribosomal biosynthesis of peptide antibiotics. *Eur. J. Biochem.* 192, 1-23.

Kleinkauf, H. and von Dohren, H. (1997), Peptide antibiotics. In "Biotechnology" (H.J. Rehm, G. Reed, A. Puhler and P. Stadler, eds.) 2nd edn. pp. 277-322. VCH, Weinheim.

Lancini, G. and Lorenzetti, R. (1993), "Biotechnology of Antibiotics and Other Bioactive Microbial metabolites, p. 236, Plenum, New York.

Martin, J.F. (1997), Biosynthesis of polyene macrolide antibiotics. *Ann. Rev. Microbiol.* 31, 13-38.

Martin, J.F., Gutierez, S. and Aparicio, J.F. (2000), Secondary metabolites. In J Lederberg (ed.) Encyclopedia of Microbiology, Vol. 4, 2nd edn., San Diego, Academic Press, New York.

Martin, J.F., Gutierrez, S. and Demain, A.L. (1997), β-Lactams. In "Fungal Biotechnology" (T. Anke, ed.), pp. 91-127. Chapman & Hall, Weinheim.

■■■■

Chapter - 24

Microbial Toxins

24.1. INTRODUCTION

A toxin, according to Dorland Medical Dictionary, is a poisonous substance produced within living cells or organisms. They can be small molecules, peptides, or proteins that can cause disease on contact with or absorption by body tissues, interacting with biological molecules such as enzymes or cellular receptors. Toxins vary greatly in their toxicity, ranging from usually minor to almost immediately deadly. Toxins are produced by a wide range of living organisms; bacteria, fungi, algae animals and plants. Toxins produced by first three groups of organisms are referred to as **microbial toxins**. In microbiology, the term 'toxin' describes products of microbial metabolism that are lethal to cells or affect them negatively, even at very low concentrations. Often this results in pathological changes in the entire body.

A. BACTERIAL TOXINS

Pathogenic bacteria employ various strategies to invade the host tissues/cells and take control of host metabolism and to obtain the nutrients and subsequent growth and proliferation. Amongst the various mechanisms developed by pathogenic bacteria to cause disease, toxins play a prominent role. Toxins can act locally or at a distance from the infectious site and are responsible for severe disease in man and animals. They damage the host via invasive and toxic attributes. Invasive bacteria injure the host through the production of extracellular enzymes and damage host tissue, or through the modulation of host response system such as upregulation of cytokine expression. In contrast, bacterial pathogens that damage the host through the action of toxins are often non-invasive and have limited capacity to disseminate inside the host. In general, both these categories of chemicals produced by bacteria are referred as 'toxins'.

At chemical level, there are two main types of toxins, polysaccharides which are associated with the cell wall of Gram-negative bacteria, and proteins, which are released from bacterial cells and may act at tissue sites removed from the site of bacterial growth. The cell associated toxins (lipopolysaccharides) are referred to as **endotoxins** and extracellular diffusible toxins are referred to as **exotoxins**. The major differences between endotoxins and exotoxins are presented in Table 24.1.

Table 24.1. Differences between exotoxins and endotoxins

Character	Exotoxin	Endotoxin
Source	Living Gram-positive and Gram-negative bacteria	Lysed Gram-negative bacteria
Location	Released from the cell, diffusible	Part of outer membrane
Chemical nature	Protein	Lipopolysaccharide
Molecular weight	Around 10kDa, filterable	50-1000KDA, not filterable
Heat stability & denaturation	Heat labile (60-80°C), on boiling get denatured	Heat stable (250°C), on boiling not denatured
Specificity	Specific to particular bacterial strain	Non-specific
Receptor on host	Usually bind to specific receptors	Specific receptors are not found
Immune reactions	Strong; highly immunogenic, stimulate the production of neutralizing antibody (antitoxin)	Weak; relatively poor immunogenicity
Enzyme activity	It has mostly enzymatic activity	It has no enzymatic activity
Conversion to toxoid	Possible	Not possible
Potency	High (1μg)	Low (> 100μg)
Detection	Many tests (precipitation, neutralization etc)	Limulus lysate assay
Location of genes	Located on plasmids or bacteriophages	Located on chromosomes
Pyrogenicity	No	Yes
Mode of action	Specific, either cytotoxin, enterotoxin or neurotoxin with defined action cells on tissues	General; fever, diarrhea, vomiting
Producing bacteria	*Staph. aureus, B. cereus, Strept. pyogenes, Vibrio cholerae* and *B. anthracis.*	*E.coli, S. typhi, Shigella, Vibrio cholerae*
Diseases	Tetanus, Diphtheria and Botulism	Menigococcemia, sepsis

24.2. ENDOTOXINS

Endotoxins are cell associated substances that are structural components of the outer membrane of Gram-negative bacteria. Thus, they refer specifically to the lipopolysaccharide (LPS) located on the outer membrane of Gram- negative bacteria. Although, the term endotoxin is mostly used synonymously with LPS, there are few endotoxins that are not related to LPS such as delta endotoxin proteins secreted by *Bacillus thuringiensis*. Despite the structural components of cell walls, soluble endotoxins may be released form growing bacteria or from cells that are lysed as a result of effective host defense mechanisms or by the activities of certain antibiotics like penicillin. LPS are also secreted as a part of normal physiological activity of membrane vesicle trafficking in the form of bacterial outer membrane vesicles (OMVs) which may also contain other virulence factors and proteins.

A single *E. coli* contains about 2 million LPS molecules per cell. Endotoxins have a high heat stability making it impossible to destroy them under regular sterilization conditions. They are amphipathic molecules that carry out a net negative charge in solution. Due to their hydrophobicity, they are likely to have strong affinities for other hydrophobic materials like plastic products used in laboratory.

The pathophysiological activity of endotoxin is associated with lipopolysaccharide. Toxicity is associated with the lipid component (Lipid A) and immunogenicity is associated with the polysaccharide components. Endotoxins are less potent and less specific in their action, since they do not act enzymatically. Endotoxins are heat stable, but certain powerful oxidizing agents such as superoxide peroxide and hypochlorite have been reported to neutralize them. Endotoxins, although antigenic cannot be converted to toxoid. Endotoxins induce fever and elicit a variety of inflammatory responses and they activate complement by the alternative (properdin) pathway, thus they become a part of the pathology of Gram-negative bacterial infections.

Lipooligopolysaccharides (LOS) are glycolipids found in the outer membrane of some types of Gram-negative bacteria such as *Neisseria* spp and *Haemophilia* spp. In fact, they are truncated parts or low molecular weight forms of LPS. Chemically LOS lack O antigen and possess only lipid A- based outer membrane-anchoring moiety, and an oligosaccharide core. The outer core oligosaccharide chain varies depending on the bacterial strain. LOS can be categorized into two forms: the high molecular weight (or smooth) possessing repeating polysaccharide O-chain, while the low molecular weight (or rough) form, lacks the O-chain but possesses a short oligosaccharide in its place. LOS play an important role in the pathogenesis of certain bacterial infections because they can act as immunostimulators and immunomodulators. Furthermore, LOS molecules are responsible for the ability of some bacterial strains to display molecular mimicry and antigenic diversity, aiding in the evasion of host immune defenses and thus contributing to the virulence of these bacterial strains.

Chemistry of Endotoxin

Most of the work to elucidate the chemical structure of endotoxin has been carried out with species of *Salmonella* and *E. coli*. LPS can be extracted from whole cells by treatment with 45% phenol at 90°C. Mild hydrolysis of LPS yields Lipid A plus polysaccharide. LPS are complex amphiphilic molecules with a molecular weight about 10kDa. There is a wide variation in chemical composition both between and among bacterial species. The general structure of LPS as shown in LPS gene clusters are highly variable between different strains, subspecies and species of bacterial pathogens of plants and animals.

LPS consists of three components or regions: Lipid A, core or R polysaccharide and O antigen or O polysaccharide (Fig. 24.1).

Lipid A: It is the basal region and lipid component of LPS. It constitutes the hydrophobic membrane anchoring region of LPS. Lipid A consists of a phosphorylated N-acetylglucosamine (NAG) dimer with 6 or 7 saturated fatty acids. Some fatty acids are attached directly to the NAG dimer and others are esterified to the 3-hydroxy fatty acids that are characteristically present. The structure of Lipid A is highly conserved among Gram-negative bacteria, and among Enterobacteriaceae lipid A is virtually constant. Its biological activity appears to depend on a peculiar conformation that is determined by glucosamine disaccharide, the phosphate groups, the acyl chains, and the KDO containing inner core.

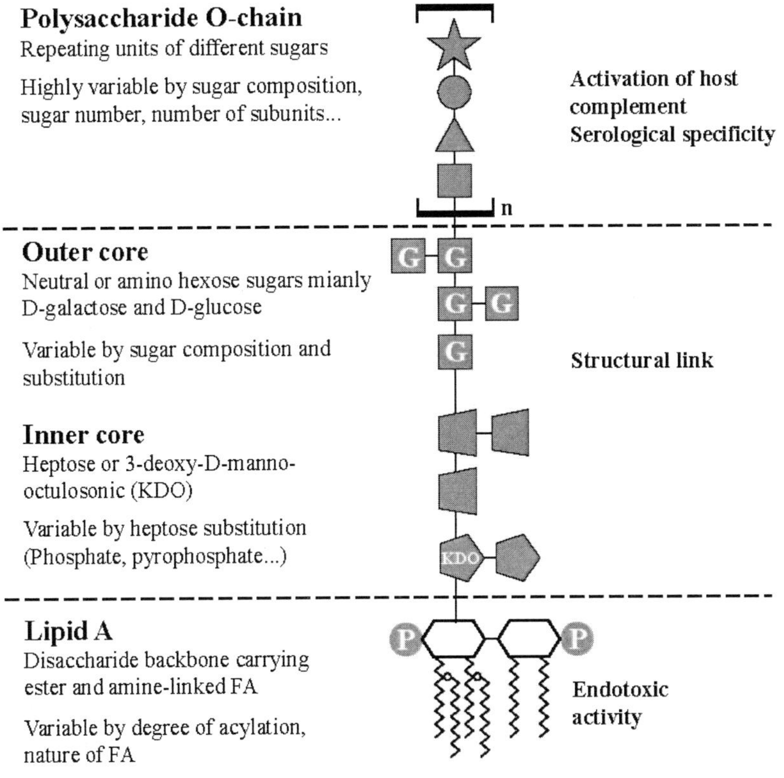

Fig. 24.1. Structural details of lipopolysaccharide

The lipid A domain is responsible for much of the toxicity of Gram-negative bacteria. When bacterial cells are lysed by the immune system, fragments of membrane containing lipid are released into the circulation causing fever, diarrhea and possible fatal endotoxic shock (sceptic shock).

Core (or **R polysaccharide):** Core connects the bottom lipid A and upper O antigen of LPS. It is attached to the 6 positions of one NAG and consists of short chain of sugars such as 2-keto-3-deoxyoctonic acid (KDO), heptose-heptose-glucose-galactose-glucose-N-acetyl glucosamine. It is interesting to note the presence of two unusual sugars *viz,* heptose and KDO in the core. KDO is unique and invariably present in LPS and hence it has been used as an indicator in assays for LPS (endotoxin). Structural variations occur within species and genera of bacteria. For example, with minor variations, the core polysaccharide is common to all members of a bacterial genus *Salmonella,* but it is structurally distinct in other genera of Gram-negative bacteria. *Salmonella, Shigella* and *Escherichia* have similar but not identical cores. The LPS cores of many bacteria also contain non-carbohydrate components, such as phosphate, amino acids and ethanolamine substituents. Core region of LPS has neither toxicity nor antigenicity property.

O-antigen (or **O polysaccharide):** O antigen comprises the outermost domain of the LPS molecule and is attached to the core oligosaccharide. It consists of repeating oligosaccharide subunits (glycans). Each subunit is made of 3–5 sugars. The individual chains vary in length ranging up to 40 repeat units. The O polysaccharide is much longer than the core oligosaccharide and bestows the hydrophilicity to the terminal domain of the LPS molecule.

A great variation exists in the composition of sugars in the O polysaccharide among species and even strains of Gram-negative bacteria. At least 20 different sugars are known to occur and many of these sugars are unique dideoxyhexoses, which occur in nature only in Gram-negative bacterial cell walls. Variations in sugar content of O polysaccharide contribute to the wide variety of antigenic types of *Salmonella* and *E. coli* and also other strains of Gram-negative species. For example, there are over 160 different O antigen structures, especially the terminal ones that confer immunological specificity of the O antigen. O antigen is exposed on the very outer surface of the bacteria and therefore, is a target for recognition of host antibodies.

The presence or absence or loss by mutation of O chains determines whether the LPS is considered 'rough' or 'smooth'. Fully length O-chains would render the LPS smooth, whereas the absence or reduction of O-chains would make the LPS rough. Bacteria with rough LPS usually have more penetrable cell membranes to hydrophobic antibiotics, since rough LPS is more hydrophobic. Virulence of a bacterial strain is related to smooth or rough LPS. For instance, in *E. coli* and *Salmonella* loss of the O antigen results in partial loss of virulence, suggesting that this portion of LPS is important during a host-parasite interaction. It is known that such 'rough' mutants are more susceptible to phagocytosis and serum bactericidal reactions.

24.2.2. LPS and Virulence of Gram-negative Bacteria

Both lipid A (the toxic component of LPS) and the polysaccharide side chains (the non-toxic but immunogenic portion of LPS) are determinants of virulence in Gram-negative bacteria.

The O antigen and virulence

As indicated earlier, the virulence of bacterial strain is attributed to intact O antigen or in other words 'smoothness'. The involvement of the polysaccharide chain in virulence is shown by the fact that small changes in the sugar sequences in the side chains of LPS result in major changes in virulence. It is assumed that one or combination of following possibilities may be involved in the expression of virulence in relation to O antigen.

i. O-specific antigens enable the organisms to adhere specifically to certain tissues especially epithelial tissues.

ii. Smooth antigens may allow the organism to escape the phagocytosis by host macrophages. On the other hand, rough mutants are more readily engulfed and destroyed by phagocytes.

iii. The hydrophilic O polysaccharides may act as water solubilizing carriers for hydrophobic toxic lipid A, which in turn facilitates the binding of LPS to the host cell surface.

iv. The O antigens may provide protection from damaging reactions with antibody and complement.

Variation in the structure of O antigen is the basis of antigenic variation among many important Gram-negative pathogens. This antigenic variation ensures the existence of multiple serotype of the bacterium, so that it is afforded multiple opportunities to infect host, even it can bypass the immune response against a different serotype.

Lipid A and virulence

Many of the physiological activities of LPS are mediated mainly by the lipid A component of LPS. Lipid A is a powerful stimulator of mammalian immune system. Endotoxins released from, or part of, multiplying cells significantly contribute to the symptoms and pathology of the disease involved. Lipid A imbedded in the outer membrane of bacterial cells exerts its toxic effects when released from multiplying cells in soluble form, or when the bacteria are lysed as as result of autolysis, complement and the membrane attack complex (MAC) ingestion and killing by phagocytes, or killing with certain types of antibiotics. Injection of living or killed Gram-negative cells or purified LPS into experimental animals causes a wide range of nonspecific pathophysiological reactions such as fever, changes in white blood cell counts, disseminated intra vascular coagulation (DIC), hypotension, shock and death. The sequence of events in pathogenesis follows a regular pattern: i. latent period, ii. physiological distress (diarrhea, prostration, shock) and iii. death. However, susceptibility varies on the dose of the endotoxin, route of administration and species of animals.

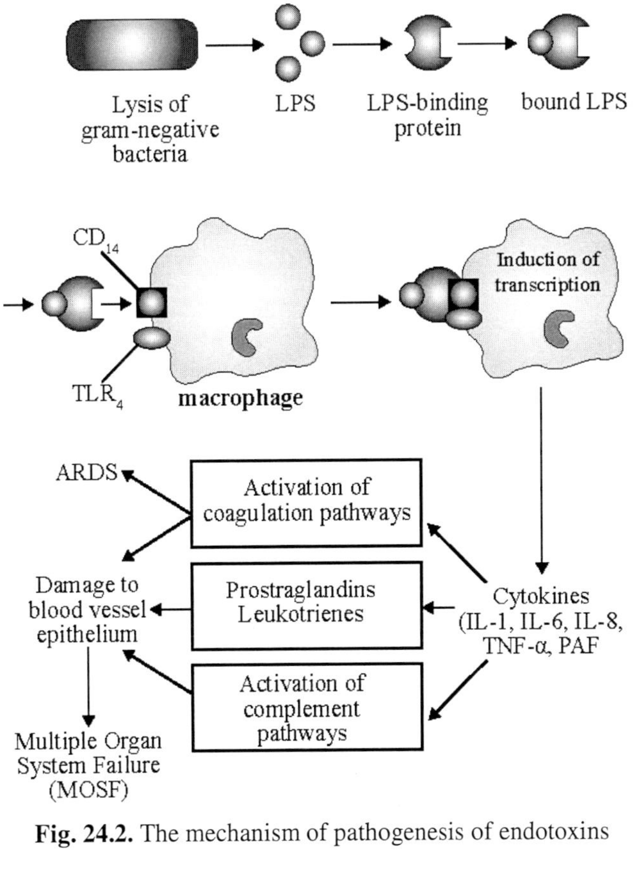

Fig. 24.2. The mechanism of pathogenesis of endotoxins

The mechanism of endotoxin pathogenesis is complex. In humans, LPS binds to a lipid binding protein (LBP) in the serum which transfers it to CD_{14} on the cell membrane (Fig. 24.2). This complex is transferred to another non-anchored protein, MD2, which associates with Toll-like-receptor-4 (TLR-4). CD_{14} and TLR4 are present on several cells of the immunological system cells including macrophages and dendritic cells. The complex consisting of LPS-CD_{14}-TLR4 and attached to cell surface triggers the signaling cascade for macrophage/endothelial cells to secrete pro-inflammatory cytokines, nitric oxide and eicosanoid, and stimulate growth of B cells. In particular with macrophages, three types of events are triggered during their interaction with LPS.

Production of cytokines, including 1L-I, 1L-6, 1L-8, tumor necrosis factor α (TNF-α) and platelet activating factor. These in turn, stimulate production of prostaglandins and leukotrienes. These are powerful mediators of inflammation and septic shock that accompanies endotoxin toxemia. LPS activates macrophages to enhanced phagocytosis and cytotoxicity. Macrophages are stimulated to produce and release lysosomal enzymes, 1L-1 and TNFα as well as other cytokines and mediators.

Activation of the complement cascade involving C3a and C5a cause histamine release leading to vasodilation and effect neutrophil chemotaxis and accumulation. The result is inflammation.

Activation of the coagulation cascade: Initial activation of Hageman factor (blood-clotting factor-XII) can activate several humoral systems resulting in :

- Acute disseminated intravascular coagulation (DIC)
- Activation of the complement alternate pathway
- Plasmin activation, and kinin activation leading to cause hypotension. The net effect is to induce inflammation, intravascular coagulation, hemorrhage and shock.
- LPS also acts as a β cell mitogen, stimulating the polyclonal differentiation and multiplication of β cells and the secretion of immunoglobulins, especially IgG and IgM.
- Known pathological effects of endotoxins are summarized in Table 24.2.

Table 24.2. Known pathological effects of endotoxins

1. **Bind to specific receptors on macrophages, B lymphocytes and other cells**
   - Stimulates release of acute phase immunoreactants and lymphokines (e.g. IFN-g, IL-1, TNF-alpha, IL-6, histamine, prostaglandins).
   - Stimulate growth of B cells (mitogenic)
2. **Fever (pyrogenicity)**
   - Any elevation of the body temperature above the normal; functions to speed up
   - reactions and limit/slow bacterial growth and multiplication.
3. **Activation of alternative complement pathway**: C3a; C5a

4. **Circulatory system effects**
   - Leukopenia (abnormal reduction in the number of leukocytes) followed by
   - leukocytosis (abnormal increase in the number of leukocytes).
   - Increased vascular permeability (vasodilation)
   - Decreased perfusion (blood flow) of blood to major organs
   - Decreased peripheral circulation.
   - Capillary leakage; microhemorrhage; formation of petechiae (round, purple lesions caused by intradermal or submucosal microvascular hemorrhaging).
   - Hypotension (low blood pressure).
5. **Effects on blood coagulation**
   - Disseminated intra vascular coagulation (DIC)
   - Activation of clotting pathway
   - Thrombosis: Formation of blood clot (thrombus) in heart or blood vessel
   - Thrombocytopenia: Abnormally low number of blood platelets.
6. **Effects on metabolic and liver functions**
7. **Decreased iron availability**
8. **Hypoglycemia: Abnormally low glucose levels**
9. **Cellular death (cytotoxicity)**
10. **Organ necrosis**
    - Sum of morphological changes indicative of cell death and caused by the progressive degradative action of enzymes.
11. **Shock**
    - Characterized by failure of the circulatory system to maintain adequate blood flow to the vital organs.
    - Symptoms include: Hypotension; weak pulse; rapid and shallow breathing; low body temperature; CNS (central nervous system) effects (e.g., nausea).
12. **Death**

24.3. EXOTOXINS

Exotoxins are heat labile proteins secreted by living bacteria during exponential growth. However, in some cases, exotoxins are only released by lysis of the bacterial cell. Exotoxins are mostly proteins, minimally polypeptides. They are invasive i.e., act at a site removed from bacterial growth, but some bacterial exotoxins act at the site of pathogen colonization and may play a role in invasion. Exotoxins are produced by a variety of bacteria including Gram-positive and Gram-negative bacteria. The production of toxin is generally specific to a bacterial species that produces the disease associated with toxin. Within the species, most strains produce the same exotoxin, while in other only a subset of strains produces an exotoxin. Usually, virulent strains of the bacterium produce the toxin, while non-virulent strains do not, and the toxin is the major determinant of virulence.

Most of the exotoxins have enzymatic activity. They cause damage to the host by destroying cells or disrupting normal cellular metabolism, and stimulate a variety of host responses. Bacterial exotoxins are strongly antigenic. *In vivo* specific antibody neutralizes the toxicity of these bacterial exotoxins (antitoxin). However, many exotoxins are so potent that they may be fatal to the host before the immune system has a chance to mount defenses against them. For this reason, antitoxin, antiserum containing antibodies, is injected to confer passive immunity. Protein exotoxins are inherently unstable. The toxin properties of most exotoxins can be inactivated by heat or chemical treatment, to produce **toxoid**. They lose the toxicity, but retain their antigenic specificity and can thus be used to produce antitoxins and, in the case of diphtheria and tetanus toxoids are used as vaccines. Toxin coding genes are located on plasmids and in some cases genes are phage-borne. For example, diphtheria toxin is coded by a temperate beta phage and production of anthrax toxin is mediated by a temperature-sensitive plasmid, pXOI. Bacterial exotoxins are the most powerful human poisons known and retain their high activity at very high dilutions. Since exotoxins are proteins, various immunological techniques such as precipitation, agglutination, ELISA, RIA can be used to detect the toxins.

Nomenclature: Exotoxins which attack a variety of cell types are called **cytotoxins,** whereas exotoxins that attack specific cell types are named according to cell type or organ they damage such as **neurotoxin, enterotoxin, leucotoxin**. Exotoxins are also named from the species, which produces them and from the disease with which they are associated (*e.g.* cholera toxin, tetanus toxin). Toxins are also named based on their activities (*e.g.* adenylate cyclase, hemolysin, pneumolysin, leucidin). Still some are given simple a letter (*e.g.* Exotoxin A of *Pseudomonas aeruginosa*).

Exotoxin producing bacteria: Both Gram-positive and Gram-negative bacteria are known to produce different exotoxins.

Gram-positive bacteria: *Staphylococcus aureus, Bacillus cereus, B. anthracis, Clostridium tetani, C. botulinum, C. perfringens, Corynebacterium diphtheriae, Streptococcus pyogenes, S. pneumoniae, L. monocytogenes, Bacillus thuringiensis.*

Gram-negative bacteria: *V. cholera, Shigella* spp., *Salmonella* spp., Enterotoxigenic *E. coli,* Enteroinvasive *E. coli,* Enterohemarrhagic *E. coli, Pseudomonas aeruginosa, Bordetella pertussis, Yersinia* spp., *Xanthomonas compestris, Aeromonas* spp., *Helicobacter pylori, Haemophilus ducreyi.*

24.3.1 Classification of Exotoxins

A large number of exotoxins have been reported from a wide range of bacteria. The unifying features of all these exotoxins are: they are proteins or peptides and secreted into the cell environment. However, they differ with each other in many respects such as producer organism, method of secretion, target organism, tissue cell or cell organelle, pathological, biochemical and immunological responses, chemical organization, domain architecture etc. As a result, different classifications are in practice. These are based on:

- Organism generating the toxin (*e.g.* botulin toxin)
- Organism susceptible to the toxin
- Method of secretion (Type I to IV)
- Host tissue susceptible to toxin (*e.g.* Enterotoxin, neurotoxin)
- Structure of toxin (*e.g.* AB5 toxin)
- Domain architecture of the toxin (*e.g.* polymorphic toxin)
- Physical properties of toxin (*e.g.* heat labile and heat stable)
- Order of identification (*e.g.* toxins A, B, C)

In accordance with the theme of this book, classification based on the mode of action is followed here:

A. Pore-forming toxins
B. Extracellular toxins acting on the cell surface and superantigens
C. Toxins acting on intracellular targets
D. Toxins directly delivered into cytoplasm

A. Pore-forming Toxins (PFTs)

Pore-forming toxins, as the name suggests, insert a transmembranous pore into a host cell membrane, thereby disrupting the selective influx and efflux of ions across the membrane. These toxins are often identified as lytic factors (lysins). Since erythrocytes have often been used to test the activity of these toxins, many of them are known as hemolysins. Pore-forming proteins possess the ability to switch from a water-soluble form to a membrane inserted pore form. Bacteria release PFTs as soluble subunits which form stable multimeric complexes on the target membranes, and translocate across lipid membranes through several mechanisms.

Over 80 different families of PFTs constitute approximately one-third of all the characterized bacterial toxins, and are found across both Gram-positive and Gram-negative bacterial species. A remarkable feature of PFTs is their ability to exist in both water-soluble and membrane inserted forms. In the soluble form, the membrane spanning portions of the PFT are commonly stabilized by interaction with the core of the molecule. This class of proteins are amphipathic and thus facilitate one part interacting with the hydrophilic cavity filled with water and the other interacting with the lipid chains or nonpolar segments of integral membrane proteins. Pore formation often involves an extensive conformational change that permits the membrane spanning residues to insert in to the lipid bilayer. In addition to conformational mobility, another key feature of many PFTs is their ability to self-assemble into dough nut shaped oligomers. Together oligomerization and conformational change can allow the formation of pores. Depending upon the toxin, the pore diameter may vary from 1 to 50nm.

Disruption of cell permeability ultimately leads to cell death. However, the early consequences of losing the permeability barrier are often the release of cytokines, activation of intracellular proteases, and sometimes induction of apoptosis.

Pore-forming toxins are broadly classified into two groups: i. large-pore-forming and ii. small-pore-forming toxins based on dimension of holes produced on the plasma membrane and the kind of interaction that they establish with the eukaryotic receptor. In addition, the pore-forming RTX family of toxins include a large group of Ca^{2+}-dependent hemolysins secreted by both Gram-positive and Gram-negative bacteria which are characterized by a conserved glycine- and aspartate rich motif of nine amino acids.

i. Large-pore-forming toxins

This class of cytolysins comprises more than 20 family members, which are generally secreted by taxonomically diverse species of Gram-positive bacteria and which have common property of binding selectively to cholesterol on the eukaryotic cell membrane. Hence, they are also known by **cholesterol-dependent cytolysins (CDC)** (Table 24.3). Large-pore-forming toxins consist of a single polypeptide chain with molecular weights ranging from 50kDa to 80kDa and are characterized by a remarkable sequence similarity. These toxins contain a common motif (ECTGLAWEWWR) located approximately 40 amino acids from the carboxy terminus. Oxidation of cysteine residue, included in this motif, abolishes the lytic activity.

Large-pore-forming toxins belong β-PFTs which insert into the membrane as β-sheets. The pores are organized in a variety of subunit numbers (7–50) and sizes (2 nm – 50 nm). These toxins are produced by *S. pyogenes, S. pneumoniae, Bacillus* spp. a variety of clostridia, including *Clostridium tetani* and *Listeria.* To date, the best characterized are listeriolysin O, an essential virulence factor of *Listeria monocytogenes* that causes meningitis and abortion; perfringolysin O (PFO), a virulence factor of *Clostridium perfringens* that causes gas gangrene, pneumolysin, the major causative agent of streptococcal pneumonia and meningitis; and streptolysin O (SLO) secreted by *S. pyogenes.*

Large-pore-forming toxins share a similar mechanism of action. Pore complex formation occurs before protein insertion into the membrane and is pH dependent. The mechanism of pore insertion includes steps that involve soluble monomer association with the plasma membrane, pre-pore monomer oligomerization, pore formation and insertion of the oligomer into the membrane. It is interesting to note that the membrane bound receptor, cholesterol plays an important role in all these steps. Once the pre-pore complex has assembled on the membrane, a conformational change occurs in which the complex inserts into the membrane due to β-barrel pore formation.

All large PFTs perform a multitude of functions in bacterial infection. These toxins disrupt plasma membranes causing cell death by necrosis and facilitating bacterial invasions, or disrupt endosomal or phagosomal membranes to release bacteria into the interior of the cell. In addition to their pore-forming properties, many large PFTs possess, pro-inflammatory properties that enhance tissue damage at the site of infection. Further, certain toxins of this group possess the ability to kill cells by alternative mechanisms. For example, in bacterial meningitis caused by *Streptococcus pneumoniae*, PLY has been shown to form pores in the mitochondria of neurons, activating cell death pathways.

Table 24.3. Some important cholesterol dependent cytolysin (CDC) toxins, their producer bacteria and effect on host

Toxin	Bacterial Species	Effect on host
Anthrolysin O (ALO)	*Baccillus anthracis*	Kills human cells neutrophils, monocytes and macrophages
Botulenolysing (BLY)	*Clostridium botulinum*	Coronary vasconstriction is triggered causing cardiac dysfunction, loading to systomic hypertension and death in rat model
Perfringolysin O (PFO)	*C. Perfringense*	Promotes dysfunctional human PMN/endothelial cell adhesion contacts and vascular leukostasis. Inhibits human PMN chemoltasis
Telanolysin (TLY)	*C. telane*	Observed lysis of rabbit lysosomes in a suspension of large grannule fraction of rabbit liver.
Listeriolysin O (LLO)	*Listeria monocytogenes*	Intracelluar release of LLO; Suppression of phagocytosis by macrophages. Induces expression of 1L-1α, 1L-I2, IFN-γ IL-8, macrophase chemotoxis protein1, adhesion molecules on the surface of human epithelial cells; Activates NF-αB
Alveolysin (ALV)	*Paenibacillus alvei*	Induces IL-8 expression in human, polymorphonuclear leucocyte, lymphocyte, monocytel and basophil cell populations
Pneumolysin (PLY)	*Streptococcus pneumonial*	Directly activates the complement eascade, incuces, IL-1β and TNFα expression from human monocytes. Inhibetion of IgG production and proliferative response from human lymphocytes
Streptolysin O (SLO)	*S. pyogenes*	Induces chemotoxis and mobility of human PMNs. Induces expression of IL-1β, IL-6 and 1L-8 and release of prostaglandin E2 from human keratinocytes. Induces expression of IL-1β and TNFα expression from human monocytes

ii. Small-pore-forming toxins

The small-pore-forming family of toxins form very small pores (1 to 1.5 nm in diameter) in the membrane, allowing the selective permeabilization of cells to solutes that have a molecular mass less than 2000Da. This group of toxins include single-component toxins, such as *S. aureus* α-toxin, hemolysin II produced by *B. cereus* and bicomponent toxins such as panton-valentine leucocidin (Luks-PV-LuKF-PV) and the ϒ-hemolysins (HlgA-HlgB and HlgC-HlgB). These proteins are secreted as water soluble monomers and assemble on the surface of susceptible cell to form heptameric transmembrane channels.

S. aureus alpha-toxin is considered as the model of oligomerizing pore-forming toxins. The alpha toxin is synthesized as a 319-amino acid precursor molecule that contains an N-terminal signal sequence of 261 amino acids. The secreted mature toxin, or protomer, is a hydrophilic molecule with a molecular weight of 33kDa mushroom-shaped heptamer comprising three distinct domains. The cap and rim domains of the heptamer are situated at the surface of the plasma membrane, while the stem domain serves as a transmembranous ion-channel through the membrane.

When acted upon by small-PFTs, the cells become permeable only to molecules smaller than the pores (up to 2000Da). Ions like Ca^{2+} can enter the cells, so that electrical permeability is established, while large molecules such as cytoplasmic enzymes are retained within the cell. At this stage, the toxin induces a number of changes in the cell, such as production of eicosanoids, activation of endonucleases, release of cytokines, and early apoptotic events. At high concentrations, the toxin can cause membrane rupture and cell lysis, thus killing the cells. The toxin causes membrane damage in a variety of cells, including erythrocytes, platelets and leucocytes

Aerolysin secreted by *Aeromonas hydrophila* and alpha toxin produced by *Clostridium septicum* bind to a family of GPI-anchored specific receptor on the surface of target cells, including the T-cell receptor RT6. Following binding, the protein oligomerizes and forms heptameric structures that insert into the cell membrane forming pores.

iii. Membrane-perturbing toxins

Delta-hemolysin or δ-toxin, a 26-amino acid peptide, secreted by *Staphylococcus aureus* does not form the membrane pores. This toxin binds nonspecifically parallel to the surface of any membrane. At high concentration, the peptide self-associates and increases the perturbation of the lipid bilayer and eventually breaks into discoidal or micellar structures. The target eukaryotic cells first become permeable to small solutes and eventually swell and lyse, releasing cell intracellular contents.

RTX toxins

These toxins belong to a large family of Ca^{2+}-dependent hemolysins. RTX toxins produced by different genera of Gram-negative bacteria include prototype *E. coli* hemolysin (HLyA) (Table 24.4). The bifunctional adenylate cyclase/hemolysin (CyaA) of *Bordetella pertussis,* the leukotoxins of *Pasteurella haemolytica* AL (LKtA) and *Actinobacillus actinomycetemcomitans* (AaLtA), and further the hemolysins of *Actinobacillus pleuropneumoniae, Actinobacillus suis, Proteus vulgaris* and *Morganella morganii.* All these toxins have in common the structural motif of several repeats of consensus sequence of nine amino acids (Leu/Ile/Phe-X-Gly-Gly/X-X-Gly-Asn/Asp-Asp-X). The number of repeated units varies from 6 to 47. Based on this genetically conserved structure, acronym RTX (repeats in toxins) has been proposed for this family of toxins. The respective units of the RTX toxins are responsible for Ca^{2+}-dependent binding to membranes. The *E. coli* hemolysin is encoded by four genes, one of which hlyA, encodes the 110kDA hemolysin, while the others are required for post-translational modification and secretion. In fact, the toxin is activated by post translational modifications, and binding to Ca^{2+} ions to the repeat units. Following binding, the N-terminal domain formed by hydrophobic and amphipathic transmembrane pores, that are small, unstable, hydrophilic, and actin specific. The toxin lyses a variety of cells including erythrocytes and human T lymphocytes. Like other small-pore-forming toxins, the initial permeabilization causes the loss of small solutes, uptake of water, and osmotic cell lysis. *E. coli* hemolysin induces apoptosis of T lymphocytes as well as the release of inflammatory lipid mediators like leukotrienes from human neutrophils and of interleukin 1b from monocytes.

Table 24.4. Some important RT toxins, their producer bacteria, mode of action and effect on host

Toxin	Bacterial species	Mode of action	Effect on host
HlyA	*E. coli*	Calcium-dependent formation of transmembrane pores	Cell lysis of erythrocytes, leukocytes, endothelial cells, and human T lymphocytes
AppA	*A. pleuropneumoniae*	Calcium-dependent formation of transmembrane pores	Lysis of erythrocytes and a variety of nucleated cells
AaltA	*A. actinomyce-temcomitans*	Calcium-dependent formation of transmembrane pores	Apoptosis
LktA	*P. hemolytica*	Calcium-dependent formation of transmembrane pores	Activity-specific vs ruminant leukocytes

Insecticidal toxins

Insecticidal proteins, also known as δ-endotoxins, include many toxins produced by species of *Bacillus thuringiensis* and that exert their toxic activity by making pores in the epithelial cell membrane of the insect midgut. The δ-endotoxins form two multigenic families, *cry* and *cyt*. Members of *cry* family, that include toxins such as Cry IA, Cry IIA, Cry III, Cry IV and Cry V, are toxic to insects of lepidoptera, diptera and coleoptera. On the other hand, members of *cyt* family are specifically lethal to the larvae of dipteran insects. The insecticidal toxins of the *Cry* family are synthesized by the bacterium as protoxins of molecular masses of 70 to 135 kDA; after ingestion by the susceptible insect, the protoxin is cleaved by gut proteases to release the active toxin of 60–70 kDa. In this form, they bind specifically and with high affinity to protein receptors channels of 10–20°A in the cell membrane.

Escherichia coli HlyE toxin

HlyE is a novel pore-forming toxin produced by *E. coli* which is completely unrelated to the *E. coli* hemolysin HlyA of RTX family. Recent studies suggest that HlyE could be the prototype of a new family of HlyE-like hemolysins specific for Gram-negative bacteria. This new class of pore-forming toxins forms cation-selective water permeable pores of 25 to 30°A in diameter. The channel formation either could be a part of mechanism for iron acquisition by the bacterial cell or may promote bacterial infection by killing immune cells and causing tissue damage.

B. Extracellular Toxins Acting on the Cell Surface and Superantigens

i. Receptor-targeting toxins

Some toxins modulate the physiological receptors of the cells which influence intracellular reaction cascades. Heat stable toxin of *E. coli* and toxins produced by staphylococci and streptococci which are often referred to as superantigens fall under this category.

Stimulation of membrane-bound guanylate cyclase: *Escherichia coli* heat stable enterotoxin

E. coli heat-stable enterotoxin (STa or STI) acts by stimulating the membrane-bound guanylate cyclase C, which is part of a signal transduction pathway normally exploited by endogenous neurohormonal regulators of intestinal salt and water transport. The heat-stable enterotoxin is a plasmid encoded peptide of 18–20 amino acids with three disulfide bridges. STa has a 50% homology with guanylin, an intestinal 15-amino acid peptide. STa binds the guanylate cyclase C (GC-C) receptor on the surface of epithelial cells of the small intestine and colon.

The initial step of intoxication is that the enterotoxin binds directly to the guanylate cyclase receptor or/and another kind of receptor. STa-bound GC-C activates protein kinase G (PKG) and protein kinase C (PKC), increasing IP3-mediated calcium to elevate extracellular GMP(cGMP). Increased cGMP activates phosphorylation of the Cl^- ion channel, cystic fibrosis transmembrane regulator, increasing the concentration of Cl^- ions in the extracellular space and causing fluid accumulation within the lumen of the intestine.

Besides STa, enterotoxigenic *E. coli* strains produce a second kind of heat-stable enterotoxin, a 48-amino acid peptide termed STb or ST II. Despite the similarity of nomenclature, STB does not show any homology with STa. STb binds sulfatide, a glycospingolipid on epithelial cells and internalization is required for intoxication. This toxin does not cause cGMP elevation as observed with STa, however, it stimulates fluid accumulation through a pertussis toxin sensitive GTP-binding regulatory protein, which has been proposed to stimulate the elevation of intracellular calcium through an unidentified ligand-gated Ca_2^+ channel. Recent studies revealed that-STb forms multimeric complexes and enhance membrane permeability, implicating a pore-forming toxin like activity.

ii. Interaction with MHC class II and T cell receptor: Bacterial superantigens

Superantigens are a class of structurally related bacterial protein toxins, 22-29kDa in size, which are extremely potent mitogens of T lymphocytes. The designation superantigen was chosen because the mechanism of T lymphocyte stimulation closely mimics that of "classical" antigens. Superantigens are produced by a variety of bacteria (Table 24.5). Established members of the superantigen family are the staphylococcal enterotoxins (SE), A, B, C1, C2, C3, D and E which are among the leading sources of food poisoning. The toxic-shock syndrome toxin (TSST-1) of *Staphylococcus aureus* induces an acute and sometimes life-threatening diseases. The pyrogenic (erythrogenic) toxin A and C of *Streptococcus pyogenes* are involved in scarlet fever. *Mycoplasma arthritidis* mitogen (MAM) causes an acute inflammatory infection.

Members of the superantigen family differ to a varying degree in their sequence homology but they show a similar three-dimensional structure. One β-pleated domain is formed by the N- and C-terminal regions which is involved in MHC and TCR binding. Diametrically opposed to this domain is a shallow cavity at the C-terminal region which most likely constitutes an additional TCR binding site.

In contrast to classical antigens, superantigens do not require proteolytic processing, but act as intact proteins. The affinity of superantigens to MHC class II in humans depends on the subtype of the human leucocyte antigen (HLA). Staphylococcal enterotoxins bind to HLA-DR and HLA-DQ but not to HLA-DP. Superantigens can bind MHC II by several mechanisms. TSST-1binds the MHC II α-chain and a region of the β-chain by extending over the presented peptide (class I). Staphylococcal and streptococcal superantigens bind the MHC II α-chain through the OB-domain and do not contact the peptide (Class II). Class IV and V superantigens bind through the C-terminal domain to the β-chain of MHC II, utilizing a coordinated zinc site as well as the presented peptide. MAM binds in a different fashion by binding the MHC II α1-helix and β1 helix, causing the dimerization of the two MHC II molecules.

Table 24.5. Some superntigen toxins, their produce bacteria, mode of action and effect on host

Toxin	Organism	Mode of action	Effect on host
SEA-SEH, TSST-1, SPEA, SPEC, SSA, SPEL, SPEM	*S. aureus* and *S. pyogenes*	Binding to MHC class II molecules and to Vβ or Vγ of T-cell receptor	T-cell activation and cytokine secretion
SMEZ	*S. pyogenes*	Binding to MHC class II molecules and to Vβ or Vγ of T-cell receptor	T-cell activation and cytokine secretion
SPEB	*S. pyogenes*	Cysteine protease, integrin binding	Proteolytic cleavage of fibronectin and vitronectin
ETA, ETB, ETD	*S. aureus*	Trypsin-like serine pro teases	T-cell proliferation, intraepidermal layer separation

Classical antigenic peptides interact with variable components of both the α-chain and β-chain of the TCR (Fig. 24.3). In this way, a small fraction of T cells is stimulated in a controlled way, with physiological amounts of cytokines being released. In contrast, the superantigen MHC II complex contact specifically, a small conserved region of the variable β-chain (V β) of the TCR, via diverse mechanisms, that include binding to the CDR2 loop of TCR V β. This interaction is sufficient to activate up to 20% of T cells. Most superantigens, however have an increased specificity to TCR which is defined by specific interactions with the TCR hypervariable areas surrounding the CDR2 loop.

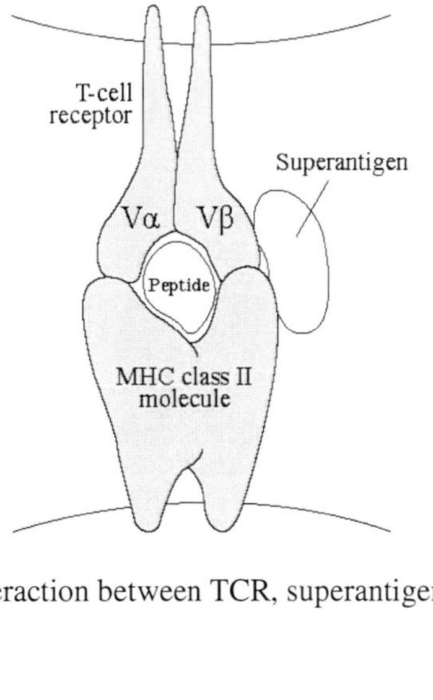

Fig. 24.3. Interaction between TCR, superantigen and MHC

Stimulation induces T cells to release interleukin-2 and interferon –Υ and subsequently tumor necrosis factor α, interleukin-1 and other monocyte mediators which are believed to play an important role in diseases, such as toxic shock syndrome induced by toxic shock syndrome toxin (TSST-1), vomiting and diarrhea caused by staphylococcal enterotoxins, and the exanthema caused by the pyrogenic streptococcal exotoxins. In addition, these toxins have also been implicated in the pathogenesis of several acute or chronic human disease states, such as Kawasaki syndrome and in other life-threatening events such as food poisoning.

The exfoliate toxins ETA and ETB of *S. aureus* have been recognized as the causative agents in staphylococcal scalded skin syndrome. Both superantigens are trypsin-like serine proteases which induce T-cell proliferation and cause intraepidermal layer separation. However, it is still a debate as to whether these two toxins should be classified as superantigens.

C. Toxins Acting on Intracellular Targets

The group of toxins with an intracellular target (A/B toxins) contains many toxins with different structures that have only one general feature in common: they are composed of two domains generally identified as A and B. A is the active portion of the toxin; it usually has enzymatic activity and can recognize and modify a target molecule within the cytosol of eukaryotic cells. B is usually the carrier for the A subunit; it binds to the receptors on the cell surface and facilitates the translocation of A across the cytoplasmic membrane. The A component is not active until it is released from the native (A+B) toxin. The functions of the A, B subunits are specific. Isolated A subunits are enzymatically active, but lack binding and cell entry capability. Similarly, isolated B subunits may bind to target cells but they are nontoxic.

A and B subunits in different bacteria are synthesized and arranged in different ways. A toxin indicated by A+B denotes that the toxin is synthesized and secreted as two separate protein subunits that interact at the target cell surface. A-B or A-5B indicates that the A and B subunits are synthesized separately, but associated by non-covalent bonds during secretion and binding domain of the target; 5B indicates that the binding domain of the proteins is composed of 5 identical subunits. On the other hand, A/B cleaved as a single polypeptide into A and B domains by proteases.

i. Attachment and entry of toxins

Attachment and entry of toxins are accomplished by two well-known mechanisms: i. Direct entry and ii. Receptor-mediated endocytosis (RME) entry.

In direct entry, the B subunit of the native (A+B) toxin binds to a specific receptor on the target cell and induces the formation of a pore in the membrane through which the A subunit is transferred into the cell cytoplasm. The specific receptors for the B subunit of toxins on target cells or tissues are usually, disialogangliosides (glycoproteins) called G-proteins on the cell membrane. In RME, the native toxin binds to the target cell and the A+B structure is taken into cell in a membrane enclosed vesicle called an endosome. H^+ ions enter the endosome lowering the internal pH which causes the A+B subunit to separate. The B subunit affects the release of the A subunit from the endosome, so that it will translocate to its target in the cell cytoplasm. The B subunit remains in the endosome and is recycled to the cell surface.

A few bacterial toxins (*e.g.* diphtheria) are known to utilize both direct entry and RME to enter host cells. On the other hand, bacterial toxins with similar enzymatic mechanisms may enter their target cells by different mechanisms. For example, the diphtheria toxin and *Pseudomonas* Exotoxin A, which have identical mechanisms of enzymatic activity, enter their host cells in slightly different ways.

Depending on their target, A/B toxins are divided into different groups, that act on protein synthesis, signal transduction, actin polymerization and trafficking within eukaryotic cells.

ii. Toxins inhibiting protein synthesis

Toxins like diphtheria toxin, *P. aeruginosa* exotoxin A and shiga toxin inhibit protein synthesis, causing rapid cell death at extremely low concentration.

Diphtheria toxin – A model for protein exotoxins with AB organization and enzymatic activity

Of all the protein exotoxins, diphtheria toxin (Dtx) produced by *Corynebacterium diphtheriae* is well studied and best known. It serves as a prototype to understand the structure and function of A-B exotoxins especially A/B exotoxins. It is produced as a single polypeptide chain with a molecular weight of 60,000 Daltons. This polypeptide is cleaved by trypsin in the presence of thiol (reducing agent) into A and B fragments or subunits. Subunit A, with a molecular weight of 21,000 Daltons, contains enzymatic activity and subunit B with a molecular weight of 39,000 Daltons, is responsible for binding to the membrane of a susceptible host. The B subunit possesses a region T (translocation) domain which inserts into the endosome membrane thus ensuring the release of the enzymatic component into the cytoplasm. In native state, *i.e.,* when the two fragments are intact the enzymatic activity of A fragment is masked by B fragment. Further, B fragment is required to enable fragment A to reach the cytoplasm of susceptible cells. The C terminal of the fragment B is hydrophilic and contains determinates that interact with specific membrane receptors on sensitive cell membranes. The specific membrane receptor for the B fragment has been shown to be a transmembranous heparin-binding protein on the susceptible cells surface.

Entry, translocation and enzymatic function: The diphtheria toxin enters its target cell by either direct entry or receptor mediated endocytosis. In either case, the first step is irreversible binding of the C-terminal hydrophilic portion of fragment B to the receptor. During RME (Fig. 24.4), the whole toxin is then taken up in an endocytic vesicle. Acidification of the endocytic vesicle allows unfolding of the A and B chains exposing the hydrophobic T-domain of the toxin. The T domain inserts into the endosome membrane translocating the A fragment into the cytoplasmic side of the membrane. There, reduction of proteolytic cleavage by trypsin releases the A chain into the cytoplasm. The A fragment is released as an extended chain but regains its active (enzymatic) globular conformation in the cytoplasm. The active A chain catalyzes the ADP ribosylation of elongation factors-2 (EF-2), thus inhibits protein synthesis.

$$\text{EF-2} + \text{NAD}^+ \longrightarrow \text{ADP-ribose diphthamide-EF2} + \text{Nicotinamide} + \text{H}^+$$

Pseudomonas Exotoxin A: Pseudomonas Exotoxin A (Exo A) is a $66k_{Da}$ single chain protein encoded by chromosomal gene. Exo A has a mechanism of action (ADP-ribosylation of EF2) identical to that of DT. However, the proteins are totally unrelated and have no primary sequence homology.

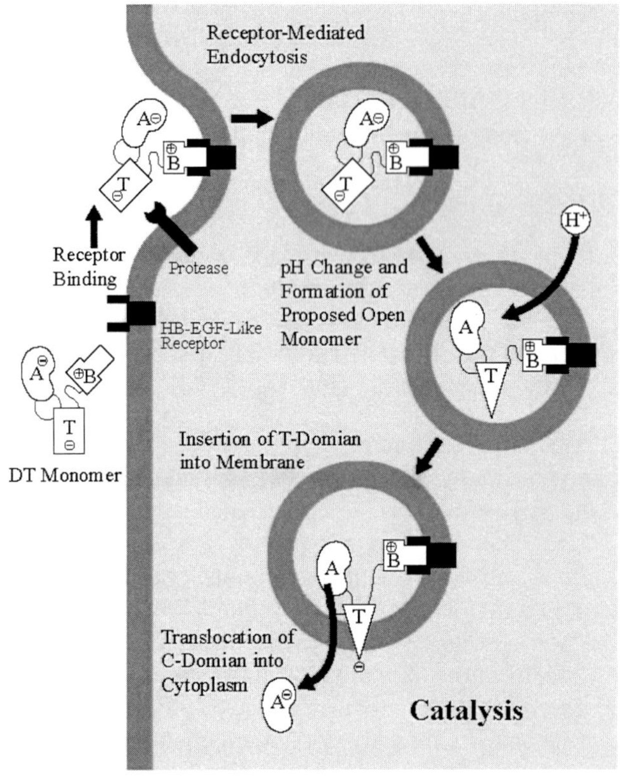

$$EF\text{-}2 + NAD^+ \longrightarrow ADP\text{-}RIBOSE\text{-}DIPTHAMIDE\text{-}EF\text{-}2 + NICOTINAMIDE + H^+$$

Fig. 24.4. Entry, translocation and enzymatic function of diphtheria toxin

Shiga toxin: Shiga toxin (also known as verotoxin) is a prototype of a number of related toxins produced by the causative agents of dysentery (*Shigella dysenteriae*) and hemorrhagic colitis (*E. coli*) producing shiga toxin types 1 and 2. Shiga toxin is a typical A/B toxin, with an enzymatically active A peptide of 35kDA that is responsible for the toxicity. This has an N-glycosidase activity, which cleaves an adenine residue from the 28S RNA, altering the function of ribosomes, which are no longer able to interact with elongation factors EF1 and EF2. This results in an inability to carry out protein synthesis and thus cell death.

iii. Toxins acting on signal transduction

Communication between and within cells is essential for any living organism. In eukaryotic cells, signals from outside stimulate receptors on the cell membrane mainly by two mechanisms: i. tyrosine phosphorylation of the cytoplasmic C-terminal part of the receptor; and ii. modification of a receptor-coupled GTP-binding protein. Bacterial toxins act all levels of signal transduction; they modify GTP binding proteins (CT, LT and PT), produce secondary messengers such as cAMP (Cya A and EF), and act on the intracellular signal cascade either by modifying small

GTP-binding proteins such as Rho, Rac and cdc42 (*Clostridium difficile* toxins A and B and *E. coli* CNF1 and CNF2) or by modifying kinases such as mitogen-activated MAPKK1 (LF).

iv. Toxins acting on G-protein

CT and *E. coli* LT-1 and LT-II: CT (Cholera toxin) is produced by *Vibrio cholerae,* the causative agent of cholera and LT-I and LT-II (heat labile) are produced by enterotoxic *E. coli* (ETEC), isolated from the humans with traveler's diarrhea. Both toxins are almost identical in structure and function, having an 80% sequence homology. These toxins are generally described as having an AB_5 composition, meaning that they contain one enzymatically active A subunit and a pentameric B subunit composed of five identical monomers. Cholera toxin is chromosomally encoded, whereas *E. coli* heat-labile toxins encoded on a plasmid.

A prerequisite for the enzymatic activity of the toxins is the proteolytic cleavage of the A subunit and the reduction of the disulfide bond to yield a catalytically A_1 fragment (192 amino acids) and an A_2 fragment (47 amino acids). A_2 mediates the binding of A_1 to the B pentamer. B protomer binds to the ganglioside receptor GM_1 on the membrane of target intestinal epithelial cells (Fig. 24.5).

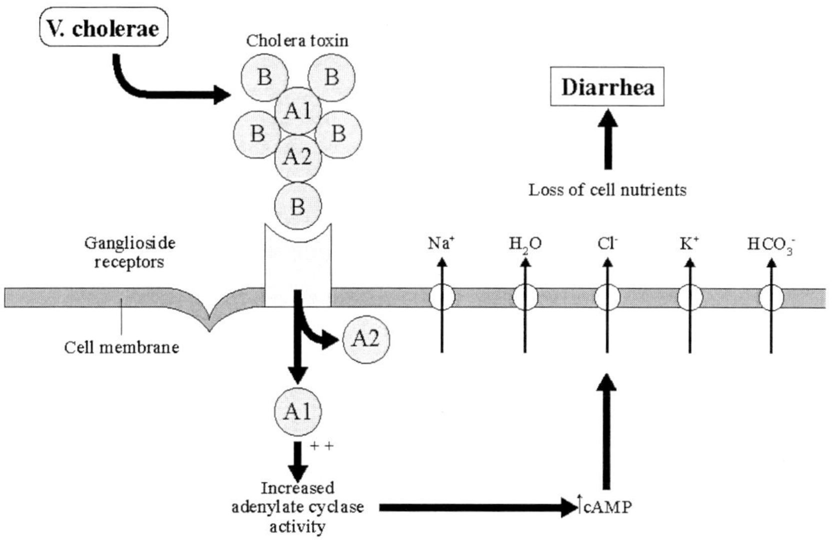

Fig. 24.5. Composition and mode of action of cholera toxin

After translocation of the enzymatically active A_1 subunit into the cytosol, ADP ribose is transferred to arginine-174 of the Gsα subunit of the G protein. ADP ribosylation results in inhibition of the GTPase activity of the Gsα protein. The Gs protein and the Gs dependent adenylate cyclase are fixed in their active statuses and cAMP is permanently produced. This inhibits the electro-neutral sodium chloride transport systems of the jejunum and the ileum and decreases water absorption. The simultaneous opening of the electrogenic chloride channels of the colon increases the intraluminal liquid in flow. These alterations of the liquid flows result in the characteristic severe diarrhea. In addition to above effects, the cholera toxin may also induce water and electrolyte secretion by stimulating prostaglandins.

Cholera-like enterotoxins are also produced by various other bacteria including *Vibrio mimicus, Salmonella, Pseudomonas, Campylobacter jejuni* and *Aeromonas hydrophila.*

Pertussis toxin: Pertussis toxin (PT) is a chromosomally encoded, 150kDa virulence factor secreted by *Bordetella pertussis*, the bacterium that causes whooping cough. The toxin is composed of five noncovalently linked subunits named S_1through S_5, which are organized into two functionally different domains called A and B. The five subunits are individually secreted to the periplasmic space, where the toxin is assembled and then released by specialized type IV secretion apparatus. The A domain, which is composed of S1 subunit, is an enzyme that shares the active-site enzymatic activity and structure with DTx, Exo A, CT and LT and intoxicates eukaryotic cells by ADP-ribosylating their GTP-binding proteins. The B domain is a nontoxic oligomer formed by subunits S_2, S_3, S_4 and S_5 in a 1:1:2:1 ratio. This domain binds to glycoproteins having a branched mannose core with attached N-acetylglucosamine.

The toxin is then internalized by receptor mediated endocytosis and follows the retrograde transport through the Golgi as in the case of CT. Golgi network is a prerequisite for cellular intoxication. Like CT and LT, pertussis toxin acts by transferring ADP-ribose to heteromeric G protein. The protein transfers the ADP-ribose group to cysteine located in the C-terminal-XCGL motif of the α-subunit of many G proteins such as G_i, Go, G_1 and G_{gust}. The consequence of ADP ribosylation is the uncoupling of G proteins from their receptors, with alteration of all signals that are transduced by them. This may have different consequences in different tissues. Some of the symptoms of whooping cough, such as the release of histamine, lymphocytosis and hyper insulinemia can be attributed to the effects of PT.

Pt is one of the main components of acellular vaccines against whooping cough. In these vaccines, the toxin is chemically detoxified (by formaldehyde or hydrogen peroxide) or genetically detoxified (by changing the two catalytic amino acids by site-directed mutagenesis).

v. Toxins generating cAMP

Adenylate cyclase: The adenylate cyclase (Cya A) or adenylate cyclase toxin (ACT) produced by *Bordetella pertussis* causes a transient increase in the intracellular levels of cAMP up to 100-1000 times the normal status. Cya A is a bifunctional protein composed of a cell-invasive and calmodulin-dependent adenylate cyclase domain (400 N-terminal residues) fused to a pore-forming hemolysin consisting of 1,306 residues. The C-terminal hemolytic domain binds the receptors and delivers the N-terminal catalytic domain into the cell, where it binds calmodulin and catalyzes high-level synthesis of the second messenger cAMP there by disrupting cellular functions. The target cells are believed to be mainly alveolar macrophages and leukocytes.

Thus, *B. pertussis* has evolved two mechanisms for elevating cAMP in the host immediately, but transiently effective adenylate cyclase toxin, on one hand, and the much slower, but persistently active pertussis toxin, on the other.

Anthrax edema factor: Edema factor (EF) (adenylate cyclase) and lethal factor (LF) are the two major exotoxins produced by *B. anthracis*. These two proteins are transported into host cells by the pore-forming PA (Anthrax protective antigen). The three subunits, encoded by a large plasmid, are synthesized and secreted independently. The protective antigen acts as a B protomer and mediates the binding to a specific, as yet unidentified protein receptor. After binding, a cell surface protease identified as furin cuts off a 20kDa fragment from the N-terminus of PA (Fig. 24.6). The remaining receptor bound PA of 63KDa (PA63) is now able to bind either EF or LF, which compete for the binding, to form either adenylate cycles toxins (PA-EF or ACT) or lethal toxin (PA-LF). These toxins receptor complex in subsequently internalized by receptor-mediated endocytosis and released into the cytosol. The binding to calmodulin then activates AC, and the cAMP level of the host cell becomes elevated. However, the major pathogenic effects of anthrax are not caused by the invasive adenylate cyclase, but by the lethal factor.

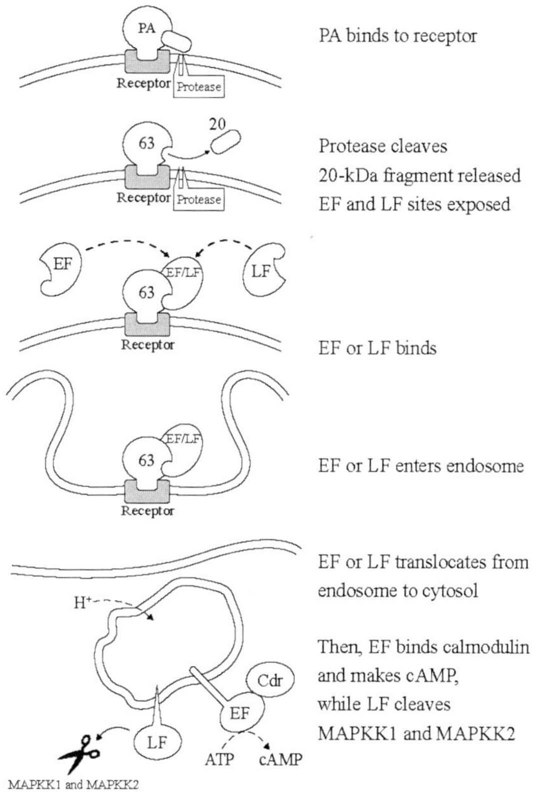

Fig. 24.6. Mode of action of anthrax toxin

vi. Toxins inactivating the cellular kinases

Anthrax lethal factor (LF), metalloprotease, released into the cytoplasm as described above, cleaves the N-terminus of the mitogen-activated protein kinases, MAPKK1 and MAPKK2. This cleavage inactivates MAPKK1 and inhibits the MAPK signal transduction pathway, a conserved pathway that controls cell proliferation, signal transduction and gene expression. The ability of LF to induce apoptosis of activated macrophages depends on the inhibition of P38MAPKs activation. Inhibition of P38 is sufficient to sensitize macrophages to activation-induced death by preventing induction of an antiapoptotic factor.

vii. Toxins inducing arrest in the G2 phase of the cell cycle

Transition of cells from G2 into mitosis requires activation of cyclin-dependent kinase $P34^{cdc2}$ by dephosphorylation. The cytolethal toxin produced by *Haemophilus ducreyi* (Hd CDT) intoxicates eukaryotic cells by causing a three-to-fivefold gradual distension and induces cell cycle arrest in the G2 phase. HdCDT induces an increased level of the tyrosine-phosphorylated from (inactive) of $P34^{cdc2}$ and hence leads to cell cycle arrest HdCDT is a complex of three proteins, CdtA, CdtB and CdtC, however CdtB is the active subunit of the CDT toxin. The role of CdtA and CdtC is still unknown.

viii. Toxins with phospholipase activity

The α-toxin is the key virulence factor produced by *Clostridium perfrigens* which is responsible for gas gangrene and is involved in the pathogenesis. This toxin is a zinc metalloenzyme that has phospholipase C (PLC) activity. The C-terminal domain of this proteinaceous toxin interacts with fatty acyl residues of phosphatidylcholine in the membrane, which is attributed for the observed hemolytic activity.

ix. Toxins modifying small GTP-binding protein: *Clostridium difficile* toxins A and B

Enterotoxin A and cytotoxin B are secreted by *C. difficile*. They bind to eukaryotic cell and are taken up by RME. Enterotoxic activity of *C. difficile* is mainly due to toxin A. Intracellularly, both toxins monoglucosylate small GTP-binding protein such as Rho, Rac and Cdc42. The site of modification is threonine 37 in Rho, and threonine 35 in Rac and Cdc42. Monoglucosylation results in breakdown of the cellular actin stress fibres. Toxin A has also been reported to induce apoptosis in T84 intestinal epithelial cells, in a time and dose dependent manner.

Cytotoxin-necrotizing factors (CNF) secreted by *E. coli* and dermonecrotic toxin (DNT) produced by *Bordetella* species exhibit similar mode of action.

x. Toxins acting on intracellular trafficking

Vital cellular processes, such as receptor mediated endocytosis and exocytosis, need specialized vesicles either to internalize the special molecules into the cell or to export molecules on to the cell surface. In some tissues, specialized vesicle transport can be used for the release to destined locations. A typical example of this is the local release of neurotransmitters at the presynaptic membrane. The key molecules involved in vesicle docking and membrane fusion, such as VAMP/Synaptobrevin, SNAP-25 and syntaxin, are specific targets of neurotoxins produced by *Clostridium tetani* and *Clostridium botulinum*, which are specific zinc-dependent proteases.

Neurotoxins of *Clostridium* spp

Clostridium botulinum and *C. tetani* secrete very potent neurotoxins, which cause botulinum and tetanus, respectively in humans and animals. The toxins of these two species and also other species such as *C. butyricum* and *C. baratii* produce toxins that dysfunct the functions of nervous system. Hence, these toxins are named as neurotoxins.

C. tetani forms a homogenous bacterial species which produces only one type of tetanus toxin (TeNT), whereas botulinum neurotoxin (BoNT)-producing strains are heterogeneous. *C.botulinum* is divided into 4 groups, which on the basis of phenotypic and genotypic parameters, correspond to different species. In addition, some strains of other species, such as *C.butyricum* and *C.baratii* can produce a related BoNT type E and F, respectively. Seven BoNT toxinotypes (A, B, C, D, E, F and G) are distinguished according to other antigenic properties. Each toxinotype is now divided into subtypes based on BoNT sequence variations, which can impact antibody binding and neutralization or affinity as well as catalytic efficiency for their substrate. BoNTs are associated with non-toxic proteins (ANTPs)to form large complexes. The function of ANTPs is still unclear, however, it is believed that they protect the neurotoxin from the acidic pH of the stomach and digestive proteases.

Structure: Both BoNT and TeNT share a common structure. They are synthesized as a precursor protein (about 150KDa), which is inactive or weakly active. The precursor which does not contain signal polypeptide is proteolytically activated in the extra bacterial medium either by *Clostridium* proteases or by digestive proteases in the intestinal lumen. The active neurotoxin consists of a light chain (L, about 50KDa) and a heavy chain (H, about 100kDa) which remain linked by a disulfide bridge. The structure of BoNTS shows three distinct domains: L-chain containing α-helices and β-pleates and including the catalytic zinc binding motif, the terminal part of the H-chain forming two unusually long and twisted α-helices, and the C-terminal part of the H-chain consisting of two distinct subdomains (H_{cn} and H_{cc}) involved in the recognition of the receptor.

The overall sequence identity at the amino acid level BoNT and TeNT ranges from 34–97%. Several domains are highly conserved which account for the common mode of action of these toxins. The central domains of L chains contain the consensus sequence (His-Glu-X-X-His) characteristic of zinc metallo-protease active site. The half N-terminal domain of the H-chain is also highly conserved and it is involved in the translocation of the L-chain in to the cytosol. Thus, a similar mechanism of internalization of the intracellular active domain into target cells is shared by all the clostridial neurotoxins. In contrast, the half C-terminal parts of H-chain, mainly the H_{cc} subdomains, are the most divergent. This accounts for the different receptors recognized by the clostridial neurotoxins.

Mode of action: BoNTs enter by oral route or are produced directly in the intestine by colonized organism and then undergo a transcytosis across the digestive mucosa. After diffusion into the extracellular fluid and blood stream circulization, BoNTs target motoneuron endings. In contrast, TeNT is formed in wounds colonized by *C. tetani*. TeNT diffuses in the extracellular fluid and can target all types of nervous endings (sensory, adrenergic neurons and motoneurons), but it is mainly retrogradely transported through the motoneurons.

Each type of BoNT and TeNT recognizes specific receptors on demyelinated terminal nerve endings, mainly through the H_{cc} Subdomaine. BoNT/A, /C/, E/, F exploit the three isoforms of the vesicle protein SV_2 as specific receptors, while BoNT/B and /G bind to synaptotagmin I or II. The GPI-anchored membrane protein Thy-1 has been proposed to act as a TeNT receptor. Gangloside-binding sites have been characterized in the Hcc subdomain. TeNT exhibits two carbohydrate binding sites, whereas BoNT/A and BoNT/B show only one. Accordingly, TeNT can bind simultaneously to two gangliosides (Fig. 24.7). BoNT/C and BoNT/D interact with gangliosides

(GD1b, GT1b) and phosphatidylethonolamine, respectively by their Hcc subdomain. The role of HCN subdomain is still unclear.

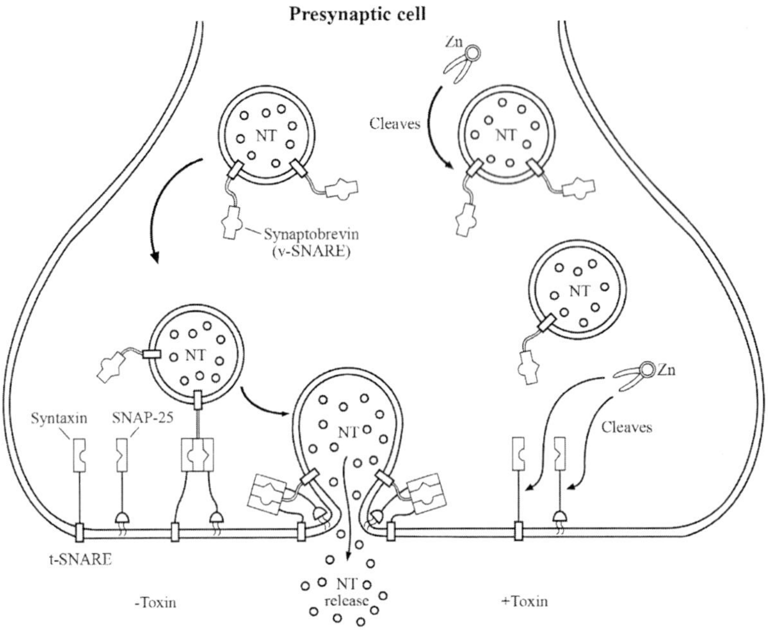

Fig. 24.7. Mode of action of neurotoxins

xi. Cleavage of VAMP/Synaptobrevin

TeNT and BoNT serotypes of B, D, F and G specifically show similar mode of action. During intoxication TeNT is internalized at junction and migrates retroaxonally within the motoneuron to the spinal cord, where it is released into the intersynaptic space located between the motoneuron and the inhibitory interneuron cell, penetrates the later, is translocated to the cytosol following acid-induced conformational change and blocks neuroexocytosis by cleaving the VAMP.

In case of BoNTs, H domain binds to receptors present in the motor neuron at the neuromuscular junction and translocate the active L form into the cytosol. Recent data show that the H-L complex is embedded in the membrane in the form of a transmembrane chaperone that acts as a dynamic structure to achieve final translocation of the L chain. L-chains of neurotoxins B, D, E and G are zinc dependent proteases that specifically cleave VAMP/Synaptobrevin.

Cleavage of SNAP-25 and Synaptobrevin: Like TeNT and BoNT B, D, F and G, L domain of BoNTA, E cleave SNAP-25 (synaptosome associated protein of 25KDa) at positions 197 (typeA) or 180 (typeE). SNAP-25 is a member of the SNARE protein family which is responsible for fusion. The cleaved SNAP-25 is unable to mediate fusion of vesicles with the host cell membrane, thus preventing the release of the neurotransmitter acetyl chlorine from axon endings. This blockage is slowly reversed as the toxin loses activity and SNARE proteins are slowly regenerated by the cell.

Type C toxin has the general H and L structure of the other botulin neurotoxins. The only difference is that it cleaves syntoxin, another protein involved in the exocytotic machinery. Type C is also unique because it can also cleave SNAP-25, thus being the only one able to cleave two substrates.

xii. Toxin acting on late endosomes/vacuoles

Helicobacter pylori, a pathogen responsible for gastritis and peptic ulcer, produces a vacuolating cytotoxin (VacA, 95KDa). This toxin causes massive growth of vacuoles with in epithelial cells. The vacuoles are coated with large quantities of Rab7, a small GTP-binding protein. The toxin induces an alteration of the endocytic pathway that results in selective swelling of the vesicles having the typical marker of late endosomes. VacA also forms water soluble oligomeric complexes and can somehow insert into lipid bilayers to anion selective channels.

xiii. Toxins acting on cytoskeleton

The cytoskeleton, a structure composed of microfilaments microtubules and intermediate filaments, controls the shape and spatial organization of the cells and is involved in cell movement, endo and exocytosis, vesicle transport, cell contact and mitosis. Bacterial toxins act mostly on microfilaments that are 7–9 nm in diameter and are built by polymerized actin. Many of the toxins affect the structure of the cytoskeleton by modifying the regulatory small G proteins, such as Ras, Rho, Rac and cdc42 which control cell shape. Many of the other bacterial toxins show indirect effect. The only toxin shown to affect directly the cytoskeleton is the C_2 toxin of *C.botulinum*, which ADP-ribosylates monomeric actin, making actin unable to polymerize.

D. Toxins Directly Delivered into Cytoplasm

Many virulent bacteria such as *Salmonella*, *Shigella* and *Yersinia* neither release the toxic proteins into the culture supernatant nor into the matrix of intracellular spaces of the host tissue. These bacteria intoxicate their hosts by using a contact dependent secretion system to inject or deliver toxic proteins into cytoplasm of eukaryotic cells. This is done by using specialized secretion systems that in Gram-negative bacteria are called type III or type IV depending on whether they use transmembrane structure like flagella or conjugative pili, respectively.

i. Toxins acting on signal transduction

Proteins acting on phosphorylation

Yersinia YpkA and YopH: Phosphorylation is central to many regulatory functions associated with the growth and proliferation of eukaryotic cells. Bacteria have developed a mechanism to interfere with these key functions in several ways. The best-known system is that of *Yersinia*, where a protein kinase YpkA and a protein tyrosine phosphate YopH are injected into the cytoplasm of eukaryotic cells by a type III secretion system to paralyze the macrophages before they can kill the bacterium.

YpkA is a ser/thr protein that also displays autophosphorylation activity. The natural eukaryotic substrates of YpkA belong to the class of small GTPases and comprise RhoA and Rac-

1, but not Cdc42. YopH acts by dephosphorylating cytoskeletal proteins, thus disrupting phosphotyrosine dependent signaling pathways necessary for phagocytosis.

EPEC Tir: A unique protein acting on signal transduction in eukaryotic cells is Tir (Translocated intimin receptors) of enteropathogenic *E. coli* (EPEC). A chaperone for Tir, called CesT is required for stability of Tir in EPEC cytoplasm. The protein mediates attachment of bacteria to the eukaryotic cells and is tyrosine phosphorylated. Following tyrosine phosphorylation, the protein mediates actin nucleation, resulting in pedestal formation and triggering tyrosine phosphorylation of additional host proteins including phospholipase C-Y.

H-Pylori Cag A: Cag A, the first bacterial protein linked to cancer, is 128 kDa protein characterized by a central region containing a repeating EPIYA motif. Once injected into the host cell, the protein is tyrosine phosphorylated at the EPIYA motif by the kinase C-Src and Lyn. The tyrosine phosphorylated CagA (Cag A-P) activates SHP-2 and inactivates C-Src, leading to cortactin dephosphorylation triggering a signal transduction cascade which results in cellular scattering proliferation. Recent data indicate that toxic activity of Cag A is due to the disruption of the apical junctional complex (AJC) of cells by interactions with ZO-1and Jam.

Proteins acting on small G proteins

SptP: *Salmonella* protein tyrosine phosphatase (SptP) is an effector protein secreted by type III secretion apparatus of *Salmonella enterica*. SptP is a modular protein composed of two functional domains; C-terminal domain and N-terminal domain. It was, recently demonstrated that the N-domain possesses strong GTPase-activating domain protein (GAP) activity for Cdc42 and Rac1.

Yopm protein of *Yersinia pestis* is believed to show similar mode of action.

ii. Proteins acting on cytoskeleton

***Salmonella enterica* serovar Typhimurium SopE:** SopE is injected into eukaryotic cells by a type III secretion system of *Salmonella* serovar *typhimurium* and it binds and activates the small G proteins Rac and Cdc42. The activation of Rac and CDc42 causes membrane ruffling and cytoskeletal rearrangements that mediate the uptake and internalization of the bacterium within the host eukaryotic cells. In addition, SopE stimulates nuclear response that induce the synthesis of proinflammatory cytokines to the induction of diarrhea.

***P. aeruginosa* Exoenzyme S:** *P. aeruginosa* exoenzyme S is a 49KDa protein with ADP-ribosylating activity that ADP-ribosylates the small G protein Ras at arginine 41. To become enzymatically active, Exo S requires an interaction with a cytoplasmic factor named factor activating exoenzymes S, or 14-3-3 protein. The toxin is injected into eukaryotic cells by type III secretion system. When the cells are intoxicated, the cytoskeleton collapses and the cell morphology changes resulting in sounding of the cells.

***C. botulinum* Exoenzyme C3:** Certain strains of *C.botulinum* produce the ADP ribosyltransferring C3 toxin (Syn.exoenzyme C3) in addition to the neurotoxins C1 and D and C2 toxin. The substrates of C3 toxin are Rho and Rho like G proteins, which are ADP-ribosylated at arginine-41. Rho is involved in the regulation of actin polymerization. If the protein is microinjected into cells or if the cells are transfected with the C3 gene under a eukaryotic promoter actin stress fibres are disrupted, the cells are rounded and arborescent protrusions are formed. Toxins with activity similar to C3 have been identified in *S. aureus* (ED1N), *Clostridium limosum, B. cereus* and *Legionella pneumophila.*

YopE: YopE, a protein encoded by the *Yersinia* pathogenicity island, is known to paralyze macrophage phagocytosis by causing actin depolymerization and disrupting the cell stress fibres, with consequent rounding of the cells and loss of cell shape. *In vivo* YopE is able to inhibit Rac but not Rho or Cdc42 regulated actin structures.

YopT: YopT is a prototype of a new family of 19 proteases with potent effects on host cells. YopT cleaves the post-translationally modified cysteine located at the C-terminal end of Rho GTPases (DKGCASS) causing the loss of the prenyl group from RhoA, Rac and Cdc42 and releasing them from the membrane. The inability of Rho to be located to the membrane causes disruption of the cytoskeleton.

SipA: SipA is a *Salmonella* protein essential for pathogenicity and exerts its function by binding F-actin finally resulting in the stabilization of F-actin.

VirA: VirA is *Shigella* effector protein, which is delivered into the host cell by a specialized type III secretion system. This protein interacts with tubulin to promote microtubule destabilization and membrane ruffling. By this mechanism, *Shigella* is able to remodel the cell surface and thus promote its entry into the host.

iii. Proteins altering inositol phosphate metabolism

SopB, a protein secreted by *Salmonella dublin* shows inositol phosphate phosphatase activity. This enzyme hydrolyzes phosphatidylinositol triphosphate (PIP_3) which is a messenger molecule that inhibits chloride secretion thus favouring fluid accumulation and diarrhea. Furthermore, SopB mediates actin cytoskeleton rearrangements and bacterial entry in a Rac-1 and Cdc42 dependent manner.

S. flexineri virulence factor IpgD is homologous to SopB, suggesting that a similar mode of virulence is also present in *Shigella.*

iv. Apoptosis inducing proteins

YopP is 30 KDa protein with cysteine protease activity that is encoded by a large plasmid of *Y. enterocolitica.* This protein is called YopJ in *Y. pseudotuberculosis* and is homologous to the Sipb protein of *Salmonella* to IpaB of *Shigella* to Avr A of *S. enterica* and to the AvrRxv protein kinases in host cells. Recent studies have shown that the mechanism of action of YopP involves the disruption of the NF-kb signaling pathway and therefore the occurrence of apoptosis in macrophages.

IpaB, a protein secreted by *Shigella* also induces macrophage apoptosis by binding to the interleukin-converting enzyme precursor (ICE or caspase I), that induces the processing to the active caspase, a well-known mediator of apoptosis. Like IpaB, SipB, a *Salmonella* invasion protein also induces apoptosis by binding ICE.

v. Cytotoxins with unknown mechanism of action

Despite the enormous amount of information available on different aspects of bacterial toxins, still there are many toxins whose mode of entry and mechanism are not known. Nevertheless, they are implicated in virulence and pathogenesis of a variety infections. Some of such toxins include: ExoU toxin produced by *Pseudomonas aeruginosa*; the zonula occludens toxins (Zot) produced by toxigenic strains of *V. cholerae*; HBL enterotoxin produced *by B. cereus*; bile salt hydrolase (BSH) elaborated by *Listeria monocytogenes.*

Novel bacterial toxins detected by genomic studies

With the advent of the genomic era, identification of bacterial factors possibly involved in virulence became easier. With the vast amount of information available on toxins, including sequence data, it is possible to proceed by homology criteria to predict novel members of important classes of bacterial toxins. Several examples exist where computer based methodologies have been instrumental in the identification of novel potential bacterial toxins in sequenced genomes. Mono-ADP-ribosyltransferases (MADPRTs) constitute a class of novel potent toxins discovered through this strategy. Followed by MADPRTs, more than 20 novel putative ADP-ribosyltransferases have been identified both in Gram-positive and Gram-negative organisms, including 5 from *Pseudomonas syringae*, five from *Burkholderia cepacea*, two from *Enterococcus faecalis* and one each from *Salmonella enterica* serovar *typhi*, *S. pyogenes*, *Mycoplasma pneumoniae*, *Streptomyces coelicolor*, *Bacillus halodurans* and *Vibrio parahaemolyticus*. Recently, a new protein toxin, *Neisseria* ADP-ribosylating enzyme (NarE) from *Neisseria meningitides* has been detected by computer analysis.

B. BACTERIOCINS

Microbes produce on extraordinary array of microbial defense systems. These include broad-spectrum classical antibiotics, metabolic by-products that include a wide range of primary and secondary metabolites, lytic agents such as lysozymes, numerous types of protein exotoxins. Most of these chemicals are discussed in detail in preceding chapters. In addition to these defense systems, almost all bacterial species produce another category of proteinaceous molecules known as bacteriocins. **Bacteriocins are proteinaceous compounds of bacterial origin that are lethal to bacteria other than the producing strain.**

The first bacteriocin discovered by Gratia (1925) was originally called principle V. It was produced by one strain of *Escherichia coli* and effective against another strain of *E. coli*. The term colicin was coined by Glatia and Fredericq (1946), whereas the term bacteriocin was used by Jacob and co-workers in 1953. The bacteriocins are characterized by :

i. Proteinaceous molecules.
ii. Exhibit bactericidal activity primarily against taxonomically closely related species.
iii. Attach to specific cell receptors.
iv. Their genetic determinants are plasmid encoded.
v. Synthesis of bacteriocins is lethal to even the producer organism.
vi. They are given the names according to the genus or species of the strain that produces them, for example, pediocins from *Pediococcus*, colicins from *E.coli* and boticins from *Clostridium botulinum*.

24.4. BACTERIOCINS vs. ANTIBIOTICS

Bacteriocins are often confused with antibiotics because of their shared characters like cidal activity and proteinaceous (at least some antibiotics). However, bacteriocins differ from traditional antibiotics in several characters. The main differences between bacteriocins and antibiotics are summarized in Table 24.6.

Table 24.6. Difference between bacteriocins and antibiotics

Characteristic	Bacteriocins	Antibiotics
Application	Food	Clinical
Synthesis	Ribosomal	Secondary metabolite
Activity	Narrow spectrum	Varying spectrum
Host cell immunity	Yes	No
Mechanism of target cell resistance or tolerance	Usually adaptation affecting cell membrane composition	Usually a genetically transferable determinant afecting different sites depending the mode of action
Interaction requirements	Sometimes docking molecules	Specific target
Mode of action	Mostly pore formation, but in a few cases possibly cell wall biosynthesis	Cell membrane or intracellular targets
Toxicity/side effects	None known	Yes

Bacteriocins have been found in all major lineages of bacteria. Every bacterial species examined to date was found to secrete bacteriocins and within a species tenor even hundreds of different kinds of bacteriocins are produced. According to Klaenhammer, 99% of all bacteria may make at least one bacteriocin and the only reason we have not isolated more is that few researchers have looked for them.

Bacteriocins can be distinguished broadly based on whether they originate from Gram-negative or Gram-positive bacteria. The difference in the structure of the bacteriocins of the two groups of bacteria appears to have resulted in the evolution of bacteriocins that are only active against similar bacteria within each group.

24.5. BACTERIOCINS PRODUCED BY GRAM-NEGATIVE BACTERIA

Many Gram-negative bacteria are known to produce bacteriocins. However, colicins and microcins produced by strains of *E. coli* and cloacin DF13 produced by *Enterobacter cloacae* are most extensively studied (Table 24.7) and their genetics, molecular structures and modes of action are well documented.

Colicins : Colicins produced by *E.coli* share certain key characteristics. Colicin gene cluster are encoded on plasmid and are composed of a colicin gene which encodes the toxin, an immunity gene which encodes a protein conferring specific immunity to the producer cell by binding to and inactivating the toxin protein; and a lysis gene, which encodes a protein involved in colicin release through lysis of the producer cell. Colicin production is mediated by the SOS regulon and is, therefore, principally produced under times of stress. Toxin productionis lethal for the target organism. This mode of targeting results in the relatively narrow phylogenetic killing range often cited for bacteriocins. Colicins are transported across the outer membrane and into the periplasm of the target cell using the bacterial transport systems, Tol or Ton. Once in the periplasm colicins can cause cell death by different mechanisms. Colicins E3,E4 and E6 and cloacin DF13 are active as RNAases. Colicin E3 and cloacin DF13 cleave 16s tRNA at a specific site which causes the termination of protein synthesis. Colicin E1, A, B, Ia, Ib, K and N form ion-permeable channels in the cytoplasmic membrane. Such channels may allow leakage of cations from the cytoplasm as well as low-molecular weight compounds such as ATP. Destruction of the electrical potential of the membrane will also occur. Colicin M inhibits the synthesis of peptidoglycan and lipopolysaccharide O antigen. In all cases, the colicin molecule needs to insert or pass through the inner membrane of the target cell. Hence, colicins, indeed all bacteriocins produced by Gram-negative bacteria are large proteins, Pore-forming colicins range in size from 449 to 629 amino acids, Nuclease bacteriocins have an even broader size range, from 178 to 777 amino acids.

Table 24.7. Bacteriocins from Gram-negative bacteria

Bacteriocin type	Molecular weight (kDa)	Producer	Examples
Colicin	4-90	*Escherichia coli*	Colicins A, B, D, E1-E7, Ia, Ib, K, L, M, N, V
Cloacin	56	*Enterobacter cloacae*	Cloacin DF13
Microcin	<1-5	*Escherichia coli*	Microcins A15, B17, C7, D15, D140, E492, H47, 15m

Colicins are classified into two distinct groups depending on whether they use the Tol pathway (Group A which includes colicins A, N, E1, K, E2-E9) or the Ton pathway (Group B which includes colicins B, Ia, Ib, D, M, V, 10, 5). A characteristic feature of colicin molecules is an organization of domains each associated with a step in killing activity; receptor binding, translocation across the outer membrane and cell death (pore formation, DNAase, RNAase etc.). It appears likely that colicins evolved by recombination of DNA fragments that encoded these domains. It is true for other bacteriocins produced by Gram-negative bacteria.

Although colicins are representative of Gram-negative bacteriocins, there are intriguing differences found within this subgroup of the bacteriocin family. *E.coli* encodes its colicins exclusively on plasmid replicons. The nuclease pyocins of *Pseudomonas aeruginosa*, which show sequence similarity to colicins and other yet uncharacterized bacteriocins are found exclusively on the chromosome. Another close relative to the colicin family, the bacteriocins of *Serratia marcescens* are found on both plasmids and chromosomes.

24.6. BACTERIOCINS OF GRAM-POSITIVE BACTERIA

Bacteriocins of Gram-positive bacteria are as abundant and even more diverse as those found in Gram-negative bacteria. They differ from Gram-negative bacteriocins in two fundamental ways. First, bacteriocin production isnot necessarily the lethal event it is for Gram-negative bacteria. This critical difference is due to the transport mechanisms. Gram-positive bacteria encode to release bacteriocin toxin. Some have evolved a bacteriocin-specific transport system, whereas others employ, the *sec*-dependent export pathway. Secondly, the Gram-positive bacteria have evolved bacteriocin-specific regulation, whereas bacteriocins of Gram-negative bacteria rely on host regulatory networks.

Klaehammer (1993) classified the bacteriocins of Gram-positive bacteria into 3 classes and four groups based on their amino acid sequences, molecular weights (kDa), structures and heat stabilities (Table 24.8). Later this classification was revised and only two classes were proposed (Colter *et al.* 2005). This classification scheme divides the bacteriocins into two distinct classes.

i. Class I bacteriocins (lantibiotics)

ii. ClassII bacteriocins

The large heat labile murein hydrolases (formerly class III bacteriocins) were moved out with a separate designation called '**Bacteriolysins**'.

Table 24.8. Classification of bacteriocins from Gram-positive bacteria

Class	Characteristics	Examples
I	Lantibiotics: small (19-37 amino acids), heat-stable	Nisin, lacticin 481, lactocin S, carnocin U149
II	Small, heat-stable nonlantibiotics (<15 kDa)	Pediocin PA-1, diplococcin, lactacin B, lactacin F, leucocin A, UAL 187, lactococcin A
III	Large, heat-labile proteins (>15 kDa)	Helveticin J, caseicin 80, lacticins A and B

i. Class I bacteriocins (Lantibiotics)

These bacteriocins are characterized by the presence of lanthionine type thioester bonds between certain side chains of amino acid residues of the protein. These modifications are introduced into the peptide post-translationally. In general, class I peptides typically have from 19 to more than 50 amino acids. These bacteriocins are further characterized by their unusual amino acids, such as lanthionine, methyl-lanthionine, dehydrobutyrine and dehydroalanine. Class I bacteriocins are further subdivided into class Ia and class Ib. Class Ia bacteriocins consist of cationic and hydrophobic peptides that form pores in target membrane and have a flexible structure compared to the more rigid class Ib. Class Ib bacteriocins, which are globular peptides, have no net charge or a net negative charge.

Nisin

Nisin is a well-known member of class Ia bacteriocins and is produced by strains of *Lactococcus lactis* subsp. *lactis*. Its structure is illustrated in Fig. 24.8. Nisin consists of 34 amino acids: however, it is initially synthesized as prenisin, consisting of a 23-amino acid leader peptide and the 34-amino acid pronisin peptide. Variants of nisin differing in one amino acid are known. Certain serine and threonine residues in the pronisin are converted to dehydroalanine and dehydrobutyrine through dehydration. Thioether bonds are then formed by reaction with the sulfhydryl groups of cysteine residues in the pronisin (lanthionine, Ala-S-Ala; 3-methylanthionine, Ala-S-Aba; Aba, aminobutyric acid). Following these chemical modificationsin the pronisin segment of prenisin, export and concomitant removal of the leader sequence yield active nisin. Normally, two nisin molecules form a functional dimer.

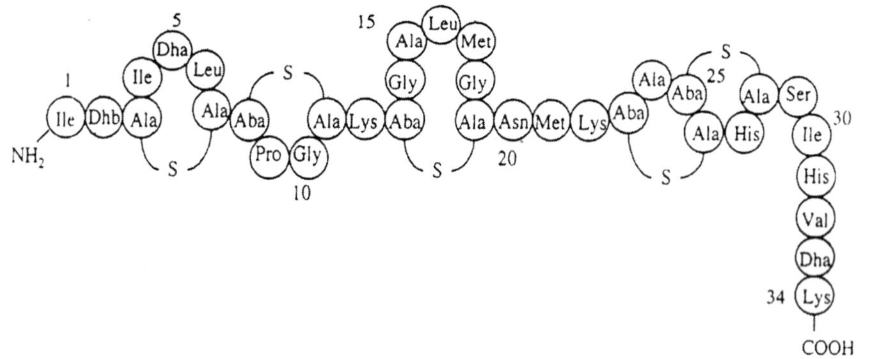

Fig. 24.8. Structure of nisin. Standard three-letter abbreviations are used for amino acids. Aba, aminobutyric acid; Dha, dehydroalanine; Dhb, dehydrobutyrin.

The gene encoding prenisin (nisA) is a part of a nisin gene cluster containing 11 genes that are responsible for nisin production (Fig. 24.9). This gene cluster is part of a conjugative transposable element that permits transfer of the nisin-producing capability to previously, nonproducing strains. Two of the 11 genes, nisR and nisk, encode products of a two-component regulatory system a response regulator and a sensor histidine kinase. Interestingly, nisinregulates its own products via this signaling system.

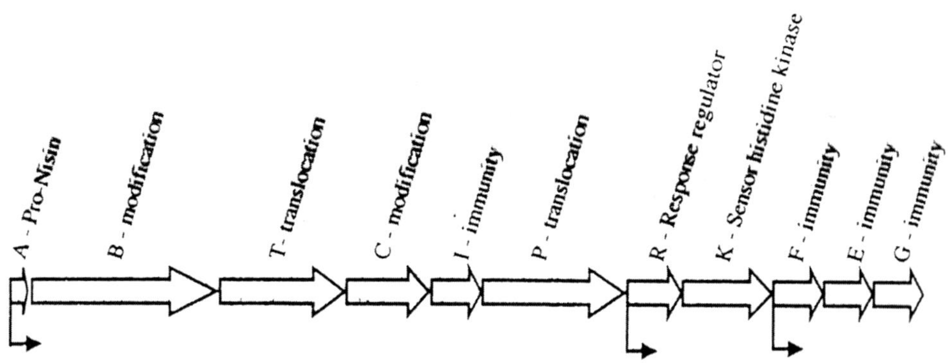

Fig. 24.9. The nisin A regulon of *L. lactis*. The organization of the genes is indicated by wide arrows. Above each arrow, the *nis* gene product and the function ascribed to the product are indicated. Promoters are marked by bent arrows.

Nisin uses lipid II, a peptidoglycan precursor as a docking molecule. Nisin and otherlantibiotics act on the energy transducing cytoplasmic membrane. The positively charged nisin binds to negatively charged components of the cell wall and membrane. Aggregation and insertion into the energized cytoplasmic membrane leads to the formation of channel. Channel formation results in the dissipation of the membrane potential and the chemical gradient across the membrane, and it ultimately leads to cell death.

Merscadidin, a class Ib bacteriocin, also uses lipid II as docking molecule, however, it inhibits peptidoglycan synthesis.

ii. Class II bacteriocins

The class II bacteriocins consist of the non-lanthionine containing peptide bacteriocins that are not subject to extensive post-translational modifications. They are low-molecular weight (<10 kDa) and heat stable peptides. In addition, a common feature of class II bacteriocins is a high content of small amino acids, such as glycine. Also, the peptides are strongly cationic with pIs in the range of 8 to 11. Related to their function as membrane active compounds, these peptides also contain hydrophobic or amphiphilic regions.

Most commonly, the genetic determinants for these bacteriocins reside on plasmids, although a few cases of chromosomal location are known. The bacteriocin genes encode preforms of the bacteriocins consisting of leader peptide, required for translocation of the bacteriocin across the cytoplasmic membrane, as well as the bacteriocin. Next to or downstream from the bacteriocin gene, a gene encoding protein conferring immunity is normally found. These immunity proteins are generally small peptides with approximately 50-150 amino acids. Regulatory system for class II bacteriocins is illustrated in Fig. 24.10. The inducer molecules, 1F, interact with the histidine kinase whichin turn transfers the phosphate group, P, to the response regulator, RR the phosphorylated protein activates the expression of several operons, including the bacteriocin operon.

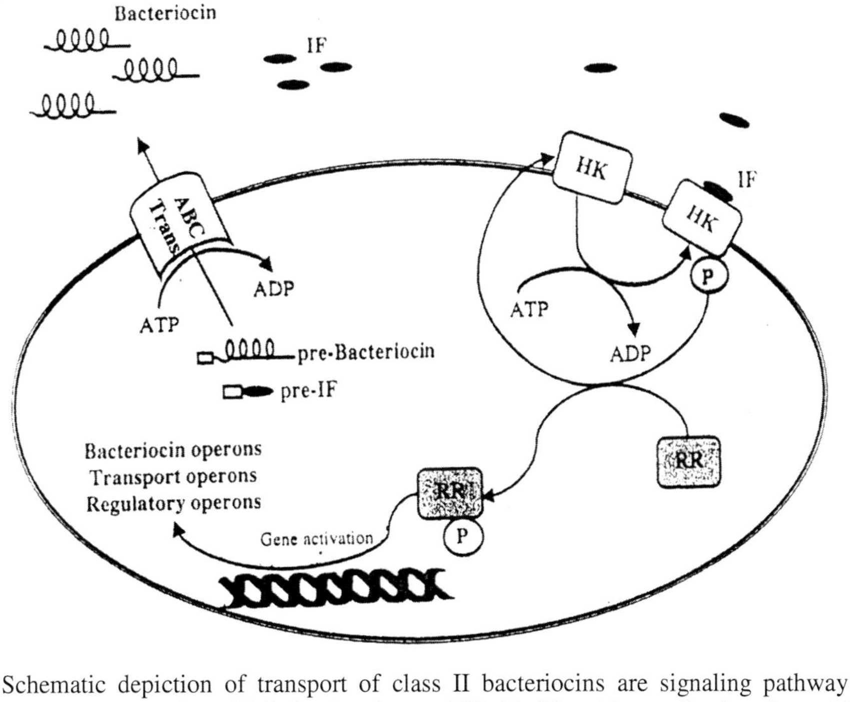

Fig. 24.10. Schematic depiction of transport of class II bacteriocins are signaling pathway leading to bacteriocin expression, IF, induction factor; HK, histidine kinase; P, phosphate; RR, response regulator; ABC Trans., ABC transporter system

Classification: Class II includes a large number of bacteriocins and is divided into four subclasses (IIa, IIb, IIc and IId).

Class IIa consists of the anti- Listeria one–peptide pediocin-like bacteriocins that have very similar amino acid sequences. Conservation between peptides is most evident in the hydrophilic, cationic N-terminal pediocin box region. The C-terminal domain is variable, hydrophobic, and/or amphiphilic. Class IIb contains the two peptide bacteriocins (e.g., lactacin F, lactococcin G and plantaricin S). These bacteriocins are unique in that they consist of two very different peptides. The two-peptide bacteriocins require the combined activity of both peptides with a mechanism of action that involves the dissipation of membrane potential, the leakage of ions and/or a decrease in intracellular ATP concentrations. Class IIc consists of the cyclic bacteriocins whose N- and C-terminal are covalently linked resulting in a cyclic structure. These bacteriocins target the cytoplasmic membrane of sensitive cells, leading to pore formation that results in loss of ions, dissipation of membrane and ultimately, cell death. The linear non-pediocin-like one-peptide bacteriocins that show no sequence similarity to the pediocin like bacteriocins are placed in class IId

Class IIa and Class IIb bacteriocins are synthesized as probacteriocins, consisting of a leader sequence (double-glycineleader) with characteristic glycine-glycine residues located at the cleavage site between the leader peptide and mature bacteriocin. Secretion of classIIa and Class IIb bacteriocins is mediated by the ATP-binding cassette (ABC) transporter (Fig. 24.11), a protein that is localized in the membrane. It consists of cytoplasmic domain with ATP-binding activity, a centralmembrane spanning domain, and a second cytoplasmic domain with proteolytic activity responsible for cleavage of the leader peptide. In case of lactococcin A, the ABC transporter is encoded by lcnC, whose product is C (Fig. 24.11). In addition, the product of *lcnD* is also involved in transport and maturation of prolactococcin A form the cytoplasm through membrane and the cell wall to the extracellular medium.

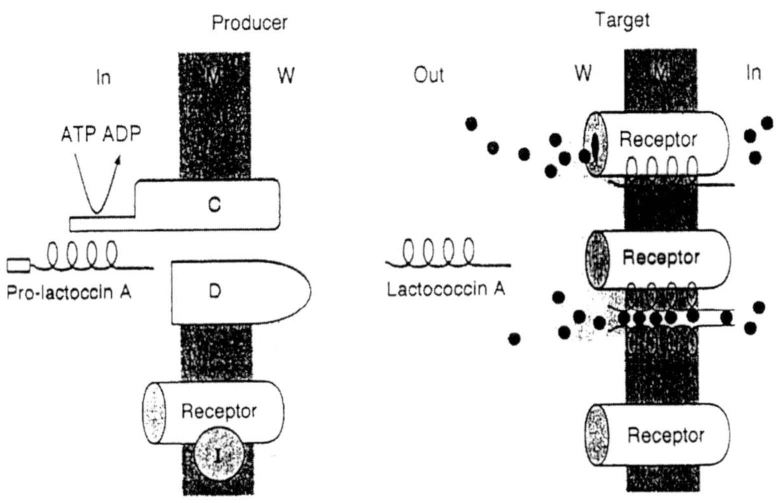

Fig. 24.11. Production of a class II bactteriocin and its effect on a target cell illustrated by the example of lactococcin A, In, inside the cytoplasmic membrane; M, cytoplasmic membrane; W, bacterial cell wall; Out, outside the cell, growth medium; C, LncC, a component of the ABC transporter system; D, LncD, a second component of the transport system; R, receptor for bacteriocin; I, immunity protein rendering receptor insensitive to lactococcin A action.

24.7. IMMUNITY AND RESISTANCE OF BACTERIOCIN PRODUCING CELLS

One interesting phenomenon of bacteriocin producing cells is the immunity towards their own bacteriocins. This phenomenon distinguishes bacteriocins from antibiotics. The genes coding for immunity proteins are in close genetic proximity to other bacteriocin structural and processing genes. The immunity of lantibiotic was initially thought to be due to an immunity gene, such as *nis*I for nisin and *Spa* I for subtilin, which code for *Nis*I and *Spa*I immunity proteins, respectively. It appears, however, that immunity to these bacteriocins is the result of the influence of several proteins, since the deletion of other genes results in altered host immunity. The phenomenon of immunity is simpler in the nonlantibiotics, class II bacteriocins. One gene codes for the immunity protein. Usually, it is a basic protein between 50 and 150 amino acid residues long that is loosely associated with the membrane. The lactococcin A immunity protein is by far the most studied one yet the basic mechanism behind the immunity is still not understood.

Although bacteriocins are not antibiotics there is concern that exposure to bacteriocins will render cells more resistant to antibiotics. Since antibiotics and bacteriocins (e.g. nisin) have different modes of action, it has been shown that exposure to nisin has no effect on the frequency of resistance of *L. monocyte genes* Scott A to ampicillin and chloromphenicol. In another study, several multi-drug resistant bacteria were subjected to upto 400 IU/ml nisin, and these organisms remained sensitive to nisin. Antibiotic resistance is usually associated with a genetic determinant, facilitating the transfer of resistance between cells, strains and species. Unlike most antibiotic resistance, bacteriocin resistance results from physiological change in the target cell membrane. Mechanismstherthan change in the cell membrane, are also reported. For example, some mutants produce an enzyme, nisinase, which degrades nisin. There is contradictory data regarding cross-resistance when bacteriocins from different classes are used.

24.8. ECOLOGICAL ROLE OF BACTERIOCINS

Undoubtedly, bacteriocins serve some function in microbial communities. These toxins play a critical role in mediating microbial population or community interactions. However, the precise role needs to be elucidated.

Bacteriocins may serve as anti-competitors enabling survival of microbial community. They may also play a defensive role and act to prohibit the invasion of other strains or species into an occupied niche or limit the advance of neighboring cells. An additional role has recently been proposed for Gram-positive bacteriocins, in which they mediate quorum sensing. It is likely that whatever roles bacteriocins play, these role changes as component of the environment, both biotic and abiotic change.

24.9. APPLICATIONS OF BACTERIOCINS

Bacteriocins and human health: The rapid rise and spread of multiresistant bacterial pathogens have forced the consideration of alternative methods of combating infection. Bacteriocins provide an alternative solution. With their relatively narrow-spectrum of killing activity, they can be considered "designer drugs" with target specific bacterial pathogens. Given the diversity of bacteriocins produced in nature, it is relatively simple task to find bacteriocins active against specific human pathogens. However, strategy is still at experimental stage.

Lactobacillus salivarius is a promising probiotic candidate frequently isolated from human, porcine and avian gastrointestinal tracts (GIT), many of which are products of unmodified bacteriocins of subclasses IIa, IIb and IId. It is well characterized bacteriocin producer and probiotic organisms.

Bacteriocins and food preservation: Bacteriocins are attractive candidates in food preservation. The main advantage over chemical preservatives is their ability to preserve without affecting the sensory qualities of the food while adhering to the demand for natural preservatives. Currently, only two bacteriocins are being used commercially as food preservatives: nisin produced by *Lactobacilluslactis* has been used commercially for 50 years and carnicyclin A a circular bacteriocin produced by *Corynebacterium maltaromaticum* UAL307 is an approved biopreservative in the U.S and Canada developed to inhibit *Listeria monocytogenes* in ready -to - eat (RTE) meat products. Circular bacteriocins, because of their stability and activity are projected as next generation biopreservatives.

Bacteriocins in hurdle technology: Hurdle technology combines different preservation methods to inhibit microbial growth. Bacteriocins often have synergies with other treatments and can be used as a hurdle to improve the safety of food. An understanding of mode of action of each individual hurdle allows the most effective combination of treatments. For example, the application of pulsed electric field (PEF), which increases the permeability of cell membranes, has been used together with nisin, which can also act at the level of the cell membrane. The growth of Gram-negative pathogens such as *E. coli* 0157: H7 and *Salmonella* can also be controlled when metal chelators such as EDTA are used in combination with nisin. EDTA disrupts the outer membrane, allowing the penetration of nisin.

Bacteriocins in cancer treatment: Research into colicins and their cytotoxic effects with human tumor cells began about 30 years ago, when colicin E3 was first reported to have significant cytocidal effects on human HeLa cells. Some promising results were observed from a colicin extracted from the *E. coli* strain HSC10 added to murein ascetic lymphatic leukemia cells. Currently, active research is being carried out in this area.

Other applications of bacteriocins: Ambicin is a nontoxic bacteriocin formula that has been shown to kill bacteria that cause dental plaque on teeth and gums. It has been evaluated for intended use in mouth wash, tooth paste, soap and other skin care applications. Ambicin is also effective in the prevention of bovine mastitis, an infection of the udder of dairy cows.

Bacteriocins have application in biotechnology as genetic markers and for maintenance of plasmid-bearing bacterial cells propagated in continuous culture. As genetic markers genes for bacteriocin synthesis and immunity can substitute for antibiotic resistance markers that are currently predominantly used to tag or identify plasmids.

C. MYCOTOXINS

Mycotoxins (*Myko*=fungus or mushroom; *Toxikon* = poison) are the low molecular weight, non-antigenic secondary metabolites produced by some filamentous fungi (moulds). These moulds mostly grow profusely on relatively dry materials like foods, feeds, fodders and sporulate heavily. When such mould infested materials are ingested by humans and livestock, toxins elaborated by these moulds incite poisonous effects resulting in pathological conditions. The pathological effects of mycotoxins, in the form of morbidity and mortality, on warm-blooded animals are called **mycotoxicosis**. Mycotoxicosis is often defined as poisoning of host following the entry of toxic substances of fungal origin into body resulting in various pathological conditions.

The mycotoxin problem received a greater attention with the report of mass death of Turky poults due to the ingestion of peanut meal infested by *Aspergillus flavus* in 1961 by Sargeant and co-workers. The death was attributed to the carcinogenic effect of aflatoxins produced by *Aspergillus flavus* present in the peanut meal. Since then a number of mycotoxicoses were sporadically reported from different parts of the world. These episodes have generated the interest in scientific community to undertake intensive studies on various aspects of mycotoxins and mycotoxicoses. Council for Agricultural Science and Technology (CAST), USA has estimated that 20,000 to 30,00,000 unique mycotoxins can exist however, only relatively a few have been well characterized. As on today, a conservative figure of 300-400 mycotoxins, produced by about 160 mould species have been characterized.

There is a great diversity among mycotoxins with respect to producer mould, metabolic pathway of synthesis, chemical nature and mode of action on the host cell. Nevertheless, the unifying features of all mycotoxins are :

i. They are secondary metabolites of filamentous fungi.

ii. Non-antigenic

iii. Macromolecule and complex molecules (of different chemical classes)

iv. Toxic to warm-blooded animals

v. Heat stable

24.10. MYCOTOXINS PRODUCING MOULDS

Mycotoxin producing fungi are ubiquitous and occur in all types of habitats especially those which are moist and rich in organic nutrients. However, those that grow on food commodities, feed and fodders are important as they cause mycotoxicoses in consumed animals including man. More than fifty genera of fungi are known to contain mycotoxigenic species. Taxonomically, most of them belong to Deuteromycetes (fungi imperfecti) especially Hyphomycetes, which are asexual states of ascomycetes. Though species of many genera are known to produce mycotoxins, most of the mycotoxigenic species belong to *Aspergillus, Penicillium* and *Fusarium*. Mycotoxigenic fungi are comparatively xerophilic and potential to produce conidial mass in abundance and need warmer conditions to elaborate these toxic metabolites. Table 24.9 summarizes the toxigenic fungi and respective toxins produced by them. It is interesting to note, from Table 24.9 that a single fungal species may produce more than one mycotoxin and, similarly many species belonging to different genera may produce the same type of mycotoxin. Thus, no specificity is linked to a fungal species and production of particular toxin.

24.11. CHEMISTRY OF MYCOTOXINS

All mycotoxins, reported so far are low molecular weight secondary metabolites. However, they are chemically diverse group of compounds: polypeptides, alkaloids, benzoquinones, xanthones, coumarins, terpenes and steroid derivatives. Structures of some of the mycotoxins are given in Fig. 24.12 in accordance with diverse chemical nature, there is a wide variation in their biosynthetic pathways (Table 24.10).

Table 24.9. Some common mycotoxins and the respective producer fungi

Mycotoxin	Fungi which produce them
Aflatoxin	*Aspergillus flavus, A. parasiticus*
Citrinin	*Aspergillus carneus, A. terreus, Penicillium citrinum, P. hirsutum, P. verrucosum*
Cyclopiazonic acid	*Aspergillus versicolor*
Deoxynivalenol (Vomitoxin)	*Fusarium graminearum, F. culmorum*
Diacetoxyscirpenol	*Fusarium moniliforme, F. equiseti*
Fumonisin B1	*Fusarium moniliforme, F. culmorum, F. avenaceum, F. nivale*
Fusarenon-x	*Fusarium nivale*
HT-2 toxin	*Fusarium moniliforme, F. culmorum, F. avenaceum, F. nivale*
Islanditoxin	*Penicillium islandicum*
Moniliformin	*Fusarium moniliforme, F. equiseti, F. oxysporum, F. culmorum, F. avenaceum, F. roseum, F. nivale*
Mycophenolic acid (MPA)	*Penicillium brevicompactum, P. roqueforti*
Nivalenol	*Fusarium moniliforme, F. equiseti, F. oxysporum, F. culmorum, F. avenaceum, F. roseum, F. nivale*
Ochratoxin-A	*Aspergillus ochraceus, Penicillium viridictum, P. nordicum*
Patulin	*Aspergillus clavatus, Penicillium expansum, Botrytis, P. roquefortii, P. claviforme, P. griseofulvum*
Penicillic acid	*Aspergillus ochraceus*
Penitrem-A	*Penicillium flavigenum*
Roridin E	*Myrothecium roridum, M. verrucaria, Dendrodochium* spp., *Cylindrocarpon* spp., *Stachybotrys* spp.
Rubratoxin	*Penicillium rubrum*
Satratoxins, F, G, H	*Stachybotrys chartarum, Trichoderma viridi*
Sterigmatocystin	*Aspergillus flavus, A. nidulans, A. versicolor, Penicillium rugulosum*
T-2 toxin	*Fusarium moniliforme, F. equiseti, F. culmorum, F. solani, F. avenaceum, F. roseum, F. nivale*
Terreic acid	*Penicillium crustosum*
Territrem-B	*Aspergillus terreus*
Trichodermin	*Trichoderma viride*
Trichothecin	*Trichothecium roseum*
Zearalenone	*Fusarium culmorum, F. graminearum, F. oxysporum, F. roseum, F. moniliforme, F. avenaceum, F. equiseti, F. nivale*

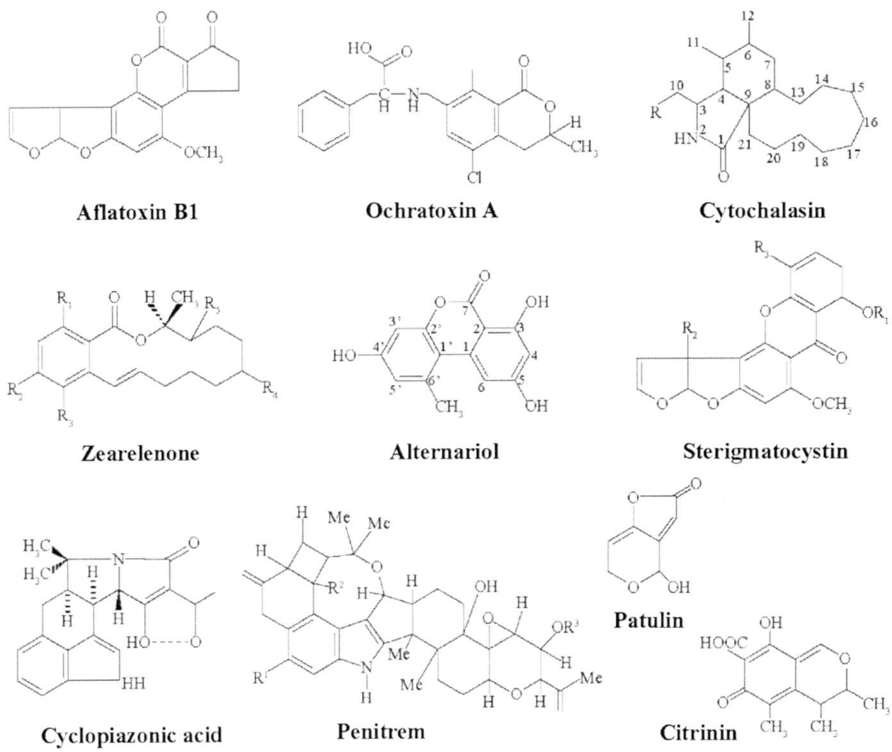

Fig. 24.12. Chemical structures of some important mycotoxins

Table 24.10. Mycotoxins and their biosynthetic categories of secondary metabolites

Biosynthetic	Representative mycotoxins
Poly ketides	Zearalenone
Tetra-	Patulin, penicillic acid, chlorflavonin
Penta-	Citrinin, ochratoxins
Hexa-	Maltoryzine
Hepta-	Viriditoxin, cytochalasins, rugulosin.
Octa-	Ergochromes
Nona-	Zearalenone, viridicatum toxin
Deca-	Aflatoxins, austocystins, erythroskyrine
Tetronic/ tertramic acid	Tenuazonic acid, cyclopiazonic acid, cytochalasins
Diketopiperazines Simple	Aspergillic acid, echinulins
Modified	Brevianamides, sporidesmins, fumitremorgons, oxaline
Peptides	Tentoxin, ergotamine, tryptoquivaline

Biosynthetic	Representative mycotoxins
C_6C_3 Products	Chlorflavonin, xanthocillin, terphenyllin
Terpenes Mono-	Viridicatum toxin
Sesqui-	Trichothecenes
Di-	Paspaline

24.12. CLASSIFICATION OF MYCOTOXINS

A large number of mycotoxins are reported and characterized through different parameters, such as producer fungi, chemical nature, metabolic pathway, mode of action at molecular level, biological effects and diseases caused. However, no accepted classification was proposed so far. Some classifications are based on the chemical nature of mycotoxins, for example, aflatoxins, ochratoxins, patulins, territrems, fumonisins, trichothecenes, zearalenone etc. Still some classifications are based on the gross biological effects induced by the toxins. The toxins known by this type of classification are carcinogenic, nephrotoxic, tremorgenic, estrogenic, cytotoxic, gastrointestinal toxins etc.

24.13. MODE OF ACTION OF MYCOTOXINS

Biochemical effects of mycotoxins result in the impairment of vital biochemical processes. The susceptible critical molecules are known as molecular receptors of the attacking toxins. The receptors are DNA, RNA, functional proteins, enzyme cofactors and membrane constituents (Table 24.11).

Table 24.11. Identified or suspected effects of the major mycotoxins and cellular and molecular mechanisms of action

Toxin	Effects	Cellular and molecular mechanisms of action
Aflatoxin B1 + M1	• Hepatotoxicity • Genotoxicity • Oncogenicity • Immunomodulation	• Formation of DNA adducts • Lipid peroxidation • Bioactivation by cytochromes P450 Conjugation to GS- transferases
Ochratoxin A	• Nephrotoxicity • Genotoxicity • Immunomodulation	• Effect on protein synthesis. • Inhibition of ATP production Detoxification by peptidases
Patulin	• Neurotoxicity • *In vitro* mutagenesis	• Indirect enzyme Inhibition
Trichothecenes (T-2, DON)	• Hematotoxicity • Immunomodulation • Skin toxicity	• Induction of apoptosis in haemopoietic progenitor cells and immune cells. • Effect on protein synthesis • Abnormal changes to immunoglobulins

Toxin	Effects	Cellular and molecular mechanisms of action
Zearalenone	• Fertility and Reproduction	• Binding to oestrogen receptors • Bioactivation by reductases Conjugation to glucuronyltransferases
Fumonisin B1	• Central nervous system damage • Hepatotoxicity • Genotoxicity • Immunomodulation	• Inhibition of ceramide synthesis • Adverse effect on the sphinganin/ sphingosin ratio • Adverse effects on the cell cycle

Source : French Food Safety Agency (AFSSA) Summary-Report, 2006

The interaction between mycotoxin and their molecular receptor may be covalent irreversible or non-covalent reversible. The specificity of a molecular receptor of mycotoxins is determined by:

i. Its affinity to the active form of toxin

ii. Its role in vital biochemical processes.

iii. The persistence of the lesion it bears.

iv. The severity of the consequences of lesion.

i. Nucleic acids as receptors: Alteration in the structure of DNA and RNA molecules by the mycotoxins impair the template activity resulting in the inhibition of synthesis of DNA, RNA and protein. In point mutations leading to synthesis of erroneous macromolecules which may result in cellular death or cell transformation to cancerous condition. The well-studied toxins in this aspect are aflatoxins.

ii. Lesion on proteins: Proteins are most common receptors in cell because of their abundance as cellular constituents and the presence of nucleophilic N, O, S hetero atoms in their functional groups. The mycotoxins alter the structures and activities of proteins. In this process, it results in conformational changes which alter or block the active sites for binding or allows the mycotoxin to compete with endogenous substrates for the active sites.

iii. Target sites: Mycotoxins usually enter the human body through ingestion. The prominent sites of interaction following oral route of a mycotoxin include gastrointestinal micro-flora, intestinal epithelial cells, portal blood, liver, bile, circulatory blood and extrahepatic organs.

The liver is obviously the primary target organ of orally administered mycotoxins. However, due to the structural diversity of mycotoxins, the other organs like kidney, intestine, blood, respiratory system, central nervous system, immune system, skin and lung may become primary target organs of some mycotoxins.

iv. Inhibition of energy production: Inhibition of cellular energy production is a major metabolic effect of mycotoxins at acute level of exposure. The molecular receptors in this mode of action are enzymes of tricarboxylic acid cycle and electron transport system in mitochondria. The immediate response is reduced oxygen consumption. Aflatoxins B1, G1 and m1 inhibit ETS between cyt. B and C or C1 in rat liver mitochondria. Luteoskyrin, emodin and the ergot mycotoxin, inhibit the site III in a way similar to cyanide. Moniliformin acts like fluoroacetate on the TCA cycle and inhibits mitochondrial pyruvate and α-ketoglutarate oxidation.

Some of mycotoxins like aflatoxin B1 also uncouple oxidative phosphorylation resulting in the immediate depletion of ATP in the cell, which may cause mitochondrial swelling and degeneration. Another way in which mycotoxins affect cellular energy production and utilization is the alteration of ATPase activity. The effect of mycotoxins on ATPase enzyme is toxin specific.

v. Inhibition of macromolecular synthesis: Inhibition of cellular biosynthesis of macromolecules is a major metabolic effect of mycotoxins at acute level of exposure, eventually leading to the cell death. Inhibition of DNA synthesis is one of the first consequences of aflatoxin B1 and also other genotoxic mycotoxins will affect the intricate transcription and translation processes of cellular protein synthesis.

Mycotoxins inhibit protein biosynthesis by modification of DNA, template activity inhibition, inhibition of RNA synthesis and maturation, inactivation of biosynthetic enzymes and impairment of amino acid transport. The various mechanisms of inhibition of protein synthesis have been observed for aflatoxin B1, ochratoxin A and trichothecenes. Inhibition of protein synthesis results in the alterations of immune responses. Mycotoxins also alter immunogenesis by inactivation of lymphocytes and phagocytes.

vi. Effect on central nervous system: Based on mode of action on central nervous system, the neurotoxic mycotoxins are of three classes.

Those causing paralysis and respiratory arrest, e.g. citreviridin and asteloxin.

Those inducing trembling in animals, e.g. penitrem A, verruculogen, gumitremorgen A

Those cause animals to vomit, e.g. trichothecenes such as vomitoxin.

The first class of toxins kill nerve cells due to disruption of energy supply by the toxins. Tremorgens act on the central nervous system by altering functional states of neurotransmitters. Vomitoxins may act on chemoreceptor trigger zone in the medulla oblongata of central nervous system.

vii. Modification of cytoskeletons: About forty fungal metabolites are known to affect the function of microtubules, microfilaments that regulate locomotion, cell division, cell membrane formation and other cellular functions. All these 40 metabolites are collectively known as 'cytochalasins'. These are the first chemicals discovered to consistently block cytoplasmic division, but not interfering with nuclear division.

viii. Diseases of animals and man: mycotoxins are reported to be casual agents of different diseases and maladies in animals and man due to the ingestion of either mould infested or mycotoxin contaminated feeds and food respectively.

24.14. BIOLOGICAL EFFECTS

Biological effects of mycotoxins are varied. These effects include from acute to chronic toxicity. Cytotoxicity, carcinogenicity, antitumor properties, mutagenicity, teratogenicity, immune-suppressive activity, antimicrobial properties, insecticidal effects and phytotoxicity are the major biological effect (Table 24.12).

Table 24.12. The spectrum of toxic effects of mycotoxins in experimental animals

Mycotoxin	**Effect**
Aflatoxins, Sterigmatocystin	Hepatotoxic
Ochratoxin A, Citrinin	Nephrotoxic
Citreoviridins, Ergot alkaloids, Penetrim A	Neurotoxic
Aflatoxins, Ochratoxin A, Trichothecenes	Immunotoxic
Penitrem A	Tremorgenic
Trichothecenes	Vomiting
Slaframine	Excessive salivation
Trichothecenes	Hemorrhagic
Zearalenone	Estrogenic
Zearalenone	Abortion
Aflatoxins, Sterigmatocystin, Ochratoxin A, Fumonosins	Carcinogenic
Zearalenone	Teratogenic
Cyclopiazonic acid (CPA)	Mutagenic

Acute and chronic toxicity: Toxicity of mycotoxins varies greatly depending on both their structure and animal species involved. Both these aspects can be well illustrated by aflatoxins. Aflatoxin B1 is the most acute toxic followed by G1, B2, G2 in order of toxicity. Some mycotoxins are target-oriented and act on specific organs of the body. For instance, aflatoxins act on liver, citrinin on kidney, penicillic acid on heart; penitrem A and other tremorgenic toxins on central nervous system, vomitoxin and some trichothecenes on digestive tract.

Cytotoxicity: Growth inhibition and mitotic suppression was observed when heteroploid human embryonic lung cells were treated with four aflatoxins. Aflatoxins also inhibit mitosis in culture of human leucocytes.

The characteristic effect of cytochalasins is the inhibition of cytokinesis and resultant failure to sustain continuous cycles of cell division. Cytochalasin D (zygosporin) showed the strongest cytotoxic effect. Trichothecenes are potent inhibitors of eukaryotic protein synthesis.

Carcinogenicity: Carcinogenic potential of a number of mycotoxins is well known in experimental animals. Aflatoxin B1 and sterigmatocystin are considered as most potent carcinogens. Sterigmatocystin which is structurally related to aflatoxin B1 is a potent hepatocarcinogen. It is believed that interaction with DNA is responsible for carcinogenicity and mutagenicity of these mycotoxins. Carcinogenicity *in vitro* has also been confirmed with other mycotoxins such as citrinin, penicillic acid, patulin, luteoskyrin and rugulosin.

Mutagenicity: Mycotxins were demonstrated to interfere with DNA and cause mutations. Aflatoxin B1, sterigmatocystin, versicolorin A, austocystins A and D and kojic acid are some of the established mutagenic mycotoxins. A wide variety of toxins such as rubroskyrin, rugulosin and crysophenol were also reported to be mutagenic.

Teratogenicity: Citrinin and aflatoxin B1 cause teratogenic effects in chick embryos with malformation primarily in the extremities. However, other changes also occurred. Ochratoxin A was also reported to be teratogenic in pregnant rats.

Antitumor properties: Several mycotoxins possess antitumor properties. The trichothecens, verrucarins A and B and roridin A were cytotoxic and inhibited the tumor in mice and walker carcinoma in rats. Inhibitory effect of diacetoxyscirpenol on the growth and experimental tumors in mice and rats was also reported. Penicillic acid and mycophenolic acid also exhibit antitumor effects.

Antimicrobial activity: Surprisingly, a variety of fungal metabolites presently known as mycotoxins were originally discovered as antibiotics. Mycophenolic acid, penicillic acid, citrinin patulin and trichothecene are a few of such examples.

The antiviral activity of mycophenolic acid was demonstrated in tissue culture against vaccinia, measles, herpes simplex and new castle diseases. Patulin inhibited the replication of mycovirus in *Penicillium stoloniferum. Bacillus thuringiensis* was found to be sensitive to kojic acid, patulin, penicillic acid, DAS, T2 toxin and rubratoxin B. Roquefortine selectively inhibited only gram-positive bacteria. Morphological changes in *Botrytis cinerea* and *Paecilomyces varioti* were induced by aspergillic acid, citrinin, griseofulvin, patulin, rugulosin and trichothecene.

Insecticidal properties: Several mycotoxins such as aflatoxins, fusaric acid, kojic acid, citrinin, rugulosin are proved to possess insecticidal properties. Insecticidal and larvicidal activities of aflatoxin B1, rubratoxin, patulin and diacetoxyscirpenol towards *Drosophila melanogaster* was also reported.

Phytotoxicity: Aflatoxins inhibited the germination of grass seeds and later at lower concentrations induced chlorophyll deficiency. Cytochalasin B acts as a thiol reagent influencing tip growth, cytoplasm streaming, organelle movement, cell wall synthesis, auxin transport etc. Permeability of plant cell membranes was damaged by patulin, citrinin and fusaric acid. T2 toxin and diacetoxyscripenol inhibit the germination of *Nicotiana sylvestris* pollen.

D. PHYCOTOXINS

Phycotoxins (*Phykos*=seaweed, *toxikon*=toxin) are potent organic compounds produced by dinoflagellates, other flagellated phytoplankton, and cyanobacteria that inhabit marine, brackish or freshwater bodies. These metabolites are in most cases, not harmful to the producer but may be toxic to either one or many members of the aquatic food web. Phycotoxins are large and complex molecules, energetically expensive to produce hence they are synthesized in small quantities. They accumulate in other organisms and can reach high concentrations during algal blooms. Most characterized phycotoxins have some economic and health impact on humans. Their potential applications exist in pharmaceuticals and cellular research. As biologically active metabolites, they may produce ecological effects at such as deciding the biogeographic distribution of phytoplankton and bloom dynamics.

The history of phycotoxins can be traced back to 1832 in which Charles Darwin attributed discoloration of ocean waters due to population explosion of dinoflagellates. In 1878 a water bloom appeared due to excessive growth of cyanobacterial algae in lake Alexandria of south Australia which was popular as poisonous Australian lake. Since then several reports appeared in different parts of world including India. Tyagi *et al* (2000) reviewed the incidence of algal blooms and toxicity of water which are responsible for health problems of animals including man.

24.15. HARMFUL ALGAL BLOOMS (HAB) AND PHYCOTOXINS

An algal bloom is a rapid increase or accumulation in the population of algae in freshwater or marine water systems, and is recognized by the discolouration in the water from their pigments. Several reasons are attributed to the development of algal blooms, however, the foremost among them especially in freshwater bodies is eutrophication. In fresh water bodies, most harmful algal blooms are formed by cyanobacteria. In oceanic water, they are caused by haptophytes, dinoflagellates, green algae, raphidophytes, euglenophytes, diatoms and cryptophytes. Algal blooms are increasingly a global concern. They can threaten human and aquatic ecosystem health, and cause major economic damage. Harmful effects include: (i) produce extremely dangerous toxins that can drastically effect food web and human health, (ii) create dead zones in the water, (iii) raise treatment cost for drinking water, (iv) effect industries that depend on clean water and (v) reduce the quality of water for recreation.

Due to their multifold adverse effects on environment and human health, algal blooms are often referred to as harmful algal blooms (HAB). HAB are completely natural phenomena which have occurred throughout recorded history but in the past two decades the public health and economic impacts have increased in frequency, intensity and geographic distribution.

Many algal bloom forming species produce potent natural toxins (phycotoxins) that can persist in the water or enter food web, leading to illness or death of aquatic animals and/or human sea food consumers. Phycotoxins associated with HAB can be broadly categorized into: (i) cyanobacteria bloom (cyanoHABs) toxins (cyanotoxins), preferentially occur in soft water reservoirs and drinking water, with direct risk to human health, (ii) marine biotoxins/marine algal toxins by dinoflagellates and diatoms occur in oceanic water which accumulate at high concentrations in various tissues of aquatic organisms such as bivalve mollusks and fish, entering food chain, threatening consumers health. Thus, phycotoxins of HAB are public health concern worldwide. Their outbreaks occur continuously. Toxins produced by HABs are associated with several syndromes including paralytic (PSP), diarrheic (DSP) amnesic (ASP) and ciguatera (shell) fish poisoning (CFP) caused by consumption of contaminated seafood. Toxins are bioaccumulated by organisms that ingest algae, and thus transmitted through the food web up to humans.

It may be remembered that all HABs are not toxic, however, still they are harmful because of other harmful effects. For example, *Noctiluca scintillans*, a marine dinoflagellate with a high biomass, but non-toxic bloom.

24.16. CLASSIFICATION AND CHEMICAL NATURE OF PHYCOTOXINS

Phycotoxins are classified in different ways such as chemical structure, symptoms of intoxication, solubility. According to chemical structure, the phycotoxins are classified into eight groups; azaspiracids (AZAs), brevetoxins (BTXs), cyclic imines, domolic acid (DA), okadaic acid (OA), pectonotoxins (PTXs), saxitoxins (STXs) and yessotoxins (YTXs) (Fig. 24.13). They belong to the chemical classes of amino acids (DA), tetrahydropurines(STXs), and polyketides (OA, AZAs, PTXs, BTXs, YTXs). Most groups of phycotoxins consist of several analogs. Phycotoxins are also classified either by symptoms of intoxication or by their solubility in water and organic solvents. Five different categories of human shell fish poisoning have been characterized; paralytic shellfish poisoning (PSP) caused by STXs, neurotoxic shellfish poisoning (NSP) caused by BTXs, diarrheic shellfish poisoning (DSP) caused by OA group of toxins, and dianophysis toxins (DTXs), amnesic shellfish poisoning (ASP) caused by DA, and azaspiracid shellfish poisoning (AZP) caused by AZAs. According to their solubility phycotoxins are classified into hydrophilic (STXs and DA) and lipophilic toxins, respectively. The group of lipophilic toxins is subdivided into four categories: the AZA group, the YTX group, the PTX group and OA group.

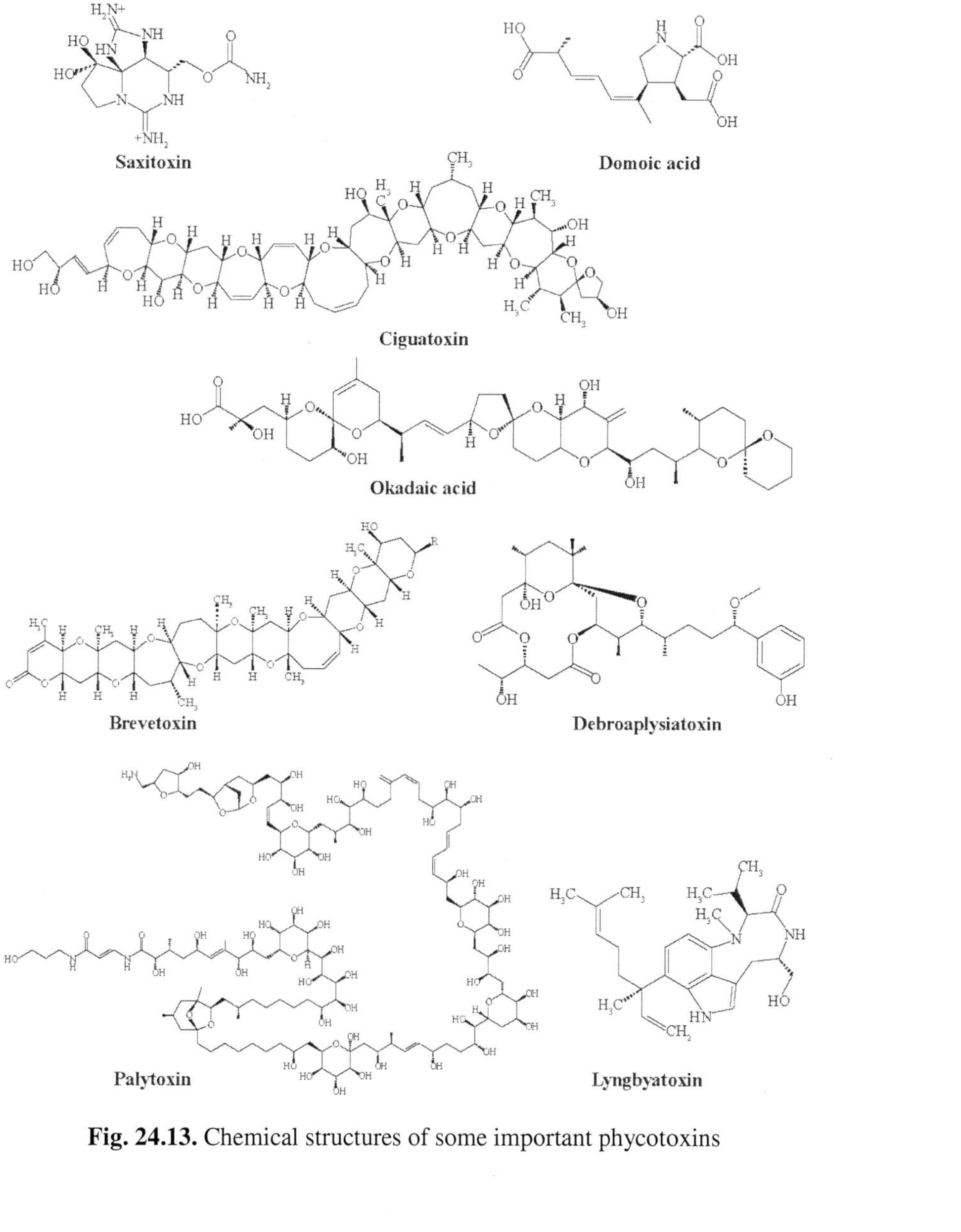

Fig. 24.13. Chemical structures of some important phycotoxins

24.17. PHYCOTOXIN PRODUCING ORGANISMS

Though about 300 species of algae are responsible for algal blooms, but only 40 species are reported to produce phycotoxins. These species belong to diverse taxa of algae and cyanobacteria. True algal species mostly belong to Dinophyceae and some belong to Raphidophyceae, Bacillariophyceae and Prymnesiophyceae. Genera and species of various taxa producing phycotoxins are presented in Table 24.13.

Table 24.13. Phycotoxin-producing organisms and the phycotoxins produced by them

Class	Toxin producing species	Toxin
Cyanobacteria	*Scytonema.* spp. *Tolypothrixcon glutinat* var. *colorata* *Lyngbya majuscua*	Scytophycins Tolytoxin Debromoaplysia toxin Majusculamides Curacins Lynbya toxins
	Anabaena spp. *Aphanizomenon* spp., *Cylindrospermipsis* spp., *Lynbya* spp., *Plantothrix* spp and *Oscillatoria*	Saxitoxins (Neo- saxitoxinsgynautoxins)
Bacillariophyceae	*Pseudo-nitzschia spp.* *p. multiseries*	Domoic acid Bacillariolides
Raphidophyceae	*Chatanella marina* *C. antigua., C.* cf. *verruculosa*	Brevetoxins
Prymnesiophyceae	*Prymnesiun parvum*	Prymnesins
Euglenoidea	*Euglena sanguinea*	Euglenophycin
Dinophyceae	*Alexandrium* spp. *Pyrodinium bahamense,* *Gymnodium catenatum*	Saxitoxins
	Gambierdiscus toxicus	Cigua toxin Gambieric acid
	Ostreopsis lemticularis	Osterotoxin
	Coolia monotis	Coolinia toxin
	Karenia brevis, K. brevi-sukata	Brevotoxin
	Protoceratium reticulatum *Lingulodinium polyedrum*	Yessotoxin
	Dinophysis spp. *Prorocentrum* spp	Okadaic acid Dinophysistoxins

Class	Toxin producing species	Toxin
	Dinopgysisfortii, D. acuta	Pectenotoxin
	Protopteridinium crassipes	Azaspir acid
	Karenia selliformis	Gymnodimine
	Alexandrium ostenfeldii	Spirolide
	Ostreopsis siamensis	Ostereocin
	Amphidrium spp.	Goniodomin
		Prorocentrolida
	Proroentrium lima	Amphidinols
	Amphidinium Spp	Amphiketide

24.18. ECOLOGICAL ADVANTAGES OF PHYCOTOXINS TO PRODUCER ORGANISMS

Phycotoxin producing organisms derive ecological benefits by producing the toxins. The effect of phycotoxins are mostly at low concentrations and subtle. At smaller scale, they may derive the benefits like protection from grazing animals, anti-microbial effects or competition, but at larger scale they have the potential to impact the biogeographic distribution of phytoplankton and bloom dynamics.

Phycotoxins may prevent grazing by mechanisms like grazer death, infertility, or deference. For example, three different species of copepods were able to distinguish between a saxitoxin producing and non-toxic *Alexandrium* spp. by chemosensory means. These three different species copepod species grazed predominantly on non-toxigenic species. This implies that saxitoxin producing *Alexandrium* spp. have an advantage over non-toxigenic dinoflagellates. The phycotoxins may also be used to ward off parasitic or algicidal heterotrophic bacteria. It has been demonstrated that acrylic acid produced by *Phaeocystis* blooms inhibited the bacteria. At very low concentrations, it serves to keep bacteria away from the phytoplankton that compete for a limited nutrient resource. It is possible that phycotoxin production is used as a method to either kill competitors or to keep other phytoplankton out of the producer's nutrients space.

24.19. MODE OF ACTION AND PHYCOTOXINS HEALTH EFFECTS

Phycotoxins belong to diverse group of chemical compounds. On par with chemical structure each one has a distinct mechanism of toxicity. Intoxication of phycotoxins can occur through oral, respiratory or dermal exposure to the toxins produced by diatoms or dinoflagellates; respiratory exposure through aerosols, exposure to soft water contaminated with cyanotoxins, through drinking water; fresh water fish, dermal exposure, e.g. bathing in contaminated lakes, or through contaminated equipment or fluids. Table 24.14 summarize the toxins described in humans, mechanism of actions and clinical syndromes. Most of the information presented in tables deals with acute exposure and effects.

Table 24.14. Cyanotoxins mechanism of action and human healthy effects

Cyanotoxins	Mechanism Action	Symptoms
Hepatotoxins		
Microcystin MCs	Inhibition of protein phosphatases type 1 and 2A Repeated low-level exposure: carcinogenesis, genotoxicity	Within 4-24h to few days: Diarrhea, vomiting, liver hemorrhage, shock, abdominal pain, jaundice, weakness, dyspnea respiratory distress, multiple organ failure
Nodularin	Inhibition of protein phosphatases type 1 and 2A	Diarrhea, vomiting, goose bumps, weakness, liver hemorrhage
Cylindrospermopsin (CYN)	Glutathione and protein synthesis as well as cytochrome P450. Repeated low-level exposure: carcinogenesis, genotoxicity	Symptoms up to several days after exposure or later. Gastroenteritis abdominal pain, vomiting, bloody diarrhea, acute liver inflammation. Liver and kidney failure, hay fever, asthma
Neurotoxins		
Anatoxin-a / Homoanatoxin-a	Nicotinic receptors: Irreversible link to the nicotinic receiver S of acetyl-choline in neuromuscular junction	Muscle twitching, cramping staggering, paralysis, convulsions, gasping, respiratory failure, death by suffocation
Anatoxin-a (S)	Irreversible inhibitor acetylcholinesterase	Salivation, muscle twitching cramping, paralysis
Saxitoxins	Neurotoxic, target the peripheral nervous system. Selective high affinity block sodium conductance in voltage-gated sodium channels, receptor site 1 at nerve fibers	Paralytic Shellfish Poison: Nausea, vomiting. peri-oral burning ataxia, drowsiness, paraesthesia, fever, tachycardia, muscular paralysis, respiratory failure, death
beta-Methylamino-L-alanine (BMAA)	Experimentally acts predominantly on motor neuron - excitotoxic through glutamate receptors	Not fully elucidated. Implicated on chronic neurodegenerative diseases
Aplysiatoxins	Potent tumor promoters Potentiation of protein kinase C (PKC)	Skin irritation, asthma like symptoms
Lyngbyatoxin	Potent tumor promoters Potentiation of protein kinase C (PKC)	Smooth muscle contraction. Skin irritation

Cyanotoxins

Cyanotoxin is the term used for the secondary metabolites from cyanobacteria, toxic to living organisms including human. They are divided according to their chemical structure and also biological effects. According to chemical structure they are divided into cyclic peptides (microcystin, nodularin), alkaloids (anatoxin-a, anatoxin-a(s), saxitoxins cylindrospermopsin, aplysiatoxin, lyngbiatoxin-a), and lipopolysccharides. Based on biological effects and target tissues affected cyanotoxins are categorized into: hepatotoxins, neurotoxins, cytotoxins, dermatotoxins and irritant toxins. Neurotoxic cayanotoxins are: anatoxin-a/homoanatoxin–a, beta–methylamino-L-alanine (BMAA). BMAA chronic, repeated exposure is linked to neuro-degenerative diseases such as amylotrophic lateral sclerosis (ALS).

Despite their harmful effects, some cyanotoxins have properties with potential beneficial applications: antibiotics, algaecides, cytotoxic, immunosuppressive agents and enzyme inhibitors. Cyanotoxins acting as protease inhibitors and on cell cycle are some of the compounds under investigation for anticancer activity.

Marine algal toxins

Marine algal toxins largely produced by dinoflagellates and diatoms are potent toxins, contaminate seawater environments, and accumulate at high concentrations in various tissues of aquatic animals such as bivalve mollusks, crustaceans, and fish, thus entering the food chain (Fig. 24.14) Consumption of contaminated fish and shellfish is one of the main causes of seafood poisoning. Marine algal toxins are an important source of seafood contamination worldwide with serious adverse impacts to human health, wildlife, the economy and the ecosystem.

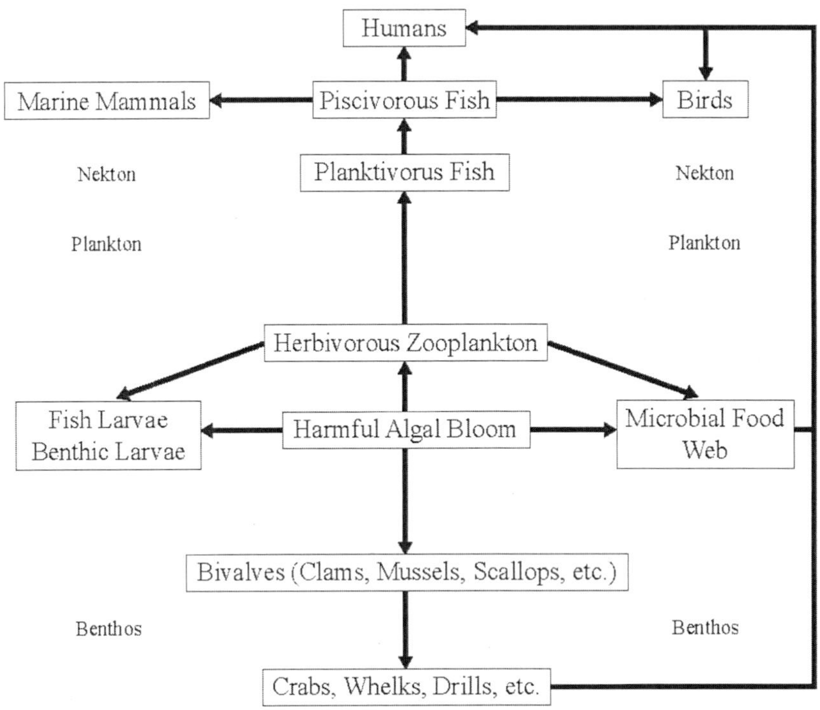

Fig. 24.14. Food chain in marine systems involving harmful algal blooms and their toxins.

The best studied marine algal toxins of seafood poisoning syndromes are diarrheic shellfish poisoning (DSP); paralytic shellfish poisoning (PSP); amnesic shellfish poisoning (ASP); neurologic shellfish poisoning (NSP); azaspiracid shellfish poisoning (AZP); ciguerata fish poisoning (CFP). Toxic syndromes associated with polytoxins (PITX) are palytoxicosis, clupeotoxism, and PITX like myotoxic compound, characterized by myalgia, rhabdomyolysis, myoglobinuria (black urine) after ingestion of cooked freshwater and brackish water fish, identified as "Half Disease" (Table 24.15).

Table 24.15. Marine algal toxins by diatoms and dinoflagellater : mechanism of action and human helath effects

Toxin group	Vector	Mechanism of Action/ Effect	Syndrome
Azaspiracids (AZAs)	Shellfish Bivalves Mollusks	In vitro effect on actin cytoskeleton cell shape and adherence. Cytoskeleton damage is not reversible after toxin withdrawal	Azaspiracids Poisoning (AZP)
Brevetoxins (BTX/PbTX)	Shellfish Bivalves Mollusks	Neurotoxic: Open sodium channels Immuno-logical effects e.g.: release of inflammatory mediators	Neurotoxic Shellfish Poisoning (NSP)
Ciguatoxins (CTXs): Pacific (P-CTX), Caribbean (C-CTX), and Indian Ocean (I-CTX)	Tropical and sub-tropical fish *e.g.:* eels, snappers, groupers, mackerels, jacks or barracudas	Neurotoxic: Opens sodium channels by binding site 5. Influx of Na, provoke action potentials, cell swell, and blebs on cell's surface	Ciguatera Fish Poisoning (CFP) Severity and type of symptoms varies with the type of CTX
Saxitoxins (STXs) and related toxins	Shellfish Bivalve Mollusks Crustaceans	Neurotoxic: Target the peripheral nervous system Selective block sodium conductance in voltage-gated sodium channels receptor site 1 at nerve fibers	Paralytic Shellfish Poisoning (PSP)
Toxin Group	Vector	Mechanism of Action/ Effect	Syndrome
Okadaic add (OA) and dynophysistoxins (DTXs)	Shellfish Bivalve Mollusks	Tumor promoter Inhibit essential serine/ threonine protein phosphatases 1(PP1) and 2A (PP2A); genotoxic, cytotoxic	Diarrheic Shellfish Poisoning (DSP)
Domoic add group	Shellfish Mollusks Crustaceans. Fish *e.g* Sardines Anchovies	Neurotoxic: Cardiatoxic: Excitatory neurotoxin; analogue of glutamate - acts through glutamate receptors. Reproductive organs	Amnesic Shellfish Poisoning (ASP)
Palytoxins (PITXs) and PITX like compounds	Shellfish Bivalve Mollusks, Crab, Fish *e.g.* Sardines Anchovies	Neurotoxic: Target the sodium potassium-ATPase pump in cell membrane; maintain the ionic gradients in cells. Membrane depolarization in excitable and non- excitable cells, contraction of muscle cell	Toxic syndromes in association with PITXs: Palytoxicosis/ Clupeotoxism, PITX like toxin is myotoxic: myalgia rhabdomyo-lysis, myoglobinuria after ingestion of cooked freshwater and brackish water fish, identified as "Half Disease"

24.20. PREVENTION AND MANAGEMENT OF PHYCOTOXIN POISONING

HAB and associated phycotoxins are a constant threat to public health, and economy, aquatic ecosystems, seafood, drinking water. Management requires a multidisciplinary action at the local and international level to manage their potentially harmful effects. Public health vigilance, monitoring, and research need to be maintained and enhanced. Despite scientific advancements phycotoxin research relating to human exposure and health consequences are sparse, while, blooms of toxigenic species have become more prevalent worldwide. At present, phycotoxin poisonings diagnosis and management is largely based on the ability of health care provider to interpret presenting clinical symptoms, collect exposure history, identify and establish the exposure event. Regulatory standards, alertness by public health organizations and primary health care providers, in regions with history of HABs, help to minimize and manage human health risks. Avoidance of non-commercially harvested shellfish not tested for phycotoxins is the best way to prevent poisoning events.

24.21. FURTHER READING

Alof, J.E. and Freer, J.R. (ed.) (1999), The comprehensive Source Book of Bacterial protein toxins. Academic Press, London, UK.

Alouf, J.E and Ladant D. (2015), The comprehensive source book of bacterial protein toxins (4th edn.) Elsevier, New York.

Anderson, D.M., Cembella, A.D. and Hallegraeff, G.M. (2012), Progress in understanding harmful algal blooms: paradigm shifts and new technologies for research, monitoring and management. *Ann. Rev. Mar Sci.* 4: 143–176.

Anonymous (2016), IOC-UNESCO taxonomic reference list of harmful microalgae.

Balfanz, J., Rautenberg, P. and Ullmann, U. (1996), Molecular mechanism of action of bacterial exotoxins *Zbl. Bakt.* 284:170–206.

Betina, V. (1989), Mycotoxins: Chemical, Biological and Environmental aspects. Elsevier Publications, Amsterdam

Blanke, S.R. (2006), Portals and pathways. Principles of bacterial toxin entry into host cells. *Microbe* 1:26–32.

Braun Volkmar *et al.* (2015), Import and export of bacterial protein toxins. *Int. J. Med. Microbiol* 305: 238–242.

Cloveland, J. *et al.* (2001), Bacteriocins: Safe, natural antimicrobial for food preservation. *Int. J. Food. Microbial.* 71: 1–20.

Eiley, M.A. and Wertz (2002), Bacteriocins: Evaluation, Ecology and Application. *Ann. Rev. Microbial.* 56: 117–137.

Galanos, C. and Freudenberg, M.A. (1993), Bacterial endotoxins: Biological properties and mechanisms of action. *Mediators of Inflammation* 2: S11–S16.

Henkel, J.S., Baldwin, M.R. and Barbieri, J.T. (2010), Toxins from bacteria EXS 100: 1–29.

Klaenhammer, T.R. (1993), Genetics of bacteriocins produce by lactic acid bacteria. FEMS Microbiology Review 12: 39–85.

Lax, A.L. (ed.) (2005), Bacterial protein toxins Cambridge University Press.

Lemichez, E. and Barbieri, J.T. (2013), General aspects and recent advances on bacterial protein toxins. Cold Spring Harb. Perspect. Med. 3: an 13573.

Llewelyn, M. and Cohen, J. (2002), Superantigens: Microbial agents that corrupt immunity. *Lancet Infect.* Dis 2; 156162.

Magan, N. and Oslen, M. (2004), Mycotoxins in food. Woodhead Publishing Ltd., Cambridge

Moayer, M. and Welch Rodney, A. (1998), Bacterial exotoxins. *Methods in Microbiology* 27:287–300.

Parker, M.W. and Feil, S.C. (2005), Pore-forming protein toxins: from structure and function progress In Biophysics and Molecular Biology 88:91–142.

Parker, M.W. and Feil, S.C. (2005), Poreforming protein toxins: From structure to function progress in Biophysics and Molecular Biology. 88: 91-142.

Pizza Mariagrazia, Masignani Vega and Rappoulin Rino (2007), Bacterial toxins. In.

Popoff, M.R. and Poulain, B. (2010), Bacterial toxins and the nervous system. Neurotoxins and multipotential toxins interacting with neuronal cells. *Toxins (Basel)* 2(4): 683–737.

Proft, T (ed.) (2013), Bacterial toxins: Genetics cellular Biology and practical applications Caister Academic Press.

Proft, T (ed.) (2009), Microbial toxins: Current Research and future trends. Caister Academic Press.

Pulido, O.M (2016), Phycotoxins and harmful algal blooms (HABs) and human poisoning: An overview. Int. Clin Pathol J 2(6):00062 DOI: 15406/1cpjl. 2016.02.00062.

Reddy, S.M., Girisham, S., Krishna Reddy, V. and Surekha, M. (2010), Mycotoxin Problem and its Management, Scientific Publishers (India), Jodhpur

Rossini G.P and Hess P (2010), Phycotoxins: chemistry mechanism of action and shellfish poisoning, EXS 100:65–122.

Todar Kenneth (2012), Online Text Book of Bacteriology. Madison Wisconsin

Toerger, R.D. *et al.* (2000), Bacteriocins. In Encyclopedia of Microbiology Vol I, (Ed. Lederberg) Academic Press.

Van der Goot F.G. (2001), Pore-forming toxins. Berlin Heidelberg: Springer Verlag.

Visciano, P *et al.* (2016), Marine biotoxins: occurrence, toxicity, regulatory limits and reference methods. *Front Microbiol* 7:1051.

Williams, J.M and Tsai, B. (2016), Intracellular trafficking of bacterial toxins. Current Opinion in Cell Biology 41:51–56.

Zanchett, G., Oliveira-Flho, E.C. (2013), Cyanobacteria and cyanotoxins: from impacts on aquatic ecosystems and human health to anticarcinogenic effects: *Toxins (basel)* 5(10): 1896–1917.

■■■■

Chapter - 25

Antimicrobials – Mode of Action and Resistance

25.1 INTRODUCTION

The study of microbial physiology makes us to understand the basic aspects of microbial life processes and the underlying mechanism of vital functions. This understanding in turn enables us to exploit the beneficial activities of microorganisms as well as minimizing or eliminating the harmful activities. A wide range of microorganisms, bacteria, fungi, yeasts, protozoa, nematodes and viruses causes life-threatening diseases in humans and other animals. These diseases not only claimed millions of lives over a long period of human history, but also changed the course of different civilizations. Not surprisingly, man discovered some of the therapeutic agents much before knowing the actual cause of disease. Search and discovery of chemical agents that can absolve the miseries of man from pathetic effects of microorganisms is one of the greatest achievements of human endeavour.

Today, there are hundreds of antimicrobial agents in our armament and still many more are in pipeline. The need of the hour is one must exercise perseverance, patience and perspiration. Scrutiny followed selection and intelligent use of these chemicals are required for effective use of these chemicals. In depth knowledge of microbial metabolic processes is essential to understand the selective mode of antimicrobial compounds targeting the microbial pathogens, while without affecting the host system. By understanding the basis of action of antimicrobial agents, we may on one hand, devise intelligent strategies for disease treatment and, on the other hand, more intimately understand subtle differences in the relationships between structure and function of pathogen and host.

With the background information provided in this book, one will be in a comfortable position to understand the selectivity and mode of action of different antimicrobials targeting the pathogenic organism. This chapter, primarily deals with the mode of action of different antimicrobial chemicals along with the physiological function of microbe affected leading to its inhibition or death. In addition, it will also bring forth the reasons behind the development of resistance to antimicrobial agents.

25.2 TERMINOLOGY

Antimicrobial

An antimicrobial is any substance of natural, semisynthetic or in synthetic origin that kills or inhibits the growth of microorganisms, but causes little or no damage to the host.

Antibiotic

An antibiotic is a low molecular weight substance produced by microorganism that at low concentration inhibits or kills other organisms. Antibiotics by strict sense, do not include antimicrobial substances that are synthetic (*e.g.* sulfonamides and quinolones) or those which come from plants (*e.g.*, quercetin and alkaloids) or animals (lysozyme).

It is important to notice that all antibiotics are antimicrobials, but not all antimicrobials are antibiotics.

Spectrum of action

A range of bacteria or other microorganisms that is affected by a certain antibiotic is expressed as its spectrum of action. Antibiotics effective against prokaryotes that kill or inhibit a wide range of Gram-positive and Gram-negative bacteria are said to be **broad spectrum** antibiotics (*e.g.*, chloramphenicol and tetracyclines). If effective against a specialized group of organisms or diseases, they are **narrow spectrum** (*e.g.*, penicillin, glycopeptides, rifamycin). If effective against a single organism or disease they are referred to as **limited spectrum** (*e.g.*, isoniazid).

Cidal and Static

Bactericidal antimicrobials are those that kill target organism. Examples of bactericidal agents include aminoglycosides, cephalosporins, penicillins and quinolones.

Bacteriostatic drugs, on the other hand, inhibit or delay bacterial growth and replication. Examples of such agents include tetracyclines, sulfonamides and macrolides.

Some antimicrobial agents can be both bacteriostatic and bactericidal depending on the dose, duration of exposure and the state of the target organism.

Selectivity

Preferential targeting of the pathogenic microorganism present in the host without affecting the host. It is because of the difference in structure and metabolic functions between the pathogen and host.

Susceptibility

If a pathogenic organism is effectively inhibited or killed by the specified dose of the antimicrobial agent, it is said to be susceptible and the phenomenon is known as susceptibility.

Antimicrobial resistance

The ability of microorganism to survive treatment with a clinical concentration of an antimicrobial agent, at which the normal population of the microorganism would be killed.

Intrinsic resistance of the host is due to the presence of wild genes which are transmitted vertically to the progeny. Acquired resistance, on the other hand, is due to stress conditions which may result in gene/chromosomal mutations or by horizontal transfer.

Analogues

A chemical compound that is structurally like another (usually a naturally occurring) compound, but differs slightly in composition. For example, sulfonamide, an analogue of paraminobenzoic acid (PABA), substitutes PABA and interferes with the normal functioning of folic acid. Similarly, base analogues interfere with the synthesis of RNA and DNA.

Chemotherapy

Treatment of diseases by chemical agents. It is aimed at killing the cells (cytotoxic effect). Chemotherapy is most commonly used in treating cancers. Pyrazinamide is one of the most important chemotherapeutic drugs in the treatment of tuberculosis.

Minimum Inhibitory Concentration (MIC)

It is the lowest concentration of an antimicrobial agent that is needed to hinder the growth of bacteria.

25.3 MECHANISMS OF ACTION OF ANTIMICROBIAL AGENTS

As noted earlier, hundreds of antimicrobials are discovered and the process of discovering novel antimicrobials is a continuous process. In addition, efforts are being made to modify the known antimicrobials to enhance their efficacy and safety. Each one of the antimicrobial agent preferentially interferes with the specific metabolic process. In other words, each one has its own mechanism by which it kills or inhibits the target pathogen, at the same time not interfering with the host metabolism. The subtle difference in metabolic processes that exists between the pathogen and host is the basis for selective action of antimicrobials.

Despite the existence of a wide range of antimicrobials, the mechanisms of action of antimicrobials can be broadly categorized into the following five types:

1. Inhibition of cell wall biosynthesis
2. Alteration of cell membrane functions
3. Inhibition of protein synthesis
4. Inhibition of nucleic acid synthesis
5. Inhibition of metabolic processes

25.3.1 Inhibition of Cell Wall Biosynthesis

Many antimicrobial agents exert their inhibitory effect on cell wall synthesis (Fig. 25.1). Since rigid walls are found in both prokaryotic and eukaryotic microbes, but are absent in animal cells, wall inhibiting agents have wide spread clinical use. A majority of the wall-affecting agents interfere with the synthesis of polymers of cell wall-murein in bacteria and chitin in fungi, which under normal conditions provide the rigidity to the cell wall. The general structure and synthesis of bacterial cell wall are discussed in Chapter 2. There, it is mentioned that murein synthesis is a compartmentalized phenomenon. Murein precursors are formed within cytoplasm at the cytoplasm–cytoplasmic membrane junction, transported through the membrane and incorporated into the growing polymer at an extra cytoplasmic membrane location. The structural complexity of murein as well as complexity of its synthesis make the process of murein synthesis readily amenable to disruption. As such, murein synthesis can be disrupted in variety of ways that include: (i) disruption of polymer formation, (ii) disruption of formation of the carbohydrate backbone, (iii) disruption of formation of the lactic acid moiety, (iv) disruption of amino acid addition to the side chains, (v) disruption of peptide bond synthesis and (vi) disruption of the function of the lipid carrier molecule, undecaprenol, required for transport of murein precursor to their growing site on the polymer. Interestingly, examples of all of the above mechanisms are known. Some important antimicrobials inhibiting cell wall biosynthesis and their specific mode of action are presented in Fig. 25.1.

Among the cell wall inhibiting antimicrobials, β-lactams and glycopeptides are important as they are widely used in veterinary and human medicine.

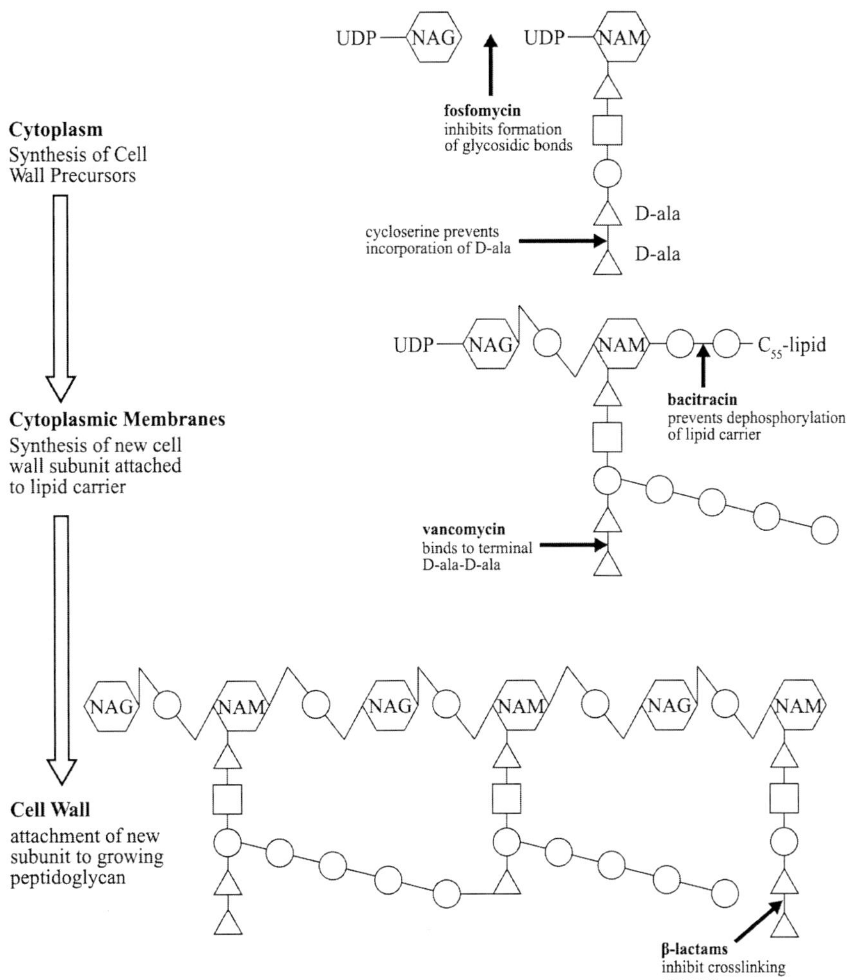

Fig. 25.1. Bacterial cell wall bisynthesis and steps blocked by different antibiotics

β-Lactams

β-lactam antibiotics are a class of broad-spectrum antibiotics, consisting of all antibiotic agents that contain a beta lactam ring in their molecular structure. They have been used safely and effectively for over 70 years. β-lactam antibiotics are analogues of D-alanyl-D-alanine – the terminal amino acid residues on the precursor NAM/NAG-peptide subunits of the nascent peptidoglycan layer. This structural similarity facilitates their binding to the active site of penicillin binding proteins (PBPs). β-lactam antibiotics include penicillins and their derivatives (penams), cephalosporins and their derivatives (cephems), carbapenems and monobactams. Structures of some commonly used β-lactams are shown in Fig. 25.2. All the β-lactams utilize a similar strategic mechanism to produce their antibacterial activity. The enzymatic targets for these antibiotics operate in the final stages of bacterial cell wall synthesis.

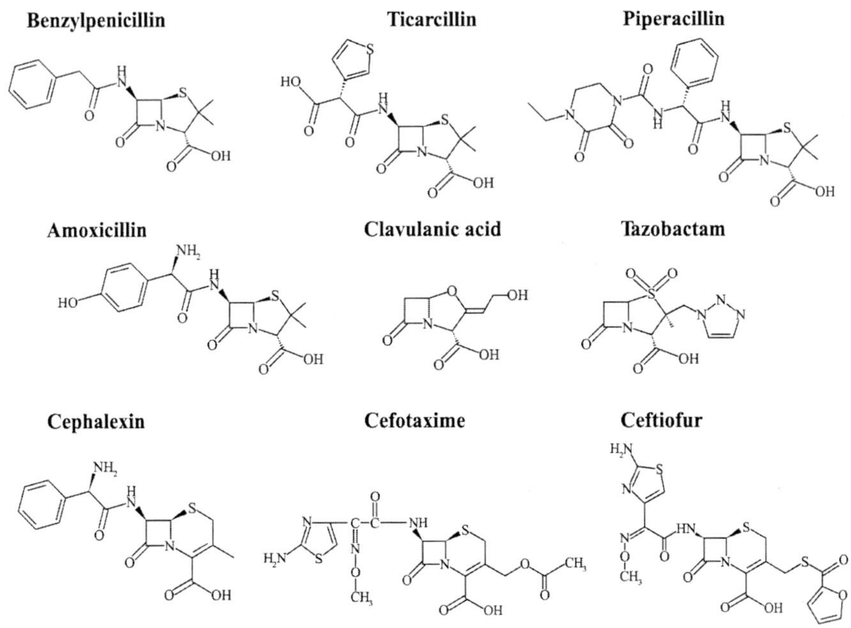

Fig. 25.2 : Representative β-lactam antibiotics

Penicillins

Penicillins are a group of closely related compounds of β-lactam antibiotics produced by *Penicillium* species. The basic structure of penicillins is 6-amino penicillanic acid (6-APA) which consists of thiazolidine ring with a condensed β-lactam ring. The 6-APA carries a variable acyl moiety in position 6. The part of the structure that makes each form of penicillin unique is the benzyl ring and the compounds that attach to it (Fig. 25.2). There are two different forms of penicillins, biosynthetic penicillins and semisynthetic penicillins. Biosynthetic or natural penicillins are produced directly from *Penicillium* molds through fermentation. Semisynthetic penicillins are derivatives of penicillin which include ampicillin, penicillin V, carbenicillin, oxacillin and methicillin. Around 50 types of penicillins are reported so far.

However, subtle differences are found in the antimicrobial action of various penicillins. An important feature of cell wall synthesis is the transpeptidation reaction, which results in the cross-linking of two glycan-linked peptide chains (Chapter 2). The enzymes that accomplish this task, the transpeptidases, are also capable of binding to penicillin or other antibiotics with the β-lactam ring. Thus, these transpeptidases are known as penicillin binding proteins (PBPs). The PBPs upon binding to penicillin are inactivated and can no longer catalyze the transpeptidase reaction *i.e.,* cross cross-linking of peptidoglycan. The cell wall continues to be formed, but is no longer cross-linked and becomes progressively weaker as the peptidoglycan backbone is laid down. In addition, the penicillin – PBP complex stimulates the release of autolysin that digests the existing wall. The result is a weakened, eventually degraded cell wall.

Cephalosporins

Cephalosporins are another group of antibiotics that contain β-lactam ring. They differ structurally from the penicillins but they have the same mode of action as the penicillins *i.e.,* they irreversibly bind to the PBPs and prevent the cross-linking of peptide glycan (Fig. 25.1). Cephalosporins are generally have a broader spectrum of antibiotic activity than the penicillins and are often more resistant to the action of enzymes that destroy β-lactam rings, the β-lactamases.

Resistance: Two major resistance mechanisms for β-lactams are operative, which vary with the organisms involved. In Gram-positive bacteria, the most common mechanism is acquisition of new PBPs that have low affinity for common β-lactams. One example of this in *Staphylococcus aureus*. The presence of the PBP2a-encoding *mec A* gene results in a PBP with low affinity for all except the recently developed β-lactams, leading to methicillin-resistant *S. aureus* (MRSA).

In Gram-negative bacteria, the major resistance mechanism is production of β-lactamases rather than modification of PBP. In addition, efflux and modification or deletion or porin function can play a role in resistance to specific β-lactams. Loss of porin function may occur in concert with β-lactamase production, which results in high levels of resistance to virtually all β-lactams.

β-Lactamases

β-lactamases are a family of enzymes produced by bacteria involved in providing bacterial resistance to β-lactam antibiotics. β-lactamases confer antibiotic resistance to bacteria by breaking the antibiotic structure. Through hydrolysis, the lactamases break the β-lactam ring open, deactivating the molecules antibacterial properties. Penicillinase, the first β-lactamase to be identified, is a specific type of β-lactamase showing specificity to penicillins, again by hydrolyzing the β-lactam ring. (Fig. 25.3)

More than 1000 unique β-lactamases have been described, with varying molecular and functional properties, based on their substrate preference and inhibitor profile. On a molecular level, β-lactamases can be divided into those that utilize serine at the active site (classes A, C and D) and those that require one or two zinc ions to facilitate β-lactam hydrolysis (Class-B). Functional grouping of these enzymes allows the correlation of molecular structure with biochemical characteristics.

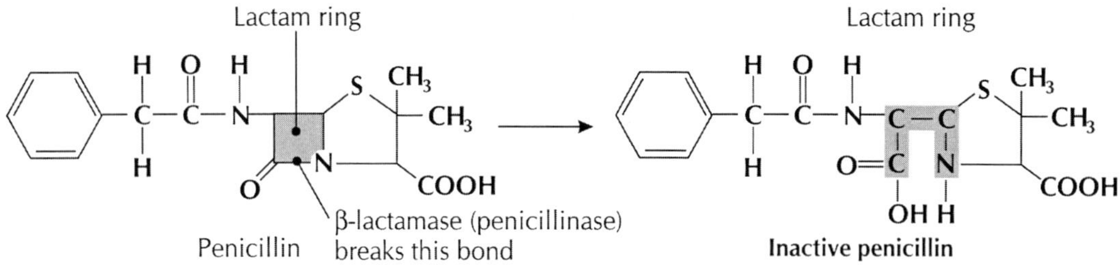

Fig. 25.3 : Mode of action of β-lactamase

Extended-Spectrum β-Lactamases (ESBLs): As bacteria developed resistance to one type of β-lactam antibiotic, new antibiotic derivatives were made by researchers. But the bacteria continually evolved and changed existing β-lactamase enzymes to break down the new compounds. The enzymes that can break down the newer derivatives are known as the extended spectrum β-lactamases, which are clinically important and transmitted horizontally among bacteria.

β-Lactamase inhibitors: Strategies for combating the β-lactamase conferred resistance have included the development of new β-lactam antibiotics that are more resistant to cleavage, and development of the class of enzyme inhibitors called β-lactamase inhibitors. β-lactamase inhibitors have little antibiotic activity of their own. However, they prevent bacterial degradation β-lactam antibiotics. Some of the β-lactamase inhibitors are: clavulanic acid, sulbactam, tazobactam, avibactam and relebactam. These drugs are co-formulated with some β-lactam antibiotics that have the same half-life.

Glycopeptides

Vancomycin was the first glycopeptide to be introduced into medical practice to treat infections caused by Gram-positive cocci. Other glycopeptides that have been introduced include teicoplanin, avoparcin and telavancin. Glycopeptides inhibit the growth of bacteria by forming a complex between antibiotic and the C-terminal D-ala-D-ala dipeptide of the nascent peptidoglycan on the outer surface of the cytoplasmic membrane (Fig. 25.1). The formation of this complex prevents the transglycosylation and transpeptidation reactions that are necessary for completion of the peptidoglycan chain, resulting in an incomplete cell wall and subsequent cell death. In addition to its interaction with D-ala-D-ala, telavancin binds to lipid-P, a cell wall precursor on the cytoplasmic side of the cell membrane. This interaction results in membrane depolarization and eventual membrane disruption, thereby providing a second mechanism that leads to cell death.

The transportation of precursors to the growing chain of intact murein is mediated by the 55-carbon lipid carrier, undecaprenol. To be active in murein precursor transport, undecaprenol must be in a monophosphorylated form. However, during addition of a murein precursor to the growing rigidity polymer, a dephosphorylated form of undecaprenol is released. Further, participation of undecaprenol in murein synthesis requires that the monophosphorylated form of undecaprenol be regenerated from the diphosphorylated form produced during addition of a building block to the growing murein polymer. Bacitracin interferes with murein synthesis by inhibiting formation of undecaprenol monophosphate from undecaprenol diphosphate, thus inhibiting ability of undecaprenol to serve as a carrier molecule (Fig. 25.1).

Resistance: Resistance to glycopeptides was initially thought to be difficult if not impossible to attain. The reason was that primary killing targets are on the outer surface of the cell membrane and thus they have no membrane barrier to overcome. These agents target the substrates rather than the enzymes. Thus, the major glycopeptide resistance mechanisms that have been identified are associated with structural modifications in the substrates for enzymes that incorporate the final amino acids in the pentapeptide precursors. It involves, for examples, the replacement of C-terminal D-ala by D-lactate or D-serine, resulting in D-ala-D-lac or D-ala-D-ser dipeptides with reduced binding affinities for a glycopeptide. Resistance to vancomycin and teicoplanin was attributed to transposon-mediated in enterococci or chromosomal mutations that may be either inducible or constitutive.

Phosphonic Acid

It is also known by fosfomycin or phosphonomycin. Fosfomycin is a simple epoxide–containing antibiotic isolated from streptomycetes and related to phosphoenolpyruvate. It has generated increased attention over the past few years due to its bactericidal activity against a wide range of microorganisms, including multidrug – resistant Gram-positive and Gram-negative bacteria. Fosfomycin inhibits the cell wall synthesizing enzyme MurA, the enol pyruvyl transferase that catalyzes the first committed step of peptidoglycan synthesis. Since this enzyme is essential for formation of the bacterial cell wall, fosfomycin inhibits the growth of a broad range of bacterial species.

Resistance: Resistance to fosfomycin is reported to involve several different mechanisms.

i. Mutations in *MurA* gene that can lead to produce an enzyme with reduced affinity for fosfomycin.
i. Increased levels of MurA can be produced, requiring higher concentration of fosfomycin for growth inhibition.
iii. Increased efflux or decreased influx due to alterations in the glpT and/or uhp transport systems.
iv. Inactivation of the antibiotic by the enzyme FosA.

25.3.2 Alterations of Membrane Functions

The cytoplasmic membrane is a critical structure for cells. In prokaryotes, apart from an agent of selective permeability, it serves a variety of additional functions. The structure and functions of prokaryotic cell membrane are explained in chapter-I. The delicate membrane can be adversely affected in many ways. Its composition may be altered, its fluidity may be changed, its organization may be destroyed or its charge may be abolished. Any one or combination of these phenomena exert a devastating effect on microbes – leakage of critical intracellular materials to the external environment, uptake of critical nutrients, disruption or destruction of enzyme and transport activities, abolition of the ability to generate energy, all of which ultimately lead to dysfunction and death of cell.

A variety of substances are known to affect the membranes of microbial cells. Some are generalized in their activities, while others are specific. Antimicrobials affecting the cell membrane include detergents, phenols, membrane-affecting antibiotics, polymyxins, polyene antibiotics, agents affecting membrane charge and ionophore antibiotics.

Detergents and Phenols

Detergents manifest their action primarily by disruption of membrane protein-lipid interactions, leading to severe membrane structural changes and leakage of small molecules from the cell interior. Additional effects of detergents include protein degradation and cell lysis by the action of autolytic enzymes.

Phenols are non-selective in their mode of action. They disrupt the membrane and inhibit enzymes. The effectiveness of phenols is frequently determined by the nature of the materials attached to the basic molecule. In general, the more lipid soluble the phenol, the more effective its antimicrobial activity.

i. Membrane-affecting antibiotics

A variety of antibiotics affect membrane function that include cyclopeptides, lipopeptides and ionophores.

Cyclopeptides

Polymyxins, the cyclopeptide class of antibiotics, produced by *Bacillus polymyxa*. Polymyxin B and colistin (polymyxin E) are commercially available. They are similar in their spectrum of activity, mechanism of action, resistance characteristics and toxicity. These antibiotics preferentially affect the membranes of Gram-negative bacteria. Polymyxins are bactericidal antibiotics that bind to lipid A of the lipopolysaccharide in the bacterial membrane resulting in membrane disintegration. Exposure to these compounds lead to leakage of 260 nm – absorbing materials and other intracellular substances. These phenomena reflect a generalized interaction with phospholipids, particularly phosphatidyl ethanolamine, with resultant generalized membrane disruption and loss of selective permeability.

The gramicidins, linear peptides that preferentially affect Gram-positive bacteria also appear to exert their effects by generalized membrane disruption.

Resistance: Resistance to polymyxins though rare, most frequently involves modifications of the outer membrane. These mechanisms may be either intrinsic or adoptive, involving numerous regulatory systems. Intrinsic resistance occurs when the phosphate groups on the lipid A anchor in the membrane lipopolysaccharide are modified in polymyxin resistant Gram-negative bacteria. Some studies have demonstrated that mutations in the PmrAB and PhoP/PhoQ two-component regulatory systems are correlated with polymyxin E resistance.

Lipopeptides

The only lipopeptide currently approved for the therapeutic use is daptomycin, a cyclic lipopeptide of high molecular weight. It has rapid bactericidal activity against Gram-positive cocci. Originally described as an inhibitor of peptidoglycan synthesis, daptomycin was later demonstrated to cause calcium-dependent membrane depolarization, which results in the cessation of macromolecular synthesis and disruption of the cellular membrane in the dying bacteria. These results were recently confirmed through transcriptional profiling studies.

Resistance: Clinical resistance to daptomycin is relatively rare, but it has been observed among occasional enterococcal and staphylococcal clinical isolates. Decreased susceptibility to daptomycin in *S.aureus* has been associated with point mutations in MprF, a lysylphosphatidylglycerol synthetase and with a nucleotide insertion in YycG, a histidine kinase.

ii. Ionophore antibiotics

Ionophore antibiotics are compounds produced by microorganisms (mainly spore forming bacteria). They are monocarboxylic polyether antibiotics. They may be monovalent (*e.g.* monensin, salinomycin) or divalent (*e.g.* lasolocid). They facilitate transport of ions across biological membranes by forming lipid – soluble complexes with mono or divalent cations. Ionophores disrupt transmembrane ion concentration gradients, required for the proper functioning and survival of microorganisms and thus have antibiotic property. Many antibiotics, particularly the macrolide antibiotics are ionophores that exhibit high affinities for Na^+ and K^+.

Ionophore antibiotics are generally extremely hydrophobic with a molecular weight greater than 500 Daltons. Consequently, they cannot penetrate the outer membrane of Gram-negative bacteria and this group of organisms is generally ionophore resistant. In contrast, Gram-positive organisms, lacking the protective outer membrane are susceptible. The ionophores are active against both aerobic and anaerobic Gram-positive organisms including *Lactobacillus* spp., *Eubacterium* spp., *Peptococcus* spp., *Peptostreptococcus* spp., *Streptococcus bovis*, as well as *Acholeplasma* and *Mycoplasma* spp. Both salinomycin and monensin have been shown to suppress *Clostridium perfringens*.

Use of ionophore antibiotics is generally known to be safe and effective at therapeutic doses in animal species like poultry birds, pigs. Their use in humans is rather rare.

Resistance: Ionophore resistance appears to be mediated by extracellular polysaccharides (glycocalyx) that excludes ionophores from the cell membrane. Genes responsible for ionophore resistance in bacteria have not been identified, but there is little evidence that ionophore resistance can be spread from one bacterium to another.

iii. Agents affecting membrane charge

Cell membrane under intact and live conditions performs energy generating functions involving a series of dependent events: electron transfer generating electrochemical gradient or proton motive force (PMF), coupling of electron transfer to phosphorylation, synthesis of ATP and ATP-ADP exchange. Membrane disrupting antimicrobial agents may affect these functions mediated through cell membrane (Table 25.1). Agents like antimycin A block electron transfer from cytochrome b to cytochrome c_1, whereas myxothiazol, rotenone, amytal, pericidin A, prevent electron transfer from Fe-S center to ubiquinone. Compounds like 2, 4-dinitrophenol, the salicylanilides and related compounds such as tetrachlorosalicylanilide (TCS), tetramethyl dipicrylamine (TMPA) and carbonyl-cyanide chlorophenylhydrazone (CCCP) manifest their action specifically by destroying the proton gradient required. Valinomycin operates in a similar manner.

Table 25.1 : Agents that interfere with oxidative phosphorylation or photophosphorylation

Type of interference	Compound*	Target/mode of action
Inhibition of electron transfer	Antimycin A	Blocks electron transfer from cytochrome *b* to cytochrome C_1
	Myxothiazol Rotenone Amytal Piericidin A	Prevent electron transfer from Fe-S center to ubiquinone
Inhibition of ATP synthase	DCMU Aurovertin	Competes with Q_B for binding site in PSII Inhibits F_1
	Oligomycin Venturicidin	Inhibit F_0 and CF_0
	DCCD	Blocks proton flow through F_0 and CF_0
Uncoupling of phosphorylation from electron transfer	FCCP, TCS DNP, TMPA CCCP	Hydrophobic proton carriers
	Valinomycin	K^+ ionophore
Inhibition of ATP-ADP exchange	Atractyloside	Inhibits adenine nucleotide translocase

*DCMU is 3-(3,4-dichlorophenyl)-l, l-dimethyl urea; DCCD, dicyclohexylcarbodiimide; FCCR cyanide-p-trifluoromethoxyphenylhydrazone; DNP 2,4-dinitrophenol.

Some agents affect ATP synthesis by selective mechanism. Aurovertins inhibit soluble ATPase (F_1) activity. Oligomycin and venturicidin bind to membrane component of F_o and CF_o ATPase and dicyclohexylcarbodiimide (DCCD) binds to the proton translocase proteins of F_o and CF_o ATPase complex.

Atractyloside interferes with the ADP – ATP exchange cycle by inhibiting adenine nucleotide translocase.

Thus, these examples illustrate the diversity of ways in which antimicrobial chemicals may inhibit membrane structure and function.

25.3.3 Inhibition of Protein Synthesis

Protein synthesis is a complex multistep process involving many enzymes as well as conformational alignment. It is one of the most fundamental biological processes by which individual cells build their specific proteins. Within this process are involved both DNA, RNA, complex cell organelles like ribosomes and an array of enzymes and precursor molecules. It is a multi-step process that includes transcription and translation whose spatial and temporal separation of reactions slightly vary in prokaryotes and eukaryotes. Major events in protein synthesis are mediated by the ribosomal particles.

The selective action of antimicrobial substances on protein synthesis reflects, to a large extent, intrinsic differences between the formation and function of 70s and 80s ribosomes and differences in the ability of substances to attach to 70s and 80s ribosomal particles. As a prelude to understand the mode of action of different antimicrobials and antibiotics, the differences between prokaryotic and eukaryotic ribosomes and protein synthesis processes are presented in Tables 25.2 and 25.3.

Table 25.2 : Differences in the constitution of ribosomes of prokaryotesc (*E. coli*) and eukaryotes (mammal)

Property	Prokaryote	Eukaryote
Smaller Subunit		
S-value	30S	36.95 (40S)
RNA	16S	18S
Proteins	21	≈30
Larger Subunit		
S-Value	50S	56.35 (60S)
RNAs	5S, 23S	5S, 5.8S, 28S
Proteins	31	49
Total RNAs	3 types (5S, 16S, 23S)	5 types (5S, 5.8S, 18S, 28S)
Total Proteins	52	≈80

Details of Tables 25.2 and 25.3 clearly indicate that there are several potential points for interference in protein synthesis processes. Nevertheless, the binding sites of the antibiotics are concentrated at and around functional hotspots of the small 30s and large 50s subunits of the bacterial ribosome (Fig. 25.4). These sites are often dominated or exclusively built by elements of the rRNA. Both the subunits of ribosomes serve as target sites for synthesis inhibiting chemical agents. Most of the natural antibiotics inhibit the peptidytransferase center (PTC) on the 50s subunits (Fig. 25.4). Reasons for the prevalence of the PTC as antibiotic target are: i. the high number of crevices allowing binding of small molecules with high affinity, ii. the PTC needs a high amount of structural flexibility and interference by drug binding.

The most common target site of the 30s subunit is the decoding center of the A site, where the large group of aminoglycosides interferes with the fidelity of aminoacyl-tRNA (aa-tRNA)-EF-TU-GTP selection. Examples of binding sites on the 30s part of the A site outside the decoding center are tetracyclines, which block stable binding of aa-tRNAs to the A-site. Fig. 25.4 summarizes the interference points of the various groups of antibiotics with ribosomal functions. Some antibiotics act directly on translational G-proteins. Antibiotics like kirromycin bind to the elongation factor EF-TU-GTP and block its switch into the GDP conformer after delivery of the aa-tRNA to the A-site. The result is that EF-TU remains on the ribosome and thus blocks protein synthesis. Another example is fusidic acid (FA) acting on the second elongation factor G (EF-G). EF-G is responsible for the translocation of ribosomes to the next codon of the mRNA after a decoding step. FA also blocks the conformational switch into GDP conformer of EF-G, which is necessary for the dissociation of this factor after it has fulfilled its ribosomal function.

Table 25.3 : Comparison of protein synthesis in prokaryotes and Eukaryote

Character	Prokaryote	Eukaryote
Overall process	Involves less proteins, comparatively simple	Involves more proteins steps are intricate
Transcription and Translation	Occur simultaneously	Spacially and temporally separated
Ribosomes	70S (30S, 50S)	80S (40S, 60S)
Ribosomal RNAs	16S, 5S, 23S	18S, 5S, 5.8S, 28S
Initiation tRNA	N-formyl methionine Met-tRNAs	Methionine Met-tRNAi
Start codon	AUG	AUG
mRNA	Polycistronic, modification absent	Monocistronic, undergoes modification
Initiation complexes	Less initiation factors	More initiation factors (eIF)
Elongation factors	EF-Tu; EFTS	EF1α-EFβy
Termination facters	Two release factors	Single release factor

Some important antibiotics inhibiting the protein synthesis, their mode of action, resistance and their spectrum of activity are presented in Fig. 25.4 and Table 25.4.

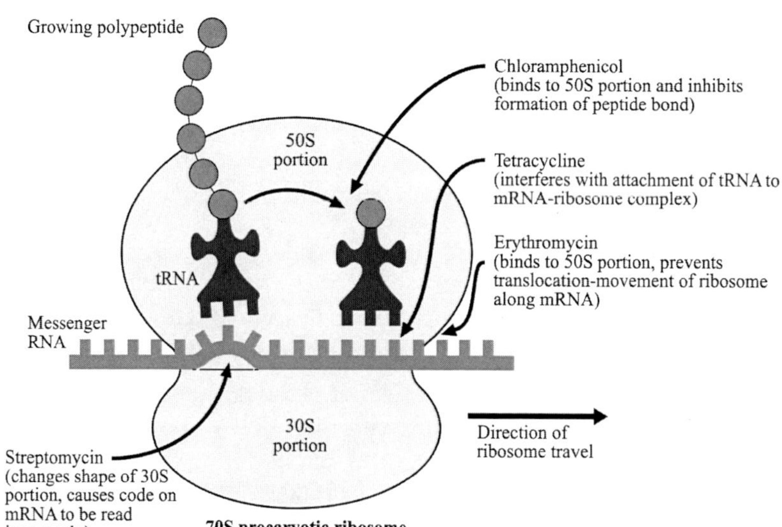

Fig. 25.4 : Interference of different antibiotics in protein biosynthesis

25.3.4 Inhibition of Nucleic Acid Synthesis

Nucleic acids are the utmost important molecules in all living systems, since they directly control the metabolic functions and thus phenotypic characters. These molecules except in rare cases, are structurally and functionally similar in all organisms. Therefore, it is difficult to selectively inhibit the nucleotide metabolism of microbe and host. Hence, the use of antimicrobials that inhibit nucleic acid synthesis is very much limited.

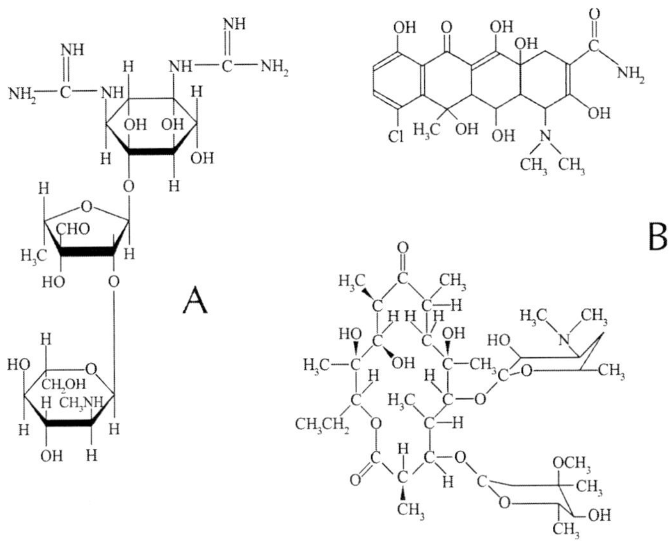

Fig. 25.5 : The structures of streptomycin (A) and tetracycline (B) Both antibiotics interact with the small subunit of the 70S ribosome and inhibit translation, but the mechanisms by which the two substances inhibit translation are different.

Many chemicals that can inhibit nucleic acid synthesis and function are reported. These chemical agents may interfere with the formation of purine, pyrimidine nucleotides, synthesis of nucleic acids, transcription or replication process.

Table 25.4 : Protein synthesis inhibitors

Drug	Mechanism of Action	Mechanism of Resistance	Spectrum of Activity
Macrolides • Erythromycin **Azalides** • Azithromycin • Clarithromycin	Inhibits formation of 50S ribosome blocking transpeptidation or translocation.	• decreased permeability of bacterial envelope to macrolide • chromosomal mutation of 50S ribosomal protein • active efflux of antibiotic - plasmid-mediated • enzymatic inactivation by esterases or phosphotransferases - plasmid mediated	broad-spectrum bacteriostatic

Drug	Mechanism of Action	Mechanism of Resistance	Spectrum of Activity
Lincosamide • Clindamycin • Lincomycin	mechanism of action similar to macrolides		bacteriostatic
Aminoglycoside • Streptomycin • Kanamycin • Tobramycin • Neomycin	primary site of action – interferes with mRNA translational accuracy at 30S ribosome → misreading and premature chain termination binds to and alters bacterial cell membrane cause leakage and disruption of cell wall	• quite common with aminoglycosides • primary mechanism - enzymatic modification of the aminoglycoside by adenylation, phosphorylation or acetylation (enzymes on plasmids or transposons, often in concert with Gram- positive beta-lactamase) • anaerobes are completely resistant b/c lack an oxygen- dependent transport system • chromosomal mutations	narrow-spectrum bactericidal
Tetracyclines • Doxycycline	binds to 30S portion of ribosome and prevents access of aminoacyl tRNA molecules of the mRNA ribosome peptide complex	• common in both Gram-positive and Gram-negative bacteria • generally, but not exclusively, plasmid- mediated • decreased uptake and increased excretion of the drug (efflux pump); resistance is conferred to all tetracyclines • associated with extensive use in animal food	broad spectrum bacteriostatic
Streptogramins combination drug that derived from pristinamycin: 30% quinupristin 70% dalfopristin	both interfere with 50S ribosomal subunit, different target site quinupristin inhibits peptide chain elongation dalfopristin interferes with peptidyl transferase	• primarily occurs by methylation of-MLS binding site; plasmid-mediated; cross-resistance occurs in that methylation of MLS site will cause resistance to all three families of protein synthesis inhibitors • also have drug modification or efflux pump (less common)	limited spectrum bactericidal • Gram-positive cocci staphylococci, streptococci, and enterococci (only *E. faecium* not *E. faecalis)*
Oxazolidinone • linezolid	binds at 50S ribosomal subunit and inhibits formation of initiation complex	• resistance by mutation of 23S RNA (50S subunit) • resistance described in enterococci and staphylococci	• bacteriostatic synthetic antibiotic • activity against streptococci, pneumococci,

Drug	Mechanism of Action	Mechanism of Resistance	Spectrum of Activity
		• does not, at present, exhibit cross-resistance with other protein synthesis inhibitors	enterococci (active against both types of enterococci species), and staphylococci
Choramphenicol	Inhibition of transpeptidation by binding to the 23S subunit of the 50S bacterial ribosome subunit	• Enhanced efflux from bacterial cell • Ribosomal protection by production of methylase enrymes • Hydrolysis by esterases • Mutation of 50S ribosomal protein	Broad spectrum Bacteriostatic
Quinupristin/ Dalfopristin	Inhibiton of protein synthesis by binding to the 50S bacterial ribosome subunit	• Production of methylase enzyme that prevents binding to target • Production of acetyltransferase that inactivates the drug • Enhanced efflux from bacterial cell	Broad spectrum Bacteriostatic
Spectinomycin	Inhibition of protein synthesis by binding to the 30S bacterial ribosome submit	• Mutations in rRNA • Modification of the drug by adenyl-transferase	Broad spectrum Bacteriostatic
Mupirocin	Inhibition of protein synthesis by binding to isoleucyl transfer - RNA synthetase	• Production of a bypass synthetase that binds mupirocen poorly	Broad spectrum Bacteriostatic
Fusidic acid	Blocks the confor-mational switch into the GDP comformer of EF-G	• Development of point mutations in *fusA* that codes for EF-G	Broad spectrum Bacteriostatic

Synthesis

Analogues of glutamine like L-azaserine and 7-diazo-5-oxo-L-norleucine (DON) inhibit phosphoribosyl formylglycineamide synthase, and hydantocidin, an aspartic acid analogue, inhibits adenyl succinate synthetase. Such inhibitions interfere with normal purine and pyrimidine formation and interconversions of among purine and pyramidine precursors.

Transcription

Interference with nucleic acid function may be physiologically manifested in a variety of ways. Direct incorporation of nucleotide base analogues may result in abnormal nucleic acid replication or in the formation of inaccurate transcripts. Rifampicin and actinomycins interfere with transcription by inhibiting the DNA dependent RNA polymerase. Selective toxicity is based on the far greater affinity for bacterial polymerases than for the equivalent host enzymes. Actinomycins, at low concentrations, interact with single stranded RNA and inhibit transcription. However, at higher concentrations, they act as intercalating agents that inhibit replication of

DNA. Resistance to rifampicin is provided by chromosomal mutations that alter the RNA polymerase target, which then has lowered affinity for rifampicin and escapes inhibition.

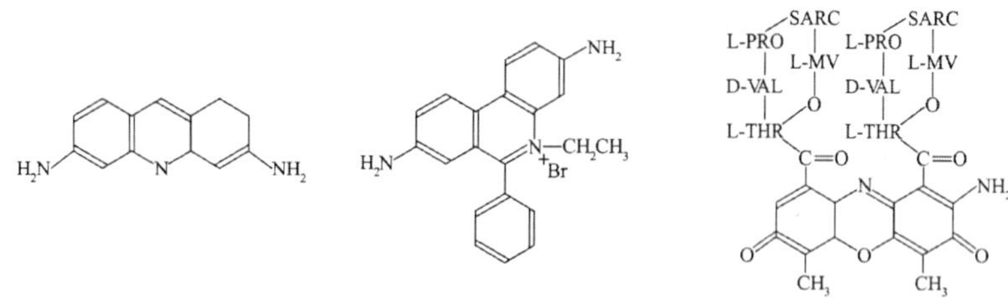

PROFLAVINE **ETHIDIUM BROMIDE** **ACTINOMYCIN D**

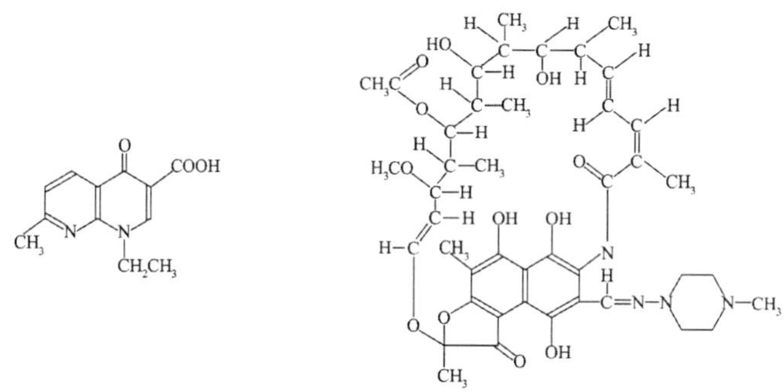

NALADIXIC ACID **RIFAMPIN**

Fig. 25.6 : The structures of agents that interfere with DNA and RNA formation and function.

Replication

Interference with DNA replication is the most physiologically devastating effect of antimicrobial chemicals. Chemical agents like proflavine, actinomycin-D and ethidium bromide are potent intercalating agents (Fig. 25.6). Their interference may be accomplished in one of three ways: (i) interference with strand separation, (ii) false residue incorporation and (iii) interference with the action of DNA polymerases particularly DNA polymerase III. Frameshift mutations, the effect of intercalating agents, would lead to formation of inaccurate transcripts.

Topological change

Intercalating agents like mitomycin form cross-links between opposite strands affecting the normal double stranded structure. This effect inhibits smooth sailing of DNA polymerase during replication.

Quinolones are a large family of broad spectrum bactericidal synthetic agents (Fig. 25.7). Nalidixic acid is the first-generation prototype, but the addition of fluorine at position 6 of the main quinolone ring has improved antibacterial activity, leading to the synthesis of many additional, more commonly used compounds. The antibacterial activity of quinolones is due to their ability to inhibit the activity of bacterial DNA gyrase (topoisomerase-II) and topoisomerase-IV. During replication of the bacterial double stranded circular DNA, DNA gyrase produces and removes supercoils in DNA ahead of replication fork to maintain the proper tension required for efficient DNA duplication. Topoisomerase-IV similarly acts to remove supercoils and to separate newly formed DNA daughter strands after replication. These enzymes act in concert to ensure that the DNA molecule has the proper conformation to efficient replication and packaging within the cell. Quinolones can interfere with these essential enzymes in bacteria while not affecting their counterparts in host cells.

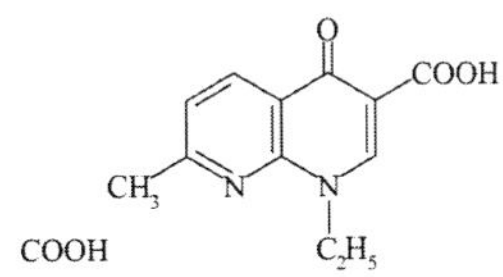

Nalidixic acid

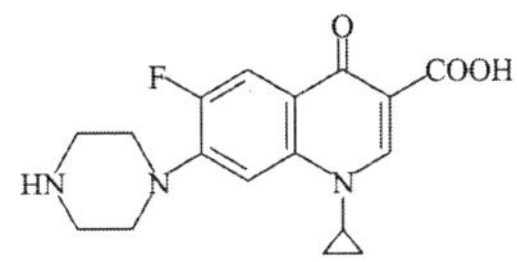

Ciprofloxacin

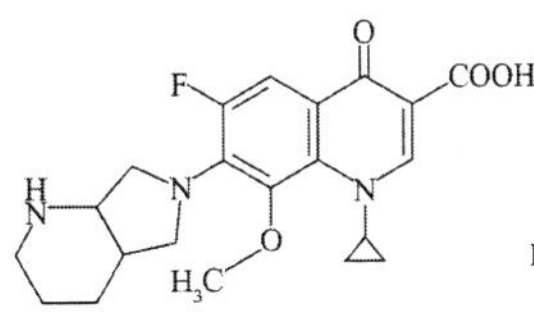

Moxifloxacin

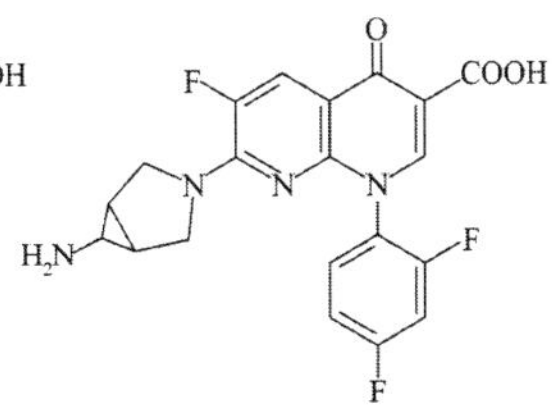

Trovafloxacin

Fig. 25.7 : Structures of some quinolone antibacterial agents

Resistance to quinolones mediated through chromosomes is exhibited in two forms: (i) mutations which change the target enzymes in a manner that affects quinolone binding, (ii) changes in cell wall permeability, resulting in decreased uptake, or by efflux. Plasmid-encoded quinolone resistance involves production of protein (termed qnr) that protects the target DNA from quinolone binding. This protein has been shown to act in concert with a plasmid encoded enzyme capable of reducing the activity of some fluoroquinolones, resulting in increased levels of quinolone resistance.

25.3.5 Inhibition of Metabolic Processes

Several commonly used antimicrobial agents inhibit bacterial metabolic pathways including those which produce precursors for nucleic acid synthesis. These antimicrobials often referred to as **antimetabolites**. A diversity of antimetabolites selectively interfere with the normal metabolic process, resulting in the termination of process or formation of an altered product. Many of these antimetabolites, which are commonly used as chemotherapeutants more often, interfere with enzyme structure and function. The physiological consequences of these effects may be varied and range from inactivation of the enzymes themselves to inactivation of other important molecules.

Analogues - Competitive inhibition

Many of the antimicrobials are structural analogues of cell constituent chemical substances. In this context, the most thoroughly documented antimicrobials are sulfonamides (*e.g.* sulfadiazine, sulfapyridine, sulfathiazole, diaminopyramidines, trimethoprim, methotrexate).

Sulfonamides

Sulfonamides are structural analogues of para-aminobenzoic acid (PABA). These compounds are synthetic bacteriostatic compounds that act in competition with PABA, for the active site of dihydropteroate synthetase, an enzyme that catalyzes an essential reaction in the synthetic pathway of tetrahydrofolic acid (THFA), which is required for the synthesis of purines and pyrimidines and therefore, for nucleic acid synthesis (Fig. 25.8). Selective toxicity depends on the fact that many bacteria synthesize THFA, whereas human cells lack this capacity and depend on an exogenous supply of folic acid. Bacteria that can use preformed folic acid are similarly unaffected by sulphonamides.

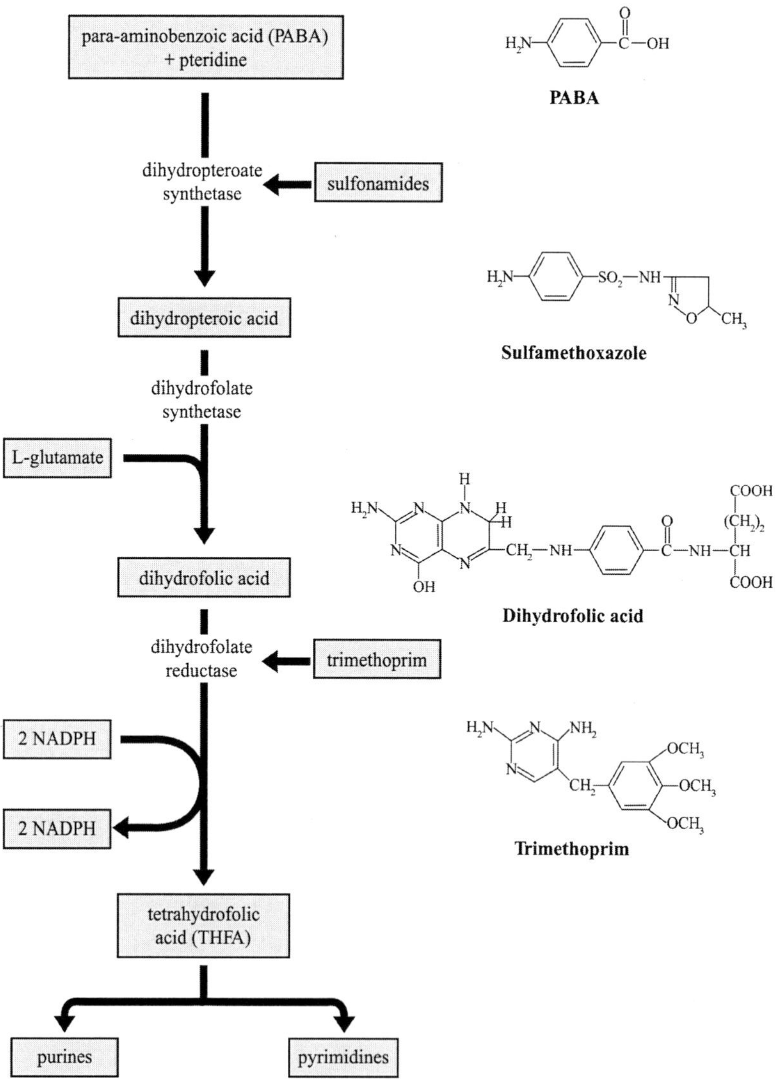

Fig. 25.8 : Mode of action of structural analogues (sulfonamlde and trimethoprim)

Resistance to sulphonamides is widespread with plasmid–mediated genes coding for an altered dihydropteroate synthetase. Though it has not totally changed its affinity, but has a greatly decreased affinity for the sulphonamides. A resistant cell, therefore, possesses two distinct enzymes: a sensitive chromosome-encoded enzyme and a resistant plasmid – encoded enzyme.

Trimethoprim

Trimethoprim is one of a group of pyrimidine-like molecules analogous in structure to the aminohydroxypyrimidine moiety of the folic acid molecule. This chemical, like sulphonamides, prevents THFA synthesis, but at a later stage by inhibiting dihydrofolate reductase. This enzyme is present in mammalian cells, as well as bacterial and protozoan cells, and the selective toxicity depends upon the far greater affinity of trimethoprim for the bacterial enzyme.

Resistance to trimethoprim is provided by plasmid-encoded dihydrofolate reductases. Plasmid-encoded dihydrofolate reductases with altered affinity for trimethoprim allow the synthesis of THFA to proceed unhindered by the presence of trimethoprim. The replacement enzymes are approximately 20,000-fold less susceptible to trimethoprim, while retaining their affinity for normal substrate.

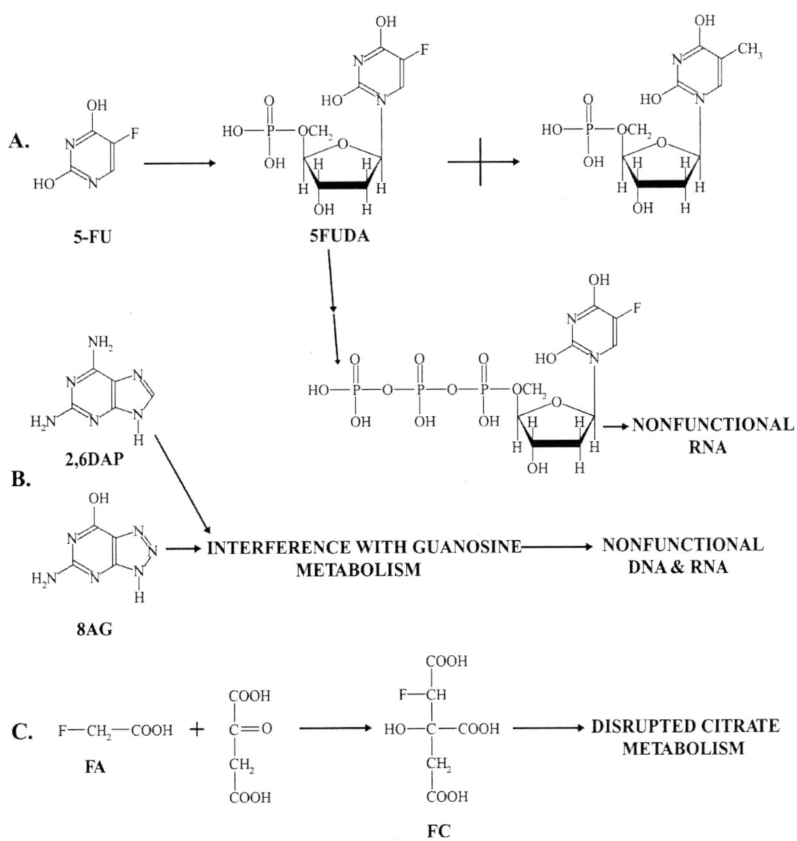

Fig. 25.9 : Modes of action, and the physiological consequences, of representatives structural analogues of normal metabolites.

Analogues-Altered products

In addition to competitive inhibition, substances structurally like normal metabolites may exert their actions in other ways. Formation of an altered product is such a mechanism (Fig. 25.9). The incorporation of halogenated derivatives of uracil (*e.g.* 5-flurouracil) into RNA exemplifies this mechanism. Due to incorporation of altered bases into the RNA molecule, the molecule functions improperly, or not at all. In addition, iodinated forms of uracil disrupt thymidine synthesis. Similarly, nonfunctional purines are formed when 2, 6 diaminopurine or 8-azaguanine are substituted for normal metabolites.

The use of fluoroacetate in place of acetate for citrate synthesis is an additional example of altered product formation because of structural analogue use. Such a procedure results in formation of fluorocitrate, a non-metabolizable substrate for aconitase. As a result, normal operation of the TCA cycle is disrupted or abolished. Mechanisms that involve incorporation of a nonfunctional structural analogue into a product are known as **lethal syntheses.**

Reverse transcriptase inhibitors

Reverse transcriptase inhibitors (RTIs) are a class of antiretroviral drugs used to treat HIV infection or AIDS and in some cases hepatitis-B. RTIs block reverse transcriptase's enzymatic function and prevent completion of synthesis of double stranded viral DNA, thus preventing HIV from multiplying.

Based on chemical structure RTIs are identified into three types (Table 25.5).

i. Nucleoside analogue reverse transcriptase inhibitors (NARTIs or **NRTIs**).

ii. Nucleotide analogue reverse transcriptase inhibitors (NtARTIs or **NtRTI**s).

iii. Non-nucleoside reverse transcriptase inhibitors (**NNRTI**s).

The antiviral effect of NRTIs and NtRTIs is essentially the same. They are analogues of the naturally occurring deoxynucleotides needed to synthesize the viral DNA and they compete with the natural deoxynucleotides for incorporation into the growing viral DNA chain. However, unlike the natural deoxynucleotides substrates, NRTIs and NtRTIs lack a 3'-hydroxyl group on the deoxyribose moiety. As a result, following incorporation of an NRTI or an NtRTI, the next incoming deoxynucleotide cannot form the next 5'-3' phosphodiester bond needed to extend the DNA chain. Thus, when an NRTI or NtRTI is incorporated viral DNA synthesis is halted, a process known as **chain termination**. All NRTIs and NtRTIs are classified as competitive substrate inhibitors. In contrast, NNRTTIs have a completely different mode of action. They block reverse transcriptase by binding directly to the enzyme (Fig 25.10). NNRTIs are not incorporated into the viral DNA like NRTIs, but instead inhibit the movement of protein domains of reverse transcriptase that are needed to carry out the process of DNA synthesis. NNRTIs are therefore classified as non-competitive inhibitors of reverse transcriptase.

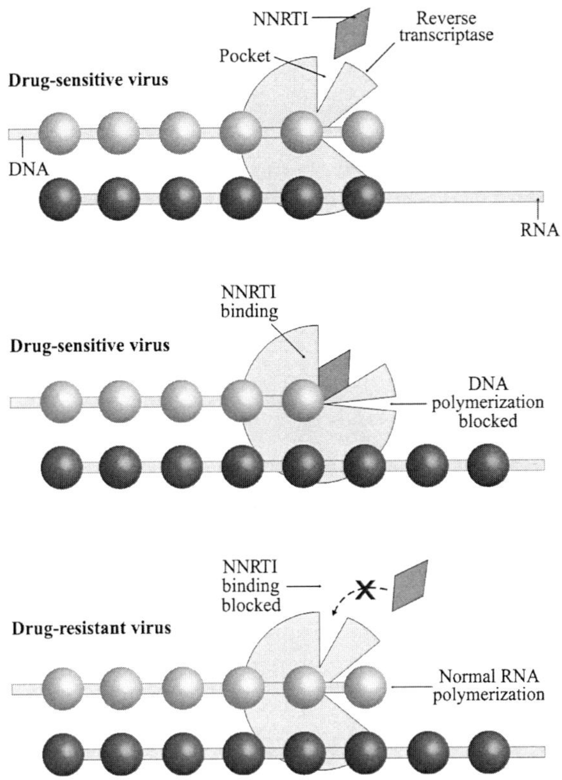

Fig. 25.10 : Blockage of action of reverse transcriptase by NNRTI and blocking of binding site in drug resistant virus

Table 25.5. Classification of reverse transcriptase inhibitors

Nucleoside analogue reverse transcriptase inhibitors (NRTIs)	Nucleotide analog reverse transcriptase inhibitors (NtRTIs)	Non-nucleoside analogue reverse transcriptase inhibitors (NNRTIs)
• Abacavir	• Tenofovir	• Nevirapine
• Didanosine	disoproxyl	• Delavirdine
• Emtricitabine	fumarate	• Efavirenz
• Lamivudine	• Adefovir	
• Stavudine		
• Zidovudine		
• Zalcitabine		
• Lamivudine		

Resistance: As NRTIs and NNRTIs are alike in their mode of action i.e., termination of DNA synthesis and HIV replication, the resistance mechanisms of the virus against these are also similar. Reverse transcriptase of HIV especially HIV-1 does not have proof-reading activity. This combined with selective pressure from the drug, leads to mutations in reverse transcriptase gene (Pol I) that makes the virus less susceptible to NTRIs and NNRTIs. Aspartate residues 110, 185, and 186 in the reverse transcriptase polymerase domain are important in the binding and incorporation of nucleotides. The side chains of residues K65, R72 and Q151 interact with the next incoming nucleotide. Also important is L74, which interacts with the template strand to position it for base pairing with the nucleotide. Mutations of these key amino acids results in reduced incorporation of the analogues.

NNRTIs do not bind to the active site of the polymerase but in a less conserved pocket near the active site in the p66 subdomain. Their binding results in a conformational change in the reverse transcriptase that distorts the positioning of the residues that bind DNA, inhibiting polymerase. Mutations in response to NNRTIs decrease the binding of the drug to this pocket.

25.4 ANTIFUNGAL AGENTS

Fungi, unlike bacteria which are prokaryotes, are eukaryotic organisms. They share many structural and functional features of eukaryotic counterparts like plants and animals. However, in few respects they differ from other eukaryotes also. Due to some unique characters of their own, fungi are placed in a separate kingdom on par with kingdoms of major groups of living organisms.

Fungi being heterotrophic organisms cause a wide range of plant diseases including crop plants. Apart from plant diseases they also cause a variety of human and diseases which include:

i. Dermatophyte infections, caused by *Microsporum, Trichophyton* and *Epidermophyton.*

ii. Systematic infections, such as coccidiomycosis, histoplasmosis and blastomycosis.

iii. Opportunistic infections, such as candidiasis, thrush, vulvaginitis, cryptococcal meningitis, aspergillosis, mucoromycosis and *Pneumocystis carinii* pneumonia.

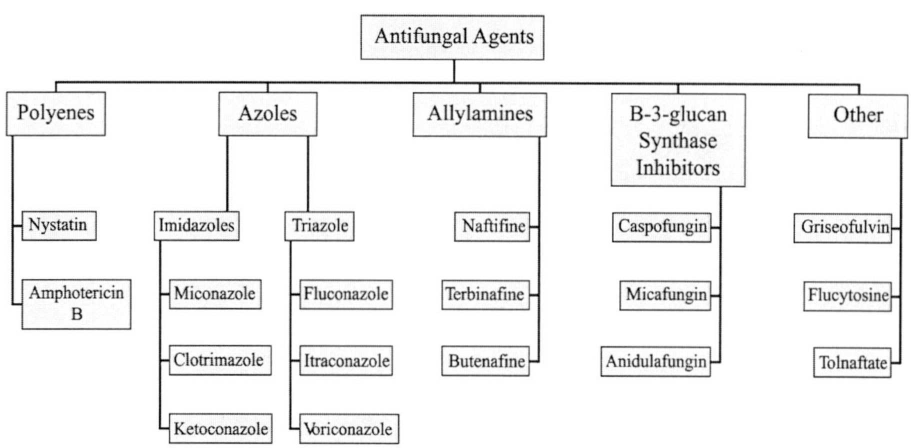

Fig. 25.11 : Classification of antifungal agents

In recent years, numerous advances have been made in developing new antifungal agents and in understanding the mechanisms existing ones. In view of discovery of many antibiotics, attempts have been made to devise an understandable classification such as chemical nature, mode of action, site of action etc. A popular classification of antifungal agents based on chemical nature is presented in Figs. 25.11 and 25.12.

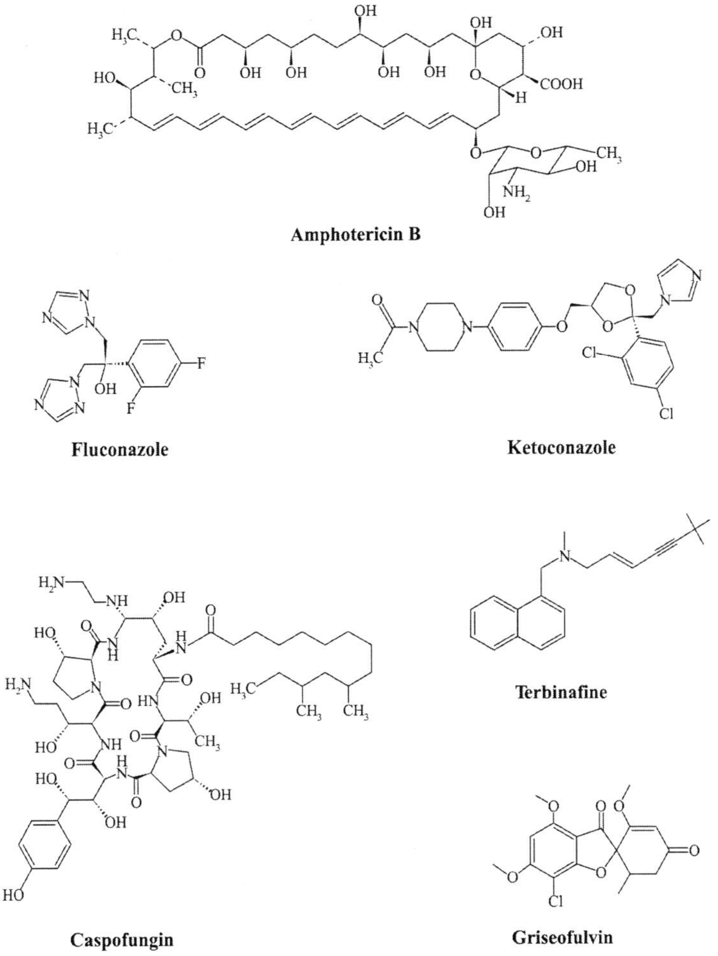

Fig. 25.12 : Chemical structures of some representative antifungal agents.

Polyenes

The polyene compounds are so named because of the alternating conjugated double bonds that constitute a part of their macrolide ring structure. The polyene antibiotics are all products of *Streptomyces* species. These antibiotics interact with sterols in cell membranes (ergosterol in fungal cells; cholesterol in human cells) to form channels through the membrane, causing the cells to become leaky. The polyene antibiotics include amphoterin B, nystatin and natamycin (pimaricin). These are broad-spectrum, mycostatic antibiotics effective against yeasts and molds.

Researches suggested that development of resistance by selection of naturally occurring resistant cells, is present in small numbers in the population. These naturally resistant cells produce modified sterols that bind to these antibiotics with lower affinity. The decreased binding of polyene was attributed to (i) a decrease in the total ergosterol content of the cell, without concomitant changes in sterol composition, (ii) replacement of some or all the polyene-binding sterols by ones which bind poorly with polyenes and (iii) reorientation, or masking of existing ergosterol, so that binding with polyene is sterically or thermodynamically less favoured.

Azoles

Azole antifungal agents are the largest class of synthetic antimycotics. They have five-membered organic rings that contain either two or three nitrogen molecules (the imidazoles and the triazoles, respectively).

Azole antifungal agents inhibit CYP 45014 α-demethylase in fungi. This enzyme is involved in the conversion of lanosterol to ergosterol. They also inhibit cytochrome p450 dependent enzymes involved in biosynthesis of cell membrane sterols. The basic nitrogen of the azole ring forms a tight bond with the heme iron of the fungal p450 preventing substrate and oxygen binding. Inhibition of the C14 α-demethylase results in accumulation of sterols still bearing a C14 methyl group, changing the exact shape and physical properties of the membrane causing permeability changes and malfunctioning of membrane imbedded proteins. They have a lower affinity for mammalian p450s. The effect is fungistatic, but may be fungicidal at lower concentrations.

Until now, there are no reports of modification of azole antimicrobials as a mechanism of resistance. Resistant strains, therefore, either exhibit a modification on the quality or quantity of target enzyme, reduced access to the target or some combination of these mechanisms. Several lines of evidence implicate a modification in the quantity or quality of 14 α-demethylase in the expression of resistance to azole antifungal agents. Other studies have implicated altered 14 α-demethylase in resistance to azoles. Over expression of 14 α-demethylase has also been implicated as a mechanism of resistance.

Allylamines

The allylamines (terbinafine, butenafine and naftifine) have been developed as a new class of ergosterol biosynthetic inhibitors that are functionally as well as chemically distinct from the other major classes of ergosterol-inhibiting antifungal agents.

Allylamines are reversible, noncompetitive inhibitors of the first step in ergosterol biosynthesis (Fig 25.13), the conversion of squalene to squalene -2,3-epoxide by squalene epoxidase. The fungal cell death is attributed primarily to the accumulation of squalene rather than to ergosterol deficiency. High levels of squalene increase membrane permeability leading to disruption of cellular organization.

Though the resistance to allylamines has been reported in few cases, the exact mechanism of resistance is still to be elucidated.

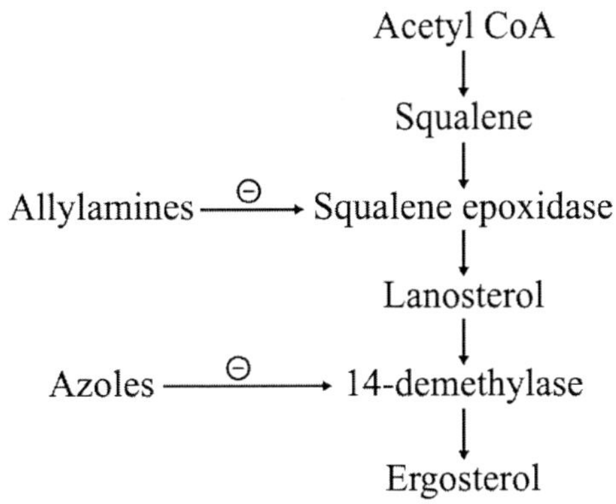

Fig. 25.13 : Mode of action of allylamines and azole antifungal drugs.

Compounds active against fungal cell walls

The fungal cell wall contains compounds such as mannan, chitin, and α- and β- glucans, that are unique to the fungal kingdom. Since these components are not found elsewhere in nature, they have been identified as possible targets that provide selective toxicity advantages. Many compounds that can affect the cell walls of fungi have been discovered. Some antifungal agents of this category have been approved for treatment and available on counters, while some other (caspofungin, anidulgafungin, micafungin, aminocandin) while some other are at the stage of clinical trials.

Inhibitors of glucan synthesis

Of the three groups of compounds (aculeacins, echinocandins and papulacandins) that are specific inhibitors of fungal β-(1-3)-glucan synthase, only echinocandins have received considerable attention. Echinocandins (caspofungin, anidulgafungin, micafungin and aminocandin) which are lipopeptides, have fungicidal activity both *in vitro* and *in vivo.*

β-glucan inhibitors act as specific noncompetitive inhibitors of β- (1,3)-glucan synthetase. Treatment of fungi with these compounds inhibits the synthesis of the structural glucan component without affecting nucleic acid or mannan synthesis. Inhibitors of glucan synthesis also have secondary effects on other components of intact cells including a reduction in the ergosterol and lanosterol content and an increase in the chitin content of the cell wall. Inhibition of β- (1,3) – glucan synthetase results in cytological and ultra structural changes in fungi characterized by growth as pseudo hyphae, thickened cell wall, and buds failing to separate from mother cells. Cells also become osmotically sensitive, with lysis being restricted largely to the growing tips of budding cells.

Though enough experimental evidence is not available, it is suggested that resistance to lipopeptide antimycotic agents is via mutations that alter the protein encoded by *FKSI,* which is the main target of the inhibitor and is presumed to be the catalytic component of the fungal cell wall glucan synthase.

Griseofulvin

Griseofulvin is an antifungal antibiotic produced by *Penicillium griseofulvum*. It is active against most dermatophytes and has been the drug of choice for chronic infections. Griseofulvin inhibits fungal cell division at metaphase by disrupting the cells mitotic spindle structure. It binds to human keratin, making it resistant to fungal infections. It may also interfere with deoxyribonucleic acid synthesis.

Resistance to griseofulvin has been reported in strains of *Trichophyton quinckeanum* and *Microsporum gypseum*. A spontaneous mutation frequency of 10^{-8} was reported for resistance to griseofulvin. The tolerance of all mutants to griseofulvin was approximately 100 times greater than in the wild strains.

Fluorocytosine

5-fluorocytosine (5FC) is a fluorinated analogue of cytosine with inhibitory activity against many yeasts. It inhibits both DNA and RNA synthesis via intra-cytoplasmic conversion to 5-fluorouracil by the enzyme cytosine deaminase. Subsequently, 5FU is converted by UMP pyrophosphorylase into 5-fluorouridylic acid (FUMP), which is phosphorylated further and incorporated into RNA, resulting in disruption of protein synthesis. 5FU is also converted to 5-fluorodeoxyuridine monophosphate, a potent inhibitor of thymidilate synthase, an enzyme involved in DNA synthesis and nuclear division. Thus, 5FC acts by interfering with pyrimidine metabolism, as well as RNA, DNA and protein synthesis in the fungal cell.

Table 25.6 : Mode of action of major groups of antifungal agents

Antifungal drug class	Examples	Mechanism of action
Polyenes	Amphotericin, Nystanin	Interacts with sterols in cell membranes forming channels that leak cellular contents
Azoles	Fluconazole, Itraconazole, Ketoconazole etc.	Inhibit ergosterol biosynthesis at the level of C14-demethylase
Allylamines	Terbinafine	Inhibit ergosterol biosynthesis at the level of squalene epoxidase
Candidins	Caspofungin, Micafungin	Inhibit glucan in fungal cell wall
Antibiotics	Griseofulvin	Inhibits mitosis (sliding of microtubules) in fungi
Antimetabolites	Flucytosine	Inhibit DNA and RNA synthesis via conversion of 5-fluorocytosine to 5-fluoricil
Thiocarbamates	Tolnaftate	Inhibit ergosterol biosynthesis at the level of squalene epoxidase

Resistance to 5FC may result from decreased uptake (loss of permease activity) or loss of enzymatic activity responsible for conversion to FUMP. Although resistance due to decreased 5 FC uptake has been found in *S. cerevisiae* and *C. glabrata,* this mechanism does not seem to be important in others. Available data suggest that blocking the formation of FUMP by loss of cytosine deaminase activity or by loss of uracil phosphoribosyl transferase (UPRTase) activity is sufficient to confer 5FC resistance. Cytosine deaminase and UPRTase constitute the pyrimidine salvage pathway and are not essential for growth under normal circumstances. Resistance is attributed to the mutational loss of one or both the above pyrimidine salvage pathway enzymes.

A summary of the mode of action of major groups of antifungal agents is presented in Table 25.6

25.5 FURTHER READING

Bhattacharjee, M.K. (2016), Antimetabolites: Antibiotics that inhibit nucleotide synthesis. In Chemistry of Antibiotics and Related drugs. Springer, Cham. https://doi.org/10.1007/978-3-319-40746-3.4

Bhattacharjee, M.K. (2016), Antibiotics that inhibit cell wall biosynthesis eBook Springer Verlag.

Bush, K. (2012), Antimicrobial agents targeting bacterial cell walls and cell membranes. *Rev. Sci. Tech. Off. int. Epiz* 31 (1) : 43 – 56.

Caldwell, D.R. (2010), Microbial Physiology and Microbial Metabolism (2nd edn.), Belmont, Calif. Star Publishing.

Corcoran *et al.* (eds.) (2012), Antibiotics III. Mechanism of antimicrobial and antitumor agents. Springer Verlag, Berlin and Heidelberg, GmBH & Co. KG.

Epand R.M. and Vogel H.J. (1999), Diversely of antimicrobial peptides and their mechanisms of action *Biochemica et Biophysica Acta* 1462 : 11 – 28.

Georgopapadakou, N.H. and Walsh T.H. (1996), Antifungal agents:chemotherapeatic targets and immunologic Strategies. Antimicrobial agents and chemotherapy 40(2) : 279 – 291

Ghanoum, M.A. and Rice L.B. (1999), Antifungal agents : Made of action mechanism of resistance and correlation to these mechanisms with bacterial resistance. *Clin. Microbiol. Review* 12(4) : 4501 – 4171.

Kong, K.F., Schneper, L. and Mathee (2009), Reta-lactam antibiotics: From antibiosis to resistance and bacteriology *APMIS* 118: 1–36.

Nikilay, R. *et al.* (2016), Ribosome assembly as anti-microbial target. *Antibiotics* 5, 18.

Walsh, C. (2000), Molecular mechanisms that confer antibacterial drug resistance. *Nature* 406 : 775 –781.

Wilson, D.N. (2009), The A-Z of bacterial translation inhibitors. *Critical Reviews in Biochemistry and Molecular Biology* 44 (6) : 393 – 433.

Wilson, D.N. (2014), Ribosome targeting antibiotics and mechanisms of bacterial resistance. *Nature Reviews Microbiology* 12: 35–48.

■■■■

Chapter - 26

Signal Transduction – Quorum Sensing

A. SIGNAL TRANSDUCTION

Microorganisms live in an ever-changing external milieu. Not surprisingly, in order to be able to compete, survive and thrive in their environment, bacteria have devised sophisticated signaling systems for eliciting adaptive responses to their environment. Motile bacteria exhibit locomotor responses to a variety stimuli, including chemicals, light, osmolarity, temperature, and electric and magnetic fields. The appearance of novel nutrients and metabolites trigger the production of transport and metabolic machinery needed to uptake and utilize them. Stress conditions of many sorts, such as antibiotics, heavy metals, turgor pressure or starvation elicit effective coping responses. Starvation counter measures can be especially elaborate. In pathogens, chemicals, temperature, or other conditions characteristic of the host environment elicit production of toxins and other virulence factors. Even symbiotic and pathogenic organisms follow chemical traits to locate their plant or animal hosts. Bacteria sense all these environmental and cytoplasmic signals through a wide range of signalling circuits and tune up their behavioural responses. The machinery behind the adaptive responses of bacteria undertakes the signalling tasks fundamental to all sensory systems: stimulus detection, signal processing, including amplification and integration of sensory inputs: and production of appropriate output responses (Fig. 26.1).

A general theme that has emerged is that a protein spanning the cytoplasmic membrane will sense an environmental signal and then interact with regulatory protein components in the cytoplasm, often through phosphorylation-dephosphorylation reactions, thereby regulating gene expression. This is called a two-component signal transduction system.

26.1. TWO COMPONENT SIGNAL SYSTEM

26.1.1. Components of Two Component Signaling System

A wide range of two component signal transduction systems enable bacteria to sense, respond and adapt to different environmental changes. The average number of two component systems in bacterial genome has been estimated at around 30 or about 1–2% of a prokaryote's genome. Although two component signaling systems are found in all domains of life, they are most common in bacteria particularly in Gram-negative and cyanobacteria. They are much less common in archaea and eukaryotes. However, they are reported in some yeasts, filamentous fungi and slime molds. They are also common in plants. Two component systems have been described as 'conspicuously absent' from metazoans.

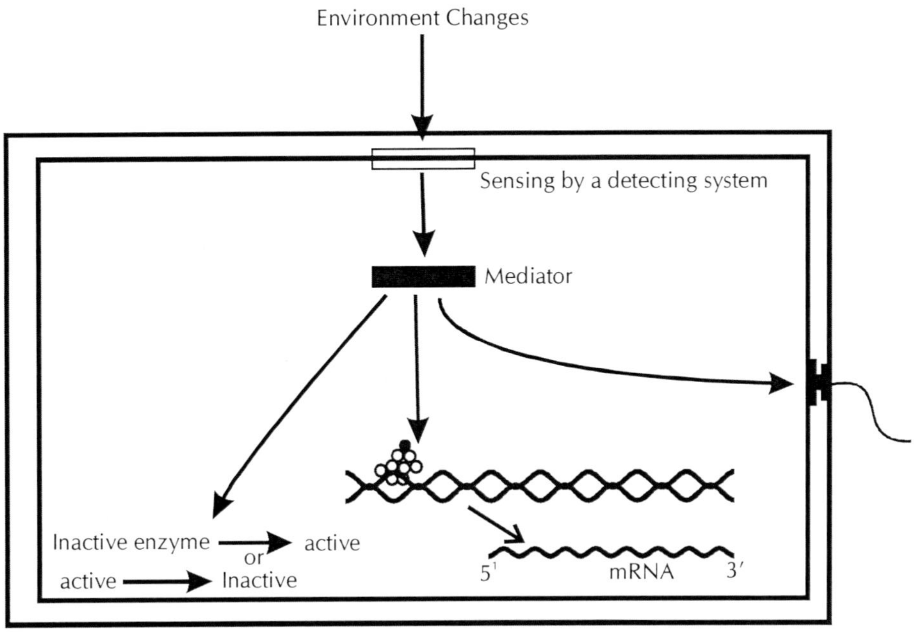

Fig. 26.1. Outlines of signal mechanism in bacteria

Two-component signaling systems have been implicated in a wide range of physiological responses. These include nitrogen assimilation, outer membrane porine synthesis, chemotaxis in *E. coli* and *S. typhimurium*, nitrogen fixation in *Klebsiella* and *Rhizobium,* sporulation in *Bacillus*, oxygen regulation of gene expression in *E. coli* and the uptake of carboxylic acids in *Rhizobium* and *Salmonella*.

The components of the two-component systems include:

i. **Histidine Kinase (HK):** It is sometimes called sensor/kinase. HKs are typically homodimeric intrinsic transmembrane proteins. They usually have an N-terminal ligand-binding domain and a C-terminal kinase ATP binding domain, but other domains may also be present. The C-terminal forms an independently folded domain that extends into the cytoplasm. It is approximately 250 amino acid residues in size and highly conserved in all members of the sensory family. The kinase domain is responsible for the auto-phosphorylation of the histidine with ATP, the phospho transfer from the kinase to an aspartate of the response regulator, and phosphotransfer from aspertyl phosphate back to ADP or to water

$$\text{HK+ATP} \longrightarrow \text{HK-P (active) + ADP}$$

Many HKs are bifunctional and possess both kinase and phosphatase activity.

ii. **Partner response regulator (RR):** Response regulators are simpler in structure than histidine kinases. These are cytosolic proteins, composed of two independently folded domains. Its N-terminal (receiver domain) is with approximately 125 conserved amino acid residues, while the C-terminal (effector domain) shows significant homology amongst certain sub-classes of regulators. The N-terminal domain of a regulator can be phosphorylated by a phosphoryl transfer from the auto kinase domain of the corresponding sensor. This phosphorylation, in turn, activates the C-terminal domain of the regulator, which then may result in either (i) stimulation of transcription of specific genes (function as transcriptional factors), (ii) regulation of rotation of flagella or (iii) modulation of enzyme activity

$$RK + HK\text{-}P \longrightarrow RR\text{-}P + HK$$

iii. Phosphatase: The phosphatase inactivates the RR-P. Phosphatase may be the histidine kinase, the response regulator or a separate protein

$$RR - P + H_2O \longrightarrow RR + Pi$$

26.1.2. Signal Transduction in Two-Component Systems

Signal transduction in two-component systems as shown in Fig. 26.2 can be conveniently thought of as occurring in three steps.

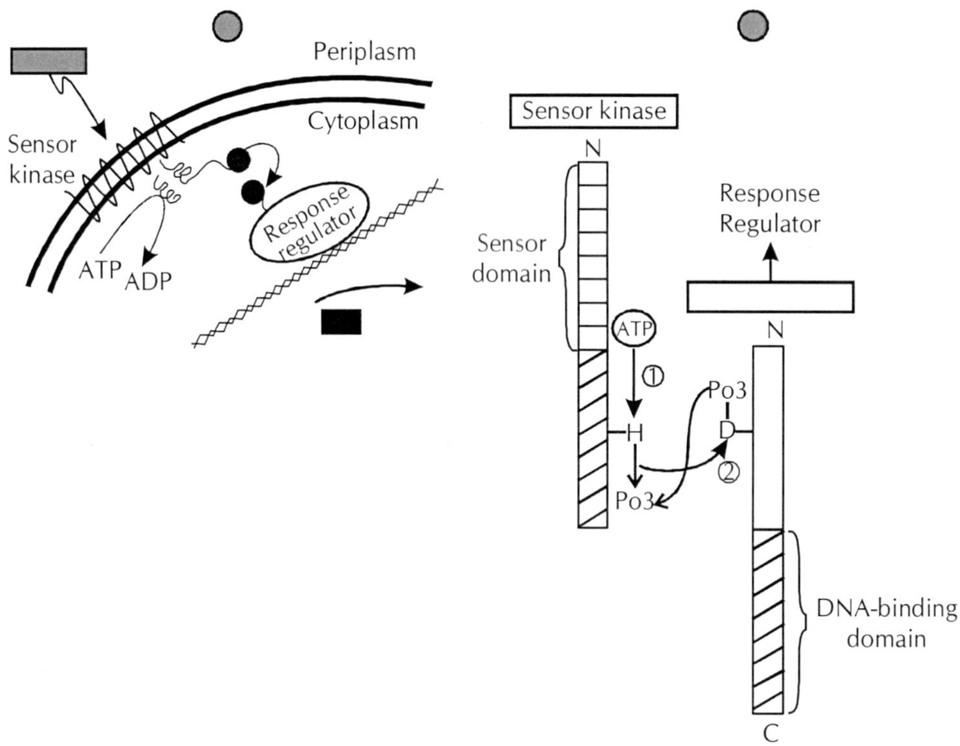

Fig. 26.2 A two-component signal system involving sensor kinase (HK) and response regulator (RR)

Step 1 : In response to a stimulus at the N-terminal domain, the histidine kinase auto phosphorylates (usually) at the C-terminal domain. The phosphoryl donor is ATP.

Step 2 : The phosphoryl group is transferred from the histidine kinase to its partner response regulator protein. All of the response regulator proteins are related in having conserved amino acid sequences in the N-terminal domain that may bind to the conserved C-terminal region of the histidine kinases.

Step 3 : The response regulator becomes activated on being phosphorylated and changes the activity of its target. The effect is usually the stimulation or repression of gene expression. Some RR proteins like NarL perform both functions. However, other response regulators may have targets other than genome. For example, in chemotaxis, the phosphorylated derivatives of the RRs Che Y and Che B change the direction of the flagella motors and the extent of methylation of the chemoreceptor proteins in the membrane, respectively.

Histidine kinases are sometimes bifunctional enzymes that can act either as kinases or as phosphatases (HK) when stimulated. When stimulation of the HK promotes phosphatase activity, it results in the dephosphorylation of the RR and hence a repression of transcription.

26.1.3. Models of Two-Component Signaling System

1. Adaptive responses by facultative anaerobes to anaerobiosis

Facultative microorganisms, such as *E. coli* and *S. typhimurium* can modify their metabolism to accommodate growth under either aerobic or anaerobic conditions. The transition between aerobic and anaerobic metabolism is accompanied by alterations in the rate, route and efficiency of pathways of electrons flow. Fig. 26.3 illustrates the basic pathways *E. coli* utilizes for aerobic versus anaerobic electron flow under aerobic conditions. *E coli* synthesizes two distinct cytochrome oxidases, cytochrome o and cytochrome d that are produced under high O_2 and low O_2 conditions respectively.

A shift from an aerobic to anaerobic atmosphere results in extensive changes in the metabolism of facultative anaerobes. Under anaerobic conditions, depending upon the availability of electron acceptors, it utilizes at least five more terminal oxidoreductases: fumarate reductase, nitrite reductase, nitrate reductase, dimethyl sulfoxide reductase and trimethylamine-N-oxide reductase. These adaptive responses have stimulated much interest. The adaptive responses to anaerobiosis are regulated by two systems (*i.e.,* the Arc system-Aerobic respiration control system) and the Fnr system (not two component).

Three systems for the regulation of gene expression by oxygen or nitrate have been characterized in *E. coli.* Two of these are responsible for regulation by oxygen, and the third mediates regulation by nitrate. They are:

i. The Arc system (two-component). This represses the transcription of several 'aerobic' genes and stimulates the transcription of the gene for cytochrome d oxidase.

ii. The Fnr system: This stimulates the transcription of several 'anaerobic' genes and represses the transcription of some 'aerobic' genes.

iii. The NarL/NarX/NarQ system (two-component). This simulates the transcription of the nitrate reductase genes and represses the transcription of other reductase genes.

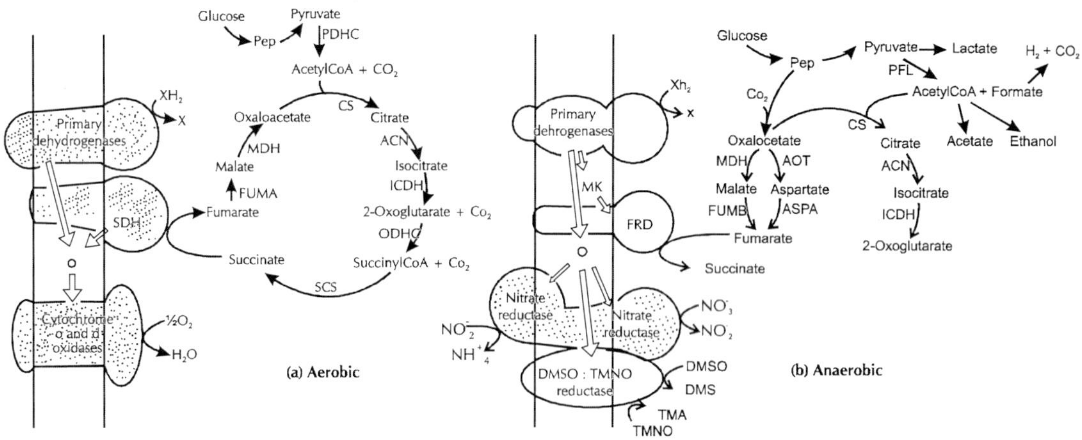

Fig. 26.3. A comparison of respiratory systems and carbon metabolism in *E. coli.*

The Arc system: The Arc system consists of histidine kinase (ArcB), and a partner response regulator (ArcC) (Fig. 26.4). Phosphorylated ArcA is responsible for:

1. Anaerobic repression of the genes for:

   Citric acid enzymes, glyoxylate cycle enzymes

   several dehydrogenases for aerobic growth (*e.g.,* pyruvate dehydrogenase, fatty acid oxidation enzymes, cytochrome *o* oxidase)

2. Induction in low oxygen of the genes for: cytochrome d oxidase and cobalamin synthesis.

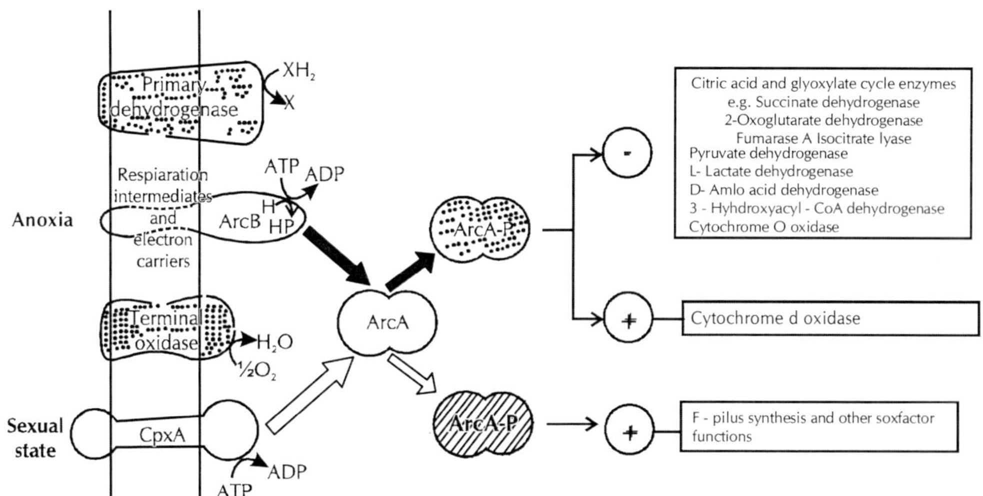

Fig. 26.4. The Arc system

The Fnr system: When *E. coli* is shifted from aerobic to anaerobic growth a number of genes are induced while others are repressed. The Fnr gene codes for a protein (Fnr) that acts as a positive regulator of transcription for many genes that are expressed only during anaerobic growth, and a repressor for certain genes that are expressed only during aerobic growth (Fig. 26.5).

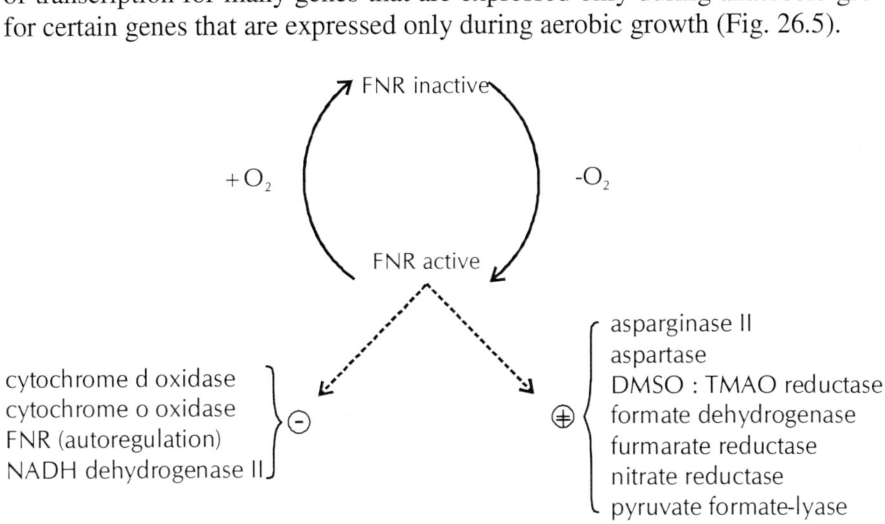

Fig. 26.5. The Fnr system

The NarL/NarX/NarQ system: The regulation of gene transcription by nitrate is mediated by NarL, NarX and NarQ. Nucleotide-derived amino acid sequences analysis and analogy with other signaling systems have placed NarL and NarX in the two-component regulatory system (Fig 26.6). NarX is a membrane bound sensor/kinase protein and NarL is a cytoplasmic regulatory protein. The gene narq was recently discovered and is thought to be a membrane bound sensor/kinase.

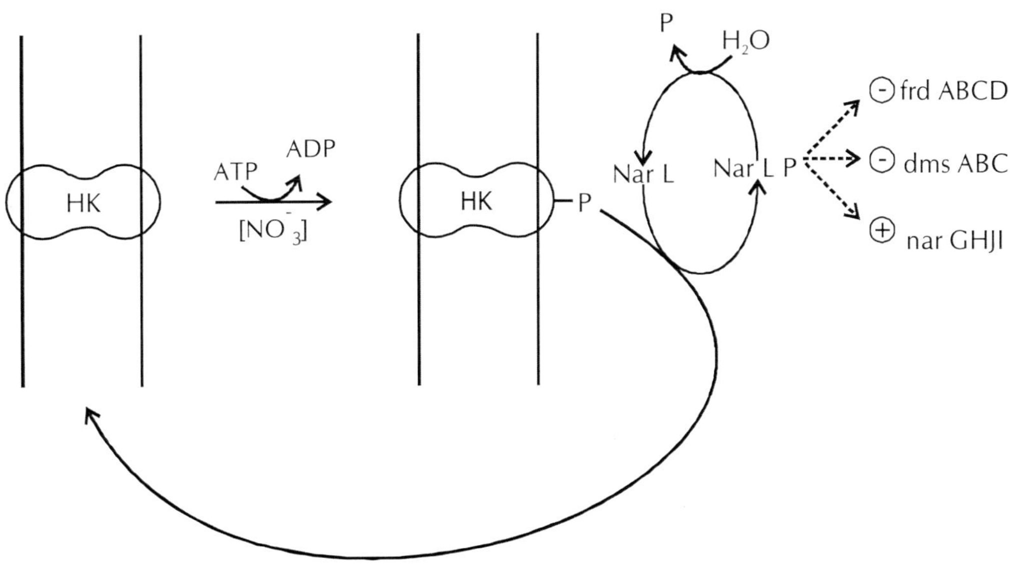

Fig. 26.6 A model for nitrate control of anaerobic gene expression

2. Chemotaxis

Chemotaxis refers to the ability of bacteria to move along a concentration gradient towards a chemical attractant (positive chemotaxis), or away from a chemical repellent (negative chemotaxis). The attractants and repellents are called ***chemo effectors.*** Bacteria measure the absolute concentration of chemo effector and then compares with the previous concentration. In the other words, bacteria measure changes in concentration over time. They actually remember the previous concentration. If the bacterium finds that it is in a higher concentration of a repellent than at a previous, it continues to move in that direction. On the other hand, if the bacterium finds that it is in a lower concentration of attractant or higher concentration of repellent than at a previous time, it moves in a randomly different direction.

Many bacteria detect chemotactic stimuli with chemoreceptor, receptor transducers or transducer proteins known as methyl-accepting chemotaxis proteins or MCP. *E. coli* has four different MCPs that mediate responses to serine (Tsr), aspartate and maltose (Tar), ribose and galactose (Trg), and dipeptides (Tap). MCPs are transmembrane proteins with periplasmic input or sensing domain and a cytoplasmic output or signaling domain. The signaling domain has no known catalytic function. Rather, it modulates the activity of the transmitter containing che A protein to elicit on chemotactic responses. In connection with chemotaxis and signal transduction, six genes namely *cheA, cheB, cheR, cheW, cheY* and *cheZ* are identified. The che proteins are cytoplasmic proteins that are part of signal transduction pathway between the attractant or repellent, and the flagellar motor switch. CheA belongs to the class of signaling proteins called histidine kinases. Chemoreceptor proteins for the chemo effector are built into bacterial cell membrane. The chemoreceptor proteins are also called receptor transducer proteins or transducer proteins. These are transmembrane proteins. The cytoplasmic domain interacts with cytoplasmic components such as the cheW and cheA proteins, whereas the periplasmic domain interacts with chemo effectors or periplasmic proteins to which the chemo effectors are bound. In *E. coli,* there are four-transducer proteins namely, Tar, Trs, Trg and Tap.

A model for signal transduction during chemotaxis in *E. coli* and *S. typhimurium* is shown in Fig. 26.7. Attractants or repellents bind to the receptor - transducer proteins in the membrane. This activates a cheW dependent change in the rate of autophosphorylation of cheA and the proteins that are phosphorylated by cheA (cheB and cheY). Attractants reduce the rate of autophosphorylation and repellents increase rate of autophosphorylation. The phosphoryl group is transferred from cheA-P to cheY and che B. CheY-P interacts with the switch proteins (Fli M, Fli N, Fli G) to cause clock-wise rotation of the motor and tumbling. Thus, repellents increase tumbling because they increase the level of che + Y-P, and attractants reduce tumbling because they reduce the level of cheY-P. CheB-P is methyl esterase whose activity results in demethylation of the receptor transducer proteins. As che B-P goes up, methylation goes down, Thus, repellents reduce the level of methylation because they increase the level of cheB-P and attractants increase methylation because they reduce the levels of cheB-P. CheB-P autophosphorylates. CheR is a methyl transferase that methylates the receptor-transducer proteins. The more highly methylated receptor-transducer protein does not transmit the chemoattractant signal to cheA. Hence adaptation to a chemoattractant is due to increased methylation of the receptor-transducer proteins. Adaptation to repellent is due to undermethylation of the receptor-transducer protein. CheZ is thought to be a phosphatase that removes the phosphate from cheY-P. The relative activities of cheA-P and cheZ determine the level cheY-P, hence the frequency of tumbling. CheW couples cheA to chemoreceptors.

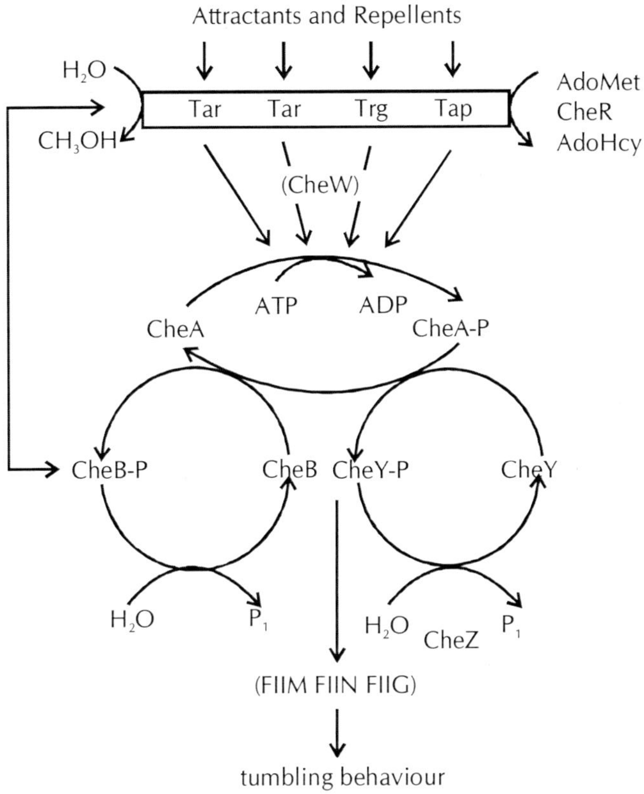

Fig. 26.7 A model of signal transduction during chemotaxis in *E. coli* and *S. typhimurium.*

3. Nitrogen assimilation: The Ntr regulon

Bacteria use inorganic and organic nitrogenous compounds as a source of nitrogen for growth. All these nitrogenous compounds are ultimately, either reduced or catabolized by the bacteria to ammonia which is assimilated via the glutamine synthetase reaction.

When the ammonia supply is adequate, the ammonia is the preferred source of nitrogen because it represses genes required for the assimilation of other nitrogenous compounds. However, when ammonia in the growth medium becomes limiting, then the genes required for ammonia assimilation from external nitrogen sources, as well as the gene for glutamine synthetase are induced. The genes that are regulated by the ammonia supply belong to the Ntr regulon. The Ntr regulon consists of many genes including the genes for glutamine synthetase, NRII (histidine kinase) and NRI (response regulator which are in an operon called ALG).

A model for signal transduction and also transcriptional regulation of gln ALG is shown in Fig. 26.8. The protein products of the operon are glutamine synthetase (GS), a histidine protein kinase/phosphatase (NRII) and a response regulator (NRI). Under ammonia-limiting conditions,

NRII phosphorylates NRI, forming NRI-P. The increased levels of NRI-P stimulate large increase in transcription rate from AP2, which uses sigma 54 RNA polymerase.

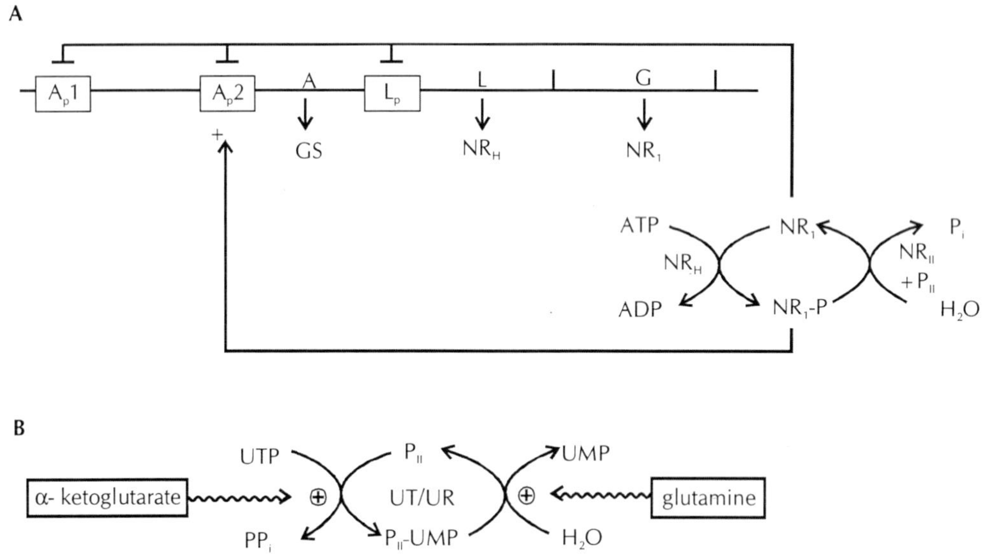

Fig. 26.8 Model for transcriptional regulation of glnALG

As levels of NRI-P increase, other Ntr operons are also stimulated. Under conditions of excess ammonia, NRII is stimulated by PII to dephosphorylate NRI-P. As a consequence, transcription from AP2 stops and sigma 70 RNA polymerase transcribes from promoters ApI and Lp at low rate. Transcription from these promoters is further regulated by repression by NRI. High glutamine (high ammonia) increases the level of PII, which stimulates the dephosphorylation of NRI-P and represses transcription whereas high α-ketoglutarate (low ammonia) stimulates the addition of UMP to PII, thus lowering the levels of PII and activates transcription (Table 26.1).

Table 26.1. Regulation of transcription of *glnALG* and glutamine synthetase

NH_3	P_{II}/P_{II}-UMP	Transcription	Glutamine synthetase
High	High	Low	Inactive
Low	Low	High	Active

4. The Pho regulon

Bacteria have evolved a signaling system to induce the formation of phosphate assimilation pathways when the supply of phosphate becomes limiting. When inorganic phosphate is in excess, it is transported into the cell by a low affinity transporter called 'Pit'. Under low inorganic phosphate conditions, *E. coli* stimulates the transcription of at least 31 genes (most of them in operons) involved in phosphate assimilation. All these genes are repressed by phosphate and are in the PHO regulon. The PHO regulon is controlled by PhoR, which appears to be a histidine kinase and response regulator, pHoB. The phosphorylated form of phoB (*i.e.,* phoB-P) activates the transcription of genes in the PHO regulon. The pho promoters use the sigma 70 RNA polymerase.

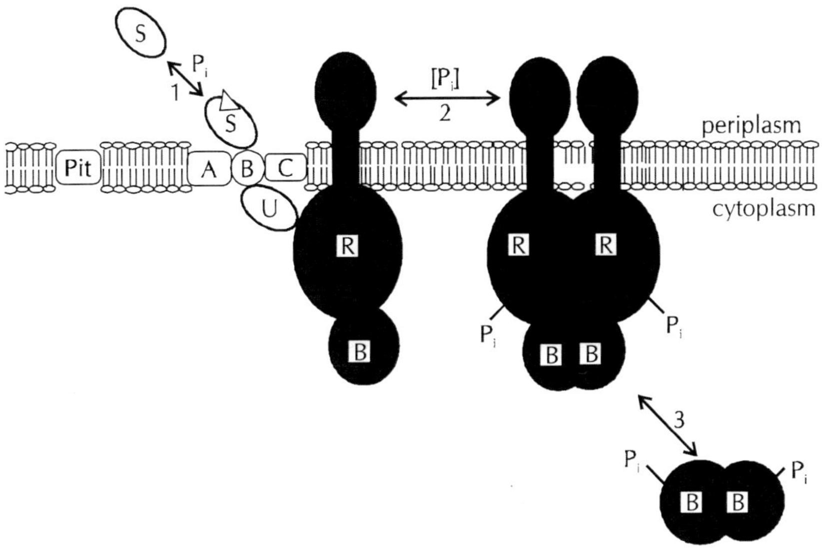

Fig. 26.9 Model for the regulation of the PHO regulon

A model for the regulation of the PHO regulon by Pi is shown in Fig. 26.9. The components required for regulation are: (i) PstS, a periplasmic Pi binding protein; (ii) PstABC, integral membrane proteins required for Pi uptake (Pst B is the permease); (iii) PhoU; (iv) PhoR and (v) PhoB. The model proposes that Pi binds to PstS and that the Pi-bound PstS binds to PstABC. A 'repressor complex' is postulated that consists of Pi bound to PstS, PstABC, PhoU and PhoR. PhoR is thought to be a histidine kinase/phosphatase bifunctional enzyme. In the 'repressor complex' PhoR is suggested to be a monomer with phosphatase activity and to maintain PhoB in its dephosphorylated (inactive state). The model further proposes that when phosphate becomes limiting, PhoR is released from the repressor complex and functions as a histidine complex as a dimer. The dimer form of PhoR autophosphorylates and activates PhoB by transferring to it the phosphoryl group. In agreement with this model, Pst and PhoU are required for repression of the PHO regulon, but not for its activation. However, it must be added that the interaction of the Pst system and PhoU with PhoR to cause Pi repression is not really understood.

5. Osmoregulation

The ability to monitor osmolarity is important to bacteria for two reasons: (i) to keep the difference in osmotic pressure inside and outside the cell becoming too great, (ii) to sense the changes in their external environment by the changes in osmolarity. One of the major mechanisms by which bacteria balance osmotic pressure across the outer membrane is by synthesizing pores to let solutes into and out of the periplasmic space. These pores are composed of outer membrane proteins called porins. To form pores, these proteins trimerize in the outer membrane. Two major porin proteins in *E. coli* are OmpC and OmpF. Pores composed of OmpC are slightly smaller than those composed of OmpF and the size of the pores can affect the rate at which solutes pass through the pores and thus confer protection under some conditions.

The signaling system that senses osmolarity and regulates expression of the ompF and ompC loci typifies the simplest of the sensory pathways based on communication modules. The signaling pathway has two components: EnvZ, an inner membrane sensor, and OmpR, a cytoplasmic response regulator (Fig. 26.10)

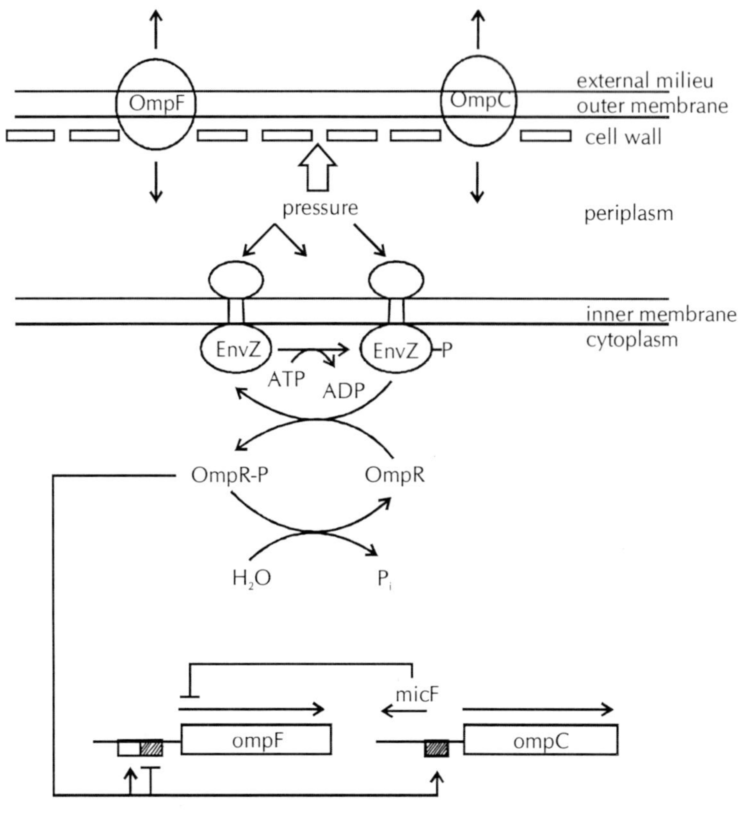

Fig. 26.10. A model to explain the osmoregulation

One model proposes that increased external osmolarity activates EnvZ, so that it phosphorylates OmpR. The model proposes that high levels of OmpR-P repress ompF and stimulate ompC by binding to low-affinity site upstream from the respective promoters, and that low levels of OmpR-P stimulate transcription of ompF promoter. Thus, when the external osmotic pressure is raised, the level of OmpR-P should increase and repress OmpF while stimulating ompC. Several lines of evidence support the model.

There appears to be an additional mode of inhibition of ompF expression. Another regulatory gene, *micF* is also stimulated by high concentrations of ompR-P and produces an RNA molecule that inhibits ompF mRNA translation. When multiple copies of *micF* are introduced on a high copy-number plasmid, there is inhibition of *in vivo* synthesis of ompF. However, deletion of the *micF* gene does not result in any change in the synthesis of ompF under high or low osmolarity conditions. Therefore, it must be concluded that there is insufficient information regarding its physiological role.

26.2. ONE-COMPONENT SIGNAL TRANSDUCTION SYSTEM

One-component signal transduction system, unlike in two-component system, contains only one component that is sensor or kinase. The response regulator is absent. However, it combines the features of the two-component system into one protein.

The ToxR system of *Vibrio cholerae*, an intestinal human pathogen, serves as an example for understanding the one-component system. The ToxR protein is a trans-membrane sensor whose C-domain specifically binds to the promoter regions of virulence-factor genes. Since it is a single protein it does not require transmitter or receiver module. The mode of signal transduction, controlled by several proteins, is under the control of ToxR regulon (Fig. 26.11).

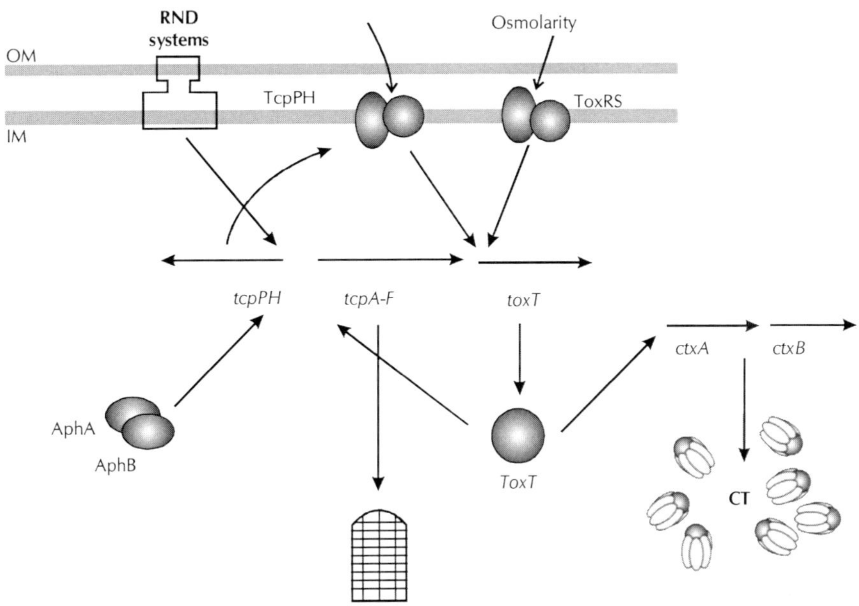

Fig. 26.11 Regulation of virulence factor production in *Vibrio cholerae*

Vibrio cholerae is a Gram-negative bacterium that is the causative agent of cholera. The virulence factors, toxin co-regulated pileus (TCP) and cholera toxin (CT) are essential for colonization of the host and enterotoxicity, respectively. These virulence factors are under the control of ToxT protein that activates transcription of the genes encoding TCP and CT. ToxT, in turn, is under the control of virulence regulatory cascade known as the ToxR regulon, which responds to environmental stimuli to ensure maximal virulence factor induction within the human intestine. The major genes in the Tox R regulon include the master regulators toxRS and tcpPH, the downstream regulator toxT, and virulence factors, the ctx AB and tcpA operons. Both the ctx AB and tcpA operons are activated primarily by the ToxT transcription factor. Expression of toxT is itself positively regulated by two sets of dimeric transcription factors Tox RS and Tox PH.

ToxR activity is greatly enhanced by the presence of another transmembrane protein, ToxS. TcpP is a homologue of ToxR and is similarly localized to the membrane with a DNA binding cytoplasmic domain; a second protein, TcpPH, enhances the stability of TcpP. Both ToxR and TcpP bind the tox P promoter. ToxR is known to regulate atleast 17 genes besides, ToxT, including the outer membrane porin genes ompU and ompT. ToxR has also been shown to directly stimulate ctxAB expression, independently of ToxT. These ToxR regulated virulence genes are collectively called the ToxR regulon, after the first discovered master regulon.

B. QUORUM SENSING

It is well known that in higher organisms, the cells in tissues interact both physically to form tissue structure and via signaling molecules, such as the well-known communication molecules (cytokines) diffusing in the intercellular space. However, for decades microbiologists believed that bacterial populations (colonies) as collections of individuals growing and behaving independently Bacterial culture was thought as a 'homogenous soup' in which individual cells live and proliferate independently of each other. More recently, it has become clear that many bacteria can communicate with one another and behave cooperatively. Bacterial population is seen as a community of interacting cells with possible differentiation. In this respect, bacterial populations can be brought in to comparison with tissues of higher organisms by their organization.

Bacteria communicate with one another using chemical signal molecules. As in higher organisms, the information supplied by these molecules is critical for synchronizing the activities of a large group of cells. In bacteria, chemical communication involves producing, releasing, detecting and responding to small hormone-like molecules termed **autoinducers** or **signal molecules**. The term autoinducer conveys that the synthesis of molecules is regulated by positive feedback. These autoinducers or signal molecules bring about a phenomenon called **autoinduction** or **quorum sensing**. Through this phenomenon bacteria monitor their own population density by sensing the levels of signal molecules whose external concentration increases as a function of increasing cell-population density. The concentration of these signal molecules increases along with the bacterial population until it rises to specific threshold and signals the bacteria that the population density has reached a critical level or quorum. The bacteria then begin expressing sets of quorum-dependent genes (Fig. 26.12 A,B). Quorum sensing has been found among both Gram-negative and Gram-positive bacteria and in bacteria living in variety ecological settings.

Quorum sensing has been observed in a variety of physiological functions like bioluminescence, sporulation, the production of virulence factors, cell divisions, antibiotics and other secondary metabolites, competence (uptake of DNA) conjugation and the formation of biofilms and growth inhibition. It also seems to be very important in many bacteria that establish symbiotic or parasitic relationships with hosts.

Quorum sensing presumably provides bacteria benefits from group activities that may be unbelievable to an individual bacterium acting alone. For example, the benefit derived from secretion of antibiotics or proteases may only occur when these exoproducts exceed a particular extra cellular concentration, and achieving this concentration is only possible through the synchronous activity of a group of cells.

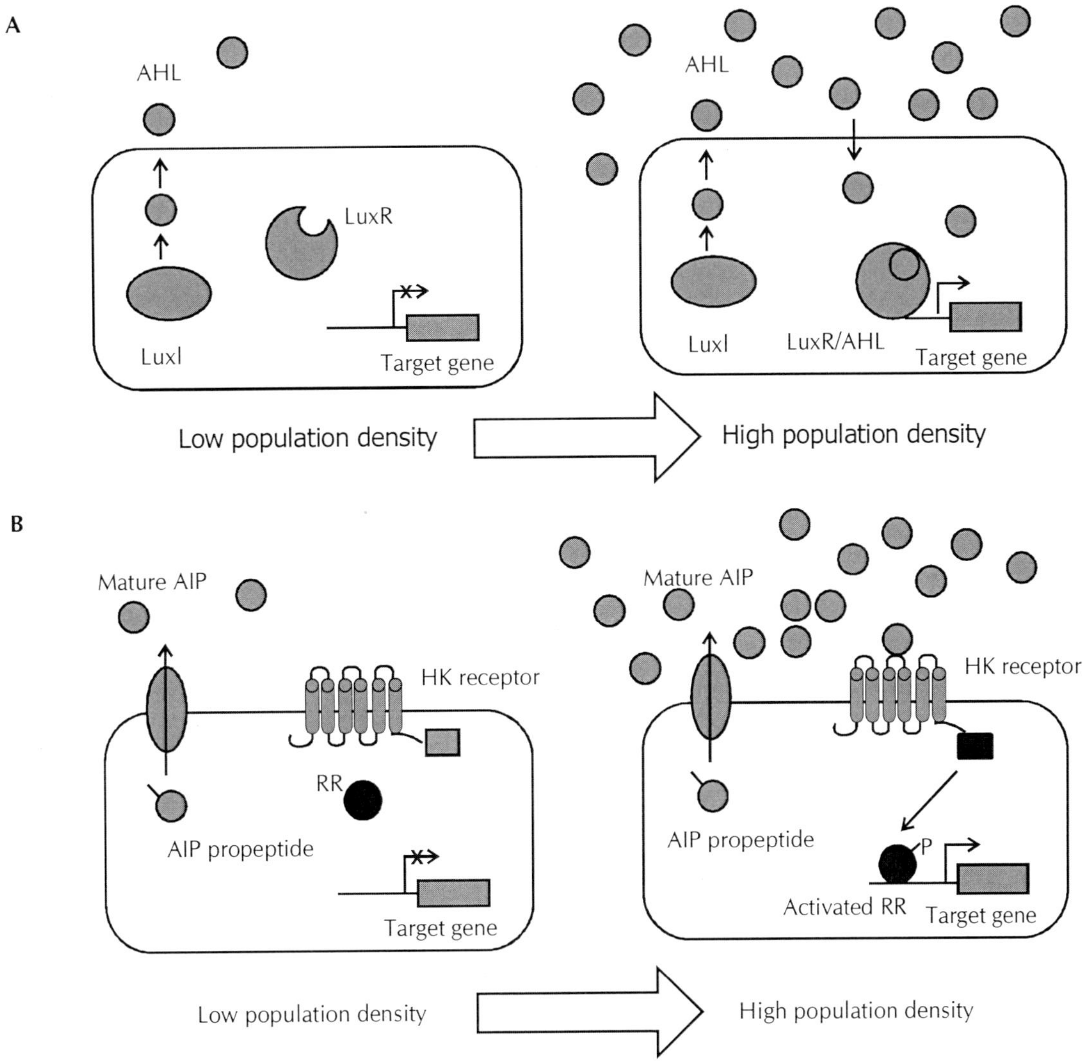

Fig. 26.12. Quorum sensing in **A.** Gram-negative, **B.** Gram-positive bacteria

26.3. AUTO INDUCERS OR SIGNAL MOLECULES

A.S. Khokhlov, in 1970s, discovered a small molecule called A-factor secreted by a streptomycete and found that when it accumulates in the medium, induces both sporulation and antibiotic production by its own cells. At that time, it was treated only as a curious phenomenon, but in the last few decades an increasing number of studies have revealed a number of communicatory molecules secreted into the medium with biological effect similar to that of A-factor. A variety of bacterial living processes appear to depend on those molecules. All these substances exhibited similar mode of action: they cause specific changes in the bacterial metabolism when they reach a critical concentration in the medium. This principle called quorum sensing involves the action of a broad spectrum of chemical substances like low molecular weight secondary metabolites, peptides, lipids and secreted proteins. These communicatory or signaling molecules of prokaryotes differ from counter parts of eukaryotes in that they: (i) are produced by the organism itself (same cell), (ii) have very low effective concentration and (iii) do not require cleavage for expression of their activity.

Signal molecules, the language of bacteria, are secreted by both Gram-negative and Gram-positive bacteria. Gram-negative cells use low molecular weight substances in particular N-acyl homoserine lactone's (AHLs) for communication, whereas Gram positive bacteria make use of either small (sometimes modified) oligopeptides or proteins. However, the difference between communication languages of Gram-positive and Gram-negatives is not always so great. Some bacteria use both languages example, *Vibrio harveyi* has one AHL based signaling system and another system using another effector, furanon, which is also secreted into the medium. It has been found that synthesis and secretion of furanons are much more widespread among bacteria than the synthesis and secretion of AHL's. Indeed, both Gram-negative and Gram-positive bacteria can synthesize furanons. Hence, it has been postulated that furanons serve as a language for interspecific communication, whereas AHL's are rather intraspecific. The furanons are presumably used in bacterial communities, such as biofilms and consortia.

26.3.1. Acyl-homoserine Lactones (AHLs)

AHL's often referred to as autoinducer 1 (Al-1) are composed of a homoserine lactone ring with an attached fatty acid chain, which can vary in length between 4 and 18 carbons, and may or may not have a keto-group in position 3 (Fig. 26.13). A diverse set of fatty acyl side chains of varying length, backbone saturation and side chain substitutions are incorporated into AHL signals. Acyl homoserine and these differences are attributed for signal specificity specificity.

AHL signal molecules are synthesized from S-adenosyl methionine and fatty acid carrier proteins by AHL synthases, LuxI and its homologues. Each LuxI protein (AHL syntheses) produces a corréct signal molecule with high fidelity. On the other hand, the structures of LuxR proteins, which function as response regulators, suggest that LuxR proteins also possess specific acyl binding pockets that allow each LuxR to bind and be activated only by its cognate signal. Hence it appears that in mixed species environments in which multiple AHL signals are present, each species can distinguish, measure, and respond only to the build of its own signal.

Most AHLs traverse membranes by diffusion and bind LuxR-like response regulators. LuxR - like response regulators (coded by LuxR gene) simultaneously act as sensors and transcription factors. This type of signal transduction system wherein signal binding domain and transcription-regulating domain are fused is referred to as one-component signal transduction system and closely related chemical inducers operate in different Gram-negative bacteria governing various processes (Table 26.2).

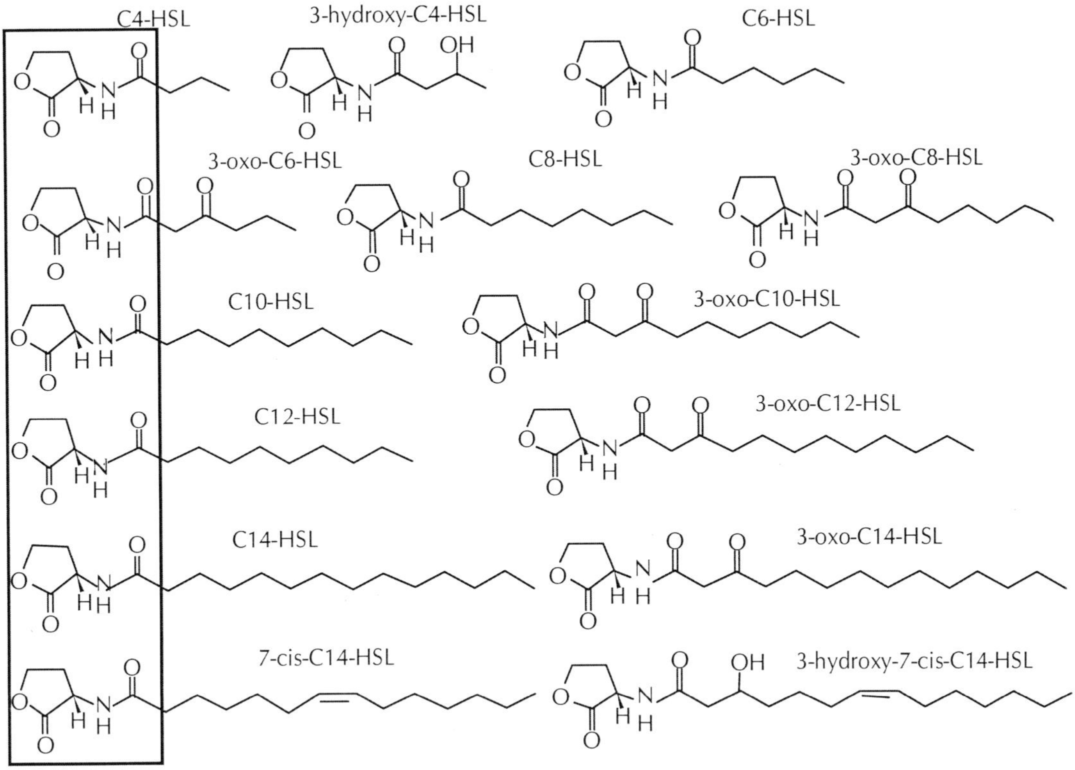

Fig. 26.13. Acylated homiserine molecule. Core component is shown in box.

Table 26.2. Organisms possessing LuxI/LuxR homologues: the regulatory proteins, the HSL autoinducers, and the regulated functions[a]

Organism	LuxI/LuxR Homologue(s)	Autoinducer Identity	Target Genes and Functions
Vibrio fischeri	LuxI/LuxR	N-(3-oxohexanoyl)-HSL	luxICDABE (bioluminescence)
Aeromonas hydrophila	AhyI/AhyR	N-butanoyl-HSL	Serine protease and metal-loprotease production
Aeromonas salmonicida	AsaI/AsaR	A-butanoyl-HSL	aspA (exoprotease)
Agrobacterium tumefaciens	TraI/TraR	N-(3-oxooctanoyl)-HSL	tra, trb (Ti plasmid conjugal transfer)
Burkholderia cepacia	CepI/CepR	N-octanoyl-HSL	Protease and siderophore production
Chromobacterium violaceum	CviI/CviR	N-hexanoyl-HSL	Violacein pigment, hydrogen cyanide, antibiotics, exoproteases and chitino- lytic enzymes

Organism	LuxI/LuxR Homologue(s)	Autoinducer Identity	Target Genes and Functions
Enterobacter agglomerans	EagI/EagR	N-(3-oxohexanoyl)-HSL	Unknown
Erwinia carotovora	ExpI/ExpR CarI/CarR	N-(3-oxohexanoyl)-HSL	(a) Exoenzyme synthesis,
			(b) Carbapenem antibiotic synthesis
Erwinia chrysanthemi	ExpI/ExpR	N-(3-oxohexanoyl)-HSL	pecS (regulator of pectinase synthesis)
Erwinia stewartii	EsaI/EsaR	N-(3-oxohexanoyl)-HSL	Capsular polysaccharide biosynthesis, virulence
Escherichia coli	?/SdiA	?	ftsQAZ (cell division), chromosome replication
Pseudomonas aereofaciens	PhzI/PhzR	N-hexanoyl-HSL	phz (phenazine antibiotic biosynthesis)
Pseudomonas aeruginosa	(a) LasI/LasR	(a) N-(3-oxodode- canoyl)-HSL	(a) lasA, lasB, aprA, toxA (exoprotease virulence factors), biofilm formation)
	(b) RhlI/RhlR	(b) N-butyryl-HSL	(b) lasB, rhlAB (rhamnolipid), rpoS (stationary phase)
Ralstonia solanacearum	SolI/SolR	N-hexanoyl-HSL, N-octanoyl-HSL	Unknown
Rhizobium etli	RaiI/RaiR	Multiple, unconfirmed	Restriction of nodule number
Rhizobium leguminosarum	(a) RhiI/RhiR	(a) N-hexanoyl-HSL	(a) rhiABC (rhizosphere genes) and stationary phase
	(b) CinI/CinR	(b) N-(3-hydroxy-7- crv-tetradecenoyl)- HSL	(b) Quorum sensing regulatory cascade
Rhodobacter sphaeroides	CerI/CerR	7,8-cis-N-(tetradecanoyl)-HSL	Prevents bacterial aggregation
Salmonella typhimurium	?/SdiA	?	rck (resistance to competence killing), ORF on Salmonella virulence plasmid
Serratia liquefaciens	SwrI/?	N-butanoyl-HSL	Swarmer cell differentiation, exoprotease
Vibrio anguillarum	VanI/VanR	N-(3-oxodecanoyl)-HSL	Unknown
Yersinia enterocolitica	YenI/YenR	N-hexanoyl-HSL, N-(3-oxohexanoyl)-HSL	Unknown

Organism	LuxI/LuxR Homologue(s)	Autoinducer Identity	Target Genes and Functions
Yersinia pseudotuberculosis	(a) YpsI/YpsR	(a) N-(3-oxohexanoyl)- HSL	Hierarchical quorum sensing cascade
	(b) YtbI/YtbR	(b) Woctanoyl-HSL	regulating bacterial aggregation and motility

26.3.2. A1-2 family

Apart from an AHL system, *V. harveyi* has parallel system whose signaling molecules are referred to as auto inducer-2 (AI-2; Fig 26.14) AL-2 is a furanosyl borate diester, whose precursor is 4, 5-di-hydroxy-2, 3-pentanedione (DPD). DPD is synthesized by LuxS from S-adenosyl methionine (SAM). Thus, SAM is a precursor both in AHL synthesis and AL-2 synthesis. It is assumed that DPD spontaneously rearranges into Al-2 when borate is available. In other species, this rearrangement does not occur, and Al-2 has a different structure.

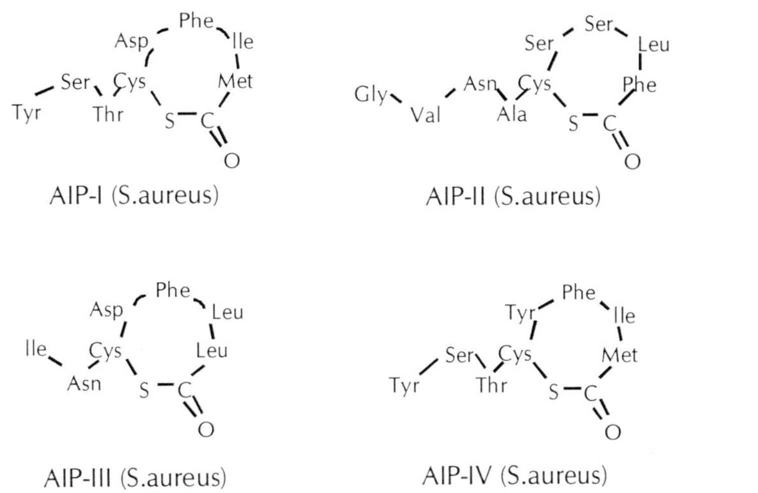

ADPITRQWGD	ComX (B.subtilis)
ERGMT	CSF (B.subtilis)
EMRLSKFFRDFILQRKK	CSP (S.pneumoniae)

Fig. 26.14. AI-2 molecules

The gene LuxS encoding the AI-2 synthase is present in roughly half of all sequenced bacterial genomes and AI-2 production has been verified in a large and diverse group of bacterial species belonging to both Gram-positive and Gram-negative groups. Similarly, AI-2 compounds were found to control gene expression in a variety of bacteria. These observations have led to the hypothesis that bacteria use AI-2 to communicate between species i.e., they serve as language for interspecific communication (AHL are rather interspecific). This non-specific language of AI-2 is supposedly used in bacterial communities, such as biofilms and consortia. Some more functions regulated by Luxs/AI-2 in pathogenic bacteria are presented in Table 26.3.

Table 26.3. Functions regulated by LuxS/AI-2 in pathogenic bacteria

Bacterium	Regulated functions
Actinobacillas actinomycetemcomitans	Leucotoxin, iron acquisition
Campylobacter jejuni	Motility
Clostridium perfringens	Toxins
Escherichia coli WELLO	Morphology, cell division
Neisseria meningitides	Bacteraemia
Phorphyromonas gingivalis	Proteases
Salmonella typhi	Biofilms
Streptococcus pyogenes	Haemolysis
Vibrio cholerae	Toxins
Vibrio harveyi	Bioluminescence, siderophores, morphology

26.3.3 Oligopeptide Autoinducers

Many Gram-positive bacteria use processed oligopeptides as signal molecules. The precursor peptides are typically between 40 and 65 amino acid residues in length. These pre-peptides are processed (cleaved) in all known cases and the resulting peptide signal molecules are typically 5 to 34 residues in length. In many cases, peptides are modified to contain lactone and thiolactone rings, lanthionines and isoprenyl groups. Representative oligopeptide autoinducers are shown in Fig. 26.15. Autoinducer peptides of *Streptomyces aureus*, commonly known as AIP are of special interest. S. *aureus* strains are classified on the basis of the sequence of their thiolactone containing AIP. At present, four different AIPs and thus four different groups of *S. aureus* are known. Each AIP inhibits activation of the virulence cascade in other three groups of *S. aureus*, while not affecting the other group's growth.

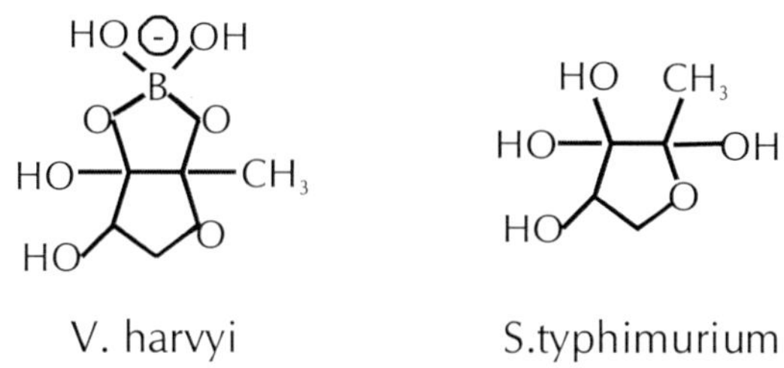

Fig. 26.15 Oligopeptide signal molecules.

Oligopeptide autoinducers do not transfer across the membrane, hence signal release is mediated by dedicated oligopeptide exporters. In most cases, concomitant with signal release is signal processing and modification. The minimum components for peptide communication, apart from the peptide signal itself, are: a membrane bound histidine kinase, and a response regulator with an aspartate phosphorylation residue, these components constitute a two-component signal transduction system, as opposed to the one-component system observed in AHL-based communication.

Peptide signals are assumed to be more flexible than the AHL signal molecules, as they do not require a specialised syntheses and can change to adapt to ecological niches by simple codon mutation. The observation that some genes encoding peptide signals are more variable than others might be an indication that this happens.

26.3.4. Other Signal Molecules

Signal molecules, other than the above described, operate in other bacteria. *Pseudomonas aeruginosa* uses quinolone molecules (PQS: 2 - heptyl - 3- hydroxy - 4- quinolone) synthesized by PqsH, which is quite hydrophobic and is exported out of the cell by a vesicular transport system analogous to the ones used by eukaryotes. *Streptomyces griseus* uses butyrolactones as autoinducers and control morphological differentiation and secondary metabolite production, fruiting body formation in the bacterium

Myxococcus xanthus is facilitated by the diffusible A signal, which seems to consist of six different amino acids generated by extracellular proteolysis, and the contact-dependent C-signal. The C-signal is a cell surface protein. Contact-dependent communication has also been observed in *E.coli*, where it is facilitated by CdiA and CdiB and leads to growth inhibition.

The list of signal molecules is by no means complete. Other QS molecules include *Bacillus subtilis* Comx, *Streptomyces coelicolor* SapB and the *Rhizobium meliloti* Nod factor. In view of vast amount of microbial diversity as revealed by metagenomic studies, it can be expected more the number of signal molecules of quorum sensing systems.

26.4. QUORUM SENSING IN GRAM-NEGATIVE BACTERIA

Quorum sensing phenomenon was observed in a large number of Gram-negative bacteria involving a variety of physiological process. However, the quorum-sensing system in bioluminescent bacteria is the best studied and understood. In fact, the first described quorum sensing system is that of the bioluminescent marine bacterium *Vibrio fischeri*, and it is considered the paradigm for quorum sensing in most Gram-negative bacteria. *V. fischeri* colonizes the light organ of the Hawaiian squid *Euphrymna scolopes*. In this organ, the bacteria grow to high cell density and induce the expression of genes required for bioluminescence. The squid uses the light provided by the bacteria for counter illumination to mask its shadow and avoid predation. The bacteria benefit because the light organ is rich in nutrients and allows proliferation in numbers unachievable in seawater.

Following the first discovery in *V. fischeri*, bioluminescence phenomenon was reported in about 30 extant bioluminescent systems, which mostly bear no evolutionary relationships with one another. Most luminous bacteria belong to three major genera-*Vibrio* (*V. fischeri, V. harveyi*), *Photobacterium* (*P. phosphoreum*) and *Photorhabdus* (*P. luminescens*). The first two of which are almost exclusively marine, whereas the last is terrestrial. All are characterized as enteric bacteria and are notable for the symbiosis in which they participate, most commonly in light organs in which the light is used by the host for some purpose.

26.4.1. Bioluminescence by Quorum Sensing: *Vibrio fischeri* Model

Bioluminescence is defined as an enzyme-catalyzed chemiluminescence, a chemical reaction in which the energy released is used to produce an intermediate or product in an electronically excited state, which then emits a photon. This phenomenon is mediated by signal molecules through quorum sensing system.

In *V. fischeri,* the species most studied, the bioluminescence is controlled by Lux operon (Lux ICDABE). The Lux operon has five structural (C, D, A, B, E) and two regulatory genes (I, R) with established functions. As shown in Fig. 26.16, these include two that code for the ∝ and β subunits of luciferase and three that code for the reductase, transferase and synthetase components of the fatty acid reductase complex responsible for aldehyde synthesis and reduction and recycling of acid product. Upstream, in the same operon, is the regulatory gene *LuxI*, and immediately adjacent but transcribed in the opposite direction is *LuxR.* Two proteins, LuxI and LuxR proteins encoded by LuxI and LuxR respectively control expression of the Lux operon. LuxI is the autoinducer synthesizer, which produces the acyl-homoserine lactone (AHL). LuxR protein, encoded by *LuxR* gene, is the cytoplasmic autoinducer receptor/DNA-binding transcriptional activator. LuxR protein using C-terminal domain binds to the cognate DNA motif palindromic Lux box where it activates transcription of the Lux ICDABE operon.

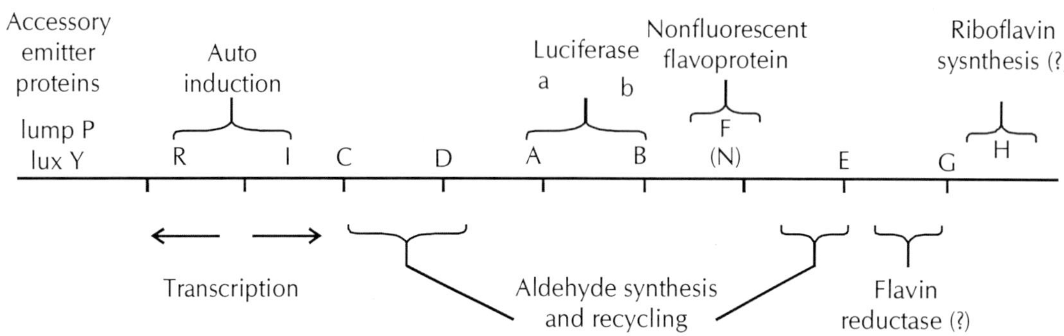

Fig. 26.16. Organization of the *lux* genes in *Vibrio fischeri*

Following production, the AHL freely diffuses in and out of the cell and increases in concentration with increasing cell density. When the signal reaches critical, threshold concentration, it is bound by LuxR and this complex activates transcription of the operon encoding luciferase (Fig. 26.17). Interestingly, the LuxR-AHL complex also induces expression of LuxI because it is encoded in the luciferase operon. This regulatory configuration floods the environment with the signals. This creates ,a positive feedback loop that causes the entire population to switch into quorum-sensing mode and produce light. A blue-green light with a maximum intensity at 490 nm then marks the bioluminescence phenotype.

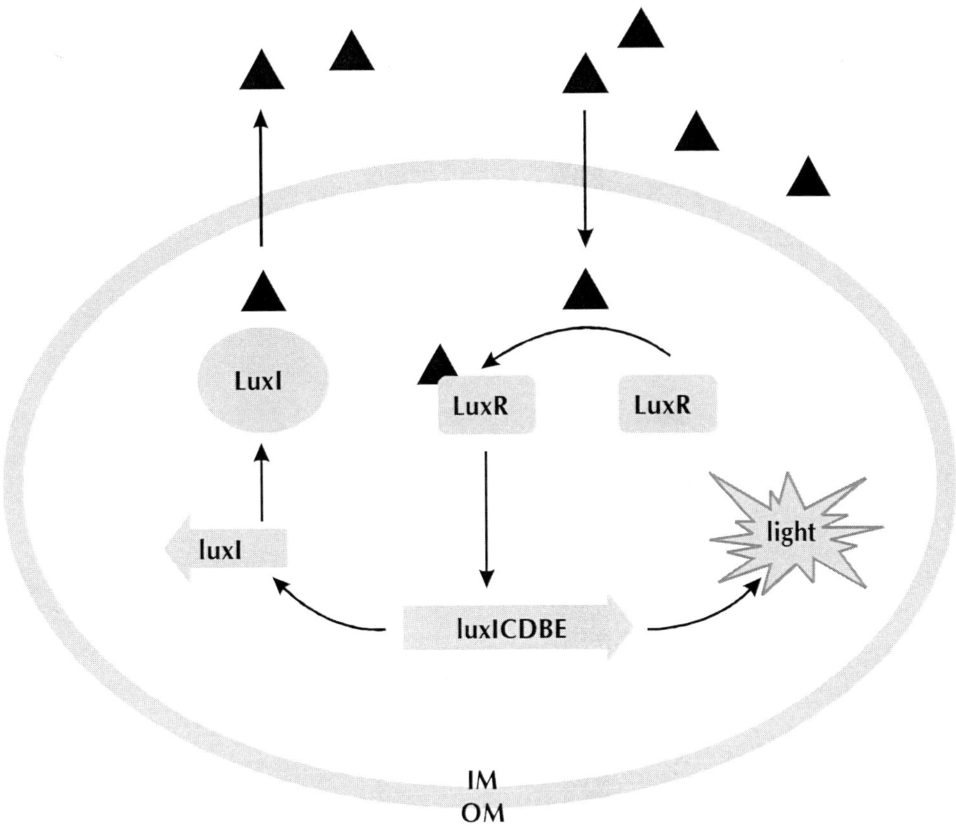

Fig. 26.17. Quorom sensing in *Vibrio fischeri*; a LuxIR signaling circuit. OM, outer membrane, IM, inner membrane.

At low cell density, the bacteria do not express bioluminescence, because the LuxR mRNA is degraded. However, at high cell density, the accumulated AI-2 molecules bind to LuxP, a periplasmic binding protein. LuxP-Al-2 binding to the LuxQ kinase triggers a dephosphorylation cascade by turning the kinases in to phosphatases. LuxQ dephosphorylates LuxU, which inturn dephosphorylates LuxO. The result is that the mRNA of the transcription factor LuxR is no longer degraded by sRNAs. This allows the translation of LuxR mRNA, production of LuxR, and expression of bioluminescence.

26.4.2. Biochemistry of Light Emission

Biochemically, light emission results from the luciferase-catalysed mixed function. Bacterial luciferase is an α-β heterodimer. It lacks metals, so prosthetic groups, and non-amino acid residues. Although the two subunits are homologous, the active site and the detailed kinetics features of the reaction are properties of the α-subunit.

In bacterial systems, flavin is the luciferin. The reaction represents a biochemical shunt of the respiratory electron transport system, carrying electrons from the level of reduced flavin ($FMNH_2$) directly to oxygen (Fig. 26.18). $FMNH_2$ reacts first with oxygen to form a linear hydroperoxide, which then reacts with long-chain fatty aldehyde to give the postulated peroxyheoniacetal intermediate. This breaks down to give long-chain acid and the intermediate hydroxyflavin in a high energy electronically excited state. Although aldehydes with chain lengths from 7 to 18 carbon atoms give light in the reaction with isolated luciferase, tetradecanal (IAC) has been identified as the naturally occurring molecule in the species studied. Interestingly, oxidation of aldehyde provides the energy, but aldehyde is not the emitter of light. The flavin is the luciferin in the bacterial system.

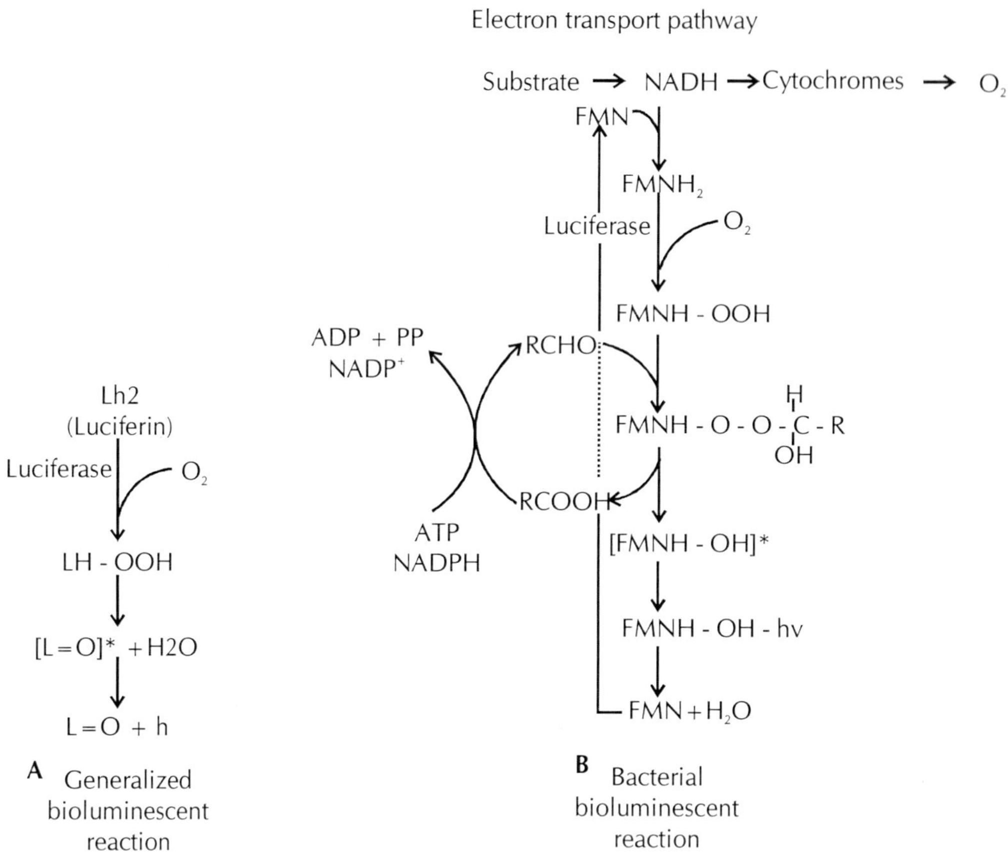

Fig. 26.18. A. Generalized reaction scheme for bioluminescent reactions. **B.** The luciferase reaction showing the components involve in the bacterial system.

The above explained module LuxI-LuxR type is considered the paradigm for quorum sensing in V. *fischeri* and most Gram-negative bacteria. It is important to note that the module of LuxI-LuxR homologs and closely related chemical inducers operate in different Gram-negative bacteria governing various processes.

In *V. fischeri* the quorum sensing for bioluminescence is brought about a single type of signal molecule (AHL). However, bacteria could communicate with multiple quorum-sensing signals. This type of quorum sensing was first observed in another Gram-negative, bioluminescent marine bacterium *Vibrio harveyi*. The *V. harveyi* quorum sensing system consists of three auto inducers and three cognate receptors functioning in parallel to channel information into a shared regulatory pathway (Fig. 26.19). Like other Gram-negative bacteria, *V. harveyi* produces one AHL signal termed HAL-1. HAL-1 binds to a membrane-bound sensor kinase (LuxN) similar to sensors in Gram-positive quorum-sensing signaling circuits. A second signal is furanosyl borate diester known as AI-2. AI-2 binds to a protein LuxP in the periplasm. The LuxP-AI-2 complex interacts with another membrane bound sensor histidine kinase, LuxQ. The third *V. harveyi* signal, an unidentified molecule termed CAI-1, is produced by the CqsA enzyme, and again this signal interacts with a membrane bound sensor histidine kinase, CqsS. All these histidine kinases (LuxN, LuxP and Lux CqsS) activated by their respective signal molecules autophosphorylate, and subsequently transfer the phosphate to the cytoplasmic protein LuxU. LuxU, in turn, passes the phosphate to the DNA binding response regulator protein LuxO.

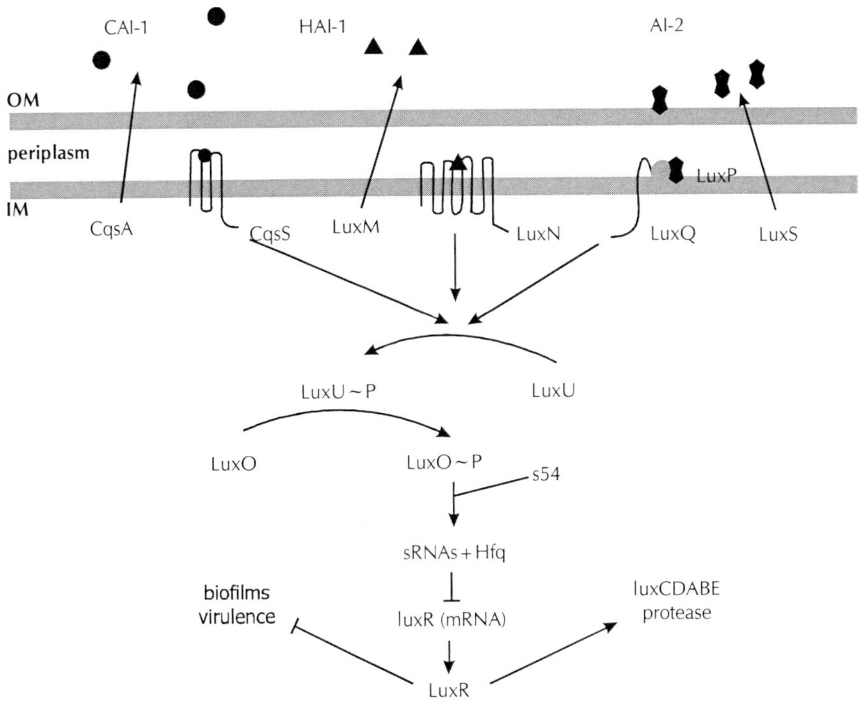

Fig. 26.19. *Vibrio harveyi* produces and responds to three distinct autoinducers. The sensory information is fed into a shared two-component response regulatory pathway.

26.5. QUORUM SENSING IN GRAM-POSITIVE BACTERIA

Gram-positive bacteria regulate a variety of phenotypes in response to increasing cell population density. Different bacterial species employ quorum sensing to regulate different phenotypes. For example, *Staphylococcus aureus* uses quorum sensing to regulate virulence; *Bacillus subtilis* and *Streptococcus pneumoniae* to develop bacterial competence; and *Enterococcus faecalis* to regulate conjugation.

Gram-positive bacteria communicate using modified oligopeptides as signal and two component type membrane-bound sensor histidine kinase as receptors. Signaling is mediated by a phosphorylation cascade that influences the acuity of a DNA-binding transcriptional regulatory protein termed response regulator. Each Gram-positive bacterium uses a signal different from that used by other bacteria and the cognate receptors are exquisitely sensitive to the signals structures. Thus, as in LuxIR systems, peptide quorum sensing circuits are understood to confer interspecies communication. Many Gram-positive bacteria communicate with multiple peptides in combination with other types of quorum-sensing signals.

A fascinating example of peptide quorum sensing exists in *Staphylococcus aureus*, which is normally a benign human communal but become a deadly pathogen upon penetration into host tissues. This organism uses a biphasic strategy to cause disease: At low cell density, the bacteria express protein factors that promote attachment and colonization, whereas at high cell density, the bacteria repress these traits and initiate secretion of toxins and proteases that are presumably required for dissemination. This switch in gene expression programs is regulated by the Agr quorum sensing system (Fig. 26.20). This system consists of an autoinducing peptide (AIP) of *Staphylococcus aureus* encoded by agrD and a two-component sensor kinase-response regulator pair, Agr C and Agr A, respectively. Agr B protein exports and adds the thiolactone ring modification to *S. aureus* AIPs. Binding of the AIP to Agr C leads to phosphorylation Agr A. Phospho Agr A induces the expression of a regulatory RNA termed RNAIII, which represses the expression of cell adhesion factors while inducting the expression of secreted factors. Activated AgrA also induces expression of the agr BDCA. This results in increased AIP levels, which ensures that the entire population switches from the lwo-cell-density to the high-cell-density state.

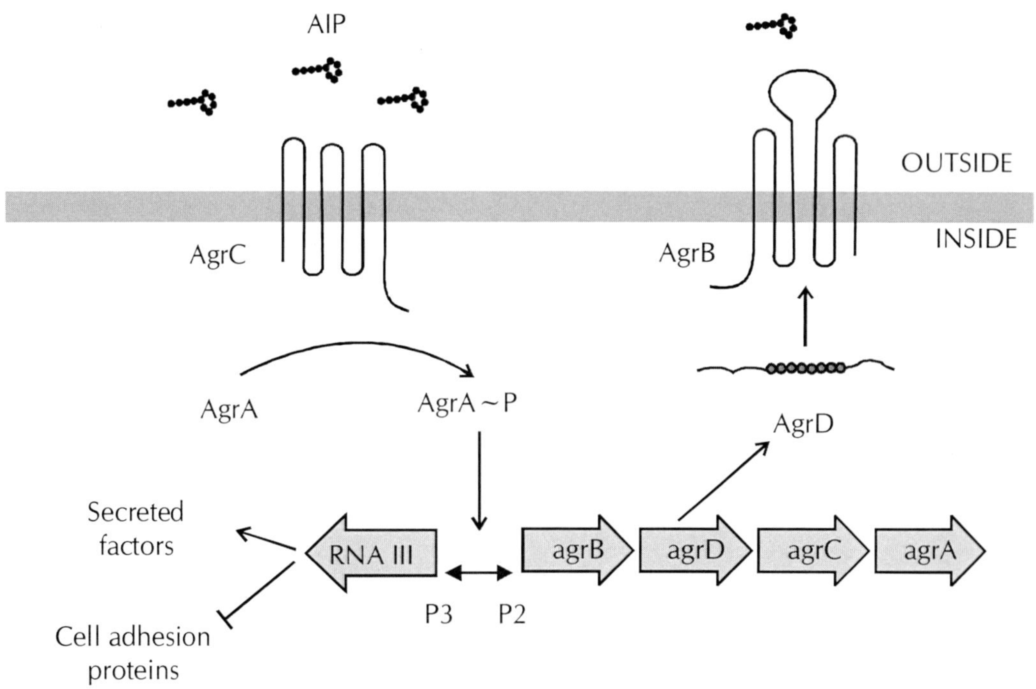

Fig. 26.20. *Quorum sensing* in *Staphylococcus aureus* using a two-component signal system and AIP as single molecule

26.6. COMPETITIVE QUORUM SENSING: *BACILLUS SUBTILIS* MODEL

Many of the quorum-sensing networks rely on multiple signals acting synergistically. However, some quorum-sensing networks are arranged such that the signals antagonize one another. Quorum sensing networks of *Bacillus subtilis* involved in competence development and sporulation are the best examples.

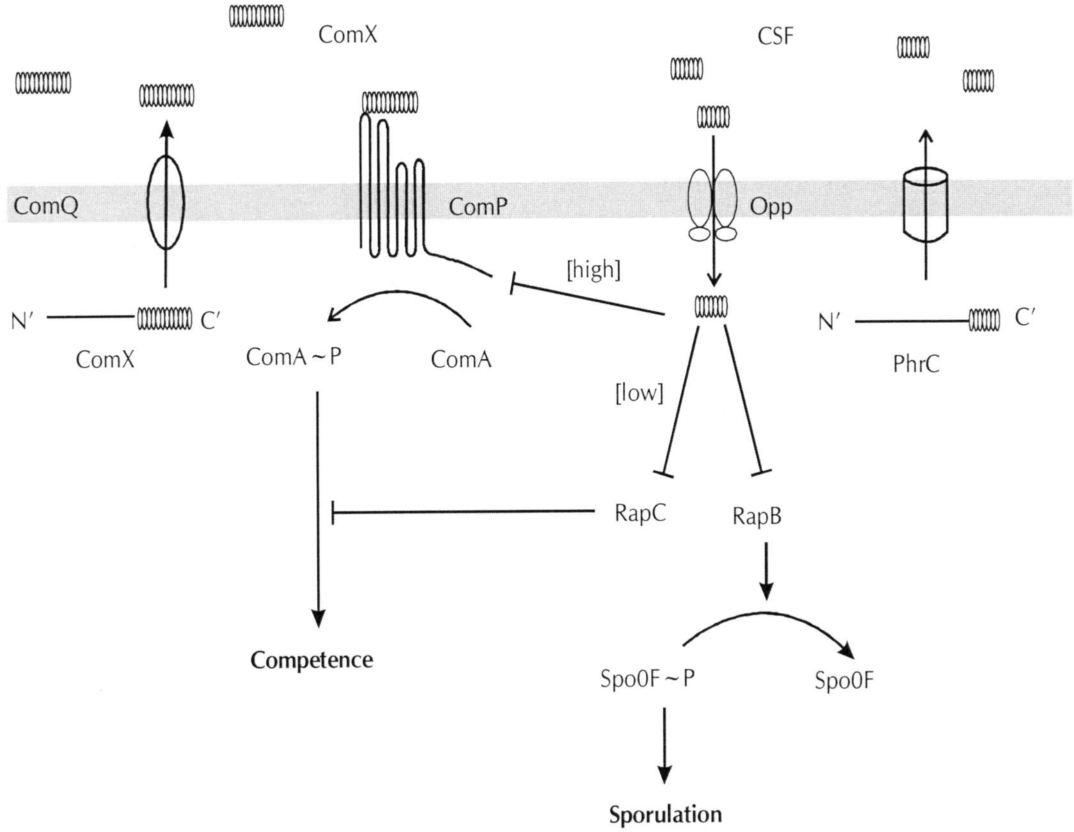

Fig. 26.21. *Bacillus subtilis* produces two autoinducting peptides that regulate two different development pathways : competence and sporulation.

B. subtilis has two autoinducing peptides functioning in a network arrangement that allows the organism to commit to one of two mutually exclusive lifestyles: competence (the ability to take up exogenous DNA) and sporulation (Fig. 26.21). ComX, a 10- amino acid peptide that is processed and secreted by ComQ, is detected by the membrane bound histidine sensor kinase ComP. ComX binding stimulates ComP to autophosphorylate and transfer phosphate to the DNA-binding response regulator ComA. Phosphorylated ComA regulates transcription of a variety of genes encoding factors required for competence development. A second oligopeptide autoinducer, competence and sporulation factor (CSF) of *B. subtilis,* encoded by the gene *phrC*, is released via the general secretory apparatus, is reinternalized through the Opp peptide transporter and acts in the cytoplasm. At low concentrations, CSF binds to a protein named RapC and disrupts Rap C binding to ComA. RapC binding to ComA inhibits competence development because DNA binding by ComA is prevented. Thus, CSF binding to RapC promotes competence development. However, at high concentrations, internalized CSF inhibits the ComP-ComA signaling cascade through an unknown mechanism, decreasing competence developments and favouring sporulation. CSF also directly promotes sporulation by inhibiting RapB mediated dephosphorylation of a response regulator named SpoOF, which in its phosphorylated state, indirectly activates genes required for sporulation.

26.7. QUORUM QUENCHING

The phenomenon of disrupting the quorum sensing is known as quorum quenching. This quorum quenching is brought about by certain enzymes degrading the signal molecules. As described earlier, cross inhibition of AIP-mediated signaling in *S. aureus* represents a clear example of a quorum-quenching mechanism because each of the four AIPs specifically inhibits quorum sensing in competing *S. aurens* groups, while not disrupting growth and other cellular functions. Many *Bacillus* species secrete an enzyme, AiiA, an AHL-acylase that cleaves the lactone rings from the acyl moieties of AHLs and renders the AHLs inactive in signal transduction. AiiA is extremely nonspecific with regard to the AHL acyl side chain, which suggests that this strategy interferes with all AHL-mediated communications between Gram-negative bacteria. Viario-*vorax paradoxus* a soil bacterium uses a different anti AHL quorum sensing strategy. Here, AHL destruction occurs via acylase (*e.g.* AiiD)-mediated lactone ring opening. Intrestingly, *V. paradoxus* uses the linearized product of the reaction as a source of carbon and nitrogen. Similarly, *Ralstonia* isolates also reported to contain AHL acylase.

In some cases, bacteria may degrade their own autoinducers, which presumably terminates quorum-sensing activities. For example, in stationary phase, *Agrobacterium tumefaciens* produces the ATTM, AHL lactonase, which can degrade the *A. tumefaciens* autoinducer. Some enteric bacteria, including *S. typhimurium* and *Escherichia coli* import AI-2 with an AI-2-specific transporter into cytoplasm wherein a series of enzymatic reactions inactivates its signaling activity. Signal molecules can also be quenched by organisms which do not produce the molecules, presumably in order to gain an advantage over communicating bacterial species in the same colonized niche. For example, *Rhodococcus* can degrade AHL signals without having any known ability to produce them. Several eukaryotic mechanisms that counteract bacterial quorum sensing have recently been reported.

Like quorum sensing, quorum quenching has also some purpose. In niches, wherein bacterial populations compete for limited resources, the ability to disrupt quorum sensing may give one bacterial species an advantage over another that relies on quorum sensing. Likewise, a host's ability to interfere with bacterial cell-cell communication may be crucial in preventing colonization by pathogenic bacteria that use quorum sensing to co-ordinate virulence. Against this information, it is presumed that analogous mechanisms exist for promoting quorum-sensing controlled behaviours when such behaviors provide benefits to organisms cohabitating with quorum-sensing bacteria.

26.8. PRACTICAL APPLICATIONS OF KNOWLEDGE OF QUORUM SENSING AND QUORUM QUENCHING

The ecological importance of quorum sensing in bacteria is well appreciated. A large number and variety of bacteria employ quorum-sensing for regulation of various phenotypes as a part of their pathogenic and symbiotic lifestyles. As such, the ability to block or promote these systems provides a powerful tool to inhibit pathogenic bacteria and enhance the symbiotic association as in case of Rhizobium-legume associations. Genetically modified plants can be used to produce AHLs to manipulate plant bacterium associations. Knowledge of quorum sensing may also provide a much-needed strategy to control multi-drug resistant strains of bacteria that employ quorum sensing for virulence and pathogenicity. It can also be exploited for control or to trigger off host-cell response, for novel antimicrobial targets and immunomodulatory agents. Therefore, it is important to take a look at the methods employed, which can enhance/block this signaling cascade.

Naturally occurring quorum-quenching processes are being tested as novel antimicrobial therapies. Researches conducted have shown some promising results. Overexpression of aiiA in tobacco and potato plants conferred resistance to *Erwinia carotovora* which requires AHL-controlled virulence factors expression to cause disease. Similarly, co-culture of *Bacillus thuringiensis* decreased *E. carotovora*-mediated plant disease in an aiiA-dependent manner. Mice treated with synthetic antagonists of *S. aureus* AIP showed resistance to infection. These observations predict that inhibition of quorum sensing offers an attractive alternative to traditional antibiotic because these strategies are not bactericidal and the occurrence of bacterial resistance therefore could be reduced.

26.9. FURTHER READINGS

Bassler, B.L.C. (1999), How bacteria talk to each other regulation of gene expression by quorum-sensing. *Current Opinion in Microbiology* 2: 282-287.

Camilli, A. and Bassler, B.L. (2006), Bacterial small-molecule signaling pathways *Science* 311(5764): 1113-1116.

Defoirdt, T. Boon, N. *et al* (2008), Quorum-sensing and quorum quenching in *Vibrio harveyi* : lessons learned from *in vivo* work. *The ISME Journal* 2: 19-26.

Dunny, G.M. and Winans, S.C. (1999), Cell-cell signaling in Bacteria, ASM Press, Washington.

Fuqus, W.C., Winans, S.C. and Greenberg E.P. (1994), Quorum-sensing in bacteria: the LuxR-LuxI family of cell density-resposive transcriptival regulators. *J. Bacterial*, 176: 269-275.

Gera Charu and Srivastava S (2006), Quorum-sensing - The phenomenon of microbial communication. *Curr. Sci* 90(5): 666-676.

Hasting, J.W. (2003), Microbial bioluminiscence. The Desk Encyclopedia of Microbiology. Elsevier Ltd.

Lukat, G.S. and J.B. Stock (1993), Response regulation in bacterial chemotaxis. *J. Cell Biochem.*, 51: 41-46.

Magasanik, B. (1993), The regulation of nitrogen utilization in enteric bacteria. *J. Cell. Biochem.*, 51: 34-40.

Miller, M.B., Basler, B.L. (2001), Quorum-sensing in bacteria *Annu. Rev. Microbiol.* 55: 165 - 199.

Parkinson, J.S. and E.C. Kofoid (1992), Communication modules in bacterial signaling proteins. *Ann. Rev. Genet.*, 26: 71-112.

Saier, M.H., Jr. (1993), Introduction: Protein phosphorylation and signal transduction in bacteria. *J. Cell. Biochem.*, 51: 1-6.

Spiro, S. and J.R. Guest (1991), Adaptive responses to oxygen limitation in *E. coli*. *TIBS*, 16: 310-314.

Stock, J.D., A.J. Ninfa and A.M. Strock (1989), Protein phosphorylation and regulation of adaptive response in bacteria. *Microbiol. Rev.*, 53: 450- 490.

Wuster, A., Madan Babu, M. (2008), Chemical molecules that regulate transcription and facilitate cell-to-cell communication. Book chapter in Wiley Encyclopaedia of Chemical Biology, Hoboken, New Jersey: John Wiley and Sons Inc.

Wanner, B.L. (1993), Gene regulation by phosphate in enteric bacteria. *J. Cell. Biochem.*, 51: 47-54.

Waters, C.M., and Barsler, B.L. (2005), Quorum sensing: cell-to-cell communication in bacteria. *Annu. Rev. Cell Dev. Biol.* 21: 319-346.

Waukan, J., and S. Forst (1992), Molecular analysis of the signaling pathway between EnvZ and OmpR in *E. coli*. *J. Bacteriol.*, 174: 1522-1527.

White David (1995), The Physiology and Biochemistry of Prokaryotes. Oxford University Press.

Whitehead, N.A., Barnard, A.M.L., Slater, H., Simpson, N.J.L. and Salmond, G.P.C. (2001), Quorum-sensing in Gram-negative bacteria. *IEMS Microbiol. Rev.* 25: 365-404.

■■■■

Chapter - 27

Endospore Formation

27.1. INTRODUCTION

A number of Gram-positive bacteria, under conditions of limited supply of C,N or P can form special, resistant, dormant structures called **endospores**. These structures are extraordinarily resistant to environmental stresses, such as heat, ultraviolet radiation, chemical disinfectants and desiccation. The main ecological roles of endospores appear to be survival in the dry state on a non-nutrient medium. Infact, some endospores have been reported to remain viable for well over 500 years. For example, endospores of *Bacillus licheniformis* and *B subtilis* are viable for 200-300 years. Due to their resistance and the fact that several species of endospore forming bacteria are dangerous pathogens, endospores are of great practical importance in food, industrial, and medical microbiology. Endospores can be examined with both light and electron microscopes. Special spore stains are used to make them clearly visible. Spores may be centrally located, subterminal or definitely terminal (Fig. 27.1). Spore position in mother cell or sporangium frequently differs among species, making it of considerable value in identification.

The endospore forming bacteria are distributed widely. They are primarily the inhabitants of soil. There are about six separate genera of bacteria that produce endospores and some of their characteristics are presented in Table 27.1.

Table 27.1. Important characteristics of spore-forming bacterial genera

Genus	G + C (mol %)	Shape	Distinguishing metabolic traits
Bacillus	33 – 66	Rods	Catalase + ve, strict aerobes
Sporolactobacillus	38 – 40	Rods	homolactic fermentation facultative anaerobes microaerophilc
Clostridium	24 – 54	Rods//filaments	strict anaerobe
Desulfotomaculum	37 – 50	Rods//filaments	sulfate reduction strict anaerobe
Sporosarcina	40 – 42	Cocci in tetrads or packets	strict aerobe
Thermoactinomyces	52 – 55	Branched filaments	strict aerobe

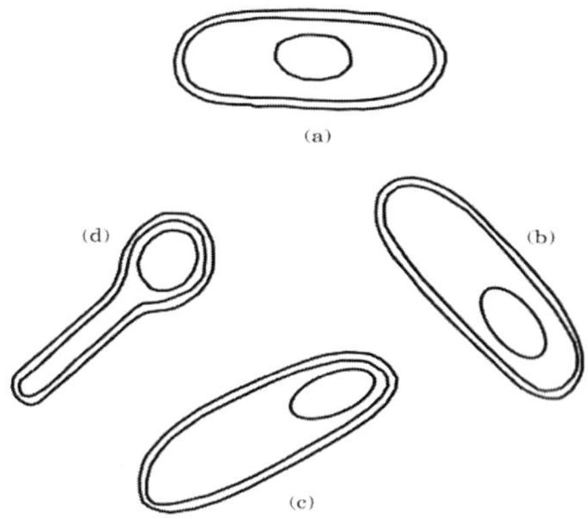

Fig. 27.1. Endospore location and size. (a) central spore (b) subterminal spore (c) terminal spore; (d) terminal spore with swollen sporangium.

27.2. ENDOSPORE STRUCTURE

Electron micrographs reveal that endospore is a complex structure (Fig. 27.2). The spore is often surrounded by a thin, delicate covering called the exosporium. A spore coat lying beneath the exosporium is composed of several protein layers, and may be fairly thick. It is impermeable and responsible for the spore's resistance to chemicals. The cortex, which may occupy as much as half the spore volume, rests beneath the spore coat. It is made of peptidoglycan that is less cross linked than that in vegetative cells. The spore cell wall (or core wall) is inside the cortex and surrounds the protoplast or core. The core has the normal cell structure such as ribosomes and a nucleoid.

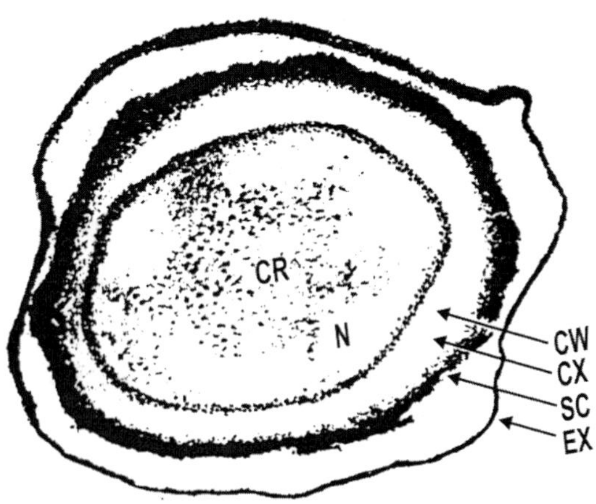

Fig. 27.2 : Electron micrograph of endospore : Ex - Exosporium, SC - Spore coat, CX-Cortex, CW- Core wall, N - nucleoid, CR- ribosomes.

Dormant endospores differ from vegetative cells not only morphologically but in physiological and biochemical aspects (Table 27.2). They contain reduced levels of a number of important metabolites, particularly high-energy compounds, such as ATP, sugar phosphates, reduced pyridine nucleotides, ribonucleotide triphosphates, and tRNA. One potential high-energy phosphate donor, 3-phosphoglycerate (PGA) is present in the dormant spores of some strains at very high concentration. These spores also contain PGA mutase, enolase and pyruvate kinase, the three enzymes required for conversion of PGA to ATP. The dormant spores also contain sufficient enzymes of glycolytic pathway and the hexose monophosphate shunt. The mechanism where by these enzymes remain inactive in the dormant spore and are activated on germination are not known with certainly.

It is still not known precisely why the endospore is resistant to heat and other lethal agents. As much as 15 per cent of the spore's dry weight consists of dipicolinic acid complexed with calcium ions. It has long been thought that dipicolinic acid was directly involved in spore heat resistance, but heat-resistant mutants lacking dipicolinic acid now have been isolated. It may be that calcium–dipicolinate complex often stabilizes spore nucleic acids. Recently specialized small, acid soluble DNA–binding proteins have been discovered in the endospore. They saturate spore DNA and protect it from heat, radiation, desiccation, and chemicals. Dehydration of the protoplast appears to be very important in heat resistance. The cortex may osmotically remove water from the protoplast, there by protecting it from both heat and radiation damage. In summary, endospore heat resistance appears to be is due to several factors; calcium–dipicolinate complex, acid soluble protein, stabilization of DNA, protoplast dehydration etc.

Table 27.2. Differences between bacterial vegetative cell and endospore.

Property	Vegetative cell	Endospore
Cell	Gram-positive	thick spore
Radiation resistance	low	high
Microscopic appearance	non – refractile	refractile
Heat resistance	low	high
Calcium presence	low	high
DPA	absent	present
Enzymatic activity	high	low
Metabolism of O_2 uptake	high	low/absent
Macromolecular synthesis of mRNA	present	absent
Resistance to chemicals & acids	low	high
Stainability by dye	stainable	stainable only with special methods
Action of lysozyme	sensitive	resistant

27.3. ENDOSPORE FORMATION

The formation of spores, called sporulation or sporogenesis, commences when growth ceases due to unfavourable conditions. Endospore formation is initiated in response to starvation of nutrients particularly carbon, nitrogen or phosphorus. The very early stages of sporulation are reversible and transfer to fresh culture medium results in resumption of vegetative growth. The sporulation seems to be initiated by the accumulation of factors that inhibit vegetative growth and depress spore genome. Sporulation is more efficient in a dense population of cell. Sporulation in *Bacillus* takes place only under aerobic conditions, while in *Clostridium,* it is strictly anaerobic. Vegetative growth and division ceases during sporulation. Several morphological and biochemical events characterize sporulation

Sporulation process has been divided in to a number of stages. In *Bacillus subtilis*, a well known endospore forming bacterium, in all seven stages have been identified treating the vegetative cell as stage O (Fig. 27.3).

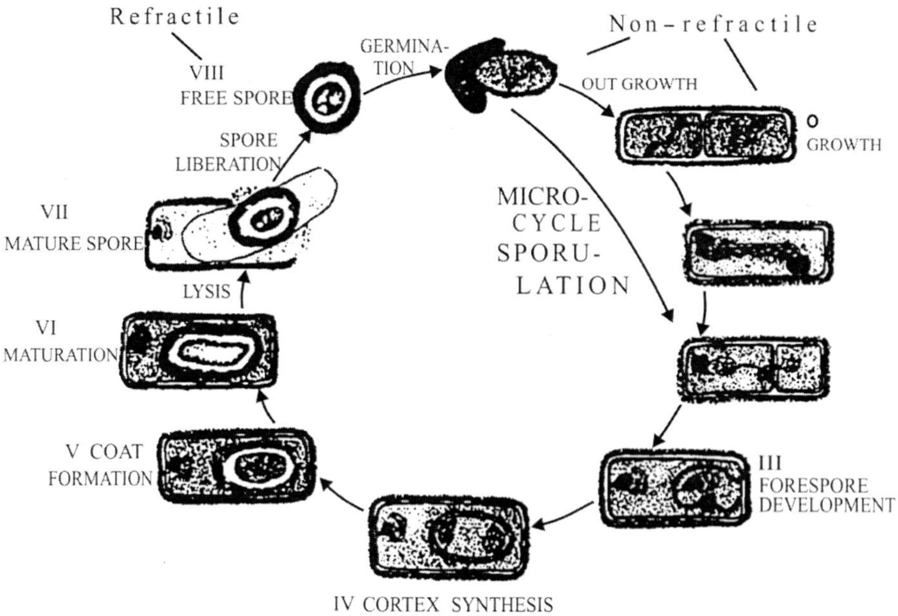

Fig 27.3. Stages in endospore formation of *Bacillus subtilis*

Stage-O

It represents the vegetative cell stage before the commencement of sporulation. Depletion of an essential metabolite limits the growth of vegetative cells and initiates sporulation.

Stage–I: Axial filament formation

The first visible change during sporulation is the fusion of the two compact nuclear bodies and redistribution of nuclear material from end to end to form an axial chromatin thread called the axial filament. The axial DNA structure could result from changes in intracellular solute and ion levels. Each chromosome may be attached to the cell membrane by a mesosome. The cell undergoes elongation as if about to undergo fission. In *Bacillus* species granules of poly-β-hydroxybutyric acid (PHB) appear. These granules represent stored food material and sporulation can be completed in distilled water. In *Clostridium* species, however, there are no food reserves and exogenous sources of food material are required for sporulation (exotrophic sporulation).

Protein turn over begins in the cell during stage 1. Synthesis of the enzymes amylase, protease, and ribosidase also take place during this stage. In many spore forming bacteria, both aerobic and anaerobic, synthesis of antimicrobial substances occurs at the onset of sporulation at this stage. Three classes of antibiotics are synthesized during stage –1 of sporulation (Table 27.3).

Table 27.3. Classes of antibiotics synthesized during stage-I of sporulation.

Class of antibiotics	Structure	Mode of action
Edeines	linear basic peptides	inhibit DNA synthesis
Bacitracins	cyclic peptides	inhibit cell wall synthesis
Gramicidin	linear and cyclic peptide	modify membrane structure and function

Stage–II: Spore septum formation

During this stage an inward growth of a double layer of the cell membrane, from the mesosome, takes place near one end of the cell. The cell is asymmetrically partitioned by the double membrane, which forms a transverse spore septum. The nuclear material divides so that each compartment contains atleast one complete chromosome. The spore septum divides the bacterial cell into a smaller and a large compartment, each containing DNA. The smaller compartment is called the forespore, while the large one is called the mother cell.

The sporulation septum has no peptidoglycan layer between the invaginating membrane in most organisms. However, in *Bacillus sphaericus*, peptidoglycan layer is observed. The mesosomes have been found to be associated with the asymmetric division of sporulation. Synthesis of alanine dehydrogenase takes place during this stage.

Stage–III: Engulfment of forespore

In normal cell division, the formation of septum is followed by the development of a transverse cell wall. In sporulation, however, the membrane of the larger cell invaginates towards the pole of the cell and engulfs the forespore. This is brought about by continued proliferation of the septal membrane and a translocation of the base of the septum towards the pole of the cell. The forespore is then enclosed by two concentric sets of membranes, its own (inner forespore membrane) and the membrane derived from the larger cell (outer forespore membrane). The outer forespore membrane has the surface polarity relative to both the plasma membrane of the mother cell and inner forespore membrane. The outer surfaces of the mother cell membrane now face each other in the double forespore membrane. This orientation is of importance in the biosynthetic activity and transport. Active transport operates in opposite direction.

The mature spore and forespore lack the biosynthetic enzymes for amino acids such as methionine and arginine. Sporulating cells take up exogenous uracil and DNA, which are transported into the forespore. The synthesis of enzymes of the TCA cycle and glyoxylate cycle as well as catalase and phosphatase takes place during this stage. Sulfolactic acid is also synthesized.

Stage – IV: Cortex synthesis

The cortex develops between the inner and the outer forespore membranes. The forespore also begins to accumulate calcium and dipicolinic acid (pyridine-2,6 dicarboxylic acid (DPA)) and as a result it becomes gray in appearance (Fig. 27.4).

At the end of the stage – III, there is a rapid increase in the accumulation of Ca ions from the sporulation medium in to the cell. The accumulated Ca ions move continuously in to the forespore compartment, so that the mature spore consists of about 9 percent of the dry weight of the spore. The shift from Ca ion efflux of vegetative bacilli to Ca ions influx during sporulation may require: synthesis of new catalytic membrane proteins or a modification of the vegetative efflux mechanism. It is proved experimentally that Ca ion is actively transported from the sporulation medium into the cytoplasm of the mother cell and then moves into the forespore by facilitated diffusion.

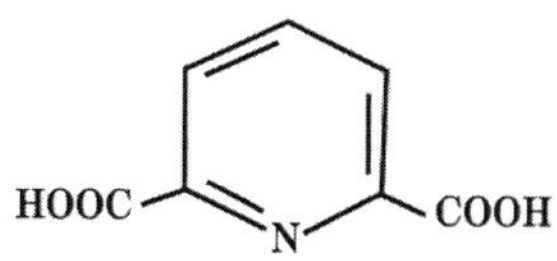

Fig. 27.4. Dipicolinic Acid.

Synthesis of DPA by the bacterial cell begins at about the same time that Ca ion uptake begins and occurs in 1:1 molar ratio with Ca ion. It is almost always extracted from spores as the calcium–DPA chelate. In sporulating cell, the total DPA is confined to the forespore compartment. Synthesis of enzymes adenosine deaminase and ribosidase take place during stage-IV. The spore becomes heat resistant following the uptake of Ca ions and synthesis of DAP; cortex specific meso diaminopimelic ligase is made. This enzyme is confined to the mother cell compartment.

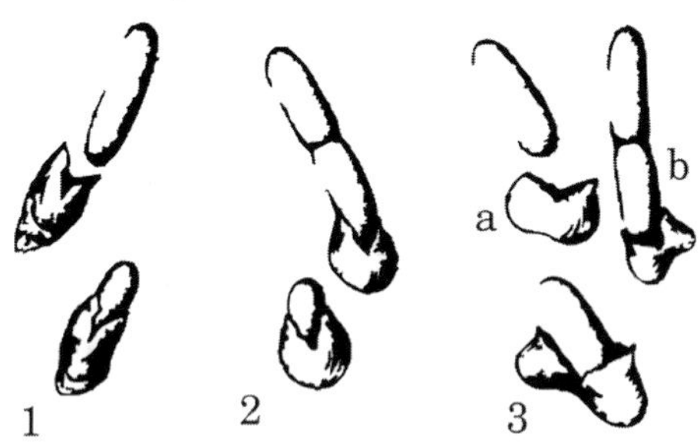

Fig. 27.5. Germinating spores. (1) Polar germination of a *Clostridium* spore with its exosporium; (2) Polar germination of a *Bacillus megaterium* spore, the spore coat remaining attached; (3) lateral germination of *Bacillus* spores; (a) *B. cereus,* (b) *B. subtilits.*

Stage – V: Early coat synthesis

Most of the proteinaceous spore coats are synthesized during stage V. The electron dense spore coats are formed outside the outer membrane surrounding the cortex. Accumulation of Ca^{+2} and

DPA continues and the forespore becomes phase white. It acquires resistance to octonal and the capacity to germinate. Large amounts of cysteine are incorporated in to the coat. In *B.cereus* group, a thin exosporium is formed around the spore coat.

Stage – VI: Maturation

The maturing spore begins to become refractile. Water is withdrawn during spore maturations. As a result the mature spore is considerably dehydrated. The cortex plays an important role in dehydration. According to the contractile cortex theory, the wastes are expelled by contraction of the loose peptidoglycan brought about by ionic cross linking with calcium ions. An alternate suggestion is that the cortex expands within the fixed volume of the coat because of loss of ionic cross-links.

Stage – VII: Spore liberation by lysis

Finally, the spore is liberated by the autolysis of the mother cell and during this period no metabolic activity is detected within the spore.

27.4. ENDOSPORE GERMINATION

Bacterial spores exhibit little or no metabolic activity and are adapted for prolonged survival under adverse conditions. Spores can remain dormant for several years. Although viability of spores decreases with the passage of time; organisms have been cultured from spores of bacilli over 300 years old. The process by which a spore is converted into a germinative cell is called germination. Three stages of germination have been distinguished,

i. Activation
ii. Germination and
iii. Outgrowth or post-germinate development

27.4.1. Activation

Some bacterial spores germinate spontaneously in a favourable medium. Others remain dormant even if placed in optimal conditions for germination. Some agents such as heat, low pH, certain chemicals such as L-alanine, adenosine, glucose and some reducing agents, may bring about activation. Heat shock or heat activation appears to be the most general mechanism for activating spores. Heating of spores in an activating fluid for 15-60 min at 65°C results in the activation of most spores. Higher temperatures of 105° to 120°C are required for thermophilic spores. The induced germinability declines if spores are returned to lower temperature for some days. Activated spores still retain their resistance to heat, refractile, resistant to staining, and still contain large amounts of dipicolinic acid.

27.4.2. Germination

Germination occurs in response to specific germinants that act as triggers, leading to rapid changes in the structure and physiology of the spore. Germination precedes the detectable changes in the spore: loss of heat resistance, ion fluxes, release of Ca^{++} and dipicolinic acid, hydrolysis of cortex peptidoglycan, rehydration of the core protoplast, and resumption of metabolic activity. Accumulation of ATP occurs within a few minutes after germination begins (150–fold increase in the first 5 min). This initial increase in ATP occurs at the expense of endogenous energy source, particularly the large PGA pool. After the exaustation of the PGA pool, energy metabolism occurs primarily via the hexose monophosphate shunt, since many enzymes of the TCA cycle are absent in germinating spores. The synthesis of RNA begins by the second minute of germination in *B. megaterium* and constitutes a major use for the ATP generated at this stage. Synthesis of RNA during the first 20 minutes of germination can be completely accomplished from nucleotides stored in the dormant spore. Chromosome replication (DNA synthesis) does not begin until late in spore germination.

Germination–specific enzymes that hydrolyze spore specific cortex peptidoglycans have been identified with studies in *B. subtilis* and *C. perfringens.* Cortex hydrolysis during germination is initiated by the attack of spore-cortex-lytic enzyme on intact spore peptidoglycan and leads to removal of cross–linkages. This step is followed by further degradation of the polysaccharide portion of the modified peptidoglycan by cortical fragment lytic enzyme.

Protein synthesis begins within the first two or three minutes of germination and can be accomplished solely from stored nitrogen reserves through the first 60 minutes of germination. The free amino acid content of dormant spores is limited but increases rapidly in germinating spores as a result of hydrolysis of spore constituents. Obviously, proteolytic activity is necessary for synthesis early in germination.

27.4.3. Out Growth

Transition from germinated spore to vegetative cell has been termed out growth (Fig. 27.5). Generally if germination occurs, it does not necessarily mean that it will be followed by out growth. The factors triggering germination donot necessarily support out growth and *vice versa.* Post-germinative out growths occur only if there are favourable conditions of nutrition and growth. Generally if a suitable nutritional environment is provided, the germinated spore will continue rapidly into the out growth process. If nutritional requirements are lacking, the process may stop altogether, or the cells may revert *i.e.,* proceed to sporulate and enter into microcycle sporulation. In the microcycle, asymmetric division, rather than symmetrical cell division, begins immediately after the elongation of the cell is initiated and the cells proceed to sporulate. However, one round of DNA replication must occur prior to reentry into the sporulation cycle.

27.5. FURTHER READING

Buchanan, C.E., A.O. Henriques and P.J. Piggot (1994), Cell wall changes during bacterial endospore formation. In J.M. Ghuysen and R. Hakenbeck (eds.) Bacterial cell wall. Elsevier, Amsterdam, The Netherlands, pp. 167-186.

Dricks, A. (1999), *Bacillus subtilis* spore coat. *Microbiol. Mol. Biol. Rev.*, 63 : 1-20.

Dricks, A., S. Roels, B. Beall, C.P. Moran and R. Losick (1994), Subcellular localization of proteins involved in the assembly of the spore coat of *Bacillus subtilis*. *Genes Dev.*, 8 : 234-244.

Errington, T. (1993), *Bacillus subtilis* sporulation: regulation of gene expression and control of morphogenesis. *Microbiol. Rev.*, 57: 1-33.

Foster, S.J. and K. Johnstone (1989), The trigger mechanism of bacterial spore germination. In: "Regulation of Prokaryotic Development" (I. Smith, R.A. Slepecky and P. Setlow (eds.), pp. 89-108. American Society of Microbiology, Washington, D.C.

Hauser, P.M. and J. Errington (1995), Characterization of cell cycle events during the onset of sporulation in *Bacillus subtilis*. *J. Bacteriol.*, 177 : 3923-3931.

Meador-Parton, J., and D.L. Popham (2000), Structural analysis of *Bacillus subtilis* spore peptidoglycan during sporulation. *J. Bacterial*, 182: 4491-4499.

Moir, A. and D.A. Smith (1999), The genetics of bacterial spore germination. *Ann. Rev. Microbiol.*, 44 : 531-553.

Piggot, P.J. (1996), Spore development in *Bacillus subtilis*. *Curr. Opin. Genet. Dev.*, 6 : 531-537.

Piggot, P.J., C.P. Moran (Jr) and P. Youngman (eds.) (1993), Regulation of Bacterial Differentiation. American Society of Microbiology, Washington D.C.

Stragier, P. and R. Losick (1996), Molecular genetics of sporulation in *Bacillus subtilis*. *Ann. Rev. Genet*, 30: 297-341.

■■■■

Glossary

Acetogens: Microorganisms that produce acetic acid.

Acetotrophic: Splitting of acetate into CH_4 and CO_2 by certain methanogens.

Acetyl-CoA (Ljungdahl-wood)pathway: A pathway of autotrophic CO_2 fixation widespread in obligate anaerobes including methanogens, homoacetogens and sulfate-reducing bacteria.

Acetylene reduction assay: Method of measuring activity of nitrogenase by substituting acetylene for the natural substrate of the enzyme, N_2. Acetylene is reduced to ethylene or ethane.

Acidophile: Organism that grows best at acidic pH values.

Actinorrhizal: Referring to an association of an actinomycete and a plant root.

Adenosine triphosphate (ATP): A compound that conserves metabolic energy in its two high-energy phosphate bonds.

Aerobes: An organism that grows in the presence of O_2: may be facultative, obligate, or microaerobic.

Aerotolerant: An anaerobe, but can tolerate low levels of O_2.

Alkalophiles: An organism that grows best at high pH.

Ammonification: Microbial conversion of nitrogen in organic compounds to ammonia (NH_3).

Amphibolic pathways: Metabolic pathways that function both catabolically and anabolically.

Amyloglucosidase: Starch digesting enzyme releasing one glucose molecule at a time.

Amylopectin: A branched polymer of D-glucose with both 1,4-glycosidic bonds and 1,6-glycosidic bonds.

Amylose: A polymer of D-glucose linked through 1,4-glycosidic bonds.

Anabolism: The biochemical processes involved in the synthesis of cell constituents from simpler molecules, usually requiring energy.

Anaerobe: An organism that grows in the absence of O_2.

Anaerobic metabolism: Metabolism that occurs in the absence of oxygen, anaerobic respiration and fermentation

Anaerobic respiration: Use of an electron acceptor other than O_2 in an electron transport-based oxidation and leading to a proton motive force.

Analogues: Structurally similar to but functionally different chemicals.

Anoxic: Absence of oxygen. Usually used in reference to a microbial habitat.

Anoxygenic photosynthesis: Use of light energy to synthesize ATP by cyclic photophosphorylation without O_2 production.

Antibiotic: A chemical compound produced by a microorganisms that inhibits the growth of or destroys other microorganism.

Antiport: Coupled movement of two solutes in opposite direction across the cell membrane.

Appendages: Structures that extend beyond the envelope of microbial cells.

Archaea: A phylogenetic domain of prokaryotes consisting of the methanogens, most extreme halophiles and hyperthermophiles and thermplasma.

Assimilatory Power: The ATP and $NADPH_2$ produced in light phase reactions, utilized to reduce CO_2 to carbohydrate.

Assimilatory reduction: The reduction of an inorganic molecule to incorporate it into organic material. No energy is made available during this process.

Autotrophs: Organism able to utilize CO_2 as a sole source of carbon.

Auxotroph: An organism that has developed a nutritional requirement through mutation contrast with a prototroph.

Axial filament (or endoflagellum): The modified flagella of spirochetes that lie within the periplasm; used for movement.

Bacteriocins: Toxic proteins produced by bacteria that kill other closely related bacteria.

Bacteriorhodopsin: A protein containing retinal that is found in the membranes of certain extremely halophilic Archaea and that is involved in light-mediated ATP synthesis.

Bacteroid: A swollen, deformed *Rhizobium* cell found in the root nodule: capable of nitrogen fixation.

Balanced growth: Microbial growth in which all cellular constituents are synthesized at constant rates relative to each other.

Barophile: An organism that lives optimally at high hydrostatic pressure.

Barotolerant: An organism able to tolerate high hydrostatic pressure, although growing better at 1 atm.

Batch culture: A closed-system of microbial culture of fixed volume, and for a definite period.

Bioenergetics: The study of energy relations to biological reactions.

Bioluminimetry: A technique for estimating the intensity of light emitted by enzyme-substrate complex: luciferase-luciferin.

Bioluminioscence: Phenomenon of emittance of light by certain animals, plants, and microorganisms.

Bioremediation: Using organisms to detoxify or eliminate toxic materials.

Biosynthestic pathways: Sequences of metabolic reactions that convert the precursor metabo-lites into building blocks for synthesis of macromolecules and other essential small molecules.

Biotechnology: The use of organisms for practical purposes.

Blue-green algae: Former designation of cyanobacteria.

Brown-rot fungi: Fungi capable of degrading only cellulose and not lignin.

Calvin-Benson cycle: The pathway through which most autotrophs incorporate CO_2 into cellular constituents.

Capsule: The diffuse outermost layer, usually carbohydrate, of many bacterial and some other microbial cells.

Carbon cycle: The cyclic conversion of carbon-containing compounds that occurs in nature.

Carboxydobacteria: The bacteria that utilize CO as an electron donor.

Carboxymethylcellulose: Soluble cellulose and it can be hydrolysed into cellobiose.

Cardinal temperatures: The minimum, optimum and maximum temperatures required for the growth of microorganisms.

Catabolism: The biochemical processes involved in the breakdown of organic or inorganic compounds, usually leading to the production of energy.

Catabolite repression: Decreased enzyme activity due to the accumulation of hydrolytic product.

Catalysis: Increase in rate of chemical reaction.

Catalyst: A substance that promotes a chemical reaction without itself being changed in the end.

Cellobiase: An enzyme attacking cellobiose releasing two glucose molecules.

Cellobiohydrolase: One of cellulases which converts insoluble cellulose to soluble cellulose.

Cellulosome: Cell bound multienzyme complex of cellulases.

Chemiosmosis: A means of generating ATP by forming an ion gradient across a membrane that drives ATP synthesis from ADP by an enzyme (ATPase) located in the membrane.

Chemoautotrophs: Microorganisms that generate ATP and reducing power from inorganic chemical reactions.

Chemolithotroph: An organism obtaining its energy from the oxidation of inorganic compounds.

Chemoorganotroph: An organism obtaining its energy from the oxidation of organic compounds.

Chemostat: A continuous culture device controlled by the concentration of a limiting nutrient and dilution rate.

Chemotaxis: The process by which cells sense certain chemicals and swim towards regions that contain optimal concentrations of them.

Chemotherapy: Treatment of diseases by chemical agents.

Chemotrophs: Microorganisms that derive the energy from chemical reactions.

Chitin: A polymer of N-acetylglucosamine, found in the cell walls of many fungi.

Chlorosomes: Cigar-shaped structures enclosed by a nonunit membrane and containing the light-harvesting bacteriochlorophyll (C, C_s, d or e) in green sulfur bacteria and in *Chloroflexus*.

Cholesterol-dependent cytolysins (CDC): These are large-pore-forming endotoxins bind selectively to cholesterol on the eukaryotic cell membrane.

Citric acid cycle: A cyclical series of reactions resulting in the conversion of acetate to CO_2 and NADH. Also called the tricarboxylic acid cycle or the Kreb's cycle.

Coenzymes: Organic molecules that are need for activation of certain enzymes.

Cofactors: Inorganic ions that certain enzymes need to be active.

Colony: A clone of cells large enough to be visible on a solid medium.

Cometabolism: The metabolic transport of a substance while a second substance serves as primary energy or carbon source.

Consortium: A two- (or more) membered bacterial culture (or natural assemblage) in which each organism benefits from the others.

Continuous culture system: A culture system with constant environmental conditions maintained through continual provision of nutrients and removal of wastes.

Cryptic growth: The invisible growth occurring during the lag phase.

Culture medium: A fluid or gelled solution, containing nutrients needed for growth of a microorganism.

Cyanobacteria: The oxygen-producing, phototrophic bacteria.

Cyclic AMP (cAMP): 3',5'-cyclic adenosine monophosphate that acts as a chemical messenger in cells.

Cyclic photophosphorylation: The process by which phototrophs generate ATP by passing an activated electron from chlorophyll through a membrane-located electron transport chain to chlorophyll.

Cyclodextrins: Cyclic oligosaccharides containing 6-12 glycopyranose units bonded through α (1-4) linkages.

Cytochrome: Iron-containing porphyrin complexed with proteins, which functions as an electron carrier in the electron transport system.

Cytoplasm: The matrix (ground substance) of a cell, composed primarily of water and protein.

de nov pathway: Synthesis of nucleotides from simple molecules, such as sugars amino acids and phosphates.

Death phase: The phase of microbial growth following the stationary phase in which cells die at exponential rate.

Defined medium: Culture medium for which the chemical composition is known because it is prepared from pure chemicals.

Denitrification: Conversion of nitrate into nitrogen gas under anoxic conditions.

Deoxyribonucleic acid (DNA): A macromolecule composed of four deoxyribonucleotides (A, G, T and C) that encodes an organism's genetic information.

Desulfurylation: The process of converting sulfur constituents of organic compound to hydrogen sulfide.

Dextrin: A product of amylopectin hydrolysis with both 1,6-glycosidic bonds and 1,4-glycosidic bonds.

Diauxic growth: A biphasic growth pattern or response in which a microorganism, when exposed to two nutrients, initially uses one of them for growth and then alters its metabolism to make use of the second.

Differential media: Media used to identify microorganism.

Disproportionation: The splitting of a chemical compound into two new compounds, one more oxidized and other reduced than the original compound.

Dissimilatory reduction: The use of a substance as an electron acceptor in energy generation. The acceptor (e.g. sulfate or nitrate) is reduced but not incorporated into organic matter during biosynthetic processes.

Doubling time: The time needed for a population to double.

Endergonic reaction: A chemical reaction requiring input of energy to proceed.

Endoglucanase: It acts on cellulose randomly decreasing the chain length.

Endospore: A differentiated cell formed within the cells of certain gram-positive bacteria that is extremely resistant to heat as well as to other harmful agents.

Endotoxin: The lipopolysaccharide component of the outer membrane of Gram-negative bacteria that is harmful to humans and other animals: most of the toxicity is mediated by lipid A.

Enrichment culture: A method of cultivating microorganisms designed to isolate a particular microorganism or type of microorganism from a large, complex natural population.

Enthalpy: The heat content of a system: in a biological system it is essentially equivalent to the total energy of the system.

Entropy: A measure of the randomness or disorder of a system: A measure of that part of the total energy in a system that is unavailable for useful work.

Enzymes: Protein (with the exception of a few RNA molecules) that catalyze specific reactions.

Equilibrium constant: The constant that is characteristic of a chemical reaction at equilibrium and is based on the relative equilibrium concentrations or activities of products and reactants.

Eukaryotes: Organisms composed of cells with internal membranes (all cellular organisms except bacteria and archaea).

Eurythermal: The microorganisms which can grow over a broad range of temperature.

Exergonic reaction: A chemical reaction that proceeds with the liberation of energy.

Exosporium: The thick, impermeable outer layer of the endospore.

Exotoxin: A toxin released by a living bacterial cell into its surroundings.

Exponential growth: (also called the logarithmic growth) The growth phase during which the number of cells in the population continues to double during the same time interval.

Exponential phase: A period during the growth cycle of a population in which growth increases at an exponential rate.

Facilitated diffusion: Movement of molecules across a membrane from regions of higher to lower concentration, mediated by proteins that permit passage of only specific molecules.

Facultative anaerobes: Organisms that use oxygen to grow when it is available but can also grow without it.

Fastidious organism: The microorganism that require numerous complex nutrients to grow. *e.g.* Lactic acid bacteria.

Fats: Esters that are solid at room temperature, formed between a molecule of glycerol and three molecules of fatty acids.

Fatty acid synthase: A multi-enzyme that catalyzes the synthesis of fatty acids.

Feedback inhibition: A decrease in the activity of the first enzyme of a pathway caused by the final product of the pathway.

Fermentation: Catabolic reactions producing ATP in which organic compounds serve as both primary electron donor and ultimate electron acceptor and ATP is produced by substrate-level phosphorylation.

Flagella: Appendages of prokaryotic and eukaryotic cells that confer motility.

Flavine adenine dinucleotide (FAD): A compound that transfers reducing power by taking up and releasing hydrogen atoms.

Flavoprotein: A protein containing a derivative of riboflavin, which functions as electron carrier in the electron transport system.

Fluid mosaic model: The proposal that proteins embedded in a phospholipid membrane move freely.

Fueling pathways: Metabolic pathways that generate precursor metabolites, ATP and reducing power.

Galacturonic acid: A basic unit of pectic acid and pectin.

Gangrene: Tissue death due to impaired blood supply.

Gas gangrene: An infection caused by *Clostridium perfringens*.

Glucoamylase: Same as amyloglucosidase.

Glucose effect: Glucose inhibiting the expression of catabolic enzymes.

Glycerate pathway: A pathway through which glyoxylate is converted to 3-phosphoglycerate.

Glycocalyx: The slimy or gummy substance that constitutes the outermost layers of the envelope of some bacteria.

Glycogen: An alpha-linked, branched chain polymer of glucose.

Glycolysis: The catabolic metabolic pathway that converts a molecule of glucose into two molecules of pyruvate.

Griseofulvin: It is an effective antifungal antibiotic.

Group translocation: Entry of a compound into a cell and simultaneous chemical alteration.

Growth rate: Measurement of how rapidly a microbial population is increasing usually expressed in doubling times per hour.

Halophile: An organism requiring salt (NaCl) for growth.

Halorhodopsin: A pigment found in halophiles.

Harmful algal blooms (HAB): HAB are formed by rapid increase or accumulation in the population of algae in fresh or marine water system.

Heat labile: Easily destroyed by heat.

Hemicellulose: An heteropolysaccharide containing both pentoses and hexoses.

Heterocyst: A differentiated cyanobacterial cell that carries out nitrogen fixation.

Heteropolysaccharide: Polysaccharide containing more than one type of monosaccharides.

Heterotroph: (Chemoorganotroph) Organisms that obtain carbon from organic compounds in their medium or diet.

High-energy bonds: Chemical bonds that require considerably less energy to break than is released when new ones form.

Homeostasis: Maintenance of constancy of internal environment.

Homopolysaccharide: A polysaccharide containing only one type of monosaccharides.

Hydrogenase: An enzyme that oxidizes NADH with the production of H_2.

Hydrophilic compounds: Compounds that dissolve in water.

Hyperthermophile: A prokaryote having a growth temperature optimum of 80°C or higher.

IMP: Inorine monophosphate, a key compound in the synthesis of nucleotides.

Inclusion bodies: In prokaryotes, visible structures within the cell other than the nuclear region and ribosomes.

Inducer exclusions: Growth inducer prevents the entry of second substratum.

Ionophore: A compound that can cause the leakage of ions across membranes.

Ionophores: An ionophore is a chemical species that reversibly binds ions. They are lipid soluble.

Isomaltose: Oligosaccharide with two molecules of glucose linked by 1,6-glycosidic bond.

Isotopes: Atoms with the same atomic number but, different atomic weight.

Joule (J): A unit of energy equal to 10^7 ergs: 1000 joules equal 1 kilojoule (KJ).

Lag phase: The period after inoculation of a population before growth begins.

Latent period: The period of time following viral infection during which no new virions are produced.

Lectins: Protein binding carbohydrates.

Leghemoglobin: A protein found in the root nodules of legumes, which scavenges oxygen.

Lichen: A symbiotic assocition between a fungus and an alga or a cyanobacterium.

Lignin: Highly stable polymer of mostly methyoxylated phenyl propanoic residues synthesized as a part of the cell wall of vascular plants.

Ligno-cellulose: Close association of cellulose with lignin.

Limiting nutrient: The scarcest nutrient in a medium that limits the growth.

Lipopolysaccharide (LPS)**:** A compound found only in the outer membrane of Gram-negative bacteria.

Lithotrophs: The microorganisms utilize inorganic substance as a source of electron H^+ donors.

Log phase: (also called the logarithmic, or exponential phase). The phase of microbial growth cycle when exponential growth occurs.

Luciferin: A reduced substance that is when acted upon by enzyme luciferase emits light.

Lyases: Pectin enzyme which breaks the pectic acid chain by introduction of double bond between C_4 and C_5.

Lysozyme: An enzyme that degrades peptidoglycon by hydrolyzing the $\beta(1\rightarrow4)$ bond that joins N-acetylmuramic acid and N-acetylglucosamine.

Magnetosomes: Small particles of Fe_3O_4 present in cells that exhibit magnetotoxins.

Magnetotaxis: Directed movement of bacterial cells by a magnetic field.

Maltose: A disaccharide consisting two molecules of glucose.

Membrane filter: A nitrocellulose membrane with holes too small for microbial cells to pass through.

Mesophile: Organism living in the temperature range near that of warm-blooded animals, and usually showing a growth temperature optimum between 25°C and 40°C.

Metabolic intermediates: Compounds formed at various steps of metabolic pathways.

Metabolism: All biochemical reactions in a cell, both anabolic and catabolic.

Methanogen: A methane-producing prokaryote: member of the archae.

Methanotroph: An organism capable of oxidizing methane.

Methylotroph: An organism capable of oxidizing organic compounds that do not contain carbon-carbon bonds: if able to oxidize CH_4, also a methanotroph.

Microaerophilic: Requiring O_2, but at a level lower than atmospheric.

Microbial antagonism: Inhibition of microbial growth by another microorganism.

Middle lamella: Amorphous substances between plant two cells which mainly consists of calcium pectate.

Minimum inhibitory concentration (MIC): It is the lowest concentration of an antimicrobial agent that is needed to hinder the growth of bacteria.

Mixotroph: An organism able to assimilate organic compounds as carbon sources while using inorganic compounds as electron donors.

Monera: The bacteria.

Mordants: Compounds that intensify staining reactions.

Mutualism: A symbiotic relationship in which both partners benefit.

Mycotoxins: Toxins produced by certain moulds and can be found in food.

Negative staining: Use of a stain that does not penetrate cells or capsules, causing them to appear bright against a dark, background.

Neurotoxins: A toxin or poison which acts on the nervous system, *e.g.*, botulinum toxin.

Neutrophile: An organism that grows best around pH 7.

Nicotinamide adenine dinucleotide (NAD): A compound that acts as a reservoir of reducing power by accepting and donating pairs of hydrogen atoms.

Nicotinamide adenine dinucleotide phosphate (NADP): A phosphorylated form of NAD that serves a similar function.

Nisin: A type of bacteriocin produced by strains of *Lactococcus lactis* subsp. *lactis*.

Nitrification: The microbiological conversion of ammonia to nitrate.

Nitrogen cycle: Conversions of nitrogen compounds that occur in nature.

Nitrogenase: The enzyme that catalyzes biological nitrogen fixation.

Nodulins: The signal molecules released by rhizobia during infection.

Nuclear envelope: A double membrane structure that encloses the eukaryotic nucleus.

Nucleoid (or nuclear region): The mass of DNA in bacterial cells not membrane-bound.

Nucleoside triphosphate: A purine or pyrimidine bonded to ribose or deoxyribose and three phosphate groups.

Nucleotide: A purine or pyrimidine bonded to ribose or deoxy-ribose and one to three phosphate groups.

Obligate aerobes: Organisms that grow only in the presence of oxygen.

Obligate anaerobes: Organisms that grow only in the absence of oxygen.

Organotrophs: Organisms that use reduced organic compounds as their electron source.

Osmophilic: Microorganism that grows best in or on media of high solute concentration.

Oxidation: A process by which a compound gives up electrons (or H atoms) and becomes oxidized.

Oxidation-reduction (redox) reaction: A pair of reactions in which one compound becomes oxidized, while another gets reduced and takes up the electrons released in the oxidation reaction.

Oxidative (electron transport) phosphorylation: The non-phototrophic production of ATP at the expense of a proton motive force formed by electron transport.

Oxygenase: An enzyme that introduces molecular oxygen into a reduced commpound; monoxygenases, dioxygenases.

Oxygenic photosynthesis: Use of light energy to synthesize ATP and NADPH by non-cyclic photophosphorylation with the production of O_2 from water.

Pasteur effect: The decrease in the rate of sugar catabolism and change to aerobic respiration that occurs when microorganisms are switched from anaerobic to aerobic condition.

Pectic acid: Demethylated polymer of pectin.

Pectin: Galacturonic acid esterified with methyl alcohol.

Pectinic acid: A short chain of galacturonic acid after removal of methyl alcohol.

Peptide: A product of protein hydrolysis with a short chain of amino acids.

Peptidoglycan: The rigid layer of the cell walls of bacteria, a thin sheet composed of N-acetyl glucosamine, N-acetylmuramic acid, and a few amino acids. Also called murein.

Periplasm: The organelle between the cytoplasmic membrane and outer membrane of Gram-negative bacteria.

Phosphorylation: Chemical addition of a phosphate group to a molecule.

Photosynthesis: The use of light energy to drive the incorporation of CO_2 into cell material.

Phycocyanin: A blue phycobiliprotein pigment used to trap light energy during photosynthesis, occur in cyanobacteria.

Phycoerythrin: A red photosynthetic phycobiliprotein pigment used to trap light energy.

Phycotoxins: Toxins produced by certain algae like dinoflagellates, diatoms, and cyanobacteria.

Pore-forming toxins: A bacterial toxin that inserts a transmembranous pore into a host cell membrane.

Porins: Protein channels in the lipopolysaccharide layers of gram-negative bacteria through which small to medium sized molecules can flow.

Porters: Membrane proteins that function to transport substances into and out of the cell.

Prochlorophyte: A prokaryotic oxygenic phototroph that contains chlorophyll a and b but lacks phycobilins.

Proton motive force: An energized state of a membrane created by expulsion of protons usually occurring through action of an electron transport chain.

Protopectin: Long chain of pectin present in young tissues.

Protoplast: A cell from which the wall has been removed.

Psychrophile: An organism able to grow at low temperatures and showing a growth temperature optimum of 15°C.

Psychrotrophs: Organisms that grow at low temperatures.

Pullulan: Branched a polymer of glucose, containing both 1,4 and 1,6 glycosidic bonds

Pullulanase: An enzyme hydrolyzing pullulan.

Quinones: Hydrophobic lipid electron carriers found in cell membranes.

Quorum quenchihng: The phenomenon of disrupting the quorum sensing.

Quorum sensing: It is the ability to detect and respond to a cell population density by gene regulation.

Recalcitrant: Resistant to microbial attack.

Reduction potential (E!): The inherent tendency, measured in volts, of a compound to act as an electron donor or an electron acceptor.

Reduction: A process by which a compound accepts electrons to become reduced.

Reductive amination: Introduction of amino group into organic acid in the presence of dehydrogenase enzyme.

Respiration: Catabolic process producing ATP in which either organic or inorganic compounds are primary electron donors and organic or inorganic compounds are ultimate electron acceptors.

Retinal: A light absorbing pigment present in halophiles.

Reverse electron transport: The enregy-dependent movement of electrons against the thermodynamic gradient to form a strong electron donor from a weaker electron donor.

Reverse transcription: A transcription process opposite to normal transcription. Synthesis of DNA from mRNA.

Root nodule: A tumor like growth on certain plant roots that contains symbiotic nitrogen-fixing bacteria.

RTX toxin (Repeat-in-toxin): These toxins belong to a large family of Ca^{++}-dependent hemolysins.

Rumen: Four chambered stomach of herbivores.

Salvage pathway: Synthesis of nucleotides from intermediates in the degradative pathways.

Selective media: Media that favour the growth of certain microorganisms over others.

Sequential induction: Each of the enzymes in a pathway in inducible by its specific substrate.

Siderophore: An iron chelator that can bind iron present at very low concentrations.

Simple stain: Single dyes used to increase contrast of a specimen.

Slime layer: A thin slimy or gummy layer that surrounds some bacterial cells.

Slime molds: Two groups of microorganisms: cellular slime molds (Acrasiease) and true slime molds (myxogastria).

Smear: A thin film spread of bacteria on a microscope slide.

Spheroplast: A spherical, osmotically sensitive cell derived from a bacterium by loss of some but not all of the rigid wall layer. If all the rigid wall layer has been completely lost, the structure is called a protoplast.

Sporulation: The process of forming spores.

Stains: Microscopy dyes used to increase contrast, storage granules: granular inclusions in the cytoplasm that hold reserve supplies of nutrient.

Standard free-energy change: The free energy change of a reaction at 1 atmosphere pressure when all reactants and products are present in their standard states; usually the temperature is 25°C.

Stationary phase: The period during the growth cycle of a population in which growth ceases.

Stenothermal: Organisms that can grow well only over a narrow temperature range.

Stickland reaction: A fermentation reaction in which one amino acid is used as a donor and another as acceptor of electrons.

Strains: Clones that are presumed or known to be genetically different.

Stringent response: The response of bacteria at molecular level to amino acid starvation.

Substrate-level phosphorylation: Synthesis of high-energy phosphate bonds through reaction of inorganic phosphate with an activated organic substrate.

Sulfur cycle: The chemical conversions of sulfur that occurs in nature.

Superantigens: These are bacterial protein toxins, which are extremely potent mitogens of T lymphocytes.

Sym. plasmid: A plasmid present in rhizobia carrying the genes necessary for symbiotic association.

Symbiosis: A relationship between two organisms.

Symport: Coupled movement of solutes in the same direction across the cell membrane.

Synchronous growth: Growth in a cell population in which all cells divide at the same time.

Synergistic action: Interaction of two organisms resulting in increased production of enzymes.

Syntrophy: A nutritional situation in which two or more organisms combine their metabolic capabilities to catabolize a substance not capable of being catabolized by either one alone.

Terminal electron acceptor: Compound at the end of an electron transport chain; *e.g.*, oxygen for aerobic respiration.

Thermal death point (TDP): The lowest temperature required to kill all microorganism in a particular liquid suspension in 10 minutes.

Thermal death time (TDT): The minimal time required to kill all microorganisms in a particular liquid suspension at a given temperature.

Thermoacidophiles: A group of archaea that grow at high temperatures and high acidic pH.

Thermophile: An organism with a growth temperature optimum between 45 and 80°C.

Trace elements: Certain inorganic elements that are essential to life but required in only minute amounts.

Transamination: Transfer of amine group from amino acid to organic acid to form an amino acid.

Transeliminase: Same as lyases.

Tricarboxylic acid (TCA) Cycle: A cychic metabolic pathway that oxidizes acetate, generates ATP and forms four precursor metabolites.

Turbidity: Cloudiness of a liquid caused by suspended particles.

Turbidostat: A continuous culture system equipped with a photocell that adjusts the flow of medium through the culture vessel so as to maintain a constant cell density or turbidity.

Turgor pressure: A cell's internal pressure resulting from a higher intracellular than extracellular osmotic strength.

Uniport: The movement of single solute molecule across the cell membrane.

Uptake hydrogenase: An enzyme present in diazotrophs which oxidizes the H_2 through oxidative phosphorylation releasing the energy.

Vacuole: A membrane-bound intracellular vesicle.

Viable count: Measurement of the living cells in a population.

Virulence: The ability of a pathogenic organism to cause the disease.

Vitamins: Nutrients required in minute quantities, primarily as precursors of enzyme cofactors.

Water activity (a_w): An expression of the relative availability of water in a substance, pure water has an a_w of 1.0.

White-rot fungi: Fungi capable of degrading lignin along with cellulose.

Wood-workman reaction: Carboxylation of pyruvic acid to form dicarboxylic acid

Xenobiotic: A completely synthetic chemical compound not naturally occurring on earth.

Xerophile: An organism adapted to growth at very low water potentials.

Xerotolerant: An organism that can tolerate and grow at low water activity (a_w).

Xylan: Other name of hemicellulose.

α-Amylase: Starch hydrolyzing enzyme which attacks 1,4-glycosidic bonds randomly resulting a short chain dextrins.

β-Amylase: Starch hydrolyzing enzyme which attacks 1,4-glycosidic bonds from non-reducing ends. It attacks alternate 1,4-glycosidic bonds resulting in the decrease of chain length by two glucose molecules (maltose).

β-hydroxyaspertate pathway: A pathway through which glycoxylate is finally converted to oxaloacetate.

β-lactoms: A group of antibiotics, chemically having a lactam ring, *e.g.*, penicillin.

β-oxidation pathway: An important pathway through which fatty acids are oxidized to acetyl CoA and CO_2.

■■■■

Subject Index